4th International Conference on Advanced Composite Materials and Manufacturing Engineering (ACMME 2017)

IOP Conference Series: Materials Science and Engineering
Volume 207

Xishuangbanna, China
20 – 21 May 2017

ISBN: 978-1-5108-4265-6
ISSN: 1757-8981

Printed from e-media with permission by:

Curran Associates, Inc.
57 Morehouse Lane
Red Hook, NY 12571

CURRAN ASSOCIATES INC. proceedings.com

Some format issues inherent in the e-media version may also appear in this print version.

Printed by Curran Associates, Inc. (2017)

Institute of Physics
Dirac House, Temple Back
Bristol BS1 6BE UK

Phone: 44 1 17 929 7481
Fax: 44 1 17 920 0979

techtracking@iop.org

Additional copies of this publication are available from:

Curran Associates, Inc.
57 Morehouse Lane
Red Hook, NY 12571 USA
Phone: 845-758-0400
Fax: 845-758-2633
Email: curran@proceedings.com
Web: www.proceedings.com

Table of contents

Volume 207

4th International Conference on Advanced Composite Materials and Manufacturing Engineering 2017

20–21 May 2017, Xishuangbanna, Yunnan, China

Accepted papers received: 1 June 2017
Published online: 15 June 2017

Preface

Composite Materials

Materials Applied in Manufacturing Engineering

Energy Materials and Environmental Materials

Engineering Materials and Material Corrosion

IOP Conf. Series: Materials Science and Engineering **207** (2017) 011001 doi:10.1088/1757-899X/207/1/011001

Preface

The organizing committee of ACMME 2017 is proud to present the proceedings of the 2017 4th International Conference on Advanced Composite Materials and Manufacturing Engineering (ACMME 2017: http://www.icacmme.org/), held on May 20-21, 2017 in Xishuangbanna, China.

ACMME 2017 intends to be a global forum for researchers and engineers to present and discuss recent innovations and new techniques related to materials engineering, in particular, advances in composite materials, materials processing engineering, functional materials and manufacturing engineering, as well as their analytical techniques and related environmental issues.

ACMME 2017 received more than 180 manuscripts, and less than 120 submissions have been accepted by our reviewers. By submitting a paper to ACMME 2017, the authors agreed to the review process and understood that papers would undergo a peer-review process. Manuscripts were reviewed by appropriately qualified experts in the field selected by the conference committee, who took detailed comments and-if the submission was accepted-the authors would submit a revised version that took into account this feedback.

All papers were reviewed using a double-blind review process: authors declared their names and affiliations in the manuscript for the reviewers to see, but reviewers did not know each other's identities, nor did the authors receive information about who had reviewed their manuscript. The committees of ACMME 2017 invested great efforts in reviewing the papers submitted to the conference and organizing the sessions to enable the participants to gain maximum benefit. We do believe your great participation will make our conference success and we would appreciate your participation.

Hopefully, all participants and other interested readers will benefit scientifically from the proceedings and also find it stimulating in the process.

With our warmest regards,

Conference Organizing Chair
Xishuangbanna, China

IOP Conf. Series: Materials Science and Engineering **207** (2017) 011002 doi:10.1088/1757-899X/207/1/011002

Committees

Conference General Chair
Prof. D. J. Fang, Wuhan University, China

Steering Committee Chairs
Prof. Kwon Myung Jang, Pukyong National University, South Korea
Prof. Q. S. Han, South-Central University for Nationalities, China

Program Committee Chairs
Prof. P. Zhao, Beijing Material Research Center, China
Prof. H. V. Lee, Universiti Sultan Zainal Abidin, Malaysia
Prof. F. L. Zhang, Shanxi University, China

Technical Committees
Prof. Y. D. Zhang, Wuhan Institute of Technology, China
Prof. G. Q. Xia, Wuhan Institute of Technology, China
Prof. X. Han, Wuhan Institute of Technology, China
Prof. Z. G. Fan, Wuhan University, China
Prof. H. Y. Qiu, Wuhan University, China
Prof. S. W. Wei, Shanxi University, China
Prof. K. Wei, Shanxi University, China
Prof. Z. X. Hou, Hainan University, China
Prof. H. L. Yan, Yunnan University, China
Prof. M. Xu, Yanshan University, China
Prof. S. D. Wei, Northeastern University, China
Prof. A. M. Qian, Beijing Material Research Center, China
Prof. L. G. Sun, Beijing Material Research Center, China
Prof. D. L. Mo, South-Central University For Nationalities, China
Prof. J. Z. Wu, South-Central University For Nationalities, China
Prof. C. K. Feng, South-Central University For Nationalities, China
Prof. S. P. Narasimha, Universiti Sultan Zainal Abidin, Malaysia
Prof. G. L. Miao, University of Science and Technology of China, China
Senior Engineer, H. Yang, China Petroleum Pipeline Engineering Co., Ltd, China
Dr. L. Y. Cai, Wuhan Institute of Technology, China
Dr. Xu Han, Hunan University, China
Dr. Danghuang Xu, Nanjing University of Aeronautics and Astronautics, China
Dr. G. Tao, Industry and Information Technology Department of Hebei Province, China

Peer review statement

All papers published in this volume of *IOP Conference Series: Materials Science and Engineering* have been peer reviewed through processes administered by the proceedings Editors. Reviews were conducted by expert referees to the professional and scientific standards expected of a proceedings journal published by IOP Publishing.

Honeycomb nano cerium oxide fabricated by vacuum drying process with sodium alginate

Guozheng Zhao[1, 2, *], Changbo Li[1], Honglin Zhang[1]

[1]Liaoning Technical University, College of Mining, Liaoning Fuxin, 123000, China
[2]Liaoning Shihua University, College of Chemistry, Chemical Engineering and Environmental Engineering, Liaoning Fushun, 113001, China

[*]Corresponding author e-mail: zhaoguozheng668@163.com

Abstract. Nano cerium oxide (CeO2) with honeycomb structure were synthesized simply and rapidly by vacuum drying method with sodium alginate as the biological template agent, Ce(NO3)3·6H2O as cerium source. The composition, aperture size, specific surface area and morphology of the prepared samples were characterized by Fourier transform infrared spectroscopy (FTIR), X-ray diffraction (XRD), N2 adsorption-desorption and scanning electron microscopy (SEM). Simultaneously, the effects on the morphology of the samples, which were caused by the drying method and the concentration of sodium alginate, were investigated. The results indicate that the prepared samples were nano CeO2 with high crystallinity and uniform dispersion, most of which had mesoporous, macroporous and honeycomb structure. The specific surface area of CeO2 is 210.0 m2/g, and the average aperture is 12.77 nm. The prepared samples can act as catalyst in the catalytic wet oxidation process for the treatment of high concentration organic wastewater, and the COD removal rate could exceed 90%.

1. Introduction

Cerium oxide (CeO2), as a typical rare earth oxide, is widely used as a promoter in three-way catalysts (TWCs) for the elimination of toxic auto-exhaust gases, oxygen permeation membrane systems, ultraviolet absorption, as well as environmental chemistry, etc.[1-5]

The different application properties of nano CeO2 are directly related to its structure and morphology. So the study on the structure and morphology of nano CeO2 attracts so much attention.[6-8] At present, the structure according to the different dimensions of the nano CeO2 can be divided into one-dimensional nano tube, nano wire and nano rod, two-dimensional nano sheet, nano disk and three-dimensional nano ball.[9, 10] The various structures of nano CeO2 are fabricated by coprecipitation, sol-gel method, hydrothermal synthesis and template method.[11-14] And biological template method for metal oxide preparation is a kind of environment-friendly and low cost method. Among biological templates that have been developed, sodium alginate is a polysaccharide, and extracted from brown algae. Alginates belong to exopolysaccharides and are non-repeating copolymers composed of β-D-mannuronic acid (M) and α-L-guluronic acid (G) which are linked by 1-4 glycosidic bonds. Since its safety and biocompatibility, abundant reserves and renewable, it is widely used in the field of membrane science, food industry, biological medicine, pollution control, etc.[15]

In this work, honeycomb nano CeO2 was fabricated by vacuum drying process with sodium alginate as template. The work focused on the influence of process parameters on the microstructure of the material. Meanwhile, the catalytic performance was examined by catalytic wet oxidation treatment with high concentration organic wastewater.

2. Experimental

2.1. Preparation of materials

All chemical reagents were of analytical grade and used without any further purification. In a typical synthesis, Ce(NO3)2·6H2O was dissolved in deionized water to produce a transparent solution of 10 mol/L. Then added 2 mL Ce(NO3)2·6H2O solution prepared into 10 mL 1% sodium alginate solution. After being stirred at room temperature, the mixture turned into a homogeneous gel, and then the gel was dried in vacuum -0.06 MPa at 30 ℃ for 5 h. Then the yellowish ceria powder with nanostructures was obtained by calcining at 600 ℃ for 2 h.

2.2. Characterization

The prepared samples were characterized by several instrumental analysis techniques. The crystallinity of the samples was measured with powder X-ray powder diffraction patterns which were obtained on a Bruker diffractometer using CuK α radiation (λ=0.1547 nm). BET measurements including surface areas and pore volumes were performed using an automatic physical static analyzer (Quantachrome, Autosorb-IQ2-MP). The samples were degassed at 300 ℃ for 3 h and characterized at liquid nitrogen temperature (77.350 K). SEM images were collected with a Hitachi SU8010 microscope. The FTIR spectra were recorded at room temperature using Agilent Technologies FTIR-660+610 spectrometer over the range of 400-4000 cm-1.

2.3. Measurement of catalytic activity

The prepared samples were used as catalyst in the experiment of catalytic wet oxidation treatment of high concentration organic wastewater. Added 0.2 g prepared samples and 1 mL H2O2 (6%) into 50 mL wastewater taken from acrylic chemical factory. The mixture was stirred at room temperature for 3 h, and then placed into microwave reactor at 100 ℃ for 40 min. After reaction, the COD value of wastewater was determined.

3. Results and discussion

3.1. Characterization analysis by XRD

Fig. 1 shows the XRD patterns of the as-made nanostructure. The results match with PDF#34-0394-CeO2 comparisons. Clear crystal lattice diffraction peaks appear at 28.5°, 33.1°, 47.6°, 56.5°, 59.3°, 69.7°, 77.1°, 79.5° and 89.0°, which indicate that the prepared sample CeO2 has high crystallinity and uniform dispersion.

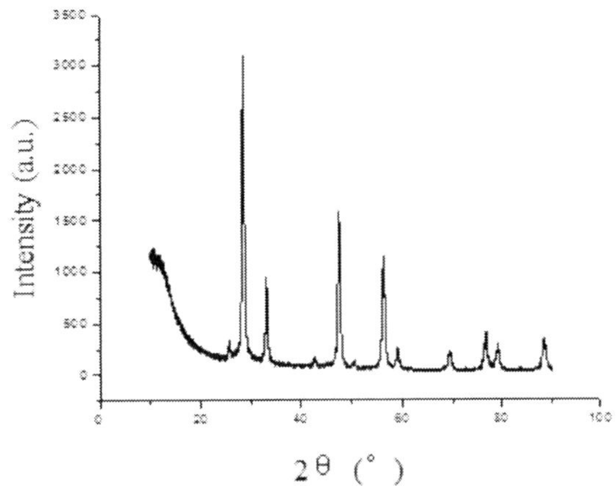

Fig.1 XRD patterns of the sample

3.2. Characterization analysis by FTIR

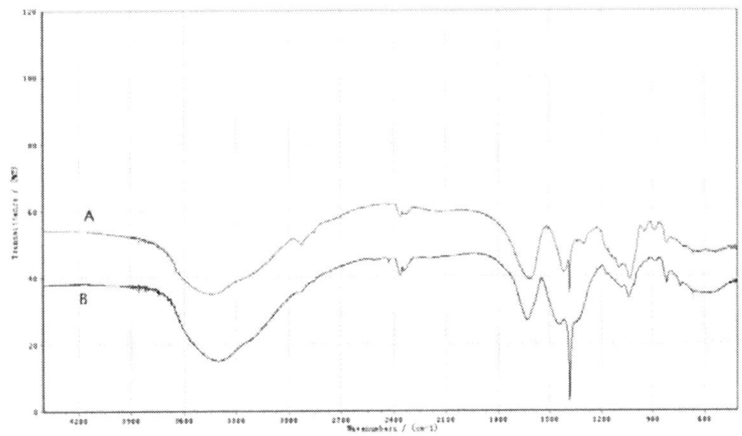

Fig.2 FTIR spectra of sample and sodium alginate (A: sodium alginate, B: sample)

Fig. 2 shows the comparisons between sodium alginate and the sample with FTIR spectra. In contrast with A, the absorption peak of B at 3446 cm-1 is the O-H bond of sodium alginate chains which shifts to low wave number direction. That is because the oxygen atom of the carbonyl group on the macromolecular chain forms an association with Ce2+, which produces chelate effect, and also the bending vibration absorption of O-H decreases. The absorption peak of B at 1417 cm-1 is decreased, which belongs to C-H bond in-plane bending vibration absorption peak. That is because the association of alginate molecular chain and the formation of network structure limit the vibration of the C-H bond, and then the absorption peak decreases. The absorption peak of B at 1350 cm-1 is enhanced which is the C=O bond stretching vibration peak. The reason lies in the role of carboxyl and Ce2+. In summary, the oxygen atoms of O-H bond and C=O bond in alginate molecular chains produces the chelate effect with Ce2+. And the formed "egg box" structure promotes the internal network structure formation of alginate.

3.3. Characterization analysis by N2 adsorption-desorption

Fig. 3 shows the N2 adsorption-desorption isotherm of CeO2. The pore characteristics of the samples can be determined by adsorption and desorption isotherms.

According to the BET classification of the adsorption isotherms, the adsorption isotherm of CeO2 is found to be close to that of the Type II adsorption isotherm (presents reverse "S" shape) [21]; the front adsorption curve rises slowly, slightly upward convex; the back end rises dramatically; even the relative pressure is close to 1.0, there is no significant adsorption saturation phenomenon presents. It is shown that there appears capillary condensation phenomenon in the N2 adsorption process, and the sample contains a certain amount of mesopores and macropores.

The adsorption and desorption isotherms of the sample adsorption isotherms do not coincide in the part of high pressure, but form adsorption loop. The type of loop reflects the characteristics and types of pore structure. Four different types of adsorption loops are recommended by International Union of Pure and Applied Chemistry (IUPAC).[22] The adsorption curve of the sample is very steep near the saturated vapor pressure, while the desorption curve is steep at medium pressure. This is close to type H3 loop recommended by the IUPAC. It indicates that the pores of prepared samples are mainly composed of nanopores, and the pore morphology is open mainly open hole such as cylinder hole with both ends opens, parallel plate hole with four sides open, and so on. The opening degree of pore is related to the rising rate of adsorption curve, the greater the pore opening, the faster the curve rise.

Calculating aperture by BJH model, the most probable aperture of CeO2 is 5.3 nm. The specific surface area and average aperture of the CeO2 obtained by BET method is 210.0 m2/g and 12.77 nm, respectively.

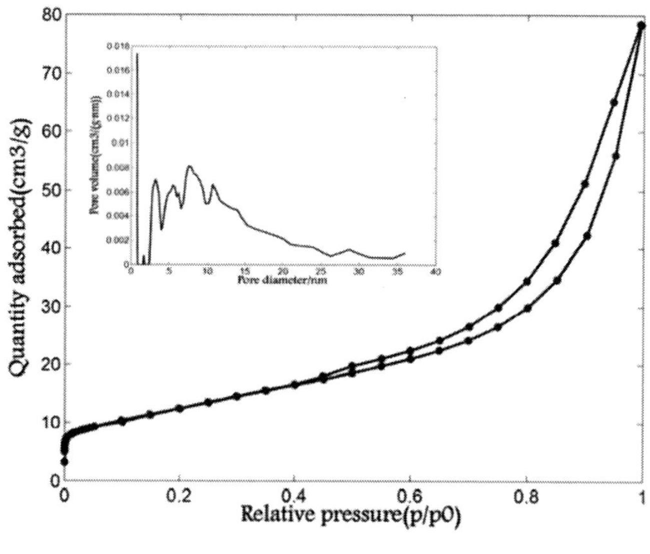

Fig.3 Nitrogen adsorption-desorption isotherms of the sample

3.4. Characterization analysis by SEM

3.4.1. Morphology of samples at different magnification. The morphology of samples was observed using magnification of 5 K and 20 K, and the results are shown in Fig. 4. The honeycomb structure can be observed clearly in figure, which is consistent with the result of N_2 adsorption-desorption isotherms.

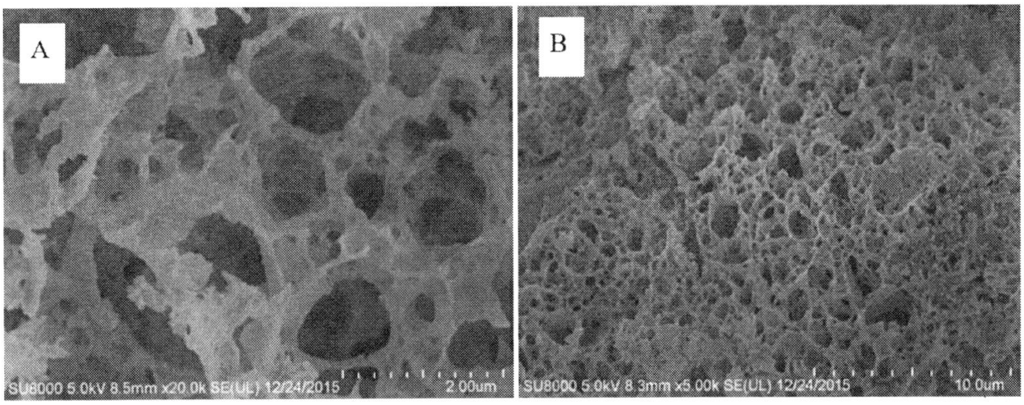

Fig.4 SEM images of the sample at different magnification (A: 20 K, B: 5 K)

3.4.2. The influence of drying methods on morphology. The samples were prepared in two ways: ordinary hot air drying (40 ℃, 48 h dehydration) and vacuum drying (-0.06MPa, 30 ℃, 5 h dehydration). The SEM scanning results of prepared samples are shown in Fig. 5. Ordinary hot air dried samples (Fig. 5 (A)) shows appearance of pore shrinkage and compact structure. This is because the material would lose water, shrink, and adhere together during ordinary hot air drying, and the more water loss, the stronger the adhesion. While the internal structure of vacuum dried samples (Fig. 5 (B)) is loose. This is because the dehydration process of material was performed in a vacuum state, the difference between internal and external pressure leads to the loose structure and easier to form honeycomb structure.

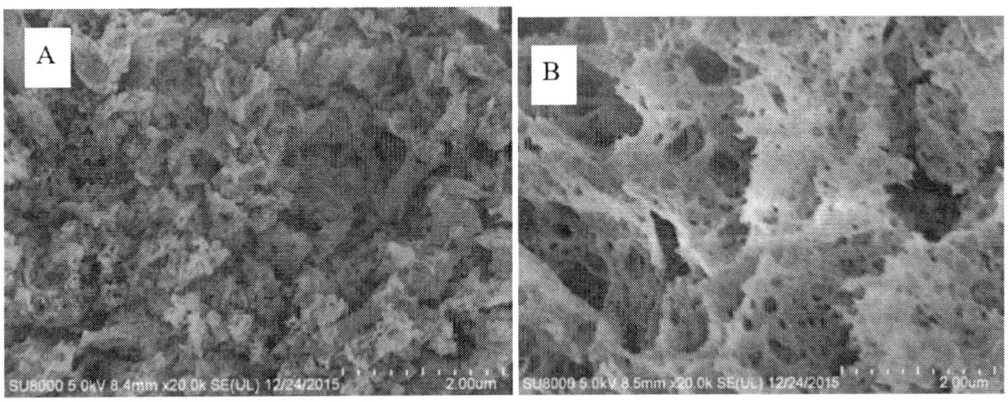

Fig.5 SEM images of the sample at different drying methods (A: Ordinary hot air drying, B: vacuum drying)

3.4.3. The influence of sodium alginate concentration on morphology. The effect of sodium alginate concentration on the morphology of the prepared samples was investigated at a concentration of 0.5%, 1.0% and 0.5 %, and the results are shown in Fig. 6, respectively. Sodium alginate solution has the characteristics of non-newtonian fluid, its viscosity increases with the increase of solution concentration. When sodium alginate concentration is low (0.5%), it is difficult to form three-dimensional structure; with the increase of the sodium alginate concentration (1%), the molar number of sodium alginate in unit volume increases, at the same time hydrogen-bonding is enhanced so that porous material presents honeycomb three-dimensional structures; when the sodium alginate dosage continued to increase (2%), the solution viscosity increases, and because of the slow migration of water and sodium alginate molecules, the resulting crystal structure has no obvious structural features.

IOP Conf. Series: Materials Science and Engineering **207** (2017) 012001 doi:10.1088/1757-899X/207/1/012001

The concentration of sodium alginate is directly related to the viscosity of the solution, and then affects the microstructure of the samples prepared.

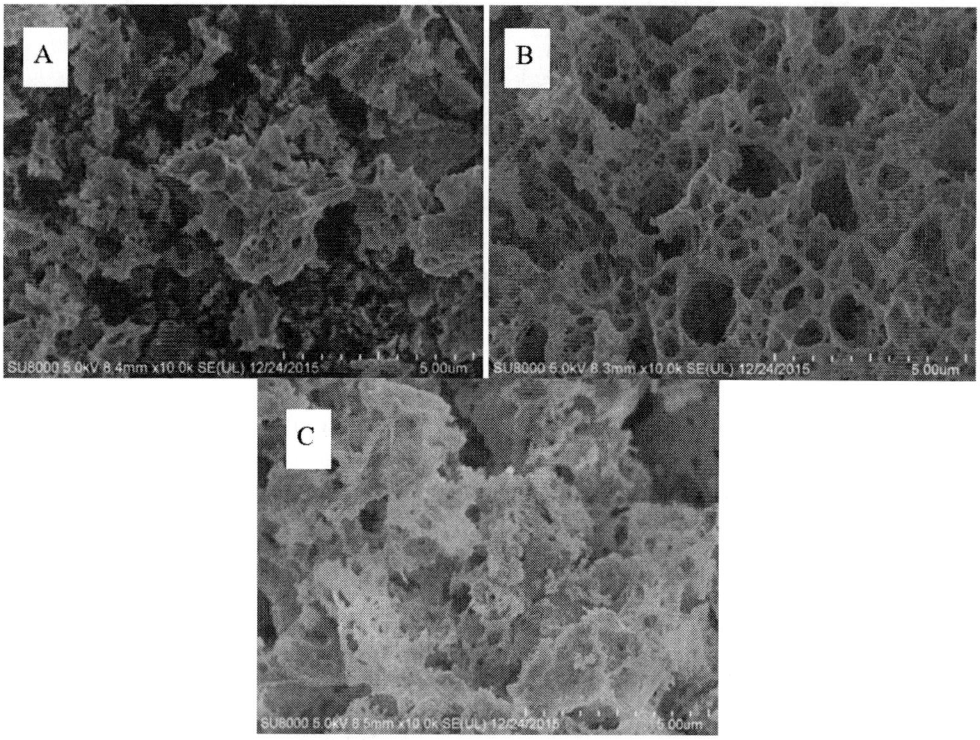

Fig.6 SEM images of the samples prepared at different sodium alginate concentration (A: 0.5%, B: 1.0%, C: 2.0%)

3.5. Catalytic performance test

1mL 6% H2O2 and 0.2 g prepared nano CeO2 were added into 50 mL acrylic chemical wastewater, and then mixed with a magnetic stirrer for 1 h in a constant temperature. Subsequently, the mixture was reacted in the microwave reactor at 100 ℃ for 40 min. The test results prove that the removal rate of COD was above 90%.

4. Conclusion

(1) It is feasible to prepare porous nano CeO2 with honeycomb structure using sodium alginate as template and under vacuum drying conditions. This method has the advantages of simple operation, mild condition, energy saving and consumption reducing. Meanwhile, compared with other templates, sodium alginate exhibits the benefits of biological materials: wide source, low price, environment friendly and so on.

(2) The drying method and the concentration of sodium alginate can affect the morphology of the samples. Vacuum drying is easier to form honeycomb structure than ordinary hot air drying, and its drying time is much shorter than the ordinary hot air drying. Too high or too low concentration of sodium alginate is not conducive to the formation of honeycomb structure. The porous nano CeO2 with honeycomb structure is obtained at the sodium alginate concentration of 1.0%.

(3) The prepared samples were applied to the treatment of high concentration organic wastewater by catalytic wet oxidation. The removal rate of COD gained from test was above 90%, which indicates that the prepared porous nano CeO2 has good catalytic performance.

References

[1] Chunwen Sun, Hong Li and Liquan Chen. Nanostructured ceria-based materials: synthesis, properties, and applications [J]. Energy & Environmental Science, 2012, 5, 8475-8505
[2] Yuan Q, Duan HH, Li LL, et al. Controlled synthesis and assembly of ceria-based nanomateria ls [J]. Colloid Interface Sci., 2009, 335: 151-167.
[3] Trovarelli A. Catalysis by Ceria and Related Materials[M]. Imperial College Press. 2002
[4] Jurriaan Beckers, Gadi Rothenberg. Sustainable selective oxidations using ceria-based materials [J]. GreenChem., 2010,12: 939-948.
[5] Peng Yang, Shanshan Yang, Zhinan Shi, Zhonghua Meng, Renxian Zhou. Deep oxidation of chlorinated VOCs over CeO2-based transition metal mixed oxide catalyst [J]. Applied Catalysis B: Environmental. 2015,162:227-235.
[6] Yu Zhang, Fei Hou, Yiwei Tan. CeO2 nanoplates with a hexagonal structure and their catalytic applications in highly selective hydrogenation of substituted nitroaromatics [J]. Chem. Commun., 2012, 48: 2391-2393.
[7] Yu Ren, Zhen Ma and Peter G. Bruce, Ordered mesoporous metal oxides: synthesis and applications [J]. Chem Soc Rev, 2012; 41(14): 4909-4927.
[8] V. K. Ivanova, O. S. Polezhaevaa, and Yu. D. Tret'yakovb. Nanocrystalline Ceria: Synthesis, Structure-Sensitive Properties, and Promising Applications [J]. Russian Journal of General Chemistry, 2010, 80(3): 604-617.
[9] R. J. La, Z. A. Hu, H. L. Li, X. L. Shang and Y. Y. Yang, Template synthesis of CeO2 ordered nanowire arrays [J], Mater. Sci. Eng., A, 2004, 368: 145–148.
[10] C. W. Sun, H. Li, Z. X. Wang, L. Q. Chen and X. J. Huang, Synthesis and characterization of polycrystalline CeO2 nanowires [J], Chem. Lett., 2004: 662–663.
[11] C. W. Sun, H. Li, H. R. Zhang, Z. X. Wang and L. Q. Chen, Controlled synthesis of CeO2 nanorods by a solvothermal method [J], Nanotechnology, 2005, 16: 1454–1463.
[12] C. C. Tang, Y. Bando, B. D. Liu and D. Golberg, Cerium oxide nanotubes prepared from cerium hydroxide nanotubes [J], Adv. Mater., 2005, 17: 3005–3009.
[13] T. Yu, B. Lim and Y. Xia, Aqueous-phase synthesis of single-crystal ceria nanosheets [J], Angew. Chem., Int. Ed., 2010, 49: 4484–4487.
[14] N. Ozer, Optical properties and electrochromic characterization of sol–gel deposited ceria films [J], Sol. Energy Mater. Sol. Cells, 2011,68: 391-400.
[15] D. Wang, Y. Kang, V. Doan-Nguyen, J. Chen, R. K€ungs, N. L. Wieder, K. Bakhmutsky, R. J. Gorte and C. B. Murray, Synthesis and oxygen storage capacity of two-dimensional ceria nanocrystals [J], Angew. Chem., Int. Ed., 2011, 50: 4378–4381.

Preparation and physiological activities of carboxymethylated derivative purified from corn bran

Linghui Zhu[1], Miaoli Fang[2], Jianjun Ma[2], Qing Mo[2,*]

[1]Zhejiang Chinese Medical University, Hangzhou, Zhejiang, China.
[2]Hangzhou Medical College, Hangzhou, Zhejiang, China.

[*]Corresponding author e-mail: moqing2006@163.com

Abstract. Two water-soluble polysaccharides extracted from corn bran were chemically modified to obtain their carboxymethylated derivatives (C-CBP1, C-CBP2). Theresults of degree of substitution and FT-IR analysis showed the carboxymethylation of polysaccharides were successful. The average molecular weight (Mw) of C-CBP1 and C-CBP2 were 368 and 263kDa, respectively. The degree of substitution (DS) of C-CBP1 and C-CBP2 were determined to be 0.44 and 0.46. The results showed that derivatives were effective in antioxidant and bile acidbinding activityin a dose dependent way. And C-CBP2 had the higher activity for hydroxyl radical, superoxide anion scavenging activities and bile acid capacity, as lower molecular weight plays a critical role in antioxidant activities and bile acid capacity. The results suggest that the carboxymethylated derivatives are potential natural antioxidant and blood fat reduce agent that can be used as drugs or functional food ingredients.

1. Introduction

Corn bran polysaccharides (CBP), derived the main component of corn bran and/or endosperm tissues, is a water-soluble polysaccharide and has significant influence on human health. In recent years, many studies have focusedon finding more valuable properties of CBP including adhesive, thickening and stabilizing, and film forming and emulsifying [1-3]. However, to the best of our knowledge, there are no available references on the antioxidant activity of the carboxymethylated corn fiber polysaccharide. Our previous study reported that carboxymethylation could enhance the solubility of pachyman [4], thus resulting in the improvement of the activities. In this case, chemical modification of carboxymethylation of corn fiber polysaccharides was carried out.

2. Materials and methods

2.1. Materialsand reagents

Corn bran was collected in Hangzhou,China. The materials were de-oil, de-starched in accordance with Yadav *et al.*[3]. All the reagents were purchased from Aladdin Industrial Co. and used without further purification.

2.2. Preparation of polysaccharide and its carboxymethylated derivatives

2.2.1. Preparation of Corn Bran polysaccharide (CBP). Corn bran polysaccharide (CBP) was extracted from the de-oiled and de-starched corn bran following the instructions provided by alkaline hydrogen peroxide procedure [3].Then, CBP was separated by the column of DEAE Sepharose Fast Flow (2.6cm×60cm) equilibrated with distilled water [5]. CBP were isolated, then concentrated, dialyzed and lyophilized, respectively.

2.2.2. Preparation of Carboxymethylated derivatives. The carboxymethylation of CBP was carried out on the previous researches [3]. The carboxymethylated derivatives were obtained.

2.3. Characterization of carboxymethylated derivatives

2.3.1. Compositional analysis. The carbohydrate content of all the CBP samples obtained above were analyzed by the phenol-sulphuric acid method [5]. Protein was determined by Coomassie Brilliant Blue method [6]. The samples were dissolved to 1mg/mL solution and scanned with an ultraviolet spectrophotometer (U-1900UV, Hitachi high-Technologies Corp., Japan).

2.3.2. Determination of carboxymethyl content. The degree of substitution (DS) of carboxymethylated derivatives was determined by neutralization titration with a few modifications [4].

2.3.3. Homogeneity and molecular weight determination. Molecular weight (Mw) and molecular size (Mn) of the purified fractions were determined by size exclusion chromatography with laser light scattering (SEC-LLS) [4].

2.3.4. Infrared spectrum analysis. Fourier transform infrared spectra (FT-IR) of the derivatives were obtained byNicolet FT-IR spectrometer (Magna-IR 760 E.S.P, Nicolet Instrument Corp., Madison, WI). Samples were ground with potassium bromide (KBr) at a ratio of 1:20 and then pressed into pellets for transformation infrared spectra measurement in a frequency range of 4000-500 cm^{-1}.

2.4. Antioxidant activity

2.4.1. 1,1-Diphenyl-2-picryhydrazyl (DPPH) radical scavenging assay. DPPH radical scavenging activity was determined as the procedure [7]. The DPPH radical-scavenging activity (%) was calculated by the following equation:

$$\text{Scavenging activity (\%)}=(1-A_1/A_0)\times 100\% \tag{1}$$

Where A_0 was the absorbance of the control (without sample) and A_1 was absorbance of the samples.

2.4.2. Hydroxyl radical scavenging activity. The hydroxyl radical scavenging activity was analyzed according to our previous researcher [4].

$$\text{Scavenging activity (\%)}=(1-(A_1-A_2)/A_0)\times 100\% \tag{2}$$

Where A_0 was the absorbance of the control (without sample) and A_1 was the absorbance in the presence of the sample, A_2 was the absorbance without sodium salicylate.

2.4.3. Superoxide radical-scavenging activity. The assay was based on the capacity of the sample to inhibit the photochemical reduction of nitroblue tetrazolium in NADH-NBT-PMS according toourpreviousresearch [4].

$$\text{Scavenging activity (\%)}=(1-A_1/A_0)\times 100\% \quad (3)$$

Where A_0 was the absorbance of the control (without sample) and A_1 was absorbance in the presence of the sample.

2.5. Bile acid binding capacity

Based on the method of our previous research [4], the effect of the samples on the bile acid capacity was investigated *in vitro*.

Statistical Analysis

The results were means ± standard deviation (SD) of determinations of triplicate cultures. The data were analyzed by one-way ANOVA and a level of $P<0.05$ was regarded as statistically significant with Duncan's new multiple ranges.

3. Result and discussion

3.1. Chemical analysis of C-CBP

Table 1. Carbohydrate content, Molecular weight (M*w*) and DS of CBP and C-CBP

	carbohydrate content (%)	Protein (%)	Mw/kDa	DS
CBP1	91.6	0.97	332	/
CBP2	93.3	1.50	187	/
C-CBP1	86.5	0.82	368	0.44
C-CBP2	84.6	1.46	263	0.46

The DS, carbohydrate content and molecular weight of CBP and its derivatives are listed in Table 1. Compared withthe native sample, not only does the carbohydrate content of CBP and C-CBP decreased significantly ($p < 0.05$) but also themolecular weight was increased, which was in accordance withthe results of Wang*et al.*[4]. The DS of the C-CBP1 and C-CBP2 was 0.44, 0.46 respectively. However, their carbohydratecontent and molecular weight listed in a converse order, whichindicated that a slight degradation might occur in the process ofC-CBP preparation.

3.2. IR Spectra of CFP and its derivatives

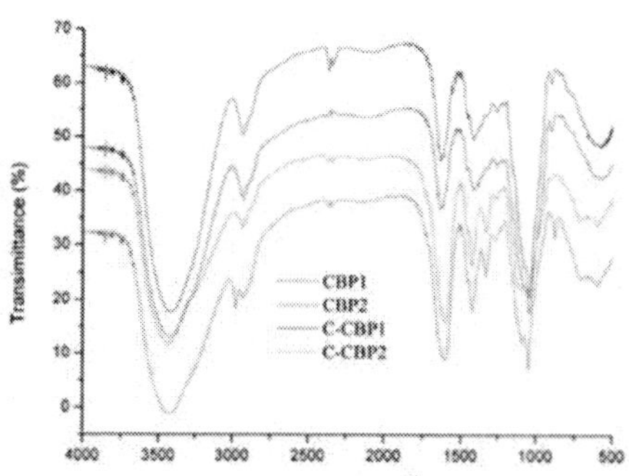

Fig.1. FT-IR spectrum of CBP1, CBP2, C-CBP1 and C-CBP2

The FT-IR spectra of the crude polysaccharide (CBP1,CBP2) and itsderivatives (C-CBP1, C-CBP2) are shown in Fig.1. The intensity of bands around 3375.76 cm^{-1} in the IRspectrum was due to the hydroxyl stretching vibration of thepolysaccharide and as expected they were broad. The absorption bands at 2920cm^{-1} were attributed to the stretching vibration of methylene group[8]. After carboxymethylation, two new bands at 1602 cm^{-1} [$\gamma_{sym}(COO^{-})$] and 1426cm^{-1} [$\gamma_{as}(COO^{-})$] were observed[8]. The results were similar to the observations reported in the carboxymethyl derivatives of Ganodermalucidum [9] and Psyllium arabinoxylan [10]. Thus, the native polysaccharide was carboxymethylated successfully without change of the molecular structure in this study.

3.3. Antioxidant activities

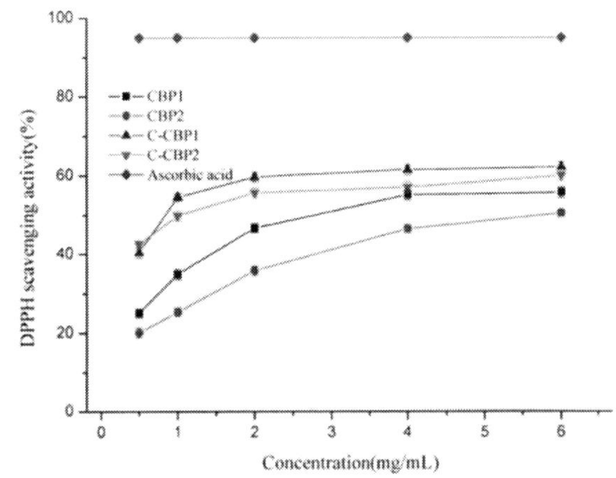

Fig.2. DPPH radical-scavenging rates of CBP1, CBP2, C-CBP1 and C-CBP2. Each point is the mean ± SD of triplicates.

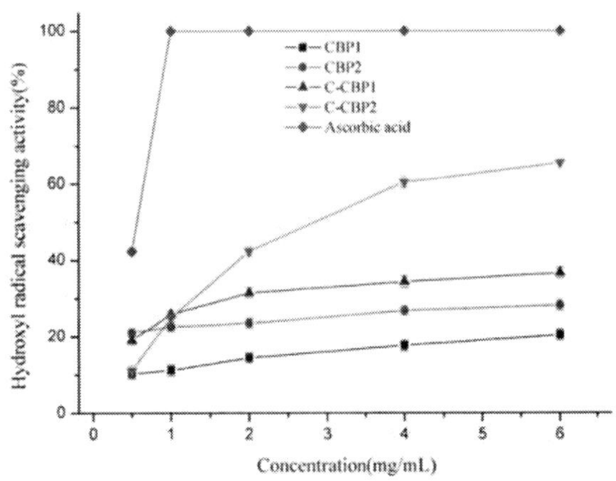

Fig.3. Hydroxyl radical-scavenging activity of CBP1, CBP2, C-CBP1 and C-CBP2. Each point is the mean ± SD of triplicates.

As shown in Fig. 2, C-CBP1 and C-CBP2 exhibited higher scavenging activitythan the native polysaccharide through etherifying hydroxyl groups with carboxymethyl groups. When the concentration of samples was 6.0mg/mL, DPPH free radical scavenging activities were 55.67±1.37%, 50.41±0.94%, 62.07±1.25%, and 59.97±1.29% for CBP1, CBP2, C-CBP1 and C-CBP2, respectively. This result demonstrated the importance of the carboxymethyl group in antioxidant ability, especiallyat a high M*w*. The carboxymethylated derivatives showed excellent scavenging activity on DPPH radical, which might be attributable to its strong hydrogen-donating ability by activating the hydrogen atom of the anomeric carbon [11].

Hydroxyl radical is considered as one of the most reactive oxygen radicals, which can easily cross cell membranes, react with most biomolecules and lead to tissue damage or cell death [13]. Fig. 4 showed that the hydroxyl radical scavenging ability of C-CBP had concentration dependencealthoughweaker than vitamin C in the same concentrations. It increasedto 36.64±1.35%, 65.47±0.42%of C-CBP1 and C-CBP2 when the concentration increased to 6 mg/mL. Our study showed that scavenging hydroxylradical ability of the carboxymethylated derivatives were improved compared native polysaccharides.This is in agreement with the results in which carboxymethylated derivatives from *Auriculariaauricular* [14]. For hydroxyl radical, there were two types of antioxidant mechanisms[9]: One mechanism was that the effect of themetal complexes led to the suppression against hydroxyl radicalgeneration, and the other was the direct clearance of the generatedhydroxyl radical. The mechanism of C-CBPs on cleaning hydroxyl radicals needed to befurther investigated.

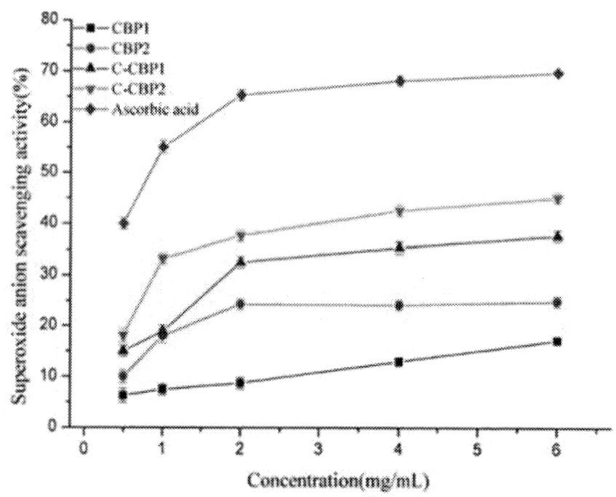

Fig.4. Superoxide anion-scavenging activity of CBP1, CBP2, C-CBP1 and C-CBP2. Each point is the mean ± SD of triplicates.

Superoxide anion radical was easily transformed from molecular oxygen, and cause oxidative damage in human body [10]. Fig.5 showed both of the native and carboxymethylated derivative had a concentration-dependent increase in the superoxide anion radical scavenging activity. In the concentration of 2.0 mg/mL, the scavenging of superoxide radicals of C-CBP1 and C-CBP2 were 32.51±1.08%, 37.67±1.08%, respectively. In the present study, comparing with the original CBP, the scavenging activity of C-CBP against superoxide anion moderately improved. The addition of electron-donating substituents probably increased radical scavengingactivity as a result of increasing electron density on the heterocyclicring of the carbons [12, 15]. After modification, carboxylic groups were introduced into the polysaccharide`s chains, which presented C-CBPwith strong electron-withstanding ability and enhanced thesuperoxide scavenging activity.

3.4. Bile acid binding capacity

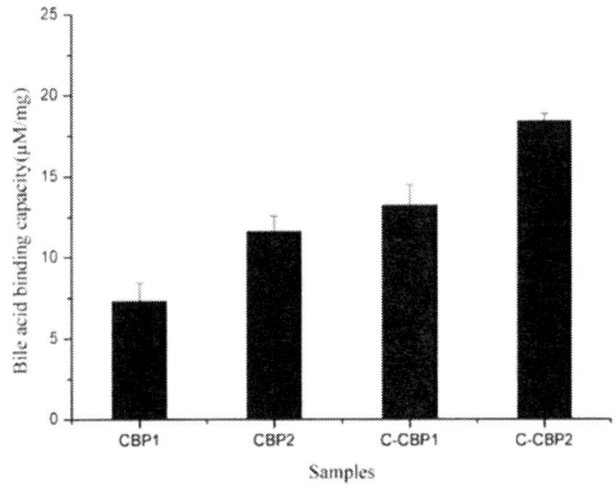

Fig.5. Bile acid binding capacity of CBP1, CBP2, C-CBP1 and C-CBP2. Each point is the mean ± SD of triplicates.

The bile acid binding capacities of the samples were presented at Fig.6. Compared with the native polysaccharide, the bile acid binding capacity became more pronounced after carboxymethylation. Particularly, the C-CBP2 showed the best capacity with18μmol/mg, indicating the molecular weightand the substituent group played an important role in the bile acid binding capacity. Therefore, better functionality of carboxymethylated polysaccharides to bind bile acids *in vitro* might be partly explained by their improved molecular weight [16].

4. Conclusion

In the present study, two water-soluble polysaccharide (CBP) extracted from corn fiber were modified successfully to obtain its carboxymethylated derivatives. In general, the carboxymethylation enhanced the antioxidant and bile acid binding activities in a dose-dependent way. Higher antioxidant activities were found for the derivatives with C-CBP2 with lowermolecular weight. The research will promote comprehensive utilization of the corn bran and further researchers need to be focused on. It is noteworthy that C-CBP should be explored as a novelpotential antioxidant after its further research on the safety for humanconsumption and antioxidant activity *in vivo*.

Acknowledgments

This work were financially supported byZhejiang Medical College fund (No. 2015XZA03) and Zhejiang Chinese Medical University fund (No. 2015ZY30).

References

[1] M.P. Yadav, M.L. Fishman, H.K. Chau, D.B. Johnston, K.B. Hicks, Molecular characteristics of corn fiber gum and their influence on its emulsifying properties. Cereal Chemistry, 2007(2): 175-180.

[2] M.P. Yadav, D.B. Johnston, K.B. Hicks, Structural characterization of corn fiber gum from coarse and fine corn fiber and a study of their emulsifying properties. Journal of Agricultural and Food Chemistry, 2007(55): 6366-6371.

[3] M.P. Yadav, N. Parris, D.B. Johnston, C.I. Onwulata, K.B. Hicks, Corn fiber gum and milk protein conjugates with improved emulsion stability. CarbohydratePolymers, 2010(2): 476-483.

[4] Y.J. Wang, Q. Mo, Z.N. Li, H.W. Lai, J. Lou, S.W. Liu, J.W. Mao, Effects of degree of carboxymethylation on physicochemical and biological properties of pachyman. International Journal of Biological Macromolecules, 2012(51): 1052-1056.
[5] X.M. Wang, Z.S. Zhang, M.X. Zhao, Carboxymethylation of polysaccharides from Tremellafuciformis for antioxidant and moisture-preserving activities. International Journal of Biological Macromolecules, 2015(72): 526-530.
[6] M. Dubois, K.A. Gilles, J.K. Hamilton, P.A. Rebers, F. Smith, Colorimetricmethod for determination of sugars and related substances.Anal.Chem. 1956(29): 350-356.
[7] Y.L. Xiang, X.Q. Xu, J. Li, Chemical properties and antioxidant activity of exopolysaccharides fractions from mycelial culture of Inonotus obliquus in a ground corn stover medium. Food Chem. 2012(134): 1899-905.
[8] H.M. Qi, Q.B. Zhang, T.T. Zhao, R.G. Hu, K. Zhang, Z.E. Li, In vitro antioxidant activity of acetylated and benzoylated derivatives of polysaccharide extracted from Ulva pertusa (Chlorophyta). Bioorg. Med. Chem. Lett, 2006(16): 2441-2445.
[9] Y. Chen, M.Y. Xie, S.P. Nie, C. Li, Y.X. Wang, Purification, composition analysis and antioxidant activity of a polysaccharide from the fruiting bodies of Ganodermaatrum. Food Chemistry, 2008(107): 231-241.
[10] A. Banerjee, N. Dasgupta, B. De, In vitro study of antioxidant activity of Syzygiumcumini fruit. Food Chemistry, 2005(90): 727-733.
[11] Z.Y. Guo, R.E. Xing, S. Liu, H.H. Yu, P.B. Wang, C.P. Li, P.P. Li, The synthesisand antioxidant activity of the Schiff bases of chitosan and carboxymethylchitosan. Bioorganic & Medicinal Chemistry Letters, 2005(15): 4600-4603.
[12] E. Tsiapali, S. Whaley, J. Kalbfleisch, H.E. Ensley, I.W. Browder, D.L. Williams, Glucans exhibit weak antioxidant activity, but stimulate macrophagefree radical activity. Free Radical Biology and Medicine, 2001.(30): 393-402.
[13] W. Liu, B.C. Zhang, Q. Wang, Z.H. Xie, W.B. Yao, X.D. Gao, L. Yu, Effects of sulfation on the physicochemical and functional properties of psyllium. Journal ofAgricultural and Food Chemistry, 2010(58): 172-179.
[14] L.Q. Yang, T. Zhao, H. Wei, M. Zhang, Y. Zou, G.H. Mao, X.Y. Wu, Carboxymethylation of polysaccharides from Auriculariaauricula and theirantioxidant activities in vitro. International Journal of Biological Macromolecules, 2011(5): 1124-1130.
[15] L.Q. Yang, T. Zhao, H. Wei, M. Zhang, Y. Zou, G.H. Mao, X.Y. Wu, Carboxymethylation of polysaccharides from Auriculariaauricula and theirantioxidant activities in vitro. International Journal of Biological Macromolecules, 2011.(5): 1124-1130.
[16] W. Liu, B.C. Zhang, Q. Wang, Z.H. Xie, W.B. Yao, X.D. Gao, L. Yu, Effects of sulfation on the physicochemical and functional properties of psyllium. Journal of Agricultural and Food Chemistry, 2010(58): 172-179.

IOP Conf. Series: Materials Science and Engineering **207** (2017) 012003 doi:10.1088/1757-899X/207/1/012003

Preparation of micelle supported magnetic hydroxylated multi-walled carbon nanotubes based DSPE for determination of PAHs

Mingyu Wang[1, 2], Shaojun Zhang[2, *], Xiao Zhang[1], Deyu Li[2]

[1]Ocean School, Yantai University, 32 Qingquan Road, 264005 Yantai, China
[2]College of Naval architecture and Marine Engineering, Shandong Jiaotong University, 1508 Hexing Road, 264310, Weihai, China

[*]Corresponding author e-mail: 15163118785@163.com

Abstract. A micelle supported Fe_3O_4 magnetic nanoparticles decorated hydroxylated multi-walled carbon nanotubes material was synthesized. The material was facilely synthesized between carbon nanotubes and Fe^{2+}. The synthesized nanomaterial served as an excellent support for micelles, exhibiting high loading capacity and selectivity. The prepared material used in dispersive solid-phase extraction (DSPE) for investigation of gaseous phase polycyclic aromatic hydrocarbons (PAHs) emitted from marine diesel engine for the first time. The application showed good response (R_2 > 0.9981) in the range of 0.02 – 1.0 μg/L, satisfactory reproducibility (variation less than ± 10%) and high precision. Limits of detection of sixteen PAHs ranged from 0.009 to 0.018 μg/L (S/N=3). The spiked recovery of proposed method (72.65-96.54 %) was 1.01 - 2.32 times higher than that of the conventional method. The enrichment factors reached to 39.65-121.32 that exhibited good enrichment ability.

1. Introduction

Polycyclic aromatic hydrocarbons (PAHs) are ubiquitous environmental pollutants that formed from both natural and anthropogenic sources with mutagenic and carcinogenic properties. Because of their hydrophobicity, PAHs and its metabolites tend to be interacted with DNA, causing malignancies and heritable genetic damage in humans [1]. Therefore, close attention has been paid to the research of PAHs, especially determination gaseous phase PAHs emitted from diesel engine for atmospheric pollutant control [2].

Directly determination of PAHs in environmental samples using analytical instrument is still challenging because of the extremely low concentrations and complex unknown interferences [3]. Therefore, pre-concentration is required. Modern pretreatment technique dispersive solid-phase extraction (DSPE) greatly simplified pre-concentration and cleaning process that overcomes the limitations of the classical techniques. Many researches focused on synthesis or modification of adsorbent nanomaterials such as nanofibers, graphene, glassy carbon, activated carbon, multi-walled carbon nanotubes (MWCNTs). Among them, MWCNTs is considered as a most widely used adsorbent materials [4].

There have no report for determination of PAHs using MWCNTs as adsorbent; the reason lies in that the selectivity of the MWCNTs is low for direct adsorption the PAHs from complex environmental samples. In order to selectively extract specific pollutants, MWCNTs can be covalently bonded with hydroxyl groups (-OH) and then packed into a cartridge as SPE mode. SPE mode overcomes the separation step of MWCNTs-OH from the solvent, but the entire process is still time consuming and using large amount of organic solvents that harmful to human body and environment. Thus, magnetite/MWCNTs-OH (MMWCNTs-OH) composite were prepared by electrostatic self-assembly as so to be easily collected by magnet thus greatly simplify the extraction process [5]. Finally, the extraction efficiency of MMWCNTs-OH composite is still need to be improved, and this problem could solved by combining micelles extraction technique [6]. In the current study, a surfactant modified magnetic hydroxylated multi-walled carbon nanotubes (MMWCNTs-OH) adsorbent was prepared and used as a platform for the formation of micelle in DSPE. Extraction of MMWCNTs-OH was performed by applying an external magnetic field. This method used in determination of gaseous phase PAHs emitted from marine diesel engine for the first time that greatly simplifies the extraction process.

2. Experimental

2.1. Synthesis of MMWCNTs-OH

MWCNTs-OH and MMWCNTs-OH were prepared according to the method of Ling and Zhang with some modification [7]. The flow chart was shown in Fig. 1. $FeCl_3 \cdot 6H_2O$ (0.060 g) and $FeCl_2 \cdot 4H_2O$ (0.020 g) were dissolved in 40 mL of deionized water. The solution was constantly stirred in a 100 mL flask. Subsequently, MWCNTs-OH (0.100 g) was added into the mixed solution followed by the slow addition of 30 mL (5%, v/v) ammonium hydroxide solution at 80 °C, and the mixture was stirred for 60 min. The whole reaction process was performed under nitrogen gas protection. Thereafter, the black precipitates formed (MMWCNTs-OH) were separated from the reaction medium via centrifugation at 2000 r min^{-1} and then washed with deionized water several times until the pH close to neutral. The MMWCNTs was collected by a magnet and dried under vacuum.

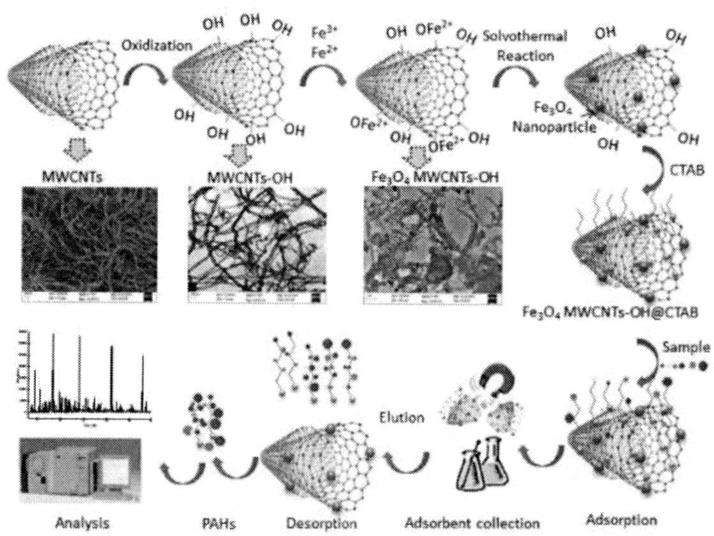

Figure 1. Scheme diagram of MMWCNTs-OH synthesis and DSPE procedures.

2.2. Sample preparation

PAHs emitted from diesel engine were collected from a RT-flex60c marine diesel engine (Qingdao Wartsila Corporation, China) according to MARPOL 73/78 annex VI. A stainless steel joint connecting tube was welded on the tail end of the exhaust pipe and connected with a brass tube. The diameter of the diesel engine exhaust pipe was 54 mm, and the internal diameter of brass tube was 5.0 mm. Each brass tube connected with a 250 mL aqueous glass containers for collection of exhaust emissions without diluting with air. The optimized aqueous solution in the containers including, sodium chloride (0.01mol/L) and 5% ethanol: *n*-hexane (3:1, v/v). Aqueous capture was collected for 10 min. The same batches of samples were used in the experiment.

In the sample extraction process, a micelle supported nanomaterial MMWCNTs-OH was prepared. Adjust the sample solution to pH3 by hydrochloric acid (HCl) firstly. Then, measured 100 mL sample solution and adding 1.0 mL MMWCNTs-OH solution (concentration 1.0 mg/mL) containing CTAB (1.0 mg/mL) to it. Extraction time was twenty minutes. A magnet was used for collecting MMWCNTs-OH material. The desorption solution using acetone at the volume of 1.0 mL. Finally, the obtained effluent was detected in a GC/MS-QP 2010 Plus.

3. Results and discussion

3.1. Characterization of MMWCNTs-OH

The SEM image of MMWCNTs-OH was shown in Fig. 1. Fe_3O_4 nanoparticles were observed evenly depositing on the surface of MWCNTs. There was no obvious aggregation observed. The diameter of Fe_3O_4 magnetic particles was about 10 nm. The SEM morphology of MMWCNTs-OH@CTAB adsorbent confirmed well decorated on the platform. This synthesis used oxidation-reduction reaction, which could not introduce impurities into the adsorbent after the reaction.

VSM was used for determination of magnetic properties of MMWCNTs-OH. As shown in Fig. 2a, the saturation magnetization of MMWCNTs-OH reached to the maximum at the value of 12.58 emu/g. The highest value was sufficient for determination of the target object. MMWCNTs-OH was rapidly dispersed after exposure to the external magnetic field. MMWCNTs-OH showed super paramagnetic that made it a good adsorbent. This phenomenon was caused by MMWCNTs-OH has abundant functional groups hydroxyl ions (-OH). The hydroxyl ions improved the proposed material better dispersion in solution. Moreover, the hydroxyl ions were susceptible to provide platform for micelles. The experiment studied the magnetic properties of MMWCNTs-OH@CTAB. After loaded with CTAB, the highest saturation magnetization was at the value of 12.19 emu/g, which was a little lower than the value of MMWCNTs-OH. But 12.19 emu/g was still sufficient for determination of object compounds in the sample solution. We could deduce from the result that MMWCNTs-OH played a more important role than MWCNTs-OH in magnetic properties of MMWCNTs-OH@CTAB.

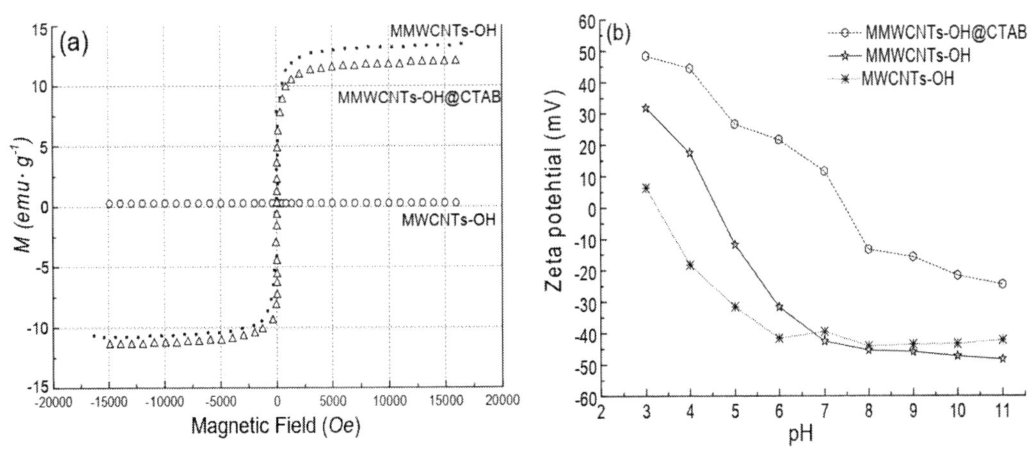

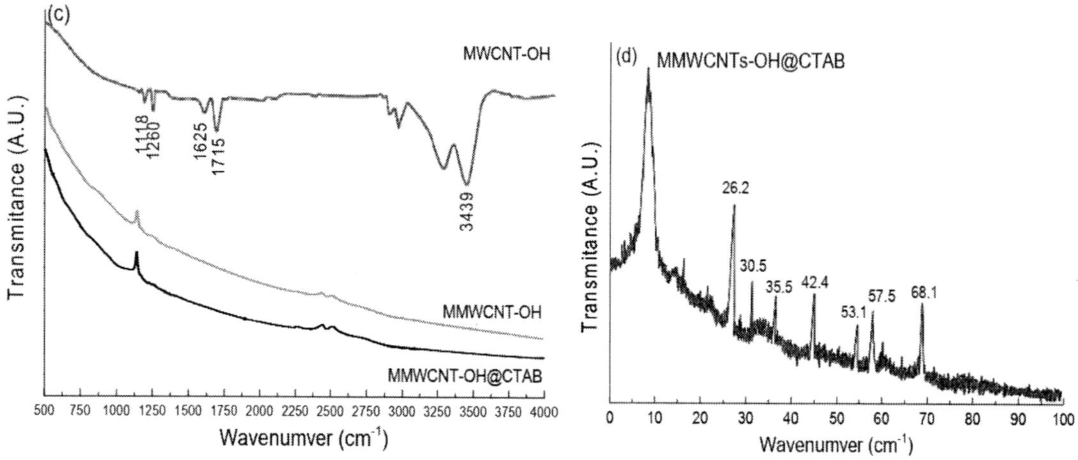

Figure 2. (a) Magnetic hysteresis loop; (b) Zeta potentials; (c) FTIR spectroscopy; (d) XRD spectroscopy of MMWCNTs-OH@CTAB sorbents

Zeta potential means the potential of shear plane, which is an important index to characterize the stability of colloidal dispersion system. As values of Zeta potential have relative with the stability of colloidal dispersion, Zeta potentials under pH 3-11 conditions were compared among MWCNTs-OH, MMWCNTs-OH and MMWCNTs-OH@CTAB (Fig. 2b). Zeta potential of MWCNTs-OH was 6.35 mV at pH3. Zeta potential reduced to -39.46 mV at pH7.0. Zeta potential did not show significant changes when solution conditions increased to pH11.0. This phenomenon indicated MWCNTs-OH dispersed well in alkaline conditions. MMWCNTs-OH was negatively charged in the solution at the range of pH5.0 to pH11.0. The highest Zeta potential of MMWCNTs-OH@CTAB solution only reached 24.52 under alkaline conditions, which indicated its unstable dispersed. But the solution was stable dispersed under acidic conditions with Zeta potential reached 48.42, Therefor, MMWCNTs-OH@CTAB in acidic conditions was used for DSPE.

Figure 2c showed the vibrational band assignments (cm^{-1}) of the MWCNTs-OH: 1118 cm^{-1} (C-O), 1260 cm^{-1} (=C-O-C-OH), 1625 cm^{-1} (=C-O-H), 1715 cm^{-1} (C=O), 3439 cm-1 (O-H). The comparison among MWCNTs, MWCNTs-OH and MMWCNTs-OH indicated that the MWCNTs can be successfully modified after the oxidation-reduction reaction, and the hydroxyl functional groups were introduced to MWCNTs-OH and MMWCNTs-OH.

Fig.2d showed the XRD patterns of MMWCNTs-OH@CTAB. Characteristic peak of MWCNTs was observed at 2 θ of 26.2°. Iron oxide mainly includes Fe_3O_4(Mn), γ -Fe_2O_3(Mh), α -Fe_2O_3(H) and FeO(OH). Among them, only Fe_3O_4 and γ - Fe_2O_3 are magnetic. Four characteristic peaks at 2 θ of 30.5°, 35.5°, 42.4°, 57.5° represented magnetic Fe_3O_4 or γ -Fe_2O_3. Peaks at 2 θ of 53.1°, 68.1°, which illustrated there still have α -Fe2O3 existed in MMWCNTs-OH@CTAB.

3.2. Method validation

The details of GC/MS mass spectrometry parameters for determination of PAHs were listed in Table 1. Table 1 summarized the results of method validation, which showed the proposed method achieved a high precision. Satisfactory correlation coefficient (R^2 > 0.9981) was obtained when PAHs concentration at the range of 0.02 – 1.0 μg/L. Values for the limit of detection (LOD, S/N≥3) and limit of quantitation (LOQ) of MMWCNTs-OH@CTAB DSPE method were in the ranges of 0.009 - 0.018 μg/L and 0.01 - 0.020 μg/L. The corresponding relative standard deviation (RSD) was in the range of 4.23-9.96%.

Table 1. Method validation of MMWCNTs-OH@CTAB as adsorbent for DSPE followed by GC/MS analysis

PAHs	Linear range (μg/L)	R^2	LOD (μg/L)	Recovery (%)	RSD (%)	variability (μg/m^3)
NAP	0.02-1.0	0.9981	0.018	72.65	6.98	12.86±2.4
ACE	0.02-1.0	0.9995	0.016	75.36	8.26	2.78±1.2
ACT	0.02-1.0	0.9990	0.018	74.23	8.26	4.85±1.8
FLU	0.02-1.0	0.9990	0.015	80.29	4.65	11.14±2.5
PHE	0.02-1.0	0.9994	0.013	82.12	9.96	4.71±1.5
ANT	0.02-1.0	0.9992	0.016	83.54	6.54	1.51±0.9
FLT	0.02-1.0	0.9998	0.010	81.06	7.58	2.56±0.8
PYR	0.02-1.0	0.9997	0.012	79.93	8.69	0.21±0.1
BaA	0.02-1.0	0.9994	0.014	84.56	5.64	9.69±2.1
CHR	0.02-1.0	0.9991	0.010	86.11	4.56	1.48±0.3
BbF	0.02-1.0	0.9995	0.009	96.54	6.36	0.55±0.2
BkF	0.02-1.0	0.9994	0.010	89.63	5.25	2.03±0.4
BaP	0.02-1.0	0.9991	0.009	89.20	6.24	0.95±0.4
IcP	0.02-1.0	0.9998	0.10	94.99	4.23	2.36±0.5
DaA	0.02-1.0	0.9994	0.10	90.64	4.26	1.60±0.1
BgP	0.02-1.0	0.9993	0.10	91.57	3.58	0.65±0.2

Repeated measurements of a group of 3 times, a total of 3 days (9 groups), and the arithmetic mean value was taken as measurement results of the group. The measurement results in the difference between maximum and minimum values as the stability of the measuring instruments in this period of time. As is shown in Table 1, the inter-day precision for 3 days showed a variation of less than ± 10%. The enrichment factors of the PAHs were in the range of 39.65-121.32. The results showed that the MMWCNTs-OH@CTAB DSPE method was reliable for detection PAHs at very low concentrations.

3.3. Application of the method to the analysis of real samples

Aqueous-capture-collection, DSPE and GC/MS method under SIM mode for an exhaust emissions sample added 16 PAHs (10 μg/L) as spiked standard samples to evaluate recoveries. The spiked recoveries were 72.65-96.54 %, and the corresponding RSD was in the range of 4.23-9.96%. (Table 1). The analysis time was 40 minutes and there were no noticeable water effects in the chromatogram. Practical analysis of PAHs emitted from exhaust pipe included 16 main PAHs, and the total amount was 62.36 μg/m3. Analysis results showed that all target PAHs dissolved in aqueous solution, even the largest one. The proposed method also captured and detected the heaviest of PAHs such as chrysene and indeno(1,2,3-cd)pyrene with ideal recovery yields. A clean separation and good chromatogram were readily achieved without the presence of sample matrix interference. Quantitative analysis of PAHs cost less than 30 min from the sample collection to detection, and the detection limits reached to nanogram grade, which illustrated the method achieved a high precision. This optimized method was also compared with traditional methods. As the collection and pretreatment time was only 21 min, it was less than 10% of the traditional methods. This demonstrated that the proposed method was comparable for collection and concentration of PAHs.

4. Conclusion

In this work, MMWCNTs-OH was synthesized with a simple procedure, and the loading capacity was high. MMWCNTs-OH can be charged and supported as anionic micelle. The MMWCNTs-OH with micelles showed higher carrying capacity than Fe3O4 magnetic nanoparticles material. The recovery of target compounds increased about 1.01 - 2.32 times than SPE method. The method was suitable for analyzing PAHs emitted from marine diesel engine. A traditional method for collection and extraction of PAHs was also performed, and the results showed that in the traditional process lost 27.52%

compounds. As the sample collection time was short from sampling to analysis, which cost less than 10% time of the traditional method so as to decrease the PAHs losses to minimum.

Acknowledgments
This research was supported by the funds of Shandong Provincical Higher Educational Science and Technology Foundation, China (J14LD04; J16LA61), Shandong Provincial Natural Science Foundation, China (ZR2014JL038), National Natural Science Foundation of China (51609131).

References
[1] C. Gao, Z.N. He, J. Li et al. Specific long non-coding RNAs response to occupational PAHs exposure in coke oven workers. Toxicology Reports. (2016) 160-166.
[2] T.P. Hu, J.Q. Zhang, C. Ye et al. Status, source and health risk assessment of polycyclic aromatic hydrocarbons (PAHs) in soil from the water-level-fluctuation zone of the Three Gorges Reservoir, China. Journal of Geochemical Exploration. 172 (2017) 20-28.
[3] Z. Nyiri, M. Novák, Z. Bodai et al. Determination of particulate phase polycyclic aromatic hydrocarbons and their nitrated and oxygenated derivatives using gas chromatography - mass spectrometry and liquid chromatography - tandem mass spectrometry. Journal of Chromatography A. 1472(2016) 88-98.
[4] M.H. Dehghani, Z.S. Niasar, M.R. Mehrnia et al. Optimizing the removal of organophosphorus pesticide malathion from water using multi-walled carbon nanotubes. Chemical Engineering Journal. 310 (2017) 22-32.
[5] C. Su. Environmental implications and applications of engineered nanoscale magnetite and its hybrid nanocomposites: A review of recent literature. Journal of Hazardous Materials, Part A, 2017, 322: 48-84.
[6] M. Hryniewicka, B. Starczewska, The usage of micellar extraction for analysis of fluvastatin in water and wastewater samples. J. Pharm. Biomed. Anal. 106 (2015) 129-135.
[7] H.F. Zhang, Y.P. Shia. Preparation of Fe3O4 nanoparticle enclosure hydroxylated multi-walled carbon nanotubes for the determination of aconitines in human serum samples. Analytica Chimica Acta. 724 (2012) 54– 60.

IOP Conf. Series: Materials Science and Engineering **207** (2017) 012004 doi:10.1088/1757-899X/207/1/012004

Controlled synthesis of α-Al_2O_3 via the hydrothermal-pyrolysis method

Zhao Li, Kunyao Wu, Jing Cao, Yongfeng Wang

School of Materials Engineering, Xi'an Aeronautical University, Xian 710077, China

Abstract. Taking aluminum sulfate and urea as the raw materials produce α-Al_2O_3 by employing the hydrothermal-pyrolysis method. The study analyzes the characterization of the products by XRD and SEM, The results indicate as follows: after 6 hours' hydrothermal reaction in the 120℃ water, with the aluminum sulfate and urea as the raw materials, spherical α-Al_2O_3 can be obtained through calcination at 1200℃.

1. Introduction

There are many crystal forms of Al_2O_3, while α-Al_2O_3 has a regular form but low reaction capacity and insoluble in acid and alkali. α-Al_2O_3 is widely applied in various fields, such as ceramics, fireproofing materials, grinding and polishing, filling materials, optics and electronics. α-Al_2O_3 in different shapes have different performances and application values. α-Al_2O_3 in fiber-shape boasts a good thermal stability, abrasive resistance and oxidative stability. Meanwhile, it enjoys advantages like high tenacity, high strength, high insulation and high dielectric constant; therefore, it has been widely used in insulation materials, fiber protection and reinforcing materials, etc. As for the tabular-shaped α-Al_2O_3, with its two-dimensional plate structure, favourable adhesive force, and remarkable shielding effect, it has been widely employed in the producing pigment, paining, phosphor powder materials, and cosmetics. The spherical α-Al_2O_3, with its good dispersibility, pressure-molding ability and firing features, has been widely used in the production of high quality ceramics· Hence, to synthesize α-Al_2O_3 powders in different shapes is of great theoretical significance and practical value to the development of the functional alumina materials.

At present, a great number of methods are available to produce α-Al_2O_3 powders in different shapes. Chen Xiaojun etc., by employing the sol-gel method, made the raw materials, aluminite powders and aluminum chloride hexahydrate, and the solvent, water, reacted at the temperature of 1200℃ and produced the fiber-shaped α-Al_2O_3. L.C. Pathak, T.B. Singh, etc., adopted the combustion method and obtained tabular-shaped α-Al_2O_3 by controlling the pH of the precursor. Zhang Guifang, Sheng Qiang, etc., taking aluminum sulfate and urea as the raw materials, produced spherical α-Al2O3 by the homogeneous precipitation method. Because of the mild and easy control condition, the hydrothermal reaction is a common synthesizing method in soft chemistry. Within the hydrothermal reaction system of soft chemistry, to synthesize α-Al2O3 in different shapes by controlling different conditions not only explores the mechanism and theory of the morphology of growth, but also has a significant meaning to the theoretical research of the crystal's growth.

This study, taking aluminum sulfate and urea as the raw materials, produced α-Al_2O_3 by the hydrothermal method through controlled synthesis. Besides, its mechanism has been explored and discussed.

2. Experiment

2.1. Sample Preparation

Prepare aluminum sulfate solution concentrated at certain degree. When the solution becomes clear and transparent, pour it into the hydrothermal reaction tank. Then put the reaction tank into the oven at a constant temperature of 120℃ for different hours. The precursor AACH can be obtained by filtering and washing the products. Then calcine the precursor at the temperature of 1200℃ for 2 hours to get the α- Al_2O_3 sample.

2.2. Sample Characterization

Conduct characterization of the crystalline structure and morphology of the sample products produced by the hydrothermal method. The crystalline structure is tested and examined by the D/Max2550VB+/PC XRD. The test condition is: CuKαtarget, tube current 50mA, tube voltage 40kV, scan step 0.02°, range of measurement between 3°~90°. The morphology is examined with the Japanese JEOLJSM-6510LV scanning electron microscope. The test conditions are as follow: accelerating voltage 20kV, object distance 11mm, beam spot size 40mm. Before testing, the sample should be dispersed in the absolute ethyl alcohol with ultrasonic wave for 10mins. When the sample is dried, spray metal for 10seconds.

3. Results and Discussion

3.1. Phase Analysis

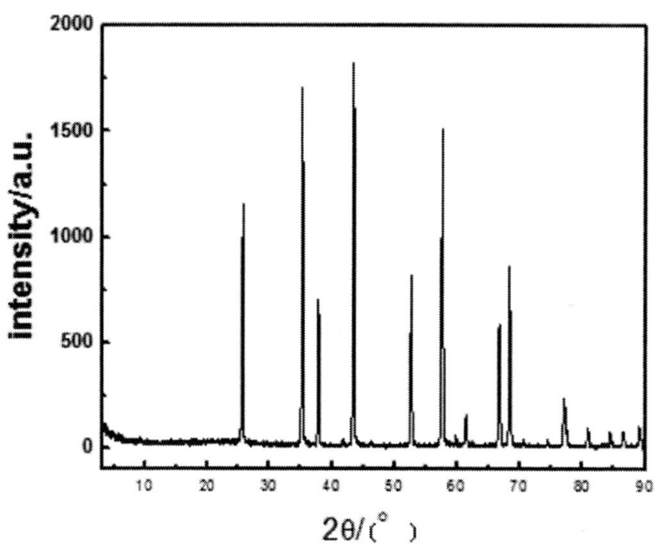

Figure 1 XRD Chart of the α-Al2O3 Powder

Figure 1 is the XRD spectrum of the products, which is obtained by calcining AACH at the temperature of 1200℃ for two hours. It is a complete match to the JCPDS standard card (NO.43-1484). All the diffraction peaks belong to α-Al_2O_3, indicating that pure phased α-Al_2O_3 with good crystalline structure has been produced.

3.2. Controlled Synthesis of the Spherical α-Al_2O_3

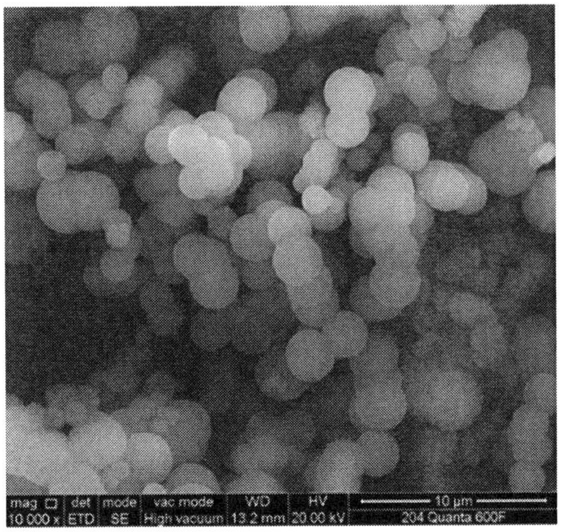

Figure 2 SEM of Spherical α-Al_2O_3

Figure 2 is the SEM picture of the products α-Al_2O_3 obtained by calining the precursors at the temperature of 1200℃ for two hours. The results suggest that the products maintain the shape of its precursor with good dispersibility; hence, it indicates that the shape of the obtained alumina powders is inherited from the shape of the precursor AACH. In the 6-hour hydrothermal reaction, due to the short reaction time span, few amount of OH- are provided by the urea in the solution, which result in a slow hydrolysis of the Al^{3+}. Meanwhile, a relative strong electrostatic attraction between SO_4^{2-} and Al^{3+} prevent the Al^{3+} and OH^- from nucleating too fast. Therefore, the course of crystallization is rather stable, assuring a full growth of every face and resulting in a solid sphere of the crystal. On the other hand, the particle size of the alumina, which is obtained by calcining the precursor from a 12-hour hydrothermal reaction, is enlarged, and its surface presents an interpenetration structure. This is resulted from the extension of the reaction time. Consequently, more OH- is formed by decomposing the urea, which makes the solution alkaline. Then the AlOOH structure is completed and the crystal generates two-dimensional tabular structure. At the same time, in order to decrease its surface energy, layers of tabular structures accumulated on the outside surface of the previous solid sphere. When the solid sphere completely disappears, the crystal grows into a hollow sphere with a circular hole in the center.

4. Conclusion

Taking aluminum sulfate and urea as the raw materials, employ the hydrothermal method to synthesize α-Al_2O_3 powders shapes by controlling the reaction temperature and reaction time. The XRD and SEM characterization results prove that the products are pure phase alumina, such as sphere. .

References

[1] R.Venkatesh, S.R.Ramanan.Effect of organic additives on the properties of sol-gel spun alumina fibres [J]. Journal of the European Ceramic Society. 2000, 20: 2543-2549
[2] B.Wilshire, F. Carreno. Deformation and damage processes during tensile creep of ceramic-fibre-reinforced ceramic-matrix composites [J].J Eur.Ceram.2000, 20: 463-472
[3] Deng Sanmao, Ye Hongqi, Su Zhou, etc.Synthesis and Properties of Flaky Oxide Powders [J]. New Chemical Materials. 2004, 32(7): 13-15
[4] Dong Yan, Jiang Jianqing, Liu Gang, etc. Preparation of Alumina as a Raw Material of

Phosphor and Controlling of Its Particle Size and Shape [J]. Journal of the Chinese Ceramic Society. 2004, 32(4): 393-397

[5] Li Shaohua, Ke Jiangjun, Yuan Fangli, etc.. Morphology Control of α-Al_2O_3 Powder Prepared by Indirect Wet Chemical Method [J]. Chinese Journal of Rare Metals. 2004, 28(4): 731-734

[6] Huang Qianjun, Chen Xiaojun, Chen Lifu. Preparation of the Polycrystalline Alumina Fubers by Sol-gel Progress [J]. Journal of Xiamen University (Natural Science). 2006, 45(3): 375-378

[7] L.C.Pathak, T.B. S.Das,etal. Effect of pH on the Combustion Synthesis of Nano-crysalline Alumina Powder [J]. Materials Letters. 2002, 57: 380-385

[8] Zhang Guifang, Shen Qiang, Zhang Lianmeng. Fabrication and Characterization of Mono-Dispersed Al_2O_3 Powders [J]. Journal of Aeronautical Materials. 2006, 26(4): 96-99

[9] Zhu H Y. Riches J D, Bany J C.γ-Alumina Nanofibers Prepared from Aluminum Hydrate with Poly (ethylene oxide) Surfactant [J]. Chem.Mater., 2002, 14(5): 2086-2093

[10] Sun X, Li J, Zhang F, et al. Synthesis of Nanocrystalline Al_2O_3 Powders from Nanometric Ammonium Aluminum Carbonate Hydroxide [J]. Journal of the American Ceramic Society, 2010, 86(8): 1321-1325.

[11] R.Venkatesh, S.R.Ramanan. Effect of organic additives on the properties of sol-gel spun Alumina fibres [J]. Journal of the European Ceramic Society, 2000, 20: 2543-2549.

[12] B.Wilshire, F.Carreno. Deformation and damage processes during tensile creep of ceramic-fibre-reinforced ceramic-matrix composites [J]. J.Eur.Ceram, 2000, 20: 463-472.

[13] A.Kebbede, J.Parai, A.H.Carim. Anisotropic grain growth in α-Al_2O_3 with SiO_2 and TiO_2 additions [J]. J.Am.Ceram. Soc, 2000, 83(11): 2845-2851.

[14] Hill R F, Danzer R. Synthesis of Aluminum oxide platelets [J]. J.Am.Ceram. Soc. 2001, 84(3): 514-520.

[15] Wu Y Q, Zhang Y F, Pezzotti G. Influence of AlF_3 and ZnF_2 on the phase transformation of gamma to alpha Alumina [J]. Mater Lett. 2002, 52(2): 366-369.

IOP Conf. Series: Materials Science and Engineering **207** (2017) 012005 doi:10.1088/1757-899X/207/1/012005

Synthesis of AuPd/g-C_3N_4 nanocomposites and their electrochemical properties

Ling Wu[1, a], Peng Xiao[2, b], Deli Jiang[2, c] and Min Chen[2, d]

[1]Department of clinical laboratory of the Affiliated Hospital of Nanjing University of TCM, Nanjing 210029, China
[2]School of Chemistry and Chemical Engineering, Jiangsu University, Zhenjiang 212013, China

[a]wling0605@163.com, [b]xp709160283@126.com, [c]dlj@ujs.edu.cn, [d]chenmin3226@sina.com

Abstract. In the present work, AuPd/g-C_3N_4 nanocomposites were prepared by the facile chemical reduction method. The TEM results show that the metal nanoparticles were uniformly deposited on the surface of the g-C_3N_4 surface. The electrochemical properties towards the sensing of H_2O_2 were tested. It was found that the AuPd/g-C_3N_4 shows the superior electrochemical performance compared with the Au/g-C_3N_4 and Pd/g-C_3N_4.

1. Introduction

In the past years, quantitative detection of H_2O_2 has received considerable attention, since H_2O_2 is an essential intermediate in biomedical, pharmaceutical, environmental protection, and electrochemistry. Many traditional techniques, such as chemiluminescence, titrimetric spectrophotometry, fluorescence and electrochemical method were used to quantitative detection of H_2O_2. Especially, non-enzymatic electrochemical method gained a great of interests among various techniques owing to its high sensitivity, low cost, easy operation, and the possibility for real-time detection. In the past decades, many nanomaterials, including metal nanoparticles, metal hexacyanoferrate, metal oxide, have been empolyed to fabricate the electrochemical H_2O_2 sensors [1, 2]. Our group has recently synthesized Ag/C nanocomposites and silver deposited porous graphitic g-C_3N_4 nanocomposites for nonenzymatic electrochemical H_2O_2 sensing which exhibited high sensitivity and low detection limit [3]. Compared with the single-component metal nanoparticles, the bimetallic nanoparticles shown superior electrochemical properties towards the detection of H_2O_2. For example, Chen and co-workers recently constructed RGO/Ag-Au/Cu_2O ternary nanocomposites electrochemical sensor for the detection of H_2O_2, and the sensor exhibited good electrocatalytic activity, with a linear response range from 0.05 to 50.75 mmol/L at -0.2 V and a sensitivity of 0.14 μA mmol/L^{-1} cm^{-2} [4]. However, up to date, there is no report on the construction of g-C_3N_4/bimetallic nanoparticle nanocomposite electrochemical sensor. In this work, AuPd/g-C_3N_4 nanocomposites were prepared by a facile chemical reduction method. The TEM results show that the metal nanoparticles were uniformly deposited on the surface of the g-C_3N_4 surface. The electrochemical properties towards the sensing of H_2O_2 were tested. It was found that the Au-Pd/g-C_3N_4 shows the superior electrochemical performance compared with the Au/g-C_3N_4 and Pd/g-C_3N_4.

2. Experimental

2.1. Preparation of AuPd/g-C_3N_4, Au/g-C_3N_4 and Pd/g-C_3N_4 samples

All chemical reagents were of analytic purity and used directly without further purification. The g-C_3N_4 nanosheets were prepared by a thermal polymerization strategy [5]. The AuPd/g-C_3N_4 was prepared by a facile chemcial reduction method. In a typical synthesis, 100 mg of g-C_3N_4 was first dissolved in 100 mL of distilled water, followed by the sonication for 30 min. Then, the 0.5 mL of $HAuCl_4$ (0.01 mol·L^{-1}) and 0.94 ml H_2PdCl_4 (0.01 mol·L^{-1}) was added into the above solution under vigorous stirring for 30 min, followed by added of 20ml $NaBH_4$ (0.15 mol·L^{-1}). Then the solution was heated to 100℃ and kept for 24 h. Then, the sample was washed with deionized water and ethanol and eventually dried under vacuum at 60 °C for 6 h. For comparison, the Au/g-C_3N_4, Pd/g-C_3N_4 sample was prepared by the simialr method in the absence of $HAuCl_4$ or H_2PdCl_4, respectively.

2.2. Characterization

The morphology of the as-prepared samples was examined by transmission electron microscopy (TEM) instrument (Tecnai G2F30 S-TWIN, Philips). The phase purity and crystal structure of the obtained samples were examined by X-ray diffraction (XRD) using D8 Advance X-ray diffraction (Bruker axs company, Germany) equipped with Cu-KR radiation (λ) 1.5406 (Å). The chemical states of the as-prepared sample were examined by X-ray photoelectron spectroscopy (XPS, PHI 5300C ESCA System) with Al Kα radiation operating at 250W.

2.3. Preparation of the modified electrode and electrochemical measurements

1.0 mg of AuPd/g-C_3N_4 composite and 10 mL of Nafion solution (5 wt%) were dispersed in 1 mL water–isopropanol mixed solvent (3 : 1 v/v) by at least 30 min sonication to form a homogeneous catalyst solution. Then 10 mL of the catalyst solution was dropped onto the surface of pretreated GCE and lefto dry at room temperature to get AuPd/g-C_3N_4-modified GCE (denoted as AuPd/g-C_3N_4/GCE). Amperometric and cyclic voltammetry (CV) experiments were carried out by a CHI 650D electrochemical analyzer (CH Instruments, Inc., Shanghai), in a home-made three-electrode electrochemical cell consisting of a twisted platinum wire as an auxiliary electrode, a saturated calomel electrode (SCE) as a reference electrode and a modified-GCE (0.07 cm^2) as a working electrode. All experiments were carried out at the ambient temperature. Prior to the experiment, the electrolyte, consisting of a solution of 0.2 M phosphate buffer solution (PBS, pH 7.4), was purged with high-purity nitrogen for at least 30 min and a nitrogen atmosphere was maintained over the solution.

3. Results and discussion

The XRD patterns of Au/g-C_3N_4, Pd/g-C_3N_4 and AuPd/g-C_3N_4 binary alloy composites were measured to investigate the crystal structure. As shown in Fig. 2, there are two distinct peaks at 13.0° and 27.4°can be indexed as the (100) and (002) peaks for graphitic materials, respectively. For the Au/g-C_3N_4 sample, due to the low content of Au metal nanoparticles, the characteristic diffraction peaks of Au cannot be observed. In contrast, the diffraction peak centered at 38.2° in the Pd/g-C_3N_4 and Au/g-C_3N_4 sample which can be ascribed to the (111) plane of Pd nanoparticles. However the peak ascribed to the (111) plane of the Pd shifted to higher degrees, probably due to the formation of alloyed nanoparticles.

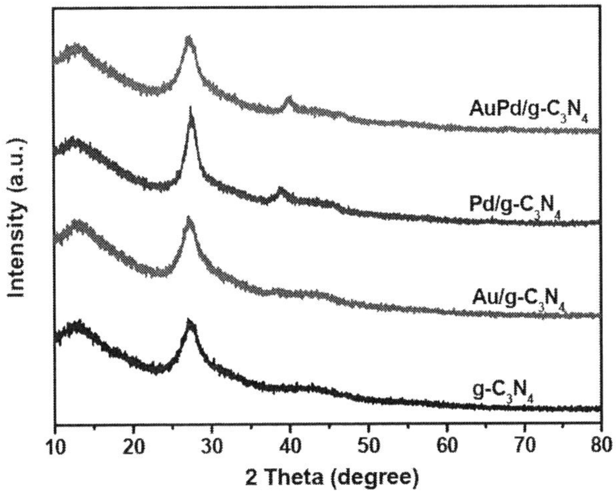

Fig. 1. XRD patterns of the as-prepared g-C_3N_4, Au/g-C_3N_4, Pd/g-C_3N_4, and AuPd/g-C_3N_4.

The morphology and microstructure of the as-prepared samples were then investigated by TEM. Fig. 2a shows that some mesopores with several tens of nanometres in size can be observed in the g-C_3N_4 sheets, which could serve as a support for the loading of metal particles in this composite system. The TEM image for the Au/g-C_3N_4 sample shows that a few of the Au nanoparticles were deposited on the surface of g-C_3N_4 sheets (Fig. 2b). Similarly, the Pd nanoparticles with 2-5 nm in size was deposited on the surface of g-C_3N_4 sheets in the Pd/g-C_3N_4 samples (Fig. 2c). For the AuPd/g-C_3N_4 sample, tiny AuPd bimetatic nanoparticles were loaded on the g-C_3N_4 nanosheet.

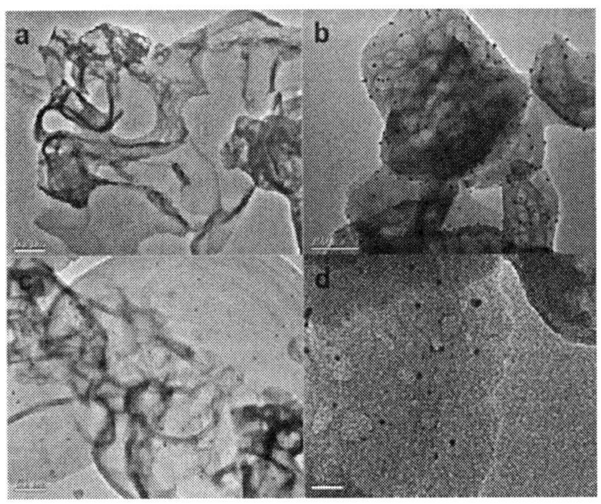

Fig. 2. Typical TEM images of (a) g-C_3N_4, (b) Au/g-C_3N_4, (c) Pd/g-C_3N_4, and (d) AuPd/g-C_3N_4.

The chemical composition of the AuPd/g-C_3N_4 composites were characterized by XPS. Fig. 3a displays the C 1s XPS spectra, and it can be ascribed to carbon and defect-containing sp^2-hybridized carbon atoms present in graphitic domains. The second peak at 287.5 eV is assigned to the sp^2-bonded carbon (the major carbon species in the g-C_3N_4 polymer) in N-containing aromatic rings (N-C=N). The high resolution N 1s spectrum of Au-Pd/g-C_3N_4 (as shown in Fig. 3b) can be also fitted into two different peaks at binding energies of 398.2 and 399.2 eV, respectively, which are attributed to sp^2 hybridized N bonded with C (C-N=C) and the bridging tertiary N (N-$(C)_3$). The Au 4f XPS signal can be deconvoluted into two peaks at around 88.0 and 84.3 eV, indicating that metallic Au^0 is the main species. As exhibited in Fig. 3d, the Pd $3d_{5/2}$ and Ag $3d_{3/2}$ peaks are located at 335.3 eV and 340.6 eV, respectively, confirming the existence of metallic Pd in the binary samples. These XPS results further confirm formation of AuPd/g-C_3N_4.

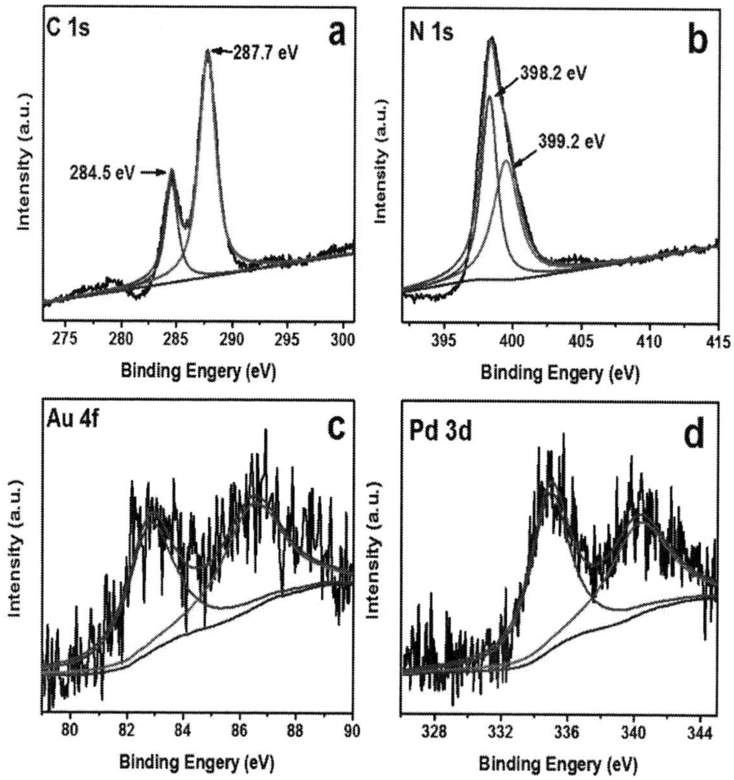

Fig. 3. XPS spectra of the AuPd/g-C_3N_4 composite: (a) C 1s, (b) N 1s, (c) Au 4f and (d) Pd 3d.

The AuPd/g-C_3N_4 binary nanocomposites are potential electrochemical sensor materials. An enzymeless H_2O_2 sensor has been constructed by direct deposition of the as-prepared catalysts on a bare GCE surface to study their electrochemical activities. Fig. 4 displays the CVs of the g-C_3N_4/GCE, Au/g-C_3N_4/GCE, Pd/g-C_3N_4/GCE, and AuPd/g-C_3N_4/GCE nanocomposites in the presence of 1 mM of H_2O_2 in N_2-saturated PBS solution (0.2 M, pH=7.4), and the scan rate is 50 mV s^{-1}. The responses of the bare g-C_3N_4/GCE, Au/g-C_3N_4/GCE and Pd/g-C_3N_4/GCE toward the reduction of H_2O_2 are insensitive. There is no obvious change in current density with the increase of electric potential. When the AuPd bimetallic nanoparticles anchored on the surface of g-C_3N_4 nanosheets, the electrode shows a sensitive response towards the reduction of H_2O_2. This might be because of the presence of alloyed metal nanoparticles with good electrical conductivity which could accelerate the electron transfer of the electrode.

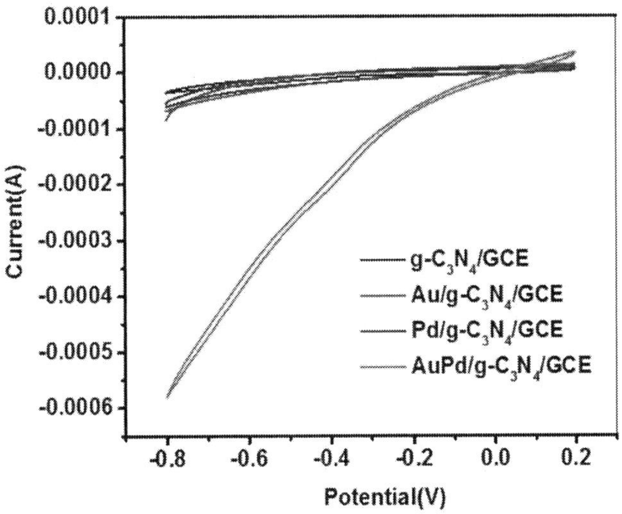

Fig. 4. CVs of different materials modified electrodes in the absence and presence of H_2O_2 for the bare g-C_3N_4/GCE, Au/g-C_3N_4/GCE, Pd/g-C_3N_4/GCE, and AuPd/g-C_3N_4/GCE in N_2-saturated 0.2 M pH 7.4 PBS with presence of H_2O_2, scan rate: 50 mV s^{-1}.

4. Summary

In summary, AuPd/g-C_3N_4 nanocomposites were synthesized by the facile chemical reduction method. The TEM results show that the metal nanoparticles were uniformly deposited on the surface of the g-C_3N_4 surface. The electrochemical properties towards the sensing of H_2O_2 were tested. It was found that the AuPd/g-C_3N_4 shows the superior electrochemical performance compared with the Au/g-C_3N_4 and Pd/g-C_3N_4.

Acknowledgements
This work was supported by the financial supports of Key Research Plan of Zhenjiang City (GY2015031 and GY2015044).

References

[1] D.L. Jiang, Y. Zhang, M.H. Huang, J.Liu, J. Wan, H.Y. Chu, M. Chen: J. Electroanal. Chem. Vol. 728 (2014), p. 26-33.
[2] S.H. Chen, R. Yuan, Y.Q. Chai, F.X. Hu: Microchim. Acta Vol. 180 (2013), p. 15-32.
[3] D.L. Jiang, Y. Zhang, H.Y. Chu, J. Liu, J. Wan, M. Chen: RSC Adv. Vol. 4 (2014), p. 16163-16171.
[4] D. Li, L.Y. Meng, P. Xiao, D.L. Jiang, S.C. Dang, M. Chen: J. Electroanal. Chem. Vol. 791 (2017), p. 23–28.
[5] D.L. Jiang, L.L. Chen, J.J. Zhu, M. Chen, W.D. Shi, J.M. Xie: Dalton Trans. Vol. 42 (2013), p. 15726-15734.

IOP Conf. Series: Materials Science and Engineering **207** (2017) 012006 doi:10.1088/1757-899X/207/1/012006

Zinc-rich $Ga_{1-x}Zn_xN_{1-x}O_x$ solid solutions with tunable composition prepared from a constant-pH coprecipitation method

Yanling Hu[1,*], Lihua Wei[1], Juan Zuo[1], Dongya Sun[1], Chunhai Jiang[1], Yongsheng Fu[2]

[1]School of Materials Science and Engineering, Fujian Provincial Key Laboratory of Functional Materials and Applications, Xiamen University of Technology, Xiamen, 361024, P. R. China
[2]Key Laboratory for Soft Chemistry and Functional Materials, Nanjing University of Science and Technology, Ministry of Education, Nanjing, 210094, P. R. China

[*]Corresponding author e-mail: huyanling@xmut.edu.cn

Abstract. Wurtzite Zinc-gallium oxynitrides ($Ga_{1-x}Zn_xN_{1-x}O_x$) has attracted a great deal of interest these years as photocatalysts capable of overall water splitting and contamination degradation under visible light irradiation. $Ga_{1-x}Zn_xN_{1-x}O_x$ particles were synthesized by nitridation of Zn/Ga/CO_3 layered double hydroxides (LDHs) using a coprecipitation method with constant pH of 8. Single-phase $Ga_{1-x}Zn_xN_{1-x}O_x$ with Zn: Ga mole ratios of 2:1, 3:1, 4:1 and 5:1 were obtained, as characterized by X-ray diffraction and scanning electron microscope. Photocatalytic activities of the obtained particles were evaluated under visible-light irradiation against the photodegradation of methylene blue (MB). The oxynitrde with a Zn:Ga mole ratio of 2:1 showed the highest photoactivity among all the samples.

1. Introduction

Zinc-gallium oxynitride solid solution ($Ga_{1-x}Zn_xN_{1-x}O_x$) with a wurtzite structure, also known as GaN:ZnO or ZnGaNO, has attracted a great deal of interest these years as photocatalysts capable of overall water splitting and contamination degradation under visible light irradiation [1-4]. $Ga_{1-x}Zn_xN_{1-x}O_x$ solid solutions have exhibited high quantum efficiency (17.3%) [5], reliable long-term durability [6], and tunable band-gap energy [7].

A number of methods have been employed to synthesize $Ga_{1-x}Zn_xN_{1-x}O_x$ solid solution by nitridation of different starting materials/precursors [8-17]. Most of these precursors require complicated processes and long time of nitridation (5-15 h) at high temperatures (600-850°C) for complete phase transformation. Zn/Ga/CO_3 LDHs, however, only need half hour to be transformed into wurtzite structure, which make them as attractive precursors to synthesize $Ga_{1-x}Zn_xN_{1-x}O_x$ photocatalyst. The molecular formula of Zn/Ga/CO_3 LDHs can be expressed as $[(Zn^{2+})_{1-x}(Ga^{3+})_x(OH)_2]^{x+}(CO_3^{2-})_{x/2}\cdot yH_2O$, which is a class of intercalation compounds consisted of positively charged brucite-like ($Mg(OH)_2$-type) layers separated by CO_3^{2-} and water molecules.

Typically, LDHs are prepared using coprecipitation methods [18]. Zn/Ga/CO_3 LDHs have been successfully prepared using a coprecipitation method with increasing pHs during the agent mixing process. [13,19]. Our previous work revealed that only a coprecipitation process with a constant pH can obtain single-phase wurtzite $Ga_{1-x}Zn_xN_{1-x}O_x$ particles with Zn:Ga of 2:1, while those with decreasing pHs produced a ZnO/$Ga_{1-x}Zn_xN_{1-x}O_x$ composite, and those with increasing pHs obtained a porous ZnGaNO/$ZnGa_2O_4$ structure.[20] In this paper, a coprecipitation method with constant pH was employed to prepare Zn/Ga/CO_3 LDH precursors with different Zn:Ga mole ratios, which would used to synthesize single-phase $Ga_{1-x}Zn_xN_{1-x}O_x$ solid solutions with tunable Zn:Ga ratios. The microstructures of the LDHs and $Ga_{1-x}Zn_xN_{1-x}O_x$ solid solutions were investigated, and the photocatalytic activities of the $Ga_{1-x}Zn_xN_{1-x}O_x$ were evaluated and compared.

2. Experimental

2.1. Preparation of ZnGaNO solid solutions

A series of metallic nitrate solutions (Solution A) were prepared with [Zn^{2+}]/[Ga^{3+}] mole ratio of 1:1, 2:1, 3:1, 4:1, and 5:1 using a fixed concentration of 0.3 mol/L for $Ga(NO_3)_3 \cdot 9H_2O$ and increased concentration of $Zn(NO_3)_2 \cdot 6H_2O$ of 0.3, 0.6, 0.9, 1.2, and 1.5 mol/L. 2 M NaOH and 1 M Na_2CO_3 were mixed to form a base solution (Solution B).

Solution A and Solution B were simultaneously added drop-wisely to a vessel containing stirred deionized water (300 mL, the initial pH of the solution was adjusted to 8 by Solution B) at such a rate that the pH of the reaction mixture was maintained at 8. The mixing process was carried out at 40 °C. After complete delivery of the Solution A, the reaction mixture was aged at 80 °C for 12 - 24 h with good stirring. The product suspension was then repeatedly centrifuged and washed with deionized water. After that, the precipitate was dried at 70 °C for 24 h. The procedure was repeated using different Solution A to obtain Zn/Ga/CO_3 LDHs with different [Zn^{2+}]/[Ga^{3+}] mole ratios, and corresponding $Ga_{1-x}Zn_xN_{1-x}O_x$ particles were obtained by heating Zn/Ga/CO_3 LDH powders at 800 °C under a mixed flow of NH_3 (150 mL/min) and CH_4 (50 mL/min) for 30 min.

2.2 Characterization

The structure of the prepared powders was identified by X-ray diffraction (XRD, Philips, PANalytical X'pert, Cu Kα radiation (λ=1.5417 Å)). The morphology of catalysts was examined using a field emission scanning electron microscope (SEM, Hitachi-S4800).

The Zn-Ga oxynitride powders were immersed in a $RuCl_3$ ethanol solution and stirred at 80°C until all the ethanol was completely evaporated. The resulting mixed powder was heated in air at 350°C for 1 h to convert the Ru species to RuO_2.

A 0.1 g sample of the 5 wt.% RuO_2-loaded catalyst powder was dispersed in 100 mL MB aqueous solution. The initial concentration of MB is 15 $mg \cdot L^{-1}$ with pH of 4.5 (adjusted by H_2SO_4 solution). The mixture was then stirred in the dark for 30 min to attain adsorption equilibrium on the surface of the catalyst. It was then stirred and irradiated under visible light from a 300W Xenon light source (PLS-SXE300C, Perfectlight, China) equipped with a cutoff filter to provide visible light (λ > 400 nm). The average light intensity was 140 mW/cm^2. The reactant solution was maintained at room temperature by a flow of cooling water during the reaction. About 5 mL of the solution was withdrawn and filtered at given time intervals to be analyzed.

3. Results and Discussions

The XRD patterns for the Zn/Ga/CO_3 LDH precursors and nitridation products with different Zn:Ga mole ratios are shown in Fig. 1(a) and 1(b), respectively. In Fig. 1(a), all five XRD patterns exhibit the major peaks that fit well to a typical LDH structure (JCPDS No. 30-1835, a=3.114 Å, c=22.41Å). There are minor impurity peaks in the background of 1:1 and 2:1 samples, indicating the presence of a small amount of $ZnGa_2O_4$ (JCPDS No. 38-1240). And there are increasing amount of additional $Zn_4CO_3(OH)_6H_2O$ (JCPDS No. 11-0287) in the order of 3:1, 4:1 and 5:1 samples. In Fig. 1(b), all five

XRD patterns have major XRD peaks corresponding to a wurtzite structure with lattice parameters close to ZnO (JCPDS No. 36-1451, a=3.249Å, c=5.207Å). 1:1 nitrided sample, however, has additional peaks indicating the presence of $ZnGa_2O_4$.

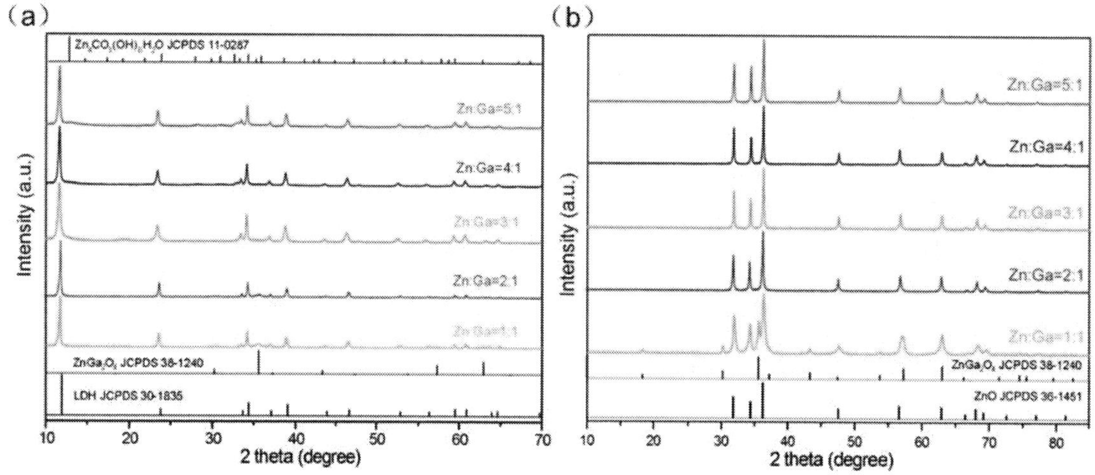

Figure 1. XRD patterns of (a) LDH precursors and (b) nitridation products with different Zn:Ga mole ratios in the solutions.

The precise lattice parameters of the wurtizte structure were calculated for each sample using the refinement function of Jade 6.0, and results were drawn in Fig. 2 in comparison with the lattice parameters of ZnO and GaN (JCPDS No. 50-0792, a=3.189Å, c=5.186Å). According to Fig. 2, it is obvious that the lattice parameters of $Ga_{1-x}Zn_xN_{1-x}O_x$ deviations from ideal solid solution behaviour. Lattice parameter *c*, especially, showed a bowing effect with increasing *x* in the solid solution. The data are consistent with the results in Ref [15], where the $Ga_{1-x}Zn_xN_{1-x}O_x$ nanoparticles were synthesized from nanocrystalline $ZnGa_2O_4$ and ZnO.

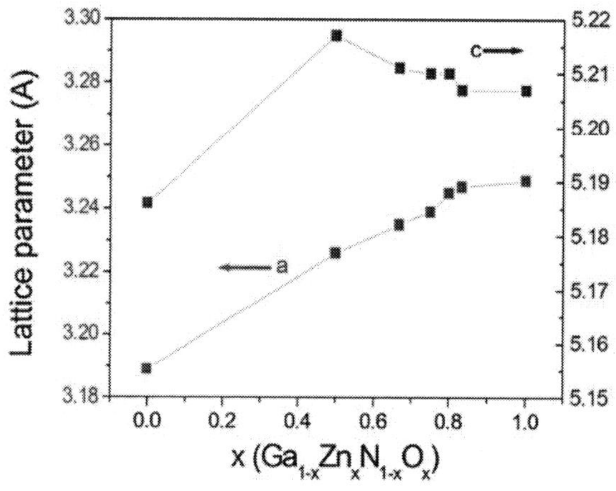

Figure 2. Lattice parameters *a* and *c* of the wurtzite $Ga_{1-x}Zn_xN_{1-x}O_x$ with different Zn:Ga mole ratios

Presented in Fig. 3 are SEM images of the LDH precursors and the corresponding nitridation products. According to Fig. 3 (a), all the precursor particles are platelets with an average diameter of ~300 nm and thickness of less than 100 nm. It is also obvious that 1:1 LDH has a large amount of nanoparticles less than 50 nm in diameter on the surface, which should be $ZnGa_2O_4$ according to the XRD results. Fig.3

(a) also show that the amount of $ZnGa_2O_4$ nanoparticles decreased in the order of 1:1 LDH, 2:1 LDH, and 3:1 LDH. For the 4:1 and 5:1 LDH, however, cloud-like impurities showed up again, which should correspond to $Zn_4CO_3(OH)_6H_2O$ according to Fig. 1(a). After nitridation, 1:1 sample remained the platelet morphology but with $ZnGa_2O_4$ nanoparticles appeared inside the oxynitride particles. Figs. 3(b) also showed that all other particles collapsed to form oxynitrides with an average diameter of 200 nm. For the 3:1, 4:1, and 5:1 oxynitrides, it is also noted that uniformity of the particle size decreased with increasing Zn:Ga ratios.

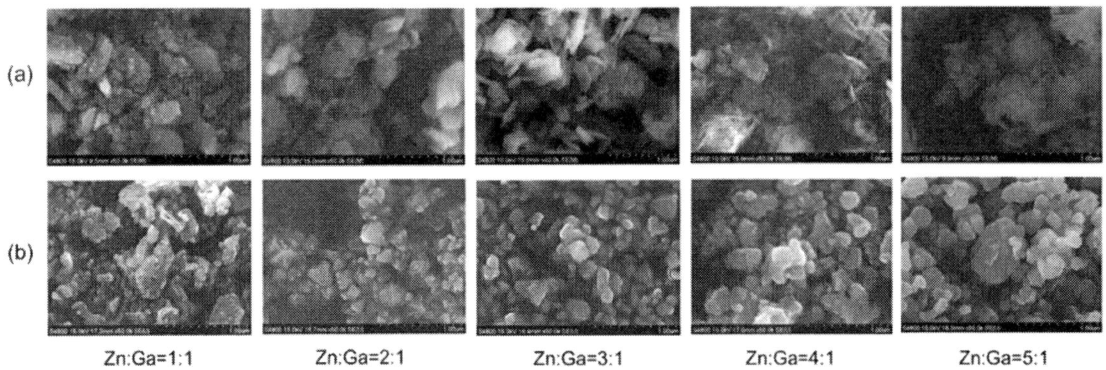

Figure 3. SEM images of (a) LDH precursors and (b) nitridation products with different Zn:Ga mole ratios in the solutions.

EDS/SEM were conducted to identify the atomic composition of metallic elements in the LDHs and oxynitrides, and the mole ratios of Zn:Ga for Solution A, LDHs, and oxynitrides were listed in Table 1 for comparison. It is clear that the Zn:Ga ratios of LDH precursors are very close to those of Solution A. Except for 3:1 oxynitride, Zn:Ga ratios of all nitridation products are slightly higher than those of LDHs, which could stem from reduction of gallium oxide by CH_4 and re-deposition of Zn on the surface of particles.

Table 1. Zn:Ga mole ratio in the solution, LDH precursors, and solid solutions

Zn:Ga (Solution A)	Zn:Ga (LDHs)	Zn:Ga (oxynitrides)
1:1	1.1:1	1.3:1
2:1	2.2:1	2.6:1
3:1	3.1:1	2.9:1
4:1	4.01:1	4.2:1
5:1	4.87:1	5.1:1

The formation of $ZnGa_2O_4$ and $Zn_4CO_3(OH)_6H_2O$ in the LDHs can be explained by Fig. 4, which gives the concentrations vs. pHs for Zn and Ga ions and their hydroxylated species in the aqueous solution. For 1:1 and 2:1 samples, there are high level of overlap between $[Zn^{2+}]$ and $Ga(OH)_4^{-1}$ curves at the pH of 8, leading to the formation of $ZnGa_2O_4$ during the aging process of LDHs. When the Zn:Ga ratio is increased, a left shift of all Zn-related curves would occur while Ga-related curves are fixed, resulting in high concentration of ZnO_2^{2-} in the aging solution, resulting in an increasing amount of $Zn_4CO_3(OH)_6H_2O$ in order of 3:1, 4:1, and 5:1 samples.

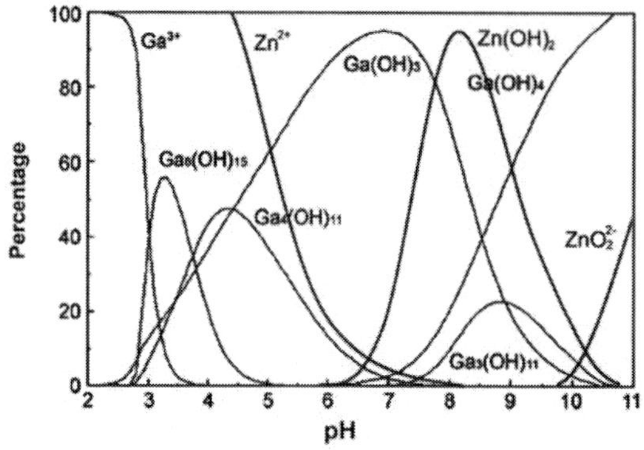

Figure 4. Schematic diagram of the distribution of Ga^{3+}, Zn^{2+} and their related hydroxylated species as a function of pH. For the curves related to Ga, the data was copied from Ref. [21] with [Ga^{3+}] of 2 M. Zinc-related curves were schematically drawn using K_{sp} of 2.09×10^{-16} and [Zn^{2+}] of 0.4 M.

Photocatalytic activities of the oxynitride particles with different Zn:Ga mole ratios were evaluated under visible-light irradiation against the photodegradation of MB. The percentage of degraded MB after two hours of visible light irradiation with presence of oxynitrdes are shown in Fig. 5. It's clear that the oxynitrde with a Zn:Ga ratio of 2:1 exhibited the highest photoactivity among all the samples. The reduced photoactivity of 1:1 oxynitride may relate to the inner $ZnGa_2O_4$ nanoparticles, which may block the movement of the photogenerated hole-electron pairs, and thereby enhance possibility of recombination. For the 3:1, 4:1, and 5:1 oxynitrides, the decreased photoactivities are likely to be ascribed to improper valence band edge levels where electrons are not able to reduce O_2 to $O_2^{-\cdot}$ radicals, which could be the major active species to degrade MB.

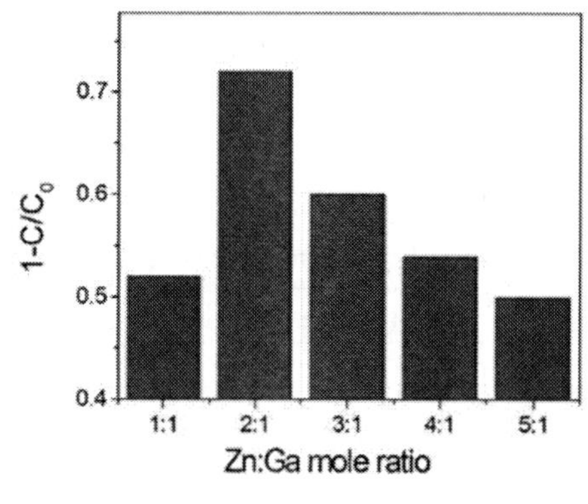

Figure 5. MB Photodegradation in the presence of 5 wt.% RuO_2-loaded $Ga_{1-x}Zn_xN_{1-x}O_x$ particles with different Zn:Ga mole ratios under visible-light irradiation ($\lambda > 400$ nm).

4. Conclusion

Zinc-gallium oxynitrides were synthesized by nitridation of $Zn/Ga/CO_3$ LDHs using a constant-pH coprecipitation method. $Zn/Ga/CO_3$ LDH platelets with few $ZnGa_2O_4$ impurity were obtained for the precursors with Zn:Ga mole ratios of 1:1 and 2:1, and LDH platelets with a small amount of $Zn_4CO_3(OH)_6H_2O$ were synthesized for the precursors with Zn:Ga mole ratios of 3:1, 4:1 and 5:1. After nitridation with NH_3 and CH_4 at high temperature at 800°C for half hour, single-phase wurtizite $Ga_{1-x}Zn_xN_{1-x}O_x$ particles were obtained with Zn:Ga mole ratios of 2:1, 3:1, 4:1, and 5:1, respectively. Nitridation of LDH precursor with Zn:Ga ratio of 1:1, however, leaded to $Ga_{1-x}Zn_xN_{1-x}O_x$ platelets with inner $ZnGa_2O_4$ nanoparticles. Photocatalytic activities of the oxynitride particles with different

Zn:Ga mole ratio were evaluated under visible-light irradiation against the photodegradation of MB. Oxynitrde particle with a Zn:Ga ratio of 2:1 showed the highest photoactivity among all the samples.

Acknowledgments
This work is supported by the National Natural Science Foundation of China (No. 51402249, No. 21303148, No. 21321062), National Basic Research Program of China (2012CB932900), and the Fundamental Research Funds for the Central Universities (No. 30916014103).

References
[1] K. Maeda, K. Teramura, T. Takata, M. Hara, N. Saito, K. Toda, Y. Inoue, H. Kobayashi, K. Domen, J. Phys. Chem. B 109 (2005) 20504-20510
[2] K. Maeda, K. Domen, Chem. Mater. 22 (2010) 612-623
[3] K. Maeda, K. Domen, J. Phys. Chem. Lett. 1 (2010) 2655-2661
[4] K. Maeda, K. Teramura, D.L. Lu, T. Takata, N. Saito, Y. Inoue, K. Domen, Nature 440 (2006) 295-295
[5] Y. Li , L. Zhu, Y. Yang, H. Song, Z. Lou, Y. Guo, Z. Ye, Small 11 (2015) 871-876.
[6] T. Ohno, L. Bai, T. Hisatomi, K. Maeda, K. Domen, J. Am. Chem. Soc. 134 (2012) 8254-8259
[7] K. Maeda, T. Takata, M. Hara, N. Saito, Y. Inoue, H. Kobayashi, K. Domen, J. Am. Chem. Soc. 127 (2005) 8286-8287
[8] C-H. Chuang, Y-G. Lu, K. Lee, J. Ciston, G. Dukovic, J. Am. Chem. Soc. 137 (2015) 6452-6455.
[9] B. Adeli, F. Taghipour, ECS J. Solid State Sci. Tech. 2 (2013) Q118-126
[10] H. Chen, L. Wang, J. Bai, J.C. Hanson, J.B. Warren, J.T. Muckerman, E. Fujita, JA Rodriguez, J. Phys. Chem. C 114 (2010) 1809-1814
[11] X. Sun, K. Maeda, M. L. Faucheur, K. Teramura, K. Domen, Appl. Catal. A-Gen. 327 (2007) 114-121
[12] W.-Q. Han, Z. Liu, H.-G. Yu, Appl. Phys. Lett. 96 (2010) 183112
[13] J. Wang, B. Huang, Z. Wang, P. Wang, H. Cheng, Z. Zheng, X. Qin, X. Zhang, Y. Dai, M.-H. Whangbo, J. Mater. Chem. 21 (2011) 4562-4567
[14] J. Wang, P. Yang, Z. Wang, B. Huang, Y. Dai, J. Alloy Compd. 652 (2015) 205-212
[15] K. Lee, B.M. Tienes, M. B. Wilker, K. J. Schnitzenbaumer, G. Dukovic, Nano Lett. 12 (2012) 3268-3272
[16] S. Yan, Z. Wang, Z. Li, Z. Zou, J. Mater. Chem. 21 (2011) 5682-5686
[17] A.A. Reinert, C. Payne, L. Wang, J. Ciston, Y. Zhu, P.G. Khalifah, Inorg. Chem. 52 (2013) 8389-8398
[18] F. Cavani, F. Trifirb, A. Vaccari, Catal. Today, 11 (1991) 173-301
[19] B. Adeli, F. Taghipour, Appl. Catal. A: Gen 521 (2016) 250-258
[20] Y.-L. Hu, Z. Wu, X. Zheng, N. Lin, Y. Yang, J. Zuo, D. Sun, C.-H. Jiang, L. Sun, C.-J. Lin, Y.-S. Fu, accepted by J. Alloys Compd.
[21] B. Hacht, Bull. Korean Chem. Soc. 29 (2008) 372-376

IOP Conf. Series: Materials Science and Engineering **207** (2017) 012007 doi:10.1088/1757-899X/207/1/012007

Synthesis of corundum-type In_2O_3 nanocrystals via hydro/solvothermal route and their photocatalytic properties

Guodong Liu*, **Weidong Wu, Chengcheng Li, Xinjin Shi, KeKe Gao, Min Li, Yuan Tian**

Department of Chemistry and Chemical Engineering, Jining University, Jining 273155, China

*Corresponding author e-mail: liugd615@163.com

Abstract. The understanding of the polymorphism of In_2O_3 has evoked recent interest in terms of fundamental scientific and technological aspects. A simple yet effective hydro/solvothermal route was introduced for the preparation of In_2O_3 nanocrystals with different phases using In(acac)3 as indium source. Through adjusting the systemic amount of water, the hydrolysis reaction of In3+ can be efficiently controlled. Moreover, the phases and morphologies of products can also be regulated. Meanwhile, the photocatalytic property of the prepared corundum-type In_2O_3 was also measured, which showed superior photocatalytic performance on the decomposition of phenol compared to that of cubic phase In_2O_3 nanocubes.

1. Introduction

Until now, nanoscale metal oxides have received great attention in a myriad of research fields such as chemical sensors, photonic devices, drug delivery carriers, field emission displays, photocatalyst, and lithium-ion batteries, owing to their unique morphologies and physicochemical properties which are different from their bulk counterparts.[1-3] As one of the most important part of new research area of functional materials, metastable materials, physical and chemical properties being different from their stable phase, are becoming ever more essential to resolve new problems. [4, 5]

As a key n-type III-VI semiconductor, Indium oxide (In_2O_3), has attracted considerable attention because of the peculiar electronic and optical properties and increasingly extensive applications in solar cells, sensors, electrocatalyst and nanoscale transistors etc.[6, 7] Normally, In_2O_3 prefers to crystallize in the form of cubic structure (C-In_2O_3), whereas metastable corundum-type In_2O_3 (H-In_2O_3) was obtained by the phase transition method under high temperature and high pressure (HTHP) conditions. [3] The understanding of the polymorphism of In_2O_3 has evoked recent interest both due to fundamental scientific and technological aspects. [8]

Here, we aimed at exploring a novel low-temperature hydrothermal & solvothermal methods to synthesize metastable phases of functional materials, and studing the formation mechanism of the metastable phases under the hydrothermal & solvothermal conditions. Among so many solution synthesis routes reported, the calcination of corresponding hydrolysis precursors (such as $In(OH)_3$ and InOOH) is necessary, which can be concluded that the amount of H_2O and hydrolyzation of In^{3+} play a crucial role in the process. So far, the influences of reactional precursors, temperature, and time, surfactants on morphologies, sizes, as well as phases of products have been deeply researched. In our previous work, the influence of H_2O in solution system was reported.[9] As a polar solvent, H_2O plays sevaral roles in hydro/solvorhermal process, which can be summarized: (i) As solvent: Reaction substances dissolved and temperature and pressure transferred medium. (ii) As reaction substance: take part in the hydrolysis reaction to form oxide or hydroxide. (iii) Influence the number of hydroxyl on the surface of obtained nanocrystals, and further affect the ir aggregation. Hence, How to use the adjustment of systemic H_2O to control solution-phase synthesis nanomaterials has an important research value. Now we use indium acetylacetones ($In(acac)_3$) to synthesis InOOH, the precursor of H-In_2O_3. In previous reports, $In(acac)_3$ was frequently used as a precursor to prepare In_2O_3 through a solvothermal process, and C–In_2O_3 rather than H–In_2O_3 was obtained.[10, 11] Herein, a simple hydro/solvothermal route to the In_2O_3 nanocrystals with different phase by using $In(acac)_3$ as indium source was introduced. Through adjusting the systemic amount of water, the In^{3+} hydrolysis reaction can be controlled. Of course, the phases and morphologies of products also can be controlled. Furthermore, the photocatalytic property of the prepared H–In_2O_3 was also studied, which shows superior photocatalytic performance on the decomposition of phenol compared to the prepared C–In_2O_3 nanocuboids.

2. Experiments

2.1 Synthesis

In a typical synthesis, 0.206 g (0.5 mmol) of $In(acac)_3$ was added to 20.00 mL of ethanol. The reaction mixture was transferred into a 30 mL inner volume stainless autoclave with Teflon liner, and different amount of water was injected into the reaction system before solvothermal reaction. The reaction was matained at 200 °C for 24 h in a furnace. The resulting products were centrifuged, and washed with ethanol for several times, and subsequently being dried in air at 60 °C. The as-prepared InOOH nanocrystals were calcined in a boat crucible at a temperature of 500 °C and maintained for 2 h in air with rate temperature of 5 °C min^{-1}.

2.2 Characterization.

X-ray diffraction (XRD) patterns of the samples were recorded on an X-ray diffractometer (Rigaku D/Max 2200PC) with a graphite monochromator and CuKα radiation (λ=0.15148 nm) in the range of 10-80° at room temperature while the voltage and electric current were held at 40 kV and 20 mA. The morphology and microstructure of the products were determined by transmission electron microscopy (TEM, JEM-100CXII) with an accelerating voltage of 80 kV and high-resolution TEM (HR-TEM, GEOL-2010) with an accelerating voltage of 200 kV.

3. Results and Discussion

Table1. Phases and Morphologies of Products obtained from different system

Volume of Mixed Solvent	H_2O (x mL)	Product	Morphology
20+x mL	0-50 μL	C-In_2O_3	Nanocubes
	0.05-0.10 mL	C-In_2O_3	Nanoparticles
	0.50-1.50 mL	InOOH	Nanoparticles
	1.50-5.00 mL	InOOH & $In(OH)_3$	Nanoparticles
	>5.00 mL	$In(OH)_3$	Nanoparticles

Through adjusting the systemic amount of water, different kinds of products were obtained. (See Table 1) The XRD pattern of the sample obtained in ethanol solvothermal route, which is denoted as sample a, can be indexed to the cubic In_2O_3 (JCPDS, No. 65-3170), no other peaks can be observed, revealing its phase-pure cubic structures. The size of sample is 23 nm calculated from the Scherrer equation. Further calculation revealed that the cell parameters of the sample is 1.0142 nm, which is in well agreement with the reported data of a=1.0140 nm (JCPDS, No. 65-3170). The data was collected using KCl as standard graph and the scan speed was one degree per minute. TEM images (Figure 2a) of the samples obtained from ethanol solvothermal route show that the nanocuboids are with smooth and regular quadrangle surface. FE-SEM images (Figure 3a) of sample a indicate that the In_2O_3 nanocrystals show as regular nanocubes with the size ca. 25 nm and thickness ca. 8-9 nm.

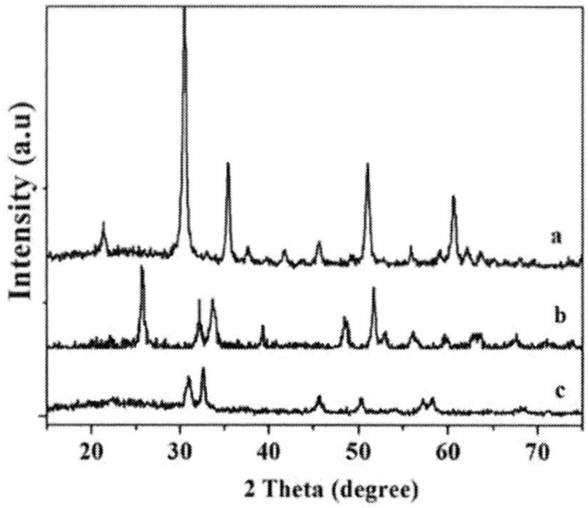

Figure 1 XRD patterns of C-In_2O_3 (a) InOOH (b) and its calcined product H-In_2O_3 (c)

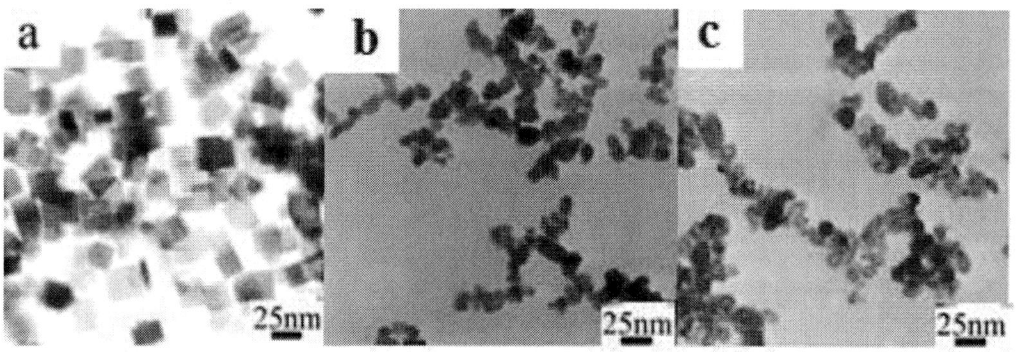

Figure 2 TEM image of C-In_2O_3 (a), InOOH (b) and its calcined product H-In_2O_3 (c)

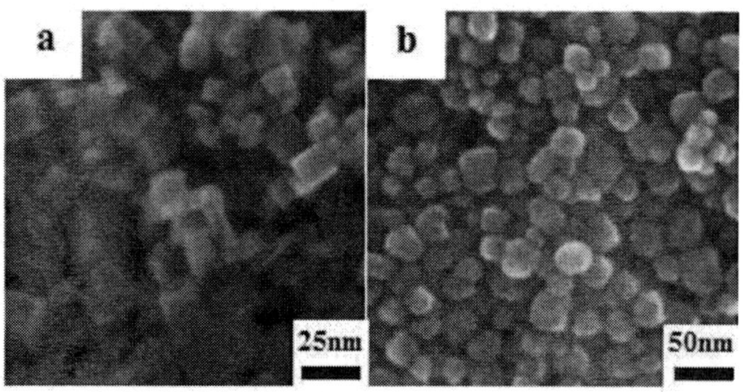

Figure 3 FE-SEM image of C-In_2O_3 (a) and H-In_2O_3 (b)

As known, the metal indium ion easily hydrolyzes to form metal hydroxide or oxide. In ethanol solvothermal reaction system, When 50 μL of H_2O water added, hydrolysis reaction between In^{3+} and H_2O were occurred to form indium hydroxide. From the XRD patterns of products obtained from different time during solvothermal procedure, we can know the generated $In(OH)_3$ from hydrolysis of indium salt further dehydrated to form C-In_2O_3 nancrystals which can be proposed as a "dissolution-recrystallization" mechanism, and the driving force is the lower solubility of the In_2O_3 compared to the $In(OH)_3$ in the system. (See Figure 4) The reaction equations are as follows:

$$In^{3+} + 3H_2O \rightarrow In(OH)_3 + 3H^+$$

$$2In(OH)_3 \rightarrow In_2O_3 + 3H_2O$$

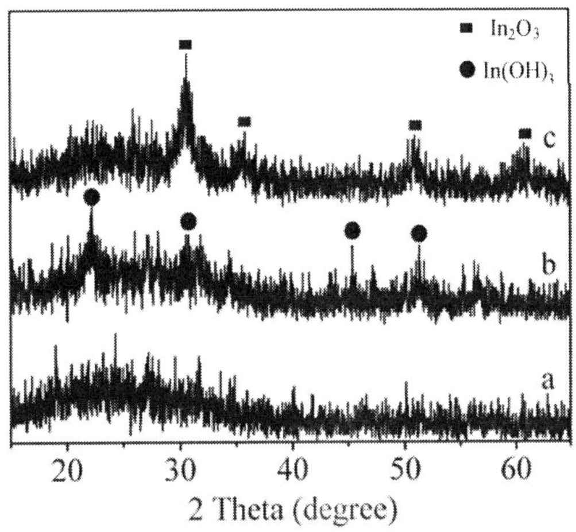

Figure 4 XRD patterns of Products obtained from different time during solvothermal procedure (When 50 μL mL H2O water added). (a)0.5 h, (b)1.0 h, (c) 3.0 h.

When 0.50-1.50 mL water was added, InOOH nanoparticles with size about 20 nm (See Figure 1b and Figure 2b) were obtained. Because the amount of water is not enough in this synstem, the dehydration reaction of $In(OH)_3$ is occurred. InOOH can be regarded as the result of part-hydrolysis of indium ion. The reaction equations are as follows:[12]

$$In^{3+} + 3H_2O \rightarrow In(OH)_3 + 3H^+$$

$$In(OH)_3 \rightarrow InOOH + H_2O$$

After calcining the as-prepared InOOH nanocrystals in a boat crucible at a temperature of 500 °C and maintained for 2 h in air, phase-pure metastable In_2O_3 nanocrystals were obtained. The XRD pattern of the product (Figure 1c) was in good agreement with that of H–In_2O_3 (JCPDS, No.73-1809). All reflections show high intensity, which reveals its high crystallinity, of which size is about 20-30 nm. (See Figure 1c, Figure 2c and Figure 2b) When more water was added, the system adequately hydrolyzed and $In(OH)_3$ nanoparticles were obtained. The reaction equations are as follows:

$$In^{3+} + 3H_2O \rightarrow In(OH)_3 + 3H^+$$

Through adjusting the systemic amount of water, the In^{3+} hydrolysis reaction can be controlled. Of course, the phases and morphologies of products also can be controlled. Herein, a simple hydro/solvothermal route to the In_2O_3 nanocrystals with different phase by using $In(acac)_3$ as indium source was introduced instead of $In(NO_3)_3$, which is not similar to our previous report.[9] $In(acac)_3$ was frequently used as a precursor to prepare In_2O_3. As a stable indium complex, $In(acac)_3$ can release indiun ion slowly under hydro/solvothernal environment in our system, where hydrolysis reaction speed can be controlled.

The IR spectra of C-In_2O_3 and H-In_2O_3 were shown above (Figure 5). The IR spectrum of C-In_2O_3 indicated that the bands around 3400 and 1630 cm^{-1} can be ascribed to the -OH vibrations of absorbed water and ethanol. The bands around 2925, 2858 and 1416 cm^{-1} can be ascribed to the C-H vibrations of methyl and methylene. The bands around 1050 cm^{-1} can be ascribed to the C-O vibrations. The bands around 500 cm^{-1} can be ascribed to the In-O vibrations of In_2O_3.[13, 14] All data shown above indicated that some acetylacetone molecular and ethanol on the surface of C-In_2O_3 through coordination bond. On the contrary, the IR spectrum of H-In_2O_3 is simple which indicted that their surface is very clean after calcining process.

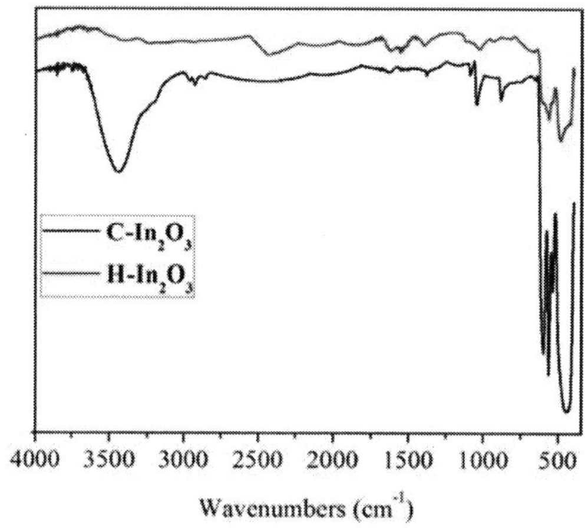

Figure 5 IR spectra of C-In_2O_3 and H-In_2O_3

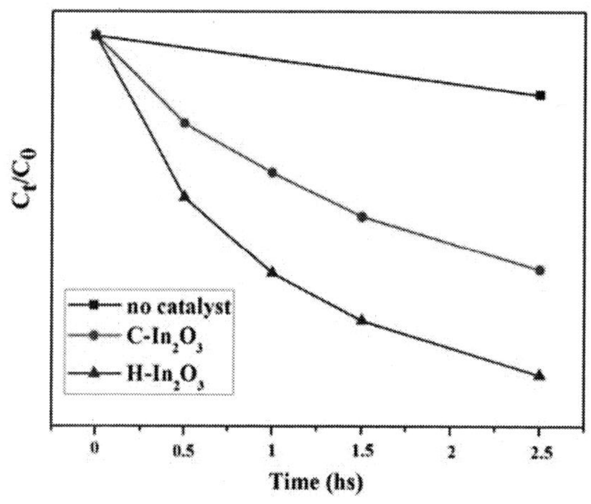

Figure 6 Photo-degradation graphs of phenol solution (1.0 mmol/L, 75.00 mL) under UV light (300-400 nm, 125 W).

Moreover, the photocatalytic performances of C–In_2O_3 and H–In_2O_3 nanocrystals were compared by the degradation of phenol under UV irradiation. As shown in Figure 6, the phenol hardly decomposes after irradiation with UV light for 2.5 h without catalyst. C-In_2O_3 exhibits a relatively poor catalytic performance with only 52% phenol decomposing, while higher catalytic activity is obtained with H–In_2O_3 as catalyst and ca.70% of the phenol decomposes under the same conditions. The higher photocatalytic activity of H–In_2O_3 compared to that of the as-prepared C–In_2O_3 might relate to their UV absorptions. Further experiments show that as a photocatalyst, H–In_2O_3 is stable and can be repeatedly used many times. N_2 adsorption experiment gives the BET surface areas of 32 and 26 m^2 g^{-1} for C–In_2O_3 and H–In_2O_3, and no adsorption of the phenol on the In_2O_3 nanoparticle's surface was detected. Then, it is proposed that the high catalytic performance of the H–In_2O_3 may relate to its high atom density of the (001) plane.[15]

4. Conclusion

In summary, the H–In_2O_3 nanoparticles with high crystallinity have been synthesized by a simple hydro/solvothermal method. The micro-difference of the amount of water added would result in the different reaction mechanisms and form different polymorphs. Moreover, the H-In_2O_3 exhibited high photocatalytic activity for the degradation of phenol.

Acknowledgements

This work is supported by National Natural Science Foundation of China (21441001), Natural Science Foundation of Shandong Province (ZR2014BM018) and Preponderant Discipline Innovative Talent Team Incubation Program of University in Shandong Province.

References

[1] Binglin Tao, Ying Zhang, Dezhi Han, Yanpeng Lia and Zifeng Yan, J. Mater. Chem. A, 2 (2014) 5455;
[2] J. Liu, Y. Zhu, J. Liang, Y. Qian, Int. J. Electrochem. Sci., 7 (2012) 5574;
[3] Won Joo Kim, Bong-Ki Min, Debabrata Pradhanc and Youngku Sohn, CrystEngComm, 17 (2015) 1189;
[4] B. García-Domene, J. A. Sans, F. J. Manjón, etc., J. Phys. Chem. C 119 (2015), 29076;
[5] Liping Gao, Fumin Ren, Zhixuan Cheng, etc., CrystEngComm, 17 (2015) 3268;
[6] Xinghua Liang, Qingqing Song, Lin Shi, Yusi Liu, etc. Int. J. Electrochem. Sci., 10 (2015) 4682;
[7] N. Pinna, G. Garnweitner, M. Antonietti and M. Niederberger, J. Am. Chem. Soc., 127 (2005) 5608;
[8] Eva-Maria Köck, Michaela Kogler, Matthias Grünbacher, etc., J. Phys. Chem. C 120 (2016) 15272;
[9] Zhuangdong Yuan, Jing Zhang, Guodong Liu, Int. J. Electrochem. Sci., 8 (2013) 1794;
[10] J. Buha, I. Djerdj and M. Niederberger, Cryst. Growth Des., 7 (2007), 113;
[11] Guodong Liu, Dairong Chen and Xiuling Jiao, CrystEngComm. 11 (2009) 1828;
[12] Yu D P, Yu S H, Zhang S Y, et al. Adv. Funct. Mater., 13 (2003) 497;
[13] Xing, X.; Chen, S.; Me, S. The workable index of infrared spectra (in Chinese), Tianjin Science and Technology Press, Tianjin, 1992;
[14] Ying, Y.; Wang, S.; Li, C.; Han, X.; Li, Z.; Li, Y. The infrared spectra of minerals (in Chinese), Science Press, Beijing, 1982.
[15] Li, Z; Xie, Z.; Zhang, Y.; Wu, L.; Wang, X.; Fu, X. J. Phys. Chem. C, 111 (2007) 18348.

Research on spontaneous activity in adult anisometropic amblyopia with regional homogeneity

Yufeng Huang* **and Yifeng Zhou**

Center of Biomedical Engineering, Department of Electronic Science and Technology, University of Science and Technology of China, Hefei, Anhui 230027, PR China

*Corresponding author e-mail: yufeng@mail.ustc.edu.cn

Abstract. Amblyopia usually occurs in early childhood and results in monocular visual impairment. The functional magnetic resonance imaging (fMRI) studies have reflected functional anomaly in amblyopia. In resting-state fMRI study, spontaneous activity changes abnormally in anisometropic amblyopia could be revealed by the regional homogeneity (ReHo). Twenty two adult anisometropic amblyopes and Twenty one normal controls participated in this fMRI study. Two sample T test was carried out to analysis ReHo within the whole brain for the inter groups. Compare with normal group, our study found that the amblyopia's ReHo mainly increased in the left frontal lobe, while decreased in the left cerebellum, the temporal lobe (left and right), and the left parietal lobe. And the ReHo values in middle and inferior temporal lobe, the prefrontal lobe, frontal lobe (positive) and parietal lobe and medial frontal gyrus (negative) could be correlated with the acuity deficit of amblyopia. The results increased in ReHo may indicate compensatory plasticity in higher vision information process, while the decreased in ReHo may reflect decreased ability in eye movement, spatial sense and visuo-motor coordination. The correlation revealed that the vision deficit may correspond to the spontaneous in certain brain area.

1. Introduction

Amblyopia, or "lazy eye", is a disorder of visual system characterized by vision in one eye is poor despite the other eye being healthy. The disorder may due to monocular deprivation in the early life and is also an important cause of unilateral visual deficit [1, 2].

Recently, neuroimaging studies have been used to analysis the functional abnormalities in amblyopia. Goodear and Hess [3, 4] report that the visual cortex is the origin deficits site of the visual pathway and the primary and extra striate vision region are responsible for the abnormalities in amblyopia. Compare with the task fMRI researches, the resting state studies need no stimulate and could value the spontaneous neuronal activity of human[5]. Regional homogeneity (ReHo) is a statistic way to evaluate the similarity between the time series of certain voxel and its nearby voxels.[6]. The ReHo has been effectively used to study lots of diseases [7-9]. For the vision area, the blind patients have higher ReHo result than the normal group in the visual cortex[10].Compared with normal vision, Lin[11] first reports that the anisometropic amblyopia shows significant ReHo values change in multiple brain regions. However the subjects have a large range of age and the ReHo results have no relationship with vision acuity deficit. Our study decreases the range of age in order to reduce the age effect and increase the number of subjects with different rank of acuity deficit, so that we can make a full understanding of the spontaneous brain activity patterns defects in amblyopia.

2. Methods and Materials

2.1 Subjects
This study contained two groups: normal controls (n = 21, mean age: 25.12 ± 1.58 years old) and anisometropic amblyopes (n = 22; mean age: 24.27 ± 2.23 years old). A comprehensive eye examination was carried out by a clinician. All participants had no strabismus and no history of eye pathology before the study (Table 1). All subjects were naive to the purpose of the experiment. This study complied with the Declaration of Helsinki and was approved by the University of Sciences and Technology of China Guidelines on Research Ethics.

Table 1. Summarized clinical details for the amblyopes

Number			22
Age (years old)	Mean ± SD		24.27 ± 2.23
	Min: Median: Max		20:24:28
Gender	*Female: Male*		5:17
Best corrected visual acuity (LogMAR)	Mean ± SD	Fellow eye	0.035 ± 0.05
		Amblyopic eye	0.581 ± 0.33
	Min: Median: Max	Fellow eye	0:0:0.1
		Amblyopic eye	0.22:0.46:1.30

2.2 Magnetic resonance imaging acquisition
Scanning was performed at the Anhui Medical University using a GE Signa HDx 3.0T whole-body with an eight-channel receive-only head coil. Each scanning session began with the acquisition of high-resolution three-dimensional T1 weighted images, which were acquired using an MP-RAGE sequence (TR = 2300 ms; TE = 2.94 ms; flip angle = 9°; 176 slices; voxel size = 1×1×1 mm^3. Functional scans were then conducted immediately thereafter. All functional scans consisted of T2*-weighted gradient-echo planar images depicting blood oxygen level-dependent (TR = 2000 ms; TE = 28 ms; flip angle = 72°; 26 slices; voxel size = 3×3×3mm^3).

2.3 Data preprocessing
All data processing were carried out using SPM8, DPARSF v2.2(http://restfmri.net) [12]. To avoid start-up magnetization transients, data that were collected in the first 10 seconds (5 × TR) were removed in each functional run. Each subject's fMRI data were slice-time corrected, motion corrected, and temporally detrended with respect to the head motion and up to order 3 polynomial drift correction. Then the corrected data were registered to the standard Montreal Neurological Institute (MNI) template and further spatially smoothed with a Gaussian kernel with FWHM of 4 mm. Data with a head movement larger than 2 mm was removed from further analysis.

3. ReHo analysis and statistical analysis
ReHo was calculated by Kendall's coefficient of concordance (KCC) and represented the similarity of the time series of the voxels[13, 14]:

$$W = \frac{\sum (R_i)^2 - n(\overline{R})^2}{\frac{1}{12}K^2(n^3 - n)} \quad (1)$$

Ri is the sum of the ith time point; K is the number of time series within a measured cluster, R is the mean of the Ri; and n is the number of ranks[6].

An Independent two-sample t-test was carried out to judge if the ReHo value were significantly different between normal and amblyopia groups. The correlation between the ReHo and the amblyopic eye's visual acuity was calculated as the Pearson correlation coefficient (ranging between -1 and 1; 2-tailed).

4. Result

There were no significant differences between the amblyopia group and controls in age (P=0.35, two-sample t-test) and gender (p=0.8, Chi-squared test). Compared with the control groups, the amblyopia showed significantly decreased in the left cerebellum, the temporal lobe (left and right), and the left parietal lobe while increased in the left frontal lobe (figure 1 and table 2).

Fig. 2 shows the correlation between the ReHo and the visual acuity of the amblyopic eye. Pearson correlation analysis showed that there was a significantly positive correlation in the middle and inferior temporal lobe, the prefrontal lobe and frontal lobe while negative correlation in parietal lobe and medial frontal gyrus. Such phenomena existed in the figure 2 and table 3 as we studied here.

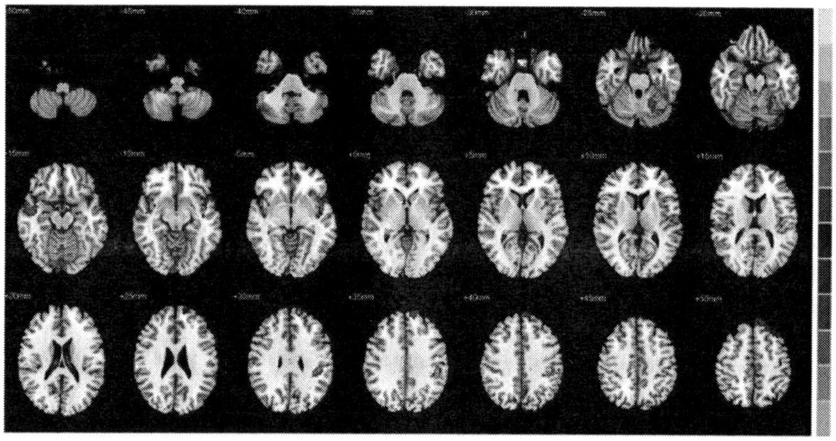

Figure1 The group different brain areas in the ReHo for the normal controls and the amblyopia groups. ($p<0.05$, Alphasim corrected, cluster size 85, 2-tailed independent two-sample t-tests).

Table2 The difference ReHo values of brain areas between normal controls and amblyopia groups ($p<0.05$, Alphasim corrected, cluster size 85, 2-tailed independent two-sample t-tests).

Brain region	Brodmann region	Cluster size	T-values	Coordinates in MNI(x,y,z)
The left Cerebellum		97	3.0799	-30, -69, -21
The right Temporal Lobe	22	98	4.297	42, 3, -6
The left Temporal Lobe	22	112	3.3108	-48, -3, -6
The left Parietal Lobe	39	85	4.6193	-48, -35, 33
The left Frontal Lobe	8	128	-3.867	-21, 36, 57

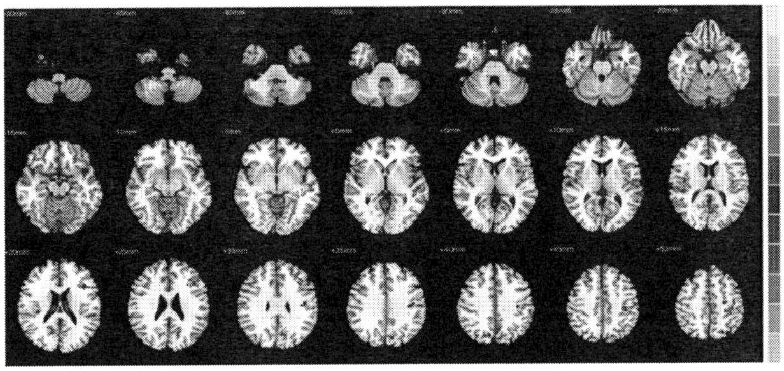

Figure 2 The significant correlate brain areas (Pearson correlation coefficients) between the ReHo and the visual acuity (logMAR) of the amblyopic eye. ($p<0.05$, Alphasim corrected, cluster size 85)

Table 3 The significant correlate brain areas between the Reho and the visual acuity (logMAR) of the amblyopic eye. (Pearson correlation coefficients, $p<0.05$, Alphasim corrected, cluster size 85)

Brain region	Brodmann region	Cluster size	T-values	Coordinates in MNI (x,y,z)
The left Temporal Lobe	20	86	0.61813	-39, -15, -30
	21	91	0.78747	-45, 6, -3
The left Frontal Lobe	10	97	0.6753	-12, 69, 15
	9	97	0.71099	-36, 24, 27
		91	0.75176	-6, 51, 51
The left Parietal Lobe	7	108	- 0.62165	-33, -30, 48
The left Frontal Lobe	9	91	0.75176	-6, 51, 51
The left Medial Frontal Gyrus	6	188	- 0.72901	-9, -12, 60

5. Discussion

The first main findings are the decrease and increase ReHo in several brain areas. The decrease value in cerebellum is believed to effect the control of the eye movement and interact with the frontal lobe[15, 16]. The temporal lobe is involved in processing sensory input into appropriate derived meaning of visual memory[17], so that the amblyopia may have difficult in this aspect. The parietal lobe integrates sensory information among various modalities, so the decrease may affect the spatial sense and visuo-motor coordination[18] for amblyopia[17].While the increase in frontal lobe may due to it involves the ability to various cognitive and perceptual tasks, so the amblyopia need more higher cognitive functional to do more work[19].

Further, the ReHo value has positive correlation with the severity of amblyopia in middle and inferior temporal lobe, the prefrontal lobe and frontal lobe. The results are similarity to the above t-test results with normal control group. Its reveal that the anisometropic amblyopia spontaneous activity defects in certain brain areas are related to acuity deficit. And ReHo is a new method to evaluate the level of visual deficit and the correlate is reasonable because the abnormal areas are basically in accordance with the T-test results. The negative correlations reflect the compensatory plasticity in parietal and medial frontal gyrus.

The findings are encouraging, while there still exist several limitations as follows: We need to enlarge the sample number in order to get a more reliable ReHo result. And we should classify the left or right eye impairments amblyopia separately. Further the strabismus amblyopia should be considered to get the more comprehensive understanding about the models of amblyopia.

Acknowledgements
Financial support: This work is supported by the National Natural Science Foundation of China (NSFC 81261120562 and 31230032 to Yifeng Zhou)

References
[1] Hess, R. and E. Howell, *The threshold contrast sensitivity function in strabismic amblyopia: evidence for a two type classification.* Vision research, 1977. 17(9): p. 1049-1055.
[2] Holmes, J.M. and M.P. Clarke, Amblyopia. The Lancet, 2006. 367(9519): p. 1343-1351.
[3] Goodyear, B.G., et al., BOLD fMRI response of early visual areas to perceived contrast in human amblyopia. Journal of Neurophysiology, 2000. 84(4): p. 1907-1913.
[4] Hess, R.F., et al., The contrast dependence of the cortical fMRI deficit in amblyopia; a selective loss at higher contrasts. Human brain mapping, 2010. 31(8): p. 1233-1248.
[5] Raichle, M.E., et al., *A default mode of brain function.* Proceedings of the National Academy of Sciences, 2001. 98(2): p. 676-682.
[6] Zang, Y., et al., *Regional homogeneity approach to fMRI data analysis.* Neuroimage, 2004. 22(1): p. 394-400.
[7] Shukla, D.K., B. Keehn, and R.A. Müller, Regional homogeneity of fMRI time series in autism spectrum disorders. Neuroscience letters, 2010. 476(1): p. 46-51.
[8] Wu, T., et al., Regional homogeneity changes in patients with Parkinson's disease. Human brain mapping, 2009. 30(5): p. 1502-1510.
[9] Wang, L., et al., Regional homogeneity of the resting-state brain activity correlates with individual intelligence. Neuroscience letters, 2011. 488(3): p. 275-278.
[10] Liu, C., et al., Increased regional homogeneity of blood oxygen level-dependent signals in occipital cortex of early blind individuals. Neuroreport, 2011. 22(4): p. 190-194.
[11] Lin, X., et al., Altered spontaneous activity in anisometropic amblyopia subjects: revealed by resting-state FMRI. PloS one, 2012. 7(8): p. e43373.
[12] Yan, C. and Y. Zang, DPARSF: a MATLAB toolbox for" pipeline" data analysis of resting-state fMRI. Frontiers in systems neuroscience, 2010. 4: p. 13.
[13] Liang, P., et al., Regional homogeneity changes in patients with neuromyelitis optica revealed by resting-state functional MRI. Clinical Neurophysiology, 2011. 122(1): p. 121-127.
[14] Kendall, M.G., Rank correlation methods. 1948.
[15] Middleton, F.A. and P.L. Strick, Cerebellar projections to the prefrontal cortex of the primate. Journal of neuroscience, 2001. 21(2): p. 700-712.
[16] Nitta, T., et al., Involvement of the cerebellar dorsal vermis in vergence eye movements in monkeys. Cerebral Cortex, 2008. 18(5): p. 1042-1057.
[17] Schacter, D.L., D.T. Gilbert, and D.M. Wegner, Psychology, 2nd ed., Worth, New York, 2010.
[18] Milner, A.D. and M.A. Goodale, Two visual systems re-viewed. Neuropsychologia, 2008. 46(3): p. 774-785.
[19] Fadiga, L. and L. Craighero, Hand *actions and speech representation in Broca's area.* Cortex, 2006. 42(4): p. 486-490.

IOP Conf. Series: Materials Science and Engineering **207** (2017) 012009 doi:10.1088/1757-899X/207/1/012009

Chemical compounds from *Chenopodium album* Linn.

Wei Zheng*

Research Center on Life Sciences and Environmental Sciences, Harbin University of Commerce, Harbin 150076, China

*Corresponding author e-mail: wei2013zheng@163.com

Abstract. Bioactive components from *Chenopodium album* Linn. were isolated and identified in this research. Light petroleum, dichloromethane and n-BuOH were firstly applied to partition the 75% EtOH extract of *Chenopodium album* Linn. which were then subjected to normal-phase silica, ODS silica gel column chromatography and semi-preparative HPLC chromatography. By the employment of NMR method in this study, chemical structures of the compounds were elucidated. Three known compounds were isolated from *Chenopodium album* Linn., and identified as Isolariciresinol 4-O-β-D-glucopyranoside (1), (7'S, 8R, 8'R)-Isolariciresinol (2) and (7'S, 8R, 8'S)-Isolariciresinol (3) by comparison of their spectral data with references. This is the first time that isolation of the compounds mentioned above from *Chenopodium album* Linn. was achieved.

1. Introduction

Chenopodium album Linn. is a native plant widely distributed around the world, which contains more than 250 species [1]. According to the reports, C. Album can be represented by 21 species in Indian. Some species are grown as edible vegetables, providing people vitamins, fibre, essential fatty acids and minerals. A few other species were cultivated for the grains gained from the plant [2]. In addition, it was also reported that C. Album naturally grow as weed in the fields of barley, wheat, gram, mustard and some other crops [3-4]. Compared with the weed of low growing, the cultivated plants are leafy and tall. The entire young plant is not only a sorts of nutritious food, but also a herbal medicine [5-6]. C. Album has been traditionally employed as a diuretic, blood purifier, hepatoprotective, sedative, antiscorbutic laxative. Pharmacological research reveals its containing anthelmintic, contraceptive and sperm immobilizing properties [7].

2. Material and Methods

2.1. Materials and Chemicals

The Chenopodium album Linn., leaves were purchased from Hebei Qixin Traditional Chinese Medicine Pellets Co., Ltd., P.R. China, and identified by Dr. Chen Ning, Research Center on Life Sciences and Environmental Sciences, Harbin University of Commerce. Organic solvents (analytical grade or HPLC grade) for the experiment were purchased from Kermel Chemical Co., (Tianjin, China).

2.2. General Experimental Procedures

In this research, tetramethylsilane was used as internal standard on a Bruck AVANCE 500 FT-NMR spectrometer to record the nuclear magnetic resonance (NMR) spectra. Silica gel (200-300 mesh)

(Marine Chemical Factory, Qingdao, China) and Octa Decylsilyl Silicion (ODS) (YMC C18, 40-70 μm) were adopted to perform column chromatography. The ODS column (YMC ODS-A, 5 μm, 250×10 mm) was applied to perform High Performance Liquid Chromatography (HPLC) separation.

2.3. Extraction and Isolation

Dried leaves of Chenopodium album Linn., (20 kg) were powdered and extracted twice times by using of 75% EtOH under reflux. A condensed extract (2.2 kg) was obtained via evaporation of the solvent under reduced pressure, and then subjected to macroporous resin and eluted with a gradient of EtOH-H2O (0%, 20%, 50%, 95%, v/v). 11 fractions were obtained after 50% fraction went through normal phase silica gel column chromatography and was eluted with a gradient of CH2Cl2-MeOH (100:0 → 0:100, v/v). Fraction 3 (CH2Cl2-MeOH, 30:1, v/v) was further separated using ODS silica gel column chromatography eluted with MeOH in water. Subfraction 3.2 and 3.4 were purified by semi-preparative HPLC to give compound 1 (14.1 mg), 2 (16.7 mg) and 3 (18.6 mg).

3. Structural Identification

Isolariciresinol 4-O-β-D-glucopyranoside (1): yellow oil. 1H-NMR (600 MHz, DMSO-d6) δH: 6.94 (1H, brs, H-2), 6.63(1H, d, J = 1.6 Hz, H-2'), 6.53 (1H, d, J = 8.1 Hz, H-5'), 6.50(1H, brs, H-5), 6.49 (1H, dd, J = 8.1, 1.6 Hz, H-6'), 4.51 (1H, d, J = 7.8 Hz, H-1''), 3.75 (1H, d, J = 8.4 Hz, H-7'), 3.70 (3H, s, 3-OMe), 3.69 (3H, s, H-3'-OMe), 3.61 (1H, m, H-6''b), 3.56 (2H, m, H-9), 3.41 (2H, m, H-9'), 3.32 (1H, m, H-6''a), 3.26 (1H, m, H-5''), 3.25 (1H, m, H-3''), 3.23 (1H, m, H-2''), 3.11 (2H, m, H-4''), 2.75 (2H, d, J = 9.6 Hz, H-7), 1.83 (1H, m, H-8'), 1.65 (1H, t, J = 9.8 Hz, H-8); 13C-NMR (125 MHz, DMSO-d6) δC: 147.2 (C-3), 146.7 (C-3'), 144.7 (C-4'), 144.1 (C-4), 136.4 (C-1'), 132.6 (C-6), 130.1 (C-1), 121.4 (C-6'), 116.5 (C-5), 115.2 (C-5'), 113.3 (C-2'), 112.1 (C-2), 100.1 (C-1''), 76.9 (C-5''), 76.6 (C-3''), 73.0 (C-2''), 68.5 (C-4''), 63.5 (C-9), 59.9 (C-6''), 59.3 (C-9'), 55.7 (C-3'-OMe), 55.6 (C-3-OMe), 45.9 (C-7'), 45.3 (C-8'), 38.0 (C-8), 32.2 (C-7).

(7'S, 8R, 8'R)-Isolariciresinol (2): yellow oil. 1H-NMR (600 MHz, DMSO-d6) δH: 6.96 (1H, brs, H-2), 6.70 (1H, d, J = 8.1 Hz, H-5'), 6.65 (1H, d, J = 1.7 Hz, H-2'), 6.50 (1H, dd, J = 8.1, 1.7 Hz, H-6'), 6.11 (1H, brs, H-5), 3.74 (1H, d, J = 10.1 Hz, H-7'), 3.73 (6H, s, H-3-OMe, -3'-OMe), 3.59(2H, dd, J = 10.6, 4.0 Hz, H-9), 3.46 (2H, m, H-9'), 2.68 (2H, dd, J= 12.7, 7.4 Hz, H-7), 1.86 (1H, m, H-8'), 1.64 (1H, t, J = 10.1 Hz, H-8); 13C-NMR (125 MHz, DMSO-d6) δC: 147.5 (C-3), 145.6 (C-3'), 144.8 (C-4'), 144.2 (C-4), 137.2 (C-1'), 132.8 (C-1), 126.8 (C-6'), 122.5 (C-6), 116.4 (C-5), 115.4 (C-5'), 113.4 (C-2'), 111.9 (C-2), 63.8 (C-9), 60.0 (C-9'), 55.8 (C-3'-OMe), 55.6 (C-3-OMe), 46.1 (C-7', -8'), 38.3 (C-8), 32.4 (C-7).

(7'S, 8R, 8'S)-Isolariciresinol (3):yellow oil. 1H-NMR (600 MHz, DMSO-d6) δH: 6.69 (1H, d, J = 1.6 Hz, H-2'), 6.61 (1H, brs, H-2), 6.55 (1H, d, J = 8.1 Hz, H-5'), 6.21 (1H, brs, H-5), 6.20 (1H, dd, J = 8.1, 1.6 Hz, H-6'), 4.08 (1H, d, J = 3.9 Hz, H-7'), 3.69 (3H, s, H-3-OMe), 3.64 (3H, s, H-3'-OMe), 3.30 (2H, m, H-9'), 3.13 (2H, m, H-9), 2.78 (2H, dd, J = 16.9, 5.5 Hz, H-7), 1.98 (1H, m, H-8'), 1.79 (1H, t, J = 10.1 Hz, H-8); 13C-NMR (125 MHz, DMSO-d6) δC: 146.5 (C-3), 146.2 (C-3'), 144.5 (C-4), 144.2 (C-4'), 134.1 (C-1'), 131.5 (C-1), 126.8 (C-6'), 122.5 (C-6), 116.1 (C-5), 114.7 (C-5'), 114.6 (C-2'), 111.7 (C-2), 63.3 (C-9), 60.6 (C-9'), 55.6 (C-3'-OMe), 55.4 (C-3-OMe), 44.1 (C-7'), 43.1 (C-8'), 38.5 (C-8), 31.8 (C-7).

4. Conclusion

In this study, Isolariciresinol 4-O-β-D-glucopyranoside (1), (7'S, 8R, 8'R)-Isolariciresinol (2) and (7'S, 8R, 8'S)-Isolariciresinol (3) from Chenopodium album Linn. were successfully isolated and identified by comparison of their spectral data with the reported reference [8-10].

Acknowledgments

This work was financially supported by the Doctoral Science Foundation of Harbin University of Commerce [Grant number 14LG06 and 13DL015], Scientific Research Fund of Heilongjiang

Provincial Education Department [Grant number 12541202] and Young Talent Project of Harbin [Grant number 2016RQQXJ127].

References

[1] Risi J, Galwey NW, In Advances in applied biololgy, in: Coaker TH (ed), The Chenopodium grains of the andes: inca crops for modern agriculture. Academic, Landon, 1984, pp. 145-216.
[2] Yadav N, Vasudeva N, Singh HS, Sharma SK, Medicinal properties of genus Chenopodium Linn, Natural Product Radiance. 6 (2007) 131-134.
[3] Khurana SC, Malik YS, Pandita ML, Herbicidal control of weeds in potato C.V. Kufribadshah, Pesticides. 20 (1986) 55-56.
[4] Bhattacharjee SK, Handbook of medicinal plant. Pointer publishers: Jaipur, 3rd Edn., 2001, p1-2.
[5] Laghari AH, Memon S, Nelofar A, Yasmin KM Determination of free phenolic acids and anti-oxidant activity of methanolic extracts obtained from fruits and leaves of Chenopodium album, FoodChem. 126 (2011) 1850-1855.
[6] Nayak DP, Swain PK, Panda OP, Pattanaik P, Srinivas B, Antimicrobial and anthelmintic evalution of chenopodium album, Int J Pharma World Res. 4 (2010) 201-215.
[7] Amrita Poonia and Ashutosh Upadhayay, Chenopodium album Linn: review of nutritive value and biological properties, J Food Sci Technol. 52 (2015) 3977-3985.
[8] Kim MH, Jiang JH, Oh MH, et al., The comparison of DPPH-scavenging capacity and antiinflammatory effects of phenolic compounds isolated from the stems of Stewartia koreana Nakai, Natural Product Research. 28 (2014) 1409-1412.
[9] Anjun Deng, Haijing Zhang, Zhihui Zhang, et al, Constituents with anti-oxidative activity from seeds of Jufeng grape, China Journal of Chinese Materia Medica. 21 (2015) 102-105.
[10] Ramdayal FD, Kiemle DJ, LaLonde RT, Directed, DDQ-promoted benzylic oxygenations of tetrahydronaphthalenes, The Journal of Organic Chemistry. 64 (1999) 4607-4609.

IOP Conf. Series: Materials Science and Engineering **207** (2017) 012010 doi:10.1088/1757-899X/207/1/012010

Preparation, characterization and properties of ZnO nanomaterials

Jiaolian Luo[1, 2], Xiaoming Zhang[1, 3], Ruxue Chen[1, 2], Xiaohui Wang[1, 3], Ji Zhu[1, 2], Xiaomin Wang[1, 2]

[1]Special and Key Laboratory of Guizhou Provincial Higher Education for Green Energy-Saving Materials. Guizhou Minzu University, Guiyang 550025, P.R. China
[2]School of materials science and engineering, Guizhou Minzu University, Guiyang 550025, P.R.China
[3]College of big data and information engineering, Guizhou University, Guizhou Guiyang 550025, China

Abstract. In this paper, using the hydrothermal synthesis method, NaOH, $Zn(NO_3)_2$, anhydrous ethanol, deionized water as raw material to prepare ZnO nanomaterial, and by X ray diffraction (XRD), scanning electron microscopy (SEM) and photoluminescence spectroscopy (PL) on the synthesis of nano materials, surface morphology and phase luminescence characterization. The results show that the nano materials synthesized for single-phase ZnO, belonging to the six wurtzite structure; material surface shaped, arranged evenly distributed, and were the top six party structure; ZnO nano materials synthesized with strong emission spectra, emission peak is located at 394nm.

1. Introduction

With the people's living standard continues to improve, the rapid development of science and technology civilization, in health, environmental protection, energy saving in the drive and drive today, nano technology has gradually become one of the research fields of high-tech fields the most promising. The arrival of the era of nano science and technology will become inevitable, and can bring great benefits to people's lives,this makes the study of nano ZnO favored by many researchers, Tao Xinyong et al[1], ZnO was synthesized by PEG assisted hydrothermal method, and the effects of sodium hydroxide concentration and reaction time on the morphology and size of were studied; Fu Sanling et al[2], Nano ZnO materials were prepared by hydrothermal method ,and the preparation characteristics and preparation mechanism were studied; Huang peng et al[3], Nano ZnO crystals were prepared by hydrothermal method, and the influence of preparation parameters on the morphology crystals was studied by changing the experimental conditions; Zhang Changlong[4], ZnO crystals were grown in large diameter autoclave by hydrothermal method, and the growth rate, morphology and optical properties of the crystals in different directions were studied; A N Baranov[5],the ZnO nanorod with diameters less than 20 nm were prepared at high temperature. Shin P K[6],the ZnO nanowires were formed on a variety of substrates by gas phase method and liquid phase method at the temperature of 200 to 350℃; Li Haili [7] used two step hydrothermal method synthesized ZnO successfully with different structures and the effects of different factors on the structure and properties of ZnO were studied systematically; Yao Chao et al [8], ZnO were prepared by sol-gel method and their photocatalytic properties were studied. Li XiuPing et al [9] ZnO were successfully prepared by homogeneous precipitation method and their photocatalytic properties were studied. In this paper, NaOH, $Zn(NO_3)_2$, anhydrous ethanol, deionized water as raw material, using the hydrothermal synthesis method, prepared ZnO nanomaterial successfully, and through XRD, SEM and PL on the synthesis of nanomaterial phase, surface morphology and luminescent properties characterization.

2. Experiment

Raw material	Annealing Temperature	Annealing Time
$Zn(NO_3)_2$+ NaOH	95℃	2h
$Zn(NO_3)_2$+ NaOH	90℃	2h
$Zn(NO_3)_2$+ NaOH	100℃	2h
$Zn(NO_3)_2$+ NaOH	95℃	4h
$Zn(NO_3)_2$+ NaOH	95℃	6h
$Zn(NO_3)_2$+ NaOH	95℃	8h

According to the experimental measurement, take an appropriate amount of zinc nitrate, adding deionized water to prepare zinc nitrate solution with a certain concentration; at the same time take an appropriate amount of sodium hydroxide (AR) was a sodium hydroxide solution with deionized water dissolving; the solution with constant temperature magnetic stirrer at room temperature stirring; washing with anhydrous ethanol pretreatment; pretreatment after drying at room temperature in 24h, then precipitated precursor to the drying oven (101-1AB) drying 10min; adding deionized water dissolving to obtain precursor into the autoclave, placed in a drying oven, in the different temperature and time of annealing; natural cooling to room temperature after filtering, the white powder into a drying oven for drying, finally produced white ZnO nanomaterial.

3. The experimental result and discussion

3.1 The XRD analysis result of sample ZnO

Figure 1 shows the XRD diffraction pattern (DY1632) of the synthesized nano ZnO materials at different temperatures and different annealing time (a). By comparing the PDF cards, the diffraction peaks of the have been identified. The 6 groups of the experimental materials are the six party wurtzite structure, the lattice constant a=3.2495, c=5.2069, and the intensity and angle of the diffraction peak is basically the same as the ZnO standard spectrum, which is a single-phase ZnO material. The figure shows: the different temperature and time of synthetic samples have strong diffraction peak, but in the hot water temperature is 95℃, the annealing time is 2h, the synthesis of diffraction peak material relative to the other 5 groups of XRD diffraction peaks will be more sharp, by contrast with the PDF card, can be clearly seen in the crystal the surface (100), (002), (101), (102), (110), (103), (200), (112), (201), (004), the diffraction angle for ZnO nanomaterial under 2 theta test out strength (CPS) for more than 12000. It can be seen that the raw material after repeated ethanol pretreatment, no extra impurity peaks, ZnO nanometer material with good crystallinity and high purity, and under the condition of annealing temperature 95℃and the annealing time 2h ,it has best crystallization properties.

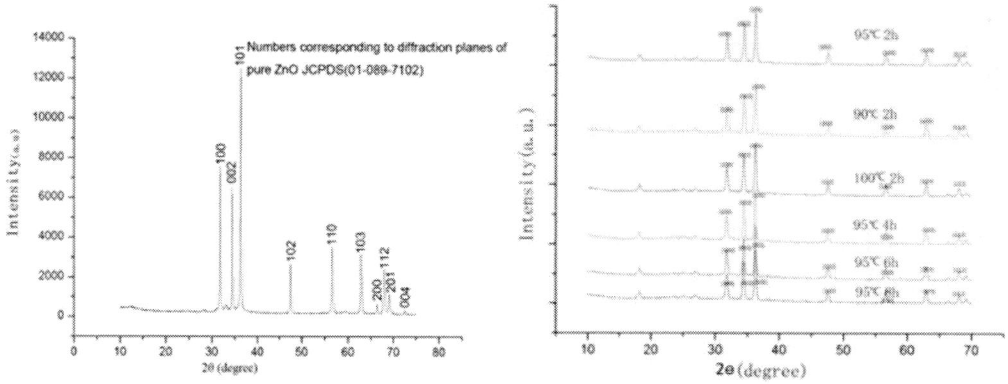

Figure 1 XRD diffraction patterns of the synthesized ZnO materials at different temperatures and different annealing time (a) and pure ZnO (b)

3.2 The SEM analysis result of sample ZnO

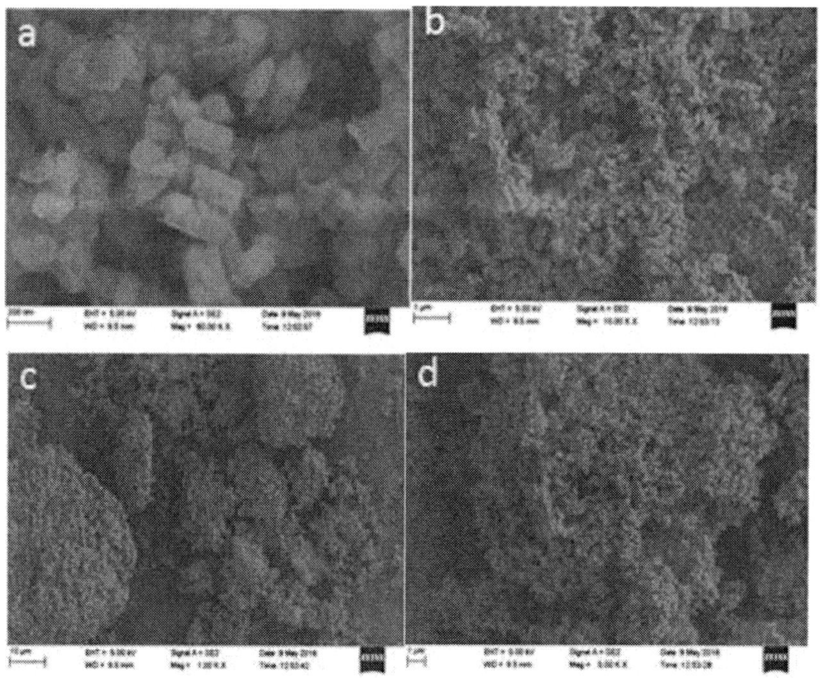

Figure 2 The SEM spectra of ZnO nano materials were prepared by annealing at 95℃ and annealing time at 2h

3.3 The exaction spectrum and emission spectrum analysis of the sample ZnO

Figure 3 is the fluorescence spectra of samples of ZnO nanomaterial (Hitachi, S4100) under the condition of water temperature 95℃, annealing time 2h, can be drawn from the figure: by the excitation of 325 nm light source, showing a wide range of emission spectrum, and the emission peak is located at 394nm, showed that the luminescence intensity of samples very strong, with good photoluminescence properties, can be used in the near ultraviolet and blue chips.

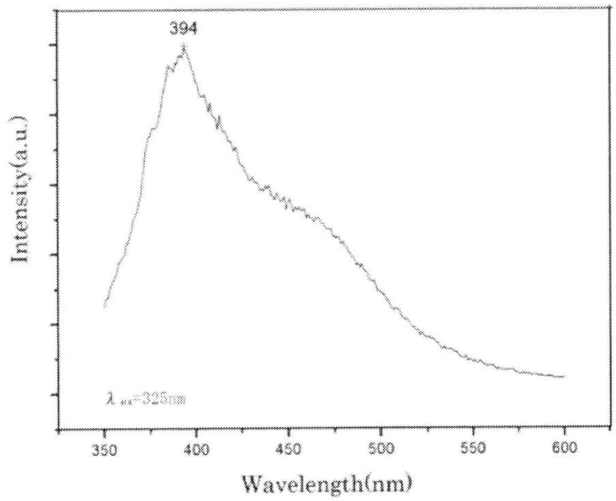

Figure 3 The luminescence property of ZnO nanomaterial were prepared by annealing at 95℃ and annealing time at 2h

4. Conclusion

This paper synthesized by hydrothermal method has relatively uniform grain size, good dispersibility, smooth surface and top ZnO nanomaterial with six party structure, and the phosphor synthesis of the crystal structure, morphology and spectral analysis, the results show that the consistent experimental synthesis of ZnO nanomaterial with the standard PDF card, it has best crystallinity and uniformity under the condition of annealing temperature is 95℃, the annealing time is 2h, while there is a small amount of nanorods, with a wide range of emission spectrum, and the emission front is located at 394 nm.

Acknowledgments

Project supported by the Special and Key Laboratory of Guizhou Provincial Higher Education for Green Energy-Saving Materials, The national college students' innovative entrepreneurship training program of the Chinese Education Department, in 2015,China(201510672001);The Science and Technology Foundation of Guizhou Province, China(The contract LH of Guizhou No.[2014]7384);The Natural science Foundation of Guizhou Province education department, China(The provincial construction projects of undergraduate course teaching, in 2015); The introduction of talent research foundation of Guizhou Minzu university, China (NO.2014(12))

References

[1] Tao X,Zhang X,Kong F,et al. Hydrothermal preparation of PEG assisted ZnO nanorods [J]. Journal of Chemistry, 2004, 62(17):1658-1662.

[2] Fu S,Cai S,Zhang X,et al. Progress in hydrothermal synthesis of ZnO crystals and nanomaterials[J]. Journal of synthetic crystals, 2006, 35(5):1016-1021.

[3] Huang P, Hydrothermal synthesis and luminescence properties of nanocrystalline ZnO[D]. Lanzhou University, 2010.

[4] Song C,Hang Y,Zhang Y,et al.Characteristics of Hydrothermally Grown ZnO Crystal, 2005, 34(6):1083-1087.

[5] Lyapina O A, Baranov A N, Panin G N, et al. Synthesis of ZnO nanotetrapods[J]. Inorganic Materials, 2008, 44(8):846-852.

[6] No I J, Shin P K. ZnO Nanowires Grown by Hydrothermal Synthesis Using Synthesis Solution Prepared with Various Preheating Time [J]. Journal of the Korean Institute of Electrical & Electronic Material Engineers, 2012, 25(6):481-485.

[7] Li H,Hydrothermal synthesis of ZnO nanomaterials and their photocatalysis properties [D], Harbin Institute of Technology,2013.
[8] Yao C, Li F, et al. Preparation and photocatalytic properties of nanometer ZnO [J], Journal of Beijing Institute of Graphic Communication, 2012, 20(2):76-78.
[9] Li X, Li X, Zhao R, Hu Y .Preparation of Nano-ZnO with Homogeneous Precipitation and Its Photocatalytic Property [J], Journal of Liaoning Shihua University, 2014(1):39-42.

IOP Conf. Series: Materials Science and Engineering **207** (2017) 012011 doi:10.1088/1757-899X/207/1/012011

Effect of vulcanization temperature and humidity on the properties of RTV silicone rubber

Xutao Wu[1], Xiuguang Li[1, a], Lu Hao[2, *], Xishan Wen[2, b], Lei Lan[2, c], Xiaoqing Yuan[2, d], Qingping Zhang[1]

[1]Power Science Research Institute of State Grid Ningxia Power Co., Yinchuan, 750001, China

[2]School of Electrical Engineering, Wuhan University, Wuhan, 430072, China

[*]Corresponding author e-mail: dqhl@whu.edu.cn, [a]13895677195@126.com, [b]xswen@whu.edu.cn, [c]lanlei@whu.edu.cn, [d]xqyuan68@whu.edu.cn

Abstract. In order to study the difference in performance of room temperature vulcanized (RTV) silicone rubber in vulcanization environment with different temperature and humidity, static contact angle method, FTIR and TG is utilized to depict the properties of hydrophobicity, transfer of hydrophobicity, functional groups and thermal stability of RTV silicone rubber. It is found that different vulcanization conditions have effects on the characteristics of RTV silicone rubber, which shows that the hydrophobicity of RTV silicone rubber changes little with the vulcanization temperature but a slight increase with the vulcanization humidity. Temperature and humidity have obvious effects on the hydrophobicity transfer ability of RTV silicone rubber, which is better when vulcanization temperature is 5°C or vulcanization humidity is 95%. From the Fourier transform infrared spectroscopy, it can be concluded that humidity and temperature of vulcanization conditions have great effect on the functional groups of silicone rubber, and vulcanization conditions also have effect on thermal stability of RTV silicone rubber. When vulcanization temperature is 5°C or vulcanization humidity is 15% or 95%, the thermal stability of silicone rubber becomes worse.

1. Introduction

With the rapid development of Chinese national economy, the demand of power is increasing, which results in a continuous increase of voltage level of power transmission equipment and the transmission capacity. The safe and stable operation of power system has been attracting more and more attention [1]. The pollution flashover has the characteristics of large area and low reclosing success rate. It may lead to blackout in a large area, which will bring huge economic loss and negative impact to the country and society [2]. Therefore, the pollution flashover problem has become one of the main factors affecting the safe and stable operation of the power grid.

Room temperature vulcanized (RTV) silicone rubber has excellent hydrophobicity and transfer of hydrophobicity, which can significantly improve the flashover voltage of porcelain and glass insulators[2].As a kind of organic insulating material, the performance of RTV silicone rubber has attracted widely attention from scholars all over the world [3, 4].

Few researches on the effect of temperature and humidity on the vulcanization properties of RTV silicone rubber were carried out [5, 6]. Some scholars have studied the effect of vulcanization process of silicone rubber on its vulcanization degree [7], but the basic physical and chemical properties have little been involved.

Dealcoholization type RTV silicone rubber is vulcanized by absorbing the moisture in the air. Small alcohol-based molecules supplied are removed and cross-linking reaction happened. The vulcanization mechanism mentioned above is shown below [8].

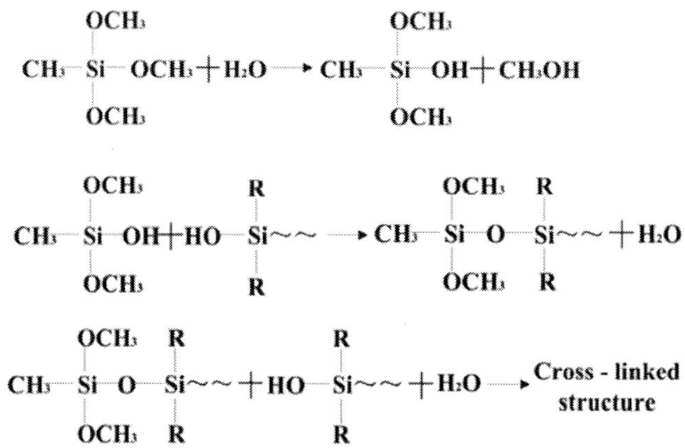

Figure 1 The vulcanization reaction mechanism of RTV silicone rubber

It can be seen from figure 1 that the moisture content in the air has a great impact on the vulcanization reaction. All chemical reactions are dependent on temperature because it affects the extent and speed of chemical reactions. Therefore, the study of temperature and humidity on the properties of RTV silicone rubber vulcanization is very meaningful.

According to DL/T 627-2012, the environmental requirements for RTV silicone rubber is under the temperature of 25°C±2°C, and the relative humidity of 40%~70% [9]. However, in the field, due to environmental constraints, construction technology and other restrictions, it is difficult to meet the vulcanization conditions. In order to solve this problem, RTV silicone rubber vulcanization was carried out under different temperature and humidity conditions. Some basic properties including hydrophobicity, transfer of hydrophobicity, FTIR and TG of the RTV silicone rubber coating were analysed, and the conclusion provided an in-depth understanding of the vulcanization mechanism of RTV silicone rubber coating.

2. Experimental Setup

The test was conducted in a programmable constant temperature and humidity chamber. A kind of commercial available RTV silicone rubber was coated on a 10 cm×10 cm square ceramic sheet, vulcanized for 48 hours under the conditions shown in Table 1, and placed for 7 days under the laboratory environment.

The vulcanization temperatures were chosen to be 5°C/25°C/65°C, in which 25°C is the normal ambient temperature. Relative humidity of 15%, 55% and 95% were chosen, in which 15% is for some certain arid areas, 55% is for the normal humidity and 95% is for the humid areas.

In this paper, we considered the influence of single factor on the vulcanization of RTV silicone rubber. The sample numbers and vulcanization conditions are shown in Table 1.

Table 1 Sample numbers and vulcanization conditions

Sample number	temperature(°C)	Relative humidity (%)
S1	25°C	55%
S2	65°C	55%
S3	5°C	55%
S4	25°C	95%
S5	25°C	15%

The hydrophobicity and transfer of hydrophobicity of RTV silicone rubber coating were studied using SL 200B contact angle measuring device. The contact angles were measured at three points and the average was plotted. In the paper, we use same method for smear treatment [3, 4]. The kieselguhr was chosen as insoluble and sodium chloride as the soluble pollutant. The ash density was 1.0 mg/cm^2 and the salt density was 0.1 mg/cm^2. After brushing and drying, the samples were placed into a plastic box for hydrophobic migration. And temperature of 25°C and humidity 70% were chosen as the migration environment. The infrared analysis was performed by the United States NICOLET company FTIR 5700 Fourier infrared spectrometer, and TG analysis instruments was the Hitachi 7300 TG analyser.

3. Effect of Different Vulcanization Temperature on Properties of RTV

Figure 2 shows that, with the increase of vulcanization temperature, the change of static contact angle is not obvious, which indicates that the vulcanization temperature has little effect on the hydrophobicity of RTV silicone rubber.

From curves of in figure 3, it can be seen that the transfer of hydrophobicity of the sample after vulcanization at 5°C is significantly better. It indicates that under 5°C vulcanization environment, the cross-linking reaction occurs incompletely, resulting in more small molecules of siloxane. With the increase of vulcanization temperature, the cross-linking reaction is further deepened.

Table 2 shows the range of the wave number corresponding to the typical characteristic peak of RTV silicone rubber [10]. From figure 4, it can be seen that the intensity of O-H peak decrease with the increase of vulcanization temperature. This indicates that with the increase of vulcanization temperature, the degree of cross-linking is deepened. The cross-linking reaction is slow at 5°C, and the degree of cross-linking is relatively light.

As shown in figure 5, with the increase of vulcanization temperature, the thermal stability of silicone rubber becomes better, and it can be explained as follows. Vulcanization reaction under 65°C is more thorough, 65°C makes some products degrade during the vulcanization process, and the final product does not contain such degradable substances.

Table 2 Functional groups and characteristic absorption peak

Functional groups	Wave number (cm^{-1})
O-H	3700~3200
C-H in Si-CH_3	2960
Si-CH_3	1280~1240
Si-O-Si	1100~1000
Si$(CH_3)_2$	840~790

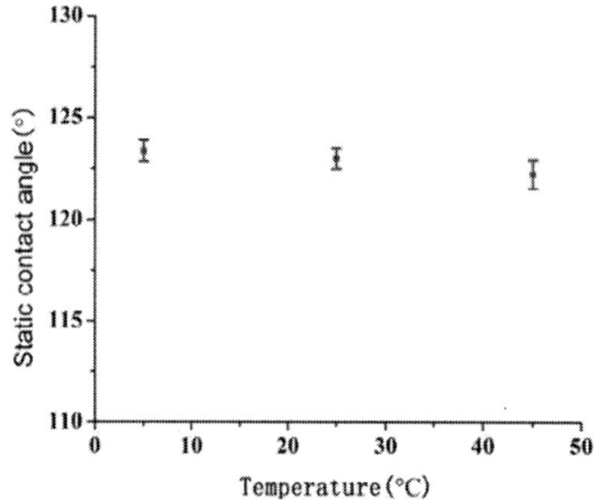

Figure 2. Hydrophobicity of RTV silicone rubber at different vulcanization temperature

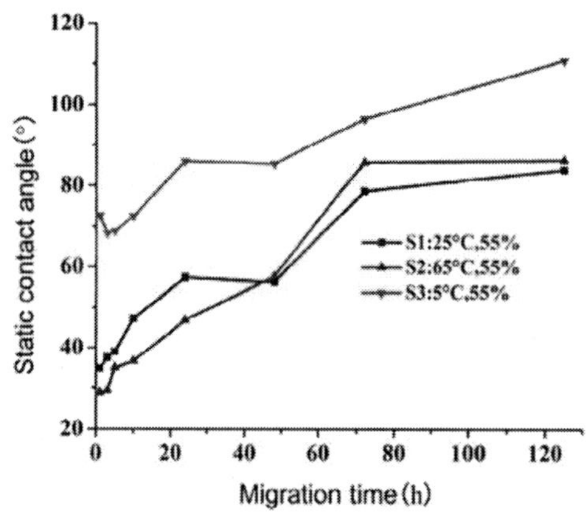

Figure 3. Hydrophobicity transfer ability of RTV silicone rubber at different vulcanization temperature

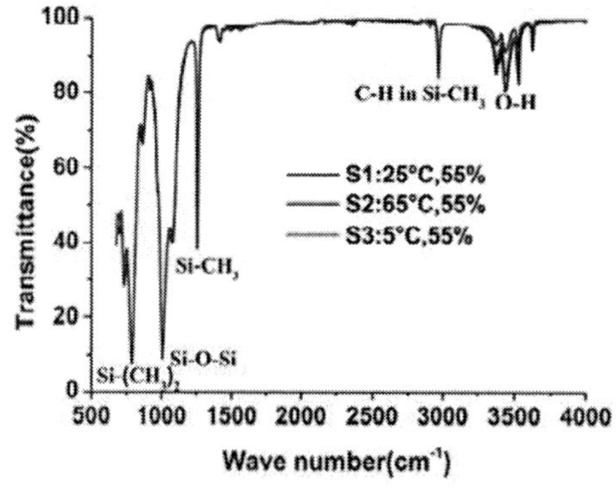

Figure 4. FTIR spectra of RTV silicone rubber at different vulcanization temperature

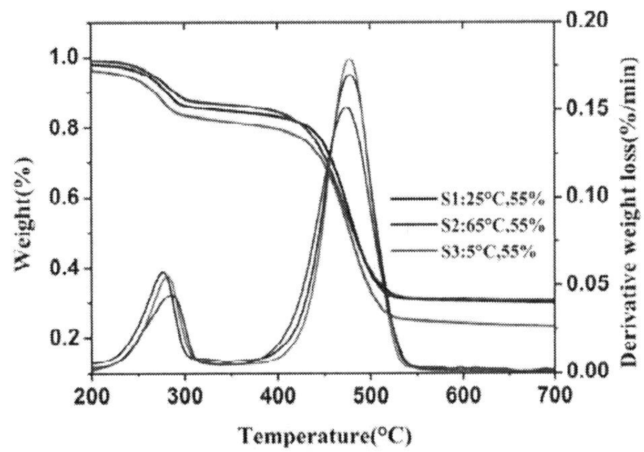

Figure 5. TG analysis of RTV silicone rubber at different vulcanization temperature

4. Effect of Different Vulcanization Humidity on Properties of RTV

Figure 6 shows that the hydrophobicity of the samples increases slightly with the increase of the vulcanization humidity, but the static contact angles differ in three cases.

As shown in figure 7, samples under high humidity vulcanization environment show a better hydrophobic migration, which means that the moisture content in the air has a great impact on the vulcanization reaction. Chemically speaking, the reaction of water and PDMS is reversible [12], when the humidity is extreme high, hydrolysis will occur and the content of PDMS in the sample will increase.

Similarly, from figure 8 we can see, Si-CH_3 peak increases with the increase of vulcanization humidity, which is in good agreement with the hydrophobicity of RTV silicone rubber. Si-O-Si peak intensity and Si-$(CH_3)_2$ peak intensity of sample in the low humidity environment are less than those of the other two, which is in good agreement with the thought that the vulcanization of the silicone rubber coating is incomplete under low humidity conditions. In the case of high humidity, Si-O-Si reacts with water due to the high content of water in the air and decompose into short chains, resulting in the generation of more PDMS[11], thus the sample have a good characteristic of hydrophobic migration.

As shown in figure 9, both high humidity and low humidity condition lead to the deterioration of the thermal stability of the silicone rubber material. Under high humidity, the moisture in the air degrade the macro molecular chains into small molecules, and decompose at lower temperatures, leading to deterioration of their thermal stability [12, 13]. In the case of low humidity condition, the cross-linking reaction is not complete. When it reaches a certain temperature, its decomposition is aggravated, and its thermal stability is deteriorated.

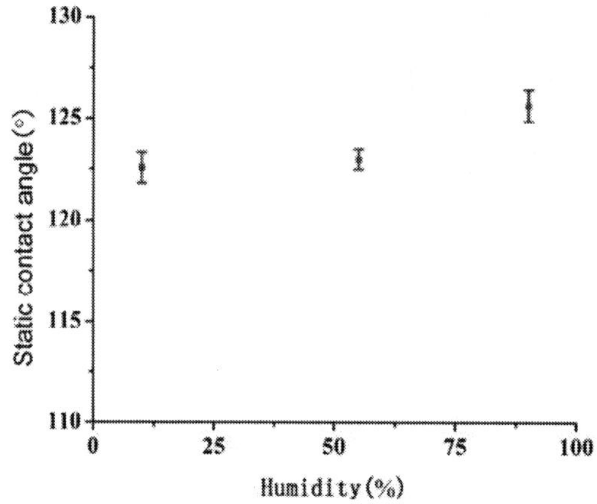

Figure 6. Hydrophobicity of RTV silicone rubber at different vulcanization humidity

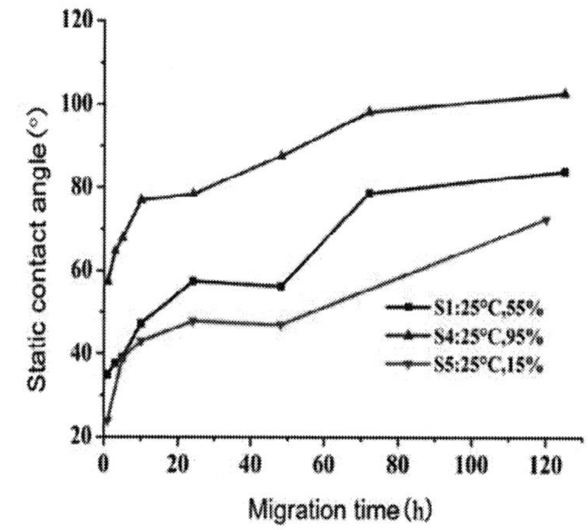

Figure 7. Hydrophobicity transfer ability of RTV silicone rubber at different vulcanization humidity

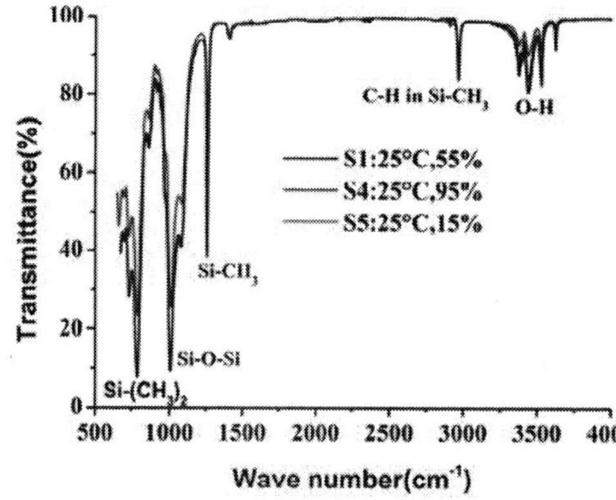

Figure 8. FTIR spectra of RTV silicone rubber at different vulcanization humidity

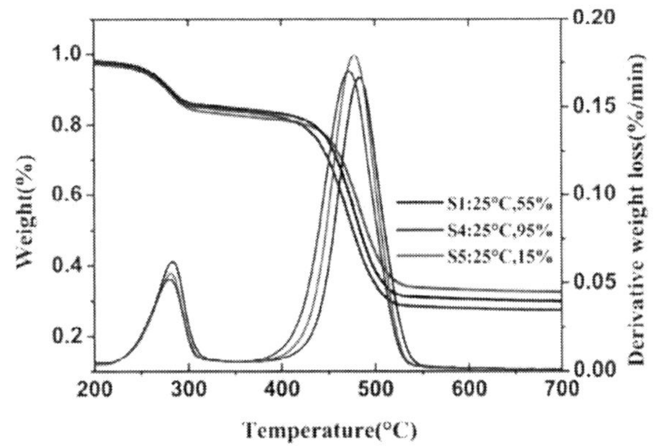

Figure 9. TG analysis of RTV silicone rubber at different vulcanization humidity

5. Conclusion

The paper study the effect of vulcanization temperature and humidity on the properties of RTV silicone rubber and make the following conclusions:

(1) Vulcanization at different temperature and humidity had little effect on the hydrophobicity of RTV silicone rubber. With the increase of vulcanization temperature, the hydrophobicity of RTV silicone rubber was not significantly affected. With the increase of vulcanization humidity, the hydrophobicity of RTV silicone rubber increased slightly. The transfer of hydrophobicity of RTV silicone rubber was affected significantly. The hydrophobicity migration of RTV silicone rubber showed a decreasing trend with the increase of vulcanization temperature. The hydrophobicity migration of RTV silicone rubber increased with the increase of vulcanization humidity.

(2) Vulcanization at different temperature and humidity have a great effect on the peak intensity of the functional groups in the RTV silicone rubber coating. When the vulcanization temperature is 65°C, the intensity of the O-H peak is significantly reduced. When the vulcanization humidity was 15%, O-H, Si-O-Si and $Si(CH_3)_2$ peak intensity are significantly reduced.

(3) The vulcanization at different temperature and humidity has an influence on the thermal stability of RTV silicone rubber. With the increase of vulcanization humidity, the thermal stability is worse, and with the increase of vulcanization temperature, the thermal stability becomes better.

(4) RTV silicone rubber vulcanization need to be paid more attention to choose the proper environment.

Acknowledgments

The authors would like to thank State Grid Corporation of China (Science and Technology Project: SGTYHT/15-JS-191) for sponsoring this study.

References

[1] Su Z and LI Q 2010 Historical Review and Summary on Measures Against Pollution Flashover Occurred in Power Grids in China J. *Power System Technology* vol 12 pp 124-130.

[2] Guan Z, Wang S, LI X, Wang L,Fan J 2000 Application and Prospect of Polymeric Outdoor Insulation in China J. *High Voltage Engineering* vol 06 pp 37-39.

[3] Jia Z, Li T, Chen C, Yang C, Chen R, Lin C 2014 Operation Characteristics of RTV Antipollution Flashover Coatings in Guangdong Region J. *High Voltage Engineering* vol 07 pp 1963-1969.

[4] Liu Y. 2015 Effects of Vulcanization Conditions on Properties of Natural Rubber for Aviation Tires *Harbin Institute of Technology*.

[5] Guo H, Peng H,Wen M,Ren W. 2013 Study on Hydrophobicity Transfer and Endurance

ofSilicone Rubber J. *Silicone Material* vol 02 pp 102-105.
[6] *DL/T 627-2012* [DL]. Beijing: China Electric Power Press, 2012.
[7] Liu C. 2011 Study on High Performance Dealcoholized RTV-1 silicone sealant *South China University of Technology.*
[8] Gorur R S. et al. 1992 Aging in silicone rubber used for outdoor Insulation J. *IEEE Trans on Power Delivery* vol 7 (2) pp 525-531.
[9] Jia Z, Gao H, Guan Z, et al. 2006 Study on hydrophobicity transfer of RTV coating base on a modification of absorption and cohesion theory J. *IEEE Transactions on Dielectrics and Electrical Insulation* vol 13(x) pp 1317-1324.
[10] Guan Z, Niu K, Wang L, Mei H, Wei X. 2012 Influence of Ambient Humidity on Hydrophobicity of Silicone Rubber Material J. *High Voltage Engineering* vol 08 p2030-2036.
[11] J. L. Goudie, M. J. Owen, T.Orbeck. 1998 A Review of Possible Degradation Mechanisms of Silicone Elastomers in High Voltage Insulation Applications. *1998 Annual Report Conference on Electrical Insulation and Dielectric Phenomena*
[12] S.M.Gubanski and A. E.Vlastos.1990 Wettability of naturally aged silicone and EPDM composite insulators J. *IEEE Trans. Power Delivery* vol 5 pp 1527-1533.
[13] Kim S H,Cherney E A,Hackam R. 1990 Loss and recovery of hydrophobicity of RTV silicone rubber insulator coatings J. *IEEE Transactions on Power Delivery* vol 5(3) pp 1491-1500.

IOP Conf. Series: Materials Science and Engineering **207** (2017) 012012 doi:10.1088/1757-899X/207/1/012012

Synthesis of Nitrogen-Doped Mesoporous Carbon for the Catalytic Oxidation of Ethylbenzene

Ruicong Wang, Yifeng Yu, Yue Zhang, Haijun Lv, Aibing Chen*

Hebei University of Science and Technology, Shijiazhuang 050018, China

*Corresponding author e-mail: chen_ab@163.com

Abstract. Nitrogen-doped ordered mesoporous carbon (NOMC) was fabricated via a simple hard-template method by functionalized ionic liquids as carbon and nitrogen source, SBA-15 as a hard-template. The obtained NOMC materials have a high nitrogen content of 5.55 %, a high surface area of 446.2 m2 g-1, and an excellent performance in catalysing oxidation of ethylbenzene. The conversion rate of ethylbenzene can be up to 84.5% and the yield of acetophenone can be up to 69.9%, the results indicated that the NOMC materials have a faster catalytic rate and a higher production of acetophenone than catalyst-free and CMK-3, due to their uniform pore size, high surface area and rich active sites in the carbon pore walls.

1. Introduction

Owing to various pore structures, high specific surface area and pore volume, mesoporous materials can be extensively used in many fields, such as purification of gases or liquids, separation of mixture, catalysis, etc. [1, 2, and 3]. With the continuous development of mesoporous carbon, mesoporous carbon materials are also capable of being utilized in a number of new filed, like flammable gas storage (hydrogen, methane), electric energy storage (supercapacitor) [4, 5]. The widely extended application areas make porous carbon have special properties, such as pore structures, surface chemical properties, crystalline order and morphologies [6, 7]. The surface chemical properties and morphologies of materials are crucial for numerous fields. Hence, preparation of highly ordered mesoporous carbon materials is the key point of many researches. In this sense, Nano casting is a very effective method for preparing ordered mesoporous materials and hard-template method is one of the typical synthetic approaches: the pores of crystalline inorganic solids (like zeolites, ordered mesoporous silica) as template are infiltrated by carbon precursors, the ordered mesoporous carbon can be obtained after carbonization to remove template [8]. As a result, carbon materials with different structures and morphologies can be made on the basis of different templates. Because SBA-15 possesses comparatively high surface are, large pore volume, uniform pore size as well as good hydrothermal stability among masses of templates, it is often applied to prepare carbon materials [9]. By using SBA-15 as hard-template and glycerin as carbon source, M. Ignaz and co-workers tuned the infiltration amount of glycerin and successfully prepared highly ordered mesoporous carbon.

However, due to the amorphous silica framework of SBA-15, the chemical activity is similar to common silica and framework is liable to collapse during high-temperature crystallization, leading to low catalytic and ion exchange capacity. In order to overcome the deficiency of carbon materials, researchers have done a large number of works to change the surface chemical properties and introduce active components. For example, introducing heteroatoms like nitrogen or boron into mesoporous carbon to replace part of carbon element can hugely change the surface structures and pore structures, enhance the hydrophilicity, affect the surface acid-base properties and provide massive active sites. The N- or B-doped porous carbon materials exhibited good performances in adsorption and catalytic reactions [8, 9]. The nitrogen doping method can be classified into in-situ doping and post processing method, and the common nitrogen sources include melamine resin, polyacrylonitrile, double acrylic amine, polyaniline, etc [10, 11, and 12]. Zhang and co-workers synthesized ordered mesoporous carbon in two-step method by using SBA-15 as hard-template, sucrose and polyaniline as carbon source and nitrogen source respectively. Because of the high nitrogen content, the product showed an excellent performance in electrochemical field, the capacitance reached 225 F g^{-1} and maintained at 95% after 2000 cycles [13]. However, conventional carbon and nitrogen precursors are detrimental to environment and residual amount is quite low. Therefore, an environmentally benign precursor with high residual amount is urgent to be developed. Ionic liquids (ILs) can be proper candidates as carbon precursors on account of the low vapour pressure, low fusion temperature and high residual carbon amount. Another significant reason is that ILs contain heteroatoms, which is equivalent to direct doping [14, 15]. Liu and co-workers have prepared N-doped mesoporous carbon utilizing SBA-15 as template, functionalized ILs as nitrogen source, exhibiting great catalytic effect in cyclization reactions [16].

Herein, we report a facile approach to synthesize N-containing ordered mesoporous carbon by using SBA-15 as hard-template and functionalized ILs as carbon and nitrogen source. The pore size of nitrogen-containing mesoporous carbon via reverse replicating SBA-15 is 3.7 nm. A series of characterizations, including SEM, XPS and XRD show the product possesses bamboo-like structure, the nitrogen content is around 5.55%. Such material demonstrates obvious catalytic activity in ethylbenzene oxidation compared with CMK-3.

2. Experimental Section.

2.1 Preparation of mesoporous silica (SBA-15).

Mesoporous silica (SBA-15) was prepared by using triblock copolymer Pluronic P123 as template, tetraethyl orthosilicate (TEOS) as silicon source. For a typical synthesis, water, hydrochloric acid (HCl) and P123 were added into a beaker. The mixed solution was stirred at 50 ° C for 30 min. After P123 was entirely dissolved, TEOS was added dropwise under vigorous stirring. Then the solution was kept stirring for 24 h. After that, the as-prepared solution was transferred into an autoclave. Hydrothermal treatment was then carried out in an oven at 100 ° C for 24 h. The white powder was obtained after filtered, washed and dried in oven. Finally, the mesoporous silica (SBA-15) was obtained after calcination at 550 ° C in air for 5 h (heating rate 1 ° C min^{-1}) to remove organic components. The reaction system possesses a molar ratio for TEOS/P123/HCl/H_2O of 1:0.017:6:138.

2.2 Preparation of nitrogen-containing ordered mesoporous carbon (NOMC).

Component A: 1.88 g (0.02 mol) of 1-vinyl imidazole and 2.18 g of ethyl bromide were added into 20 ml of methanol. After reaction at 40 ° C under nitrogen for 24 h, the product was evaporated and dried to a constant weight in a vacuum oven.

Component B: 1 g of SBA-15 was dissolved in ethanol solution, which is followed by adding 0.03 g of azodiisobutyronitrile, 3 ml of N, N-dimethylformamide and 0.07 g of nickel nitrate. The mixed solution was stirred under room temperature until ethanol was completely evaporated.

Component B was added into component A and ethanol was added to form a clear solution. Then, the reaction was carried out at 60 ° C under nitrogen for 6 h. The product was evaporated, dried to a constant weight in oven and transferred into tube furnace, calcined at 600 ° C under nitrogen atmosphere for 3 h. The NOMC was finally obtained after etching silica template and washed with acid to remove nickel.

2.3 Characterization.
Wide angle X-ray diffraction (XRD) patterns and Small angle X-ray diffraction patterns were collected using a Rigaku D/MAX 2550 diffract meter (Cu K radiation, λ = 0.15406 nm), operated at 40 kV and 40 mA. The d-spacing values were calculated by the formula $d = 2\pi/q$. Scanning electron microscopy (SEM) images were taken on a Hitachi S-4800 microscope. Transmission electron microscopy (TEM) measurements were conducted on a JEM-2010 F microscope (JEOL, Japan). Nitrogen sorption isotherms were measured at 77 K with a Micromeritics Tristar 3020 analyser. The Brunarer-Emmett-Teller (BET) method was utilized to calculate the specific surface area. The pore volumes and pore size distributions were obtained from the adsorption branches of isotherms by using Barrett-Joyner-Halenda (BJH) method, and the total pore volumes were estimated from the adsorbed amount at a relative pressure P/P_0 of 0.997. X-ray photoelectron spectroscopy (XPS) was recorded by using AXIS ULTRA DLD spectrometer with Al Kα radiation.

2.4 Catalytic measurement.
Ethylbenzene catalytic oxidation was operated in glass reactor under high pressure, equipped with thermostatic water bath, refluxing and magnetic stirring device. 10 mg of catalyst, 1 moll of ethylbenzene and 3 moll of tart-butyl hydro peroxide (TBHP) were added into 3 ml of water, the mixture was remained at 70 ° C for 24 h. Then, 0.05 g of n-dodecanal and 6 ml of dichloromethane were added respectively for extraction. The content of mixture was measured by using Agilent GC 7820A.

3. Results and discussion.

The NOMC was synthesized by using ILs as nitrogen source and SBA-15 as hard-template. From Figure 1a, it can be seen that NOMC possesses bamboo-like morphology with uniform size, indicating NOMC is successfully copied from SBA-15. In Figure 1b, the length of NOMC is about 600~1000 nm, and aggregation or fragmentation is not observed in a large area. TEM images (Figure 1c and d) show that NOMC have intact ordered mesoporous without cracked parts, further demonstrating successful replication of pore structure of original template.

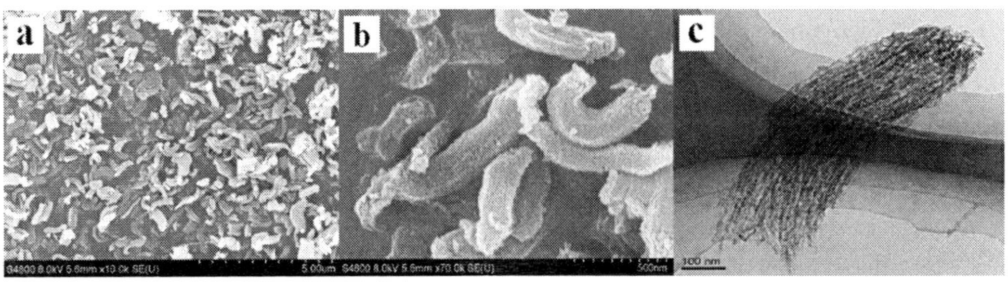

Figure 1. SEM images (a, b) and TEM images (c, d) of nitrogen-containing ordered mesoporous carbon (NOMC).

In Figure 2a, two peaks at 0.7 and 1.5 can be obviously observed, which can ben indexed to (100) and (110) facets, respectively. The space structure matched with two-dimensional hexagonal mesostructured further confirms NOMC is replicated from silica template [11]. The wide-angle X-ray diffraction pattern (Figure 2b) reveals that two peaks at 23· And 44· Correspond to (002) and (100) facets of graphite, indicating mesoporous carbon is graphitized to some extent [12]. The metal nickel is able to graphitize carbon at high temperature, resulting in graphitized structure after removal of SBA-15 and nickel [13].

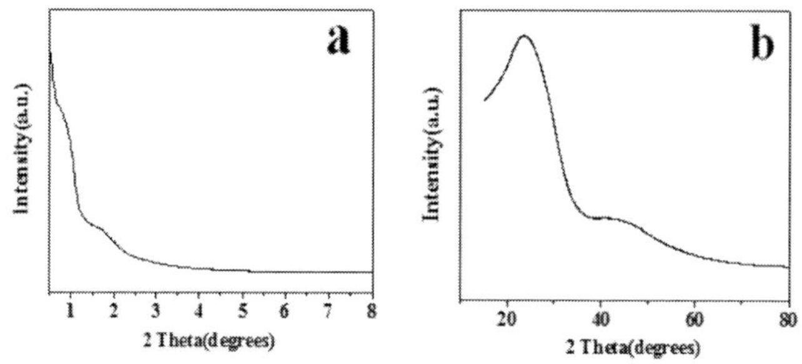

Figure 2. (a) Small-angle X-ray diffraction (SAXS) and (b) Wide-angle X-ray diffraction (XRD) patterns of NOMC.

The nitrogen adsorption-desorption isotherms (Figure 3a) reveals characteristic type corves with distinguishable hysteresis loop at relative pressure P/P_0 = 0.45-0.9, reflecting the mesostructured of NOMC is well matched with TEM and XRD results [14, 15]. The pore size distribution calculated by Barrett-Joyner-Halenda (BJH) model is cantered at 3.7 nm, which agrees well with wall thickness of SBA-15, indicating NOMC is templated from template SBA-15 [16]. The specific surface area and pore volume of NOMC are calculated to be 446.3 m^2/g and 0.77 cm^3/g, respectively.

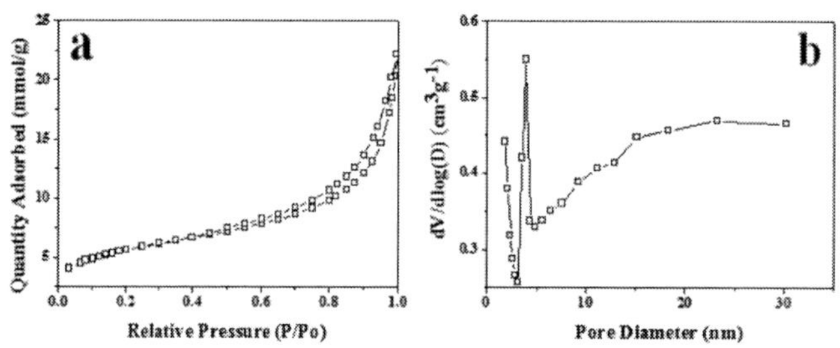

Figure 3. (a) Nitrogen adsorption-desorption isotherms and (b) pore size distribution of NOMC.

The X-ray photoelectron spectroscopy (XPS) spectrum of N1s shows three peaks. The peak at 398.1 eV corresponds to sp2 hybrid bonds or pyridine-N in lamellar graphite. The peak at 400.3 eV is matched with pyrrole-N of nitrile functional groups linked with carbon. The peak at 401.4 eV belongs to graphite-N [11, 13, and 17]. The rearrangement of nitrogen atom during high temperature carbonization leads to insertion of N atom into carbon framework. The literature reported that graphite-N is capable of facilitating C-H activation. The XPS spectrum shows the surface content of NOMC is 5.55% and graphite-N is 27.6%, which benefits catalytic oxidation of ethylbenzene in liquid phase.

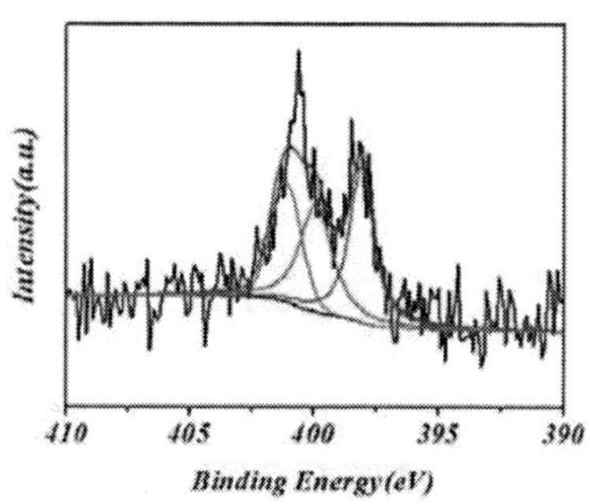

Figure 4. XPS core-level spectra of N 1s of NOMC.

In order to evaluate the catalytic performance of NOMC, we use NOMC as catalyst, tert-butyl hydro peroxide (TBHP) as oxidant to catalyze the oxidation of ethylbenzene in liquid phase. The yield of acetophenone and the conversion rate of ethylbenzene are used to evaluate the catalytic performance in this test (Table 1). The conversion rate and yield of acetyl benzene are 35% and 6% without any catalyst, revealing the autocatalysis of ethylbenzene. When CMK-3 is used as catalyst, the conversion rate and yield rise to 67.2% and 22.3%, showing mesoporous structure is favor to catalytic oxidation.

To further explore the impact factors of ethylbenzene oxidation, the catalytic abilities of CMK-3 and NOMC are carried on research. Table 1 shows the catalytic performance of NOMC (with conversion rate of 84.5% and yield of 69.9%) is far better than that of CMK-3. However, the surface area of CMK-3 is much higher than NOMC, indicating surface area of catalyst has little effect on oxidation of ethylbenzene, which is identical to report [18]. During the oxidation process of ethylbenzene to acetyl benzene, the N atom derived from NOMC is not directly involved in the catalytic reaction. Instead, N atom can change the electronic structure of adjacent C atom, and the graphite-N promotes C-H activation, resulting in considerable enhancement of catalytic performance.

Table 1. Catalytic performance of ethylbenzene oxidation with different catalysts.

Sample	BET (m^2/g)	Pore size (nm)	Pore volume (cm^3/g)	N content (XPS)	Conversion rate (%)	Yield (%)	Reaction rate ($\mu mol\ m^{-2}h^{-1}$)
No catalyst	/	/	/	/	35	6	/
NOMC	446.2	3.7	0.77	5.55	84.5	69.9	65.3
CMK-3	1271.7	3.6	0.58	/	67.2	22.3	1.14

4. Conclusion

Nitrogen-containing ordered mesoporous carbon (NOMC) was fabricated via a hard-template method by functionalized ionic liquids as carbon and nitrogen source, SBA-15 as hard-template and nickel nitrate as catalyst. NOMC samples have a high nitrogen content of 5.55 %, a high surface area of 446.2 $m^2\ g^{-1}$ as well as uniform pore size, indicating NOMC was successfully replicated from template SBA-15 and ILs were impregnated into mesoporous. The conversion rate of ethylbenzene can be up to 84.5% and the yield of acetophenone can be up to 69.9%, which are far more than CMK-3. Besides, this work has extended the application fields of metal-free catalysts.

Acknowledgments

This work was financially supported by xxx fund. This work was financially supported by Hebei Natural Science Foundation (B2015208109), Hebei Training Program for Talent Project (A201500117), Beijing National Laboratory for Molecular Sciences (20140120), Five Platform Open Fund Projects of Hebei University of Science and Technology (2015PT37), Hebei One Hundred-Excellent Innovative Talent Program (III) (SLRC2017034), Hebei Science and Technology Project (17214304D).

References

[1] Y.Y Li, K.K Han, W.G Lin, Fabrication of a new MgO/C sorbent for CO2 capture at elevated temperature, J Mater Chem A. 1 (2013) 12919–12925.

[2] Z.Z Xiao, F.F Peng, X.L Li, Surface modification of malonic acid-catalyzed carbon xerogels and their high performance for adsorption of Cu (II) ions, Surf Interface Anal. 45 (2013) 1869–1877.

[3] S Wang, W.C Li, G.P Hao, Temperature-Programmed Precise Control over the Sizes of Carbon Nanospheres Based on Benzoxazine Chemistry, J Am Chem Soc, 133 (2011) 15304–15307.

[4] Y Xiao, H.W Dong, C Long, Melaleuca bark based porous carbons for hydrogen storage,

International Journal of Hydrogen Energy. 39 (2014) 11661-11667.

[5] S.S Feng, W Li, J.X Wang, Hydrothermal synthesis of ordered mesoporous carbons from biomass-derived precursor for electrochemical capacitors, Nanoscale. 6 (2014) 14657-14661.

[6] Y Gu, Z.G Xiong, W.A Abdulla, A new approach to preparing porous carbons with controllable pore structure and morphology, Chem Comm. 50 (2014) 14824-14827.

[7] H.N Wang, X Rong, Han L, Controlled synthesis of hexagonal mesostructured silica and macroporous ordered siliceous foams for VOCs adsorption, RSC Adv. 5 (2015) 5695–5703.

[8] T Kyotani, Z Ma, A Tomita. Template synthesis of novel porous carbons using various types of zeolites, Carbon. 41 (2013) 1451–1459.

[9] M Ignat, C.J Van Oers, J Vernimmen, Textural property tuning of ordered mesoporous carbon obtained by glycerol conversion using SBA-15 silica as template, Carbon. 48 (2010) 1609–1618.

[10] J Wei, D.D Zhou, Z.K Sun, A Controllable Synthesis of Rich Nitrogen-Doped Ordered Mesoporous Carbon for CO2 Capture and Supercapacitors, Adv Funct Mater. 23 (2013) 2322–2328.

[11] J Wang, H.Y Liu, X.M Gu, Synthesis of nitrogen-containing ordered mesoporous carbon as a metal-free catalyst for selective oxidation of ethylbenzene, Chem Comm. 50 (2014) 9182-9184.

[12] M Li, J.M Xue. Integrated Synthesis of Nitrogen-Doped Mesoporous Carbon from Melamine Resins with Superior Performance in Supercapacitors, J Phys Chem C. 118 (2014) 2507-2517.

[13] N.N Liu, L.W Yin, C.X Wang, Adjusting the texture and nitrogen content of ordered mesoporous nitrogen-doped carbon materials prepared using SBA-15 silica as a template, Carbon. 48 (2010) 3579–3591.

[14] J.P Paraknowitsch, J Zhang, D.S Su, Ionic Liquids as Precursors for Nitrogen-Doped Graphitic Carbon, Adv Mater. 22 (2010) 87–92.

[15] J.P Paraknowitsch, A Thomas, M Antonietti, A detailed view on the polycondensation of ionic liquid monomers towards nitrogen doped carbon materials, J Mater Chem. 20 (2010) 6746–6758.

[16] Y Liu, J.J Peng, S.R Zhai, Synthesis of Ionic Liquid Functionalized SBA-15 Mesoporous Materials as Heterogeneous Catalyst toward Knoevenagel Condensation under Solvent-Free Conditions, Eur J Inorg Chem. 15 (2006) 2947–2949.

[17] Y.T Hu, H.J Liu, Q.Q Ke, Effects of nitrogen doping on supercapacitor performance of a mesoporous carbon electrode produced by a hydrothermal soft-templating process, J Mater Chem A. 2 (2014) 11753-11758.

[18] J Wang, H.Y Liu, X.M Gu, Synthesis of nitrogen-containing ordered mesoporous carbon as a metal-free catalyst for selective oxidation of ethylbenzene, Chem Commun. 50 (2014) 9182-9184.

Study on mechanical properties and microstructure of $Cu_{20}Fe_{80}$ alloy using the mechanical alloying

Junwei Fu[1], Suhua Yin[1], Zhongqi Dong[1,*], Yang Wang[2], Yanxia Liu[1]

[1]Engineering Research Institute, Hebei College of Industry and Technology, Shijiazhuang 050091, China
[2]School of Materials Science and Engineering, Hebei University of Science and Technology, Shijiazhuang, Hebei, China 050000, China

*Corresponding author e-mail: dongzhongqi@126.com

Abstract. It was prepared that $Cu_{20}Fe_{80}$ alloys using the mechanical alloying and cold pressing and the microwave sintering. The microstructure and the phase composition of $Cu_{20}Fe_{80}$ alloy were analyzed by SEM and XRD, respectively. The density and hardness of the alloys were measured. The evolution rule of microwave of $Cu_{20}Fe_{80}$ alloy was investigated during the microwave sintering. The results have shown that the powder milling structure of $Cu_{20}Fe_{80}$ alloy is a flake layer which is refined and the mechanical alloying enhanced with increasing milling time; The microstructure of the $Cu_{20}Fe_{80}$ alloy powder blank cold pressed is lamellar which gradually become dense, the density increases and the formability is improved with the increase of cold compressive strength; The microstructure of the $Cu_{20}Fe_{80}$ alloys sintering was lamellar that the voids decreased and the grain boundaries were gradually obvious and the density increased and the hardness increased with the increase of cold compression strength and the milling time the best milling time was 4h..

1. Introduction

Iron copper base powder metallurgy self-lubricating bearing, the addition of iron powder to save Cu elements, can greatly reduce the cost of raw materials [1-3].But the traditional bearing mixed iron copper powder, the compressive strength, hardness is based on copper based materials on the increased slightly, poor corrosion resistance; due to non-uniform iron copper powder mixing, resulting in hard phase distribution is not uniform, and the composition and microstructure in the sintering process of segregation. The inhomogeneity caused the performance of the bearing in the operation noise increased and shortened service life [4-7]. In this paper, the preparation method of the combination of mechanical alloying, cold pressing and microwave sintering is proposed. The main research is to find out a method to make the Cu-Fe alloy powder sintered uniformly, so as to obtain a good structure.

2. Experimental Methods

The purity of raw materials (Cu alloy particle size 3~5 m Fe (selection) and the size of -200 mesh) is 99.99%, according to the chemical composition (atomic ratio) configuration of $Cu_{20}Fe_{80}$ alloy, the sample weight is about 5-10 grams. The raw powder was put into the QM-1SP type planetary high-energy ball mill, and was protected by anhydrous ethanol. The alloy was milled by 2H, 4H and 6h in high energy ball mill. $Cu_{20}Fe_{80}$ alloy was placed in a vacuum drying chamber after grinding, vacuum to 2 * 10^{-2}Pa, temperature 318K, drying 2~3h. Drying after using BT-9300S laser particle size distribution instrument to determine the particle size of $Cu_{20}Fe_{80}$ alloy powder. DNS100 electronic universal testing machine was used to make the $Cu_{20}Fe_{80}$ alloy powder in the self-made mold using 30Mpa, 50Mpa, 100Mpa, 200Mpa, 300Mpa, 400Mpa, 500Mpa, cold pressing. The cold pressing sintering under control $Cu_{20}Fe_{80}$ sample loading platform of the central vacuum microwave MKX-T3-1 sintering machine, sintering time 20min, temperature from room temperature to 950 degrees C, sintering by use of two-color infrared thermometer MRISBSF alloy temperature monitoring. The microstructure and phase composition of the alloy were analyzed by VEGA3SBH scanning electron microscope and scanning electron microscope. X ray diffraction determination of powder samples by Cu-K target in X - ray diffraction spectrum, crystal structure analysis of samples.

3. Experimental results and discussion

3.1. Mechanical alloying Cu20Fe80 powder microstructure analysis

Figure 1 (a) and (b) Cu, Fe powder morphology of raw materials, Cu, Fe are small and multi - grain, Fe particles larger. Figure 2 (a) and (b) $Cu_{20}Fe_{80}$ mechanical alloying 2h, morphology of 4H and 6h, and milling powder, changed CuFe powder, $Cu_{20}Fe_{80}$ powder after the mechanical alloying are mostly lamellar particles of lamellar particles and fine fraction (Figure 3) the surface layer, and flaky particles have fractured shape, the edge is sharp teeth marks in the shape of powder surface with small particles.

From Figure 3 can also be found in particle size distribution appeared in a short two normal distribution peaks, the first wave appeared in the range of 0~2.20 selection m, the peak is very low powder content 0~2.20 selection of m is low, the second peak appeared in the range of 6.5~130 m is tested, particle concentration distribution in between 6.5~130 m selection. It can be seen from the figure of powder grinding 6h, particle size ratio between 6.5~130 tested m was higher than that of 2h, 4h, the vast majority of powder refinement to about 35 m. selection. The mechanical alloying of $Cu_{20}Fe_{80}$ powder was analyzed by means of component surface scan (Figure 2b). It was found that the powder was rich in Cu and rich in Fe, and the rich Fe layer was larger than that of the rich Cu layer. Figure 4 is a mechanical alloying 2h, X ray diffraction $Cu_{20}Fe_{80}$ powder 4H and 6h after the spectrum, Cu, Fe diffraction peaks do not overlap, but the Cu peak with increasing milling time prolonged, the Fe peak with the milling time increased, indicating that Fe and Cu gradually formed a compulsory solid solution, mechanical alloying the degree with the grinding time.

Fig.1 The morphology of powder of (a) Cu and (b) Fe

Fig.2 The morphology of $Cu_{20}Fe_{80}$ powders by mechanical alloying (a) 2h (b) 4h (c) 6h

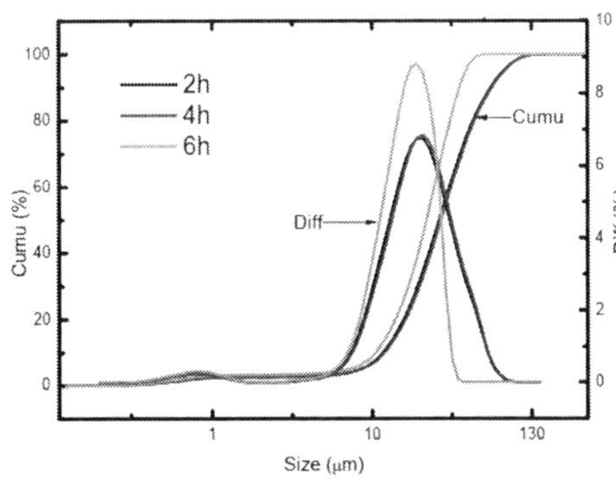

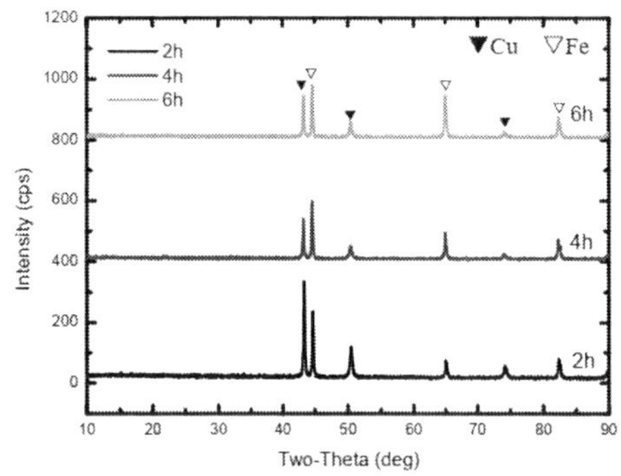

Fig.3 Particle size distribution of $Cu_{20}Fe_{80}$ powders by mechanical alloying

Fig.4 XRD patterns of powders $Cu_{20}Fe_{80}$ powders by mechanical alloying

3.2. Microstructure analysis of Cu20Fe80 alloy powder by cold pressing mechanical alloying

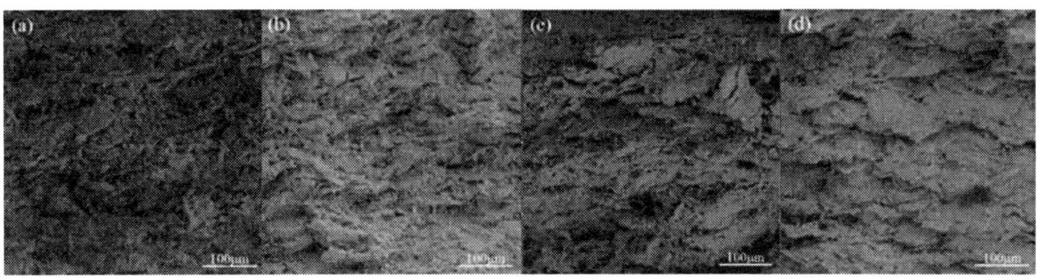

Fig.5 The fracture morphology of $Cu_{20}Fe_{80}$old pressed green (a)200Mpa (b) 300Mpa(c) 400Mpa (d) 500Mpa

The two element composite powder Cu/Fe in respectively in 30Mpa~500Mpa were cold pressed under pressing after fracture morphologies of SEM samples as shown in Figure 5 (a) ~ (d), because 30Mpa, 50Mpa, 100Mpa under the pressure of sample is too loose, so do not do in fracture analysis. It can be seen from the fracture surface of Cu/Fe composite powder that the void fraction decreases with the increase of cold compressive strength, and the fracture surface is layered, and the debris after fracture becomes granular. This is because the metal powder pressing with the increase of pressure and deformation of powder powder to enhance the binding force between reason, in low strength cold pressing (<100Mpa) deformation of powder should be elastic deformation smaller contact surface powder particles, powder particles contact shear stress at the lower, which leads to the bond strength between particles is low, in doing experiments broken fracture phenomenon; when the cold pressure increasing intensity (>200Mpa), in the process of deformation of powder particles at the contact surface due to welding and meshing force increase of the viscosity and compact strength cold compacts increase and because of the stress of static cold compact vertical stress is greater than the side deformation of powder particles in the vertical direction is higher than that of the side, therefore, pressing pressed tissue shows a lamellar. From the edge of fault, it shows that the plastic deformation of Cu/Fe two composite powder compacts is higher.

3.3. Microstructure and properties of $Cu_{20}Fe_{80}$ alloy during microwave sintering

Figure 6 (a) ~ (g) is the microstructure of $Cu_{20}Fe_{80}$ alloy under different cold compressive strength, and Figure 7 is the surface scanning energy spectrum analysis when the cold compressive strength is 400Mpa, in which the black phase is rich in Fe phase and the gray phase is rich in Cu phase. Table 1 is the density and hardness of $Cu_{20}Fe_{80}$ alloy with different cold pressing strength. In 30Mpa, the relative density of $Cu_{20}Fe_{80}$ alloy is 75.58%, the hardness is 21.5MPa, the thick layer of flaky black and grey hue phase composition, layer distribution is not uniform, the surface contains a large number of pores (see figure 6a-6c). The relative density and hardness of Cu20Fe80 alloy increased with the increase of cold pressing strength (50Mpa~500Mpa). The distribution of the alloy sheet was more uniform, and the surface pores were greatly reduced (see figure 6d-6g).

Table 1 The performance of $Cu_{20}Fe_{80}$ alloy

pressure	30MPa	50MPa	100MPa	200MPa	300MPa	400MPa	500MPa
2hdensity%	75.58%	78.01%	82.47%	84.85%	89.63%	91.80%	93.44%
4hdensity%	73.54%	77%	82.33%	86.16%	90.42%	92.26%	92.68%
6hdensity%	73.10%	76.84%	82.01%	86.21%	90.11%	91.85%	92.54%
2hdensityMPa	21.5	21.5	43.5	48.3	51.8	55	60.3
4hdensityMPa	23.2	22.8	43.3	48.5	56.5	59.5	62
6hdensityMPa	22	24.3	43	44.6	56.6	60.5	61.3

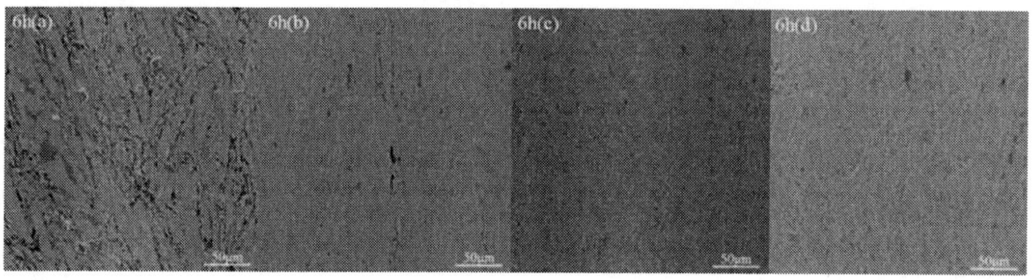

Fig.6 The microstructure of $Cu_{20}Fe_{80}$ alloy at cold compressive strength (a) 30MPa (b) 200MPa (c) 300MPa (d) 400Mpa

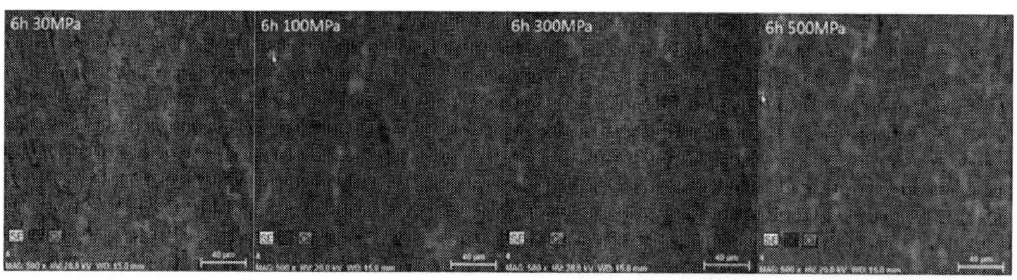

Fig.7 The energy spectrum analysis of $Cu_{20}Fe_{80}$ alloy after ball milling 6h

4. Conclusion

The Cu20Fe80 alloy was prepared by mechanical alloying and cold pressing microwave sintering. The microstructure and structure of the samples were investigated by SEM and XRD, and the results were as follows:

(1) With the increase of cold compressive strength, the $Cu_{20}Fe_{80}$ alloy powder compacts from loose to compact;

(2) With the increase of cold compressive strength, the porosity of $Cu_{20}Fe_{80}$ alloy decreases, the density and hardness increase, and the microstructure is more uniform.

Acknowledgments

Thanks to the Hebei Provincial Department of science and technology key projects (ZD2015006) and project 2015YF02 funding.

References

[1] Su Wan Wan, Jiao Minghua, Wang Yuliang, Li Lingxin, Yin Yanguo. Influence of Cu on Microstructure and Tribological Properties of High - Boron Iron Based Bearing Materials [J]. Bearing, 2015,(1): 32-36.

[2] Jia Chengchang. Sintering Metal OillessMetal World [J]. Bearing, 2011,22(1):28-34.

[3] Fu Hanguang, Hu Hong. Progess of research on high boron wear resistant cast alloys [J]. Special Summary, 2005(3) :32-37.

[4] Tong An, Fei Qin. Effects of the intermetallic compound microstructure on the tensile behavior of Sn3.0Ag0.5Cu/Cu solder joint under various strain rates [J]. Microelectronics Reliability, 2014 (54) : 932-938.

[5] Liu Y, Li B H, Li J, et al. Effect of Titanium on the ductilization of Fe-B alloys with high boron content [J]. Materials Letters, 2010, 64: 1299-1301.

[6] X.L. Yan , M. Lin. J.Y.Wang. Equilibrium and kinetic surface segregations in Cu–Sn thin films [J]. Appl Phys A, 2013, (113) : 423–430.

Effect of La_2O_3 on microstructure and properties of $Cu_{20}Fe_{80}$ alloy by microwave sintering

Junwei Fu[1], Suhua Yin[1], Zhongqi Dong[1,*], Yang Wang[2], Yanxia Liu[1]

[1]Engineering Research Institute, Hebei College of Industry and Technology, Shijiazhuang 050091, China
[2]School of Materials Science and Engineering, Hebei University of Science and Technology, Shijiazhuang, Hebei, China 050000, China

*Corresponding author e-mail: dongzhongqi@126.com

Abstract. It was prepared that $Cu_{20}Fe_{80}$ alloys using the mechanical alloying and cold pressing and the microwave sintering. The microstructure and the phase composition of $Cu_{20}Fe_{80}$ alloy were analyzed by SEM and XRD, respectively. The density and hardness of the alloys were measured. The effect of La_2O_3 on microstructure and properties of $Cu_{20}Fe_{80}$ alloy was investigated during the microwave sintering. The results have shown that the powder milling structure of $Cu_{20}Fe_{80}$ alloy is a flake layer which is refined and the mechanical alloying enhanced with increasing La_2O_3 content; The microstructure of the $Cu_{20}Fe_{80}$ alloy powder blank cold pressed is lamellar which gradually become dense, the density increases and the formability is improved with the increase of La_2O_3 content; The microstructure of the $Cu_{20}Fe_{80}$ alloys sintering was lamellar that the voids and the density and the hardness change with the increase of La_2O_3 content and the best La_2O_3 content was 0.2%.

1. Introduction

Iron - Copper - based powder metallurgy bearing material, which can save non-ferrous metals and reduce the cost of raw materials [1-3]. But the traditional bearing mixed iron copper powder, the compressive strength, hardness is based on copper based materials on the increased slightly, poor corrosion resistance, and will inevitably generate and microstructure segregation caused by inhomogeneity of hard phase and the performance of the resulting running noise increases and shorten the service life of the bearing, the in the large-scale automation production process, is not conducive to ensure the consistency and stability of the product performance[4-7]. In this paper, La_2O_3 was added in the preparation of Cu-Fe alloy, and the effect of La_2O_3 on the microstructure and properties of Cu-Fe alloy during mechanical alloying and cold pressing microwave sintering was studied.

2. Experimental Methods

The purity of raw materials (Cu alloy particle size 3~5 m Fe (tested), mesh size -200) and La_2O_3 99.99% (atomic ratio), according to the chemical composition of $Cu_{20}Fe_{80}$ alloy configuration, the sample weight is about 5-10 grams. The raw material powder was put into QM-1SP type planetary high energy ball mill and was protected by anhydrous ethanol. The alloy was grinded by 6h in high energy ball mill. $Cu_{20}Fe_{80}$ alloy is placed in a vacuum drying oven after grinding, vacuum to 2 * 10^{-2}Pa, temperature 318K, drying 2~3h. After drying, the particle size of $Cu_{20}Fe_{80}$ alloy powder was determined by BT-9300S laser particle size distribution instrument. DNS100 electronic universal testing machine was used to make cold pressing of $Cu_{20}Fe_{80}$ alloy powder in 200Mpa. The cold pressing sintering under control $Cu_{20}Fe_{80}$sample loading platform of the central vacuum microwave MKX-T3-1 sintering machine, sintering time 20min, temperature from room temperature to 950 degrees C, sintering by use of two-color infrared thermometer MRISBSF alloy temperature monitoring. Comes with VEGA3SBH scanning electron microscope and scanning electron microscope energy spectrum analysis and phase composition of alloy microstructure. X ray diffraction determination of powder samples by Cu-K target in X - ray diffraction spectrum, crystal structure analysis of samples.

3. Experimental results and discussion

3.1. Mechanical alloying Cu20Fe80 powder microstructure analysis

Figure 1 (a), (b) and (c) were Cu, Fe and La_2O_3 powder morphology, Cu, Fe and La_2O_3 were small and multi grain, Fe particles larger. Figure 2 (a) ~ (d) for the morphology of $Cu_{20}Fe_{80}$ mechanical alloying 6h, and milling powder, changed CuFe powder, $Cu_{20}Fe_{80}$ powder after the mechanical alloying are mostly lamellar particles of lamellar particles and fine fraction (Figure 3), and the surface layer flake particles have the fracture shape, the edge is sharp teeth marks in the shape of powder surface with small particles. It can be seen from the figure of powder grinding 6h containing 0.6% La_2O_3, the size ratio between the 6.5~130 selection m was higher than that of the content of 0 and 0.2%, the vast majority of powder refinement to about 35 m selection.

The mechanical alloying of $Cu_{20}Fe_{80}$ powder was analyzed by means of component surface scan (Figure 2d). It was found that the powder was rich in Cu and rich in Fe, and the rich Fe layer was larger than that of the rich Cu layer. Figure 4 is the X ray diffraction $Cu_{20}Fe_{80}$ powder containing 0~1.0% La_2O_3 mechanical alloying of 6h spectrum, Cu, Fe diffraction peaks do not overlap, but increased the content of La_2O_3 Cu peaks with the diffraction peak first increased and then decreased, with the increase of Fe peak La_2O_3 content with the first decreased and then increased, in La_2O_3 content of 0.2% for the transition point, showed that the La_2O_3 content in formation forced solid solution of Fe and Cu, have enhanced first and then weakened.

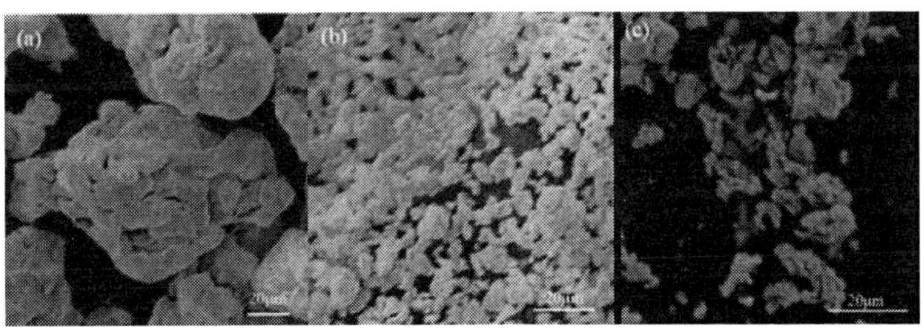

Fig.1 The morphology of powder of (a) Cu and (b) Fe (c) La_2O_3

Fig.2 The morphology of $Cu_{20}Fe_{80}$ powders by mechanical alloying (a) 0.0% (b) 0.2% (c) 0.8% SEM and (b) Surface scan

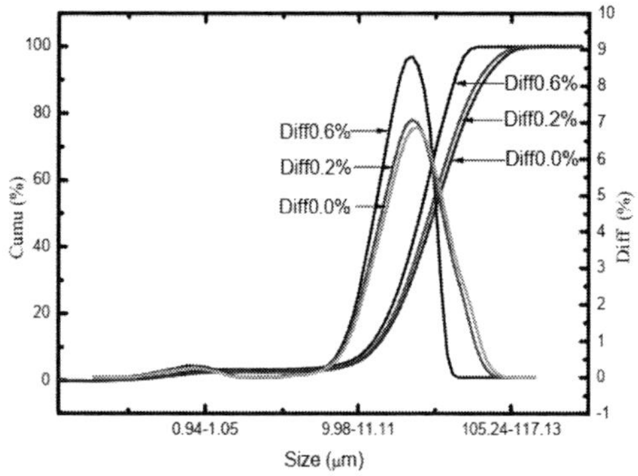

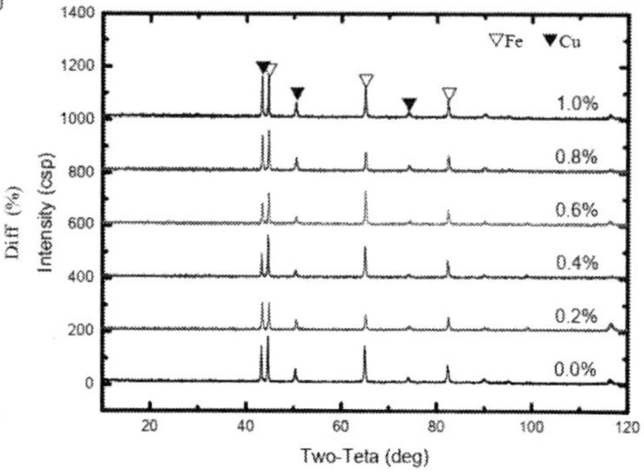

Fig.3 Particle size distribution of $Cu_{20}Fe_{80}$ Powders by mechanical alloying **Fig.4** XRD patterns of powders $Cu_{20}Fe_{80}$ powders by mechanical alloying

3.2. Microstructure analysis of Cu20Fe80 alloy powder by cold pressing mechanical alloying

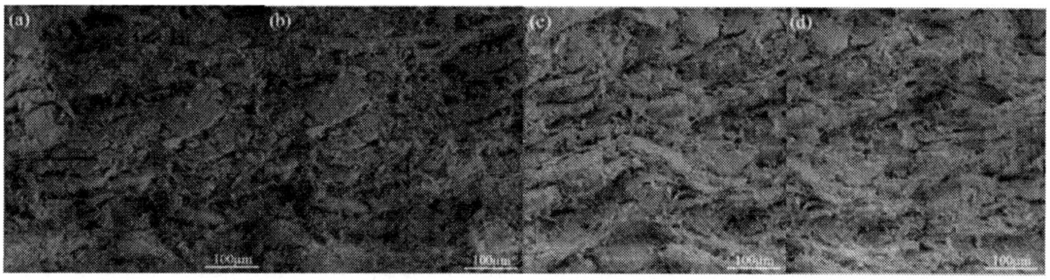

Fig.5 The fracture morphology of $Cu_{60}Cr_{40}$ cold pressed green (a) 0.0% (b) 0.2% (c) 0.4% (d) 0.6%

Cu/Fe two composite powder in the 200Mpa and La_2O_3 content of 0~1.0% were respectively subjected to cold pressing under the pressure of the fracture surface of the SEM morphology as shown in Figure 5 (a) ~ (d). It can be seen from the fracture surface of the Cu/Fe composite powder that the fracture surface is layered, and the debris after the fracture becomes granular. This is due to the increase of the pressure and the deformation of the powder and the increase of the binding force between the powders, the structure of the pressed compacts is lamellar. From the edge of fault, it shows that the plastic deformation of Cu/Fe two composite powder compacts is higher.

3.3. *Microstructure and properties of Cu20Fe80 alloy during microwave sintering*

Figure 6 (a) ~ (d) is the microstructure of La_2O_3 and 0~1.0% content of cold pressing strength of 200Mpa $Cu_{20}Fe_{80}$ alloy, Figure 7 is a ball milling 6h spectrum of $Cu_{20}Fe_{80}$ alloy surface can scan the contents of 0~1.0% and La_2O_3 cold compressive strength 200Mpa, in which the red phase is Fe rich phase, the green phase is Cu rich phase blue, La dispersed in iron matrix. Table 1 is the density and hardness of $Cu_{20}Fe_{80}$alloy with La_2O_3 content of 0~1.0% and cold pressing strength of 200Mpa. With the increase of La_2O_3 content, the density and hardness increased first and then decreased, and reached the maximum at 0.2%.Microstructure of $Cu_{20}Fe_{80}$ alloy by thicker lamellar black and grey hue in the La_2O_3 phase, the content is more than 0.4% layer distribution is not uniform, the surface contains a large number of pores (see figure 6c-d). With the increase of La_2O_3 content of $Cu_{20}Fe_{80}$ alloy relative density and hardness decrease.

Table 1 The performance of $Cu_{20}Fe_{80}$ alloy

Adding LaO_3	0.0%	0.2%	0.4%	0.6%
6hdensity%	86.21%	92.35%	90.87%	88.77%
Rockwell hardness	44.6	52	49.5	50.5

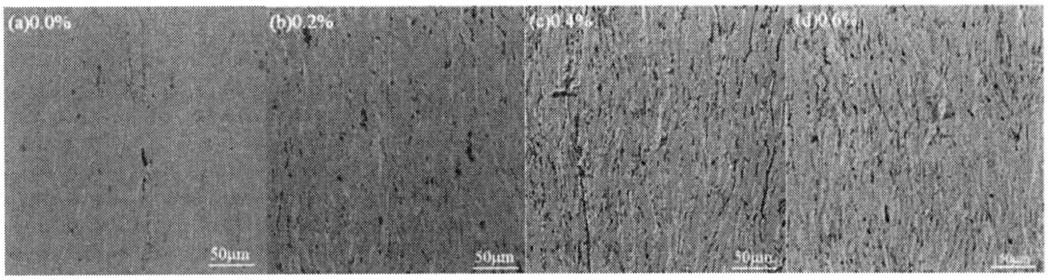

Fig.6 The microstructure of $Cu_{20}Fe_{80}$ alloy (a)0.0% (b)0.2% (c)0.4% (d)0.6%

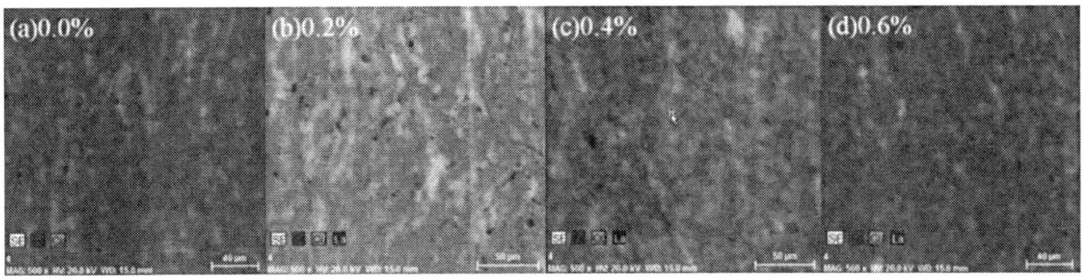

Fig.7 The energy spectrum analysis of $Cu_{20}Fe_{80}$ alloy after ball milling 6h

4. Conclusion

$Cu_{20}Fe_{80}$ alloy was prepared by mechanical alloying and cold pressing microwave sintering. The effects of La_2O_3 on Microstructure and properties of $Cu_{20}Fe_{80}$ alloy were investigated by SEM and XRD:

(1) with the increase of La_2O_3 content, the mechanical alloying of $Cu_{20}Fe_{80}$ alloy composite powder was enhanced, and the density and hardness increased first and then decreased, while the strength and density of the alloy were the best;

(2) the role of La_2O_3 in the $Cu_{20}Fe_{80}$ alloy, when the content is low can refine microstructure of alloy performance increase, but when the content is high, easy accumulation of rare earth oxides in the grain boundary, thus worsening the continuity of the matrix, the properties of the material is reduced.

Acknowledgments

Thanks to the Hebei Provincial Department of science and technology key projects (ZD2015006) and project 2015YF02 funding

References

[1] Su Wan Wan, Jiao Minghua, Wang Yuliang, Li Lingxin, Yin Yanguo. Influence of Cu on Microstructure and Tribological Properties of High - Boron Iron Based Bearing Materials [J]. Bearing, 2015,(1): 32.

[2] Jia Chengchang. Sintering Metal OillessMetal World [J]. Bearing, 2011,22(1):28-34.

[3] Fu Hanguang, Hu Hong. Progess of research on high boron wear resistant cast alloys[J]. Special Summary, 2005(3) :32.

[4] Tong An, Fei Qin. Effects of the intermetallic compound microstructure on the tensile behavior of Sn3.0Ag0.5Cu/Cu solder joint under various strain rates[J]. Microelectronics Reliability,

2014 (54) : 932.
[5] Liu Y, Li B H, Li J, et al. Effect of Titanium on the ductilization of Fe-B alloys with high boron content [J]. Materials Letters, 2010, 64: 1299.
[6] X.L. Yan , M. Lin. J.Y.Wang. Equilibrium and kinetic surface segregations in Cu–Sn thin films [J]. Appl Phys A, 2013, (113) : 423.

IOP Conf. Series: Materials Science and Engineering **207** (2017) 012015 doi:10.1088/1757-899X/207/1/012015

Preparation and characteristics of Aziridine-Benzoguanamine-Urea-Melanmine-Formaldehyde resins

Wenhuan Li[1], Jie Xu[1], Xingwei He[1], Jingjie Zhang[1], Enmin Zong[2], Xiaohuan Liu[1] and Shenyuan Fu[1,*]

[1]School of Engineering, and National Engineering and Technology Research Center of Wood-based Resources Comprehensive Utilization, Zhejiang Agriculture and Forestry University, Hangzhou Lin'an 311300, PR China
[2]Zhejiang Provincial Key Laboratory of Plant Evolutionary Ecology and Conservation, Taizhou University, Taizhou 318000, China

[*]Corresponding author e-mail: fshenyuan@sina.com

Abstract. In this study, the effect of Aziridine (AZ) addition on the properties of the BG-MF and BG-U-MF resin, such as the mechanical strength, tensile strength, moisture adsorption, and limited oxygen index, were studied in details. The best performance was obtained for modified resins with 2 wt. % added AZ. The structure characteristics of the modified MF resins were investigated using Fourier transform infrared spectroscopy (FTIR) and X-ray diffraction (XRD). It was found that the characteristic absorption peak of the hydroxyl group and the ring of the three triazine group was significantly reduced, which indicated that the addition of aziridine was beneficial to the curing reaction. It could be seen that the effect of the addition of aziridine on the crystallinity of BG15% modified resin was not obvious, but the crystallinity of BG15%+U20% modified resin was reduced. The tensile strength and elongation at break increased after the addition of 2% aziridine, and resistance to burning and boiling water were also improved.

1. Introduction

Melamine formaldehyde resin (MF) is derived from the polycondensation of melamine and formaldehyde molecules, which is a triazine-rich heteroatom polymer with high content of nitrogen1. Presently, MF is widely used in many fields. However, the high degree of crosslinking density in cured MF resin makes it inherently brittle, which resulted in limiting its application. Thus, the modification of melamine resin (MF) has become a research hotspot [1-7].

At present, the research on the modification of melamine resin (MF) has been reported. Xu et al [6] reported that Melamine-formaldehyde resin (MF)/polyvinyl alcohol (PVA) composite fibers were prepared by wet spinning, the experiments results shown that the obtained MF/PVA composite fiber shows a sea-island phase structure, and the best comprehensive properties: the breaking strength and breaking elongation are 4.29 cN/dtex, 13.55%, respectively. Yari et al [7] reported that an acrylic polyol/melamine hardener was modified with a POSS macromolecule, and POSS could enhance toughness. Hang et al [8] reported that using benzoguanamine (BG) as a substitute for partial melamine to synthesize modified resin to improve the heat resistance and toughness.

Aziridine (AZ) is transparent liquid crosslinker. It is a compound containing three membered rings, and three membered ring molecules is an unstable structure, which shows high chemical crosslinking reaction (Fig 1). In this paper, the BG-MF and BG-U-MF resin were both modified by AZ, and the impregnated kraft paper laminate were prepared by using the obtained modified resins as adhesive. The effect of AZ addition on the properties of the BG-MF and BG-U-MF resin, such as the structure characteristics, mechanical strength, tensile strength, moisture adsorption, and limited oxygen index, were studied in details.

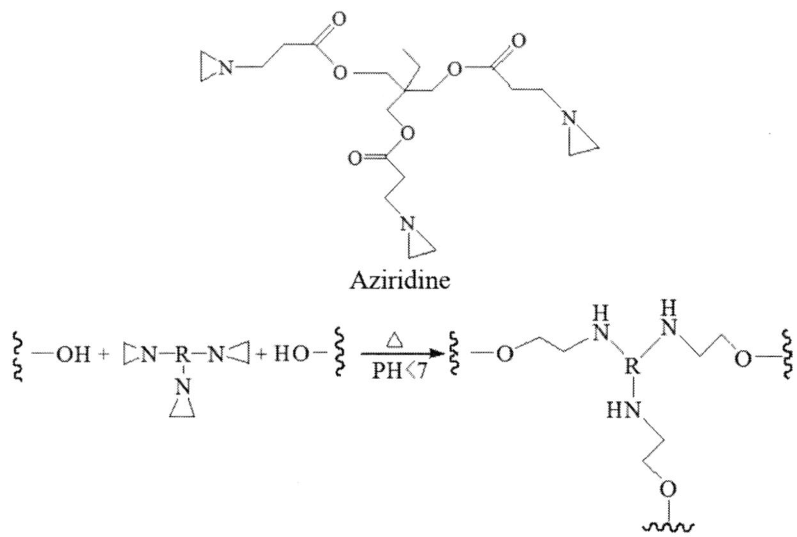

Figure 1. The mechanism of modification by aziridine (AZ)

2. Materials and methods

2.1. Materials

Formaldehyde, melamine, urea, and kraft paper were purchased from Zhejiang linglong decoration material Co. Ltd., benzoguanamine was obtained from Wuhan Organic Chemical Co. Ltd., sodium hydroxide was obtained from Sinopharm Chemical Reagent Co. Ltd.; aziridine was purchased from Dongguan xuyi hua Chemical Co. ltd.

2.2. Characterization

2.2.1. Fourier Transform Infrared (FTIR) Analysis. The modified resin was cured at 120 ℃, and then dried in vacuum oven to a constant weight. The FTIR spectra of samples are recorded using a Model Perkin Elmer 1100 series operating in the region of 4000-400 cm^{-1} using KBr pellet technique.

2.2.2. X-Ray Diffractomer (XRD) Analysis. XRD spectra of the samples were obtained from a Rigaku D/max-RA power diffraction-meter using Cu Kα radiation in the 2θ range of 5°-60°.

2.3 Preparation of modified MF resins

A typical procedure was as followed, certain formaldehyde and melamine were both added into a condensation reflux device three flasks. The pH value was adjusted to about 9.0 by sodium hydroxide solution, and then the mixture was stirred until completely dissolved. After that, the certain amount of benzoguanamine was added to the mixture solution, and the reaction mixture was heat to 85 ℃ and maintaining for a certain reaction time. The colorless transparent liquid copolymer resin was obtained, with a solid content of 61.4% and a pH value of 8.8. The content of aziridine was 2%.

A certain amount of 37% formaldehyde solution, urea and melamine were added to a condensation reflux device three flasks. The pH value was adjusted to about 8.5 by sodium hydroxide solution, and the reaction mixture was heat to 60 ℃ and maintaining for a certain reaction time. The rest of the melamine, benzoguanamine, and formaldehyde were added to the mixture solution, and the reaction mixture was heat to 85 ℃ and maintaining for a certain reaction time. The colorless transparent liquid copolymer resin was obtained, with a solid content of 57.2% and a pH value of 8.5. The content of aziridine was 2%.

2.4. Preparation and properties of impregnated paper laminated board

A typical procedure was as followed, each laminated board was made up of three pieces of kraft paper, and the amount of glue was 80%, and then dried at 60 ℃ by hot air circulation, which make the adhesive paper uniform dried. Hot-pressing condition was as followed: Temperature 140-150 ℃, pressure 6-7 MPa, time 30 min.

The appearance of the obtained hot pressed laminated plate was smooth, uniform color and gloss. The process conditions of tensile properties (GB/T 7911-2013): The tensile mechanical properties of laminated plate were tested by electronic universal testing machine. The temperature was 25 ℃, the loading rate was 5mm/min, and the number of each sample was not less than 5. Boiling water resistance test (GB/T 7911-2013): each sample is not less than 3. Determination of combustion performance (GB/T 2406.2-2009): The top ignition method and each sample is not less than 5.

3. Results and discussion

3.1. FTIR and XRD analysis

FTIR spectra were usually employed to identify the structural change of the sample. As shown in Fig 2, the band at near 3350 cm^{-1} was linked to the -OH and –NH- hydroxyl groups. The stretching band appeared at 1010 cm^{-1} was ascribed to the vibration absorption peak C-O and N-H, and the peak at 813 cm^{-1} ascribed to thiotriazinone surface of outer ring vibration absorption peak. From the results, it was concluded that existence of aziridine could accelerate curing reaction of BG15% and BG15%+U20%.

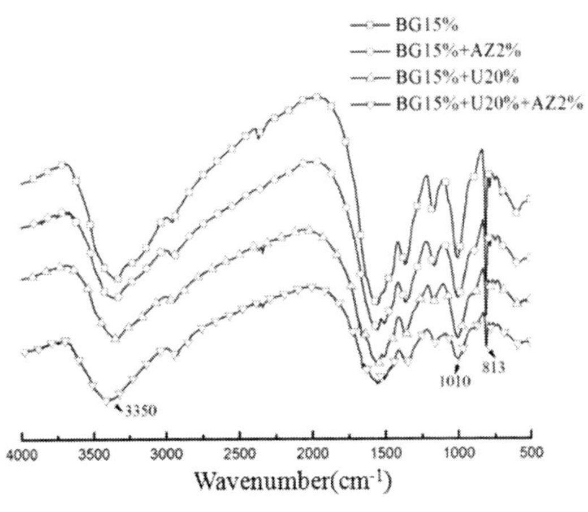

Figure 2. FT-IR of modified MF resin

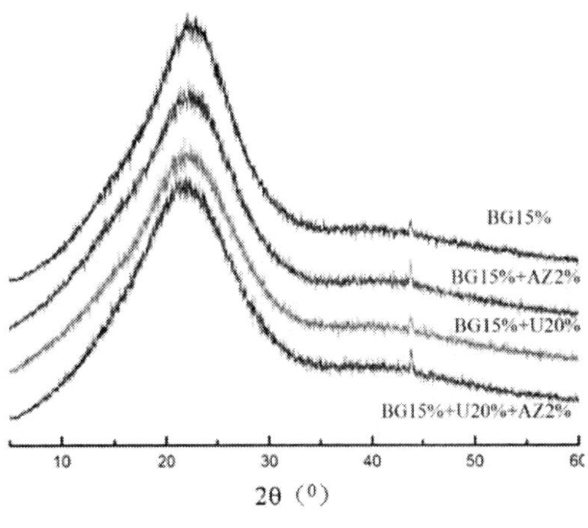

Figure 3. XRD of modified MF resin

As shown in Figure 3, there is no obvious change in the X-ray diffraction pattern of BG15% modified MF resin, and the diffraction peaks appear at 20 and 43 degrees. The relative crystallinity of BG15% modified MF was 39.45% by calculation of crystallinity. But the crystallinity of modified resin by aziridine was changed to 40.20%, this was maybe because that the new cross-linked structure was formed when the aziridine was added. By adding AZ to BG15% and U20% resin, the peak angle was slightly shifted and narrowed. According to the calculation of crystallinity, the relative crystallinity of urea modified MF was 40.78%, and the crystallinity was reduced by the modification of aziridine. This was because that the addition of aziridine molecules could destroy the regularity of the molecular chain of urea formaldehyde resin.

3.2. Mechanical of properties modified MF resin

Table 1 shown effect of aziridine on the tensile mechanical properties of modified MF resin. As shown in Table 1, In comparison with the MF resin, the tensile strength of BG15% modified MF resin increased, but the elastic modulus decreased. Meanwhile, the elongation at break increased to 3.24%. This may be due to the existence of the the large molecule of the triazine ring. The tensile strength and elastic modulus of MF modified by BG15% and U20% were increased after adding the aziridine, and the tensile strength increased from 86Mpa to 90Mpa, while the elongation at break was not obviously changed.

Table 1. Effect of Aziridine on the tensile mechanical properties of modified MF resin

Samples	Tensile strength (Mpa)	Elasticity modulus (Mpa)	Elongation at break (%)
MF	91.2	11767	2.54
BG15%	99.2	13618	3.10
BG15%+AZ2%	99.8	12593	3.24
BG15%+U20%	86.0	12280	3.47
BG15%+U30%+AZ2%	90.0	12370	3.49

3.3. Oxygen index, thickness swell water adsorption properties of modified MF resin

Fig 4 shown that effect of aziridine on oxygen index of MF modified resin laminated panel (BG15% and BG15% + U20%). The results are shown in Fig 3. There was no obvious change on the oxygen index of BG15% modified MF resin when the addition of 2% aziridine. But the oxygen index was increased to 34.9 when the aziridine added into the BG15% + U20% resin.

Fig 5 and Fig 6 shown that effect of aziridine on thickness swell and water absorption of modified MF resin laminated panel. As shown in Fig 4, it was clearly seen that the thickness swell ratio of BG15% resin was decreased from 7.72% to 7.6% and the water absorption rate was decreased from 8.73% to 8.6% after 2% aziridine adding. Meanwhile, the thickness swell ratio of BG15%+U20% resin was decreased to 7.6% and the water absorption rate decreased to 8.6% after 2% aziridine adding. It was found that the thickness swelling rate and water absorption rate were both decreased.

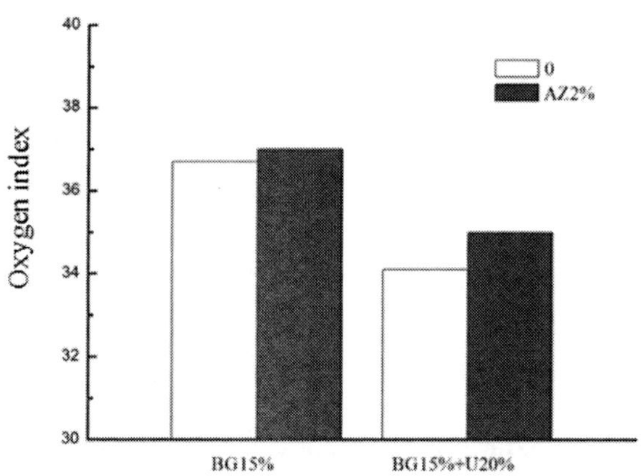

Figure 4. The oxygen index of laminated panel

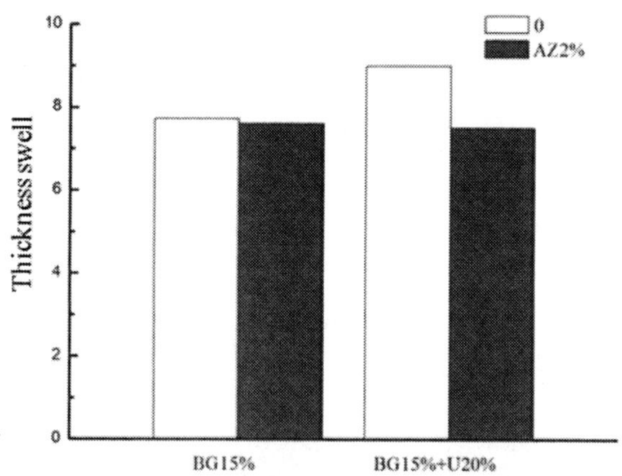

Figure 5. The thickness swell of laminated Panel

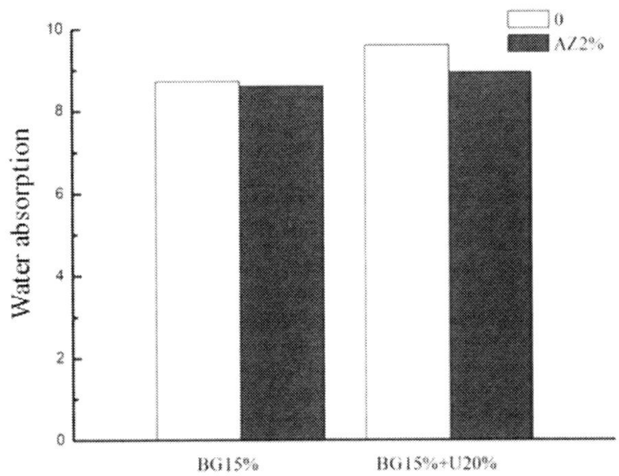

Figure 6. The water absorption of laminated panel

4. Conclusion

(1) Based on the results of FTIR analysis, it was found that the characteristic absorption peak of the hydroxyl group and the ring of the three triazine group were significantly reduced under the same curing condition, which indicated that the addition of aziridine was beneficial to the curing reaction.

(2) Based on the results of XRD analysis, it could be seen that the effect of the addition of aziridine on the crystallinity of BG15% modified resin was not obvious, but the crystallinity of BG15%+U20% modified resin was reduced.

(3) Based on the results of mechanical, oxygen index, thickness swell and water absorption properties of laminated panel, the tensile strength and elongation at break increased after the addition of 2% aziridine, and resistance to burning and boiling water were also improved.

Acknowledgments

Xiaohuan Liu acknowledges the financial support from the Program for key Science and Technology Team of Zhejiang Province with Grant No. 2013TD17. Shenyuan Fu acknowledges the financial support from the Program for Zhejiang Provincial Natural Science Foundation of China under Grant No. LZ16C160001.

References

[1] L. Huang, C. Zhao, Y. Yao, Y. You, Z. Wang, C. Wu, Y. Sun, J. Tian, J. Liu , Z. Zou, International Journal of Hydrogen Energy 41(2016) 11090-11098.
[2] Y. Shen, J. Gu, H. Tan, S. Lv ,Y. Zhang, Construction and Building Materials 120(2016)104-111.
[3] L. Licea-Jiménez, P. Y. Henrio, A. Lund, T. M. Laurie, S. A. Pérez-García, L. Nyborg, H. Hassander, H. Bertilsson , R. W. Rychwalski, Composites Science and Technology 67(2007)844-854.
[4] Y. Liu, X. Zhao, L. Ye, Industrial & Engineering Chemistry Research 55 (2016) 8743-8750.
[5] S. Mou, Y. Lu, Y. Jiang, Applied Surface Science 384(2016) 258-262.
[6] W. Xu, Yu, C., Zhao, X., Xu, J., & Jiang, M, Journal of Applied Polymer Science 133, 2016.
[7] H. Yari, M. Mohseni , M. Messori, Polymer 63 (2015)19-29.
[8] S.Z. Hang , B.H. Ping, Journal of Nanjing of Institute of Technology 6 (2008)39–44.

IOP Conf. Series: Materials Science and Engineering **207** (2017) 012016 doi:10.1088/1757-899X/207/1/012016

Synthesis of $MnFe_2O_4$ hollow microspheres via a sacrificial templates approach

Yang Li[1], Jun Ni[2], Canying Zhang[2] and Zhaoguo Meng[1,*]

[1]College of Electromechanical Engineering, Qingdao University of Science and Technology, Qingdao, China
[2]College of Materials Science and Engineering, Qingdao University of Science and Technology, Qingdao, China.

*Corresponding author e-mail: zhaoguom@qust.edu.cn

Abstract. Manganese ferrite ($MnFe_2O_4$) hollow microspheres were synthesized through a sacrificial template approach using manganese carbonate ($MnCO_3$) microspheres as sacrificial templates. Different from the literatures, the synthesis of $MnFe_2O_4$ hollow spheres was achieved through a one-pot hydrothermal process without subsequent calcination process. The temperature and reaction time of the hydrothermal process were set to be 180 ℃ and 8 h. The formation mechanism of the $MnFe_2O_4$ hollow microspheres was also investigated. Results indicate that the average diameter of the as-synthesized $MnFe_2O_4$ hollow microspheres is 1.2 μm. The shells of the hollow microspheres are composed of nanoparticles. During hydrothermal process, Mn2+ ions diffused outward and resulted in shell materials of $MnFe_2O_4$, thus $MnCO_3$ templates were consumed and $MnFe_2O_4$ hollow microspheres were obtained.

1. Introduction

Magnetic hollow structured ferrites are very attractive due to their well-defined interior voids, low density, large surface area, stability in fluids, and surface permeability. These properties make them much suitable for microencapsulation, drug delivery, magnetic separation [1-3], etc. As an important category of magnetic hollow structured ferrites, spinel structured ferrites (MFe_2O_4, M =Zn, Mg, Co, Ni, Cd) hollow spheres have been reported to be synthesized through a two-steps template method [4, 5], where microspheres such as carbonaceous saccharide microspheres and carbon microspheres are used as templates. Precursor of MFe_2O_4 is then deposited on the surfaces of the templates to form core-shell structures that will be subsequently calcinated to remove the templates and obtain hollow microspheres. Besides the above method, sacrificial template approach is a one-pot template method for hollow structure synthesis, where the template is actually one of the reactants of the synthesis reaction for shell materials and will be consumed in the process [6]. However, the synthesis of MFe_2O_4 hollow microspheres via a sacrificial template approach in wet chemistry systems is seldom reported. So it would be interesting to apply sacrificial template approach to the synthesis of MFe_2O_4 hollow microspheres.

Herein, we report the synthesis of $MnFe_2O_4$ hollow microspheres via a sacrificial templates approach using $MnCO_3$ microspheres as sacrificial templates. It will be demonstrated that $MnCO_3$ templates can be consumed and transformed to MFe_2O_4 shell materials during hydrothermal process, thus the hollow microspheres are obtained. $MnFe_2O_4$ is chosen because it is one of the most important magnetic materials and is widely used in electronic applications [7], as contrast-enhancement agents in magnetic resonance imaging (MRI) technology [8], and as magnetic recording media [9], etc.

2. Experimental

All chemicals involved in the experiment were analytical grade and used without further purification.

Preparation of $MnCO_3$ Templates: The typical procedure to prepare the template of $MnCO_3$ microspheres can be referred to Fei's work in the literature [10]. NH_4HCO_3 (10 mmol) and $MnSO_4 \cdot H_2O$ (1 mmol) were dissolved respectively in 70 mL water. Ethanol (21 mL) was then added to the $MnSO_4$ solution with stirring. When the two solutions were mixed and maintained at room temperature for 3 h, the $MnCO_3$ particles were precipitated and suspended in this water-ethanol mixed solvent. After being separated from the reaction mixture by vacuum filtration, washed several times with distilled water and dried at room temperature in a vacuum oven (ca. 0.1 MPa) for 6 h, the template of $MnCO_3$ microspheres were obtained.

Preparation of $MnFe_2O_4$ hollow spheres: In a typical procedure, appropriate $FeSO_4 \cdot 7H_2O$ (2 mmol) was dissolved in water (40 mL), and then the as-prepared $MnCO_3$ particles were dispersed in the $FeSO_4$ solution by ultrasonic treatment for 5 min. Then the obtained mixture were transferred into 60 mL stainless steel autoclave and heated at 180 ℃ for 8 h, followed by cooling to room temperature naturally. After the hydrothermal reaction, the black solid was centrifuged, and sequentially washed with distilled water repeatedly to remove the residual reactant. Then, the product was dried in air at 60 ℃ for 6 h before characterization.

Characterizations of samples: The XRD patterns were recorded on a powder X-ray diffractometer (Rigaku D/max-2500/PC) equipped with Cu $K_{\alpha 1}$ radiation source (λ=1.5406 Å) at a step width of 0.02°. The transmission electron microscopy (TEM) images were captured on a JEM-2000EX transmission electron microscopy. The scanning electron microscopy (SEM) images were taken with a JEOL-JSM6700F scanning electron microscope.

3. Results and discussion

The X-ray diffraction (XRD) pattern of the as-synthesized $MnCO_3$ microspheres (see Figure 1a) shows sharp diffractions characteristic of simple cubic $MnCO_3$ (JCPDS card No. 44-1472). No other impurities were detected in the synthesized products. SEM image shows that the $MnCO_3$ crystals have a solid spherical morphology (Figure 1b). The sizes of the microspheres, typically ranges from 1.2 to 1.7 μm, with an average size of about 1.4 μm and standard deviation of 0.3 μm. High magnification image of a single microsphere (the insert in Figure 1b) shows that the surface of the as-prepared $MnCO_3$ microsphere is composed of triangular edges and corners. A BET specific area of 6.7 m^3/g has been measured, suggesting that the $MnCO_3$ microspheres are solid inside.

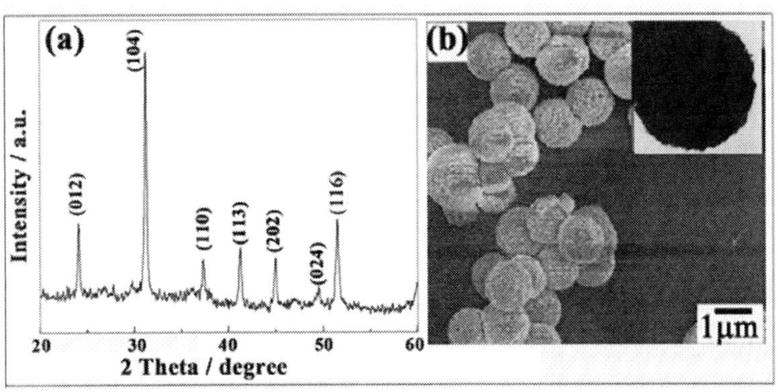

Figure 1. XRD patterns (a) and SEM images (b) of the MnCO3 templates. The inset in (b) is a high magnifications TEM image of a single microsphere.

The X-ray diffraction pattern of the resulted products after hydrothermal treatment is presented as Figure 2a. All diffraction peaks can be readily indexed to cubic $MnFe_2O_4$ (JCPDS card No.73-1964) without impurity phases detected. SEM image (Figure 2b) of the as-synthesized $MnFe_2O_4$ shows that the sample is composed of microspheres with an average diameter of about 1.2 μm. A high magnification SEM image of the microsphere surface inserted in Figure 2b shows that the microsphere is composed of nanoparticles with average particles size of about 70 nm. TEM image (Figure 2c) reveals the hollow structure of the microspheres. The hollow structure can also be confirmed by the SEM image of a broken hollow sphere inserted in Figure 2c. It can be found from the cross section that the shell of the hollow microsphere is composed of nanoparticles with size typically ranges from 50 nm to 80 nm. The results indicate that $MnFe_2O_4$ hollow microspheres have been synthesized. The magnetic hysteresis curve of the $MnFe_2O_4$ hollow microspheres (Figure 2d) shows ferromagnetic behavior with saturation magnetization (Ms), remanent magnetization (Mr), and coercivity (Hc) values of the $MnFe_2O_4$ hollow microspheres are 33.96 emu/g, 3.38 emu/g and 85 Oe, respectively (based on the magnified image of the central part of Figure 2d). The relatively lower Ms value of the $MnFe_2O_4$ hollow microspheres can be contributed to the small grain size of the nanoparticles constitute the hollow microspheres.

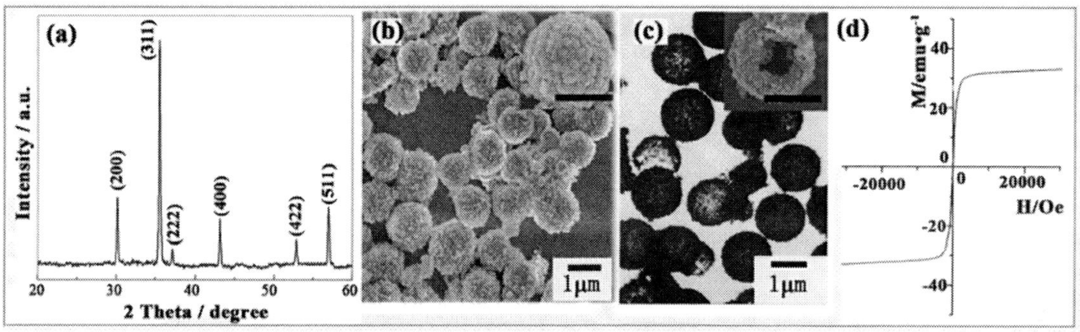

Figure 2. XRD pattern (a), SEM images (b), TEM images (c) and Magnetic hysteresis curve (d) of the MnFe2O4 hollow microsphere. The scale bars in the two insets stand for 1μm.

To study the phase and structure transitions from $MnCO_3$ microsphere to $MnFe_2O_4$ hollow microsphere, products obtained at different stages of the hydrothermal process were collected and subjected to XRD and TEM analysis. Figure 3 presents the XRD patterns and TEM images of microspheres obtained at different stages of the hydrothermal process. The microspheres obtained at 5 h are mainly composed of solid cores and loose outer layers. The corresponding XRD pattern reveals a combination of $MnCO_3$ and $MnFe_2O_4$. It can be deduced that the solid cores are $MnCO_3$ templates and the loose outer layers are newly formed $MnFe_2O_4$ that are transformed from the $MnCO_3$ templates. In this case, an interface between $MnCO_3$ template and $MnFe_2O_4$ outer layer existed in the microsphere. Driven by the concentration gradient, Mn^{2+} ions diffused outward and Fe^{3+} ions diffused inward. Reactions occurred at the interface and resulted in $MnFe_2O_4$ shell materials. The mechanism details can be proposed as follow.

When dissolved in water, $FeSO_4$ will undergo oxidation and hydrolysis reactions which can be described as equation (1). Oxygen in equation (1) comes from the dissolved O_2 in water. Then Fe^{3+} resulted from reaction (1) will partly undergo further hydrolysis reaction (equation (2)) resulting in slightly decrease of pH value. It should be noted that part of Fe^{2+} in the present work might undergo reaction (1) and even reaction (2) before hydrothermal treatment, as the solution was found to have turned yellow before transferred into the autoclave. The rest of Fe^{2+} might be transformed to Fe^{3+} during hydrothermal process. As pH value decreased at the beginning of hydrothermal process, $MnCO_3$ dissolved according to reaction (3). However, reaction (3) occurred at very low speed and

IOP Conf. Series: Materials Science and Engineering **207** (2017) 012016 doi:10.1088/1757-899X/207/1/012016

only on the surface of the $MnCO_3$ templates, because the pH value would not be very low in the present experiment. In fact, the pH value of the resulted mixture after hydrothermal process was measured to be around 4.0. The speed of reaction (3) was so low that the resulted Mn^{2+} could not diffuse from the surface of the $MnCO_3$ templates before reacted with Fe^{3+} to form $MnFe_2O_4$ (reaction (4))at elevated temperature and pressure (saturated vapor pressure of water at 180 ℃ is about 1.0 MPa). It should be emphasized that reaction (4) can only occur at elevated temperature and pressure. Under ambient conditions, precursors (frequently complex hydroxide) will be formed instead, and then subsequent calcinations are required to get $MnFe_2O_4$. The driven force of reaction (4) comes from the elevated temperature and pressure that favour the formation of $MnFe_2O_4$ because $MnFe_2O_4$ is more stable than the precursors at elevated temperature and pressure. After the outmost layer of $MnCO_3$ templates had been transformed to $MnFe_2O_4$, reaction (3) and subsequently reaction (4) continued towards the centre of $MnCO_3$ templates. It should be noted that $MnFe_2O_4$ was formed as nanoparticles with tunnel in between, thus ions and molecules could diffuse in and out. During this period, the early formed $Fe(OH)_3$ dissolved at lower pH value (around 4.0), as reaction (5) describes. Fe^{3+} resulted from reaction (5) joined reaction (4) and was then consumed.

$$12\ Fe^{2+} + 3\ O_2 + 6\ H_2O \rightarrow 8\ Fe^{3+} + 4\ Fe(OH)_3 \qquad (1)$$

$$Fe^{3+} + 3\ H_2O \rightarrow Fe(OH)_3 + 3\ H^+ \qquad (2)$$

$$MnCO_3 + 2\ H^+ \rightarrow Mn^{2+} + CO_2 + H_2O \qquad (3)$$

$$Mn^{2+} + 2\ Fe^{3+} + 4\ H_2O \rightarrow MnFe_2O_4 + 8\ H^+ \qquad (4)$$

$$Fe(OH)_3 + 3\ H^+ \rightarrow Fe^{3+} + 3\ H_2O \qquad (5)$$

As time prolonged, more and more Mn^{2+} ions diffused outward leaving some caves at the center of the microsphere. Such caves can be observed in the TEM image of the microspheres obtained at 6 h. The microspheres obtained at 7 h are mainly hollow structures with a few individual small cores inside. The corresponding XRD pattern shows a dominant $MnFe_2O_4$ phase with minor peak of the main diffraction of $MnCO_3$. Thus the individual small cores can be explained as the residual of $MnCO_3$ templates. In the hollow microspheres obtained at 8 h, the individual small cores of $MnCO_3$ have totally disappeared. The corresponding XRD pattern can be indexed to $MnFe_2O_4$ phase without any diffraction of $MnCO_3$ detected. The sacrificial templates of $MnCO_3$ microspheres have been completely transformed to $MnFe_2O_4$ hollow microspheres. Theoretically, the diameters of the $MnFe_2O_4$ hollow microspheres should be slightly larger than that of the $MnCO_3$ microspheres. However, due to the loose structure of the outer layers, some $MnFe_2O_4$ nanoparticles dropped from the surface and resulted in similar diameters of the $MnFe_2O_4$ hollow microspheres.

IOP Conf. Series: Materials Science and Engineering **207** (2017) 012016 doi:10.1088/1757-899X/207/1/012016

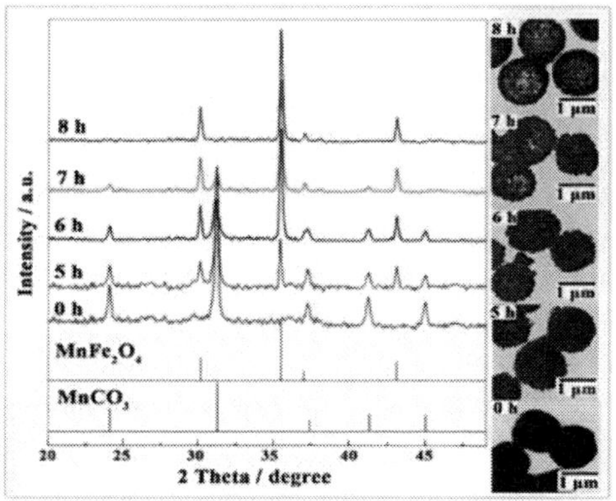

Figure 3. Conversion from $MnCO_3$ microspheres towards $MnFe_2O_4$ hollow microspheres under hydrothermal condition at 180 ℃. XRD patterns (left) and TEM images (right) of products collected at different stages of hydrothermal process.

Hydrothermal temperature is critical to the formation of $MnFe_2O_4$. In order to determine the appropriate hydrothermal reaction temperature, the hydrothermal temperature was set to be 140 ℃, 160 ℃, and 180 ℃, respectively. The hydrothermal reaction time was prolonged to 12 h thus the reaction would complete. Other conditions were identical to the typical procedure. Figure 4 shows the X-ray diffraction patterns of products obtained at different hydrothermal temperatures. The results support the idea that $MnFe_2O_4$ should be synthesized at elevated temperature and pressure. The products obtained at 180 ℃ is $MnFe_2O_4$ without detectable impurity, while the products obtained at 140 ℃ and 160 ℃ contain considerable amount of complex impurities. It seems that 180 ℃ is the appropriate hydrothermal temperature in the present work.

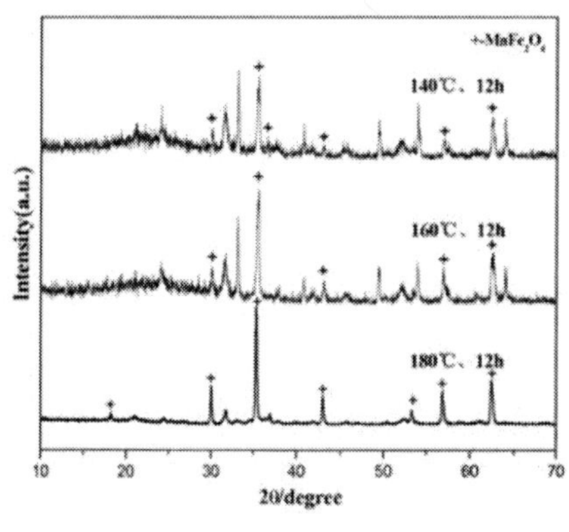

Figure 4. The XRD patterns of the samples obtained at different hydrothermal temperatures

The molar ratio of $MnCO_3$ to $FeSO_4$ is also important and should be controlled severely as 1:2, otherwise diffraction peaks of impurities will appear in the XRD patterns. Figure 5a shows the X-ray diffraction pattern of the sample synthesized with excessive Mn^{2+}. Surplus $MnCO_3$ exists in the products when there is excessive Mn^{2+}. Figure 5b shows the X-ray diffraction pattern of the sample synthesized with excessive Fe^{2+}. The pattern peaks reveal that the hematite Fe_2O_3 has been synthesized together with $MnFe_2O_4$ when Fe^{2+} is excessive.

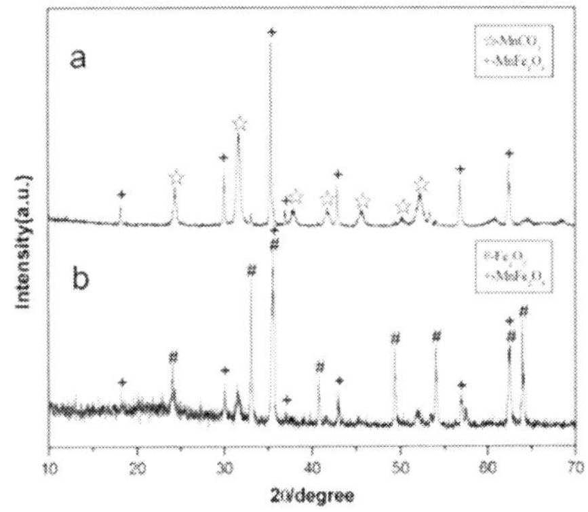

Figure 5 XRD patterns of the samples obtained with (a) excessive Mn^{2+}; (b) excessive Fe^{2+}

4. Conclusion

In summery, we report a sacrificial templates approach for the synthesis of $MnFe_2O_4$ hollow microspheres using $MnCO_3$ as sacrificial template. We demonstrate that $MnCO_3$ microspheres can be transformed to $MnFe_2O_4$ hollow microspheres through hydrothermal process at elevated temperature and pressure. Hydrothermal treatment at 180 ℃ for 8 h is the appropriate condition for the formation of $MnFe_2O_4$ hollow microspheres. The as-synthesized $MnFe_2O_4$ hollow microspheres exhibit ferromagnetic behaviour with slightly lower saturation magnetization value than their solid counterparts.

Acknowledgments

This work was supported by the National Natural Science Foundation of China (51172117, 51472134), the Natural Science Foundation of Shandong Province (ZR2013EMM003).

References

[1] L.J. Berchmans, R.k. Selvan, P.N.S. Kumar, C.O. Augustin, Structural and electrical properties of $Ni1_{-x}Mg_xFe_2O_4$ synthesized by citrate gel process. J Magn Magn Mater 2004; 279: 103-110.

[2] M. Giersig, M. Hilgendorff, Magnetic nanoparticle superstructures. Eur J Inorg Chem 2005; 18: 3571-3583.

[3] D.G. Shchukin, I.L. Radtchenko, G.B. Sukhorukov, Micron-Scale Hollow Polyelectrolyte Capsules with Nano-Sized Magnetic Fe_3O_4 Inside. Mater Lett 2003; 57: 1743-1747.

[4] Z. Li, X. Lai, H. Wang, D. Mao, C. Xing, D. Wang, General Synthesis of Homogeneous Hollow Core−Shell Ferrite Microspheres. J Phys Chem C 2009; 113: 2792-2797

[5] J. Gu, S. Li, M. Ju, E. Wang, In situ carbon template-based strategy to fabricate ferrite hollow spheres and their magnetic property. J Cryst Growth 2011; 320: 46-51.

[6] L. Wang, F. Tang, K. Ozawa, Z. Chen, A. Mukherj, Y. Zhu, et al. A General Single-Source Route for the Preparation of Hollow Nanoporous Metal Oxide Structures. Angew Chem Int Ed 2009; 48: 7048-7051.

[7] M. Sugimoto, The past, present and future of ferrites. J Am Ceram Soc 1999; 82: 269-80.

[8] B. Gillot, M. Laarj, S. Kacim, Reactivity towards oxygen and cation distribution of manganese ironspinel $Mn_{3-x}Fe_xO_4(0\leq x\leq 3)$ fine powders studied by thermogravimetry and IR spectroscopy. J Mater Chem 1997; 7: 827-831.

[9] D. Zhang, X. Zhang, X. Ni, J. Song, H. Zheng, Low-temperature fabrication of $MnFe_2O_4$ octahedrons: Magnetic and electrochemical properties. Chem Phys Lett 2006; 426: 120-123.

J. Fei, Y. Cui, X. Yan, W. Qi, Y. Yang, K. Wang, Q. He, J. Li, Controlled Preparation of MnO_2 Hierarchical Hollow Nanostructures and Their Application in Water Treatment. Advanced Materials 2010, 20:452-456.

IOP Conf. Series: Materials Science and Engineering **207** (2017) 012017 doi:10.1088/1757-899X/207/1/012017

The Research Progress of Targeted Drug Delivery Systems

Jiayin Zhan, Xizi Liang Ting, Junjie Zhu*

School of Chemistry and Chemical Engineering, Nanjing University, Nanjing 210023, China

*Corresponding author e-mail: jjzhu@nju.edu.cn

Abstract. Targeted drug delivery system (DDS) means to selectively transport drugs to targeted tissues, organs, and cells through a variety of drugs carrier. It is usually designed to improve the pharmacological and therapeutic properties of conventional drugs and to overcome problems such as limited solubility, drug aggregation, poor bio distribution and lack of selectivity, controlling drug release carrier and to reduce normal tissue damage. With the characteristics of nontoxic and biodegradable, it can increase the retention of drug in lesion site and the permeability, improve the concentration of the drug in lesion site.at present, there are some kinds of DDS using at test phase, such as slow controlled release drug delivery system, targeted drug delivery systems, transdermal drug delivery system, adhesion dosing system and so on. This paper makes a review for DDS.

1. Introduction

DDS is refers to the use of new technology and biological polymer materials technology to delivery gene or drugs to the designated the body parts according to the clinical need time and dose. it can be designed to minimize drug degradation, increase bioavailability, allow targeting to specific cells and reduce the total amount of drug needed, it can be controlled release of drugs to decreasing toxicity and harmful side-effects [1].the idea of DDS drug research and development is to use the new technology and new materials to improve curative effect, reduce the effect of patients with adverse reaction and easy to use by changing the drug pharmacokinetic model and delivery channels. DDS is the application and development in pharmacy by the modern science and technology, it has become the theme of the modern pharmacy innovation and development.

Targeted drug research began in the 1970s, its main form is for injection, with the development, the current trend of DDS drug research and development is the use of liposome, lipid, protein, fat, microspheres and a biodegradable polymers as drug carrier. Targeted drugs originally is mainly used in cancer treatment, to maximize the efficacy of a drug is of prime importance during the choice of the delivery system. Among the currently available delivery systems, which include liposomes, emulsions, polymeric micelles and micro-particles, carbon nanotubes (CNTs) , and so on. The following we will illustrate the characteristic of all kinds of DDS respectively.

2. Targeted drug delivery system

2.1 Liposomes

Because the design of viral vectors is relatively expensive and might be toxic to normal organs, a variety of Nano vectors have been explored as possible gene therapy vehicles. Liposomes, especially nano-liposomes is primarily lipid bilayer membrane. Phospholipid elimination in the blood is relatively slow, as pharmaceutical package can be released slowly by buried liposomes, which thus effectively prolongs action time of drugs [2]. Liposomes is safe and non-toxic, targeted modification advantages are more apparent [3], the newest research about bio reducible lipid nanoparticles obtain good results [4]. Researchers combining bio reducible lipid (8-O14B) nanoparticles with negatively supercharged Cre recombinase or anionic Cas9: single-guide (sg) RNA complexes drives the electrostatic assembly of nanoparticles that mediate potent pro-tein delivery and genome editing, in order to across the

Mammalian cell membrane, including escape from endosomes, developing delivery vehicles to transport active protein to their intracellular target site is thus essential to advance protein-based genome editing. Another [5] research functional targeting epirubicin liposomes for transferring drugs across the blood-brain barrier (BBB) to treat glioblastoma Pro. Zhao [6] used the lipid-mixing method to make Liposomes were developed in order to treat bacterial infections in cystic fibrosis with regard to the enhanced bactericidal activity of entrapped antibiotics released through their fusion with bacterial membranes.to overcome the acute and growing problem of antibiotic resistance of bacteria to conventional antibiotics which made it imperative to develop new liposome formulations for antibiotics.

2.2 Carbon Materials

Nanoscale carbon materials usually including carbon nanotubes97, graphene98 / graphene-oxide99, and nano diamond Targeting and selectivity during cancer cell destruction has also been reported through molecular surface functionalization of single-walled carbon nanohorns(SWCNTs)both in vitro and in vivo. The use of nanoscale materials has attracted considerable attention. Nanoscale materials have been demonstrated to offer a variety of therapeutic and diagnosic capabilities. It usually has a diameter between 10 nm and 1000 nm. The drug is usually made from natural polymer material and has been fully applied to treatment course of tumor, diabetes and vascular disease. It can effectively extend action time of drug, then effectively improve clinical effect of the drug and minimize toxic and side effects of the drug [7]. Frequency of use of nano preparation technology can provide better development space for medical use and effectively promote long-term development of nano biotechnology. Compare with other materials, Carbon can be much cheaper, but the chirality and diameter of nanotubes, which severely impact the physical and chemical properties, they are difficult to control [8].

2.3 Metallic nanomaterials

Metallic nanomaterials including gold and silver nanocrystals, and Nano rods. Have been shown to generate localized hyper thermal heating through the absorption of incident optical radiation and surface Plasmon relaxation to treat the disease. Affected by the size and shape of the particle and are susceptible to drift. One problem with drug delivery is heterogeneity within the tumor due to irregular blood vessel architecture, hindered diffusion from a dense intercellular matrix. Its advantage is low-cost, easily synthesized nanoparticles and the particles are also hown to have a high thermal stability [9] .PEG-DSPE coating may be related to better absorption, based on the stability and a pharmacokinetic improvement in the blood circulation time. This method will lead to enhanced permeation for nanoparticles to across the vascular endothelium and achieve improved accumulation in the tumor [10].

2.4. Semiconducting nanomaterials
Semiconducting nanomaterials have also received a significant amount of recent attention for hyper thermal theranostics, their structures typically allow for substantial penetration of electromagnetic fields throughout the internal volume of the particle. Yet these materials are often composed of toxic metals. The following research of semiconductors as PTT agents will certainly require the study of biocompatible materials in the future.

2.5. Iron Oxide Nanocrystals
The difference between the magnetic particles and their metallic and semiconducting counterparts is the mechanism by which the particle is heated. Briefly, magnetic hyperthermia is achieved by applying external alternating magnetic fields to cause the magnetic particles to heat through hysteresis loss (Néel relaxation) or induced eddy currents [11]. Nanoscale materials have been demonstrated to offer a variety of therapeutic and diagnostic (theranostic) capabilities.

2.6. Nano micelles
Polymeric Nano micelles are formed by amphiphilic polymers with distinct hydrophobic and hydrophilic segments. The polymer self-assemble to form micelles in aqueous solution. The mechanism of drug release from Nano micelles is dependent on the nature and strength of interactions between core-forming polymer and drug molecules, micelle stability [12].

2.7. Carbon nanotubes
Carbon nanotubes (CNT), have recently been studied as novel and versatile drug and gene delivery vehicles. When CNT are suitably functionalized, they can interact with various cell types and are taken up by endocytosis. Some anti-cancer drugs cisplatin, methotrexate, the antifungal compound amphotericin B and doxorubicin have been delivered by Centrist advantage is pro-tecting the cargo from metabolism or degradation, or through targeted delivery that can reduce side effects about the Nano toxicology of the CNT and their potentially damaging effects on the environment [13] . Shorter multi-walled CNTs (MWCNTs, i.e., 1 μm) have been reported to penetrate the cell membrane more efficiently than the longer CNTs, which can inhibit their uptake by self-arranging into a coiled or bundled shape. Cai et al. [14] made a theoretical analysis of penetration pathways. They showed that amphiphilic nanotubes can penetrate through artificial lipid bilayers via an endocytosis pathway. Different types of nucleic acids such as micro-RNA (miRNA), small-interfering (siRNA) and plasmid DNA (pDNA) can be bound to CNTs and transferred into mammalian cells. Oral administration of peptides has problems with enzymatic degradation and poor uptake from the gut, but CNTs were proposed to overcome these limitations [15].

2.8. Mesoporous silica materials
Mesoporous silica nanoparticles (MSNPs) is uniformly sized, porous and dispersible nanoparticles using colloidal chemistry and evaporation-induced self-assembly. This kind of material enables the loading of diverse cargos and cargo combinations at levels exceeding those of other common drug delivery carriers such as liposomes or polymer conjugates. This is because non-covalent electrostatic, hydrogen-bonding and van der Waals interactions of the cargo with the MSNP internal surface cause preferential adsorption of cargo to the MSNP[16], SNPs within reconfigurable supported lipid bilayers to develop new classes of responsive Nano carriers that actively interact with the target cell.it usually requires biocompatibility and low toxicity and can be used to follow bio distribution, cancer cell targeting efficiency, internalization pathways, cytotoxicity, and the progress of therapy. There have been widely differing reports concerning the toxicity of MSNP and amorphous silica in genera[l] [17-18]

3. Conclusion
Because of the curative effect and smaller side effect, the role of targeted drugs will more and more be taken seriously, with the acceleration development of research on mechanism of drug targeting, targeted drug applications continue to expand in more field.

References
[1] Bennett E. Smith, Paden B. Roder, Xuezhe Zhou,et al. Nanoscale materials for hyperthermal theranostics. Nanoscale. 2015 , 7(16): 7115–7126.
[2] Wen S, Liu H, Cai H, et al. Targeted and pH-responsive delivery of doxorubicin to cancer cells using multifunctional dendrimer-modified multi-walled carbon nanotubes. Adv healthc Mater. 2013; 2(9):1267–1276.
[3] Aili, D., Mager, M., Roche, D. & Stevens, M. M. Hybrid nanoparticle-liposomedetection of phospholipase activity. Nano Lett .2011,11:1401–1405.
[4] Ming Wang, John A. Zuris,Fantao Meng,et al. Efficient delivery of genome-editing proteins using bioreducible lipid nanoparticles PNAS .2016, 113(11) : 2868–2873
[5] Cheng-Xiang Zhang, Wei-Yu Zhao, Lei Liu,et al.A nanostructure of functional targeting epirubicin liposomes dually modified with aminophenyl glucose and cyclic pentapeptide used for brain glioblastoma treatment. Oncotarget, 2015,6(32): 32681-32700
[6] Wang ,Yufan Ma1, Hayssam Khalil. Fusion between fluid liposomes and intact bacteria: study of driving parameters and in vitro bactericidal efficacy. International Journal of Nanomedicine 2016,8: 4025–4036
[7] Carolina Garrido, Carrie A Simpson, Noelle P Dahl,et al.Gold nanoparticles to improve HIV drug delivery. Future Med. Chem. 2015,7(9), 1097–1107
[8] Jain RK, Stylianopoulos T. Delivering nanomedicine to solid tumors. Nature Reviews Clinical Oncology. 2010; 7(11):653–664.
[9] Zhang Z, Wang J, Chen C. Gold nanorods based platforms for light-mediated theranostics.Theranostics. 2013; 3(3):223–238.
[10] Jing Zhang, Yingchong Chen , Xiang Li ,et al. The influence of different long-circulating materials on the pharmacokinetics of liposomal vincristine sulfate. International Journal of Nanomedicine. 2016:11 :4187–4197
[11] Gupta AK, Naregalkar RR, Vaidya VD,et al. Recent advances on surface engineering of magnetic iron oxide nanoparticles and their biomedical applications. Nanomedicine2007; 2(1):23–39.
[12] Ravi D. Vaishya, Varun Khurana, Sulabh Patel,et al.Controlled Ocular Drug Delivery with Nanomicelles. Wiley Interdiscip Rev Nanomed Nanobiotechnol. 2014 September ; 6(5): 422–437.
[13] Mahdi Karimi, Navid Solati, Amir Ghasemi, et al. Carbon nanotubes part II: a remarkable carrier for drug and gene delivery. Expert Opin Drug Deliv. 2015 July ; 12(7): 1089–1105.
[14] Cai D, Mataraza JM, Qin Z-H, et al. Highly efficient molecular delivery into mammalian cells using carbon nanotube spearing. Nat Methods. 2005; 2(6):449–454.
[15] Pattinson SW, Ranganathan V, Murakami HK, et al. Nitrogen-induced catalyst restructuring for epitaxial growth of multiwalled carbon nanotubes. ACS Nano. 2012,6(9):7723-7730.
[16] Tarn D, Ashley CE, Xue M,et al. Mesoporous Silica Nanoparticle Nanocarriers – Biofunctionality and Biocompatibility Acc Chem Res. 2013,19; 46(3): 792–801.
[17] Lin Y-S, Haynes CL. Impacts of Mesoporous Silica Nanoparticle Size, Pore Ordering, and Pore Integrity on Hemolytic Activity. Journal of the American Chemical Society. 2010; 132:4834–4842.
[18] Nicholas A. W. Bell and Ulrich F. Keyser Specific Protein Detection Using Designed DNA Carriers and Nanopores J. Am. Chem. Soc. 2015, 137, 2035-2041.

IOP Conf. Series: Materials Science and Engineering **207** (2017) 012018 doi:10.1088/1757-899X/207/1/012018

The preparation and detection of high purity silicon tetrachloride with optical fibres level

Ye Wan[1], Xiong Zhao[1] Shuhu Guo[1,*] Dazhou Yan[2] Dian Yang[1,] Lei Wang[1]

[1]China Silicon Corporation Ltd. No. 1 MuDan Avenue, City of Luoyang, Henan province, 471023;

[2]China Enfi Engineering Corporation No.12 Fuxing Avenue, Beijing of China. 100038.

*Corresponding author e-mail:guoshuhu@enfi.com.cn

Abstract This article showed a technology for producing high purity silicon tetrachloride with optical fiber level, which using the method of adsorption unite distillation, whose material belong to by-product of polysilicon system. This technology could remove chlorosilane, metal impurities and impurities containing hydrogen effectively. Test's results showed that most indexes can meet or exceed the standards of Evonic, indicating that product produced by this technology can meet the demand of processing optical fiber preforms what is OVD level completely. This technology has the advantages of simple process, continuous operation, and large capacity and so on, and industrialization can be realized.

1. Introduction

Today, Modified Siemens Process has already been one main process for producing polysilicon, however, it can also generate too much $SiCl_4$ in the manufacturing process. For example, if one company produced 10000t polysilicon, it would also produce 200000t $SiCl_4$ [1]. How to recycle $SiCl_4$ has been one key for influencing development of polysilicon company. Most polysilicon companies usually recycle $SiCl_4$ into process system to produce trichlorosilane, adopting the method of cold hydrogenation. With the forward study of $SiCl_4$, it has found that $SiCl_4$ can also be used for manufacturing many high value-added silica-based materials, such as fumed silica, optical fiber perform, ethyl silicate [2, 3, and 4].

High purity $SiCl_4$ is an important material for manufacturing optical fiber performs, having been one indispensable materials for optical communication industry. However, being limited by domestic purification technology, 80-90% $SiCl_4$ for optical fiber has to import [5]. So, this article designed one purification technology for manufacturing high purity $SiCl_4$, using co-product $SiCl_4$ from polysilicon company. Test's results showed that index did reach the international advanced level.

2. Experiment

2.1. Preparing high purity $SiCl_4$

Firstly, technology took impure $SiCl_4$ into crude distillation column (stainless steel) to remove low and high boiling substances continuously. Low purity $SiCl_4$ went into 1# tower to remove low boiling substances, such as trichlorosilane and dichlorosilane. Those low boiling substances went out through tower top, and intermediate product located at tower bottom would go into 2#tower to remove high boiling substances containing B, P elements in $SiCl_4$. At the top of 2#tower, the mass fraction of $SiCl_4$ would be not lower than 99.99%. Then, technology made use of adsorption column (stainless steel) to remove impurities containing hydrogen in system. After that, intermediate product went into distillation column to remove low and high boiling substances again, and prepared ultrapure $SiCl_4$ in the end. The preparing process was shown in Fig.1.

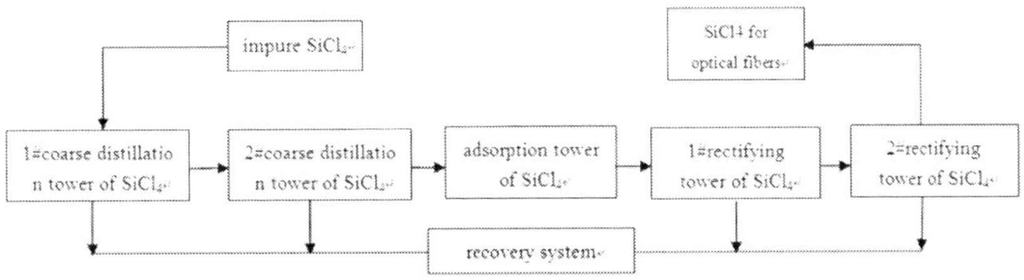

Figure 1. The preparing process of high purity SiCl4 for optical fiber.

At the section of crude distillation, it need to pump high purity N_2 into crude distillation column. By controlling temperature, the low and high boiling substances could be transferred with a certain proportion, the temperature of column bottom should be controlled at 75-80℃, of tower top should be controlled at 65-70℃. At the phase of absorption, it made use of globular activated carbon as adsorbent to remove metal impurities and impurities containing hydrogen, adsorbent temperature should be controlled at 20-40℃.

At the phase of distillation, $SiCl_4$ after adsorbed was transported into effective distillation column, at the protection of high pure N_2. Like crude distillation, low and high boiling substances should be removed with one certain proportion through controlling temperature, temperature at column bottom should be controlled at 75-80℃, at tower top should be controlled at 65-70℃, and the reflux ratio should be not less than10:1.

2.2. Analysis and Determination

Using G-C (7890A, Agilent) tested the content of chlorosilane impurities in $SiCl_4$ prepared by this technology; using ICP-MS (7700, Agilent) tested the content of metals impurities; using FT-IR (Is5, Thermos Fisher) tested the content of impurities with hydrogen. Technology also designed one online sampling system, which can sample and test online to realize the isolation between detecting system and surrounding, guaranteeing veracity of test results.

3. Results and Discussion

3.1. Coarse Rectifying

Fig. 2 was results of component test of crude $SiCl_4$ before or after coarse distillation, particular dates were showed in Table 1. It could found that $SiHCl_3$ was wiped off effectively, after coarse distillation.

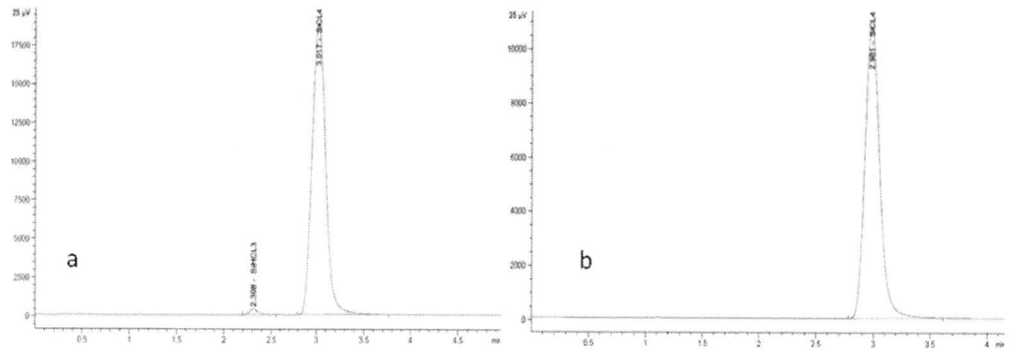

Figure 2. The component of SiCl4 through coarse distillation: (a) before, (b) after.

Table 1. The component of impure SiCl4.

component	$SiCl_4$ (%)	$SiHCl_3$ (%)	SiH_2Cl_2 (%)
impure $SiCl_4$	98.5	1.5	0
after coarse distillation	>99.99	<0.01	0

3.2. Absorbing

Table 2 was the test of metal impurities in $SiCl_4$ before and after adsorption, results showed that each metal impurity was dropped sharply after adsorption.

Table 2. Content of metal impurities before and after adsorption (ng·g-1).

sample	Al	Fe	Cu	Cr	Mn	Ni	V	Co	Zn
before	12.32	27.69	0.10	3.83	0.14	0.26	0.06	0.05	0.20
after	1.25	2.46	0.05	0.74	0.09	0.11	0.05	0.01	0.06

The basic principle of solid adsorption is based on the chemical bond polarity of each component in the compounds is different [6]. $SiCl_4$ is a symmetrical molecule without electric dipole moment, but, nearly all impurities are asymmetric molecule with substantial electric dipole moment, for example $AlCl_3$, $FeCl_3$, PCl_3 and BCl_3, those both can form chemical bond easily, and can be adsorbed easily.

3.3. Rectificating

Fig. 3(a) was FT-TR of impurities contain hydrogen in $SiCl_4$ before distillation. Results showed that the infrared transmittance of main compounds impurities contain hydrogen were still low, after being rough distillated and adsorbed, although the content of metal impurities has already as low as 5 ppb.

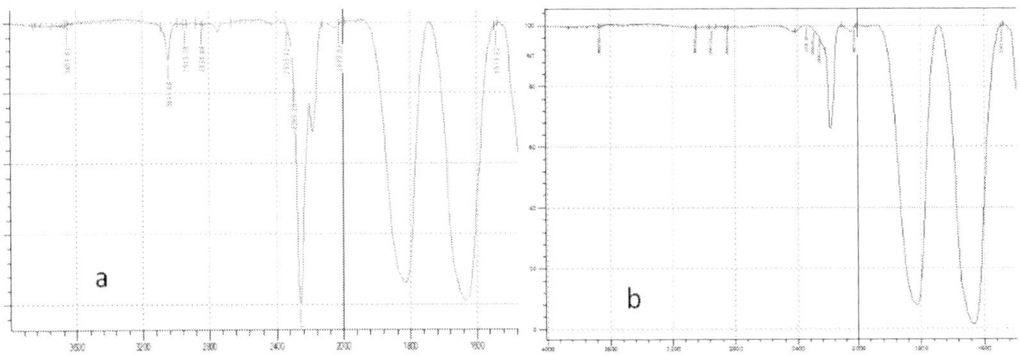

Figure 3. The infrared transmittance of hydrogen compounds in SiCl4 before and after distillation.

Fig. 3(b) was FT-TR of $SiCl_4$ after distillation, it can be found that infrared transmittance of compounds contain hydrogen increased generally at 3045cm^{-1}, 2338cm^{-1}, 2295cm^{-1}, even some perssad were up to 99%, after further distillation.

Distillation is the method which is used most widely to separate liquid mixtures. It has advantages of simple operation, automation, continuous production, large output and so on. It has already been used to manufacture high purity $SiCl_4$ for optical fiber [7, 8]. Experiment proved that distillation can remove hydrogen compounds effectively, if the distillation flow and parameters can get reasonable layout.

3.4. Quality Comparing

Table 3 was a comparison about content of metal impurities in $SiCl_4$ between this technology and YFOC or Evonic. Results showed that metal impurities' content in $SiCl_4$ prepared by this technology were all lower than YFOC, main metal impurities was inferior to Evonic. But the control for Al, Fe and Cr was not so better, it perhaps because that impure $SiCl_4$ used in this technology comes from polysilicon system, and those impurities mainly came from industrial Si. Meanwhile, Stainless Steel tube were used totally in technology system as pipeline welding and installation, unsmooth surface could cause chemical micro-pollution, which could improve the content of Cr in product. This problem can be dealed with the method of polishing treatment [9].

Table 3. Metal impurity of sample compare with YFOC and Evonic(ng·g-1)

sample	Al	Fe	Cu	Cr	Mn	Ni	V	Co	Zn
YFOC	1.00	3.00	1.00	1.00	0.50	1.00	1.00	1.00	1.00
Evonic	0.50	1.50	0.25	0.50	0.25	0.50	0.50	0.50	0.50
Sample	0.85	2.46	0.05	0.74	0.09	0.11	0.05	0.01	0.06

Table 4 was the comparison about content of impurities containing hydrogen in $SiCl_4$ prepared by this technology and standard of Evonic. It could be found that the infrared transmittance of other hydrogen compounds impurities except Si (CH_3) Cl_3 both meet the standard of Evonic. This is for that Si (CH_3) Cl_3 can form one azeotrope with $SiCl_4$, because of their nearly same boiling point; and it need too much plate number and reflux ratio to wipe off, which is a limit for the use of distillation operation for preparing very pure $SiCl_4$ with optical fiber level. For Si (CH_3) Cl_3 impurities, it can be removed through photo chlorination or low temperature plasma method.

Table 4. Content of hydrogen compounds compares with the standard of Evonic.

hydrogen compounds	wavelength	standard of Evonic	sample
$Si(OH)Cl_3$	3666cm^{-1}	≥90%	99.36%
-CH	3100-3020cm^{-1}	≥99%	99.75%
-CH	2970-2925cm^{-1}	≥95%	99.70%
HCl	2860-2830cm^{-1}	≥95%	99.86%
CO_2	2338 cm^{-1}	≥95%	99.29%
$Si(NCO)_4$	2295 cm^{-1}	≥97%	100.00%
$SiHCl_3$	2257 cm^{-1}	≥99%	100.00%
$Si(CH_3)Cl_3$	2023 cm^{-1}	≥99%	97.93%
Si_2OCl_6	1540 cm^{-1}	≥99%	99.49%

4. Conclusion

This study prepared high purity $SiCl_4$ with the method of distillation unite adsorption, results showed that contents of main metals and hydrogen compounds impurities are better than the standard of Evonic, which can meet the demand of optical fiber perform with OVD grade.

Comparing with Evonic, metal impurity containing Al and Fe are a little higher. It indicates that company should control the quality of raw material strictly, and remove it through distillation as far as possible, before starting to produce very pure $SiCl_4$. For higher content of Cr, it should handle with the pipeline selection and welding technology.

It is too hard to remove Si (CH_3) Cl_3 with traditional distillation technology for its nearly same boiling points with $SiCl_4$. But, it still can improve the quality of products further through the method of photo chlorination for removing Si (CH_3) Cl_3, which can meet the requirement for optical fiber performs with PCVD grade.

The technology has the advantages of simple process, continuous operation, and large capacity and so on. It can realize industrialization, and make a good foundation for the localization of $SiCl_4$ with optical fiber grade.

Acknowledgments

This work was financially supported by Talents in science and technology Innovation plan of Henan province.

References

[1] PAN Hongxing, ZENG Yalong, ZHU Ming, et al. Research Progress and Application of Hydrogenation Technology of Silicon Tetrachloride[J]. Materials Review, 2013, 27(s1): 154-156.
[2] Zhao Yun, Dan Jianming, Hong Chenglin. Application research of silicon tetrachloride from polysilicon enterprises[J]. Inorganic Chemicals Industry, 2014, 46(7):11-15.
[3] HU Kai-da, WANG Xiao-jun.Simulation and optimization of fiber grade silicon tetrachloride purification[J]. Modern Chemical Industry, 2013, 33(7):112-113.
[4] Cong Jinsheng, Xia Jumin. A Study on the synthesis of silic-oxide polymers by using the By-product of silicon tetrachloride from the preparation of polycrystalline silicon[J]. Journal of Hebei University of Technology, 1995(2):71-75.
[5] XU Hao-min, HUANG Guo-qiang. Research progress in refining techniques of SiCl4 for optical fiber [J]. Modern Chemical Industry, 2015(7):27-31.
[6] Liu Xiuxi, Lin Mingxi, Xue Chengshan. Properties, fabrication and applications of high purity

gases [M]. Beijing, Electronic Industry Press, 1997.
[7] ZHOU Ling-ying, XIANG Wen-qiang. Study on Four-Grade Rectification of Industrial Silicon Tetrachloride [J]. Contemporary Chemical Industry, 2016, 45(3):599-601.
[8] Zhou Xiang, Zhou Jun. Method for producing high purity silicon tetrachloride for optical fibers: CN, 1554581A [P]. 2004-12-15.
[9] Li Den-sin. Overview of Electronic Grade High Purity Gas Delivery System [J]. Equipment for Electronic Products Manufacturing, 2005, 34(2):56-58.

IOP Conf. Series: Materials Science and Engineering **207** (2017) 012019 doi:10.1088/1757-899X/207/1/012019

Effect of Cobalt Content on Microstructures and Wear Resistance of Tungsten Carbide–Cobalt-Cemented Carbides Fabricated by Spark Plasma Sintering

Renqi Ma[1], Shaodong Ju[1,*], Haiyan Chen[2] and Chenyang Shu[2]

[1]China National Offshore Oil Corporation Energy Technology Drilling & Production Company, Tianjin 300452, China

[2]College of Ocean Science and Engineering, Shanghai Maritime University, Shanghai 201306, China

*Corresponding author e-mail: jushd2@cnooc.com.cn

Abstract. We researched tungsten carbide–cobalt (WC–Co)-cemented carbides without binder and with little binder and sintered by spark plasma sintering with 0 wt. %, 0.2 wt. %, 0.5 wt. % and 0.8 wt. % Co at different temperatures. The microstructures, hardness, fracture toughness and wear resistance of the WC–Co-cemented carbides were investigated by scanning electron microscopy and mechanical property tests. The wear resistance of WC–Co-cemented carbides with little to no binder is acceptable and it decreases with increase in Co content. There is no significant difference in wear resistance at 0 wt. % and 0.2 wt. % Co content. By analysing the friction and wear mechanism, it can be concluded that because the Co phase was extruded first, the stability of the WC hard base was disrupted, the WC phase desquamated and abrasion occurred. WC-cemented carbide with 0.2 wt. % Co is the most suitable material. The best results were achieved at 1680°C with 2333 HV30 and 6.82 MPa.m1/2

1. Introduction

Tungsten carbide–cobalt (WC–Co)-cemented carbides are used widely for offshore oil exploitation engineering, in the mechanical industry and in aerospace engineering because of their high hardness, fine red hardness, suitable wear resistance and fracture toughness[1-3]. A significant improvement in mechanical properties can be achieved in cemented carbides with a finer grain size. Developments in the cemented carbide industry have led to nanomaterial additions to improve the mechanical properties of WC–Co-cemented carbides [4]. WC–Co-cemented carbides are composed mainly of a metallic ceramic and a cobalt binder and the Co content affects the hardness, fracture toughness and wear resistance of WC–Co-cemented carbides [5]. Therefore, WC-cemented carbides (Co content of less than 1wt. %) with no binder and little binder show a significant improvement in mechanical properties.

The optimum sintering method for WC–Co-cemented carbides is spark plasma sintering (SPS), which enables a powder compact to be sintered by Joule heat and high-pulsed electric current through the compact. In recent years, SPS has been used widely as a sintering method for different materials, including WC–Co-cemented carbides [6, 7]. WC-cemented carbides with little binder and as fabricated by SPS are compact and uniform; they have fine mechanical properties and an excellent wear resistance.

We have studied the microstructure, hardness, fracture toughness and wear resistance of WC-cemented carbides as sintered by SPS with 0 wt. %, 0.2 wt. %, 0.5 wt. % and 0.8 wt. % Co contents at different temperatures. Four specimens were investigated and compared to establish the optimum Co addition to the WC powder and the wear mechanism of the WC–Co-cemented carbides with little binder.

2. Experimental procedure

2.1. Materials

Table 1 lists the main characteristics of the WC powder (shown in Fig. 1a) and Co powder (shown in Fig. 1b) that were used as raw materials. Table 2 describes the nominal composition of the cemented carbides. The WC–Co composite powders were prepared using a wet ball-milled method in a high-energy planetary ball mill in which the milling ball was WC media of 5 mm and the liquid medium was ethanol. The ball-to-powder ratio was 5:1 and the rotational speed was 400 rpm. When milling was completed, the powder mixture was placed in a drying oven for 2 h at 70°C, after which the dried powder was sieved twice, using a 100-mesh (149-μm) sieve.

The sieved powders were poured into a graphite mold and placed in an SPS apparatus. Sintering was performed by applying a pulsed electric current to the SPS system. Table 3 shows the technological parameters used in the SPS process.

Table 1. Properties of powders used in the experiment

Powder	Chemical composition (wt. %)	D_{50}/μm(producers' data)	Purity	Supplier
WC	6.11 Ct; 0.04 Cf; 0.30 VC; 0.5 Cr_3C_2; WC balance	2.41	99.9%	Jinlu Tungsten Co. Ltd., China
Co	-	0.8	99.9%	Nanjing Hanrui Co. Ltd., China

Table 2. Nominal composition of alloys (mass %)

Specimen Name	Co (Wt. %)	WC-Cr_3C_2-VC (Wt. %)	Theoretical density(g/cm^3)
WC-0Co	0	100	15.556
WC-0.2Co	0.2	99.8	15.543
WC-0.5Co	0.5	99.5	15.523
WC-0.8Co	0.8	99.2	15.503

Table 3. Parameters used in SPS process

Specimen Name	Temperature (℃)	Pressure (MPa)	Holding time (min)	Heating rate (℃/min)
WC-0Co	1700	50	5	200
WC-0.2Co	1680	50	5	200
WC-0.5Co	1600	50	5	200
WC-0.8Co	1500	50	5	200

2.2. Characterization of WC–Co-cemented carbides

The microscopic morphology of the WC–Co composite powders was characterized by scanning electron microscopy (SEM). Phase analysis was implemented by using PANalyticalX'Pert PRO X-ray diffraction (XRD) at 2θ angles from 20° to 80°. The WC–Co-cemented carbide morphology was studied by using a JSM 7500F field emission SEM. The specimen density was measured by immersion in water using the Archimedes principle, to calculate the relative density of WC–Co-cemented carbides with different Co contents. The hardness and fracture toughness were measured using the Vickers hardness and fracture toughness tester or the Vickers indenter under a constant load of 5 kg.

For accuracy, specimens were polished to a near-mirror finish and were checked for face parallelism before testing. All mechanical property tests were conducted three times and the arithmetic mean value from the three tests was calculated.

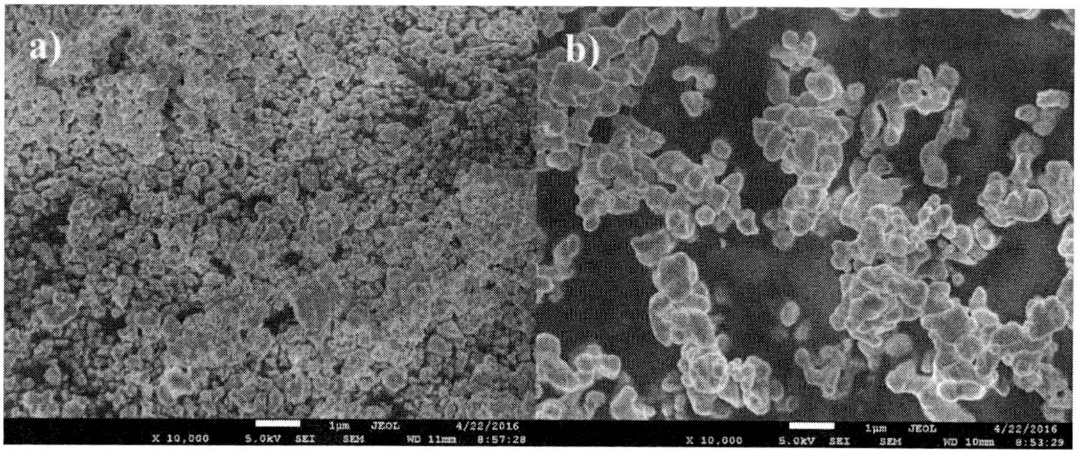

Figure1. SEM images of (a) WC powder, (b) Co powder.

2.3. Friction and wear test

According to the ASTM G133-05 standard, reciprocating wear tests of WC–Co-cemented carbides were evaluated by using a Tripoli micro-tribemate in a ball-on-flat method. Four wear test specimens were ground to a finish of 1000#siC abrasive papers, and were cut into specimens of (15 × 15 × 10 mm). Before and after each test, the surface was cleaned by ultrasound in acetone solution, and dried immediately with hot air. WC balls (diameter: 8.0 mm, hardness: 90.5 HRA) were used as counter bodies. Room-temperature dry tests (2 h) were performed under a normal load of 60 N, an amplitude of 5 mm and a frequency of 5 Hz.

After friction testing, the worn surfaces were studied by SEM and the wear track was analyzed by using white-light interferometry. The specific wear rate was studied as a measure of wear volume loss and total sliding distance, which was calculated as the ratio of wear volume loss over the normal load multiplied by the total sliding distance (mm^3/NM).

3. Results and discussion

3.1. Characterization and microstructures of WC–Co-cemented carbides

Figure 1 shows the morphological detail of the WC and Co powders. Both powders are uniform and fine, which yields a high surface energy for the solid WC particles and a fine sintering performance. XRD patterns of WC–Co-cemented carbides with different Co contents are shown in Fig. 2. WC phase peaks are clear, but Co phases are difficult to locate, possibly because very little Co was added, so they are relatively weak and therefore, are covered by the WC peak.

Figure 3 shows the SEM images of the WC-cemented carbides that contain 0 wt. %, 0.2 wt. %, 0.5 wt. % and 0.8 wt. % Co. At their respective temperatures and pressures, the four specimens exhibit good crystal structures, clear grain boundaries and a small and uniform grain size. The cohesive phases with different Co contents can sinter homogeneous and compact material specimens. As the Co content increases, the WC–Co-cemented carbide grain microstructure becomes more uniform and

IOP Conf. Series: Materials Science and Engineering **207** (2017) 012019 doi:10.1088/1757-899X/207/1/012019

compact; however, when the content exceeds 0.2 wt. %, the grain size variation decreases, and no change is apparent between the final two specimens, WC-0.5Co and WC-0.8Co. Therefore, further addition of Co to the system will not lead to any further improvement in performance of the cemented carbides.

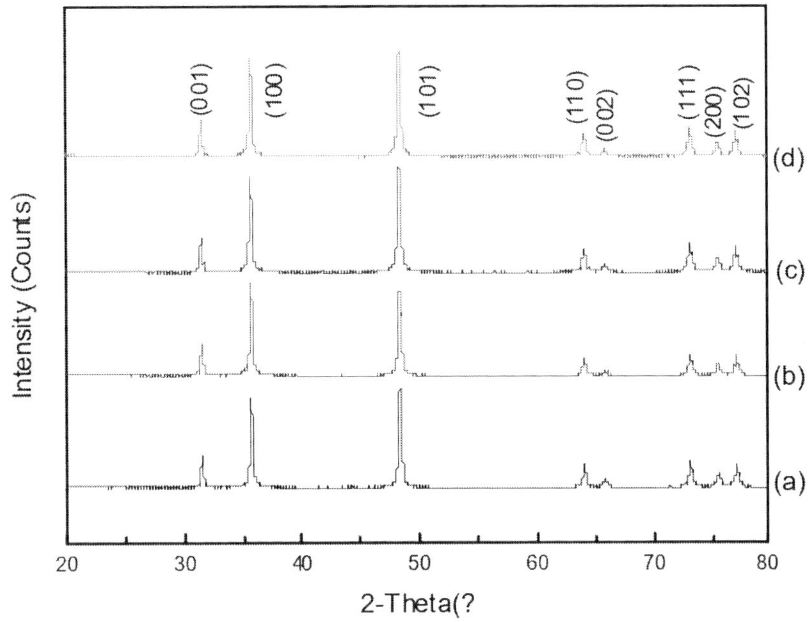

Figure 2. XRD patterns of WC–Co-cemented carbides with Co content of (a) 0 wt. %, (b) 0.2 wt. %, (c) 0.5 wt. %, (d) 0.8 wt. %.

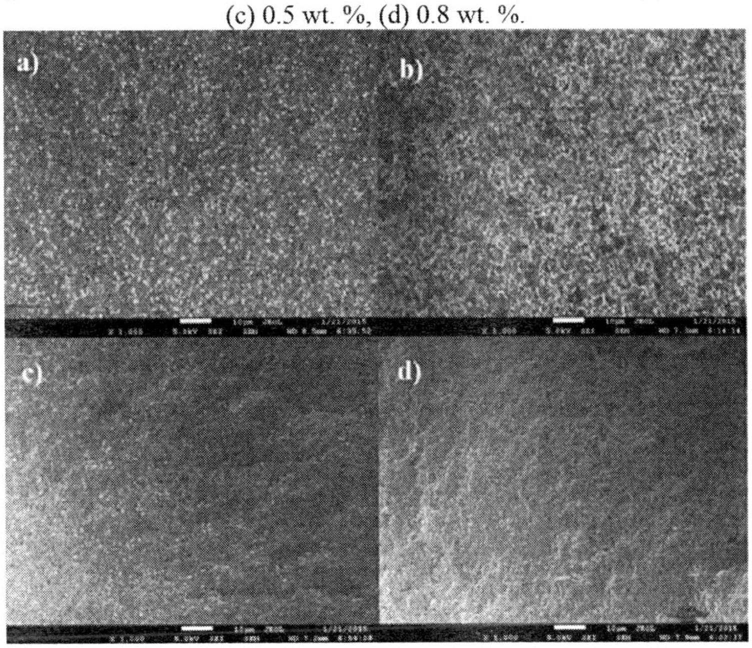

Figure 3. Microstructures of WC–Co-cemented carbides with Co content of (a) 0 wt. %, (b) 0.2 wt. %, (c) 0.5 wt. %, (d) 0.8 wt. %.

3.2. Mechanical properties

The variation in density and the porosity of the WC–Co-cemented carbides with different Co contents is shown in Fig. 4, as calculated using the Archimedes principle. According to the experimental results, the specimen density decreases with increasing Co content. The density of specimen WC-0Co is 15.498 g/cm^3, whereas that of WC-0.8Co is lowest at 15.298 g/cm^3. This occurs because the density of Co in the specimen is less than that of the WC. However, the actual density of each group differs little because the Co content in each cemented carbide is low. The porosity of the cemented carbides increases with increase in Co content, and the specimen compactness decreases accordingly. Porosity increases when the Co content increases to 0.50 wt. %, which may affect some mechanical properties adversely.

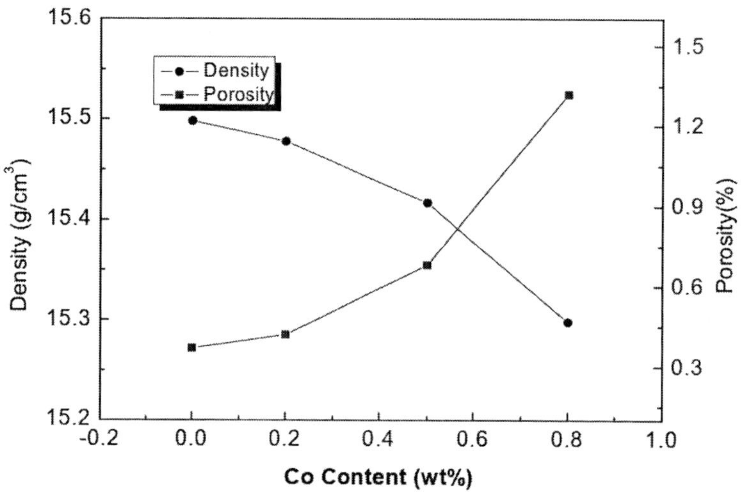

Figure 4. Dependence of density and porosity on Co content

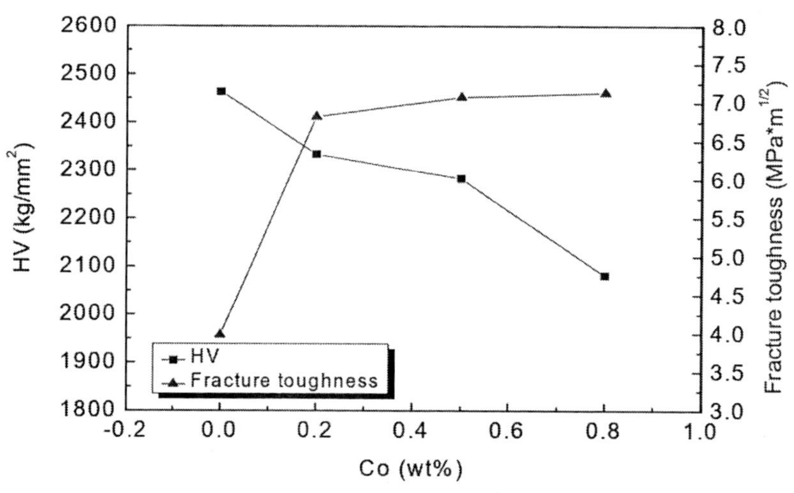

Figure 5. Dependence of Vickers hardness and fracture toughness on Co content

Hardness and fracture toughness are the two most important mechanical properties of cemented carbides [8-10]. Figure 5 shows that with an increase in Co content, the specimen hardness decreased gradually, which shows that specimens with a greater Co content have a smaller Vickers hardness.

This occurs mainly because the Vickers hardness reflects the indentation-resistance ability of the WC–Co-cemented carbide, which is associated with the hard and binder phase contents and the WC solubility in the Co phase.

Conversely, the fracture toughness of the WC–Co-cemented carbides improves significantly with increasing Co content, which means that the fracture toughness of the four specimens increases with decreasing Vickers hardness. The fracture toughness of WC-0Co is lower than that of the other three specimens. The fracture toughness of WC-0Co is close to 3.98 $MPa.m^{1/2}$ and the fracture toughness's of the other three specimens are similar in value. The main reason for this result is that the fracture toughness of the WC–Co-cemented carbides is affected by the WC grain size and the Co binding-phase content. When cracks propagate, the Co binding phase can still connect the cracks and apply a tensile stress to prevent crack expansion in the region near the crack tip. Therefore, Co addition can improve the fracture toughness of WC–Co-cemented carbides with little binder.

Therefore, WC-0.2Co exhibits a good microstructure and fine mechanical properties compared with the other three specimens, which affects the practical application of cemented carbide.

3.3. Wear properties and wear mechanism

During each test, the coefficient of friction was measured under the same conditions, and its variation was analysed after 4000 s of testing to remove the influence of the running-in period. Figure 6 shows the variation in friction coefficient of the WC–Co-cemented carbides with different Co contents. The friction coefficient of each specimens increases with increasing Co content. The average friction coefficient of the WC–Co-cemented carbides without Co is ~0.16, and a higher friction coefficient of approximately 0.21 was observed for WC–Co-cemented carbides with 0.8 wt. % Co. With increase in Co content, a higher specimen friction coefficient may result from the lower hardness. WC–Co-cemented carbides with 0.8 wt. % Co exhibited the smallest average grain size, which causes higher applied-stress propagating dislocations through the material and results in the highest friction coefficient.

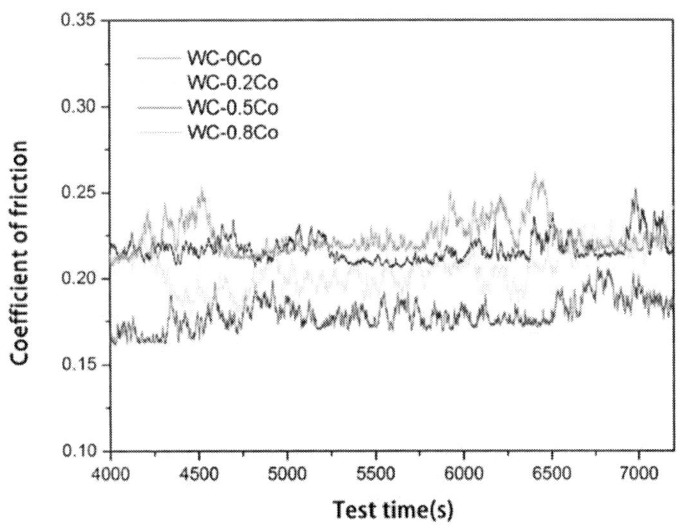

Figure 6. Relationship of frictional coefficient and time for each group of specimens

The mean values of the volume wear rate of WC–Co-cemented carbides with different Co contents are shown in Fig. 7. The volume wear rate of WC–Co-cemented carbides without Co is $3.13.10^{-8}$ mm^3/(NM) and that of the other specimens was $3.43.10^{-8}$ mm^3/(NM), $5.21.10^{-8}$ mm^3/(NM) and $6.22{\cdot}10^{-8}$

mm^3/(NM). When the Co content was 0 wt. % and 0.2 wt. %, there was no significant difference in wear resistance. The volume wear rate of the specimens increases with increasing Co content and this increase is distinct when the Co content exceeds 0.2 wt. %. Therefore, when the Co content exceeds 0.2wt%, the WC–Co-cemented carbide wear resistance decreases significantly.

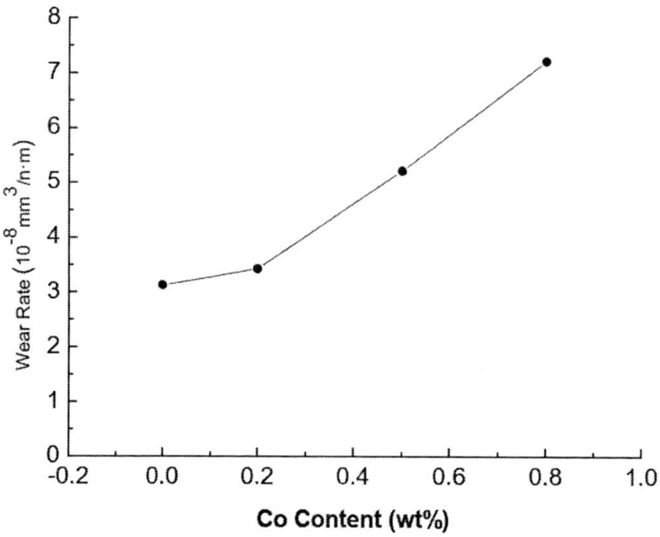

Figure 7. Volume wear rate for each specimen

Figure 8 shows the two-dimensional non-contact surface mapping of the wear tracks. The width and depth increase with increase in Co weight percentage. Excessive Co addition will reduce the material wear resistance. Figure 9 shows micrographs of the worn surface from a low to a high concentration. Some holes exist as caused by grain peeling in each worn surface. WC-0Co and WC-0.2Co have less exfoliation pits, whereas WC-0.5Co and WC-0.8Co show an obvious flaking on the surface. Peeling on the WC-0.8Co surface is serious with a large peeled area.

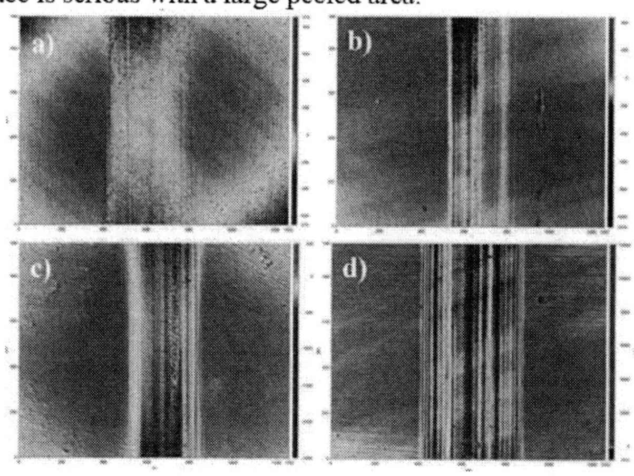

Figure 8. Two-dimensional non-contact surface mapping of wear tracks with Co content of (a) 0 wt. %, (b) 0.2 wt. %, (c) 0.5 wt. %, (d) 0.8 wt. %.

WC–Co-cemented carbides are comprised mainly of a metallic ceramic and a cobalt binder. The WC phase, as the hard phase, possesses a high hardness and results in deformation, whereas the Co phase is relatively soft and prefers to deformation plastically under a heavy load. Because the Co phase was extruded first, the WC hard base stability was reduced. When the contact stress exceeds its fracture toughness limit, WC grains are broken and peel off, which creates a peeling pit on the worn surface. The fine abrasive debris consists of pulverized detached WC particles that form a micro-cutting function in the sample surface. The peeling and micro-cutting phenomenon contribute to the material wear and tear. WC-0Co and WC-0.2Co specimens exhibit a lower volume wear rate and a stronger wear resistance. For WC-0.5Co and WC-0.8Co, the stability of the WC phase is affected by the increase in plastic deformation of the Co phase. Surface peeling is relatively obvious, and agrees with an increase in volume wear rate and the relatively weak wear resistance. Hence, for WC-0Co and WC-0.2Co, the wear mechanism is abrasive wear because of the low Co binder. For WC-0.5Co and WC-0.8Co, the main wear mechanism is adhesive wear by plastic deformation of the Co phase.

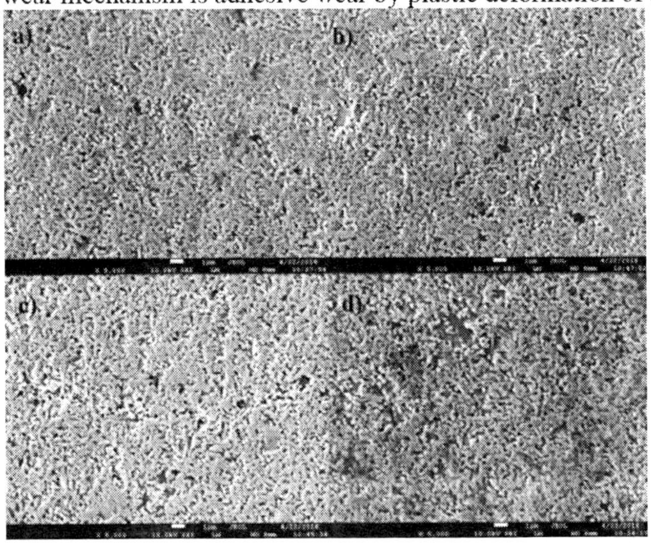

Figure 9. SEM micrographs of worn surface of WC–Co-cemented carbides with Co content of (a) 0 wt. %, (b) 0.2 wt. %, (c) 0.5 wt. %, (d) 0.8 wt. %.

4. Conclusion

WC–Co-cemented carbides are comprised mainly of a metallic ceramic and a cobalt binder, and the Co contents affect the hardness, fracture toughness and wear resistance of WC–Co-cemented carbides. The hardness and wear resistance of WC–Co-cemented carbides decrease as the Co content increases. However, the fracture toughness is improved as the Co content increases, which ensures a certain impact resistance of the WC–Co-cemented carbides. By analyzing the friction and wear mechanism, it can be concluded that because the Co phase was extruded first, the WC hard base stability was reduced, the WC phase desquamated and abrasion occurred.

Therefore, Co addition should be controlled within a certain range and a 0.2 wt. % Co sample possesses a high hardness, good fracture toughness and wear resistance. Sintering is by SPS at 50 MPa, a maximum of 1680°C and a holding time of 5 min. The Vickers hardness of the sintered carbide reached 2333 HV_{30}, and the fracture toughness achieved 6.82 $MPa \cdot m^{1/2}$.

Acknowledgments

This work was supported by the Project of China National Offshore Oil Corporation (CNOOC-KJ125ZDXM12LTD03NFGC2014-06).

References

[1] J. Watts, G. Hilmas, Crack deflection in tungsten carbide based laminates, Int. J.Refract. Met.

Hard Mater. 24 (2006) 222–228.

[2] X.L. Shi, G.Q. Zhao, X.L. Duan, R.Z. Yuan, H.H. Lin, Mechanical properties, phases andmicrostructure of ultrafine hardmetals prepared by WC-6.29Co nanocrystalline composite powder, Mater. Sci. Eng. 392 (2005) 335–339.

[3] P. Kindermann, P. Schlund, H.G. Sockel, M. Herr, W. Heinrich, K. Gorting, U.Schleinkofer, High-tempererature fatigue of cemented carbides under cyclic loads,Int. J. Refract. Met. Hard Mater. 17 (1999) 55–68.

[4] Tamara AleksandrovFabijanić,Željko Alar, DankoĆorić, Influence of consolidation process and sintering temperature onmicrostructure and mechanical properties of near nano- andnano-structured WC-Co cemented carbides, Int. Journal of Refractory Metals and Hard Materials 54 (2016) 82–89.

[5] Z. Huang, et al., Effect of Cu on the microstructures and properties of WC-6Co cemented carbides fabricated by SPS, Int J Refract Met Hard Mater (2016).

[6] Shixian Zhao, Xiaoyan Song, Jiuxing Zhang, Xuemei Liu, Effects of scale combination and contact condition of raw powders on SPS sintered near-nanocrystalline WC–Co alloy, Materials Science and Engineering: A. 473 (2008) 323–329

[7] Y. Jiang, J.F. Yang, R. Liu, X.P. Wang, Q.F. Fang, Oxidation and corrosion resistance of WC coated tungsten fabricated by SPS carburization, Journal of Nuclear Materials. 450 (2014) 75-80.

[8] K. Jia, T.E. Fischer, B. Gallois, Microstructure, hardness and toughness of nanostructured and conventional WC-Co composites, Nanostruct. Mater., 10 (5) (1998), pp. 875–891.

[9] TaharLaoui, Abbas Saeed Hakeem, KachallaAbdullahi, NouriSaheb, Nasse Al-Aqelli, Effect of consolidation mechanism on the properties on nanostructure WC-6, 9, 12 wt% cemented carbides, Proceeding of ASME 2011 International Mechanical Engineering Congress and Exposition, Nano and Micro Materials, Devices and Systems; Microsystems Integration, vol. 11, , ASME, Denver, Colorado, USA (November 11–17, 2011), pp. 667–674.

[10] X.L.Shi,G.Q.Zhao,X.L.Duan,Z.Xiong,H.Yang,Theeffectoftungstenbufferlayeronthe stability of diamond with tungsten carbide-cobalt nanocomposite powder during spark plasma sintering, Diam. Relat. Mater. 15 (2006) 1643–1649.

Synthesis and electrochemical properties of Co_3O_4 nanoparticles by hydrothermal method at different temperatures

Qiuyan Duan [1, 2], **Haiyan Chen** [1,*]

[1]School of Material Science and Engineering, Shanghai Maritime University, Shanghai, China
[2]Department of Thermal Science and Energy Engineering, University of Science and Technology of China, Hefei, China

*Corresponding author e-mail: hychen@shmtu.edu.cn

Abstract. In this work, Co_3O_4 nanoparticles were synthesized by hydrothermal method at different temperatures. The synthesized Co_3O_4 nanoparticles were characterized using transmission electron microscopy (TEM), X-ray diffraction (XRD), X-ray photoelectron spectrometer (XPS) and nitrogen adsorption-desorption. The Co_3O_4 nanoparticles prepared at the hydrothermal temperature of 140 ℃ and at the annealing temperature of 350 ℃ have a shorter crystal spacing distance associated to the (220) crystal planes, larger BET surface area and more bivalent cobalt on the surface than these of Co_3O_4 nanoparticles prepared at higher hydrothermal temperature or higher annealing temperature. The supercapacitor performances of synthesized Co_3O_4 nanoparticles were analysed by cyclic voltammograms (CVs), Galvano static charge/discharge (GCD) and the electrochemical impedance spectroscopy (EIS) in 6 M KOH aqueous electrolyte solution. The Co_3O_4 nanoparticles produced at lower temperatures exhibit good pseudo capacitance behaviour. Also, owning to the low hydrothermal temperature and the annealing temperature of Co_3O_4 nanoparticles, they can be relatively low cost in favour of a promising candidate for electrochemical supercapacitors.

1. Introduction

Given rises in environmental pollution and rapid consumption of traditional fossil fuel energy resources, it is extremely urgent to explore more efficient, clean, and low-cost energy conversion/storage resources with high energy and power capacity [1-3]. Supercapacitors have attracted considerable attentions for their high power density, long cycle life and fast charge-discharge capability [4-6]. The core of supercapacitors is the electrode materials, which directly dominates the performance of energy storage. Transition metal oxides such as MnO_2, NiO, RuO_2 and Co_3O_4 have been widely investigated for their pseudo capacitance performances in KOH aqueous electrolyte solution [7-9]. By comparison, Co_3O_4 has a high theoretical specific capacitance (3650 F/g), low toxicity and great redox ability [10]. Different methods were applied to synthesize oxide Co_3O_4, such as thermal decomposition of solid phase, chemical spray pyrolysis, chemical vapor deposition and traditional sol-gel method [11-15]. However, the size distribution of Co_3O_4 nanocrystal made by the above methods is wide. Hydrothermal synthesis is an efficient technique for preparing fine powders by chemical reaction with high temperature and high pressure in solution [16].

In this present work, Co_3O_4 nanoparticles were prepared by hydrothermal method at different temperatures. The main point is to explore how temperatures have an influence on the crystal structures and electrochemical performance of cobalt oxide. In the experiment, we chose 140℃ and 200℃ as hydrothermal temperatures, and 350℃ and 450℃ as annealing temperatures. The crystal structures of prepared Co_3O_4 nanoparticles were characterized by TEM, XRD, XPS and N_2 adsorption-desorption. The capacitance performances were studied by CVs, EIS and GCD.

2. Experimental

2.1. Synthesis of Co_3O_4 nanoparticles

All of the chemicals were of analytical grade and used without further management, which were got from Alfa Assar (China). 0.03 M Co $(NO_3)_2$ $6H_2O$ and 0.26 M CO $(NH_2)_2$ were dissolved in 210 ml distilled water. The mixture was stirred by magnetic stirring apparatus until a transparent solution was obtained and then was transferred to two Teflon-lined stainless steel autoclaves (100 ml capacity) evenly. The autoclaves were filled with distilled water up to 80% of the total capacity, which were then sealed and heated at 140℃ and 200℃ for 3h respectively, marked as A and B. After the autoclaves were cooled down to room temperature in air naturally, the precipitate was collected and washed with distilled water and absolute ethanol 3 times in turn by centrifugation (5000 r/min). The precursor was dried in a vacuum oven at 80℃ for 8h. Finally, the powder A was annealed at 350℃ and 450℃ respectively, and the powder B was annealed at 450℃ for 2h in the air with a slow heating rate of 4℃ per minute to obtain the desired Co_3O_4 nanoparticles. The three samples were marked as 140 ℃ - 350 ℃, 140 ℃ - 450 ℃, 200 ℃ - 450 ℃.

2.2. Materials characterization

The morphology and nanostructure were characterized using transmission electron microscopy (TEM, Tecnai G2 F20 S-TWIN, operated at 200kV). The Brunner-Emmet-Teller (BET) surface area was measured by nitrogen adsorption-desorption. The total pore volume and mean pore size were calculated by the Barrett-Joyner-Halenda (BHJ) method. The crystal structures were studied using Analytical Pert PRO X-ray diffraction. The surface elemental analysis was performed using an X-ray photoelectron spectrometer (XPS, Thermos ESCALAB 250XI). The mass of the electroactive materials was weighed with a microbalance (Mettle Toledo) with an accuracy of 0.01 mg.

2.3. Electrochemical measurements

To measure the electrochemical performance, electrodes were fabricated by the mixture of active materials, acetylene black and polyvinylidene fluoride (PVDF) in a weight ratio of 8:1:1 with N-methylpyrrolidone (NMP). In this experiment, the mass of active material is 20 mg in each electrode. Then the slurry was evenly pressed onto the nickel foam (10×10 mm^2), which was used as working electrodes. The working electrode was dried at 60℃ for 6h and then pressed under 6MPa for 1 min. Platinum foil was used as counter and saturated calomel electrode (SCE) as reference electrode. The electrolyte is 6M KOH aqueous solution. Cyclic voltammograms (CVs) were measured using IviumnStat electrochemical workstation between -0.2V and 0.5V. The electrochemical impedance spectroscopy (EIS) and Galvano static charge/discharge were measured in the electrochemical workstation (Auto lab PTN320N). The frequency of EIS ranged from 0.01Hz to 100 KHz at the open circuit potential. Charge/discharge cycles of the sample were measured between 0.01V to 0.35V at different current densities. All electrochemical measurements were carried out at room temperature.

3. Results and discussion

3.1. Characterization of materials

The XRD patterns of the products in Fig.1 show clear diffraction peaks were well indexed to (220), (311), (222), (400), (422), (511) and (440) planes of crystalline cubic Co_3O_4 (JCPDF card no. 01-074-1657; space group:Fd-3m), respectively. No obvious diffraction peaks of other impurity phases were checked out, indicating the successful preparation of pure Co_3O_4 by this hydrothermal method. Based on the highest diffraction peak (311 crystal plane), the average crystallite size of these samples are 20.2 nm, 26.9 nm and 156.9 nm from up to down in the XRD patterns, which were calculated by the Scherer formula[17]. Furthermore, as either the hydrothermal temperature or annealing temperature increased, the diffraction peak intensity was enhanced and the peak width was decreased, which mean the crystalline of Co_3O_4 nanoparticles become well. As shown in the figure, the crystallinity of 140 ℃-350 ℃ is relatively lower.

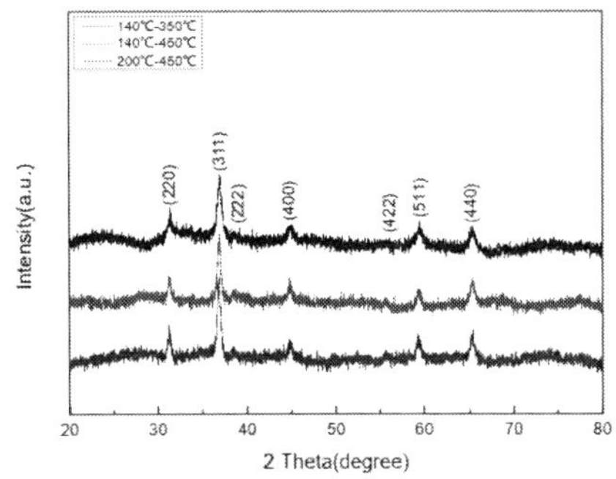

Figure 1. XRD patterns of Co_3O_4 nanoparticles at different temperature

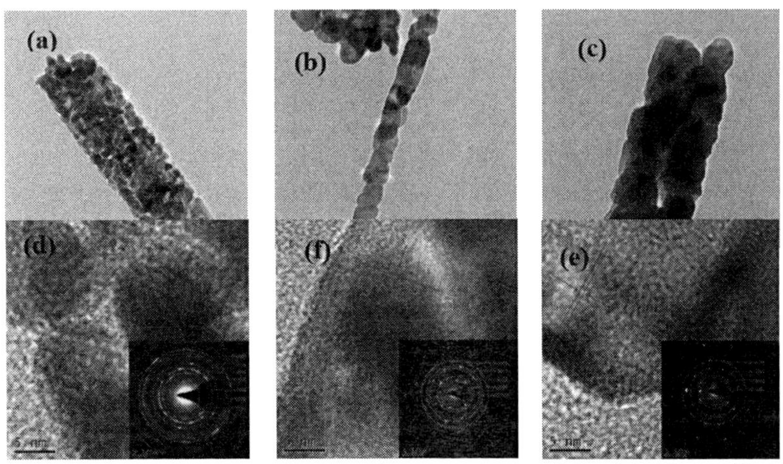

Figure 2. TEM images and HRTEM images of Co_3O_4 nanoparticles (a), (d) 140℃-350℃ (b), (e) 140℃-450℃ (c), (f) 200℃-450℃

The typical TEM and HRTEM images of different Co_3O_4 nanoparticles are displayed in the Fig.2. The TEM images (a), (b), (c) clearly demonstrate the diameter of gains become bigger as the temperature increased, which are about 10nm, 30nm and 50nm respectively in sequence. From HRTEM images (d), (e), (f), the lattice fringes with a spacing distance of 0.278nm, 0.285nm and 0.290nm are associated to the (220) crystal planes of Co_3O_4 nanoparticles. The corresponding selected area electron diffraction (SAED) pattern inset in the images (b), (d), (f) demonstrates that the oxide is

Co_3O_4 phase and the diffraction rings are mainly indexed as (440), (422), (311) and (220) respectively, which are consistent with results obtained from XRD analysis. The shorter spacing distance is beneficial for fast diffusion of the redox phase, which makes the composites display a high electrochemical capacitance [18].

To further investigate the chemical composition of these samples, X-ray photoelectron spectroscopy (XPS) was carried out. As shown in Fig.3.(a), (b), (c), the Co 2p spectrum indicate a low energy and a high energy band at 779.9 eV and 795.3 eV, corresponding to Co $2p_{3/2}$ and Co $2p_{1/2}$, with two satellites at 788.8 eV and 805.2 eV, and these are the characteristic of Co_3O_4 phases [19]. The Co $2p_{3/2}$ peak of 779.9 eV can be further disassembled into two fitting peaks at 779.7 eV and 780.7 eV, and the Co $2p_{1/2}$ peak of 795.3 eV can be further disassembled into two fitting peaks at 794.8 eV and 796.1 eV. The peaks with the binding energy of 779.9 eV and 794.8 eV can be appointed to Co^{3+}, while the peaks at 780.7 eV and 796.1 eV can be appointed to Co^{2+}. As the temperature increased, the area ratio of trivalent cobalt against bivalent cobalt on the surface is 35.1%, 25.8% and 27.68%, indicating bivalent cobalt was predominant and some trivalent cobalt was transformed to bivalent cobalt. Furthermore, the high-resolution O 1s spectrum in Fig. 3 (d), (f), (e) can be further disassembled into two fitting peaks at 530.0 eV and 531.4 eV, confirming the presence of lattice oxygen and surface oxygen on the surface[20].

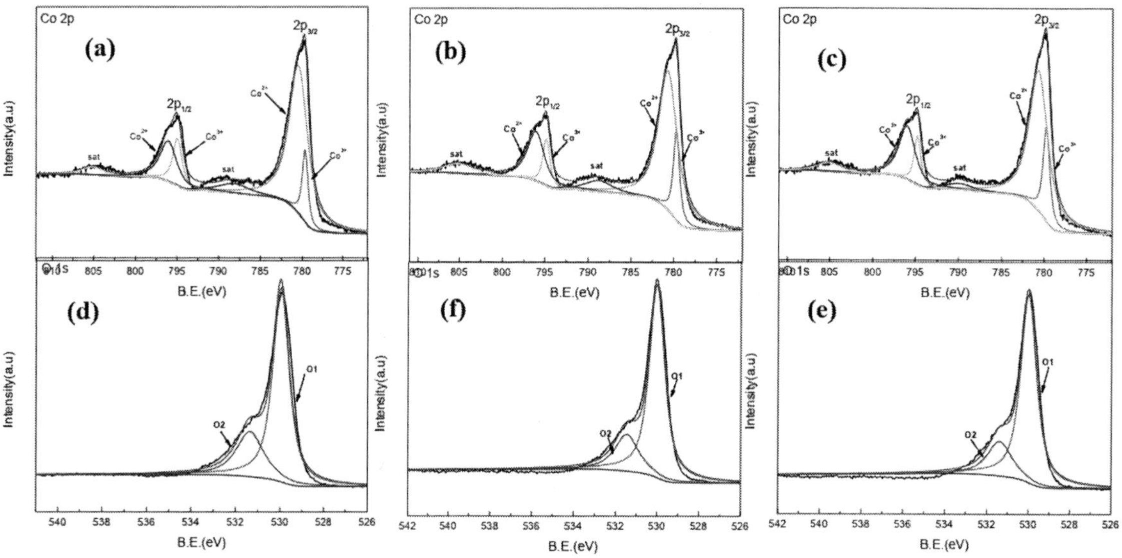

Figure 3. XPS spectra of Co 2p of the Co_3O_4 nanoparticles and O 1s of the Co_3O_4 nanoparticles (a), (d)140℃-350℃ (b), (f) 140℃-450℃ (c), (e) 200℃-450℃

The BET surface area obtained from five-point BET method is 68.142 m^2/g, 24.248 m^2/g, 21.516 m^2/g in turn. The value of 140℃-350℃ is relatively higher than another two values and most of the values reported earlier in literature (31.3 m^2/g, 43.9 m^2/g, 59 m^2/g) [18, 21]. The pore size (diameter) distribution calculated from the Barrett-Joyner-Halenda (BJH) method shows average size pore of 2.455 nm, 2.187 nm and 2.189 nm successively in Co_3O_4 nanoparticles. Similarly, the value of 140℃-350℃ Co_3O_4 nanoparticles is bigger relatively. Comparing with annealing temperature, hydrothermal temperatures have a greater influence on the BET surface area, total pore volume and mean pore size. Big pore volumes will provide short diffusion paths for electrons and ions and high surface areas can provide more active sites, which further promote electrochemical reactions. These features can particularly improve pseudo capacitive performance.

Table 1. The BET surface area, total pore volume and mean pore size of these samples

Sample	SBET (m^2/g)	VT (cm^3/g)	Mean pore size(nm)
140℃-350℃	68.142	0.0288	2.455
140℃-450℃	24.248	0.009	2.187
200℃-450℃	21.516	0.008	2.189

3.2. Electrochemical characteristic of Co3O_4 nanoparticles

A three-electrode installation was applied to examine the pseudo capacitive performances of these Co_3O_4 nanoparticles for supercapacitor electrode in 6 M KOH. Fig.4a shows the cyclic voltammograms (CVs) of 140℃-350℃ Co_3O_4 at different scan rates with a potential window from -0.2 V to 0.5 V (vs. SCE). Two major redox peaks appeared in all CV curves, indicating the capacitance characteristic was mainly governed by pseudo capacitive reactions. With increasing sweep rates, the integrated CV area of 140℃-350℃ Co_3O_4 electrode gradually increased, accompanied with redox peaks moving to more positive and negative potentials. However, the peak shape of CV curves was not deformed in any degree even though the scan rate was as high as 50mV/s, indicating the high electrochemical reversibility of the material [22].

Fig.4b shows CV curves of Co_3O_4 nanoparticles at different temperature with the scan rate of 5 mV/s. Two pairs of redox peaks showed up in CV curves. In the redox process, two sequential reactions take place as follows [23]:

$$Co_3O_4 + OH^- + H_2O \leftrightarrow 3\ CoOOH + e^- \quad (1)$$

$$CoOOH + OH^- \leftrightarrow CoO_2 + H_2O + e^- \quad (2)$$

Evidently, the area of the curve of 140℃-350℃ Co_3O_4 electrode is larger than these of 140℃-450℃ Co_3O_4 electrode and 200℃-450℃ Co_3O_4 electrode, suggesting the 140℃-350℃ Co_3O_4 electrode presents better capacitive properties on comparing with two other electrodes. The specific capacitance can be calculated using the CVs curves as follows [24]:

$$Cm = \frac{\int Idv}{mvV} \quad (3)$$

Where Cm (F/g) is the specific capacitance of the electrode, I (A) is the discharge current, V (V) is the potential, v (mV/s) is the scan rate, and m (g) is the mass of the electroactive materials in the electrodes. Fig.4c shows the specific capacitance of these samples at different scan rates. The specific capacitance decreased with the increase of scan rate. This behavior was due to the fact that the high scan rate can restrict the ion to all the pore of electrode [25, 26]. In the literature [27], it reported that only outer surface of Co_3O_4 nanoparticles can be used to storage charges. Specific capacitance of 140℃-350℃ Co_3O_4 nanoparticles is 222, 206, 203, 160, 116 and 76 F/g at the scan rate of 5, 10, 20, 50, 100 and 200 mV/s, which is higher than those of anther Co_3O_4 nanoparticles at all the same scan rates. The result is good consistent with speculation at BET results, manifesting the 140℃-350℃ Co_3O_4 electrode presents better capacitive properties in comparison with two other electrodes again.

Fig.4d compares the Galvano static charge/discharge curves for these various Co_3O_4 nanoparticles at the same current density of 0.01A/s. The obtained curves with fairly a triangular shape did not show a double-layer capacitor rather pseudo capacitance behaviors, which inferred the characteristic of charge storage mechanism [28]. Meanwhile, the charge and discharge times of 140℃-450℃ Co3O4 nanoparticles and 200℃-450℃ Co_3O_4 nanoparticles are shorter than that of 140℃-350℃ Co_3O_4 nanoparticles evidently. According to the equation as follows [29]:

$$Cm = \frac{I \times t}{\Delta V \times m} \quad (4)$$

where Cm (F/g) is the specific capacitance of the electrode, I (A) is the discharge current, ΔV (V) is the window potential, m (g) is the mass of the electroactive materials in the electrodes, the specific capacitance of 140℃-350℃ Co_3O_4 nanoparticles is relatively larger than these of two other samples. Meanwhile, regarding discharge curves, IR drops of 140℃-350℃ Co_3O_4 nanoparticles (0.0117 Ω) is relatively shorter than these of 140℃-450℃ Co3O4 nanoparticles (0.0143Ω) and 140℃-350℃ Co_3O_4 nanoparticles (0.0131Ω), suggesting a relatively smaller internal resistance of 140℃-350℃ Co_3O_4 nanoparticles. We speculate that the high specific capacitance is due to relatively low resistance derived from the shorter crystal spacing distance. EIS was carried out to further evaluate the specific capacitance of these Co_3O_4.

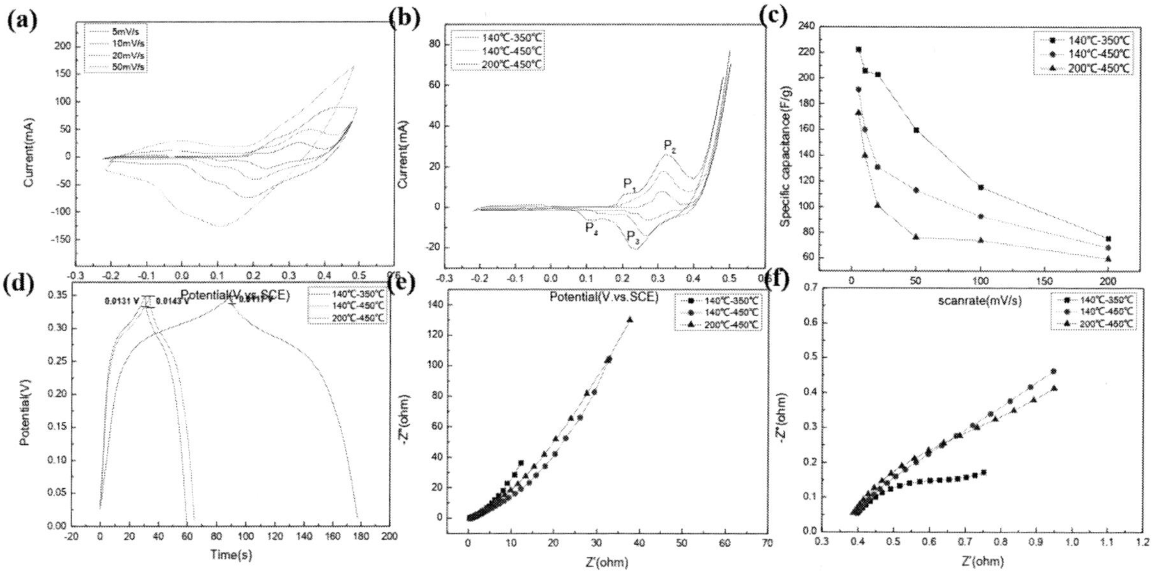

Figure 4. (a)CV curves of 140℃-350℃ Co_3O_4 at different temperatures with the scan rate of 5, 10, 20 and 50 mV/s. (b) CV curves of Co_3O_4 at different temperatures with the scan rate of 5 mV/s. (c) Plots of the specific capacitance against the scan rates. (d) The charge and discharge curves of different Co_3O_4 at a constant current density of 10mA/s. (e) Nyquist plots of different Co_3O_4 before cyclic voltammograms. (f) The enlarged part of (e)

The Nyquist curves of the obtained Co_3O_4 nanoparticles exhibit a semi-circle in the high frequency region, a straight line in the low frequency region, and a short Warburg section between the semi-circle and the slope of the line in the Fig.4e. The semi-circle in the high frequency region on the real axis represents the charge-transfer resistance, which demonstrates the electrolyte ions to the electroactive materials and the conductivity of the electrodes. The straight line ascends at the frequency region at an angle of almost 45° exhibits Warburg resistance", indicating the ion diffusion resistance of the charged ions of Co_3O_4. The straight line in the low frequency region displays the ideal capacitance behavior nature [30, 31]. The 140℃-350℃ Co_3O_4 nanoparticles has a smaller charge-transfer resistance value in the high frequency region and a more vertical line in the low frequency region than another two Co_3O_4, suggesting the 140℃-350℃ Co_3O_4 nanoparticles has a better capacitive behavior. The result is in great corresponding to the previous sections in the IR drops and gives an explanation for the higher specific capacitance of the 140℃-350℃ Co_3O_4 nanoparticles.

Above all, the 140℃-350℃ Co_3O_4 nanoparticles have a better capacitive behavior, which can be explained in terms with shorter crystal spacing distance, larger BET surface area and more bivalent cobalt on the surface.

4. Conclusion

In summary, Co_3O_4 was prepared by hydrothermal method at different temperatures successfully. The morphology, surface area, crystal structures and surface elemental analysis were studied in detail with the support of TEM, N_2 adsorption-desorption, XRD and XPS in sequence. The 140℃-350℃ Co_3O_4 has a shorter crystal spacing distance indexed to the (220) crystal planes, larger BET surface area and more bivalent cobalt on the surface. In addition, the 140℃-350℃ Co_3O_4 exhibits excellent pseudo capacitance behavior. The specific capacitance of it was higher in comparing with 140℃-450℃ Co_3O_4 and 200 ℃ -450 ℃ Co_3O_4 in 6 M KOH aqueous solution according to the results of cyclic voltammograms (CVs), electrochemical impedance spectroscopy (EIS) and Galvano static charge/discharge, which was in good agreement with their crystal structure. Due to low hydrothermal temperature and annealing temperature of 140℃-350℃ Co_3O_4, it can be relatively low cost to be benefit of a potential electrode for supercapacitor application.

Acknowledgments

This work was financially supported by Science & Technology Program of Shanghai Maritime University (No. 20130448).

References

[1] G. Wang, L. Zhang, J. Zhang, A review of electrode materials for electrochemical supercapacitors, Chem. Soc. Rev.41(2012) 797-828.

[2] H. Wang, H. Dai, Strongly coupled inorganic-nano-carbon hybrid materials for energy storage, Chem. Soc. Rev. 42 (2013) 3088-3113.

[3] Y. Luo, J. Jiang, W. Zhou, H. Yang, X. Qi, H. Zhang, H. J. Fan, D. Y. W. Yu, C. M. Li, T. Yu, Seed-assisted sy nthesis of highly ordered TiO2@α-Fe2O3 core/shell arrays on carbon textiles for lithium-ion battery applications, Energy. Environ. Sci. 5(2012) 6559-6566.

[4] C. Liu, F. Li, L.P. Ma, H.M. Cheng, Advanced materials for energy storage, Adv. Mater. 22 (2010) E28–E62.

[5] Z. Chen, Y. Qin, D. Weng, Q. Xiao, Y. Peng, X. Wang, H. Li, F. Wei, Y. Lu, Design and synthesis of hierarchical nanowire composites for electrochemical energy storage, Adv. Funct. Mater. 19 (2009) 3420–3426.

[6] A. Manthiram, A. Vadivel Murugan, A. Sarkar, T. Muraliganth, Nanostructured electrode materials for electrochemical energy storage and conversion, Energy Environ. Sci. 1 (2008) 621–638.

[7] F. Shi, L. Li, X. L. Wang, C. D. Gu, J. P. Tu, Metal oxide/hydroxide-based materials for supercapacitors. RSC Adv 2014; 4: 41910–21.

[8] H. Jiang, J. Ma, C. Li, Mesoporous carbon incorporated metal oxide nanomaterials as supercapacitor electrodes. Adv Mater 2012; 24: 4197–202.

[9] M. Zhi, C. Xiang, J. Li, M. Li, N. Wu, Nanostructured carbon–metal oxide composite electrodes for supercapacitors: a review. Nanoscale 2013; 5: 72–88.

[10] M. Deng, F. Huang, I. Sun, W.T. Tsai, J. Chang, An entirely electrochemical preparation of a nano-structured cobalt oxide electrode with superior redox activity, Nanotechnology 20 (2009) 175602.

[11] M. R. Tarasevich, B. N. Efremov, Electrodes of Conductive Metallic Oxides, Part A, Elsevier,

Amsterdam, 1980, Chapter 5;

[12] R.N. Singh, J.F. Koenig, G. Poillerat, P.J. Chartier, Electrochem. Soc.137 (1990) 1480;

[13] M. Hamdani, J.F. Koenig, P.J. Chartier, Appl. Electrochem. 18 (1988) 568;

[14] C.S. Cheng, M. Serizawa, H. Sakata, T. Hirayama, Mater. Chem. Phys. 53 (1998) 255;

[15] M.E. Baydi, G. Poillerat, J.L. Rehspringer, J.L. Gautier, J.F. Koenig, P.J. Chartier, Solid State Chem 109 (1994) 281.

[16] L.J.Cote, A.S.Teja, A.P.Wilkinson, Z.J.Zhang, Continuous hydrothermal synthesis and crystallization of magnetic oxide nanoparticles [J]. J Mater Res, 2002, 17(9): 2410-2416.

[17] H.P. Klug, L.E. Alexander, X-ray Diffraction Procedures: for Polycrystalline and Amorphous Materials, second ed., Wiley, New York, 1974.

[18] J.L.Lv, W.L.Guo, T.X.Liang, Synthesis of Co3O4@CoMoO4 core–shell architectures nanocomposites as high-performance supercapacitor electrode, Journal of Electroanalytical Chemistry 783 (2016) 250–257

[19] X. Xiang, L. Zhang, H.I. Hima, F. Li, D.G. Evans, Co-based catalysts from Co/Fe/Al ayered double hydroxides for preparation of carbon nanotubes, Appl. Clay Sci.42 (2009) 405–409.

[20] J.C.Dupin, D. Gonbeau, I. Martin-Litas, P. Vinatier, A. Levasseur, Lithium intercalation/deintercalation in transition metal oxides investigated by X-ray photoelectron spectroscopy, J. Electron Spectrosc. Relat. Phenom. 120 (2001)55–65

[21] D.Feng, X.Q.Zuo, Q.Yang, G.Li, Z.L.Ding, M.Z.Wu, Y.Q.M ,S.W.Jin, K.R.Zhu, Facile hydrothermal reduction synthesis of porous Co3O4 nanosheets@RGO nanocomposite and applied as a supercapacitor electrode with enhanced specific capacitance and excellent cycle stability, Electrochimica Acta 222 (2016) 976–982

[22] J. W. Lee, T. Ahn, J. H. Kim, J. M. Ko, J. D. Kim, Nanosheets based mesoporous NiO microspherical structures via facile and template-free method for high performance supercapacitors, Electrochim. Acta 56 (2011) 4849–4857.]

[23] W. Liu, X. Li, M. Zhu, X. He, High-performance all-solid state asymmetric supercapacitor based on Co3O4 nanowires and carbon aerogel, J. Power Sources 282 (2015) 179–186

[24] Adhikari MP, Adhikari R, Shrestha RG, Rajendran R, Adhikari L, Bairi P,et al. Nanoporous activated carbons derived from agro-waste corncob for enhanced electrochemical and sensing performance. Bull Chem Soc Jpn 2015;88:1108–15.

[25] R.B. Rakhi, W. Chen, D. Cha, H.N. Alshareef, Substrate dependent selforganization of mesoporous cobalt oxide nanowires with remarkable pseudocapacitance, Nano Lett. 12 (2012) 2559–2567.

[26] D. S. Achilleos, T. A. Hatton, Surface design and engineering of hierarchical hybrid nanostructures for asymmetric supercapacitors with improved electrochemical performance, J. Colloid Interf. Sci. 447 (2015) 282–301.

[27] K.Y. Zhang, H. Chen, X. Wang, D.L. Guo, C.G. Hu, S.X. Wang, J.L. Sun, Q. Leng,Synthesis and structure determination of potassium copper selenide nanowires and solid-state supercapacitor application, J. Power Sources 268 (2014) 522–532.

[28] G.S. Gund, D.P. Dubalb, B.H. Patil, S.S. Shinde, C.D. Lokhande, Enhanced activity of chemically synthesized hybrid graphene oxide/Mn3O4 composite for high performance supercapacitors, Electrochim. Acta 92 (2013) 205–215

[29] Gopalakrishnan M, Srikesh G, Mohan A, Arivazhagan V, In-situ synthesis of Co3O4/graphite nanocomposite for high-performance supercapacitor electrode applications, Applied Surface Science 403 (2017) 578–583.

[30] P. L. Taberna, P. Simon, J. F. Fauvarque, Electrochemical characteristics and impedance spectroscopy studies of carbon/carbon supercapacitor, J.Electrochem. Soc. 150 (2003) A292–300

[31] Y. Wang, Z. Shi, Y. Q. Huang, Y. F. Ma, C. Y. Wang, M. M. Chen, Y. Y. Chen,Supercapacitor devices based on graphene materials, J. Phys. Chem. C113 (2009) 13103–13107.

IOP Conf. Series: Materials Science and Engineering **207** (2017) 012021 doi:10.1088/1757-899X/207/1/012021

Study on absorption of Cd^{2+} and Pb^{2+} by bamboo shoots superfine powder modified at high temperature

Lina Ma[1, 2], Yiting Zhu[1, 2], Xiaobing Huang[1], Mao Lin[3], Lijing Lin[1,*] and Yicheng Wang[4]

[1]Agricultural Products Processing Research Institute, Chinese Academy of Tropical Agricultural Sciences, Zhanjiang, China
[2]College of Food Science and Technology, Huazhong Agricultural University, Wuhan, China
[3]Guizhiu Institute of Integrated Agricultural Development, Guizhou Academy of Agricultural Science Jinxinshequ, Xiao he District, Guiyang, 550006, China
[4]South Subtropical Crops Research Institute Chinese Academy of Tropical Agricultural Sciences, Zhanjiang, China

*Corresponding author e-mail: 49031788@qq.com

Abstract. In order to find better heavy metal absorption material, this modification by heating bamboo shoots superfine powder at 50 °C with 5, 30 and 90 min. The effect on the absorption of Pb^{2+} and Cd^{2+} was determined through scanning electron microscopy (SEM) and infrared spectra (FTIR). The results showed that the heating time of 30 min, its absorption of Cd^{2+} and Pb^{2+} was best, and the highest absorption Cd^{2+} and Pb^{2+} were 56.46% and 74.18%, respectively. The heating time for 5 min and 90 min, absorption rate of the kinds of ions was lower than those modified.

1. Introduction

With the rapid development of industry, smelting industry, pesticide, dye and so on every year to produce a large number of industrial waste-water with heavy metal ions [1]. Heavy metal pollution has become one of the most serious problems in our country, the river water quality pollution rate is as high as 80.10% [2]. Due to Pb^{2+} and Cd^{2+} is easy to be absorbed by human and cause a lot of diseases [3-5]. Conventional method of removing heavy metals [6], such as chemical precipitation, membrane separation and ion exchange cannot well solve the problem of heavy metal pollution. Heavy metal pollution has seriously harmed to our health and life safety, looking for efficient and low-cost the adsorption material of heavy metal is urgent.

Superfine grinding technology is a process of material crushing to micron grade [7], as a new type of high technical processing methods, it will be applied to food processing. Studies have shown that material by superfine grinding processing could make food has delicate taste and the high absorption rate. It makes materials good solubility, absorption and liquidity. This will greatly promote the development of food industry [8]. Superfine grinding technology damages fiber acicular structure of material surface, increase the content of soluble dietary fiber and carbohydrate, reduce the content of insoluble dietary fiber, but protein, fat and other nutrients content is no significant change[9].

Bamboo shoots is the tender part of the bamboo, it is a branch of *Gramineae Bambusoideae* the general [10]. Bamboo shoots enrich dietary fiber, protein, amino acids and multifarious mineral elements [11]. Bamboo shoots are planted widely in our country, the existing area about 5.38 million hectares is the highest in the world, and it is growing at an annual rate of 100000 hectares [12]. Rich bamboo resources cause bamboo shoots produce a large amount of waste in the process of machining. According to statistics, the annual output of bamboo shoot and bamboo pu top waste about 1.60 million ton [13]. Superfine grinding is nondestructive processing way, the application of bamboo shoots processing also is less, the study of absorption properties of heavy metals and bamboo shoots superfine powder there has been no relevant report. This research through studying the high temperature of modified superfine powder on the absorption of Pb^{2+} and Cd^{2+}, provide a basis for the development of efficient low-cost absorbent of heavy metals.

2. Experimental

2.1. Materials

Moao bamboo shoots was purchased from Fengnong food Co., Ltd. in Fujian. All chemicals used were of analytical grade.

2.2. The bamboo shoots of superfine powder preparation

After picking the shelled, cut into parts, ground up soaking, rinsing, boiled, drying treatment, fresh bamboo shoots will become the dried bamboo shoot, it's moisture content is under 10.00%. Dried bamboo shoot become the bamboo shoots of coarse powder by the universal mill, and further grinding using the vibrating crusher at 15 °C low temperature under the condition of grinding 40 min after the bamboo shoots of the superfine powder, stored in a dry environment, set aside.

2.3. Heat treatment

Heating the bamboo shoots superfine powder under 50 °C, respectively, 5, 30, 90 min, freeze drying the sample after processing.

2.4. SEM analysis

Microstructural images were obtained from scanning electron microscope (XL-30, Dutch Philips electronics optical Co., and Ltd) at 20-2 low magnification and an accelerating voltage of 0.2-3.0 kV to eliminate the samples charge.

2.5. FTIR analysis

A small amount of sample are taken, some dried KBr is added, mixture is grounded into powder uniform in the mortar powder, then pressured into small rounds in tablet press, and FTIR is set scanning range 4000 - 400 cm^{-1} and 0.1 cm^{-1} of scanning step length, scanning number of 32, is set in the air. Each sample deduct background before scanning. The small round piece put in infrared spectrometer scanning to obtain the infrared spectra of the sample.

3. Results and discussion

3.1. Absorption of heavy metals

The figure. 1 showed Cd^{2+} (a) and Pb^{2+} (b) absorption of the bamboo shoots superfine powder after heat treatment. It can be seen that Cd^{2+} absorption rate of untreated bamboo shoots superfine powder is about 30.00%, and the absorption of Pb^{2+} rate can reach 50.00%. After heating 30 min at 50 °C , Cd^{2+} and Pb^{2+} absorption effect of the bamboo shoots superfine powder is best, and the highest absorption Cd^{2+} and Pb^{2+} were 56.46% and 74.18%, respectively. The absorption increased significantly. After heating 5 min and 90 min at 50 °C, two kinds of ion absorption rate is lower than the untreated absorption rate. Therefore, the heating 30 min can enhance Cd^{2+} and Pb^{2+} absorption ability of bamboo shoots superfine powder appropriately. At the beginning of the absorption is lower. With the absorption time, absorption also increased. This was due to heavy metal ions and the bamboo shoots superfine powder cannot contact incompletely in the beginning, with the extension of absorption time, absorbent and heavy metal ions can reach good contact, then getting absorption saturation state.

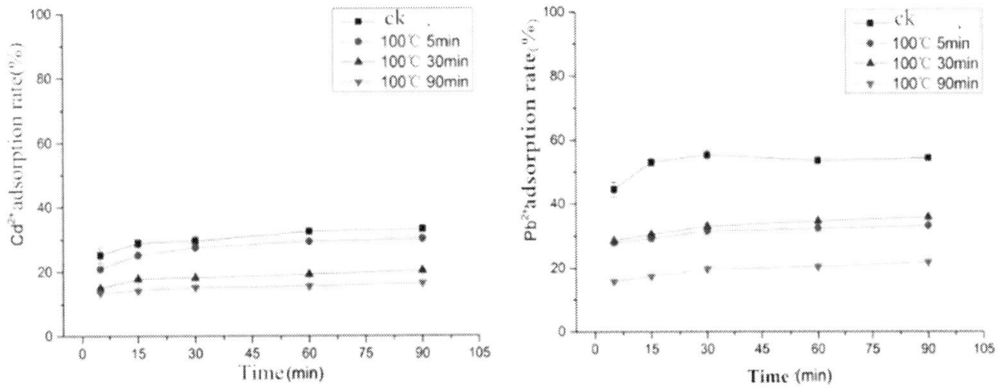

(a) (b)

Figure 1. Absorption rate of Cd^{2+} (a) and Pb^{2+} (b) by bamboo shoots superfine powder at atmospheric 50 °C.

3.2. SEM

The figure. 2 showed SEM picture of the bamboo shoots superfine powder after heating treatment, figure a, b, c is morphology structure of the bamboo shoots superfine powder after heating 5 min, 30 min and 90 min, respectively. Dimensional administrative levels of powder particles is enhanced after heating for 30 min, fold and whole is increased. It is advantageous to the absorption. The powder particle surface fold and the hole are increased after heating 5 min and 90 min, but the administrative levels is poorer, the absorption effect is not obvious, and this result is consistent with the absorption rate.

(a) (b) (c)

Figure 2. SEM images of bamboo shoots superfine powder at atmospheric 50 °C (a, 5 min; b, 30 min; c, 90 min)

3.3. FTIR

The figure. 3 showed the infrared spectrometer of the absorption before and after heating treatment of the bamboo shoots superfine powder. And a, b, c is control, 50 °C heating but no absorption, 50 °C heating and absorption of the bamboo shoots superfine powder, respectively. As can be seen there are a number of peaks in 3445.02 ~ 3239.33 cm^{-1} of untreated bamboo shoots superfine powder. They are stretching vibration peak of -OH, intramolecular hydrogen bond and intermolecular hydrogen bond.

There is a peak only in 3370.71 cm^{-1} after heating treatment, the peak only is -OH stretching vibration peak. The reason may be breaking hydrogen bonds in the bamboo shoots make -OH multiply after heating. And there is a number of peaks in 1113.94~1048.61 cm^{-1} of untreated bamboo shoots superfine powder, it is C-N stretching vibration peak. And there is a peak only in 1038.57 cm^{-1} after heating treatment that is heating treatment can make change C-N. The bamboo shoots superfine powder by heating treatment absorbed heavy metal ions, it' s peaks occurred red shift of different level in 3370.71 cm^{-1} -OH stretching vibration peak, 1651.15 cm^{-1} C=O stretching vibration peak, 1156.37 cm^{-1} -OH stretching vibration peak and 1153.75 cm^{-1} C-N stretching vibration peak, respectively. Strength of the peak is abate, this shows that absorption process -OH, C=O and C-N participate in the coordination reaction.

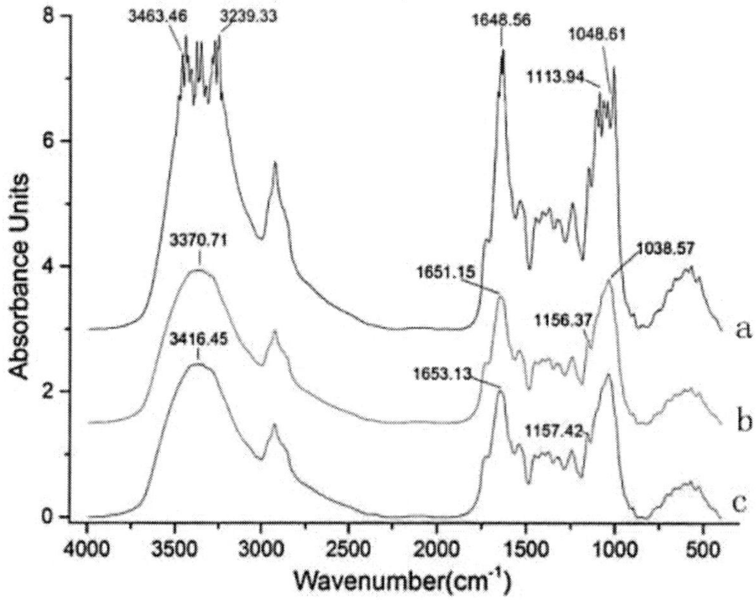

Figure 3. FTIR spectra for bamboo shoots superfine powder before treated and after absorption at atmospheric pressure and 50 °C

4. Conclusion

The results showed that absorption of Cd^{2+} and Pb^{2+} could obviously increase of bamboo shoots superfine powder after heating treatment of 30 min. SEM showed that the surface of the bamboo shoots superfine powder was destroyed and structure level got enhanced. FTIR showed hydrogen bond rupture, -OH increase, and the coordination reaction involved in chemical absorption are mainly -OH, C=O and C-N, the rest is physical absorption. Thus the heating 50 °C for 30 min made bamboo shoots superfine powder on the absorption of Pb^{2+} and Cd^{2+} effect is better. It has the great potential in removing heavy metal ions as a biological absorbent.

Acknowledgments

This research was supported by the Central Public-interest Scientific Institution Basal Research Fund for Chinese Academy of Tropical Agricultural Sciences (1630062014019), the Province Natural Science Foundation of Hainan (20163114), the Special Scientific Research Fund of Agricultural Public Welfare Profession of China (201503142-14), and the Special Fund for Guizhou Province Science and Technology Project ([2016]3035).

References

[1] Z. X. Yao, X. D. Xin, W. Si. The research progress of heavy metal detection method, J. 17 (2011) 29-30.
[2] X. K. Zheng, A. H. Lu. Heavy metal pollution present situation of the soil and the method of prevention and cure, J. Soil and Environmental Sciences. 11 (2002) 79-84.
[3] S. X. Zhou, F. D. Ge. Environmental protection. Beijing: China Environmental Science Press, J. 16(2011).
[4] X. Y. Lan, C. W. Fu, M. Huang. Atomic fluorescence determination of cadmium content in edible vegetable oil, J. Modern Food Science and Technology. 27 (2011).
[5] J. Meng, C. N. Su. Heavy metal Pb, Cd contamination analysis of Zhengzhou city environment, J. Journal of Zhengzhou University of Light Industry. 28 (2013) 52-54.
[6] X. Q. Zhao, Q. K. Xiang. Formaldehyde modified tea leaves the absorption study of chromium in waste water, J. Jiangsu Agricultural Sciences. 40 (2012) 372-374.
[7] Y. H. Xie, S. X. Wang, J. X. Xie. The present situation of the application of superfine grinding technology and the development tendency, J. China Powder Science and Technology. 15 (2009) 64-67.
[8] Y. T. Zhu, X. B.Huang, L. J .Lin. Superfine grinding effect on microstructure and nutrient composition of MAO bamboo shoots, J. Chinese Journal of Tropical Crops. 36 (2015) 937-941.
[9] G. J. Chen, L. S. Wang, F. S. Zhang. Effect of superfine grinding on physicochemical properties of mulberry leaf powder, J. Agricultural Engineerin. 31 (2015) 307-314.
[10] S. L. Zhu, N. X. Ma, M. Y. Fu. China's bamboo plant disposition, J. Beijing: China Forestry Publishing House. (1994).
[11] RFRI. Bamboo as food and medicine, J. Report of Rain Forest Research Institute. (2008).
[12] L. Yang, L. Z. Ding, Z. L. Cai. Four kinds of plants of MAO bamboo shoot forest land soil heavy metal pollution repair action research, J. Journal of Northwest Forestry University. 26 (2011) 14-17.
[13] F. Zhang. Preliminary Study of Fuel Ethanol by Bamboo Shoots Residual. (2013).

IOP Conf. Series: Materials Science and Engineering **207** (2017) 012022 doi:10.1088/1757-899X/207/1/012022

Model development of work roll wear in hot strip mill

Ziying Liu[1, *], Yingping Guan[2] and Fengqin Wang[1]

[1]Shougang Research Institute of Technology, Beijing 100043, China
[2]Graduate School of Yanshan University, Qinhuangdao 066004, China

[*]Corresponding author e-mail: zoonlzy@163.com

Abstract. This paper, based on the analysis of the main factors(specific roll force, mean roll surface temperature, irregular edge wear and contact arc length) affecting roll wear, designed a new work roll wear model, the test data shows that the model can more accurately reflect the work roll wear, can be on-line prediction of work roll wear. The roll wear curve, including constant wear and irregular edge wear, presents a box shape, and the reasons also are showed in this paper. The top roll wear and bottom roll wear in the same mill are inconsistent, and the reasons are also analysed in this paper. Results show that the construction of the work roll mathematical model accords with the general law of work roll wear and tear; it can more accurately forecast roll wear online.

1. Introduction

In the hot rolling, physical models for rolling mills have been intensively developed in recent years. Work roll wear not only worsens the strip shape, and also declines the performance of mills on shape control. As many factors affect work roll wear in the hot rolling and these factors influence each other, quantifying the impact of these factors on roll wear is important and difficulty [1, 2]. According to a literature study presented in a PhD thesis by Meng 1994 at least 182 wear equations have been published [3]. Referring to Meng, available equations are so confusing that just few designers can use any of them to predict product lifetime with confidence [4]. Thus it is not strange that the author has found just a few models of interest in the year of 2003. However the works by Archard [5, 6] shall especially be mentioned. Most models are empirical and only few influential parameters are taken into account.

In this paper, the main aim of the work is to find an empirical work roll wear model structure suitable for hot strip roll finishing mills. On the basis of this structure a new model is developed that takes more influential parameters into consideration than the presently used on-line model. The equation includes the influence of specific roll force, mean roll surface temperature, irregular edge wear, the number of roll rotation and contact arc length. The roll wear curve, including constant wear and irregular edge wear, presents a box shape, and the reasons also are showed in this paper. Analyzed the influence factors of roller wear is inconsistent in the same roll. Results show that the construction of the work roll mathematical model accords with the general law of work roll wear and tear, it can more accurately forecast roll wear online.

2. Mill Configuration

Fig.1 shows the arrangement of typical tandem hot mill. Slabs for strip rolling are heated to a temperature of 1250 °C in walking beam or pusher-type reheating furnaces. After descaling, the slabs are rolled in a roughing mill to a thickness of about 20 to 30 mm using 5 or 7 passes. The width of the work piece is adjusted in a vertical edger.

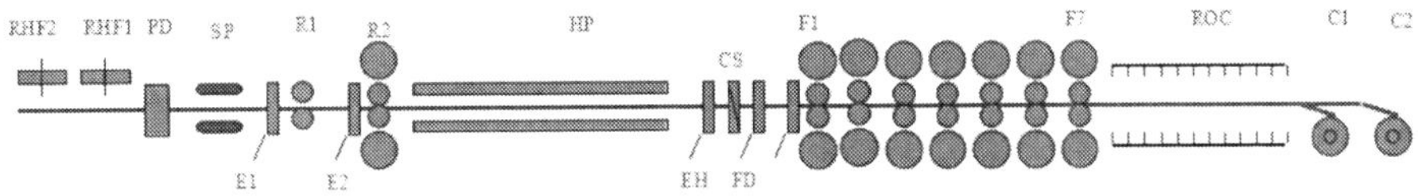

Figure 1. Arrangement of tandem hot mill rolling line and control unit

After that the work piece is rolled down to a strip of about 1.5-16.0 mm thickness, using a four-stand computer controlled finishing mill. The transverse thickness profile of the strip is maintained by controlling the shape and axial displacement of the work-rolls. The control system is based on mathematical models. It provides real time data in various rolling stages and controls the process. From the last stand, the strip continues to the cooling zone where it is chilled from about 800 °C to about 600 °C. Then the strip is coiled. The maximum weight of a coil is approximately 30 tones. If required, the strip can be skin pass rolled at room temperature to improve its properties.

3. Model Development

A flow diagram for the wear model is shown in Fig. 2. One of the important factors affecting roll wear is the roll force. To develop a wear model, it is therefore important to develop a roll force model, which calculates the rolling force at each finishing stand during rolling. Strip temperature, strip width, strip thickness, roll diameter and chemical composition are the factors to calculate the roll force. Except the roll force, mean roll surface temperature, irregular edge wear, contact arc length and work roll revolutions also influence the wear of work roll.

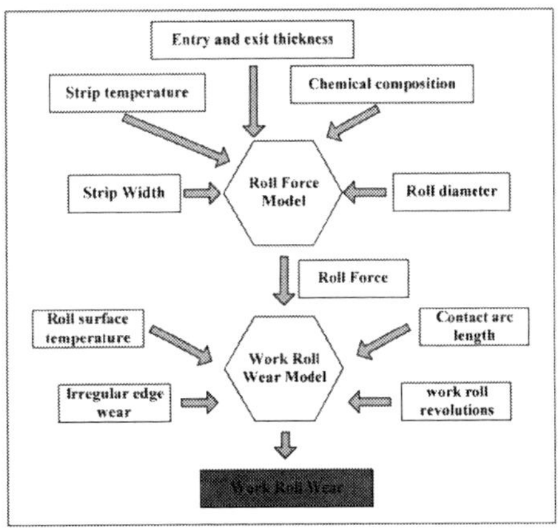

Figure 2. Flow chart of work roll wear model

The work roll wear function is a high order polynomial function as the wear rate at the strip edges is more pronounced than at the strip centre. The roll wear term is a profile over the strip width zone of the work rolls calculated from the number of work roll revolutions made. The number of work roll revolutions N_i is defined as:

$$N_i = \frac{1000\int_0^t V_{pi}dt}{2\pi R_{Wi}} \tag{1}$$

Where, V_{pi} is the stand work roll peripheral speed, m/s; t is the time since last calculation, s; R_{Wi} is the work roll radius, m. In calculating the wear profile, the work roll is segmented into thin discs across the work roll face length. The work roll wear is calculated for each disc. For (2 N_w+1) discs across the strip width, the dimensionless disc position X_j is:

$$X_j = \frac{j}{N_W}, -N_W \le j \le N_W \tag{2}$$

The work roll incremental wear due to the last N_i roll revolutions Δy_{Wij} is defined as:

$$\Delta y_{Wij} = -\left(k_{wW1i} + k_{wW2i}P_i^{k_{wW3i}}\right)\left(1 + k_{wW4i}X_j^{k_{wW5i}}\right)\left(1 + k_{wW6i}\,\theta_{Ri}\right)\left(1 + k_{wW7i}L\right) \tag{3}$$

Where, P_i is the specific roll force, KN/mm; θ_{Ri} is the mean roll surface temperature, ℃; L is the contact arc length, mm; k_{wWli} is the related parameters.

That is, the wear is a function of specific roll force, mean roll surface temperature and contact arc length. The total wear for each disc j and stand i is obtained by summing the individual roll wear increments.

$$y_{Wij} = \sum_0^{N_i} \Delta y_{Wij} \tag{4}$$

4. Experimental Procedures

Using the above mathematical model, in a programming language C++ offline simulation program, calculating the work roll wear of a rolling cycle, and the actual situation of the roll wear are measured with high precision grinding machine. The calculation scheme of work roll wear is shown in Fig.3.

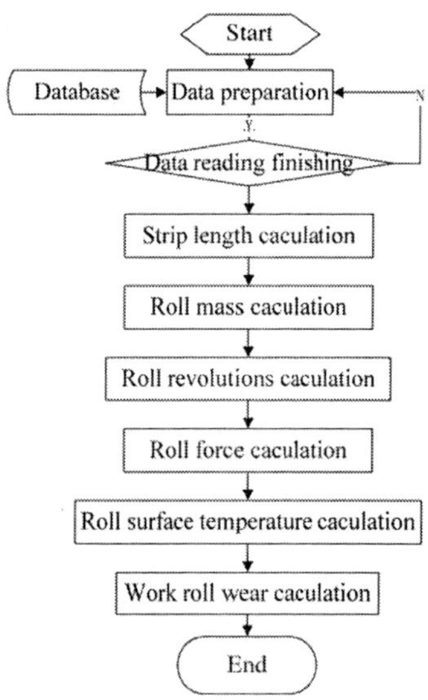

Figure 3. Model calculation scheme

The measuring direction is along the roll driven side to operation side, and the roll is divided into 21 sections, using the actual production data of Qian-Gang 2250mm HSM in March 2016, after a rolling cycle the F4 work roll wear calculation results are shown in Fig.4. From Fig.4, we can see the deviation between calculated value and the actual value is not big, so the model can be used for on-line prediction of roll wear. The actual roll wear contour approximate box, including the middle constant wear and the edge irregular wear, the reason is the roll pressure distribution difference at various points. Both sides roll wear relatively central roll wear is bigger; this is because the strip edge temperature is lower and the existence of edge thinning. We also can see that the mean value of top roll wear is greater than the mean value of bottom roll wear in the same mill. The area and pressure between strip and top roll is bigger because of mill loop, so the top work roll wear is serious, and the oxidation iron retention on the strip surface also increases the top roll wear.

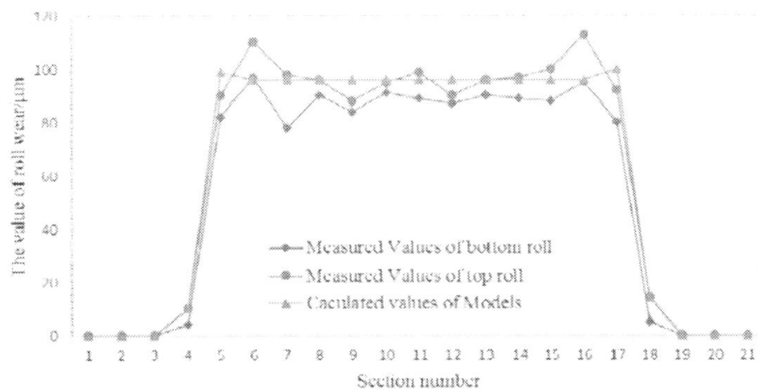

Figure 4. The calculated roll wear and measured roll wear of top work roll

5. Conclusion

A wear model was developed to predict the roll wear profile at different stands of a finishing mill. The wear behavior of finishing mill (stands F4) work rolls was analyzed experimentally. A box type wear pattern with sloping wall was observed. Comparatively higher irregularity was observed for stand F4. Agreement between the predicted and actual wear values for the top and bottom rolls for all four schedules was in general within 6.5%. The computational time required for the overall simulation was within reasonable limits. This model can therefore be used as an efficient online roll wear prediction method for the finishing mill stands of a hot strip mill.

Acknowledgments
This work was financially supported by Beijing postdoctoral research fund 2015 ZZ-152.

References

[1] S.K.Choudhary, D.Mazumdar, A.Ghosh, Mathematical Modeling of Heat Transfer Phenomena in Continuous Casting of Steel, ISIJ international. 1993, 33:764.
[2] Sauer W, Thermal camber model for hot and cold rolling, Steel Research. 1996, 67(1):18 -21.
[3] Meng HC, Wear Modeling Evaluation and Categorization of Wear Models, USA: University of Michigan. 1994.
[4] Meng H C, Ludman K C, Wear Models and Predictive Equations: Their Form and Content, Wear. 1995, 181-183(2): 443-457.
[5] Archard JF, Contact and Rubbing of Flat Surfaces, Appl. Phys. 1953, 24: 981-988.
[6] Archard JF, Rowntree RA, The Temperature of Rubbing Bodies. II. The Distribution of temperatures, Wear. 1998, 128(1): 1-17.

IOP Conf. Series: Materials Science and Engineering **207** (2017) 012023 doi:10.1088/1757-899X/207/1/012023

Experimental study on twin-roll strip casting of Cu-9Ni-6Sn alloy

Delin Tang[1], Xianghua Liu[1], Zhaofei Wang[2,*], Fengde Wang[2]

[1]Research Academy, Northeastern University, Shenyang, 110819, China
[2]Yantai Wanlong Vacuum Metallurgy Co., Ltd, Yantai, 264006, China

[*]Corresponding author e-mail: wzf2558@163.com

Abstract. A strip billet of Cu-9Ni-6Sn alloy was prepared by vacuum melting and twin-roll casting while another billet with the same chemical composition was prepared by vacuum melting and general casting. Component distribution and microstructure of the two kinds of copper alloy from different processing procedures were analysed by metallographic examination and SEM. The results showed that there were massive dendritic crystals with obvious segregation of tin in the sample prepared by general casting. The sample processed by twin-roll casting process had more fine grains with homogenous distributed tin compared to that prepared by general casting. The results demonstrated that the fine grains with homogenous distributed tin were caused by the excellent cooling ability of twin-roll casting. The extremely fast cooling speed retained the homogenous component distribution in molten condition and there was no time for grains to grow up. The experiment indicated twin-roll casting was a feasible process to produce Cu-9Ni-6Sn alloy.

1. Introduction

Cu-9Ni-6Sn alloy is one of the potential materials to supplant beryllium copper alloys because of its high strength, high hardness, good stress relaxation, excellent resistance to stress-corrosion and little time-deformationl [1]. Cu-9Ni-6Sn alloys are widely used to manufacture relay, switch, spring, installation parts, and optical instrument and so on [2]. Cu-9Ni-6Sn alloys caused wide public concern since the 1970s. Schwartz [3] found Cu-9Ni-6Sn alloy formed the mixed organization, which is alternatively constituted by tin-rich areas and tin-poor areas with Spional decomposition of α phase in threshold of desolations of supersaturated solid solution at 300~450℃. Schwartz observed the Spional decomposition by TEM for the first time. Plewes [4] made Cu-9Ni-6Sn alloy huge deformation and distinctly improved its plastic and mechanical property.

However, Cu-9Ni-6Sn alloy is difficult to be produced by melting and casting in traditional method because of the serious segregation of tin in the process of solidification [5-7]. The segregation of tin affect the mechanical properties of Cu-9Ni-6Sn alloy and make it difficult to industrialization. The methods of rapid solidification, spray deposition and powder forming have been applied to solve this problem, but achieved little due to high-cost, complex equipment and difficulties in industrial production. In this paper, twin-roll strip casting was used to produce Cu-9Ni-6Sn alloy [8-10]. Twin-roll strip casting is a process which produces the strip products directly from molten metal by the two counter rotating rolls as casting die. The microstructures and composition distribution of both samples processed by twin-roll strip casting and general casting were determined by OM and SEM.

2. Experimental procedure

2.1 Twin-roll strip casting

In the twin-roll strip casting experiment, copper, nickel and tin with mass percent of 85%, 9%, and 6% were put into vacuum furnace in proper order. Molten alloy were filled into the roll gap after 20 minutes of heating preservation between 1150~1250℃. The twin-roll strip casting system was shown in Figure 1.

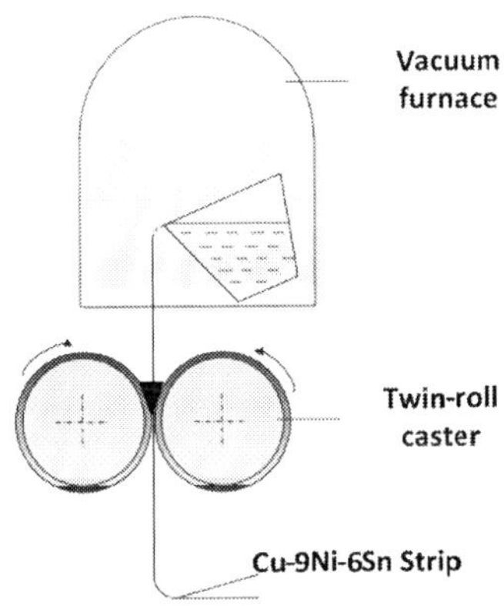

Figure 1. Twin-roll strip casting diagram

Twin-roll strip casting experiment was carried out on experimental twin-roll casting mill with the roller diameter of 500mm. Process parameters obtained from the experiment were summarized as follows: Linear speed was 37.8m/min, roll-casting force was 24kN, and the gap between two rollers was 1.5mm. After the experiment, Cu-9Ni-6Sn strip with width of 110mm and thickness of 3.1mm was produced from the exit of the twin-roll casting mill as shown in Figure 2. The strip had good surface quality. There were no obvious cracks on the surface of strip. After measurement, the strip was with good flatness. The difference gauge of same plate was less than 0.01mm.

Figure 2. Twin-roll casting strip

2.2 General casting

General casting experiment was also carried out in vacuum melting furnace. The same raw materials with that used in twin-roll strip casting experiment were put into vacuum furnace in proper order. Raw materials were heated up to 1250℃ and completed the solidification process in ingot mold. The cross profile of casting rod was shown in Figure 3.

Figure 3. The cross profile of casting rod

3. Results and discussion

Both Cu-9Ni-6Sn alloy samples by general casting and twin-roll strip casting were conducted OM analysis and the results are shown in Figure 4. The sample prepared by general casting showed obvious dendritic microstructure, and the primary dendrite arm spacing is about 0.5mm; The sample processed by twin-roll casting showed evident layering phenomenon: fine equiaxed grains near the surface of strip (less than 0.1mm beneath the surface); big columnar grains (between 0.1 to 0.4mm beneath the surface) and fine equiaxed grains near the surface of strip (more than 0.4mm beneath the surface). The grain size of fine equiaxed grains in surface and center of the strip is about 25μm and the grain size of big columnar grains is about 100μm.

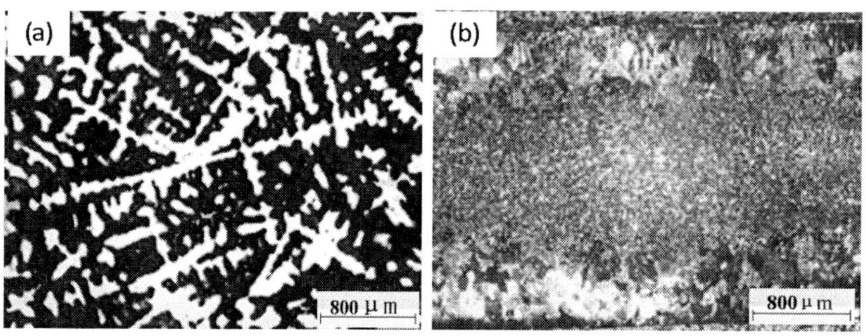

(a) Casting microstructure (b) Twin-roll casting microstructure

Figure.4 OM images of casting and twin-roll casting microstructure

In order to define the components and contents of the samples, SEM analysis was conducted. Figure 5 showed the component distribution of tin by twin-roll strip casting. Figure 5 (a) and Figure 5 (b) respectively showed the components and contents in the center and surface of the sample processed by twin-roll strip casting. The colors on both images are homogenous distributed. That means tins distribute uniformly both in the center and surface in the samples. The component and content of the three regions above-mentioned in Figure 5 is shown in Table1.

Table 1. The component and content of three test region

Test region	Cu, wt%	Ni, wt%	Sn, wt%
1	83.92	9.47	6.61
2	83.53	9.55	6.92

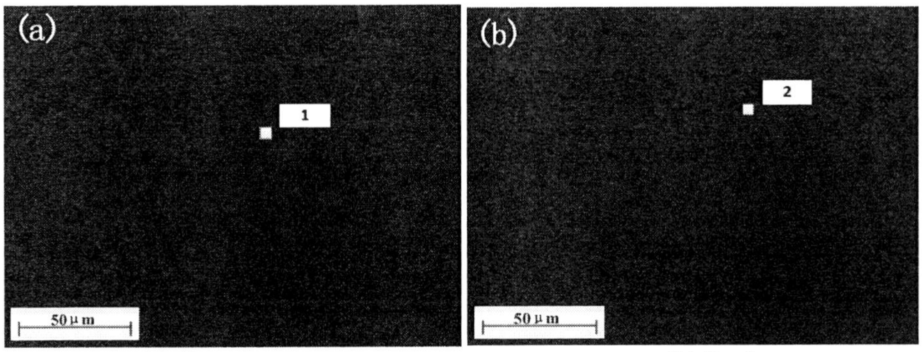

(a) Center (b) Surface

Figure. 5 SEM images of component scanning of sample by-twin-roll casting

The SEM analysis result of the sample processed by general casting is as shown in Figure 6. Three parts could be sharply separated by the shape and color of the image. In Figure 6, the first part is shown as "3" which is dull-red and without boundary; the second part is shown as "4", which is lighter than "3"and without boundary; the third part is shown as "5", which is bright red and with irregular shaped boundary.

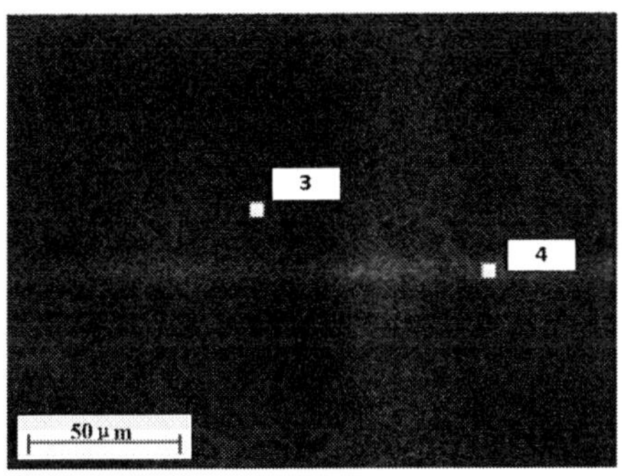

Figrue. 6 SEM images of component scanning of sample by general casting

The component and content of the three regions mentioned above in Figure 6 is shown in Table 2. Tin in the sample processed by general casting mainly distributes in the regions marked with 3and 4 in Figure 6.

Table 2. The component and content of three test region

Test region	Cu, wt%	Ni, wt%	Sn, wt%
3	84.14	11.48	5.38
4	76.08	12.41	11.51

From the above table, the content of tin in region 3 is less than that in region 4. Region 4 had more tin content than matrix. The tin content in region3 is 5.38%, which is the close to the tin mass percent in Cu-9Ni-6Sn. There was obviously segregation of tin in region4.

According to the OM and SEM result, the microstructure of Cu-9Ni-6Sn alloy produced by general casting was made up of coarse dendrites with segregation of tin. The heterogeneous distribution of tin formed tin-poor areas and tin-rich areas. The tin content in tin-rich areas was 7 times more than that in tin-poor areas; The microstructure of Cu-9Ni-6Sn alloy produced by twin-roll strip casting was made up of fine equiaxed grains with grain size of 25~100μm. The alloy composition was close to the set value. There was no obvious segregation of tin.

The reasons why twin-roll casting process produced fine grains and homogeneous composition are as following:

Twin-roll strip casting process had strong cooling ability. The cooling speed could reach 10^4 ℃/s in the solidification process. The extremely fast cooling speed retained the homogenous component distribution in molten condition and there was no time for grains to grow up. Coarse grains could be crushed into fine grains under the influence of rolling force.

4. Conclusion

The microstructure of Cu-9Ni-6Sn alloy produced by general casting was made up of coarse dendrites with segregation of tin. The tin content in tin-rich areas was 7 times more than that in tin-poor areas; the same alloy produced by twin-roll casting was made up of several layers grains with different sizes. The size of fine equiaxed grains in the surface and center was about 25μm. The strip billet had homogeneous composition. The mass fraction of tin was about 6.2%.

The experimental results showed that twin-roll casting process was more feasible to produce Cu-9Ni-6Sn alloy strip billet. Twin-roll casting process could effectively avoid the segregation of tin because of its ultra-fast cooling speed. The appropriate reduction in twin-roll casting could crush the coarse grains into fine grains and improve the mechanical properties of the strip billet.

Acknowledgments
Corresponding author: Zhaofei Wang (Email: wzf2558@163.com)

References

[1] Y.H.Wang, M.P.Wang, B.Hong, General description of Cu-9Ni-6Sn alloy, J. Materials Review. 18 (2004) 33~35
[2] S.L.Yang, W.B.Xie, A review on the research and applications of Cu-Ni-Sn alloys, J. Shanghai Nonferrous Metals. 33(2012)41~45
[3] L.H. Schwartz, Spinodal decomposition in a Cu-9Ni-6Sn alloy, J. Acta Met, 22(1974) 601~608
[4] J. T. Plewes, Spinodal decomposition and discontinuous precipitation in a Cu-15Ni-8Sn, J. Metal Trans A, ,6(1975): 537~543
[5] L.Luo, M.Luo, Y.F.Hu, Processing technology, microstructure and Performance of Cu-9.8Ni-2.2Sn Alloy, J. Hot Working Technology, 41(2012) 23~25
[6] L.H.Zhang, Influence of Mn and Cr addition on structure and properties of Cu-9Ni-6Sn Alloy, J. Shanghai Nonferrous Metals,16(1995) 220~228
[7] Y.Liu, Y.H.Luo, L.J.Wang, Research and development trend of Cu-15Ni-8Sn alloy, J. Metallic Functional Materials. 20(2013) 52~56
[8] F.Huang, H.S.Di, Numerical simulation of twin-roll casting magnesium alloy strip process, J. Journal of Northeastern University (Natural Science). 13(2015) 3~8
[9] Z.Y.Zhang, F.Q.Zhang, F.L.Wang, Development of twin-roll caster technology to produce special alloy thin strip, J. Research on Iron and Steel. 39(2011) 57~62
[10] X.Yan, H.L.Wang, Z.S.Lei, Development of twin-roll thin strip continuous casting technology, J. Shanxi Metallurgy. 103(2006) 4~6

IOP Conf. Series: Materials Science and Engineering **207** (2017) 012024 doi:10.1088/1757-899X/207/1/012024

Development of Shear Thickening Material

Kaijie Song[1], Wentao Huang[2] Hongjing Wan[1,*] and Hongjun Huang[1]

[1]Department of Physics and Chemistry, Ordnance Engineering College, Shijiazhuang
[2]School of Materials Science and Engineering, Tiedao University, Shijiazhuang

*Corresponding author e-mail: 1003307358@qq.com

Abstract. As a new kind of functional material, excellent mechanical properties make Shear Thickening Material one of the most important materials in the field of protection against external forces. According to different raw material and preparation technology, several kinds of shear thickening systems were classified and analysed, and the corresponding mechanism was analysed. On the basis, the effects of disperse phase, high polymer weight and composition on the thickening properties were discussed. The application field of shear thickening materials was reviewed, and the direction of potential research value was pointed out.

1. Introduction

As new materials, shear thickening materials include shear thickening gels and shear thickening liquids, the shear thickening liquid was developed by the U.S. Army Research Laboratory and the State University of Delaware synthetic research center in 2000, in order to create a new generation of military battle suit. This STF is a mixed fluid by the dispersion of Nano particles in polymer or long chain polymer formed, where nanoparticles can free movement, so STF is liquid; when subjected to external shear stress rapidly, the viscosity of the system increased, the area turns to kind of solid state, great loss of external stress energy. When the external force is removed, the system state gradually restored, which exhibited reversible properties [1]. In this paper, the mechanical properties of shear thickening materials in different systems, the corresponding thickening mechanism and influencing factors are reviewed.

In this paper, the mechanical properties of shear thickening materials in different systems, the corresponding thickening mechanism and influencing factors are reviewed.

2. Nanoparticle / polymer suspension system

This kind of shear thickening fluid is a kind of suspension system with special properties. It is composed of the dispersed phase and the dispersed medium. At present, the research on the preparation and properties of this kind of system plays an important role in the field of shear thickening materials. The relationship between shear thickening effect and particle size of dispersed phase was studied through experiment to control fluid thickening effect conducted by Marzano[2] et al, the steady and dynamic analysis of rheological properties of STF are carried out by LGG [3] respectively, and the viscoelastic transformation rule of STF is obtained.

2.1. Shear thickening mechanism

There are two main theories about the shear thickening mechanism of nanoparticle / polymer system. One is the principle of ODT proposed by Hoffman et al through a series of light scattering experiments, the principle for the transformation mechanism of particles from "order" to "disorder"; the second is clusters formation mechanism proposed Brady according to the Stokes Ian Dynamic experiment, which holds that micro assembly "clusters" generated in fluid under the action of shear stress, which results in the viscosity increasing [3]. Later, some scholars have found that the particles with disordered distribution can also produce shear thickening effect, so the theory of particle cluster is more reasonable. Marzano et al [4] found there does exist a "particle cluster" formation in the shear thickening transition through small angle neutron scattering (SANS) experiments [5];

2.2. Preparation and characterization

The most commonly used methods for the preparation of nanoparticle / polymer shear thickening systems are ball milling and ultrasonic method. The properties of SiO_2/PEG system prepared by different processes were studied by Xu Supping et al [6].

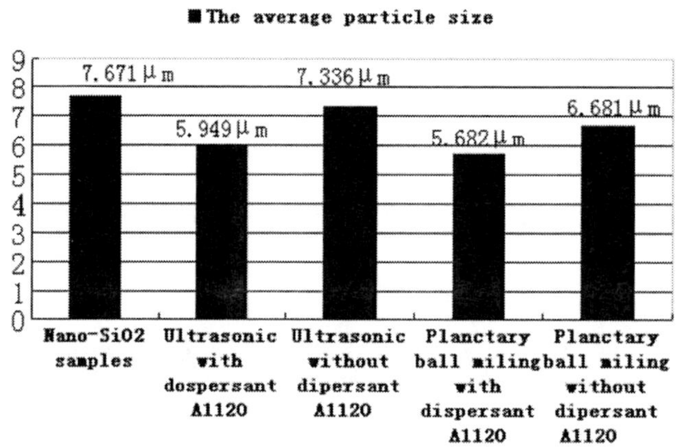

Figure 1. Dispersion of Nano- SiO2/PEG200 system with different preparation methods

The average particle size of Nano SiO_2 prepared by the two different methods in Figure 1 is used to determine the size of the dispersion. Conclusion: (1) the system prepared by planetary ball milling method has better dispersion; (2) the dispersion of the dispersant A1120 is better than that of the system. By consulting the relevant materials, the author finds that the reason lies in the formation of hydrogen bond [7] on the surface of dispersant and dispersed phase -OH, which reduces the probability of aggregation.

The microstructure of STF nanoparticles / polymer system is tested by means of rheological optical experiment [8] and neutron scattering method [9]. Rhea optical experiments is based on the relations between tensor and optical pressure during particles suspension process, hereinto light dichroism is used to identify the effects of thermodynamics and rheology on particle motion, its advantages are mainly embodied in the no external mechanical force, the result is a viscoelastic zero modification; small angle neutron scattering method by observing the changes of STF radiation "to describe the formation of clusters. The rheological characterization technology is used less fully in China at present, mainly because of the limited equipment, and the research technique is not mature. In the next step of the study, we will attempt to characterize the micro rheological properties of the shear thickening system by diffusion spectroscopy.

2.3. Influence factors

2.3.1. Properties of dispersed phase. The influence of dispersed phase particles on the rheological properties of STF includes mainly two kinds of solid content. Lee et al [10] measured the steady-state rheological behaviour of SiO2 (PEG) / SiO2 system with SiO2 volume fraction of 57% and 62%, respectively. The results show that the higher the content of the particles, the higher the critical thickening point.

2.3.2. Properties of dispersion media. The viscosity, polarity, molecular weight and steric hindrance of the dispersed medium have some influence on the performance of STF. Wu Qiumei [11] studied on several different dispersion medium STF, according to the analysis, due to the long molecular chains of PEG, so there exist large steric hindrance, which contributes to the formation of clusters and the shear thickening liquid preparation.

3. Polymer solution system

3.1. Several kinds of polymers with shear thickening properties

3.1.1 Hydrophobic ally associating polymer Among STF systems, the most common and most representative is hydrophobic associating polymer. Near or below C* (The boundary concentration of fluid and colloid), the hydrophobic groups associate together, which leads to the formation of intramolecular or intermolecular aggregation, the polymer mainly based on intramolecular association. In the shear flow field, with the increase of shear rate, the polymer chain will be stretched, which leads to the formation of the association between macromolecules.

3.1.2 High molecular in bad solvent the paper learned that some bad solvent, although the lack of hydrophobic groups, there is also a shear thickening phenomenon, bad solvent with large molecular greatly, such as the ten ethanol solution of hydrogenated naphthalene solution, polystyrene polyiso butylene and poly ethylene oxide. The mechanism of shear thickening is usually considered to be a thickening phenomenon due to the change of the structure induced by collision, In recent years, Brown's kinematic mechanics model has proved that the polymer chains in the inferior solvents collide or form higher order structures to produce shear thickening effect [12]. Briscoe he [13] found the shear viscosity rises very slowly, and the amplitude is also very limited through the study of rheological behavior of Hydrolysis degree of 20% to 40% high molecular weight polyacrylamide (PAM) in aqueous solution of glycerol. In this paper, the author considers that the principle of shear thickening of high molecular weight polymer is the hydrogen bond interaction between polymer chains.

3.2. Shear thickening mechanism

According to the relevant information provided by the shear thickening theory model, the mechanism of polymer shear thickening fluid can be roughly summarized as the following three kinds from different angles [14,15]: (1) shear induced crosslinking; (2) extensional shear induced non-Gauss chain; (3) the formation of space network structure. From the micro level, it is explained as polymer chain interaction transfer to interaction between polymer chain [16,17]. The polymer chains are stretched in shear flow, with the further increase of the shear rate, this effect will be destroyed, and the polymer chain interaction will be gradually replaced by the interaction between the polymer chains of different molecular chains by the strong interaction of strong association together. Therefore, the macroscopic behaviour is the increase of fluid viscosity, i.e., shear thickening behaviour.

3.3. Influencing factors

3.3.1 Concentration of polymer solution the effect of concentration on polymer shear thickening fluid is especially prominent. With the critical concentration of C* as the marker, the shear effect is different. According to the author's understanding and understanding of the shear thickening material, the higher the concentration, the longer the chain density, the less the space between the chain and the chain.

3.3.2 Polymer composition currently what is mostly studied as additives of nanoparticles / polymer system is associating polymer, such as Gong Xinlong [18] put PEG4000, PEG6000 and PEG10000 to use as additives of long chain polymer to discuss the effect additive of concentration on the rheological properties, results show that the system is the critical point of thickening advance with the addition of long chain molecules concentration. On the basis of mechanism analysis, the long chain polymer is adsorbed on the surface of Nano particles, and the flocculation resulted in the increase of the viscosity of the system. Similarly, the hydrophobic group has a significant effect on the rheological behaviour of hydrophobic ally associating polymer [19].

4. Major application areas

4.1 Liquid Armor

In recent years, shear thickening materials are mainly used in software protection field, especially in the military field. According to the characteristics that shear thickening fluid rapidly harden while received a force to resist the external force, and restored the soft state the moment external force removed , foreign scientists since the 90s last century studied STF as a function of individual protective armor. Professor Wanger of material research center of the State University of Delaware made composite fabric through the collocation of different composition shear thickening fluid and Kevlar fiber, namely "liquid armour" [20] the stab resistance of the which was tested .Wanger also had different infiltration of dispersed phase particles of fabrics characterized by SEM, the results show that the infiltration of Kevlar fiber SiO_2 particles can be better than PMMA particles, with nano SiO2 particles as the dispersed phase combination of shear thickening fluid and fiber fabric can absorb more energy, so pullet proof performance better.

4.2 Shock absorption field

According to the characteristics of shear thickening fluid in the presence of external impact force, it is widely used in the field of shock absorption. The shear thickening material is used in the shock absorber of the military vehicle equipment or the shock absorber of the machine tool in the military industry, which is not only convenient to carry, but also to reduce the loss of equipment in the process of transportation[21].

5. Summary and Outlook

Due to its excellent protective ability and light characteristics, the application prospect of the shear thickening material in the fields of national defines, military industry, automobile, transportation, transportation, construction and sports is bright. However, the domestic and foreign research shows that the higher standard of preparation of shear thickening of composite materials, in order to ensure the minimum critical shear stress, and more efficient combination of the shear thickening material as a function of phase and matrix phase, so the preparation process needs to be further improved, by means of comprehensive diversification, for example as well as the rheological characteristics of oscillating scanning study on STF by

IOP Conf. Series: Materials Science and Engineering **207** (2017) 012024 doi:10.1088/1757-899X/207/1/012024

steady-state scanning and frequency of STF, in view of the strain rate sensitivity [22]It is significant to study its response under shock loading mechanics will provide guidance for the application of the material in the door. At present, there is still a gap between the domestic shear thickening materials and that in foreign countries relative to the protection technology, which requires academia and industry hand in hand, so as to achieve constant progress day by day in this field. In addition, the author is currently not yet access to application at home and abroad about the shear thickening material in the foaming material, as a commonly used dual-use carrier material, strength of foam material needs to be improved, in consideration of the properties of shear thickening materials. An application technology breakthrough will achieve in shear thickening material field, if a kind of a foamed material with shear thickening effect can be developed.

References

[1] Xu J, Jamieson A M, Wang S Q, et al. Shear thickening and time-dependent rheological behavior in aqueous polyacrylic ester dispersions [J]. Journal of colloid and interface science, 1996, V182 (1): 172-178.

[2] Maranzano BJ, Wagner NJ. The effects of particle size on reversible shear thickening of concentrated colloidal dispersions[J]. Journal of Chemical Physics, 2001, 114(23): 10 514-10 527.

[3] LEE Y S,Wanger N J. Dynamic properties of shearthickening colloidal suspensions[J]. Rheologica Acta, 2003,42(3) : 199-208.

[4] JF Brady,G Bossis.STOKESIAN DYNAMICS[J].Annual Reviews,1988,20:111-157.

[5] Newstein M C, Wang H, Balsara N P, et al. Microstructural changes in a colloidal liquid in the shearthinning and shear thickening regimes [J]. Journal of Chemical Physics, 1999, V111 (10): 4827-4838.

[6] Maranzano B J, Wagner N J. Flow small angle neutron scattering measurements of colloidaldispersion microstructure evolution through the shear thickening transition [J]. Journal of ChemicalPhysics, 2002, V117 (22): 10291-10302.

[7] Eric D,Stab resistance of shear thickening fluid(STF)-kevlar composites for body armor applications[Z],Proceedings of the 24th Army Science Conference.2004.

[8] Bender J W,Wanger N J.Colloid and Interface Science,1995,172(1):171-184.

[9] Lee Y S,Wanger N J.Ind Eng Chem Res,2006,45:7015-7024.

[10] . WQM,RJM,HBY et al.Influence of dispersion medium and temperature on the rheological propertie of SiO2 dispersion.[J].Journal of Central South University,2006,37(5):862-866

[11] Kishbaugh A j,Mchugh A J.Rheologica Acta,2011,32(1):9-24

[12] Briscoe B,Lyckham P,Zhu S.Rheologic Acta,2011,38(3):224-234.

[13] Ahn K H,Osaki K. J Non-Newtonian Fluid Mech,1994,55(3):215-227.

[14] Ahn K H,Osaki K. J Non-Newtonian Fluid Mech,1995,56(3):267-288.

[15] Van Egmond J W.Curr Opin Colloid Interface Sci,1998,3(4):385-390.

[16] Mahfuz H,Clements F,Rangari V,et al.Enhanced stab resistance of armor composites with functionalized silica nanoparticles[J].Journal of Applied Physics,2009,V105(6):064307..

[17] Xu Yulei,Gong Xinlong,PengChao,et al.Shear Thickening Fluids Based on Additives with Different Concentrations and Molecular Chain Lengths[J].Chinese Journal of Chemical Physics,2010,23:342-346

[18] QYL,WJH,ZBK et al.Modification Effects of Amphiphilic Comb-like Polysiloxane Containing Polyther Side Chains on the PVDF Membranes Prepared via Phase Inversion Process.[J]. Polymeric Sinic,2007,12(1):151~168.

[19] Kalman D P,Wanger N J.Microstructure of Shear thickening Concentrated Suspensions Determined by Flow-USANS[J].Rheologica Acta,2009,48:897-908.

[20] Kalman D P,WWetzel E D,Wanger N J,et al.Polymer Dispersion Based Thickening Fluid-

fabrics for Protective Applocations[C]//Proceedings of SAMPE.Baltimore MD,2007:1-9.
[21] Brown E,Forman N A, Orellana C S,et al. Generality of shear thickening in dense suspensions[J]. Nature Materials. 2010, 9(3):220-224.
[22] Yildiz S.Synthesis and rheological behavior of shear thickening fluids(STFs) for liquid armour applications [D]. Turkey:Izmir Institute of Technology,2013.

The Synthesis and Characterization of Tetrakis [(p – amino phenoxy) methyl] methane

Yongli Peng[1], Qian Zou[2,*]

[1]School of Materials Science and Engineering, Wuhan Institute of Technology, Wuhan, China

[2]School of Materials Science and Engineering, Wuhan Institute of Technology, Wuhan, China

*Corresponding author e-mail: 854167473@qq.com

Abstract. In order to solve the shortcomings of the cured epoxy resin poor toughness, this paper proceeded from the structural design of curing agent to synthesize a special curing agent tetrakis [(p-aminophenoxy) methyl] methane which containing both Benzene ring and amino group. A Symmetric compound of tetrakis [(p - acetamidophenoxy) methyl] methane was prepared by using simple and easy to get pentaerythritoltetratosylate and acetaminophen for raw materials, after Williamson etherification reaction intermediates for synthesis of a symmetrical structure of the compound tetrakis [(p-acetamido phenoxy) methyl] methane, then hydrolysed under acidic conditions it can be tetrakis [(p-amino phenoxy) methyl] methane. The influence of reaction time, reaction temperature and reactant ratio to production yield of tetrakis [(p - acetamidophenoxy) methyl] methane was studied by orthogonal experiment of three factors and three levels, and get the optimal process parameters: the reaction time: 16 h, the reaction temperature: 170 ℃, reactant ratio, 1:5.The Structure of tetrakis [(p - acetamidophenoxy) methyl] methane and tetrakis [(p-amino phenoxy) methyl] methane were characterized by infrared and 1H-NMR.

1. Introduction

Pentaerythritol and its derivatives is a compound with symmetrical structure, this special structure makes these compounds to construct complex molecules and supramolecular compounds starting point, in particular, it offers attractive Intermolecular interaction of multiple sites that can be connected to form compounds in complex networks of the Association Center[1-3]. Pentaerythritol and its derivatives have a wide range of applications in biology and medicine, materials science and catalysts. Pentaerythritol tetra phenyl ether is a pentaerythritol as the core of the four-arm compound, thanks to the special structure of such compounds, more and more researchers introduce special functional groups from the molecular design and synthesis series of different physical and chemical properties of macromolecules to meet the needs of industrial applications[4,5] , In order to solve the poor toughness of epoxy resin cured products, this paper, from the design of curing agent structure, synthesizes a special curing agent tetrakis [(p-aminophenoxy) methyl] methane which containing both benzene ring and amino group[6,7]. Through check out large literature, the synthesis method of tetrakis [(p-aminophenoxy) methyl] methane has had reported, p-nitro phenol and pentaerythritol tetra benzene sulfonate were first reacted under basic conditions to give tetrakis [p- nitrophenoxy) methyl] methane, and then tetrakis [(p-nitrophenoxy) methyl] methane is catalytically hydrogenated to give tetrakis [(p-aminophenoxy) methyl] methane, but from the aspect of safety, economy, simplicity of operation, this synthesis is not so satisfactory[8-10]. Therefore, in this paper, acetaminophen is used instead of p-nitro phenol and pentaerythritol tetra benzene sulfonate to obtain tetrakis [(p-acetamidophenoxy) methyl] methane, and then hydrolysed under the condition of raced to get tetrakis [(p-aminophenoxy) methyl] methane[11,12]. This method is more simple and safe than the former, while the synthesis of intermediate tetrakis [(p-acetamidophenoxy) methyl] methane is particularly important, In this paper, orthogonal experiments were carried out to study the optimum conditions for the preparation of tetrakis [(p-acetamidophenoxy) methyl] methane with pentaerythritol tetra benzenesulfonate and acetaminophen[13,14].

2. Experimental

2.1 Reagents and Instruments

Pentaerythritol, Sino pharm Group Chemical Reagent Co., Ltd., chemical pure; benzene sulfonyl chloride, Tianjin Daomao Chemical Reagent Factory, Analytical pure; Pyridine, Tianjin Daomao Chemical Reagent Factory, Analytical Pure; Acetaminophen, Shanghai Aladdin Limited Liability Company, chemical pure; ethanol, Tianjin Kemei Ou Chemical Reagent Co., Ltd., analytical pure; sodium hydroxide, Tianjin Bodi Chemical Co., Ltd., analytical pure.

DF-101S collector temperature heating magnetic stirrer, Gongyi City to China Instrument Co., Ltd.; XT5 micro-melting point detector, Beijing Branch instrument electro-optical instrument factory; Fourier transform infrared spectrometer, Perkin Elmer; nuclear magnetic resonance spectrometer, Germany Bruker.

2.2 Synthetic Principle of Tetrakis [(p-aminophenoxy) methyl] methane

The synthesis of intermediate tetrakis [(p-acetyl-amino phenoxy) methyl] methane is similar to the Williamson ether synthesis reaction principles, the Williamson etherification reaction is the substitution of the hydrogen atom of the hydroxyl group in the alcohol or phenolic molecule by an alkyl or aryl group Etherification or phenolic etherification. The reaction of tetrakis [(p-acetamidophenoxy) methyl] methane is carried out by reacting the phenol with a salt under alkaline conditions to form a phenate anion, salts anion as the nucleophile and the alkyl halide or sulfonate SN2 Nucleophilic substitution reactions occurred. Tetrakis [(p-acetyl-amino phenoxy) methyl] methane was hydrolysed under acidic conditions to give tetrakis [(p-aminophenoxy) methyl] methane, and the total amount of tetrakis the reaction is shown in Figure 1:

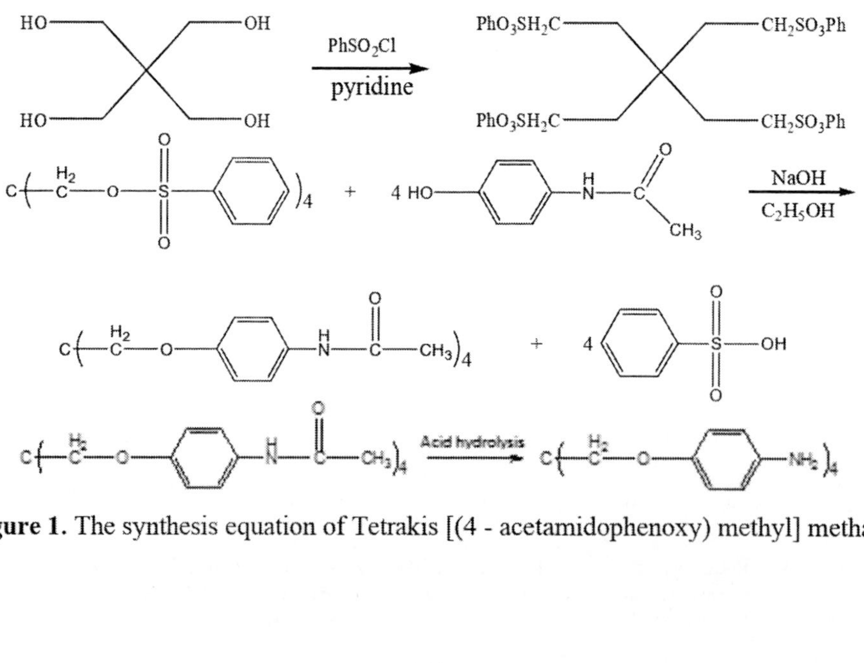

Figure 1. The synthesis equation of Tetrakis [(4 - acetamidophenoxy) methyl] methane

2.3 Synthesis of Pentaerythritol Tetra benzene Sulfonate
Into a 250ml round-bottomed flask were added 4.36g Pentaerythritol, 20 mL pyridine, 20 ml of benzene sulfonyl chloride was continuously added dropwise while stirring, the dropping rate was controlled and the reaction temperature was controlled below 35 ° C, after completion of the dropwise addition, the temperature was raised to 40 ° C and the reaction was stirred for 3 h. After the reaction, the product of the three-necked flask was poured into a large beaker containing 30 mL hydrochloric acid, 33 mL ice water and 67 mL methanol mixed solution in a timely manner, stirred and allowed to stand for sedimentation and filtration. Then configure a volume of acetone and ethanol mixed solution to filter crystals obtained from recrystallization, where v (acetone): v (alcohol-free) =1:4, continued to filter, oven drying at 70 ℃, get 19.3 g of white powder, yields 85.3%, melting point 99-101℃.

2.4 Synthesis of Tetrakis [(p - acetamidophenoxy) methyl] methane
In the 250 mL flask three by adding a certain amount of acetaminophen and sodiumhydroxide, ethanol as the solvent of the measured amount of dry, In a 250 mL three-necked flask were added a certain amount of acetaminophen and sodium hydroxide, dry ethanol was measured amount as the solvent. The reaction mixture was stirred and heated for 1h at 60 ° C, then add the appropriate amount of pentaerythritol tetra benzenesulfonate, heated to 150 ° C above the reaction for a certain time.

2.5 Synthesis of Tetrakis [(p - aminophenoxy) methyl] methane
The prepared tetrakis (p-acetamidophenoxy) methyl] methylene crystals were poured into a three-necked flask, 20 ml of glacial acetic acid and 10 ml of 8 mol / L hydrochloric acid were added, the reaction mixture was heated 100 ° C under stirring and refluxing for 10 h. When the reaction is complete, remove device, cooling the solution to room temperature, configure the 10mol/L concentration of Nao solution, the solution is titrated until the solution is neutral. Stand to precipitation filter and then get tetrakis [(p-amino phenoxy) methyl] methane.

2.6 Tetrakis [(p-acetyl-amino phenoxy) methyl] methane Analysis Method
Melting point testing: XT-5 binocular microscope melting point Tester, according to JJG 701-2 008 "melting point Analyser verification regulation of" determination of melting point.

FT-IR: with potassium bromide Tablet method, Fourier transform infrared spectrometer Perkin Elmer (USA) was tested on, scanning range of 400~4 000 cm-1.

Nuclear testing: using deuterated DMSO as solvent and TMS as internal standard, tested on Agilent 400 MR nuclear magnetic resonance spectrometer analysis.

3. Analysis and Discussion

3.1 FT-IR Analysis of Tetrakis [(p-acetyl amino phenoxy) methyl] methane
Figure 2 is an infrared spectrum of tetrakis [(p-acetamidophenoxy) methyl] methane in which 3 297 cm-1 is attributed to the stretching vibration of NH, and 1 659 cm-1 is attributed to the stretching vibration of C = O, Thereby indicating the presence of the amide bond. 1 609 cm-1 and 1 512 cm-1 attributed to the skeleton vibration absorption peak of the benzene ring proved the presence of benzene ring. 2 959 cm-1, 2 922 cm-1 and 2 868 cm-1, belonging to the telomeric vibration of methyl methylene; 1 232 cm-1 is the characteristic absorption peak of Ar-O; 831 cm-1 P-Substitution of the benzene ring. By comparing the spectra with the groups in the structural formula of tetrakis [(p-acetamidophenoxy) methyl] methane, it is basically determined that the functional groups represented by the characteristic peaks on the spectrum are tetrakis [(p-acetamidophenoxy Base) methyl] methane.

IOP Conf. Series: Materials Science and Engineering **207** (2017) 012025 doi:10.1088/1757-899X/207/1/012025

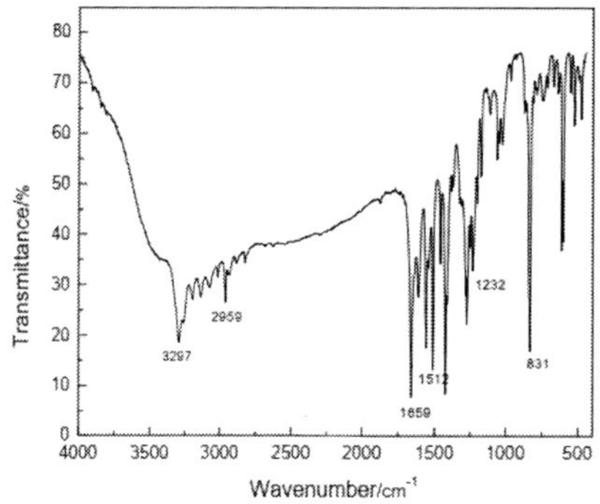

Figure 2. The infrared figure of Tetrakis [(4 - acetamidophenoxy) methyl] methane

3.2 Analysis of Nuclear Magnetic Spectrometry of Tetrakis [(p - Acetamidophenoxy) methyl] Methane

Figure 3 is a Tetrakis [(p-acetyl-amino phenoxy) methyl] hydrogen nuclear magnetic resonance spectra of methane, molecule solvents using deuterated dimethyl sulfoxide (DMSO), analysis is shown in table 1, The integrated area ratios of absorption peaks 1 to 5 coincide with the number of H at the corresponding positions in the molecular structure of tetrakis [(p-acetamidophenoxy) methyl] methane, and there are 1H NMR (400 MHz, dmso) δ 9.73 (s (P, 1H), 7.41 (d, J = 8.8 Hz, 2H), 6.87 (d, J = 8.9 Hz, 2H), 4.17 (s, 2H), 1.96 (s, 3H)which Prove that the resulting material is tetrakis [(p-acetamidophenoxy) methyl] methane.

Table 1. The chemical shifts of H atoms of Tetrakis [(4 - acetamidophenoxy) methyl] methane in 1H-NMR

Serial	δ	Number of H	Peak area
1	9.73	1	0.29
2	7.41	2	0.59
3	6.87	2	0.59
4	4.17	2	0.54
5	1.98	3	1.00

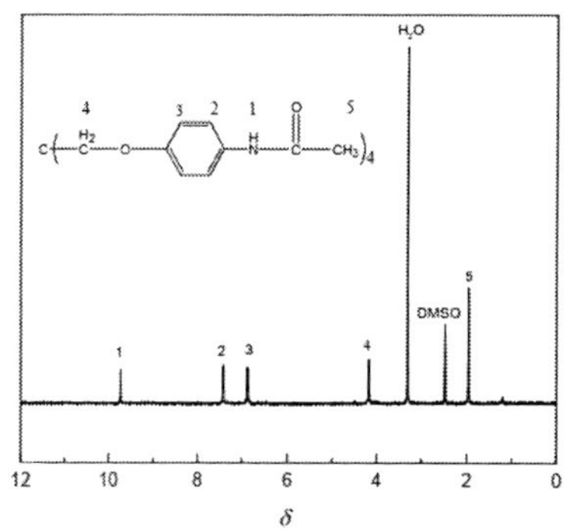

Figure 3. 1H-NMR of Tetrakis [(4 - acetamidophenoxy) methyl] methane

3.3 FT-IR Characterization of Tetrakis [(p-aminophenoxy) methyl] methane

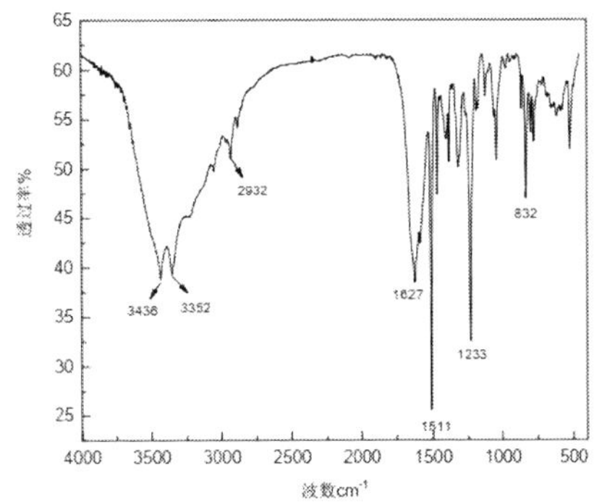

Figure 4. The infrared figure of Tetrakis [(4-aminophenoxy) methyl] methane

The characteristic peaks are shown in the following table:

Table 2. The characteristic peaks of Tetrakis [(4-aminophenoxy) methyl] methane in FI-IR

absorb position(cm-1)	Attribution statement
3436, 3352	-NH2-stretching vibration
1627	N-H in-plane bending vibration
1585,1511	Benzene ring C = C stretching vibration
1233	Ar-O
2934, 2883, 1384	-CH2-stretching vibration
832	p-substitution of the benzene ring

The absorption peak of primary amine NH stretching vibration at 3436, 3352 cm-1, the in-plane bending vibration absorption peak of NH on primary amine at 1627cm-1; 2934, 2883 and 1384cm-1 are attributed to the stretching vibration absorption peak of methylene CH, And 1585, 1511 is the skeleton vibration peak of benzene ring, 832cm-1 is the para substitution of benzene ring, 1233cm-1 is Ar-O vibration absorption peak. It can be seen that the characteristic peaks of tetrakis [(p-aminophenoxy) methyl] methane can be found in the infrared map.

3.4 Nuclear Characterization of Tetrakis [(p - aminophenoxy) methyl] methane

Figure 5 shows the NMR spectra of the tetrakis [(p-aminophenoxy) methyl] methane molecules, solvent used deuterium generation two methyl Asia sulfone (DMSO), analysis as table 2-9 by shows, absorption peak 1 to 4 of points area ratio and tetrakis [(p-aminophenoxy) methyl] methane molecular structure in the corresponding location Shang h of number anastomosis, proved proceeds material for tetrakis [(p-aminophenoxy) methyl] methane.

Table 3. The chemical shifts of H atoms of Tetrakis [(4 - aminophenoxy) methyl] methane in 1H-NMR

Serial number	1	2	3	4	5	6
Chemical shift	4.59	6.43	6.65	4.03	3.33	2.48
Number of H	2	2	2	2	water	DeuteriumDMSO
Peak area	1.00	0.98	1.03	1.05		

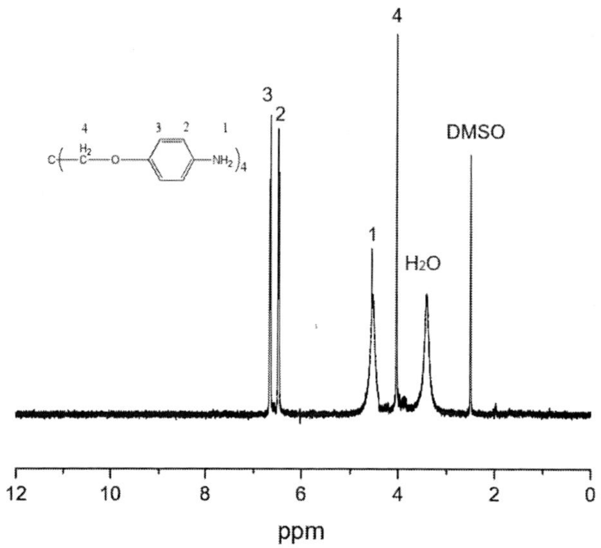

Figure 5. 1H-NMR of Tetrakis [(4 - aminophenoxy) methyl] methane

3.5 Discussion on Influencing Factors of Synthesis Reaction

Orthogonal experiment is a highly efficient, fast and economical research multi-factor and multi-level design method. As the synthesis of intermediates has a decisive effect on the yield of the final product, the yield of tetrakis (p-acetamidophenoxy) methyl] methane is taken as an index by orthogonal test. The reaction time, Reaction temperature, and reaction product feed ratio of three factors are preferred, the choice of L9 (33) for orthogonal experiments, the impact of factors were shown in Table 1.

Table 4. The levers of impact factors to the synthesis reaction of Tetrakis [(4 - acetamidophenoxy) methyl] methane

level	Reaction time /h	Reaction temperature /℃	n(pentaerythritoltetratosylate):n (acetaminophen)
1	12	160	1:4
2	16	170	1:5
3	20	180	1:6

Orthogonal table was shown below:

Table 5. The preparation process of Tetrakis [(4 - acetamidophenoxy) methyl] methane by orthogonal design

Serial number	Reaction time /h	Reaction temperature /℃	n(pentaerythritoltetratosylate): n(acetaminophen)	Yield/%
1	12	160	1:4	55.2
2	12	170	1:5	66.3
3	12	180	1:6	64.6
4	16	160	1:5	74.2
5	16	170	1:6	71.8
6	16	180	1:4	57.5
7	20	160	1:6	66.8

8	20	170	1:4	62.3
9	20	180	1:5	63.6
K1	186.1	196.2	175.0	
K2	203.5	200.4	204.1	
K3	192.7	185.7	203.2	
k1	62.01	65.4	58.3	
k2	67.8	66.8	68.0	
k3	64.2	61.9	67.7	
R	5.7	4.9	9.7	
Primary and secondary order			C>A>B	
Excellent level	A2	B2	C2	
Excellent combination			A2 B2 C2	

From the above calculation and analysis, we can see that the reaction ratio has the greatest influence on the experiment, which is the main influencing factor and the reaction temperature is not important factor. The optimal level is A2B2C2, that is, the reaction time is 16 h, the reaction temperature is 170 ℃, the ratio of reactants was 1: 5, and the reaction was carried out according to the optimum synthesis conditions. The yield of the product was 76.2%, which was higher than that of the fourth group, so that the reaction of tetrakis (p-acetamidophenoxy) methyl] methane was The optimum reaction time was 16 h, the reaction temperature was 170 ℃, and the reaction ratio was 1: 5.

4. Conclusion

(1) The experiment was carried out by L9 (33) orthogonal experiment. The results showed that the reaction ratio had the greatest effect on the synthesis of tetrakis (p-acetamidophenoxy) methyl] methane, and the reaction temperature was unimportant The optimum conditions for the synthesis of tetrakis [(p - acetamidophenoxy) methyl] methane were as follows: reaction temperature 170 ℃, reaction molar ratio 1: 5, reaction time 16 h, the yield was 76.2%.

(2) The FT-IR and NMR spectra of the obtained product were verified, and it was proved that the final product was tetrakis (p-aminophenoxy) methyl] methane, which indicated that the synthetic route was feasible.

Acknowledgements

Project Fund: Supported by the National Natural Science Foundation of China (21376182) Author: Zou Qian, Master graduate. E-mail: 854167473@qq.com Corresponding author: Pengyong Li, Ph.D., professor E-mail: PYL68@163.com

References

[1] DOMINIC L, THIERRY M, JAMES D W. Molecular Tectonics. Porous hydrogen-bonded networks built from derivatives of pentaerythrityltetraphenyl ether [J]. Organic Chemistry. 2004, 69(6): 1776-1787

[2] WEI Y, SHAO Y. The improved synthesis of pentaerythritylTetramine [J].Energetic Materials, 2002, 10(2): 49-52.

[3] PENG Y L, LUO C, CHEN B G. Synthesis and structure characterization of 1, 3-dibromo-2, 2-bis (bromomethyl) propane [J]. Journal of Wuhan Institute of Technology, 2013, 35(7): 37-42.

[4] TANG X D, ZHANG Q Z, ZHOU Q F. Application of pentaerythritol and its derivatives in the synthesis of dendrimers [J]. Organic Chemistry, 2004, 24(6): 585-590.

[5] ZHANG H C, HaAO a Y, DU G Y, etal. Derivatives of pentaerythritol as the matrix of sugar

families with sugars dendritic molecules and application [J]. Organic Chemistry, 2008(9): 1515-1522.

[6] YANG Y F, TANG G P, CHENG H F, etal. Research progress of toughening epoxy resin [J]. Materials Review, 2010, 24(8): 85-88.

[7] LI J R. Synthesis, curing behavior and propertries of a heat resistant curing agent with flexible unit [D]. Beijing University of Chemical Technology, 2013.

[8] DOMINIC L, THIERRY M, ERIC D, etal. Molecular Tectonics. Hydrogen-Bonded Networks Built from Tetra- and Hex anilines [J]. Crystal Growth & Design, 2005, 5(4): 1451-1456.

[9] LI G H. The Position of Different Substituents Star of Synthesis and Properties of Liquid Crystal Compounds [D]. Northeastern University, 2010.

[10] BACKER H J, DIJKEN G. Pentaerythritol thioethers [J]. View Issue, 1932,51(3) :289-293.

[11] PETER A, HENDERSON and CORRIE T I. Methylene-linked liquid crystal dimers and the twist-bend nematic phase [J]. Liquid Crystals, 2011, 12(38):1407–1414.

[12] WALTER A J, MICHAEL H. On nitro- and aminophenoxyacetic acids [J]. Chemical Society, 1917, 39(10):2188-2224.

[13] Ashraf A, ABBAS, ADEL S G, etal. A convenient synthesis of thiamacrocyclic dilactams [J]. Heteroatom Chemistry, 2007, 18(3): 249-254.

[14] Wang M L, LEE Z F. Synthesis of novel multisite phase-transfer catalysts and their applications to the catalyzed reaction of bisphenol a and allyl bromide [J]. Industrial & Engineering Chemistry Research, 2006, 45(14):4918-4926.

[15] SUNGGAK K, CHUNG K N, YANG S B, etal. Direct synthesis of ethers via zinc chloride mediated etherification of alcohols in dice chloromethane [J]. Journal of Organic Chemistry. 1987, 52(17): 3917-3919.

IOP Conf. Series: Materials Science and Engineering **207** (2017) 012026 doi:10.1088/1757-899X/207/1/012026

Debonding defect detection of metal and composite bonding structures

Yuke Shang[1*], Xinjun Wang[2], Ping Li[3], Nengjun Yang[4]

[1]Xian high technology institution, Xian, China
[2]Xian high technology institution, Xian, China
[3]Xian high technology institution, Xian, China
[4]Xian high technology institution, Xian, China

*Corresponding author e-mail: 454040697@qq.com

Abstract. Metal and composite bonding structures are widely used in many industrial fields. But in the material production and processing and bond- ing process, the adhered surface is not clean, is between the two surface bonding with adhesive is not strict, the existence of gas, dust and other rea- sons, the adhesive structure foaming, deboning, bonding strength decrease- e, durability and reliability reduce etc. In this paper, a kind of typical metal and composite bonding structure was studied by electromagnetic ultrasonic testing. The results show that the results are obvious. The electromagnetic ultrasonic can be used to test the adhesion of metal and composite materials.

1. Introduction

With the rapid development of industrial economy and science and technology, bonding structure of metal and composite materials have been widely used in many industrial fields, such as automobile shock damping device of rubber bonding lining layer, engine braking device in nonmetallic friction materials and metal components of the adhesive, oil and gas pipelines and anti-corrosion materials bonding [1]. But in the process of material production and processing and bonding, the adhered surface is not clean, is between the two surface bonding with adhesive is not strict, the existence of gas, dust and other reasons, makes the bonding structure of bubbles, deboning, bond strength decreased, durability and reduced reliability[2]. Application of electromagnetic ultrasonic testing technology to detect the adhesion of metal and composite materials, without coupling agent, more convenient, the results are also better[3].

2. Electromagnetic ultrasonic working principle and Detection of shell and adiabatic layer deboning principle

2.1. Electromagnetic ultrasonic working principle

EMAT is the core tool of electromagnetic ultrasonic testing technology. The EMAT consists of three parts - a magnet that provides a bias magnetic field, a coil that generates a pulsed magnetic field, and the measured pattern of exciting and propagating ultrasonic waves in the interior, as shown in figure1.

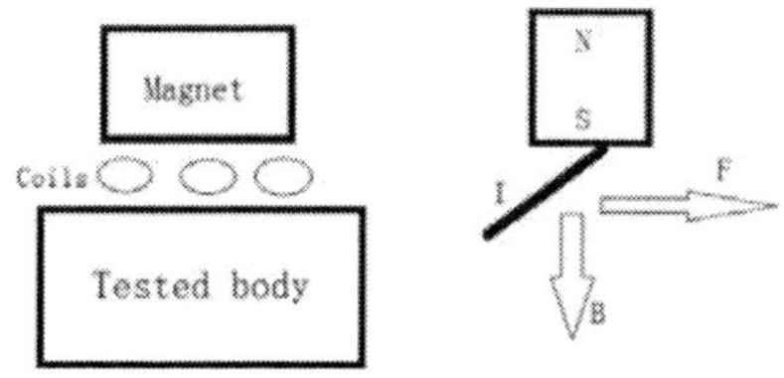

Figure 1. Basic structure of EMAT.

The process of coupling of electromagnetic field, force field and sound field and so on, which constitute the working principle of the electromagnetic transducer. The excitation coil in the bias field and high frequency pulse interaction around the radiation induced dynamic electromagnetic field; and in the dynamic electromagnetic field under the surface of the measured body feeling stimulating pulsed eddy current equal and opposite with the birth; interaction of the vortex in the influence of bias magnetic field, dynamic electromagnetic field and other physical processes under a stress field, the measured particle surface generated by vibration, acoustic form spread out, which is the principle of ultrasonic excitation[4-8].

In the process of EMAT energy conversion, the force field which is measured by the measured body is generally composed of Lorenz force, magnetic force and magnetic force, and the force field is formed by the interaction of the three force fields [4].

2.2 Detection of shell and adiabatic layer deboning principle

As shown in Figure 2, ultrasonic propagation in the measured body, when acoustic wave I_0 reaches the surface of the shell and adiabatic layer, without bubbles, cracks and other deboning defects, most sound waves pass through the first interface, reflecting very little acoustic waves. Therefore the received signal energy is low; if it is in the deboning interface, because of the large acoustic impedance of air, acoustic I_0 almost total reflection, the energy of receiving echo signal is obviously higher than that of good bonding quality. The echo generated by the multiple reflection of the ultrasonic wave in the interior of the measured body has strict repeatability, the waveform characteristic is obvious, and the echo energy is used to judge whether the adhesive surface of the metal and the nonmetal layer is free from sticking[2,9].

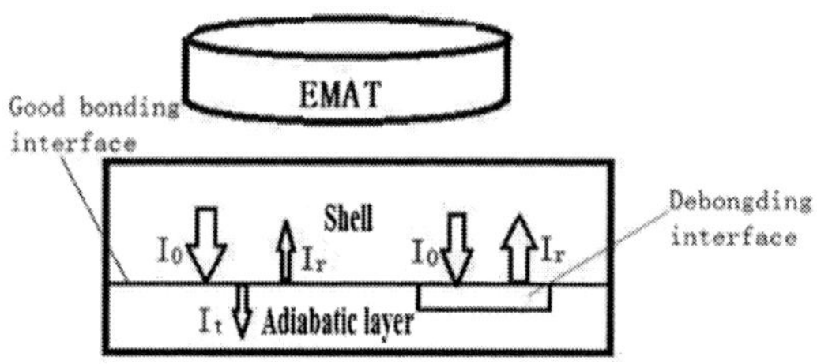

Figure 2. The propagation of sound waves in different interfaces.

The main information source of the deboning judgment is the received electromagnetic echo signal, which carries the characteristic information of the deboning. As shown in Figure 3, for a typical off stick echo signal.

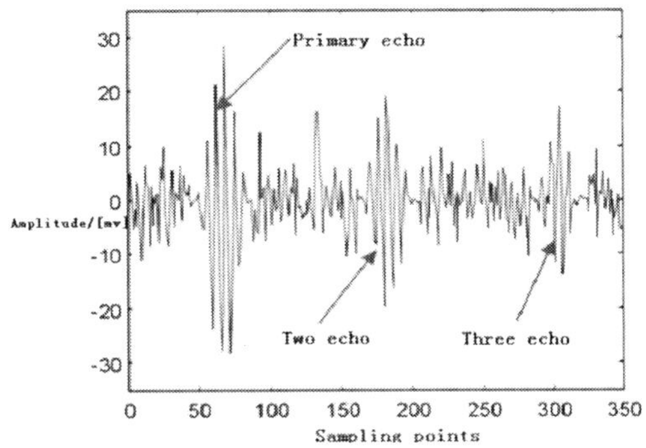

Figure 3. Typical echo signal.

It can be seen that the first echo is very obvious, but second times and third times after the echo signal is not so obvious, disturbed by the noise signal. How to deal with the signal to obtain the ideal result is also the key to detect the first interface bonding. In this paper, we select wavelet threshold to process the echo signal. The application of wavelet analysis of acoustic emission reduction technology, accurate and fast noise reduction, extraction of useful information. Select Sym8 wavelet, as shown in Figure 4.

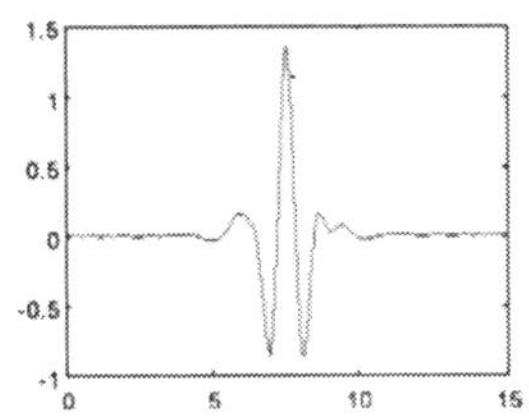

Figure 4. Sym8 wavelet function waveform.

Wavelet noise reduction of the basic processes are as follows figure 5:

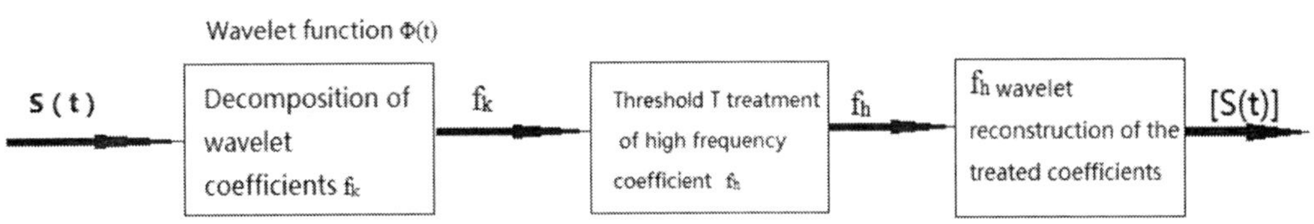

Figure 5. Wavelet noise reduction of the basic processes.

Sym8 wavelet function is used to deal with the original signal, and the result is shown in Figure 6:

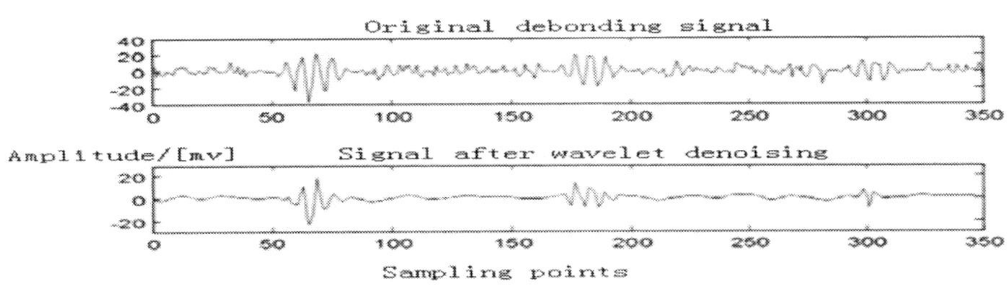

Figure 6. Sym8 wavelet processing results.

It can be seen that the Sym8 wavelet is used to deal with the signal relatively smooth, the feature information is retained more, and it is easier to identify the characteristics of deboning.

3. Test case

The detection of bonding between metal and composite material on the inner wall of rocket shell is taken as an example. We are in the bonding quality good place and artificial deboning place detection. The detected echo signal is displayed on the oscilloscope after wavelet threshold processing. Good bonding area detection and Detection of bad adhesion of the rocket shell. In the detection of bad adhesion of the rocket shell at the echo, as shown in Figure 7 received signal, we can see a number of equidistant and decreasing amplitude of echoes, through two adjacent echo time difference of rocket shell thickness can be calculated from viscosity of about 4mm. The echo signal received by the good bonding area is shown in figure 8, the signal is similar to the bad adhesion, and only all signals appear to have an overall attenuation. In contrast to the first echo amplitude of the two detection signal, the peak attenuation of the first peak value of the non-sticking point is found to be about 30%.This is because the adhesive material of the non-sticking point has a certain absorption attenuation of the ultrasonic wave.

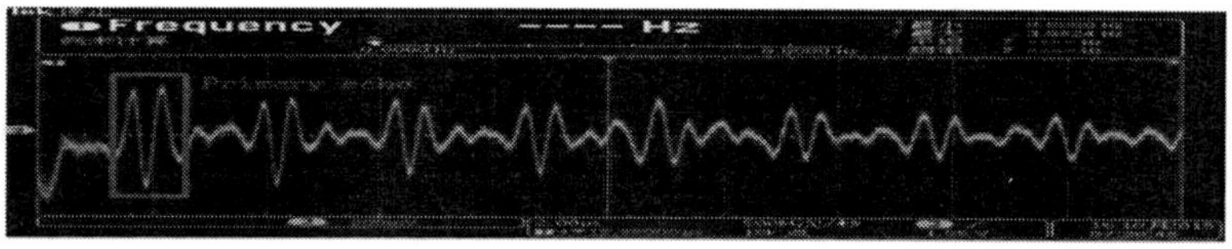

Figure 7. The echo signal received by the deboning area (An echo V pp is about 1040[mv])

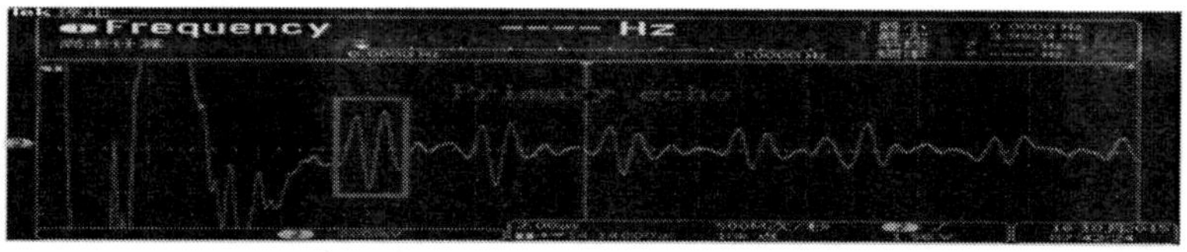

Figure 8. The echo signal received by the good bonding area (An echo V pp is about 760[mv])

Because the electromagnetic ultrasound does not need coupling agent, we directly use electromagnetic ultrasonic probe to detect the shell. The whole process is short, detection results directly in the oscilloscope display, the signal is obvious, very easy to distinguish. Electromagnetic ultrasonic can be used to measure the deboning of the first interface of solid rocket.

4. Conclusion

The adhesion of a typical metal and composite material, i.e., the solid rocket motor case and the bonding structure of the thermal insulation layer, was measured. The first peak of the signal at the deboning site was reduced by 30%. Can rapidly and accurately judge the position of sticking out. Electromagnetic ultrasound can be used to detect the bonding quality of metal and composite materials.

Acknowledgments

This work was financially supported by National Natural Science Foundation of China fund.

References

[1] Li Fang. Identification of cladding state of solid rocket motor,D. Beijing: Beijing University of Aeronautics and Astronautics, 2001.

[2] "national defense science and Technology Industrial NDT personnel qualification and certification training materials" Review Committee. Ultrasonic testing,Beijing: Mechanical Industry Press,2005,pp.3-8.

[3] Huang Songling, Wang Shen, Zhao Wei. The guided wave theory and application,M.Beijing: Tsinghua University press,2013,pp.1-2.
[4] Lei Kang, Steve Dixon, Kaican Wang, Jingmin Dai. Enhancement of signal amplitude of surface wave EMATs based on 3-D simulation analysis and orthogonal test method,J. NDT&E International.59(2013)11-17.
[5] Cao Jianhai, Yan Gong Biao, Han Xiaofeng. Principle and application of electromagnetic ultrasonic thickness measurement: a new ultrasonic thickness measurement.2002,36 Journal of Zhejiang University (Engineering Edition),J. 36(2002) 88-91.
[6] S.E. Burrows, Y. Fan, S. Dixon. High temperature thickness measurements of stainless steel and low carbon steel using electromagnetic acoustic transducers,J. NDT&E International.68(2014)73-77.
[7] Lei Huaming. Research on electromagnetic ultrasonic transducer mechanism and its application in pipeline inspection on. PhD thesis. Shanghai: Shanghai Jiao Tong University(2005)
[8] Shujuan Wang, Zhichao Li, Pengzhan Li, Xin Liu, Guofu Zhai. Numerical and experimental evaluation of the receiving performance of meander-line coil EMATs,J.Nondestructive Testing and Evaluation.29(2014)269-282.

IOP Conf. Series: Materials Science and Engineering **207** (2017) 012027 doi:10.1088/1757-899X/207/1/012027

Study on Synthesis of Thoreau-modified 3, 5-Dimethyl-Thioltoluenediamine Used as Epoxy Resin Curing Agent and Its Performance

Yongli Peng [1], Wenzheng Xiao [2,*]

[1]College of Materials, Wuhan Institute of Technology, Wuhan, China
[2]College of Materials, Wuhan Institute of Technology, Wuhan, China

*Corresponding author:308441213@qq.com

Abstract. A novel curing agent Thoreau modified 3, 5-Dimethyl-thioltoluenediamine was synthesized and its molecular structure was characterized by FTIR and DSC. The curing kinetics of a high toughness and low volume shrinkage ratio epoxy system (modified DMTDA/DGEBA) was studied by differential scanning calorimetry (DSC) under noni so thermal conditions. The data were fitted to an order model and autocatalytic model respectively. The results indicate that in order model deviates significantly from experimental data. Malik's method was used to prove that the curing kinetics of the system concerned follow single-step autocatalytic model, and a "single-point model-free "approach was employed to calculate meaningful kinetic parameters. The DSC curves derived from autocatalytic model gave satisfactory agreement with that of experiment in the range 5K/min~25K/min. As the heating rate increased, the predicted DSC curves deviated from experimental curves, and the total exothermic enthalpy declined owing to the transition of competition relationship between kinetics control and diffusion control.

1. Introduction

Epoxy resin is one of the most important thermosetting resins. The cured epoxy resins are of high crosslinked network that often exhibit high tensile strength and modulus [1], excellent chemical and corrosion resistance [2], and good thermal stability [3]. However, the ordinary epoxy resin cannot crosslink alone without curing agents or accelerant. In order to convert the resins to practical materials, it is necessary to add a curing agent. Most of the epoxy curing agents in common use are ploy functional amines [4, 5], tertiary amines, anhydrides and other latent curing agents.

Now with theorem can significantly reduce the curing temperature of epoxy resin, to synthesis a novel curing agent by theorem-polyamine condensation reaction, the study using theorem modified liquid aromatic amine, the expected good mechanical properties of the resin system.

In this paper, 3, 5-Dimethyl-thioltoluenediamine (DMTDA) was used to replace the solid aromatic amines, and the amount of active hydrogen in theorem was more than that of solid urea and polyamine. , After the reaction can improve the reactivity of DMTDA, make it better and more fully with the epoxy resin reaction. The final preparation of low toxicity, faster curing, and epoxy resin easy to mix evenly epoxy curing agent.

IOP Conf. Series: Materials Science and Engineering **207** (2017) 012027 doi:10.1088/1757-899X/207/1/012027

The thermodynamic curves of the modified DMTDA / epoxy resin curing system were obtained by non-isothermal DSC scanning [6, 7]. The kinetic parameters of the modified aromatic amine-epoxy resin were obtained by using the n-level model and the autocatalytic model respectively. Contrast to determine its kinetic equation. Finally, the curing process parameters of the resin system were obtained by the measured data and extrapolation method.

2. Experimental Section

2.1. The *Synthesis of* Theorem-*modified 3, 5-Dimethyl-thioltoluenediamine*

To a 100 ml clean three-necked flask equipped with an exhaust gas treatment unit were 0.15mol theorem and 0.6mol DMTDA which were added and nitrogen was introduced. Oil bath heating system slowly increased to 135℃ temperature, the speed of stirring, constant temperature reaction for 8h, then stop heating and stirring, cooling to room temperature, after pumping the liquid obtained by adding slowly into the beaker containing toluene solution, After stirring, into the separator funnel, to be placed after the separation, the upper layer of liquid in the beaker, after the rotary evaporation, the dark yellow oily liquid. Theorem modified polyamine reaction principle shown in Figure 1 ($n \leq 3$):

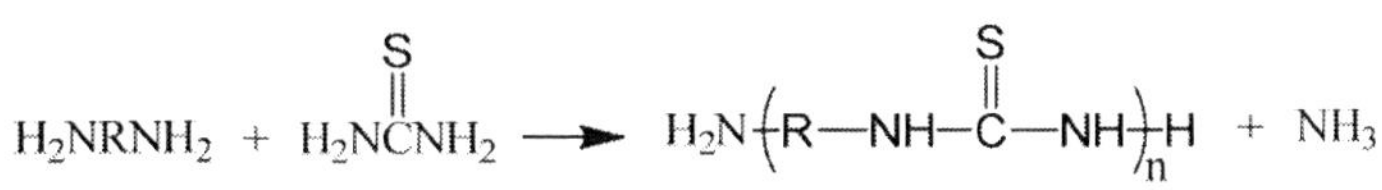

Figure.1 The synthesis reaction equation of theorem modified polyamine

2.2 Study on Curing Kinetics of Modified DMTDA / Epoxy Resin System

2.2.1 Determination of Curing Parameters of Modified DMTDA / Epoxy Resin by DSC The prepared modified DMTDA/epoxy resin(DGEBA) system was vacuum degassed at 40℃ and weighed 3~5mg for DSC analysis. The temperature was measured by 5K / min, 10K / min, 15K / min and 20K / min respectively.

The curing reaction heat (ΔH) was determined by integrating the area between the DSC curve and the baseline in units of kJ/mol.

3. Result and Discussion

3.1 FT-IR Analysis of Modified DMTDA

Figure 2 shows the FT-IR characterization of the product after separation and purification of theorem modified DMTDA at 130 ℃, 7h, nDMTDA: nthiourea = 4: 1.

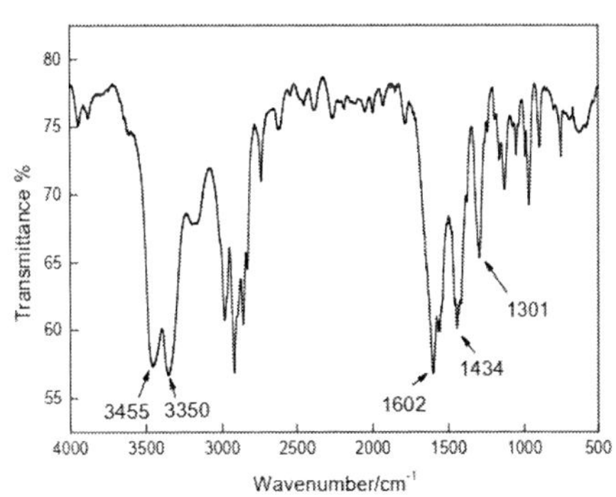

Figure.2 The infrared figure of theorem modified 3, 5-Dimethyl-thioltoluenediamine

The peaks at 3455 cm-1, 3350 cm-1 and 1602 cm-1 can be confirmed as the stretching and bending vibration peaks of -NH2-, and the presence of primary amino groups ensures that the substance remains as an epoxy resin curing agent use. 1301 cm-1 is the stretching vibration peak of -C=S- in the -CS-N- unit, and 1434 cm-1 is the stretching vibration peak of -CN- in the -CS-N- unit, which indicates that the theorem modified DMTDA.

3.2 Study on Curing Kinetics of Modified DMTDA / Epoxy Resin System

3.2.1 Study on Curing Kinetics of Resin System Based on n - level Dynamic Model Most of the kinetics of resin curing reaction using the n-level dynamic model in the image model:

$$\frac{d\alpha}{dt} = k_0(1-\alpha)^n \tag{1}$$

Where α is the curing degree; n is the reaction order; k0 is the reaction rate constant which related by the temperature.

The kinetic parameters of the curing reaction, such as the activation energy and the reaction order, play an important role in understanding the curing reaction kinetics. In general, the Kissinger and Crane equations can be used to process the DSC data obtained at different heating rates. The kinetic parameters such as Ea. n, A and so on can be obtained, and the n-level kinetic model equation of the curing reaction can be determined.

Kissinger equation:

$$\ln(\frac{\beta}{T^2}) = \ln(\frac{RT}{Eg(\alpha)}) - \frac{E}{RT} \tag{2}$$

Where β is the heating rate; Tp is the peak temperature; Ea is the apparent activation energy.

Fig. 3 shows the DSC curves at different heating rates. Table 1 shows the exothermic enthalpy ΔH, the initial temperature Ti, the peak temperature Tp and the termination temperature Tf at different heating rates.

At different heating rates, the relationship between ln (β/T2) and (1/Tp) is shown in Fig. Linear fit can be obtained slope, that is, -Ea/R intercept is ln (AR/Ea).

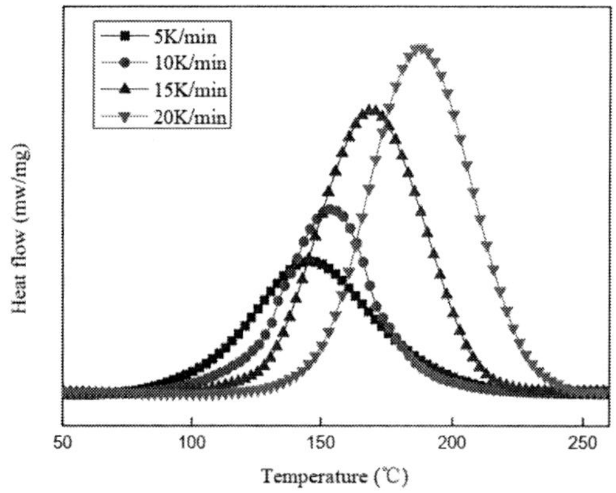

Figure.3 DSC scans of modified DMTDA/DGEBA at different heating rates

Table 1 characteristic parameters of the different reaction systems

$\beta/(K\cdot min^{-1})$	T_i/℃	T_p/℃	T_f/℃	$\Delta H/(J\cdot g^{-1})$
5	95.6	145.4	202	20.72
10	103.5	153.8	208.3	21.47
15	117.3	167.3	217.6	38.15
20	142.4	187.7	232.5	46.40

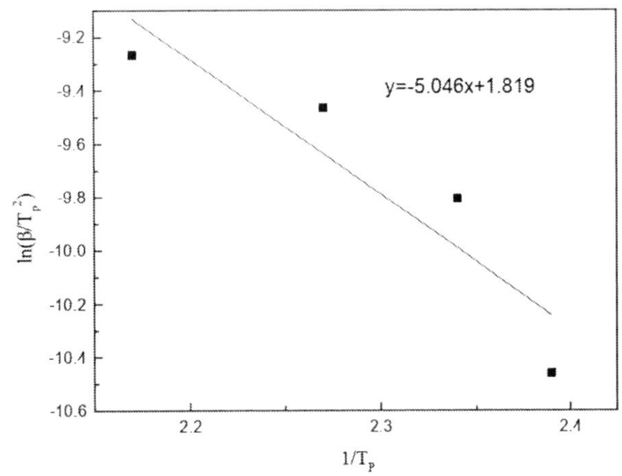

Figure.4 the Kissinger equation of DEGBA/modified DMTDA

From Fig. 4, we can get the equation y =-5.046x+1.819, and the apparent activation energy of DGEAB/modified DMTDA system is Ea = 41.9KJ/mol according to the slope.

Crane equation:

$$\frac{d(\ln\beta)}{d(1/T_p)} = -\left(\frac{E_a}{nR} + 2T_p\right) \tag{3}$$

When $\frac{E_a}{nR} \geq 2T_p$, the equation can convert to $\frac{d(\ln\beta)}{d(1/T_p)} = -\frac{E_a}{nR}$ .At different heating rates, the relationship between lnβ and 1 / Tp is shown in Fig.5, and the linear slope is linear.

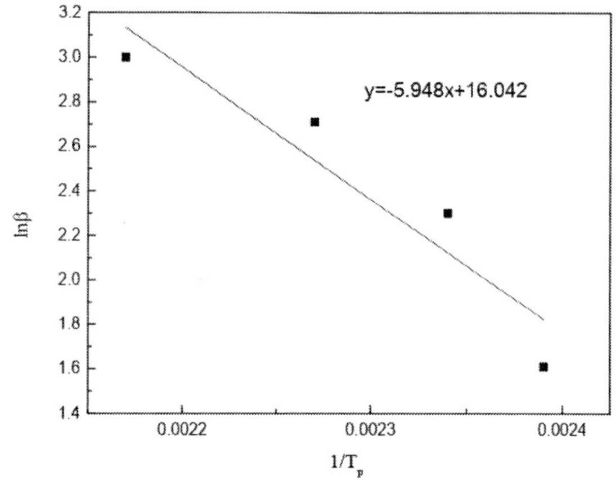

Figure.5 the Crane equation of DEGBA/modified DMTDA

From Fig. 5, we can get that the straight line equation is y = -5.948x + 16.042, and the reaction order of DGEAB/modified DMTDA system is 0.84 according to the slope.

3.2.2Study on Curing Kinetics of Resin System by Autocatalytic Model The autocatalytic model is generally used to study isothermal curing kinetics due to its simultaneous calculation of the complexity of multiple parameters:

In order to further study the curing behavior of DGEBA/modified DMTDA system, the self-catalyzed model is used to determine the appropriate reaction kinetic function. In this paper, that is based on the definition of function $y(\alpha)$ and $z(\alpha)$ to determine $f(\alpha)$.

$$y(\alpha) = (\frac{d\alpha}{dt})\exp(\chi) \tag{4}$$

$$z(\alpha) = \pi(\chi)(\frac{d\alpha}{dt})\frac{T}{\beta} \tag{5}$$

In the formulas (3-10) and (3-11), χ express E_a / RT it is said that β the heating rate $\pi(\chi)$ is the temperature integral, and the four-level approximation of Senum-Yang is used to represent the temperature integral.

$$\pi(\chi) \approx \frac{\chi^3 + 18\chi^2 + 88\chi + 96}{\chi^4 + 20\chi^3 + 120\chi^2 + 240\chi + 120} \tag{6}$$

Figure 3-6 is the normalization of DGEAB / modified DMTDA system $y(\alpha)$, $z(\alpha)$ and the reaction rate of the reaction with the conversion rate of the relationship between the heating rate of

10K / min-1, $y(\alpha)$, $z(\alpha)$ and curve of the peak Respectively, αp, αM, αp∞, detailed parameters as shown in the table.

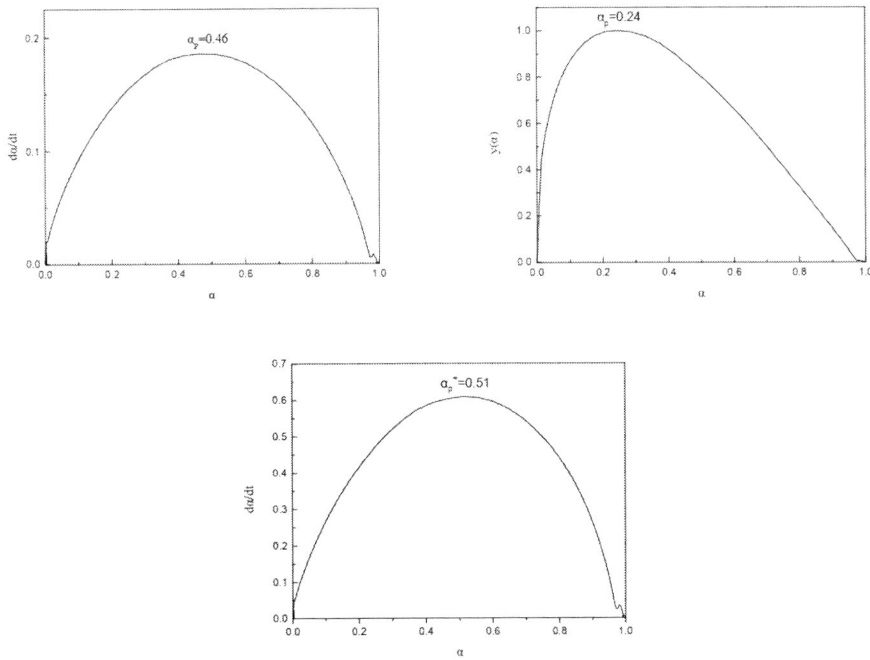

Figure.6 Variation of experimental reaction ($d\alpha / dt$), normalized $y(\alpha)$, and $z(\alpha)$ versus fractional conversion for DGEBA/modified DMTDA (10K·min-1)

Table 2 Peak values of $d\alpha / dt$, $y(\alpha)$, and $z(\alpha)$, along with calculated kinetic parameters

Heating rate/(K·min^{-1})	α_p	α_M	α_p^{∞}	n	m	lnA
5	0.47	0.20	0.51	1.57	1.39	11.52
10	0.46	0.21	0.51	1.50	1.28	12.01
15	0.47	0.27	0.52	1.17	1.04	11.69
20	0.51	0.29	0.49	1.22	1.27	11.68
Average value				1.37	1.25	11.73

It can be seen from Table 2 that 0 <αM <αp∞, and αp∞ ≠ 0.632, according to the shape, the eigenvalue determines the kinetic mechanism function closest to the SB (m, n) model.

$$\frac{d\alpha}{dt} = Ae^{-\chi}\alpha^{m}(1-\alpha)^{n} \tag{7}$$

Where m, n is the reaction order.
It can convert to,

$$\ln[(\frac{d\alpha}{dt})\exp(\chi)] = \ln A + n\ln[\alpha^{m/n}(1-\alpha)] \tag{8}$$

Where m/n is equal to αM / (1 - αM), so that in the range of 0.2 ≤ α ≤ 0.9, $\ln[(\frac{d\alpha}{dt})\exp(\chi)]$ is plotted with $\ln A + n\ln[\alpha^{m/n}(1-\alpha)]$, and the slope is n, so that the m value can be calculated and the straight line Intercept is lnA, the results shown in Table 2, Figure 7 is when the heating rate of 5K · min-1, the system $\ln[(\frac{d\alpha}{dt})\exp(\chi)]$ with $\ln A + n\ln[\alpha^{m/n}(1-\alpha)]$ changes.

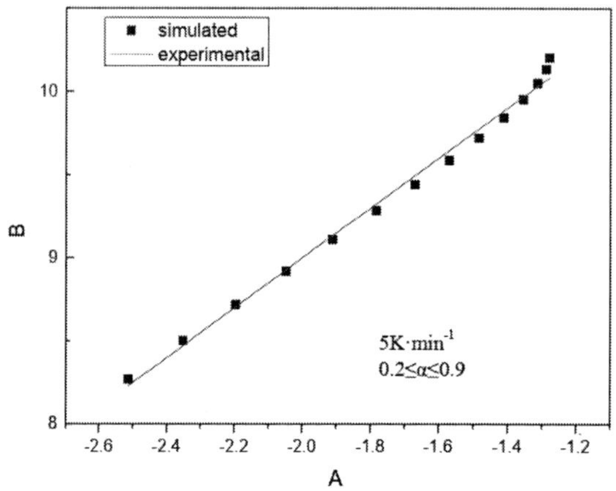

Figure. 7 variation of $\ln[(\frac{d\alpha}{dt})\exp(\chi)]$ versus $\ln A + n\ln[\alpha^{m/n}(1-\alpha)]$ for DGEBA/modified DMTDA (5K·min-1)

The results of DEGBA / modified DMTDA were as follows:

$$\frac{d\alpha}{dt} = \beta\frac{d\alpha}{dT} = 1.24\times10^{6}\exp\ (\frac{-41900}{RT})\ \alpha^{1.25}(1-\alpha)^{1.37}$$

The initial conditions were as follows: T = 0 ℃, the curing degree α0, and the calculation results shown in Figure 8.

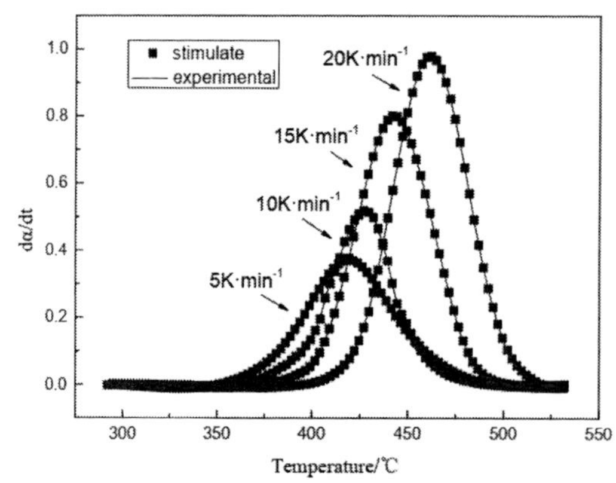

Fiugre.8 comparison rates of experimental and predicated for DGEBA/modified DMTDA from SB (m, n) model

Figure 8 shows the experimental curve and model calculation curve at different heating rates. As shown in the figure, the model curve agrees well with the experimental curve. However, with the increase of the heating rate, the experimental curve gradually deviates from the model calculation curve. This is because the curing rate of the epoxy resin system depends on the activity and activity of the functional group. The reaction rate between the reaction groups in the initial reaction is much greater than the rate of intermolecular collision, and the curing reaction is controlled by the chemical activity of the reactive groups. With the progress of the reaction, the molecular chain is growing until the system reaches the gel. The crosslinking degree of the further polymerization system is increasing, and the movement of the active group is more and more restricted, and the curing reaction is changed to the diffusion control. With the increase of the heating rate, the initial temperature of the curing reaction increased, the thermal motion of the molecule increased, the time of gelation was gradually shortened, and the influence of the diffusion control on the system became more and more obvious.

In general, the model can be used to describe the curing behavior of modified DMTDA / epoxy resin system in the temperature range of 5K / min ~ 25K / min. This curing model can be used to predict the time required for the system to reach a certain degree of cure at a certain temperature, or to achieve the desired temperature for a predetermined period of time.

4. Conclusion

Increasing the reactivity of DMTDA by polyamine-theorem condensation reaction, effectively reducing its curing temperature as epoxy resin curing agent, determining the product by FT-IR characterization, which provides the modification of DMTDA A new idea. The self-catalyzed model can describe the curing behavior of the modified DMTDA / epoxy resin system, and the model of the ϕ-T curve is calculated by the non-isothermal kinetics. The self-catalyzed model can be used to describe the curing behavior of the modified DMTDA / Epoxy resin system. Values and experimental values are basically consistent. Which provides a theoretical basis for the application of the system as a high-performance resin material and optimizing the curing process.

Acknowledgments

Project Fund: Supported by the National Natural Science Foundation of China (21376182)
Author: Xiao Wenzheng, Master graduate. E-mail: 308441213@qq.com
Corresponding author: Pengyong Li, Ph.D., professor E-mail: PYL68@163.com

References

[1]. S.L.Kim,M.D. Skibo,J.A. Manson, R.W.Hertzberg,J. Janiszewski, Tensile, impact and fatigue behavior of an amine-cured expoy resin, Polym. Eng. Sci. 18(14)(1978)1093-1100

[2]. D. Paul Schweinsberg. G.A. George, A.K. Nanayakkara, D.A. Steinert, The protective action of epoxy resins and curing agents-inhbitive effects on the aqueous acid corrosion of iron and steel, Corros. Sci. 28(1)(1988)33-42

[3]. M. Shau, T. Wang, Syntheses, structure, reactivity, and thermal properties of new cyclic phosphine oxide epoxy resins cured by diamines, J. Polym .Sci. Part A: Polym. Chem. 34 (3) (1996) 387–396

[4]. J. Wan, Z. Bu, C. Xu, B. Li, H. Fan, Preparation, curing kinetics, and properties of a novel low-volatile starlike aliphatic-polyamine curing agent for epoxy resins, Chem. Eng. J. 171 (1) (2011) 357–367.

[5]. J.M. Laza, C.A. Julian, E. Larrier, M. Rodriguez, L.M. Leon, Thermal scanning rheometer analysis of curing kinetic of an epoxy resin: 2. An amine as curing agent, Polymer 40 (1) (1999) 35–45.

[6]. Gan Li, Tian Jinli. Preparation and curing kinetics of hydration epoxy/DDS system [J]. Polymer Materials Sciences and Engineering, 2011, 27(9): 161-163

[7]. Zhou Hongjun, Yin Guoqiang, Lin Xuan. Curing kinetics of epoxy/activated nano-Al2O3 composites [J]. CIESC Journal. 2011,62(5): 1749-1755

The synthesis and characterization of Bis- (biguanide-p-phenyl) methane

Yongli Peng[a], Zhenglin Pi[*]

School of Materials Science and Engineering, Wuhan Institute of Technology, Wuhan, China

[*]Corresponding author e-mail: 329201036@qq.com, [a]PYL68@163.com

Abstract. To decrease the curing temperature of epoxy resin with dicyandiamide, a new kind of modified dicyandiamide curing agent was prepared with daimio diphenyl methane and dicyandiamide as raw materials, distilled water as solvent and hydrochloric acid as a catalyst. The optimum preparation technology of bis-(biguanide-p-phenyl) methane was studied by orthogonal experiments of three factors and three levels. The results show that the yield of bis-(biguanide-p-phenyl) methane reaches to 85.12% at reaction time of 4h, reaction temperature of 120 oC, mole ratio of 1:2.15. The structures of the compounds were analysed and confirmed by the Infrared spectroscopy and Hydrogen nuclear magnetic spectroscopy. The product is pure without by-products by this way and the modified dicyandiamide had higher reactivity than unmodified dicyandiamide.

1. Introduction

Dicyandiamide is one of the epoxy resin latent curing agents, it is widely used and has many excellent physical and chemical properties. But dicyandiamide is poor compatibility with epoxy resin, when it is used as an epoxy resin curing agent alone, the curing temperature reaches 160 to 180℃ which limit the development and application of dicyandiamide as latent curing agent for epoxy resin. So it has become a research hotspot if the curing temperature of dicyandiamide can be reduced to low temperature and it would not change its latent and the use of performance. At present, there have been a large number of studies to reduce the curing temperature of dicyandiamide, two methods are usually used. One of these methods is adding accelerator to the dicyandiamide, but this method is easy to damage the use of dicyandiamide performance, another method is Chemical modification of dicyandiamide.

The strong absorption of the cyano group reduces the electron cloud density of the amino group in the dicyandiamide molecules is the reason which will make curing temperature of dicyandiamide is too high. So it can reduces the curing temperature and maintaining its latent by using the method of Chemical modification of dicyandiamide.

2. Experimental

2.1 Reagents and Instruments

Dicyandiamide chemical pure, Sino pharm Group; diamine diphenyl methane, analytical grade, Sino pharm Group; hydrochloric acid (analytical grade), Xinyang chemical Reagent Factory.

DZF-6250 vacuum oven: Suzhou Sanqing Instrument Co., Ltd.; ZKYY intelligent oil bath pot: Gongyi City to China Instrument Co., Ltd.; TX5 micro-melting point detector,Hebei Deco Machinery Technology Co., Ltd.; Nicolet700 Fourier Infrared Spectrometer,American Thermos Fisher Scientific, Potassium Bromide. Agilent 400MR Nuclear Magnetic Resonance Spectrometer, Agilent Technologies Co., Ltd., Nuclear Magnetic Spectroscopy, Solvent DMSO.

2.2 Synthetic Principle of Bis-(biguanide-p-phenyl) methanee

The synthesis route of Bis-(biguanide-p-phenyl) methane is shown in Fig. 1, and the diamine dibenzylmethane and dicyandiamide in the molar ratio of 1: 2.15 are respectively mixed, and the amount of diaminodiphenylmethane Hydrochloric acid as the catalyst, and add the appropriate amount of water as solvent, through the argon protection, open the reflux condenser, stirring temperature and temperature control at 120 ℃ under the conditions of reaction for about 4 h after the reaction of the product after pumping water washing 3~4 times, after purification in vacuum drying to obtain a light yellow high purity Bis-(biguanide p-phenyl) methane, calculate the yield, measuring the melting point.

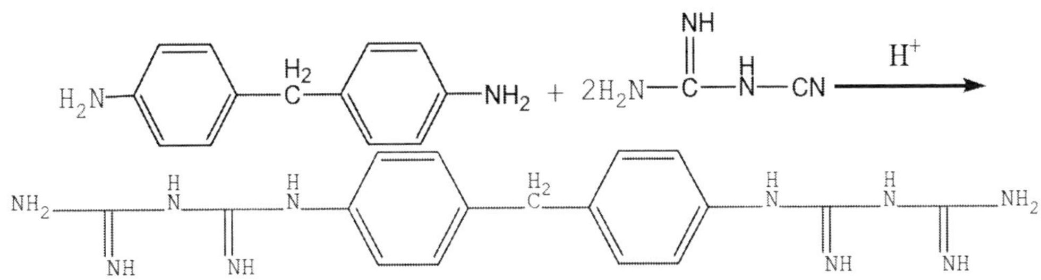

Fig. 1 Synthesis route for the bis- (biguanide-p-phenyl) methane

3. Analysis and Discussion

3.1 Discussion on Influencing Factors of Synthesis Reaction

Orthogonal experiment was used to optimize the yield of Bis-(biguanidino-p-phenyl) methane under different reaction times, reaction temperature and reactant ratio (A), reaction temperature (B) and feed mass ratio (C) were selected as the influencing factors. The orthogonal experiment was carried out by using L_9 (3^3) and the influencing factors were studied under the condition of average temperature of 25℃ ~ 30℃. The level is shown in Table 1.

Tab. 1 Factor level chart of L9 (33) orthogonal experiment

NO.	reaction time/h	reaction temperature/℃	n(DDM):n(Dicy)
1	3	110	1:2
2	4	120	1:2.15
3	5	130	1:2.3

According to Table 1, orthogonal test table L_9 (3^3) was selected and the yield and the melting range of the product of bis-(biguanidino-p-phenyl) methane were measured.

Tab. 2 Result of the L9 (33) orthogonal experiment

No.	reaction temperature/℃	reaction time/h	n(DDM):n(Dicy)	yield/%	melting range /℃
1	110	3	1:2.0	71.3	112.3~113.5
2	110	4	1:2.15	78.7	112.0~113.1
3	110	5	1:2.3	69.6	111.8~112.6
4	120	3	1:2.3	81.8	111.6~112.2
5	120	4	1:2.15	84.4	111.4~111.6
6	120	5	1:2.0	80.2	111.8~112.3
7	130	3	1:2.15	79.9	111.5~112.2
8	130	4	1:2.3	80.5	111.6~112.6
9	130	5	1:2.0	75.4	112.4~113.3
K1	219.6	233.0	231.1		
K2	246.4	243.6	238.8		
K3	235.8	225.2	231.9		
k1	73.2	77.7	77.0		
k2	82.1	81.2	79.6		
k3	78.6	75.1	77.3		
R	8.9	6.1	2.6		
Primary and secondary order		A>B>C			
Excellent level	A2	B2	C2		
Excellent combination	A2B2C2				

It can be seen from tab 2 that the reaction time and the reaction temperature are the main factors, and the optimal level is A_2B_2 according to the average value, otherwise, the feed ratio is the secondary factor, Select range between 1: 2.0~1: 2.3. From the above orthogonal table shows that the highest product yield is the fifth group of experiments. According to the theory of optimal synthesis conditions, the product yield is 85.12% when exaction time is 4 h, the reaction temperature is 120℃, the feed ratio is 1: 2.15.Significantly higher than the experimental yield of group 5.Therefore, the optimum synthesis process of bis-(biguanidino - p-phenyl) methane is reaction time 4 h, reaction temperature 120 ℃, catalyst dosage 4%, feed ratio 1: 2.15.

3.2 FT-IR Analysis of bis- (biguanide-p-phenyl) methane

Figure 2 is the infrared spectrum of the reaction product of diaminodiphenylmethane and dicyandiamide. By comparing the standard IR images of dicyandiamide and the bis-(bi guanidine-p-phenyl) methane, it show that Dicyandiamide has a characteristic absorption peak of cyano (-C = N) at 2214 ~ 2168 cm^{-1},but bis-(biguanidino-p-phenyl) methane does not have any absorption peaks in this wavelength range, at 1624~1511 cm^{-1} there was a significant absorption peak (-C = N).This indicates that when the diaminodiphenylmethane reacts with dicyandiamide, the cyano group (-C = N) is consumed and reacted completely to produce -C=N.The bis-(biguanidino-p-phenyl) methane molecule contains a C_{aryl}-N bond, so that its absorption peak at 1383 cm^{-1} has its stretching vibration. At the same time, the characteristic absorption peaks of benzene ring to substitute at 809 cm^{-1} were obtained. From the above characteristics of the absorption peak can be seen that the reaction product is Bis-(biguanide-p-phenyl) methane.

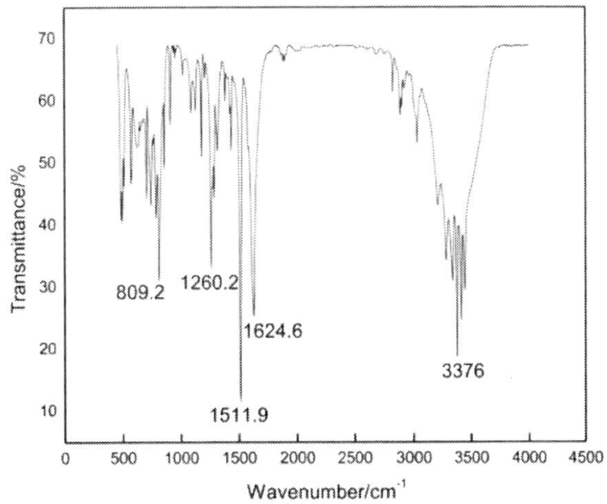

Fig. 2 The infrared figure of bis- (biguanide-p-phenyl) methane

3.3 Analysis of Nuclear Magnetic Spectrometry of bis- (biguanide-p-phenyl) methane

Figure 3 is a nuclear magnetic resonance spectrum of the bis-(biguanidino-p-phenyl) methane molecule using deuterated dimethyl sulfoxide (DMSO) as shown in Table 3. The integrated area ratios of the absorption peaks 1 to 7 were the number of H in the corresponding position of the bis-(bisguanidino-p-phenyl) methane molecule is consistent with that of bis-(bisguanidino-p-phenyl) methane.

Tab. 3 the characteristic parameters of bis- (biguanide-p-phenyl) methane in 1H NMR

No.	chemical shift	number of H	peak area
1	6.46	2	1.84
2	4.80	1	0.92
3	5.14	1	0.93
4	8.77	1	0.91
5	6.80	1	0.86
6	7.48	1	0.85
7	3.35	2	1.86

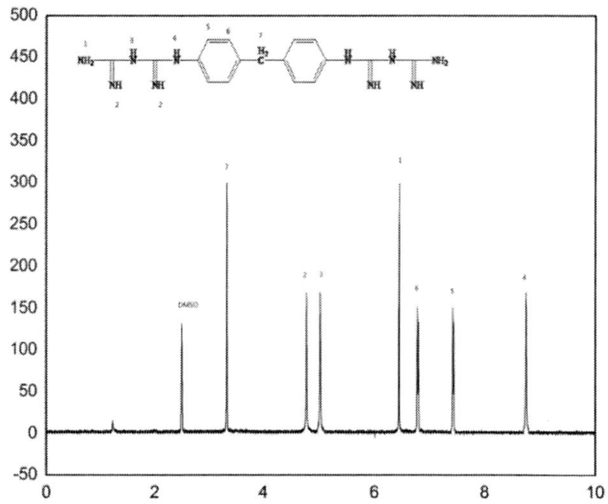

Fig. 3 1H NMR spectrum of bis-(biguanide-p-phenyl) methane

4. Conclusion

(1)The optimum synthesis process of Bis- (biguanide-p-phenyl) methane was obtained by orthogonal experiment L_9 (3^3).The ratio of diamine dibenzylamine: dicyandiamide: concentrated hydrochloric acid is 2.15:1: 2.15; reaction temperature is 120 ℃; reaction time is 4 h, the yield is 85.12%.

(2)The FTIR of the obtained product was observed at 1624~1511 cm^{-1} in the spectrum, and the absorption peak (-C = N) and the bis-(biguanide p-phenyl) methane appeared at 809 cm^{-1} ^{1}H-NMR (400 MHz, DMSO) δ 6.46 (s, 2H), 4.80 (s, 1H), 5.14 (s, 1H), ^{1}H-NMR (400 MHz, DMSO)), 8.77 (s, 1H), 6.80 (d, J = 8.7 Hz, 1H), 7.48 (d, J = 8.9 Hz, 1H), 3.35 (s, 2H). It was verified that the final product was Bis- (biguanidino-p-phenyl) methane, indicating that the synthetic route of this material was feasible.

Acknowledgements
Project Fund: Supported by the National Natural Science Foundation of China (21376182)
Author: Yongli Peng, Ph.D., professor. E-mail: PYL68@163.com
Corresponding author: Zhenglin Pi, Master graduate, E-mail: 329201036@qq.com

References

[1] PAN J P, FAN H P, LI Z L. The progress in research on the moderate temperature latent curing agent for epoxy adhesives [J]. Chemistry and Adhesion, 2010, 32(6): 45-49+54.
[2] SONG Y T, DANG M Y, LI S. Study on synthesis and properties of dicyandiamide curing agent modified by aniline[J]. Chemistry and Adhesion, 2012, 34(1): 4-6+19.
[3] WANG Y C, LIN S D, TU Y Y, et al. Synthesis and properties of a novel bis- phenylbiguanide as latent-curing agent for epoxy resin [J]. Fine Chemicals, 2016, 33(2): 224-230.
[4] TANG R, YANG H W, ZHOU J P, et al. Research on the behavior of dicyandiamide modified by bisphenol-a type phenolic aldehyde curing epoxy resin [J]. Surface Technology, 2014, 43(6): 144-148.
[5] FENG L B, WANG Y L, HAO X Z, et al. Study on behavior of modified dicyandiamide curing epoxy resin [J]. Materials Review, 2012(14): 59-62.
[6] LEI J Y, DU X D, ZHNG H L, et al. Study on latent curing agent of modified dicyandiamide for epoxy resin and its application [J]. Chemical Technology Market, 2006, 29(2): 36-40.
[7] CHENG X L, SUN F Y, BA S H. Synthesis and properties of dicyandiamide curing agent

modified with o-toluidine for epoxy resin[J]. Adhesion, 2014(12): 60-63.
[8] LI W, CHENG X L, ZHANG Z H. Preparation and performance of furfural modified dicyandiamide modified [J]. Adhesion, 2016(10): 42-45.
[9] CHENG X L, GUO X W, YANG Y L. Researches of benzaldehyde-modified dicyandiamide as curing agent for epoxy resin [J]. Adhesion, 2015(3): 43-46.
[10] BAO X H, WEI C Z, WAN Y D. Synthesis of 1-o-(tolyl) iguanid[J]. Coatings Technology & Abstracts, 2016, 37(5): 20-23.
[11] WANG Q, SHI T J, YU F. Synthesis and properties of novel latent curing agent [J]. Journal of Chemical Industry and Engineering, 2014, 65(9): 712-717.
[12] JIAO J, LAN L W, NING R C. A new epoxy resin system cured by modified dicyandiamide [J]. Journal of Northwestern Polytechnic University, 1999, 17(4): 619~623.
[13] JIAO J, LAN L W. Study on properties of epoxy/modified dicyandiamide/accelerant [J]. Fiber Reinforced Plastics/Composites, 1999(4): 14~17.
[14] WANG Q, SHI T J, ZHANG Y, et al. latent curing of polybiguanide/epoxy resin [J]. Journal of Chemical Industry and Engineering, 2015, 66(1): 464-470.
[15] CHEN L X, ZHANG H L, LEI J Y. Research progress on latent curing agent for epoxy resin [J]. New Chemical Materials, 2004, 32(7): 29-32.

IOP Conf. Series: Materials Science and Engineering **207** (2017) 012029 doi:10.1088/1757-899X/207/1/012029

Characterization of konjac glucomannan-gelatin IPN physical hydrogel scaffold

Xiliang Chen[*], Qinghua Chen, Tingting Yan and Jinkun Liu

School of Materials Science & Engineering, Kunming University of Science and Technology, Kunming, China

*Corresponding author e-mail: tsenhey@qq.com

Abstract. A novel IPN hydrogel scaffold is prepared by freeze-drying method, in which konjac galactomannan (KGM) and gelatin are physically crosslinked respectively. This scaffold is thermostable, and the structure of this scaffold is analysed by scanning electron microscope, Fourier transform infrared spectrum, and X-ray diffraction method. The FT-IR results show that hydrogen bonds are formed between KGM and gelatin molecules, which hinder the formation of their respective crosslinking. This is consistent with the XRD results that the crystallinity gets lower in the IPN gels compared with pure gelatin and KGM gels. The morphologies of freeze-dried hydrogels are observed by SEM and the mechanical properties of the scaffolds are tested to analyse the relationship between the structures and properties. Although this novel IPN hydrogel is physical gel, it shows rubber-like performance as chemical gels. And it is nontoxic, so it can be used as the scaffold for cartilage tissue engineering that embedded in human bodies.

1. Introduction

Gelatin is a kind of polypeptide derived from collagen hydrolysis, and is widely used in food, drug delivery, and tissue engineering [1]. When temperature decreases below 35 °C, coil-helix transition of gelatin molecules will happen in solution, which causes the sol-gel transition [2]. The structure of formed triple helices is identical with that of collagen [3]. Triple helices are formed by self-assembly of gelatin molecules due to the action of hydrogen bonds [4]. When temperature increases above 35 °C, triple helices will dissolve in water, and the gel transforms into the sol. The body temperature of human is above 35 °C. Hence, the physical gel of gelatin is unstable as a kind of scaffold embedded in human bodies. Therefore, chemical crosslinks are involved in gelatin gels to resist melting at body temperature. The most common crosslinking agent is glutaraldehyde [5]. But glutaraldehyde is toxic [6]. Hence, genipin becomes an alternative recently, which has low toxicity [7].

Many polymers are compounded with gelatin to form new hydrogels for the purpose of improving the hydrogel properties. The additive includes natural polymers and synthetic polymers. Chitosan, hyaluronic acid, alginate and agar are the commonly used natural polymers [8-11]. Synthetic polymers such as polyvinyl alcohol and poly (ethylene glycol) are also used commonly [12-14]. Konjac galactomannan (KGM) is a kind of polysaccharide derived from amorphophallus konjac plants [15]. When heated in alkaline solution, deacetylation will happen on KGM, and the KGM molecules are crosslinked simultaneously [16, 17]. These cross linkages are thermal stable, will not damage even heated [18]. KGM is non-toxic, and with good biocompatibility, so KGM hydrogels are used in food, pharmaceutical and biological tissue engineering [18].

In this study, gelatin and KGM are crosslinked respectively to form interpenetrating polymer network (IPN). The schematic is shown in Fig. 1. The physical properties and morphology of the hydrogels are studied by SEM, XRD, FT-IR and mechanical testing machine. The results show that hydrogen bondings are formed between KGM and gelatin, which hinder the formation of their own crosslinking. The structural properties also affect the mechanical properties of the hydrogels.

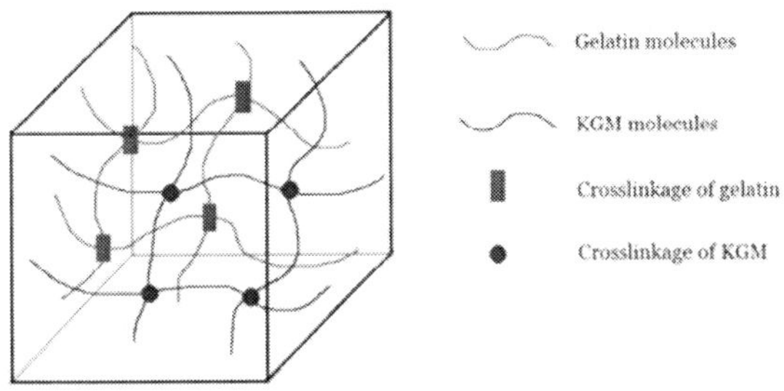

Figure 1. Schematic of prepared KGM-gelatin IPN hydrogel

2. Experimental

2.1 Materials

All the chemical reagents used were obtained from commercial sources in China. The gelatin was purchased from Sino pharm Chemical Reagent Co., Ltd, Shanghai, China. The KGM was purchased from Shiyan Huaxianzi Konjac Productions Co., LTD, Hubei, and China. The content of galactomannan is more than 95%.

2.2 Preparation of hydrogels

Three kinds of hydrogels were designed and prepared for the contrast. The ingredients of the hydrogels are: 2.0 g KGM; 1.0 g KGM and 1.0 g gelatin; 2.0 g gelatin. A certain amount of gelatin was soaked in 50 mL deionized water for at least 2 h, for making gelatin swell sufficiently. Then the water was heated to 45 °C and stirred simultaneously to make gelatin dissolved completely. After that, 0.1 mL ammonium hydroxide with the concentration of 25% was added to the solution, and then, the mixed solution was stirred evenly. KGM was added to the solution subsequently with severely stirring until the solution became homogeneous. Then the temperatures of the solutions were raised to 70 °C and kept for 24 h. After that, the samples were transferred to the fridge with the temperature of 4 °C and kept for 24 h, and then the hydrogels were completely prepared. The dry samples were prepared by freeze-drying method at the temperature of -40 °C.

2.3 Characterization

2.3.1 Morphology observations

The surface morphologies and internal structures of freeze-dried hydrogels were observed by a scanning electron microscope (SEM Quanta 200, FEI, USA).

The samples were freeze-dried at -40 °C to maintain the porous structure without any collapse.

The apertures of pores were measured by geometrical observation on screen.

2.3.2 FT-IR analysis

Fourier transform infrared spectrum experiments were performed on Bruker TENSOR27 infrared spectrometer (Germany), 4000~400 cm-1, Sweep frequency of 10 times, and Spectral resolution of 4 cm-1.

2.3.3 X-ray diffraction studies
Wide-angle X-ray diffraction patterns were recorded on an X-ray diffractometer (D/MAX-3B, Rigaku Denki, Japan) in the range of 2θ=5~90° with Cu kα radiation (λ=1.5406×10-10 m) at 40 kV and 30 mA.

2.3.4 Mechanical tests
Scaffolds were made to 10 mm × 10 mm × 10 mm cubes for mechanical tests. The tests were performed on HY-0230 computer controlled electronic universal testing machine (Shanghai HengYi Precision Instrument Co., Ltd., China).

3. Results and discussion

3.1 FT-IR analysis
In order to confirm the reactions among the functional groups, FT-IR spectra of gelatin, KGM, and composite of gelatin and KGM freeze-dried gels are shown in Fig. 2. The characteristic absorption bands of mannose in the KGM (curve a) appear at 876 and 810 cm^{-1}, and the peak at 895 cm^{-1} in the KGM corresponds to β-glucosidic bonds. These absorption peaks mentioned above also appear in the composite of gelatin and KGM (curve b). Hence, it can be concluded that the primary structures of KGM keep intact in the composite. Moreover, it can be seen from curves a and b that the peaks at around 2854 and 1740 cm^{-1} which represent acetyl groups do not appear. This is because the ammonia solution is weakly alkaline, which made deacetylation happened on KGM. The FT-IR spectrum of gelatin (curve c) showed strong absorption bands at 3330 cm-1 (N-H stretching), 1655 cm^{-1} (amide Ⅰ, C=O stretching), 1547 cm^{-1} (amide Ⅱ, N-H bending), and 1240 cm^{-1} (amide Ⅲ, C-N stretching). Theses mentioned absorption bands can also be found in curve b. Therefore, it can be concluded that the primary structures of gelatin keep intact in the composite gel.

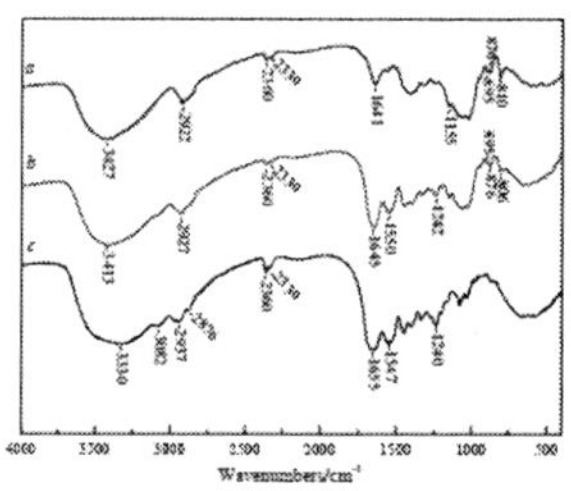

Figure 2. FT-IR spectra of KGM (a), composite of gelatin and KGM (b), and gelatin (c) freeze-dried gels

It can be seen from Fig. 2 that the location and intensity of some bands on curve a, b and c are different. The absorption band at 3427 cm^{-1} on curve a corresponds to O-H stretching vibration, but it appears at 3413 cm^{-1} on curve b. It shows significant red shift of O-H band in the composite gel compared with the KGM gel. The reason is analyzed as follows. KGM has a large number of hydroxyl groups, which can easily form hydrogen bonding with KGM molecules and gelatin molecules. There are a huge number of peptide bonds in gelatin molecules, and the six atoms which form the peptide bond are in one plane. Therein, C-N bond has the characteristics of double bond, and cannot rotate freely. The triple helices formed in gelatin gels are caused by hydrogen bonds, which occur among carbonyl groups (C=O) on the amino acid residues and amino groups (N-H) on the nearby amino acid residues. The hydrogen bonds have the properties of saturation and directivity, and the bond angels are around 180° in most cases. Hence, it can be concluded that the hydrogen bonds formed between KGM and gelatin molecules attribute to the action between the hydroxyl groups of KGM and carbonyl groups of gelatin. It can be seen from Fig. 2 that the absorption band of amide Ⅰ on curve b shifts to

1645 cm^{-1} compared with 1655 cm^{-1} on curve a, which also indicates that the carbonyl groups of gelatin interact with the hydroxyl groups of KGM.

3.2 X-ray diffraction studies
XRD patterns of KGM (curve a), composite of gelatin and KGM (curve b), and gelatin (curve c) gels are shown in Fig. 3. It can be seen in curve a that the crystalline peaks and dispersion peaks of amorphous phase stack up together. It has been found that the addition of ammonia to KGM solutions is required in order to remove the acetyl groups (deacetylate KGM) and induce the formation of a hydrated gel-forming presudo-fibrillar precipitate [17]. The gelation of the KGM molecules possibly occurred through the formation of a network structure supported by hydrogen bonds. Hence, it was considered the acetyl groups associated with KGM inhibited gel formation [17]. The presence of mild alkali in a KGM solution cleaved the acetyl groups and resulted in heat-stable elastic gels, which retained their structure under various conditions. The electron diffraction diagram of KGM gel illustrated the mannan Ⅱ-type crystallization [17].

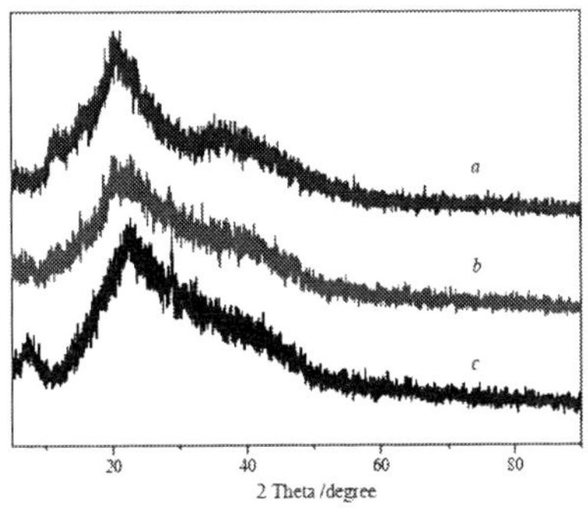

Figure 3. X-ray diffraction patterns of KGM (a), gelatin-KGM (b), and gelatin (c) freeze-dried gels

3.3 Mechanical tests
The elastic moduli were measured when the compressive deformation was 10%, which were respectively 13.336 kPa of KGM gels, 2.813 kPa of composite gels and 21.648 kPa of gelatin gels. Even the gelatin gels exhibit the highest elastic modulus among the three kinds of hydrogels, the thermal-reversible characteristic limited their applications. The gelatin gels fractured when the compressive strain reached 38%, and the corresponding compressive strength was 11.03 kPa. The KGM and the composite hydrogels did not fracture even the compressive strain reached 70%, and the elastic modulus increased with the increase of compressive strain.

It was found that the elastic modulus reduced a great value of the composite hydrogels compared with KGM gels. The reason is analyzed as follows. Generally, the moduli of polymer gels depend on the crosslinking density and crosslinking types. KGM and gelatin gels are both physical gels, and the crosslinking zones are formed by hydrogen bonds. The entanglement of KGM and gelatin molecular chains and the association between gelatin and KGM molecules by hydrogen bonds hindered the formation of respective crystallization of KGM and gelatin. Therefore, the crosslinking density decreased and the crosslinking zones shortened, which decreased the moduli of the composite hydrogel.

3.4 Morphology observations
The SEM images were obtained to characterize the microstructure of the freeze-dried hydrogels and are presented in Fig. 4.

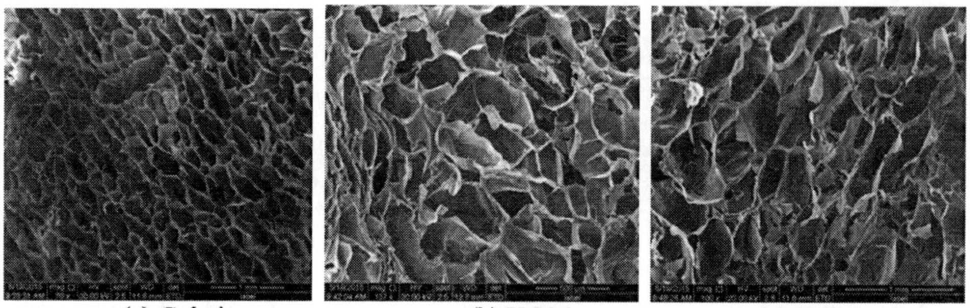

(a) Gelatin (b) KGM (c) KGM-gelatin

Figure 4. Microstructures of the freeze-dried gelatin (a), KGM (b), and gelatin-KGM IPN (c) hydrogels.

This suggests that the hydrogel matrices are porous, with a three-dimensional interconnected microstructure by virtue of the freeze-drying step with the pores being the result of ice crystal formation. The apertures in the gelatin matrices are around 200~300 μm wide estimated from the micrographs. In contrast, the apertures in KGM and composite gels are not uniform as that in gelatin gels, and spread from around 300 μm to 500 μm. It is seen that gelatin gels have the strongest walls which make the shapes of the pores uniform. This is due to the materials characteristics. As discussed in section 3.3, gelatin gels have better stiffness and worse toughness compared with KGM gels. The stiffness of the structure can maintain its shape during the freeze-drying process. While the KGM gels are the opposite. They have good elasticity and toughness, the walls of the pores are not stiff as that of gelatin gels, which causes the irregular pores in the freeze-drying process. The composite gels also have irregular pores, but are better than that of KGM gels. It is considered that the gelatin networks in the composite gel play a supporting role.

4. Conclusions

Novel IPN hydrogels were prepared by physically crosslinking gelatin and KGM respectively. Gelatin physical hydrogel will melt to sol at body temperature, but the IPN hydrogel can maintain its structure under various conditions, although it is still physical gel. This should be attributed to the KGM networks formed in the IPN hydrogels.

Hydrogen bond was formed between the carbonyl group of gelatin and hydroxyl group of KGM, which is analyzed by FT-IR method. The entanglement of molecular chains and the association between gelatin and KGM molecules by hydrogen bonds hindered the formation of their respective crystallization, which can be proved by XRD patterns. And this directly caused the crosslinking density decreased and the crosslinking zones shortened in the IPN hydrogel. So that the compressive modulus of IPN hydrogel was lower than that of KGM hydrogel. Nonetheless, the toughness of the IPN hydrogel was much higher than that of gelatin hydrogel.

This novel IPN hydrogel is a kind of physical gel, but it shows rubber-like performance as chemical gels. And it is nontoxic, so it can be used as the scaffold for cartilage tissue engineering that embedded in human bodies.

Acknowledgements

This work was financially supported by the National Natural Science Foundation of China (31260228) Program.

References

[1] Thumsing S, Supaphol P. Gelatin microspheres containing multi-walled carbon nanotubes: preparation, characterization, and use as a drug carrier. Chiang Mai Journal of Science. 2014; 41: 1287-1300.

[2] Guo L, Colby R H, Lusignan C P, Whitesides T H. Kinetics of triple helix formation in semidilute gelatin solutions. Macromolecules, 2003; 36: 9999-10008.

[3] Flory P J, Weaver E S. Helix-coil transitions in dilute aqueous collagen solutions. Journal of the American Chemical Society, 1960; 82: 4518-4525.
[4] Hauschka P V, Harrington W F. Collagen structure in solution. III. Effect of cross-links on thermal stability and refolding kinetics. Biochemistry, 1970; 9: 3734-3745.
[5] Bigi A, Cojazzi G, Panzavolta S, Rubini K, Roveri N. Mechanical and thermal properties of gelatin films at different degrees of glutaraldehyde crosslinking. Biomaterials, 2001; 22: 763-768.
[6] Fu Y T, Sheu S Y, Chen Y S, Chen K Y, Yao C H. Porous gelatin/tricalcium phosphate/genipin composites containing lumbrokinase for bone repair. Bone, 2015; 78: 15-22.
[7] Michael F B, Yiu-Fai N, Paul D A P. Mechanism and kinetics of the crosslinking reaction between biopolymers containing primary amine groups and genipin. Journal of Polymer Science: Part A: Polymer Chemistry, 2003; 41: 3941-3953.
[8] Gaetani R, Feyen D A M, Verhage V, Slaats R, Messina E, Christman K L, Giacomello A, Doevendans P A F M, Sluijter J P G. Epicardial application of cardiac progenitor cells in a 3D-printed gelatin/hyaluronic acid patch preserves cardiac function after myocardial infarction. Biomaterials, 2015; 61: 339-348.
[9] Luo Y X, Lode A, Akkineni A R, Gelinsky M. Concentrated gelatin/alginate composites for fabrication of predesigned scaffolds with a favorable cell response by 3D plotting. RSC Advances, 2015; 5: 43480-43488.
[10] Rivadeneira J, Di Virgilio A L, Audisio M C, Boccaccini A R, Gorustovich A A. Evaluation of the antibacterial effects of vancomycin hydrochloride released from agar-gelatin-bioactive glass composites. Biomedical Materials, 2015; 10: 015011.
[11] Merkle V, Zeng L K, Teng W B, Slepian M, Wu X Y. Gelatin shells strengthen polyvinyl alcohol core-shell nanofibers. Polymer, 2013; 54: 6003-6007.
[12] Xu K D, Cantu D A, Fu Y, Kim J, Zheng X X, Hematti P, Kao W J. Thiol-ene Michael-type formation of gelatin/poly(ethylene glycol) biomatrices for three-dimensional mesenchymal stromal/stem cell administration to cutaneous wounds. Acta Biomaterialia, 2013; 9: 8802-8814.
[13] Kaul G, Amiji M. Tumor-targeted gene delivery using poly(ethylene glycol)-modified gelatin nanoparticles: In vitro and in vivo studies. Pharmaceutical Research, 2005; 22: 951-961.
[14] Gao S J, Nishinari K. Effect of deacetylation rate on gelation kinetics of konjac glucomannan. Colloids and Surfaces B-Biointerfaces, 2004; 38: 241-249.
[15] Chen J, Li J, Li B. Identification of molecular driving forces involved in the gelation of konjac glucomannan: Effect of degree of deacetylation on hydrophobic association. Carbohydrate Polymers, 2011; 86: 865-871.
[16] Pan Z D, Meng J J, Wang Y M. Effect of alkalis on deacetylation of konjac glucomannan in mechano-chemical treatment. Particuology, 2011; 9: 265-269.
[17] Dave V, McCarthy S P. Review of konjac glucomannan. Journal of Environmental Polymer Degradation, 1997; 5: 237-241.
[18] Marcano J, Hernando I, Fiszman S. In vitro measurements of intragastric rheological properties and their relationships with the potential satiating capacity of cheese pies with konjac glucomannan. Food Hydrocolloid, 2015; 51: 16-22.

Synthesis and gas sensor application of nanostructure Cr_2O_3 hollow spheres

Mingming Zhao[1], Huichao Zhu[1,*], Jiaqi Zhang[1], Meihua Li[1,2] and Zhaoyun Cai[1]

[1]School of Electronic Science and Technology, Dalian University of Technology, Dalian 116024, China
[2]City Institute, Dalian University of Technology, Dalian 116600, China

*Corresponding author e-mail: zhuhuichao@dlut.edu.cn

Abstract. This paper prepared single phase Cr_2O_3 hollow spheres by the template method. The templates was carbon spheres prepared by glucose hydro-thermal method. The Cr_2O_3 hollow spheres were confirmed and characterized by TEM and N2 adsorption-desorption. Then we investigated the gas sensor application of nanostructure Cr_2O_3 hollow spheres. The gas sensor properties of the Cr_2O_3 hollow spheres to some toxic flammable gases were investigated by resistance response method. The result showed that the sensor was extremely sensitive to the temperature, the resistance of sensor was increased linearly with the concentration of reductive gases varies from 50 to 250 ppm at optimum operating temperature (300℃). The response and recovery of the sensor was fast with the time to reach 70% of the total resistance change was 16s and 7s.

1. Introduction

Hazardous and toxic gases from car and industrial exhausts are polluting the environment [1]. It is of great importance to detect these toxic gases. The sensing materials had been the focus of the research for decades. Some n-type metal oxide semiconductor, such as Zn O [2], Sn O_2 [3], WO_3 [4], Fe_2O_3 [5], had been used as sensing materials. There were also many p-type metal oxide semiconductor, including Ni O [6], Mn O_2 [7], Cu O [8], had been researched to be used for gas sensor because of their features such as ease of fabrication, high sensitivity.

As an important wide gap p-type metal oxide semiconductor, Cr_2O_3 has many outstanding properties, such as catalysis, electrochemical, magnetic, optoelectronic, mechanical and gas sensing properties [9]. So, Cr_2O_3 have been seemed to be a perspective material used for gas sensor. In this paper, a novel template method was used to synthesize Cr_2O_3 hollow spheres, which could provide big surface to volume ratio. It had be proved by the our N_2 adsorption-desorption. Then the sensor which had high sensitivity and fast response/recovery time based on the Cr_2O_3 hollow spheres powder was fabricated.

We use the resistance response method to investigate the sensor by the instrument Agilent B1500A Semiconductor Device Analyzer and special gas distribute system. The data showed that the resistance of sensor had close linear relationship to the gas concentration in a wide range from 50ppm to 250ppm at the optimum operating temperature 300℃. Also, the result showed the gas sensor had fast response/recovery time and high sensitivity.

2. Experimental

2.1 Preparation of carbon microspheres
In a typical procedure, 9g Glucose Anhydrous was dissolved in a certain amount of deionized water to produce a resulting solution 50mL with the concentration of 1mol/L. Then the solution was transferred to a Teflon-lined autoclave and maintained at 180℃ for 8h then cool it to room temperature. The purple brown products generated from the former process were then centrifuged and rinsed for at least 10 times with deionized water and ethanol to remove other organic impurities. Finally, a brown powder was generated from the cleaned solution after dried at 80℃ for 12 h [10].

2.2Preparation of Cr_2O_3 hollow spheres
For the preparation of the Cr_2O_3 hollow spheres powders, 0.5g of carbon microspheres obtained from 2.1 and 8g of $Cr(NO_3)_3.9H_2O$ were dispersed in 40mL ethanol by ultra-sonication for 1h. After aged for 24h, the solution was rinsed with ethanol for 2 times and dried at 60℃ for 12h. Next, the dried powder was calculated at 500℃ for 2h in air with the temperature rising speed of 2℃/min. Now, the powders of Cr_2O_3 hollow spheres were obtained.

2.3Characterizations
The prepared materials were characterized by transmission electron microscope (TEM: TF30, FEI USA). The element analysis was realized by EDX (TF30, FEI USA). The surface to volume ratio was investigated by N_2 adsorption-desorption.

2.4Preparation of gas sensor
Sensors based on Cr_2O_3 hollow spheres were prepared as follows. First, 0.1g of the Cr_2O_3 hollow spheres was added into a mixed solution of Terminal and ethyl cellulose. After grinding for 1h and aged for 2h at room temperature, the resulted mixture was coated on an alumina substrate with eight pairs of gold interdigital electrodes. The width and clearance of the electrodes were 0.25mm. Then the resulted mixture with the interdigital electrodes was treated with two step annealing processes. First, they were heated to 60℃ to remove the water, and then annealed at 500℃ for 2h in air to remove the organic binders from the paste. Now, the Cr_2O_3 hollow spheres based sensor was fabricated. The sample need to be heated at 400℃ in air for 1h before test. [11]

2.5Gas sensing measurement
The gas sensing measurements were performed in a dynamic testing system, which has been reported in our former study [11]. A tubular oven was applied to achieve a precise working temperature for the sensor, and the concentration of target gas was controlled by four mass-flow controllers (MFC). An Agilent semiconductor analyzer (B1500A) was used to record and display the instantaneous variation of sensor resistance. Response of gas sensor (S) is defined: S = Rg/Ra, Ra and Rg are sensor resistance in air and target gas, respectively. The response time and recovery time were defined as the time to reach 70% of the total resistance change.

3. Results and discussions

3.1 Materials characterization
TEM was adopted to analyze micro topography of the Cr_2O_3 hollow spheres powder. The images were showed in Figure 1(a)-(b). The sample presents porous hollow spheres with rough surfaces and an average diameter distribution 10-20nm. HRTEM images of Cr_2O_3 were shown in Figure 1(c)-(d), which confirmed the high crystalline nature with a lattice spacing of 0.354nm and 0.267nm were corresponding to (012) and (110) of Cr_2O_3[12,13]. The formation mechanism of carbon nanoparticles and Cr_2O_3 hollow spheres were explained elsewhere[9], As shown in Figure 1(e), We assume that the formation of the carbon spheres firstly involves the dehydration of the carbohydrate and carbonization

of the so-formed organic compounds, whose surface was hydrophilic and has a distribution of -OH and -C=O groups. In the step of 2.2, Cr^{3+} and its resulting nanoparticles were predominantly positioned near the hydrophilic shell of the carbon particles, which yields an increase in the diameter of the carbon spheres. After annealing, the carbon spheres were removed as carbon dioxide to the air, Cr^{3+} was oxidized and Cr_2O_3 hollow spheres are obtained. The rough surface of the sample provides high surface to volume ratio and a great number of active sites for gas adsorption. All these morphology and microstructure results are approved this material will have a good gas sensing property. The EDX spectra of Cr_2O_3 is shown in Figure 1(f), which indicates that the samples are composed of Cr and O elements, and further proves the impurity doping of Cr_2O_3.

The inner porous architecture and the specific surface area of the Cr_2O_3 porous microspheres are further characterized by nitrogen adsorption technique. Figure 2(a) shows the N_2 adsorption/desorption isotherm curves and Figure 2(b) shows the corresponding BJH pore size distribution of Cr_2O_3 hollow spheres. It could be seen from Figure 2(a) that the isotherms belong to a type-IV with a H3 type hysteresis loop (according to IUPAC classification) and exhibit no limited adsorption at high P/P0. As shown in Figure 2(b), the pore size distribution covered a range of 1~20 nm and showed a mesoporous structure. The sample showed a relatively high BET surface area of 49.0365 m^2/g, which indicated a kind of large textural porosity. And therefore, there might have some potential for the enhancement of gas sensing performances.

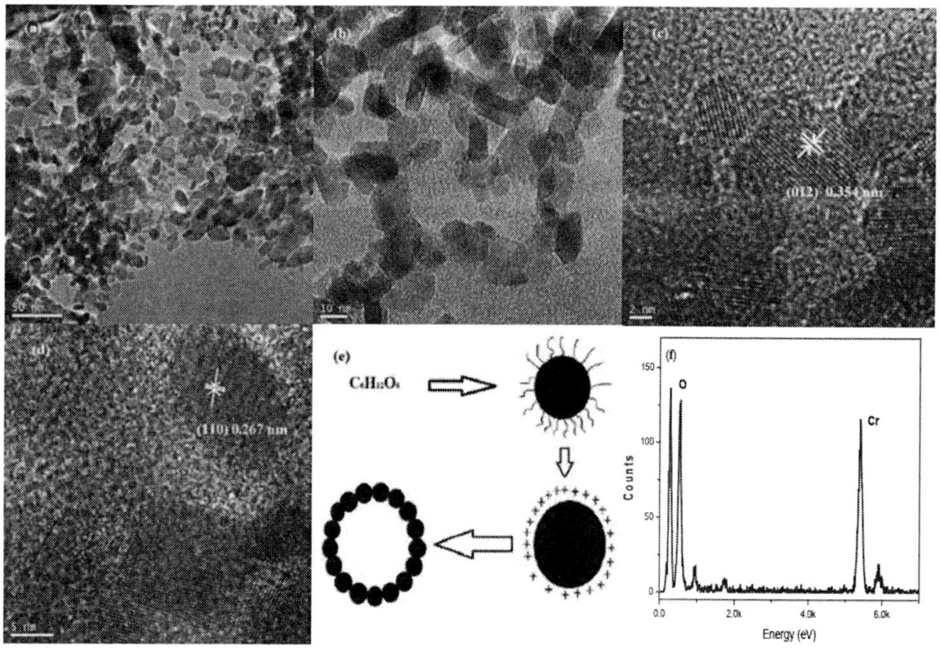

Figure 1. (a)-(b) TEM images of Cr_2O_3; (c)-(d) HRTEM image of Cr_2O_3; (e)The formation model of the Cr_2O_3 hollow spheres; (f)The EDX spectra of Cr_2O_3

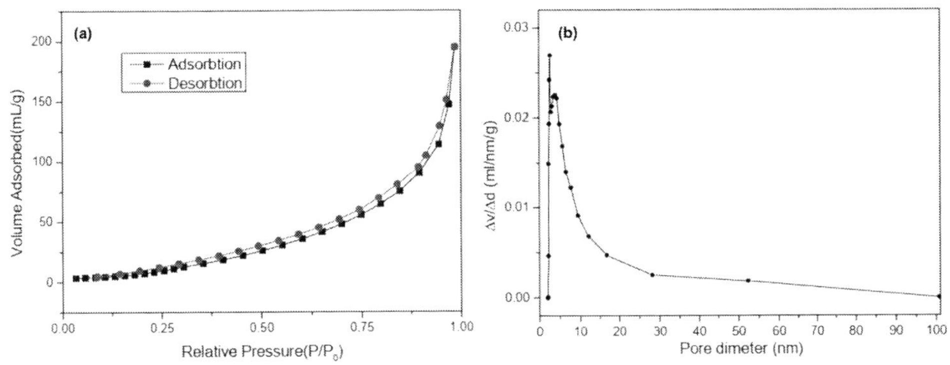

Figure 2. (a)N2 adsorption and desorption isotherm curve; (b) Pore size distribution curve of Cr_2O_3.

3.2 Gas sensing properties

It is well-known that the response of gas sensors relies greatly on the working temperature. The response of the sensor on increasing concentration 50ppm to 250ppm of the C_2H_5OH vapor, was investigated to determine the optimum operating temperature and the response and recovery time. The test result and test condition were shown in Figure 3(a). When a certain concentration of the target gas flow into the tube, the resistance of sensor increased, which indicates typical p-type behavior [14]. The sensor response amplitude to C_2H_5OH remarkably increases with concentration. It can be seen from Figure 3(b) that the response of the sensor increases almost linearly with C_2H_5OH concentration within the range of 50-250ppm in each certain temperature. The sensor also showed best response to the C_2H_5OH at 300℃. The response and recovery time for the sensor to 50 ppm C_2H_5OH was found to be about 16s and 7s seen from Figure 3(c).

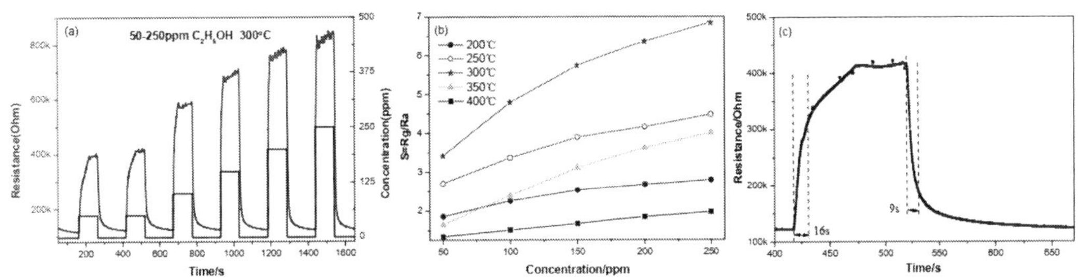

Figure 3. (a) The response of sensor to C_2H_5OH with the concentration range was 50 to 250ppm at 300℃; (b) The response to the C2H5OH at different temperature; (c) The response/recovery time for the sensor.

In order to comprehensively investigate the gas sensor performance to various gases at 300℃, 3 kinds of VOC gases, such as butyl-acetate, iso-propyl alcohol and methoxy-propanol, were employed to be the target gases. According to the experimental results that were showed in Figure 4(a)-(d), our prepared sensors showed improved properties to either C_2H_5OH or butyl-acetate, especially the latter. It could be seen that the relation between the response and the concentration of target gas was explicit linear. Compared with others, the response to iso-propyl was higher. This could explained by the reductive of the gases. Iso-propyl reacts more easily with the pre-adsorbed oxygen ions due to higher activity.

IOP Conf. Series: Materials Science and Engineering **207** (2017) 012030 doi:10.1088/1757-899X/207/1/012030

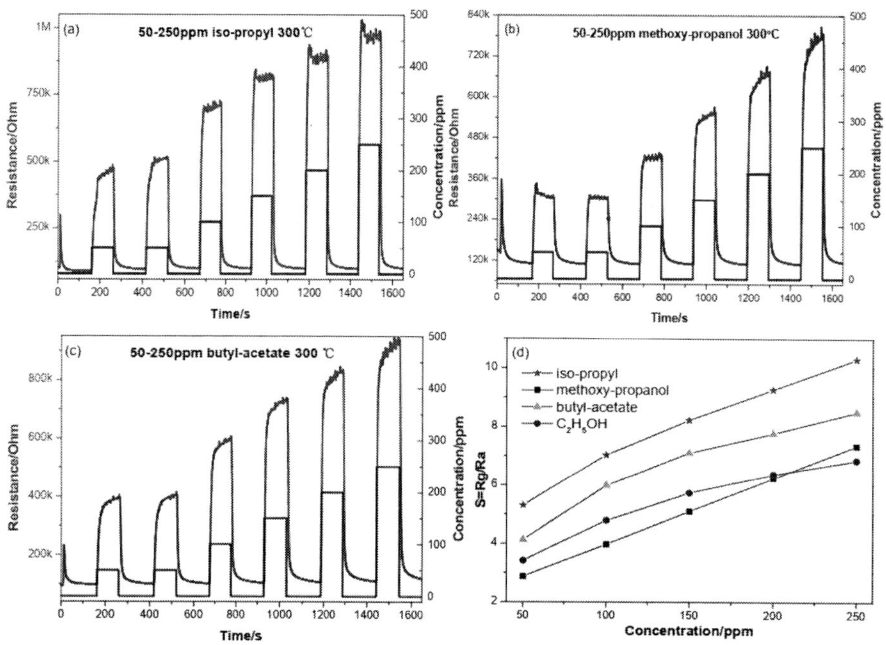

Figure 4. (a) Iso-propyl, (b) Methoxy-propanol, (c) Butyl-acetate test results at optimum operating temperature (300℃), (d) Compare of the response to different gas

3.3 Gas sensing mechanism

As we all know, the Cr_2O_3 is p-type metal oxide semiconductor. The gas sensing mechanism could be similar with that of other p-type metal oxide semiconductor, such as CuO. Gas molecules adsorbed on the surface of Cr_2O_3 particles have remarkable tuning effects on the electrical properties due to their high surface to volume ratio. The sensing property of Cr_2O_3 hollow spheres originates from the interaction between the target gases and the pre-adsorbed O^- and O^{2-} on the Cr_2O_3 surfaces. When the Cr_2O_3 was exposed to the air, the charge exchange interactions between Cr_2O_3 and O_2 as following (1)-(4) did happen and generated the O^- and/or O^{2-} on the surface of the Cr_2O_3. When such Cr_2O_3 is exposed to reductive molecules, taking ethanol as example, the adsorbed reductive molecules interact with the pre-adsorbed O^- and O^{2-} as shown in (5)-(7). The reactions between the reductive molecules and the pre-adsorbed O^- or O^{2-} release free electrons and neutralize the holes (8), the majority carriers in p-type Cr_2O_3. This compensation results in a decrease in the holes in Cr_2O_3, and consequently, an increase in sensor resistance. Note that the oxygen molecules are continuously supplied from the dilute gas (synthetic air) and are adsorbed on the Cr_2O_3 surface while the interaction with reductive molecules to desorption from the surface, with the trend towards an equilibrium state. When the concentration of reductive target gas is decreased, more oxygen molecules in air will adsorb on the surface of Cr_2O_3, and the capture of electrons through the processes indicated in (1)-(4) will reduce the sensor resistance towards the initial stable surface state of Cr_2O_3.[15,16]

$$O_2\,(gas) \rightarrow O_2\,(adsorbed) \qquad (1)$$

$$O_2\,(adsorbed) + e^- \rightarrow O_2^-\,(adsorbed) \qquad (2)$$

$$O_2^-\,(adsorbed) + e^- \rightarrow 2O^-\,(adsorbed) \qquad (3)$$

$$O^{-} \text{ (Adsorbed)} + e^{-} \rightarrow O^{2-} \text{ (adsorbed)} \quad (4)$$

$$C_2H_6O \text{ (gas)} \rightarrow C_2H_6O \text{ (adsorbed)} \quad (5)$$

$$C_2H_6O \text{ (adsorbed)} + 6O^{-}\text{(adsorbed)} \rightarrow 2CO_2 + 3H_2O + 6e^{-} \quad (6)$$

$$C_2H_6O \text{ (adsorbed)} + 6O^{2-}\text{(adsorbed)} \rightarrow 2CO_2 + 3H_2O + 12e^{-} \quad (7)$$

$$e^{-} + \text{hole} \rightarrow \text{NULL} \quad (8)$$

Conclusions

In summary, gas sensor based on template synthesis Cr_2O_3 hollow spheres have been successfully synthesized. A simple and practical test method for gas sensors with practical shelves is realized. 4 types of target gases were investigated. The optimum operating temperature was 300℃. The sensor have high sensitivity to reductive gases and great selectivity to the butyl-acetate. When reductive gases such as C_2H_5OH flow through the sensor, the resistance of sensor was increased. The sensitivity of the sensor was close linear correlated with the concentration of target gas. The response and recovery time was 16s and 7s.

Acknowledgments

The authors thank The National Natural Science Foundation of China (61107028, 61376115, and 61131004) and International Science & Technology Cooperation Program of China (No. 2015DFR10970) for financial support. It was also supported by the Fundamental Research Funds for the Central Universities (Grant No. DUT16LAB16).

References

[1] D.N. Suryawanshi, D.R. Patil, L.A. Patil, Sensors and Actuators B 134 (2008) 579-584.
[2] A.S.Kornyushchenko, Ahalapitiya H.Jayatissa, V.V. Natalich, V.I.Perekrestov,Thin Solid Films 604:48-54 (2016).
[3] Imad H. Kadhim, H. Abu Hassan. J Mater Sci: Mater Electron 27:4356-4362 (2016) DOI: 10.1007/s10854-016-4304-0.
[4] R. Faleh, M. Othman, S. Gomri, K. Aguir, A. Kachouri, IEEE SENSORS JOURNAL, VOL.16, (2016) 3123-3130.
[5] H.M. Yang, S.Y. Ma n, G.J. Yang, W.X. Jin, T.T. Wang, X.H. Jiang, W.Q. Li Materials Letters 169 73-76 (2016).
[6] Po-Cheng Chou, Huey-Ing Chen, I-Ping Liu, Chun-Chia Chen, Jian-Kai Liou,Kai-Siang Hsu, and Wen-Chau Liu , IEEE SENSORS JOURNAL, VOL. 15, 3711-3715 (2015).
[7] J. Xiao, P. Liu, Y. Liang, H. B. Li, and G. W. Yang, Journal of Applied Physics 114, 073513 (2013); doi: 10.1063/1.4819218
[8] Oleg Lupan, Vasile Postica,Vasilii Cretu,NiklasWolff,Viola Duppel,Lorenz Kienle, Rainer Adelung, Phys. Status Solidi RRL 10, No. 3, 260-266 (2016) / DOI 10.1002/pssr.201510414
[9] Li Sheng, Li Feng-Li, Zhou Shao-Min, Wang Peng, Cheng Ke, Du Zu-Liang, Chin. Phys. B Vol 18 No 9, 1674-1056 (2009).
[10] Lifang He, Yong Jia ,Fanli Meng ,Minqiang Li , Jinhuai Liu, J Mater Sci 44:4326-4333 (2009)

DOI 10.1007/s10853-009-3645-y
[11] Shufeng Xia, Huichao Zhu, Haitao Cai, Jiaqi Zhang, Jun Yua, Zhen'an Tang, RSC Adv, 4, 57975 (2014) DOI: 10.1039/c4ra09083g
[12] Ling-Yan Jiang, Sen Xin, Xing-Long Wu, Hong Li, Yu-Guo Guo, Li-Jun Wan , J. Mater. Chem., 20, 7565-7569 (2010) DOI: 10.1039/c0jm01027h
[13] F. Farzaneh, M. Najafi, Journal of Sciences, Islamic Republic of Iran 22(4): 329-333 (2011)
[14] David E. Williams, Sensors and Actuators B 57 1-16 (1999).
[15] Suman Pokhrel, C.E. Simion, V. Quemener, N. Barsan, U. Weimar ,Sensors and Actuators B 133 78-83 (2008).
[16] Nguyen Duc Hoa, Sea Yong An, Nguyen Quoc Dung, Nguyen Van Quy, Dojin Kim, Sensors and Actuators B 146 239-244 (2010).

IOP Conf. Series: Materials Science and Engineering **207** (2017) 012031 doi:10.1088/1757-899X/207/1/012031

Studies on the preparation of Caro's acid by ultrasonic enhanced electrochemistry

Linbo Li, Zeli Yu*, Tao Hong, Zhao Fang, Jishi Peng and Zhao Yang

Xi'an University of Architecture and Technology, Xi'an 710055, P. R. China

*Corresponding author e-mail: 1206764982@qq.com

Abstract. Ultrasonic cavitation effects can generate hydroxyl radicals and high energy, which is widely applied in the field of oxidation currently. Ultrasound-enhanced electrochemical is used to prepare Caro's acid, which improves the generate rate of Caro's acid. In this article, the influences of ultrasonic frequency and ultrasonic power on the electrolysis voltage, electrolyte temperature, electrolyte concentration and the concentration of additive in the process of electrochemical preparation of Caro's acid was studied. And the optimal production conditions were determined. The research results showed that ultrasonic can significantly improve the production of Caro's acid and the product can increase by about 20 g/L under the best condition.

1. Introduction

The study on ultrasonic-induced electrochemical reactions has developed rapidly, which was widely used in water treatment [1-2], ore decomposition [3], electroplating [4] and preparation of nano-materials [5-8] and other fields currently. Zhuping [9] and her co-workers had studied treatment of non-leaching refractory gold ores with electrochemical antioxidant under two conditions including ultrasonic and ultrasonic condition. The results showed that ultrasound can reduce the reaction time and increase leaching rate by 35.84%.

Electrolytic oxidation technology [10-11], which is a green and efficient chemical process, is widely used in wastewater treatment [12].with the aid of ultrasonic; the electrochemical process can enhance the mass transfer process, reduce the electrolytic voltage and reduce the energy consumption in ultrasonic environment. Moreover, ultrasonic cavitation can produce a micro environment of high temperature and high pressure, which can produce a lot of hydroxyl radicals with high oxidizing ability and improve oxidation efficiency. Domestic and foreign scholars mainly applied ultrasonic electrochemical oxidation to wastewater treatment. For example, Zhao Guohua [13] from Tongji University combined boron doped electrode with ultrasound to treat phenol and phthalic acid. The results showed that hydroxyl radical produced during electrolytic process under ultrasonic environment. Furthermore, the oxidation capacity increased and the oxidation peak increased by 32%, the degradation efficiency was significantly improved. Meanwhile the energy consumption was greatly reduced.

In this study, ultrasonic chemical and electrochemical were combined together for the preparation of Caro's acid in a low concentration of sulfuric acid. The results indicated that the anode oxidation rate, current efficiency and mass transfer rate were improved greatly with a reduced energy consumption. The problems existing in the preparation of Caro's acid by electrolysis have been solved fundamentally, which lays the foundation for industrial production.

2. Experimental part

2.1. Experimental principle

2.1.1. Electrolysis principle. Under the action of electric field force, the anions and cations move to cathode and anode respectively in electrolyte solution. Finally, the hydrogen sulfate is oxidized to Caro's acid at the anode and the hydrogen ion is reduced to hydrogen in the cathode. The main reaction and side reaction were as following:

$$HSO_4^- + H_2O - 2e \rightarrow HSO_5^- + 2H^+ \quad \varepsilon^\theta = 2.303V \text{ (Main anode reaction)} \quad (1)$$

$$\varepsilon = 2.303 - 0.0591pH + 0.0296\lg a_{HSO_5^-}/a_{HSO_4^-}$$

$$2H_2O - 4e \rightarrow 4H^+ + O_2 \uparrow \quad \varepsilon^\theta = 1.230V \text{ (Side anode reaction)} \quad (2)$$

$$\varepsilon = 1.2290 - 0.0591pH + 0.0148\lg P_{O_2} \text{ (Cathodic reaction)} \quad (3)$$

$$2H^+ + 2e \rightarrow H_2 \uparrow \quad \varepsilon^\theta = 0.000V$$

2.1.2. Principle of Ultrasonic Enhanced Electrochemistry. When electrolysis process was carried out under ultrasonic environment, ultrasonic cavitation can make electrolyte solution generate hydroxyl, hydrogen peroxide and hydrogen free radicals as shown in formula (4) ~ (8)[14]. When the cavitation bubble is been compressed, the vapor in the bubble is decomposed into hydroxyl and hydrogen free radicals. In the absence of oxidized objects, H2O2 is the main product of ultrasonic decomposition. When the air is involved in the reaction, hydrogen free radicals and oxygen were combined rapidly to form hydrogen peroxide free radicals. Hydrogen peroxide produced by ultrasonic cavitation reacted with HSO4- in solution to form Caro's acid, which was shown as formula (9).

$$H_2O \rightarrow \cdot OH + \cdot H \quad (4)$$

$$\cdot OH + \cdot OH \rightarrow H_2O_2 \quad (5)$$

$$\cdot H + O_2 \rightarrow \cdot HO_2 \quad (6)$$

$$H_2O \cdot + H \cdot \rightarrow H_2O_2 \quad (7)$$

$$H_2O \cdot + H_2O \rightarrow H_2O_2 + O_2 \quad (8)$$

$$HSO_4^- + H_2O_2 \rightarrow HSO_5^- + H_2O \quad (9)$$

2.2. *Experimental Method*

2.2.1. Experimental Device. Diagram of ultrasonic enhanced electrolytic equipment, as shown in Figure 1.

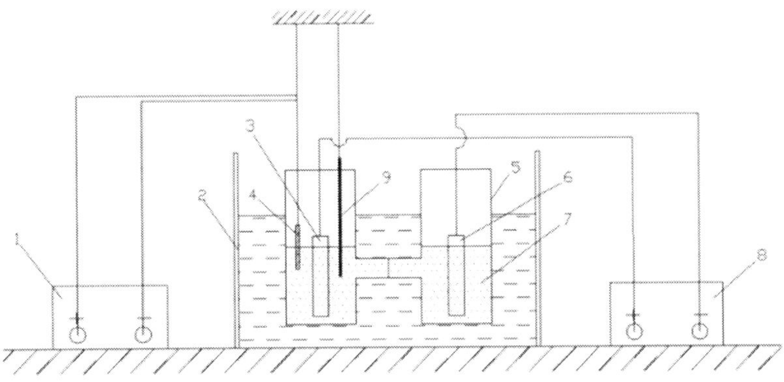

Figure 1. Diagram of ultrasonic enhanced electrolytic equipment.
(1-the ultrasonic generator, 2-low temperature thermostatic bath, 3-platinum plated titanium anode, 4-transducer probe, 5- electrolytic tank, 6- Graphite cathode, 7- electrolyte, 8- DC power supply, 9-thermometer.)

2.2.2. Analysis Method. The potassium dichromate to oxidation-reduction titration. Due to the strong oxidation of persulfate, ferrous ions can be oxidized to ferric ions. So excessive ferrous sulfate solution can be added to the solution, in which part of ferrous ions were oxidized by per sulfate and residual ferrous ions were oxidized by potassium dichromate. According to the amount of ferrous ions oxidized by potassium dichromate oxidation, the amount of ferrous ions oxidized by persulfate ions was calculated. So as the concentration of sulfuric was obtained in the solution.

2.2.3. Experimental reagents. 98% concentrated sulfuric acid, phosphoric acid, potassium dichromate, ammonium ferrous sulfate, sodium diphenylamine sulfonate, ammonium thiocyanate, etc.

3. Results and Discussion

3.1. *The Influences of Ultrasonic Frequency and Time on Electro oxidation*

In ultrasonic environment, the hydrogen free radicals and hydrogen peroxide free radicals can be generated owing to the cavitation effect in water solution, which can enhance the oxidation of the solution and increase the synthesis rate of Caro's acid. Ultrasonic can also improve the ion activity in solution, accelerate ion migration rate and reduce concentration polarization. Therefore, the effects of ultrasonic and ultrasonic time on the electrochemical oxidation process should be investigated.

Experimental conditions: the voltage of electrolysis is 6.0V, the concentration of electrolyte is 400g/L, the concentration of ammonium thiocyanate is 0.5g/L, the ultrasonic power is 150W, the temperature of reaction is 15℃ and the proper reaction time is 4 hours while a sample was taken every 15 min for testing. The influences on three conditions for electrolysis process were investigated and the three conditions included absence of ultrasonic, different ultrasonic frequency including 20 kHz and 40 kHz as well as ultrasonic time. The results are shown in Fig. 2.

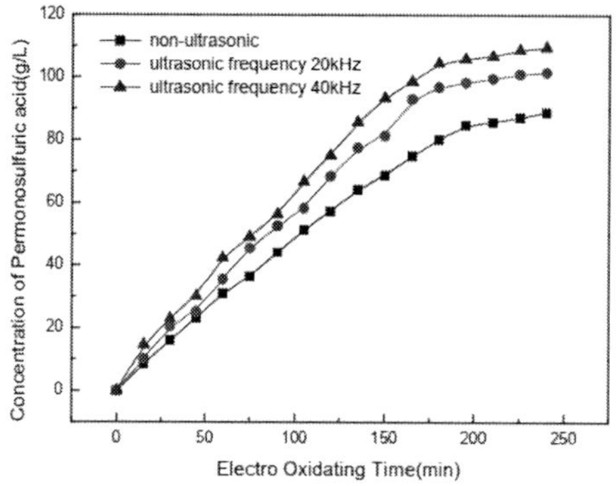

Figure 2. The influences of ultrasound frequency and time on the concentration of Caro's acid generation in the process of electrolysis

As shown in Fig. 2, the concentration of Caro's acid showed a growing trend as time went on. While the generation rate of oxidant decreased when electrolyte time was longer than 3 h, this is because Caro's acid is not stable and hydrolyzed easily and produce sulphuric acid and hydrogen peroxide. While electrolysis process was an exothermic reaction. The longer the time, the higher the temperature, which led to decomposition of Caro's acid. As a result, the optimal time was 3 h. The presence of ultrasonic promoted the generation rate of oxidant. At the same time, oxidant concentration increased when the ultrasonic power was increased. This was mainly because the acoustic cavitation effect helped generate hydrogen free radicals and hydrogen peroxide free radicals, accelerated ion mobility, increased oxidant generation rate. Based on the experiment results, ultrasonic power should be 40 kHz.

3.2. The Influence of Ultrasonic Power on Electro oxidation Process

Ultrasonic power is one of the important factors that effect of cavitation effect, which directly affects the reaction temperature and reaction rate. It is very important to investigate the influence of ultrasonic power on electro oxidation process.

Experimental conditions: the voltage of electrolysis is 6.0V, the concentration of electrolyte is 400g/L, the concentration of ammonium thiocyanate is 0.5g/L, the temperature of reaction is 15℃ and the optimal reaction time is 3hours. The experimental results are shown as Fig. 3.

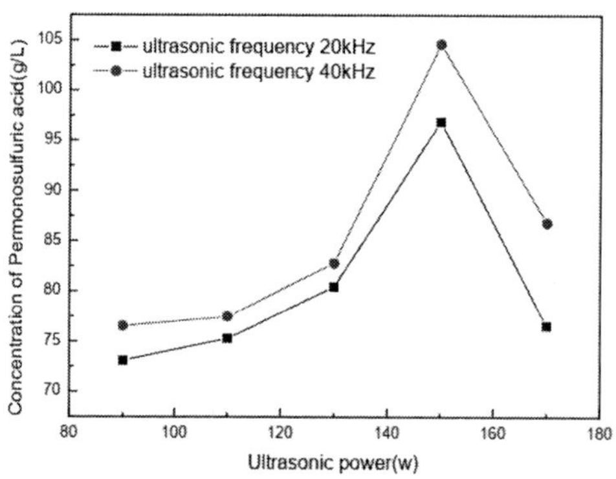

Figure 3. The influences of ultrasound power on the concentration of Caro's acid generation in the process of electrolysis

As shown in Fig. 3, when ultrasonic frequency was a constant, the generation rate of Caro's acid showed a trend of increasing firstly and then decreasing with the increasing of ultrasonic power. This is because the increase of ultrasonic power to speed up the rate of the anodic reaction, but if the ultrasonic power was too large then most of the electrical energy was converted into heat, promoting the decomposition of Caro's acid and reducing the rate of oxidation. When the ultrasonic frequency was 20 kHz and 40 kHz, the generation rate of Caro's acid reached its peak value when ultrasonic power was 150W. Therefore, the appropriate ultrasonic power is 150W.

3.3. The Influences of Ultrasonic and Electrolysis Voltage

The electrolysis voltage is composed of theoretical voltage and over potential, and the over potential is caused by the polarization of the electrode. The presence of ultrasonic for electrolysis can help reduce concentration polarization and enhance the diffusion mass transfer so as to improve the anode electrode potential and anodic oxygen evolution over potential, inhibited side reactions and improve the formation rate of Caro's acid. The experimental results are shown in Table 1.

Table 1. The concentration of Caro's acid under different electrolysis voltage and ultrasonic frequency

decomposition voltage /V		5.0	5.5	6.0	6.5	7.0
The concentration of Caro's acid /g·L^{-1}	non-ultrasonic	64.36	75.86	80.19	95.77	102.46
	20kHz	60.63	64.28	96.89	103.32	80.45
	40kHz	63.85	66.09	104.62	86.29	72.28

Experimental conditions: the concentration of electrolyte is 400g/L, the concentration of ammonium thiocyanate is 0.5g/L, the ultrasonic power is 150W, the temperature of reaction is 15℃ and the optimal reaction time is 3 hours. The experimental results are shown in Figure 4.

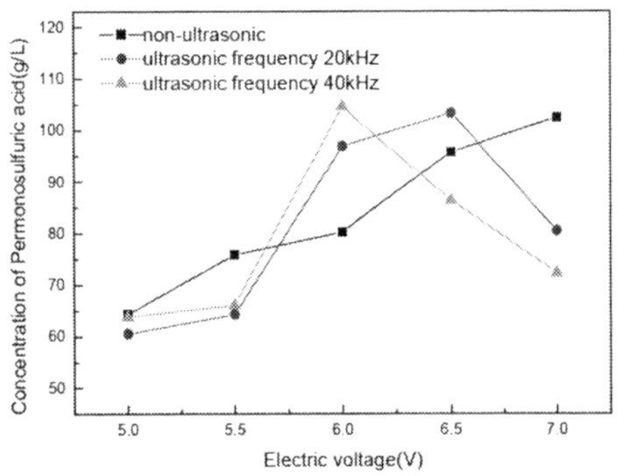

Figure 4. The influences of ultrasonic and voltage on the concentration of Caro's acid generation in the process of electrolysis

As shown in Fig. 4, the concentration of the Caro's acid was the highest when the ultrasonic voltage was 7.0V. When the ultrasonic frequency was increased, the peak value of the concentration Caro's acid shifted to the left, which indicated that the presence of ultrasound could reduce the voltage of the cell. After combining with ultrasound, the concentration of oxidant first increases and then decreases with the increase of electrolysis voltage. This is because the voltage is too high, which led to the increasing of anode over-potential and the energy consumption and the decomposition of Caro's acid was accelerated.

3.4. The Influences of Ultrasonic Frequency and Additive

After consulting the literature, [15-16] it is impossible to avoid the anode adverse reaction in electrolysis process. Therefore oxygen inhibitors must be added. In this study, ammonium thiocyanate was chosen as the oxygen inhibitor and ultrasonic frequency as well as different concentration of oxygen inhibitors were investigated. The experimental results are shown in Table 2.

Table 2. The concentration of Caro's acid under different concentrations of additive and ultrasonic frequency

Additive concentration / g·L-1		0	0.1	0.3	0.5	0.7	0.9
The concentration	non-ultrasonic	20.15	24.88	66.8	80.19	63.41	53.69
of Caro's acid	20kHz	28.78	42.54	69.26	96.89	68.19	56.21
/g·L-1	40kHz	30.35	46.65	77.98	104.62	82.68	65.89

Experimental conditions: the concentration of electrolyte is 400g/L, the voltage of electrolysis is 6.0V, the temperature of reaction is 15℃, the ultrasonic power is 150W and the optimal reaction time is 3 hours. The experimental results are shown in Fig. 5.

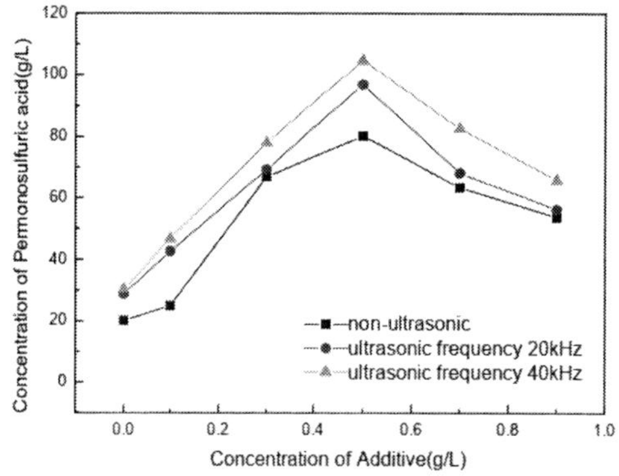

Figure 5. The influences of ultrasonic and additives on the concentration of a sulfuric acid generation in the process of electrolysis

As shown in Fig. 5 and Table 2, additives have great influence on the synthesis of Caro's acid for the reason that the addition of additives led to the formation of SCN- adsorption layer on the anode surface, which greatly improved the oxygen precipitation over potential effectively and reduced electrolytic energy consumption [17]. What we can see from Fig.4 was that the formation rate of Caro's acid firstly increases and then decreases when the additive concentration increases. When the concentration of additive is 0.5g/L, the concentration of Caro's acid reached the maximum value. The concentration of additive shouldn't be too high. The result shows that the anode adsorption layer was too thick when the concentration of additives was too high, which hindered the oxidation of Caro's acid. Ultrasound can reduce the concentration gradient of the solution around the electrode and the concentration polarization of the solution so as to improve the yield of Caro's acid. The higher ultrasonic frequency leads to more Caro's acid, but the overall trend was unchanged. Accordingly, the optimal concentration of additive should be 0.5 g/L.

3.5. The Influence of Ultrasonic and Electrolyte Concentration

Ultrasound can enhance diffusion and mass transfer as well as decrease the concentration polarization of solution. Different concentration of Caro's acid solutions have different effects on the generation rate of Caro's acid in the presence of ultrasonic. The experiment results are shown in Table 3.

IOP Conf. Series: Materials Science and Engineering **207** (2017) 012031 doi:10.1088/1757-899X/207/1/012031

Table 3. The concentration of Caro's acid under different concentration of sulfuric acid and ultrasonic frequency

Concentration of sulfuric acid / g·L-1		200	250	300	350	400	450
The concentration of Caro's acid /g·L-1	20kHz	42.75	45.61	59.34	72.19	96.89	81.34
	40kHz	49.25	60.65	72.3	75.38	104.62	97.01

Experimental conditions: the voltage of electrolysis is 6.0V, the concentration of ammonium thiocyanate is 0.5g/L, the temperature of reaction is about 15℃, the ultrasonic power is 150W and the optimal reaction time is 3 hours. The experimental results are shown in Fig. 6.

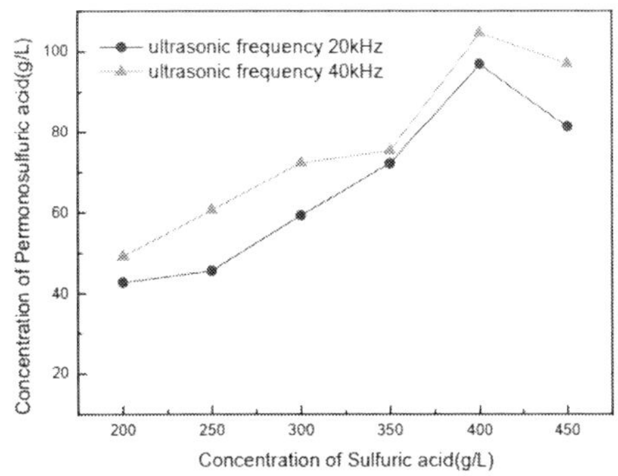

Figure 6. The influences of ultrasonic and concentration of electrolyte on the concentration of Caro's acid generation in the process of electrolysis

It can be seen that the concentration of Caro's acid firstly increases and then decreases with the increase of the concentration of sulfuric acid from Fig.6. When the concentration of sulfuric acid is 400g/L, the concentration of Caro's acid reached its maximum value. When the concentration of sulfuric acid was too low, HSO_4^- obtained from the decomposition of sulfuric acid was too little and the reaction intensity was low. When the concentration of sulfuric acid was high, the solution viscosity increased and it is not conducive to mass transfer. As a result, the ultrasonic cavitation will be weaken and ultimately lead to a low Caro's acid production rate. The higher the ultrasonic frequency was, the higher the formation rate of Caro's acid was.

3.6. The Effects of Ultrasound and Electrolyte Temperature

The addition of electrolytic process will produced a large number of cavitation bubble and cavitation bubble rupture will release heat, which led to the increasing of temperature for the solution. Thereby strengthening the ion transport and electrochemical reaction, but hydrolysis of Caro's acid was also accelerated. On the contrary, reducing temperature can inhibit hydrolysis of Caro's acid. Temperature should be reasonably control after adding ultrasound.

Experimental conditions: the voltage of electrolysis is 6.0V, the concentration of electrolyte is 400g/L, the concentration of ammonium thiocyanate is 0.5g/L, the ultrasonic power is 150W and the proper reaction time is 3 hours, the experimental results are shown in Fig. 7.

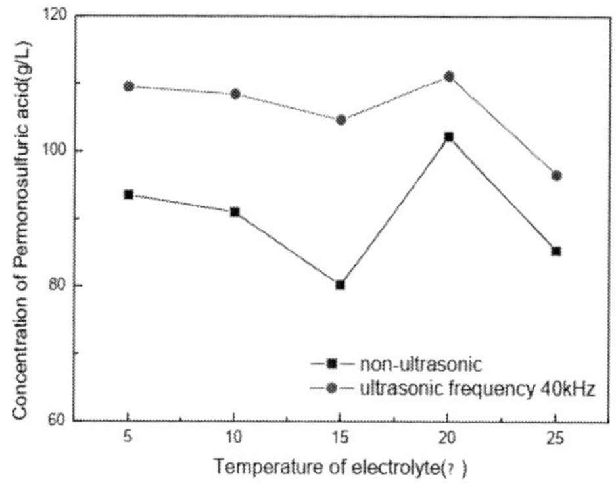

Figure7. The influences of ultrasonic and temperature of electrolyte on the concentration of Caro's acid generation in the process of electrolysis

It can be seen that the temperature had a great influence on the rate of formation of the oxidant from Fig. 7. As the temperature of the electrolyte increases, the concentration of Caro's acid firstly decreased and then increased again and then decreases. This was mainly because the increasing temperature led to increasing hydrolysis of Caro's acid and the decreasing current efficiency. The concentration of Caro's acid increased about 20 g/L, but the overall trend was unchanged. Therefore, the temperature of electrolyte was chosen to be 20℃.

4. Conclusions

(1) In the process of preparing Caro's acid by electrolysis, adding additives can increase the over potential of oxygen precipitation and restrain the occurrence of oxygen evolution side reaction effectively.

(2) Electrolysis in ultrasonic environment can accelerate the ion migration rate, reduce the tank voltage, increase the anodic oxidation rate and promote the synthesis of over Caro's acid. Therefore the purpose of strengthening production could be achieved.

(3) The experiment results showed that maximum concentration of Caro's acid can reach 104.62g/L under the condition of using 400g/L sulfuric acid as electrolyte, electrolytic voltage control in 6.0V, ammonium thiocyanate concentration is 0.5g/L, the temperature controlled at 20℃, the ultrasonic frequency is 40kHz, ultrasonic power is 150W, optimal time is 3 h.

Acknowledgments
This work was financially supported by The National Natural Science fund.
Project No. E042203.

References

[1] Bing Chen. Sonochemical Degradation of Organic Pollutants in Aqueous Solution[D]. Shandong Institute of Light Industry,2010

[2] Guanlei Qiao. The study on Ultrasonic Degradation of ciprofloxacin[D]. Zhengzhou University, 2013

[3] Jiangang Fu, Lijuan Jia, Hong Zhong.Study on design of electrolytic cell system for ultrasonic electro-oxidation of molybdenum concentrate[J].Rare Metals and Cemented Carbides, 2007,35(3),5~9.

[4] Jingquan Cheng, Suwei Yao. Application of ultrasound in electrochemistry[J].Plating and Finishing,2005,27(1),16-19

[5] Qiang Yang, Jianfeng Huang. Developments and applications of sonochemical method in preparation of nanomaterials[J].Chemical Industry and Engineering Progress,2010,29(6), 1091-1101

[6] Zhang Kai, Fu Qiang, Fan Jinghui, etal. Preparation of Ag/PS composite particles by dispersion polymerization under ultrasonic irradiation[J].Materials Letters, 2005,59(28),3682-3686
[7] CHEN Bing-di, PENG Cheng-xin,CUI Zheng. Ultrasonic synthesis of CoO/graphene nanohybrids as high performance anode materials for lithium-ion batteries[J].Transactions of Nonferrous Metals Society of China,2012,22(10):2517-2522.
[8] CHEN Deng-bin, ZHAO Yu-tao,etal. Microstructure and mechanism of in-situ Al2O3(p)/Al nano-composites synthesized by sonochemistry melt reaction[J].Transactions of Nonferrous Metals Society of China,2012,22(1):36-41
[9] Ping Zhu, Kunfang Li, Ming Zhou. Study of in situ leaching gold from refractory gold ores with method of ultrasonic-aided electro-chlorination[J]. Rare Metal Materials and Engineering, 2009,38(6),1091-1094
[10] WEI Qi-feng, ZHANG Qi-xiu.Electro-oxidation of Ce(Ⅲ) in mixed rare earths sulfurate solution by changing current density[J].Transactions of Nonferrous Metals Society of China,2004,14(11):1963-1969.
[11] WEI Qi-feng, ZHANG Qi-xiu,XIA Wen-tang. Oxidation of cerium(Ⅲ) to cerium(Ⅳ) and simultaneous deposition of copper powder by anion-exchange membrane coupling electrolysis[J].Transactions of Nonferrous Metals Society of China, 2005, 36(3):0396-0402.
[12] Jian Yang, YunTao Wu, Meiyan Xing.Treatment of Refractory Anthraquinone Dyeing Wastewater by Micro-Electrolysis and Fenton Oxidation Process[J]. Journal of TongJi University,2005, 33(12)
[13] Guohua Zhao, Junxia Gao, Shihao Shen.Ultrasound Enhanced Electrochemical Oxidation of Phenol and Phthalic Acid on Boron-Doped Diamond Electrode[J].Journal of Hazardous Materials, 2009,172(2),1076-1081
[14] Zhu CP, He SC, Shan ML, Feng R and Xu JY. Recent advance of application and research of ultrasonic technique in wastewater treatment[J].Computers and Applied Chemistry, 2005,22(1),28~32
[15] Guanghong Zhang,Yanlin Chen. Study on Electrolytic Method for Preparing Ammonium Persulfate[J].Guangzhou Chemical Industry, 2012,40(19),19-20
[16] Jianhong Zhao, Peiyong Sun, etc. Study on the Process for Electrosynthesis of Ammonium Persulfate[J].Journal of Zhengzhou University,2006, 27(1),109-112
[17] Yufeng Cong, Guang Xiao, etc. Electrochemical Preparation of Ammonium Persulfate by Dilute Solution of Ammonium Sulfate[J].Journal of Fushun Petroleum institute,2001, 21(4),29-31

IOP Conf. Series: Materials Science and Engineering **207** (2017) 012032 doi:10.1088/1757-899X/207/1/012032

Low temperature sintering and nonlinear dielectric properties of Li_2O doped $Ba_{0.6}Sr_{0.4}TiO_3$ ceramics derived from the citrate method

Xiaofei Zhang*, Xiangyi Deng, Yuting Ren

School of Mathematics and Physics, Hubei Ploytechnic University, Huangshi, China

*Corresponding author e-mail: zhangxiaofei@whut.edu.cn

Abstract. A nominal composition of $Ba_{0.6}Sr_{0.4}TiO_3$ + 0.5 wt. % Li_2O ceramic specimen was prepared from the citrate method. The addition of Li_2O was found to be effective to reduce the sintering temperature of $Ba_{0.6}Sr_{0.4}TiO_3$ ceramics, which was attributed to the formation of transient liquid phase. The ceramic specimen sintered at 900 ℃ showed a fine-grained microstructure with considerable densification and attained the relative density of 97%. The diffusion of Li_2O into $Ba_{0.6}Sr_{0.4}TiO_3$ lattice was found to have an evident effect on the dielectric properties of the ceramics. The specimen exhibited a dielectric constant of 1997 and a dielectric loss 0.67% at 10 kHz and zero bias-field together with a tenability of 27.8% and a FOM of 41 at 10 kHz and 30 kV/cm.

1. Introduction

In the past decades, the dielectric nonlinearity of barium strontium titanate ($Ba_{1-x}Sr_xTiO_3$) under bias electric field has been the subject of extensive investigations [1, 2]. The outstanding dielectric nonlinearity of $Ba_{1-x}Sr_xTiO_3$ makes it a promising candidate material for electrically tunable microwave devices, such as filters, resonators, delay lines, variable capacitors and phase shifters [3]. Miniaturization and integration of the electrical devices are the currently developed trend, which necessitates the low temperature sintering of the dielectric materials. At present the sintering temperature of $Ba_{1-x}Sr_xTiO_3$ ceramics prepared from the conventional solid state method is approximately 1400 ℃. Adopting superfine starting powders with high reactivity is a viable approach to enhance the sinter ability and reduce the sintering temperature of BST materials. In our previous work, we have proved the feasibility and made the sintering temperature decreased to be 1250 ℃ [4]. Adding sintering aids, such as glass frits and oxide additives, is an effective employed strategy to overcome the problem. By adding glass frits to the functional ceramics, the sintering temperature can be lowered drastically. However their dielectric properties were usually degraded by a large margin [5-6]. A small amount of lithium-based sintering aids such as LiF, Li_2O and Li_2CO_3 were reported to decrease the sintering temperature without sacrificing the electric properties of $BaTiO_3$-based materials [7-11]. While non-homogeneity is always the main problem for Li_2O doped BST materials produced by the conventional solid state method. In addition, fine grained ceramics prepared by chemical solution methods favor to obtain moderate dielectric constants, which contributes to the device matching.

The citrate method is essentially a polymeric precursor method, using citric acid and water as the complex agent and solvent, respectively. In this work, we produced a nominal composition of $Ba_{0.6}Sr_{0.4}TiO_3$ + 0.5 wt. % Li_2O (abbreviated as BSTL) ceramics via the citrate precursor method. The structure and nonlinear dielectric properties of the ceramic specimens were investigated.

2. Experimental

All the raw materials used in this work were commercially available reagent grade chemicals (Sino pharm Chemical Reagent Co., Ltd.). BSTL powders were synthesized by a citrate method using reagent grade Ba $(NO_3)_2$, Sr $(NO_3)_2$, $LiNO_3$, tetra butyl titivate and citric acid as starting materials. Tetra butyl titivate was first dissolved in a citric acid solution and various nitrates were then added, followed by stirring to yield a transparent aqueous solution. The mole ratio of citric acid to the total metal cation content was 1.5. A final doping level of 0.5 wt. % Li_2O was achieved by adding a corresponding molar fraction of $LiNO_3$. The precursor solution was heated to form a foam-like solid precursor. The solid precursor was pulverized and calcined at 600 ℃ for 1 h in air. The calcined powder was uniaxial pressed under a pressure of 300 MPa into discs of 13 mm diameter and 1 mm thickness. The compacted discs were sintered at 900 ℃ for 2 h in air. Details of the synthesis and preparation processes have been described elsewhere [4].

The phase purity of the calcined powder and crystal structure of the ceramic specimens were examined by a Philips X' pert PBO X-ray diffractometer using Cu Ka radiation. The morphology of the calcined powder was observed at a Hitachi S-4700 field emission scanning electron microscope (FESEM). The microstructure of the powders and ceramic specimens was investigated at a Jeol JSM-5610LV scanning electron microscope (SEM). The dilatometry measurement of the compacted powder was conducted by a Nietzsche DIL 402C dilatometer at a heating rate of 5 ℃/min between 20 and 1100 ℃ in air. The density of the ceramic specimens was measured by the Archimedes method with ethyl alcohol as the medium. The relative densities of the ceramic specimens were determined from the measured results and the theoretical densities calculated based on the X-ray diffraction (XRD) data of $Ba_{0.6}Sr_{0.4}TiO_3$ ceramics using the Jade software. The ceramic specimens were painted with silver paste on both surfaces as electrodes for measuring electric properties. The temperature dependence of the dielectric constant and dielectric loss was measured by a HP4294 impedance analyzer and a JYT-800L environmental chamber between -60 and 110 ℃ at 1, 10 and 100 kHz, respectively. The nonlinear dielectric properties were measured at room temperature by a TH2818 impedance meter at 10 kHz under external bias electric fields rising from zero to 30 kV/cm.

3. Results and discussion

Figure 1 shows the XRD pattern of BSTL powder calcined at 600 ℃. A pure perovskite structure was identified for the powder, indicating that a small amount of Li_2O didn't give rise to impurity phase at the temperature.

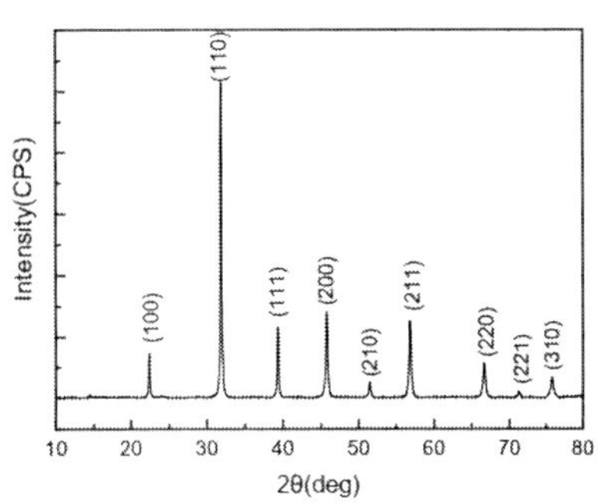

Figure 1. XRD pattern of BSTL powder calcined at 600 ℃

IOP Conf. Series: Materials Science and Engineering **207** (2017) 012032 doi:10.1088/1757-899X/207/1/012032

Figure 2 shows SEM and FESEM images of BSTL powder calcined at 600 ℃. The SEM image (Fig. 2a) offered a panoramic view of the powder morphology, displaying fine and relatively uniform. The FESEM image (Fig. 2b) further revealed that the size of preliminary particles was 30–40 nm and there exist slight agglomerations (around 150 nm) of the preliminary particles.

Figure 2. (a) SEM and (b) FESEM images of BSTL powder

Figure 3 shows the dilatometric curves of the BSTL compacted powder. One can see that an onset temperature of shrinkage occurred at around 780 ℃ and a maximum shrinkage rate at 800 ℃. The shrinkage behavior tended to be saturated after 1100 ℃. The onset temperature of shrinkage have been reported to be located at higher temperature (1060 ℃) for $Ba_{0.6}Sr_{0.4}TiO_3$ powder in our previous research [4]. The dilatometry results hint that the BSTL powders could be sintered to reasonable densification levels at relatively low temperatures. The sintering temperature for BSTL was reduced, which should be attributed to be the formation of liquid phase in the sintering process [8, 12].

Figure 4 shows the XRD pattern of BSTL ceramic specimen. A pure perovskite structure with a cubic symmetry was identified for the ceramic specimen. The result indicates that chemical reaction between $Ba_{0.6}Sr_{0.4}TiO_3$ and Li_2O phase appears to be insignificant. The inset figure shows the SEM image of the specimen. It can be observed that the ceramic specimen has a fine-grained microstructure with considerable densification. The mean grain size of the specimen was estimated to be about 0.4 μm via an image analysis using the Image-Pro Plus software. The Archimedes measurement indicated that the ceramic specimen attained about 97% of the theoretical density. The relative density of the ceramic specimen is roughly comparable with those of $Ba_{0.6}Sr_{0.4}TiO_3$ ceramics prepared by the conventional solid-state method at sintering temperature of 1400 ℃ or higher temperatures [13]. The comparison suggests Li_2O contributed to the sintering of the ceramic specimen, which is considered to be the formation of transient liquid phase. In addition, the explicit grain boundaries were observed to be quadrilateral, which should be attributed to the precipitation of BSTL phase from the eutectic liquid phase.

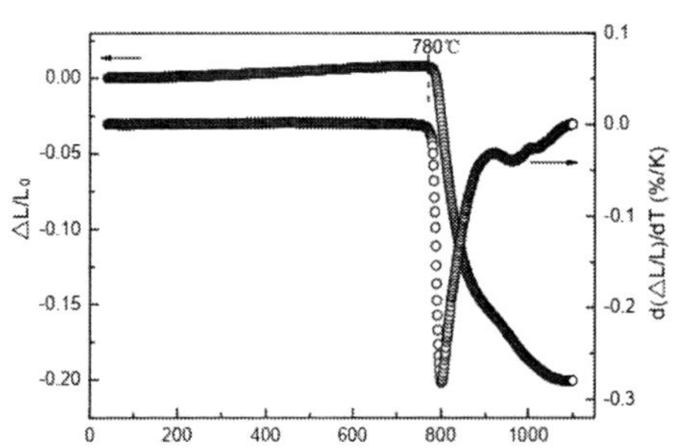

Figure 3. Dilatometry curves of the BSTL compacted powder

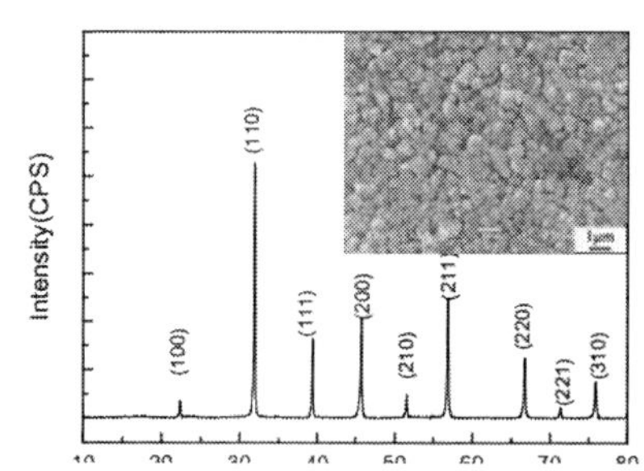

Figure 4. XRD pattern of BSTL ceramic specimen. The inset is SEM image of the specimen

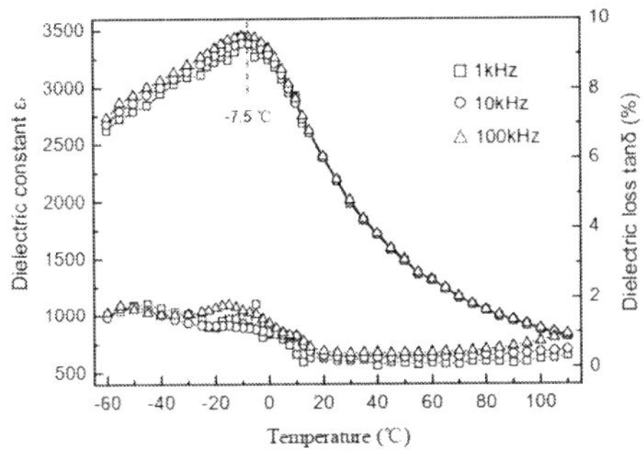

Figure 5. Temperature dependence of dielectric constant and dielectric loss at different frequencies for BSTL ceramic specimen

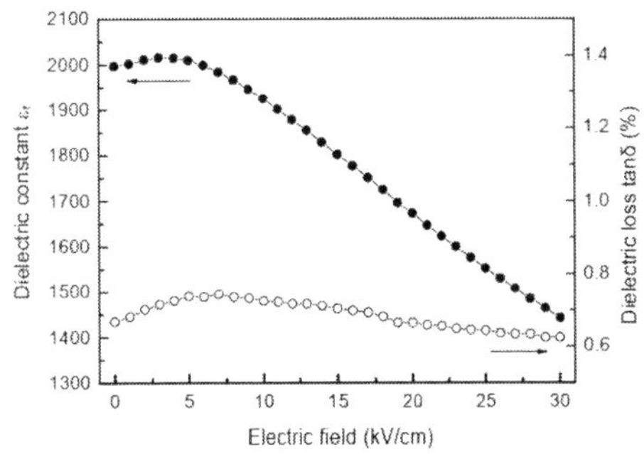

Figure 6. Bias electric field dependence of dielectric constant and dielectric loss for BSTL ceramic specimen

Figure 5 shows the temperature dependence of dielectric constant and dielectric loss at different frequencies for BST ceramic specimen. A slightly diffused peak of dielectric constant was observed at around -7.5 ℃, with a rather obscure frequency dispersion. A detectable peak of dielectric loss was also accompanied near the temperature. As well-known, the dielectric anomaly can be ascribed to a ferroelectric-Para electric phase transition. The dielectric behavior in Figure 5 is similar to previously reported results for $Ba_{0.6}Sr_{0.4}TiO_3$ ceramics [14, 15]. The temperature of dielectric constant maximum (T_m) for BSTL specimen decreased compared to $Ba_{0.6}Sr_{0.4}TiO_3$ ceramics in the literatures, which can be explained by the diffusion of lithium ions into the $Ba_{0.6}Sr_{0.4}TiO_3$ lattice.

Figure 6 shows the dielectric constant and dielectric loss as a function of bias electric field for BSTL ceramic specimen. Generally speaking, the dielectric constant decreased with bias electric field, indicating the typical feature of nonlinear dielectrics. A careful inspection indicates a rather flat variation of the dielectric constant with electric field within 5 kV/cm and a rapid decline at the higher electric fields. Correspondingly, the dielectric loss increased with electric filed through a maximum at 5 kV/cm and then tended to decrease slightly. The variation of the dielectric properties within low electric field was considered to be related with polar micro-regions [15]. The formation of polar micro-regions was believed to be caused partially by the substitution of Li^+ for Ti^{4+} in the perovskite structure [16].

The specimen exhibited a dielectric constant of 1997 and a dielectric loss of 0.67% at 10 kHz and zero bias-field. The tenability was determined as the percentage of dielectric constant change under 30 kV/cm. The figure of merits (FOM, defined as tenability/tanδ) were determined from the results of the tenability and dielectric loss. The specimen attained a tenability of 27.8% and a FOM of 41 at 10 kHz and 30 kV/cm.

4. Conclusion

Superfine and uniform powders of $Ba_{0.6}Sr_{0.4}TiO_3$ + 0.5 wt. % Li_2O with a pure perovskite phase was derived from the citrate method. The sintering temperature of the resulting ceramic specimen was reduced to 900 ℃. The ceramic specimen showed a fine-grained microstructure and relative density of 97%. The variation of the dielectric constant and dielectric loss as a function of temperature and bias electric field was found to be related with the diffusion of Li^+ into $Ba_{0.6}Sr_{0.4}TiO_3$ lattice. The ceramic specimen showed a dielectric constant of 1997 and a dielectric loss 0.67% at 10 kHz and zero bias-field together with a tenability of 27.8% and a FOM of 41 at 10 kHz and 30 kV/cm.

Acknowledgments
This work was supported by Science and Technology Research Project of Hubei Provincial Department of Education (No. B2016277), Talent Project of Hubei Polytechnic University (No. 12xjz02R) and the National Natural Science Foundation of China (No. 51372077).

References

[1] A.K. Tagatose, V.O. Sherman, K.F. Astafiev, J. Denktash, N. Setter, Ferroelectric Materials for Microwave Tunable Applications, Journal of Electro ceramics 11 (2003) 5–66
[2] L.B. Kong, S. Li, T.S. Zhang, J.W. Zhai, F.Y.C. Boey, J. Ma, Electrically tunable dielectric materials and strategies to improve their performances, Progress in Materials Science 55 (2010) 840–893
[3] C. Ang, Z. Yu, dc electric-field dependence of the dielectric constant in polar dielectrics: Multipolarization mechanism model, Physical Review B 69 (2004) 174109-1-8
[4] X.F. Zhang, Q. Xu, Y.H. Huang, H.X. Liu, D.P. Huang, F. Zhang, Low-temperature synthesis of superfine barium strontium titanate powder by the citrate method, Ceramics International 36 (2010) 1405-1409.
[5] D. Zhang, T.W. Button, V.O. Sherman, A.K. Tagantsev, T. Price, D. Iddles, Effects of glass additions on the microstructure and dielectric properties of barium strontium titanate (BST) ceramics, Journal of the European Ceramic Society 30 (2010) 407–412
[6] H.T. Jiang, J.W. Zhai, J.J. Zhang, X. Yao, Dielectric Properties and Low-Temperature Sintering of the Ba0.6Sr0.4TiO3 Ceramics with B2O3/CuO Additions, International Journal of Applied Ceramics Technology 10 (2013) 873–878
[7] J.M. Haussonne, G. Desgardin, PH. Bajolet, B. Raveau, Barium Titanate Perovskite Sintered with Lithium Fluoride, Journal of the American Ceramic Society 66 (1983) 801-807
[8] M. Valant, D. Suvorov, Low-temperature sintering of (Ba0.6Sr0.4) TiO3, Journal of the American Ceramic Society 87 (2004) 1222-1226.
[9] V.K. Pal kuru, J. Peräntie, M. Komulainen, T. Tick, H. Jantunen, Tunable microwave devices using low-sintering-temperature screen-printed barium strontium titanate (BST) thick films, Journal of the European Ceramic Society 30 (2010) 389–394
[10] Y.S. Ham, S.W. Yun, J.H. Koh, Analysis on the temperature dependent structural properties of Li2CO3 doped (Ba, Sr) TiO3 ceramics, Journal of Electro ceramics 26 (2011) 32–36.
[11] S.H. Kim, J.H. Koh, The microwave properties of Li doped 0.7(Ba,Sr)TiO3–0.3MgO thick film interdigital capacitors on the alumina substrates, Microelectronic Engineering 87 (2010) 79–82.
[12] L. Zhao, B.P. Zhang, P.F. Zhou, L.F. Zhu, J.F. Li, Effect of Li2O addition on sintering and piezoelectric properties of (Ba, Ca)(Ti, Sn)O3 lead-free piezo ceramics, Journal of the European Ceramic Society 35 (2015) 533–540
[13] B. Su, J.E. Holmes, B.L. Chen, T.W. Button, Processing Effects on the Microstructure and Dielectric Properties of Barium Strontium Titanate (BST) Ceramics, J. Electro Ceram. 9 (2002) 111-116.
[14] A. Feteira, D.C. Sinclair, I.M. Rainey, Y. Somiya, M.T. Lanagan, BaTiO3-Based Ceramics for Tunable Microwave Applications, J. Am. Ceram. Soc. 87 (2004) 1082-1087.
[15] Q. Xu, X.F. Zhang, Y.H. Huang, W. Chen, H.X. Liu, M. Chen, B.H. Kim, Sinter ability and nonlinear dielectric properties of Ba0.6Sr0.4TiO3 derived from a citrate method, Journal of Alloys and Compounds 485 (2009) L16–L20
[16] L. Zhao, B.P. Zhang, P.F. Zhou, L.F. Zhu, N. Wang, Piezoelectric and ferroelectric properties of (Ba, Ca)(Ti, Sn)O3 lead-free ceramics sintered with Li2O additives: Analysis of point defects and phase structures, Ceramics International 42 (2016) 1086–1093

Effects of Nd and Homogenizing on Cast Microstructure and Mechanical Properties of ZM21 Alloy

Quan Li [1, a], Xinwei She[1, b], Anke Du[1, c], Jian Peng[1, 2, d], Xianquan Jiang [1, e]

[1] Chongqing Academy of Science and Technology, Chongqing 401123, China
[2] College of Materials Science and Engineering, Chongqing University, Chongqing 400044, China

[a]email:156723583@qq.com, [b]email:1203657212@qq.com, [c]email: 88645177@qq.com, [d]email: jpeng@cqu.edu.cn, [e]email:574754298@qq.com

Abstract. In this paper, by adding different amounts of magnesium in ZM21 alloy we look into the influence that Trace Nd has on the microstructure and mechanical properties of magnesium alloy cast ZM21 and how homogenizing effects the mechanism of its organization and performance. The results showed that the addition of rare earth Nd doesn't make the refined grains smaller but make them larger. The hot compressive deformation resistance of ZM21 as-cast alloy increases significantly with the increase of Nd content.

1. Introduction

In the research of magnesium alloy, Nd is mainly added as the primary element to improve the mechanical properties of alloy at room and high temperature [1-2] by generating Mg-Nd dual strengthening phase. There are also some research to add Nd as micro alloy elements to magnesium alloy. For example, we add Nd to Mg-Zn alloy, which mainly generates the Mg-Zn-Nd ternary alloy phase, to strengthen the effect to the alloy. However, there are just a few research about the effect on the plasticity of magnesium alloy. It is necessary to do more research on it so as to extend the application of Nd in the magnesium alloy and provide basic research for developing high plastic deformation of magnesium alloys on the basis [3-5]. In this paper, by adding different amounts of magnesium in ZM21 alloy we look into the influence that Trace Nd has on the microstructure and mechanical properties of magnesium alloy cast ZM21 and how homogenizing effects the mechanism of its organization and performance.

2. Experiments

Raw materials used in the experiment of alloy for pure Mg (mass fraction 99.98%), Zn (mass fraction 99.7%), Mg-34%Mn and Mg-20%Nd alloy, with low carbon steel crucible in 60 kw resistance furnace refining, 5 flux as refining agent and coating agent, by water-cooled semi continuous casting system of Φ92 mm ingot casting, melting and pouring process with CO_2 and SF_6 gas mixture for their own protection. By adding ZM21 magnesium alloy to the successive Nd intermediate alloy, we get Mg-20%Nd for ZM21 magnesium alloy as substrate. The alloy has different contents of Nd, whose components through laser spectrometer detecting are as shown in table 1.

IOP Conf. Series: Materials Science and Engineering **207** (2017) 012033 doi:10.1088/1757-899X/207/1/012033

Give the sample 420℃x 10 h homogenization annealing treatment in 12 kw wind circulation box and put it in the corrosion of picric acid. Then observe it with the OPTEC MDS organization metallographic microscope. Conduct the deformation resistance test on Gleeble1500D,

Table 1. Chemical component of the alloy (wt. %)

Alloy Number	Alloy Chemical component			
	Zn	Mn	Nd	Mg
1#	1.93	0.86	—	other
2#	1.97	0.88	0.07	other
3#	1.97	0.87	0.22	other
4#	1.99	0.87	0.35	other
5#	1.94	0.88	0.47	other

hot compress temperature is 400℃, the heating rate 3℃/s, insulation 5 min, the compression rate is $0.5s^{-1}$. Give the tensile test at room temperature on the new CMT- 5105 electronic universal material testing machine, the tensile rate of 4 mm/min.

3. Test results and analysis

As the pictures shown in figure1, figure 1(a) ~ (e) are the alloy as-cast dendritic structure, which shows that 1#~5# alloy has compounds in the in tar dendritic segregation. 1# alloy has less compounds, most of which are granular and there is a small amount of short strip. After adding trace rare earth Nd, the compounds in 2# alloy increase and become bigger, and has intermittent strip. With the increase of Nd content, strip compounds increase gradually and become gradually continuous. 3#~4# alloy formed more fine mesh compounds in a row. We also found that the spacing between secondary dendrite in 1#~3# alloy is similar, but the secondary dendrite spacing of 4# alloy is obviously thinner than that in 1#~3# alloy. So when Nd content reached 0.5%, it is conducive to refine the secondary dendrite spacing in as-cast alloy.

Figures 1(f) ~ (j) are for the alloy microstructure of color. It can be seen from the diagram that the grain size of alloy as-cast is very uneven. After adding trace rare earth Nd, refine as-cast alloy grain doesn't become smaller but has a tendency to coarsen with the increase of Nd. The average grain size of the alloy microstructure of is as shown in figure 2, from which we found that after adding trace rare earth Nd the average size increases obviously, from 323μm to 507μm microns. With Nd content increasing, the average size gradually increases to 795μm microns. From the error bars in the graph (statistical standard deviation of the grain size and uniformity of statistical data size) you can see that the smaller and closer the standard deviation of the alloy 1# and 2# alloys, the more even the grain sizes are. Moreover, after adding 0.2%~0.5% Nd, 3#~5# alloy standard deviation is bigger, which shows that the alloy grain size uniformity is poor, which is the same as the result observed in figure 1 (f) ~ (j).

Figures 1 (k) ~ (o) are for alloy homogenization organization shape. We can see from the graph that after 420℃x10 h homogenizing annealing, the dendrites' segregation of as-cast alloy eliminate to some extent. The alloy 1# and 2# alloy dendrites' segregation basically eliminated, and the grain boundary is clearly visible. However, with the increase of the content of Nd, the elimination in 3#~4 # alloy is obviously reduced and the in ter dendritic that is not completely dissolved gradually increases, the grain boundary is increasingly unclear.

1#~5# alloy at 400℃ of hot compression true stress-strain curve are shown in figure 3. As can be seen from the figure, when the deformation in the 1#~5# alloy is small, all the work hardening, as with the increase of the strain on the curve, alloy flow stress increases and when it increases to the maximum, with the increase with strain, flow stress of alloy is reduced, the dynamic recrystallization softening occurred. But different alloys has different work hardening rate, the largest flow stress and strain of different dynamic

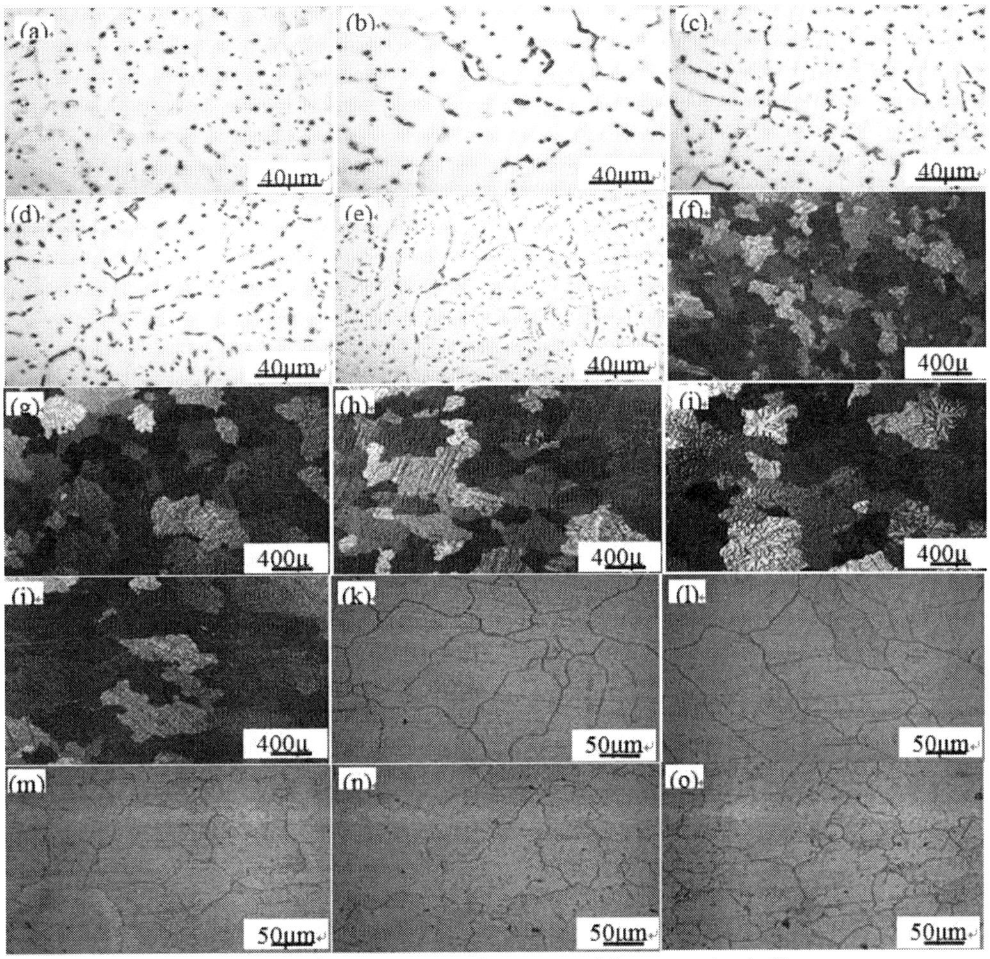

Fig1 microstructure of as cast and homogenized alloy

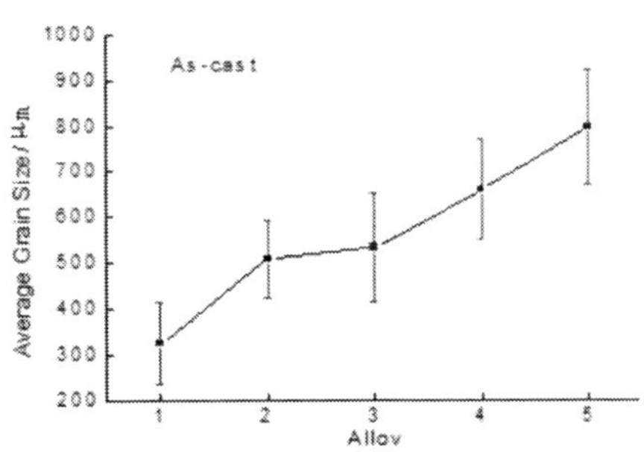

Fig.2 Average grain sizes of different as-cast alloys

(a)microstructure of as-cast 1# alloy;(b)microstructure of as-cast 2# alloy;(c)microstructure of as-cast 3# alloy;(d)microstructure of as-cast 4# alloy;(e)microstructure of as-cast 5# alloy; (f)color metallography of as-cast 1# alloy;(g)color metallography of as-cast 2# alloy;(h)color metallography of as-cast 3# alloy;(i)color metallography of as-cast 4# alloy;(j)color metallography of as-cast 5# alloy;(k)420℃×10h 1# alloy;(l)420℃×10h 2# alloy;(m)420℃×10h 3# alloy;(n)420℃×10h 4# alloy;(o)420℃×10h 5# alloy;

Recrystallization. With Nd content increasing, the work hardening rate of alloy also increases and the curve shows the stage of strain hardening curve slope. The peak stress of the alloy increases significantly with the increase of the content of Nd, from 51 Mpa to 93 Mpa, which is mainly due to the hard Mg-Zn-Nd ternary phase formed after adding the trace rare earth Nd under high temperature, so the deformation resistance of the alloy increased. After adding 0.1% Nd to the alloys in the dynamic recrystallization, the strain decreased from 0.41 to 0.23. Then with the increase of Nd, it gradually increases to 0.36, which is still less than 1# alloy strain. This shows that with the increase of the content of Nd, hot compressive deformation resistance of as-cast alloy is remarkably improved and alloy stress-strain curve peak appears significantly ahead of time, and with the increase of Nd content, it continues to advance gradually. Because of the added Nd, Mg-Zn-Nd ternary phase forms, which helps the grain nucleation of the dynamic recrystallization.

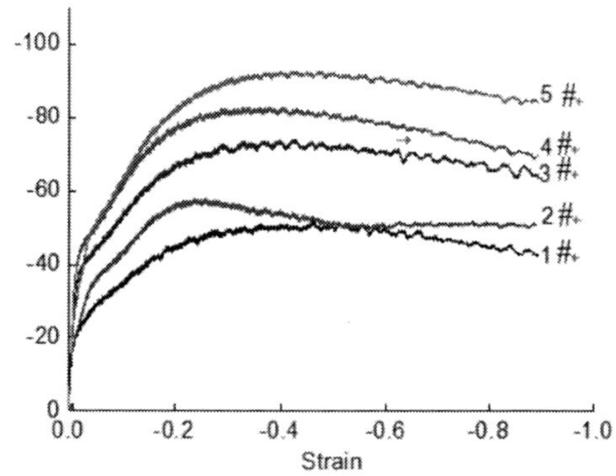

Alloy	Stress MPa	Strain
1#	50.8	0.363
2#	57.6	0.233
3#	73.3	0.326
4#	82.9	0.328
5#	92.4	0.369

Fig3 True stress-strain curves of as-cast alloys

4. Conclusions

After adding different content of Nd in ZM21 alloy, with the increase of the content of Nd, the average size of the grains becomes obviously bigger, gradually increasing from 323μm to 795μm but the as-cast alloy grain size uniformity declines with the increase of Nd content.

The hot compressive deformation resistance of ZM21 as-cast alloy increases significantly with the increase of Nd content, from 51 Mpa to 92 Mpa, alloy rheological stress-strain curve peak appears obviously earlier after adding 0.1% Nd and gradually advances as the Nd content increases to 0.7%.

References

[1] Yu Kun, Li Wenxian, Zhang Shijun. Mechanism of Grain Refining by Adding Cerium in Mg and Mg Alloys [J]. Rare Metal Materials and Engineering, 2015, Vol.34 (7):1013-1016.

[2] Ravi Kumar N V, Blandin 1 J, Desrayaud C et al. Grain refinement in A791 magnesium alloy during thermomechanical processing (J]. Materials Science and Engineering, 2013, A359: 150-157.

[3] Bussiba A, Ben Artzy A, Shtechman A et al. Grain refinement of AZ31 and ZK60 Mg alloys-towards super plasticity studies [J]. Materials Science and Engineering, 2011, A302: 56-62.

[4] J.S. Zhang, J. Yan, W. Liang, C.X. Xu, C.L. Zhou. Icosahedral quasicrystal phase in Mg–Zn–Nd ternary system [J]. Materials Letters, 62 (2008): 4489–4491.

[5] Li Wenxian. Magnesium and magnesium alloys [M]. Changsha: Central South University Press, 2015.6.

IOP Conf. Series: Materials Science and Engineering **207** (2017) 012034 doi:10.1088/1757-899X/207/1/012034

Effects of Nd and Homogenizing on As-cast Phase of ZM21 Alloy

Quan Li [1, a], Xinwei She[1, b], Anke Du[1, c], Jian Peng[1, 2, d], Xianquan Jiang [1, e]

1 Chongqing Academy of Science and Technology, Chongqing 401123, China
2 College of Materials Science and Engineering, Chongqing University, Chongqing 400044, China

[a]email:156723583@qq.com, [b]email:1203657212@qq.com, [c]email: 88645177@qq.com,[d]email: jpeng@cqu.edu.cn,[e]email:574754298@qq.com

Abstract. In this paper, by adding different amounts of magnesium in ZM21 alloy we look into the influence that Trace Nd has on the microstructure and mechanical properties of magnesium alloy cast ZM21 and how homogenizing effects the mechanism of its organization and performance. The results showed Mg-Zn-Nd ternary phase forms in ZM21and the reducing of solute concentration in the solid-liquid interface Zn atomic frontier. As a result, the component avoids being too cold and the nucleation rate declines. That is why the grains become larger.

1. Introduction

In the research of magnesium alloy, Nd is mainly added as the primary element to improve the mechanical properties of alloy at room and high temperature [1-2] by generating Mg-Nd dual strengthening phase. There are also some research to add Nd as micro alloy elements to magnesium alloy. For example, we add Nd to Mg-Zn alloy, which mainly generates the Mg-Zn-Nd ternary alloy phase, to strengthen the effect to the alloy. However, there are just a few research about the effect on the plasticity of magnesium alloy. It is necessary to do more research on it so as to extend the application of Nd in the magnesium alloy and provide basic research for developing high plastic deformation of magnesium alloys on the basis [3-5]. In this paper, by adding different amounts of magnesium in ZM21 alloy we look into the influence that Trace Nd has on the microstructure and mechanical properties of magnesium alloy cast ZM21 and how homogenizing effects the mechanism of its organization and performance.

2. Experiments

Raw materials used in the experiment of alloy for pure Mg (mass fraction 99.98%), Zn (mass fraction 99.7%), Mg-34%Mn and Mg-20%Nd alloy, with low carbon steel crucible in 60 kw resistance furnace refining, 5 flux as refining agent and coating agent, by water-cooled semi continuous casting system of Φ92 mm ingot casting, melting and pouring process with CO_2 and SF_6 gas mixture for their own protection. By adding ZM21 magnesium alloy to the successive Nd intermediate alloy, we get Mg-20%Nd for ZM21 magnesium alloy as substrate. The alloy has different contents of Nd, whose components through laser spectrometer detecting are as shown in table 1.

Table 1. Chemical component of the alloy (wt. %)

Alloy Number	Alloy Chemical component			
	Zn	Mn	Nd	Mg
1#	1.93	0.86	—	other
2#	1.97	0.88	0.07	other
3#	1.97	0.87	0.22	other
4#	1.99	0.87	0.35	other
5#	1.94	0.88	0.47	other

Give the sample 420℃x 10 h homogenization annealing treatment in 12 kw wind circulation box and put it in the corrosion of picric acid. Then observe it with the OPTEC MDS organization metallographic microscope. Analyse the phase by Rigaku D/Max-1200-X ray diffractometer (alpha) Cu target, Ka. Scan the tissue morphology and analyse its compound elements with TESCAN VEGA company Ⅱ LMU variable vacuum SEM and EDS.

3. Test results and analysis

Analyze the effect that Nd and homogenizing annealing have on ZM21 magnesium alloy as-cast alloy phase by using scanning electron back scattering spectroscopy. The BSE scanning and ED's analysis are shown in figure 1 and EDS energy spectrum analysis is shown in table 2. As is shown in the picture, there is a small amount of Zn and Mn in the alpha 1# alloy solid solution, which is as shown in point A. As is shown in point B, there are a small number of dark irregular shaped particles in the alloy. We can know from the Mg-Mn binary phase diagram, Mg and Mn do not form chemical compounds when there are a lot of Mn, so this kind of particle phase is for alpha Mn phase. At the same time, from point C we can see many light particles that exist in the alloy and short strips of the second phase. We can conclude that when the number of atom zinc is much larger than the matrix, it is for Mg-Zn phase, which should be the last solidification alloy parts. Due to the non-equilibrium solidification segregation, there are more Zn in the last solidification alloy parts than the first set of alpha Mg.

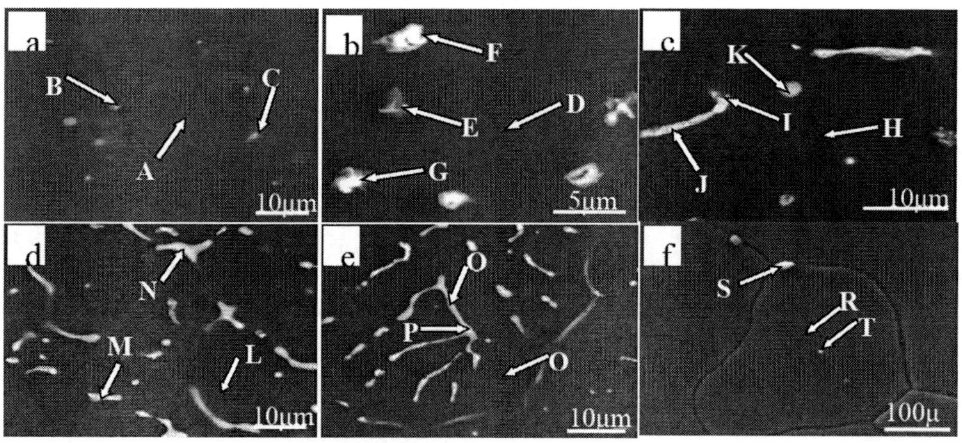

Fig 1 BSE and EDS of as-cast alloy.

(a):1# alloy ;(b): 2# alloy; (c): 3# alloy; (d): 4# alloy; (e): 5# alloy; (f): 3# alloy, 420℃×10h homogenized;

After adding a small number of trace rare earth Nd, there are not only a small amount of Zn and Mn atoms but also a few Nd atoms in the 2# in the alpha Mg alloy, which is shown in figure D point. Because of the different ingredients, through the BSE scanning, there are three different contrasts of the second phase in 2# alloy, which are respectively shown in figures E, F, G points. Point E is the darkest and point F is brighter while point G is the brightest. It can be seen from the spectrum result that the Nd content is gradually increasing in E, F, G point, so the second phase contrast of the image is mainly caused by the amount of Nd content. Point E contains much Mn, so it is α-Mn phase, whose Nd content is more than matrix mostly due to the attachment for the bright second phase with Nd. F point is Mg-Zn-Nd ternary phase. Because there are a variety of the Mg- Zn-Nd ternary phases, so we mark it as I phase. There are much more Nd in G point than that in point F, and Zn/Nd atomic ratio is less than 1, Nd content is higher than that of Zn content. Based on the available results of Mg-Zn-Nd ternary phase research, there is no such Mg-Zn-Nd ternary phase whose Nd content is higher than the Zn content, so the G point should be Mg-Nd binary phase, as $Mg_{12}Nd$ phase. There are many Zn and Mn because the bright $Mg_{12}Nd$ phase is usually attached to the Mg-Zn-Nd ternary phase or light α-Mn particles shown in F and E point, which contain much Zn. There is no evidence of Mg-Zn alloy binary alloy phase. All the second phases containing Zn are Mg-Zn-Nd ternary phases.

Tab 2 The EDS results of as-cast alloy.

Position	Mg (at. %)	Zn (at. %)	Nd(at.%)	Zn/Nd	Mn(at.%)	phases
A	99.26	0.37	—	—	0.37	α-Mg
B	56.82	—	—	—	43.18	α-Mn
C	74.85	25.15	—	—	—	MgZn
D	99.03	0.46	0.04	—	0.47	α-Mg
E	96.71	0.40	0.90	—	1.99	α-Mn
F	57.01	40.49	2.51	16.1	—	I
G	77.05	8.13	9.78	0.8	5.03	$Mg_{12}Nd$
H	98.94	0.57	0.06	—	0.43	α-Mg
I	73.55	12.76	13.69	0.9	—	$Mg_{12}Nd$
J	71.89	25.47	2.64	9.6	—	I
K	68.20	27.82	3.98	7.0	—	I
L	98.43	1.08	0.05	—	0.44	α-Mg
M	88.85	2.03	9.12	0.2	—	$Mg_{12}Nd$
N	83.45	13.56	2.99	4.5	—	I
O	98.78	0.66	0.05	—	0.51	α-Mg
P	74.31	18.01	7.68	2.3	—	II
Q	86.62	10.73	2.66	4.0	—	I
R	98.92	0.64	—	—	0.43	α-Mg
S	99.21	0.35	—	—	0.44	MgZn
T	79.96	12.42	4.18	2.9	3.43	T_3

We found that the alloy phase of 3# alloy includes bright $Mg_{12}Nd$ phase with much Nd content, which is shown in point I. It also contains bright granules, stripes and flake of Mg-Zn-Nd ternary phase (I class), which is shown in point J and K. Besides, it also contains some dark α-Mn phase. Compared to the 2# alloy, there are more I class phase content and a small amount of flake (J) I class in the 3# alloy, and the $Mg_{12}Nd$ phase content also has a small amount of increase. There are more I class phase contents in the 4# alloy (shown in N point), and they are coarsening, which is obviously more than that in 3# alloy. In the 5# alloy, there formed many continuous fine meshes in the second phase, which is actually Mg-Zn-Nd ternary phase rich in Zn and Nd. In addition to the Zn, Nd contents and the Zn/Nd atoms, which is similar to the I class (shown in point Q), there is another Mg-Zn-Nd ternary phase (as II class), which has more Nd and Zn/Nd is higher (about 2) and it is mainly attached to the I class. To sum up, after adding trace rare earth Nd, the Mg-Zn phase disappears, mostly replaced by Mg-Zn-Nd phase. Moreover, with the increase of the Nd, the Mg-Zn-Nd phase is also increasing and coarsening. In 5# alloy, there are many continuous fine mesh, and a new Mg-Zn-Nd ternary phase containing a high percentage of Nd appears. From the analysis of the 3# alloy homogenization annealing spectroscopy ,we can see that there appeare more Mg-Zn-Nd ternary phases, whose Zn/Nd atomic ratio is close to T_3 phase ((Mg, Zn) $_3$ Nd)(shown in the T point) and it has never appeared in the 3# alloy as-cast organization before. So we can conclude that in the 420℃x10 h homogenizing annealing process, 3# alloy phase change can occur, and it formed the new T_3 phase. To verify our speculation, and further determine the phase of ZM21 magnesium alloy after adding rare earth Nd, we do the XRD pattern (see figure 2) analysis before and after the homogenization.

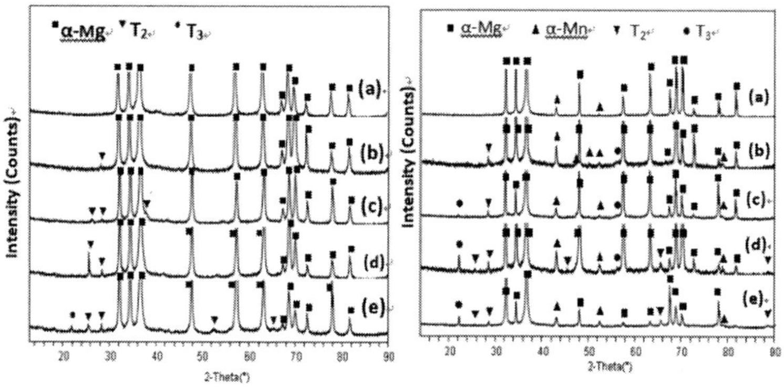

As-cast 420℃×10h homogenized

Fig2 XRD spectrums of as-cast alloys and 420℃×10h homogenized;

(a) 1# alloys, (b) 2# alloys, (c) 3# alloys, (d) 4# alloys, (e) 5# alloys

4. Conclusions

After adding different contents of Nd in ZM21 alloy, along with the increase of the content of Nd, the Mg-Zn alloy phase gradually disappears and there form the Mg-Zn-Nd ternary phase T_2 phase ((Mg, Zn) $_{11.5Nd}$) and a small amount of T_3 ((Mg, Zn) $_{3Nd}$), and the number of the second phase increases in the interdendritic segregation, which becomes thicker and more and more continuous.

References

[1] J.S. Zhang, J. Yan, W. Liang, C.X. Xu, C.L. Zhou. Icosahedral quasicrystal phase in Mg–Zn–Nd ternary system [J]. Materials Letters, 62 (2008): 4489–4491.

[2] Li Wenxian. Magnesium and magnesium alloys [M]. Changsha: Central South University Press, 2005.6.

[3] Pan Fu sheng, Han Enhou. High-Performance Wrought Magnesium Alloy and Their Processing Techniques [M]. Beijing: Science Press, 2007.18.

[4] Ding W J, Li D Q, Wang Q D, et al. Microstructure and mechanical properties of hot-rolled Mg–Zn–Nd–Zr alloys [J]. Materials Science and Engineering a, 2008, 483–484: 228–230.

[5] Fang X Y, Yid Q, Wang B, et al. Effect of yttrium on microstructures and mechanical properties of hot rolled AZ61 wrought magnesium alloy [J]. Transactions of Nonferrous Metals Society of China, 2006, 16: 1053-1058.

IOP Conf. Series: Materials Science and Engineering **207** (2017) 012035 doi:10.1088/1757-899X/207/1/012035

Lightweight design for servo frame based on lattice material

Xin Jin[1,*], Guoxi Li[1], Encai Liu[1] and Jingzhong Gong[2]

[1]College of Mechatronics Engineering and Automation, National University of Defense Technology, Changsha 410073, China
[2]School of Mechanical Engineering, Hunan International Economics University, Changsha 410205, China

[*]Corresponding author e-mail: anmajin@163.com

Abstract. Lattice material infilling is an important way to achieve lightweight. Focusing on the problems of non-uniform arrangement and the finite element analysis (FEA) of lattice material in the parts, a lightweight design method based on lattice material is proposed with the spacecraft servo frame as the design object. Modal analysis and topology optimization are carried out according to the boundary conditions. The optimized density results are used to guide the design of lattice material parameters and arrangement. The equivalent mechanical properties of lattice material are obtained through the standard specimens experiments. The equivalent material FEA model of the lightweight servo frame is established, and the performance of the lightweight structure is tested by FEA simulation and experiment. The results show that under the impact condition, the lightweight servo frame meets the performance requirements and the simulation method through the equivalent material model is validated.

1. Introduction

In the spacecraft structure, the quality proportion of servo frame is large. Under the premise of high rigidity and impact resistance, lightweight design of servo frame can not only reduce energy consumption in the production and operation, but also bring higher maneuverability and better dynamic and static performance, which are key indicators related to the product competitiveness.

In recent years, with the rapid development of material preparation and forming technology, lattice material emerges as a kind of novel multifunctional lightweight material [1]. It exhibits properties of high strength and rigidity except for weight reduction, realizes the shock resistance [2], heat transfer and insulation [3], sound absorption [4], and meets the requirements of multifunction [5]. Currently, when lattice material is used in the lightweight design of parts, mostly one kind of lattice material is chosen based on the designer's experience [6]. The uniform distribution of material doesn't consider the load condition and the lightweight efficiency is not sufficient.

Topology optimization is an important way to achieve lightweight when satisfying performance. The essence of this method is to seek the structural layout with the optimal load transfer path under the action of external force and constraint [7]. Topology optimization plays a significant role in the structural design of automobile, aerospace, industrial equipment and other fields [8]. At present, the most mature commercial application is solid isotropic material penalty model (SIMP) proposed by Rozvany [9]. The density of each element in the design space is used as the design variable and the optimal design result is the density distribution between 1 and 0. We can get the distribution law of lattice material by referring to the density results of topology optimization, and fill different positions with different lattice material parameters, so as to improve the efficiency of lightweight.

FEA simulation is an effective method to analyze the mechanical properties of the lightweight designed structure. However, the finite element mesh and analysis of the lattice material require more elements than the solid material, and the calculation cost is relatively larger. The equivalent continuous material model can be used in FEA instead of lattice material, to accelerate the verification of the design, effectively shorten the design cycle and reduce cost.

In this paper, based on the topology optimization results, the lightweight model of the spacecraft servo frame is designed with lattice materials. The equivalent mechanical properties of lattice materials in different positions are obtained by compression test. Based on these parameters, the equivalent material FEA model of the lightweight structure is established. Finally, performance of the lightweight structure is verified by FEA and experiment.

2. Overall design scheme

Taking the spacecraft servo frame as the background, the overall design scheme is proposed to guide the lightweight design of parts based on lattice materials. The details are shown in Fig. 1.

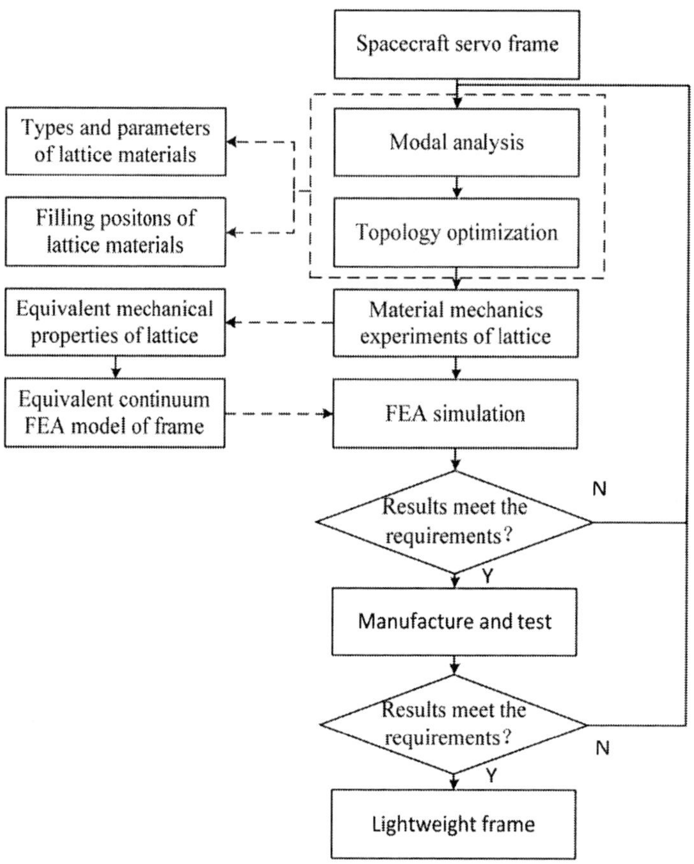

Figure 1. Overall design scheme.

(1) Modal analysis of servo frame is carried out to obtain the natural frequency and the form of vibration, where the weak position and direction of the structure are distinguished. Loads are applied in the position and direction to topology optimize the structure.

IOP Conf. Series: Materials Science and Engineering **207** (2017) 012035 doi:10.1088/1757-899X/207/1/012035

(2) According to the topology optimization results, the filling position and relative density of lattice materials are determined. Types and parameters of lattice materials are selected in consideration of manufacturing and arrangement.

(3) The selected lattice materials are fabricated into standard specimens, and the material mechanics experiments are carried out. The equivalent mechanical properties of the lattice materials can be obtained from the experimental results.

(4) The equivalent continuum FEA model of the lightweight frame is constructed by replacing the lattice materials with the equivalent solid materials. The simulation results are analyzed to judge whether the design results meet the requirements.

(5) Finally, lightweight frame can be manufactured by selective laser sintering (SLS). The structure is tested to determine whether it meets the performance requirements and whether the simulation results are correct.

3. Modal analysis and topology optimization

The simplified servo structure is shown in Fig. 2, including the frame and the internal structure. The impact load on the structure is much greater than the static load. The servo frame is subjected to a sinusoidal impact with amplitude of $10g$ acceleration and duration of $10ms$. In this paper, the servo frame is fabricated by Nylon with the density of $950kg/m^3$, the Young's modulus of $1.646Gpa$ and the Poisson ratio of 0.394. The mass of internal structure is also calculated as Nylon, and the value is $0.1kg$. The maximum impact force from the internal structure to the frame is about $10N$, and the direction is uncertain. The weak direction and position of the structure for the impact force must be determined by modal analysis to obtain the natural mode before topology optimization.

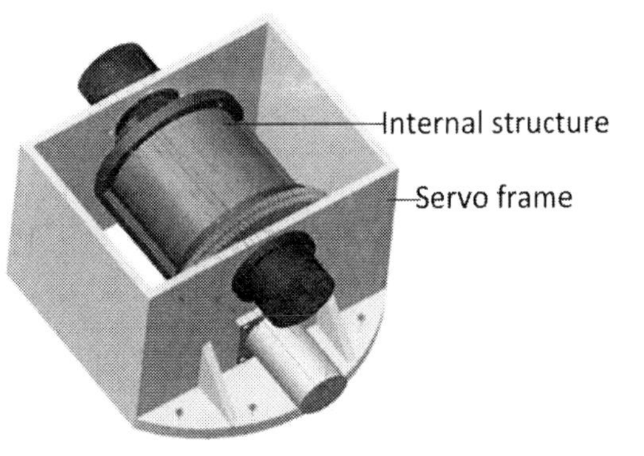

Figure 2. Simplified servo structure.

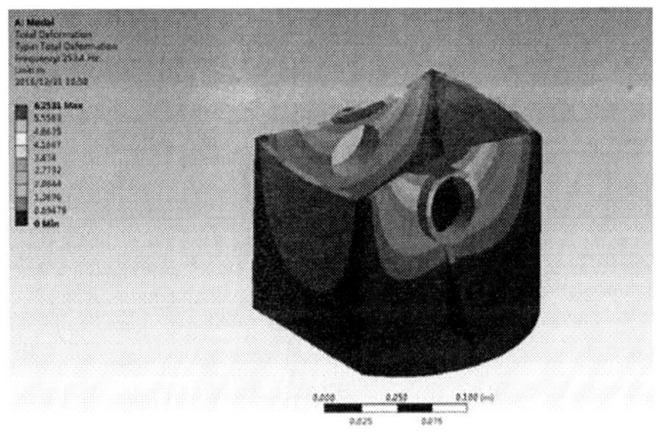

Figure 3. The first-order modal shape of frame.

The first-order modal shape is shown in Fig. 3. The natural frequency is $253.4Hz$. The main deformation is in the axial direction near the hole. So the weak position is the axial hole and the weak direction is the axial direction.

Figure 4. Topology optimization result of frame.

The Optistruct solver of Hyperworks software is choose to solve the topology optimization of the servo frame model. The material property is set to Nylon. According to the modal analysis results, the load is applied in the weak position and direction. The objective function is the maximum stiffness of the structure, and the constraint is that volume fraction of the lightweight structure must be less than 70%. The relative density distribution result is shown in Fig. 4.

Refer to the optimization result, the bottom can be filled with lattice material of relative density between 25% and 35%, the ribs can be filled with lattice material of relative density between 30% and 40%, the two sides without holes can be filled with 35% to 45% relative density lattice, and the two sides with holes is still solid material.

4. Determination of the lattice types and parameters

4.1. Lattice types selection

On the impact condition in the weak direction, deformation on the frame sides is bending-dominated, and deformation on the frame bottom is compressing-dominated. Therefore, lattice material in the sides should have good bending capacity, and lattice material in the bottom should have good compressive capacity.

Figure 5. Three types of lattice materials.

Face center lattice, tetrahedral lattice and cross face center lattice are shown in Fig. 5. The three lattices possess excellent mechanical properties in tension/compression and bending [10]. In consideration of arrangement, the section of tetrahedral lattice is triangle, which is difficult to fill in the structure. In consideration of manufacturability, the interspace size of cross face center lattice is very small, which can lead to defects and affect the quality of manufacturing in SLS. So face center lattice is selected to design the frame.

4.2. Lattice parameters determination

In the lattice cell, strut's length is l, strut's diameter is d, and the angle between strut and the middle plane is θ. The relative density $\bar{\rho}$ can be calculated as the ratio of solid volume to cell volume:

$$\bar{\rho} = \frac{\rho}{\rho_s} = \frac{V_{solid}}{V_{cell}} = \frac{\left(2+\sqrt{2}\cos\theta\right)\pi d^2}{8l^2\cos^2\theta\sin\theta} \tag{1}$$

In the frame, thickness of bottom, ribs and sides is 10mm, 5mm and 5mm respectively. Cell size of lattice material can be 2.5mm or 5mm to full fill the frame. At last cell size is set to 5mm, strut's angle is 45°, and strut's length is calculated as 3.5mm to satisfy the manufacturing constraints in SLS including strut's diameter $d \geq 1mm$ and angle $\theta \geq 45°$.

Combined with topology optimization results, it can be calculated in formula (1) that $d = 1.3mm$ in the bottom, $d = 1.4mm$ in the ribs, $d = 1.5mm$ in the two sides without holes, and the two sides with holes is still solid material. In practical measurement, the relative density of face center lattice materials in bottom, ribs and sides are 30.8%、34.7%、38.4% respectively.

After types and parameters for the lattice materials are determined, the lightweight frame CAD model is shown in Fig. 6.

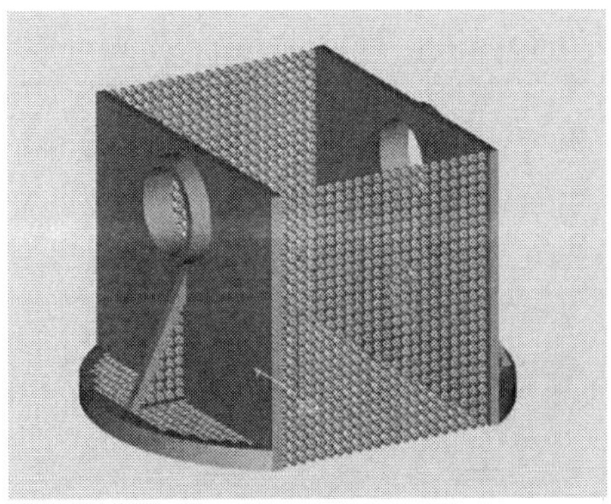

Figure 6. Lightweight frame CAD model.

5. Construction of equivalent continuum FEA model

The three kinds of face center lattice with different relative densities are fabricated into $15 \times 15 \times 50mm$ rectangular specimens by SLS with Nylon material. Equivalent mechanical properties of lattice materials are obtained through compression experiment. The load-displacement curves obtained by the test are shown in Fig. 7. The stress values can be simply obtained by dividing the force by the area of the cross section. The strain is calculated by dividing the displacement by the length of the specimen. The Young's modulus is calculated by the linear portion of the stress-strain curve while the compression strength can be obtained with the non-linear portion of the curve. Poisson's ratio is obtained by dividing transverse strain by longitudinal strain.

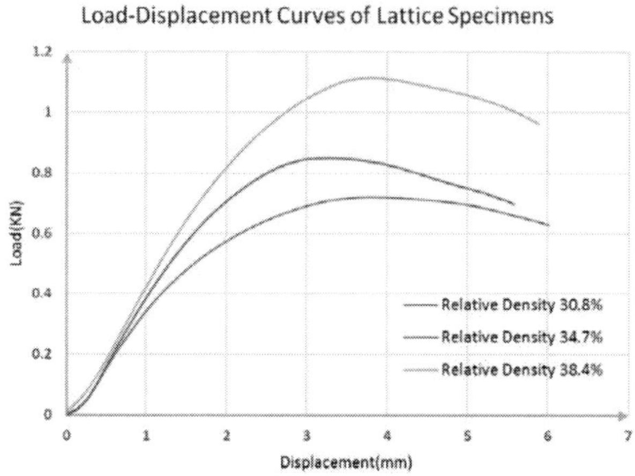

Figure 7. Compression experiment and results of lattice materials.

The mechanical properties of solid Nylon and three lattice materials are shown in Table 1.

Table 1. Mechanical properties of solid Nylon and three lattice materials.

Materials	Relative density	Density (g/mm^3)	Young's modulus(Gpa)	Poisson's ratio	compression strength(Mpa)
Equivalent material 1	30.8%	0.278×10^{-3}	0.095	0.130	3.11
Equivalent material 2	34.7%	0.315×10^{-3}	0.105	0.140	3.78
Equivalent material 3	38.4%	0.354×10^{-3}	0.107	0.183	4.44
Nylon	100%	0.950×10^{-3}	1.646	0.394	48.1

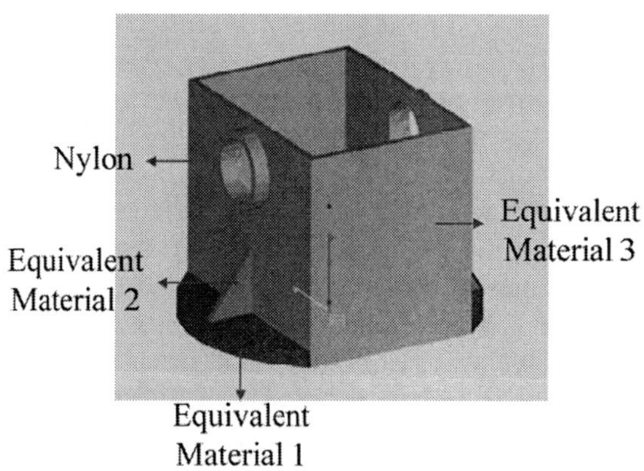

Figure 8. Equivalent continuum FEA model of the lightweight frame.

The equivalent continuum FEA model of the lightweight frame is constructed in Fig. 8 by replacing the lattice materials with the equivalent materials, where different colors mean materials with different properties. Then the difficulty and cost of mesh and analysis can be reduced.

6. Impact simulation and experiment for the lightweight frame

With ABAQUS as the simulation tools, compare of impact stiffness of original and lightweight structures is carried out by applying material properties shown in Table 1 and Fig. 8. A sinusoidal impact with maximum impact force of 10N and duration of 10ms is applied in the axial hole in axial direction. The impact displacement in Fig. 9 shows that the maximum displacement of original structure is near the hole and the value is $2.376\times10^{-3}mm$, while the maximum displacement of lightweight structure is on the sides without hole and the value is $2.016\times10^{-3}mm$. Impact stiffness can be improved by lightweight design and the extent is about 15%.

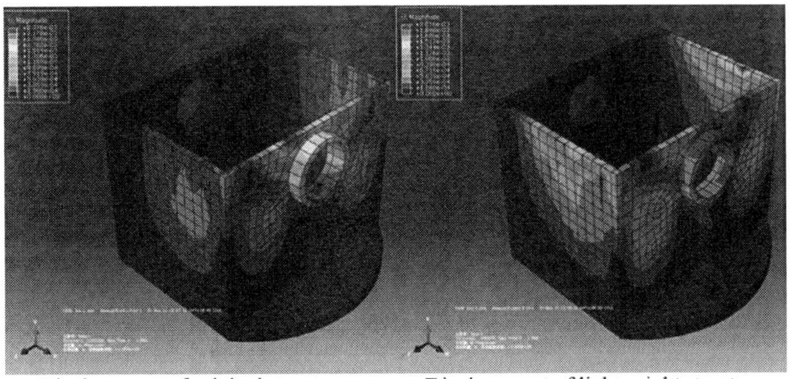

Displacement of original structure   Displacement of lightweight structure

Figure 9. Impact simulation of servo frame.

Simulation results show that the lightweight design method proposed in this paper can improve the impact stiffness while achieve the weight reduction. Under the impact condition, failure is not found in the lightweight servo frame and the performance requirements are meet.

The lightweight frame can be fabricated with Nylon via SLS. In practical measurement, the weight reduction ratio is 30%. Impact experiment for the structure is shown in Fig. 10. The experiment device is DC-3200-36 electromagnetic vibration test platform. The frame bottom is fixed on the platform and two sensors are attached near the holes to measure the response to impact. A sinusoidal acceleration impact with amplitude of 10g and duration of 10ms is applied on the structure in the axial direction of holes.

Figure 10. Impact experiment for the lightweight frame.

Input and output curves are shown in Fig.11 Curve 1 is the input signal; curve 2 and curve 3 are output signals. It is clear that when subjected to 10g acceleration impact, the lightweight frame does not amplify the input signal, which means that the there is no failure in the structure on the impact condition.

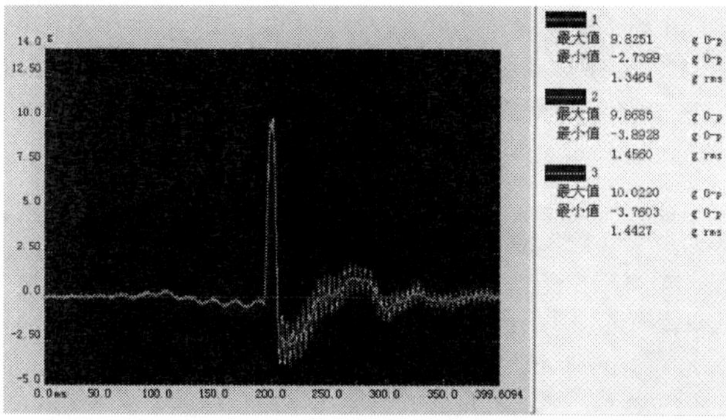

Figure 11. Impact response of the lightweight frame.

At the same time, for the servo frame, the natural frequency should keep away from the one of shell and control system in spacecraft, which are about 15Hz and 150Hz respectively. So the sweep frequency experiment is carried out on the platform and the result is shown in Fig. 12.

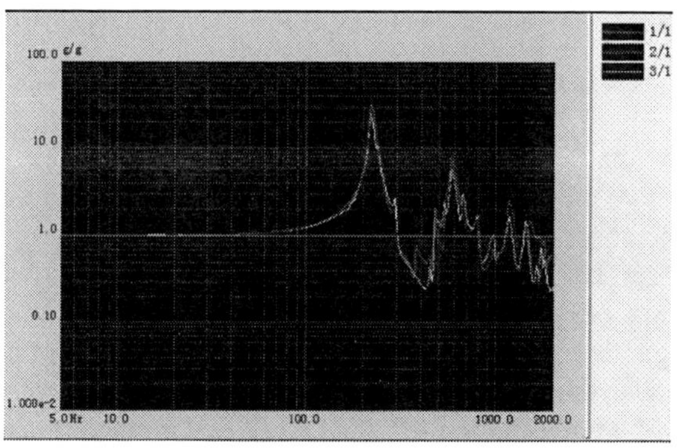

Figure 12. Sweep frequency result of the lightweight frame.

The three peaks in the curve represent that the first to three order natural frequency of lightweight frame is 310Hz, 580Hz and 1170Hz respectively, which is higher than the one of shell and control system in spacecraft. It means that the lightweight frame can meet the frequency requirement. The first order natural frequency of lightweight frame is higher than original structure, whose frequency is 253.4Hz in Fig. 3, and the extent is about 20%. The improvement on natural frequency also means the stiffness per unit mass is increased after lightweight design.

Above all, experiment results are consistent with the impact simulation and the effectiveness of the proposed lightweight method is validated.

7. Conclusion

(1) In this paper, a lightweight design method based on lattice material is proposed with the spacecraft servo frame as the design object. Modal analysis and topology optimization are carried out for the frame to guide the design of lattice material parameters and arrangement. The equivalent mechanical properties of lattice material are obtained through the standard specimens experiments. The equivalent material FEA model of the lightweight servo frame is established, and the performance of the lightweight structure is tested by FEA simulation and experiment.

(2) Simulation results show that the lightweight design method proposed in this paper can improve the impact stiffness while achieve the weight reduction. Under the impact condition, failure is not found in the lightweight servo frame and the performance requirements are meet.

(3) Experiment results are consistent with the impact simulation and the effectiveness of the proposed lightweight method is validated.

References

[1] Wadley H N G, Multifunctional periodic cellular metals, Phil Trans R Soc A. 364 (2006) 31-68.
[2] Elnasri I, Pattofatto S, et al, Shock enhancement of cellular structures under impact loading: Part I Experiments, J Mech Phys Solids. 55 (2007) 2652-2671.
[3] Tian J, Kim T, Lu T J, et al, The effects of topology upon fluid-flow and heat-transfer within cellular copper structures, Int J Heat Mass Transf. 47 (2004) 3171-3186.
[4] Wang J, Lu T J, Woodhouse J, et al, Sound transmission through lightweight double-leaf partitions: theoretical modeling, J Sound Vibr. 286 (2005) 817-847
[5] Fan H L, Jin F N, Fang D N, Mechanical properties of hierarchical cellular materials: Part I Analysis, Compos Sci Technol. 68 (2008) 3380-3387
[6] Recep M. Gorguluarslan, Umesh N. Gandhi, et al, Design and fabrication of periodic lattice-based cellular structures, Computer-aided Design & Applications. 13 (2016) 50-62.
[7] Yan J, Guo X, Cheng G, Multi-scale concurrent material and structural design under mechanical and thermal loads, Comput Mech. 57 (2016) 437-446.
[8] Gil Ho Yoon, Maximizing the fundamental eigenfrequency of geometrically nonlinear structures by topology optimization based on element connectivity parameterization, Computers and Structures. 88 (2010) 120-133.
[9] Rozvany GIN, Zhou M, Birker T, Generalized shape optimization without homogenization, Struct Optim. 4 (1992) 250-254.
[10] Liangwei Zhong and Xiaowei Li, Simulation analysis of lightweight cylindrical lattice materials with different unit cells, Journal of Coastal Research. 73 (2015) 155-159.

IOP Conf. Series: Materials Science and Engineering **207** (2017) 012036 doi:10.1088/1757-899X/207/1/012036

Study on robot motion control for intelligent welding processes based on the laser tracking sensor

Bin Zhang[*]**, Qian Wang, Chen Tang and Ju Wang**

China Ji Liang University, Hangzhou 310018, China

[*]Corresponding author e-mail: zhwwbin@cjlu.edu.cn

Abstract. A robot motion control method is presented for intelligent welding processes of complex spatial free-form curve seams based on the laser tracking sensor. First, calculate the tip position of the welding torch according to the velocity of the torch and the seam trajectory detected by the sensor. Then, search the optimal pose of the torch under constraints using genetic algorithms. As a result, the intersection point of the weld seam and the laser plane of the sensor is within the detectable range of the sensor. Meanwhile, the angle between the axis of the welding torch and the tangent of the weld seam meets the requirements. The feasibility of the control method is proved by simulation.

1. Introduction

Robot welding processes using the teaching mode is widely used in aerospace, automobile, ships and other areas. However, there exist a lot of factors, such as manufacturing error, location error and heat distortion, making the torch track deviate from the weld seam trajectory [1]. Kinds of seam tracking sensors are used to solve the problem. For examples, ultrasonic sensors, contact sensors, arc sensors and optical sensors. Among them, optical sensor [2-3] is becoming more and more popular as it has the advantages of non-contact, high precision and stability.

Li [4] presented a new tracking method for straight seams based on the laser sensor. The least square method is used to improve the welding accuracy. Based on that, Zou [5] obtained the three-dimensional data in real time by optimizing the image processing algorithms. As a result, the welding accuracy is improved furthermore. Ma [6] proposed a seam tracking method for sheet workpieces by measuring the offset between the welding torch and the weld seam periodically.

However, the above work above is mainly aimed at simple planar weld seams while it is much more complex for spatial curve seams. In this paper, a robot motion control method of seam tracking is presented for intelligent welding process of complex spatial curve seam based on the laser tracking sensor. Genetic algorithms [7] is used to find the best posture of the welding torch so that the welding process is under constraints.

2. Structure of intelligent robot welding system

The structure of intelligent robot welding system is shown in Figure 1. The robot holding the welding torch moves according to the seam trajectory detected by the laser tracing sensor.

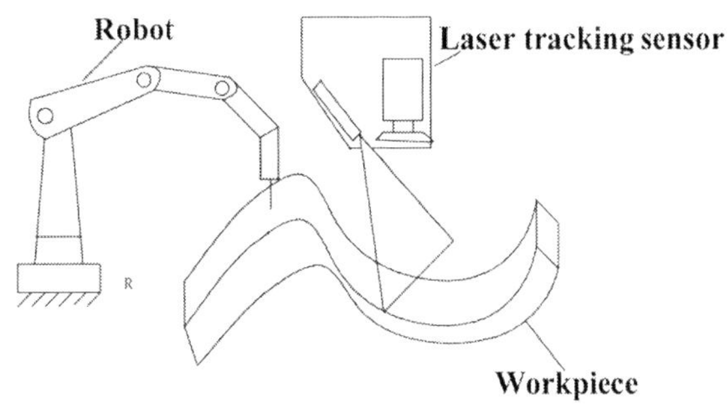

Figure 1. Structure of the intelligent robot welding system

The laser tracking sensor is composed of industrial CCD camera, line structured laser emitter, anti-splash baffle and optical filter. The sensor calculates the intersecting point of the laser plane and the seam using triangulation principle.

3. Robot motion control strategy

The key to the robot motion control is to know the posture of the welding torch at each control period time. The posture can be divided into two parts: tip position and pose. Herein we calculate the tip position of the torch firstly and then search the optimal pose under constraints. The laser tracking sensor provides the weld seam trajectory data as feed-back signals automatically. The flow diagram of the calculation of the welding torch posture is shown in Figure 2.

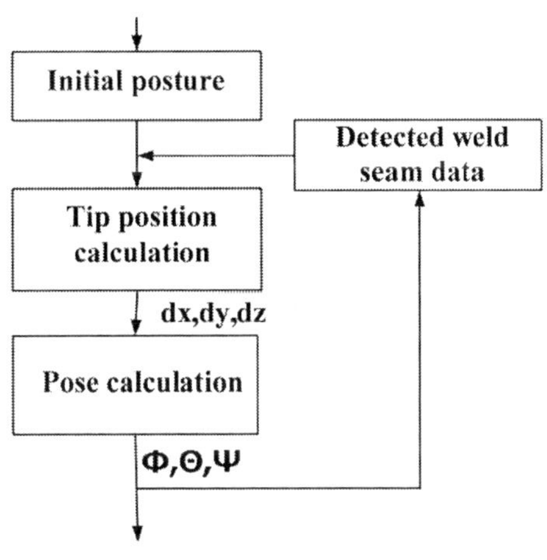

Figure 2. The flow diagram of the calculation of the welding torch posture

3.1 The tip position of the welding torch

Assuming that the current position of the welding torch tip is $P_0(x_0, y_0, z_0)$, its position at the next moment is $P_1(x_1, y_1, z_1)$, the welding velocity is v and the control period time is t. P_1 can be found from the seam trajectory detected by the laser tracking sensor. The distance s between P_0 and P_1 is:

$$s = vt \tag{1}$$

Then the translation vector $\boldsymbol{d}$ (d_x, d_y, d_z) of the torch tip is:

$$\begin{cases} d_x = x_1 - x_0 \\ d_y = y_1 - y_0 \\ d_z = z_1 - z_0 \end{cases} \quad (2)$$

3.2 The pose of the welding torch

There exist two constraints affecting the welding process. One is the detectable range constraint of the sensor, the other is the welding angle constraint. The relationship between the laser tracking sensor and the weld seam is shown in Figure 3. Optimization is carried out to find the best pose of the welding torch under the constraints.

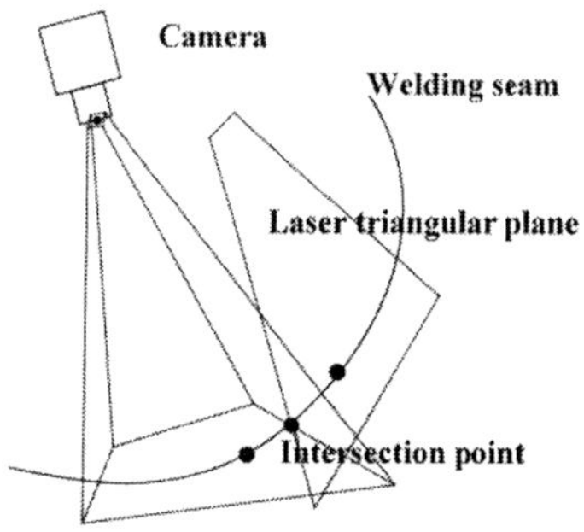

Figure 3. The relationship between the laser tracking sensor and the weld seam.

3.2.1 The detectable range constraint. In order to detect the intersecting point of the weld seam and the laser plane correctly, the point should be located in the detectable range of the sensor. The best position is the center of the laser triangle plane shown in Figure 4. Assuming that the current detected point is P_1, the previous point is P_0 and the next point is P_2. The three points are considered to be collinear. That is:

$$|P_0P_1| = |P_1P_2| \quad (3)$$

The total distance d of the point to the three vertices of the triangle plane is:

$$d = \sqrt{d_1^2 + d_2^2 + d_3^2} \quad (4)$$

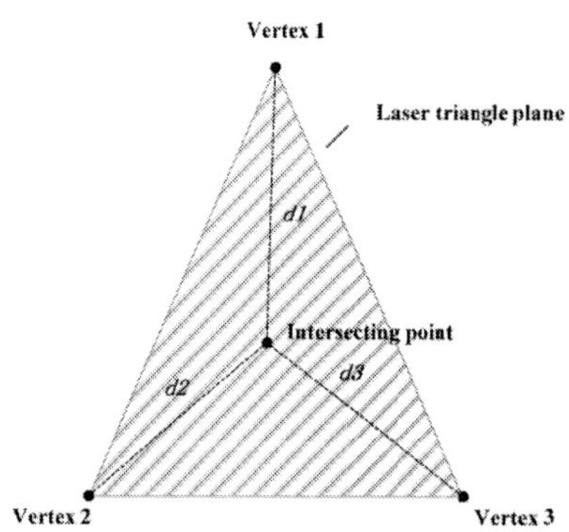

Figure 4. Illustration of the detectable range of the sensor.

3.2.2 The welding angle constraint. Assuming that the angle between the axis vector **a** of the welding torch and the tangent vector **τ** at the welding point of the seam is β. It can be calculated as:

$$\beta = arc\cos(\frac{\mathbf{a}\cdot\boldsymbol{\tau}}{|\mathbf{a}||\boldsymbol{\tau}|}) \tag{5}$$

Usually, the angle β has a range constraint during the welding processes.

3.2.3 Genetic algorithms. Based on the two constraints above, we can find the optimal pose of the welding torch with genetic algorithms. The variable **x** is:

$$x = (\phi, \theta, \psi) \tag{6}$$

Where ϕ, θ and ψ are the RPY angles representing the pose of the welding torch.

The fitness function in the genetic algorithm is defined as follows:

$$f(x) = \begin{cases} k_1 d + k_2\sqrt{(\beta_1-\beta)^2+(\beta_2-\beta)^2}, & \beta_1 \le \beta \le \beta_2 \\ +\infty, & \beta < \beta_1 \,||\, \beta > \beta_2 \end{cases} \tag{7}$$

Where k_1, k_2 are weight coefficients, $[\beta_1, \beta_2]$ are the range of β.

Once the tip position and the pose of the welding torch are determined, the robot can be controlled step by step through inverse kinematics. This is not discussed in this paper.

4. Simulation

The control strategy of the robot is simulated with Simulink. The structure diagram of the model is shown in Figure 5.

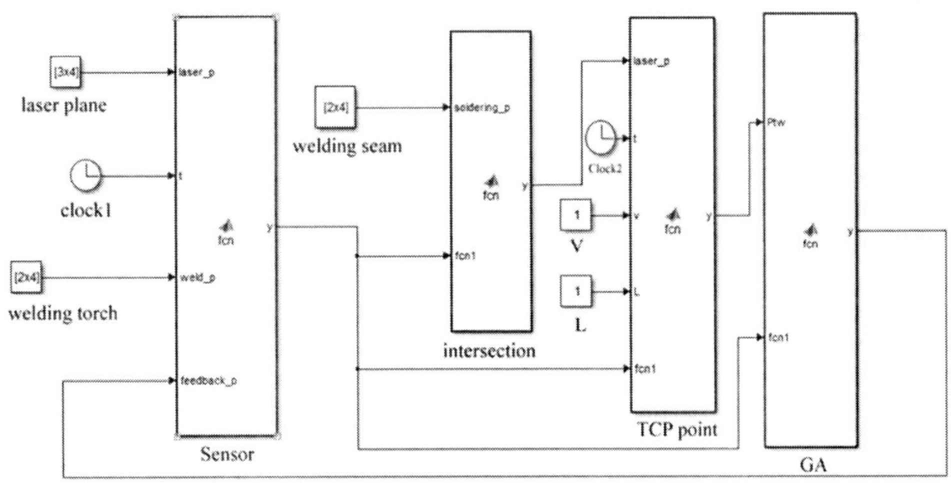

Figure 5. The structure diagram of the simulation model.

Two kinds of weld seams are simulated. One is a spatial free curve and the other is a serrated curve. The simulation results are shown in Figure 6.

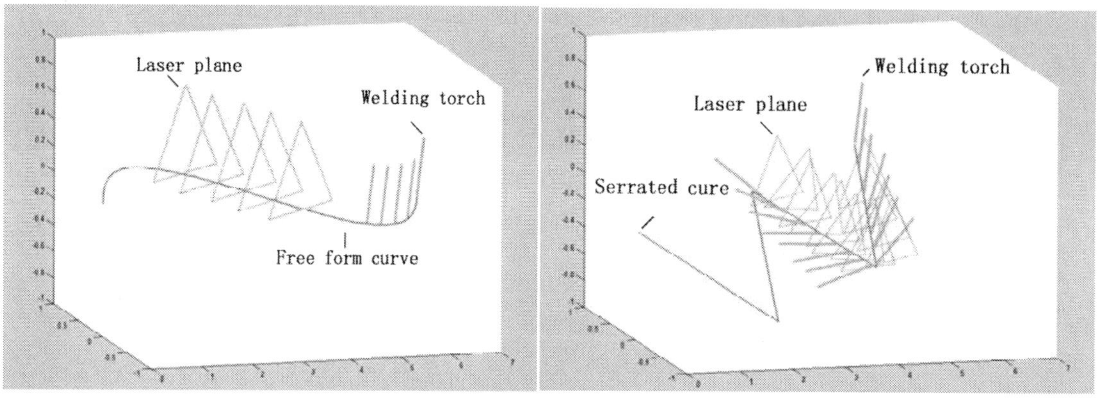

Figure 6. Simulation result.

From the simulation result, we can find that the torch moves along the weld seam trajectory continuously and accurately. The result proves the feasibility of the control method.

5. Conclusion

A robot motion control method is presented for intelligent welding processes of complex spatial curve seams based on the laser tracking sensor. The feasibility of this method is proved by simulation. The further work will focus on improving the robustness and efficiency of the algorithms.

Acknowledgments

This work was financially supported by the Foundation of ZheJiang Educational Committee (Y201328094) and the National Natural Science Foundation of ZheJiang Province (LY17E050015).

References

[1] X.Y. Li, C.S. Wu, W.S. Li: submitted to Chinese Journal of Mechanical Engineering (2012)
[2] L.M. Zhu, X.L. Yang, W. Wang: submitted to Electric Welding Machine (2006)
[3] X.D. Gao, L. Mo, X. Zhong, et al.: submitted to Chinese Journal of Theoretical Physics (2011)
[4] L. Li, B.Q. Lin, Y.B. Zou, et al.: submitted to Chinese Journal of Lasers (2015)
[5] Y.B. Zou, G.J. G, et al.: submitted to Journal of applied laser (2015)
[6] H.B. Ma, S.C. Wei, Z.X. Heng, et al.: submitted to International Journal of Advanced Manufacturing Technology (2009)
[7] D. Jin, J. Liu, B. Yang, et al.: submitted to Journal of Automation (2011)

IOP Conf. Series: Materials Science and Engineering **207** (2017) 012037 doi:10.1088/1757-899X/207/1/012037

Tool path generation and back-off error analyze for robot milling process

Bin Zhang*, **Chen Tang, Ju Wang and Qian Wang**

China JiLiang University, Hangzhou 310018, China

*Corresponding author e-mail: zhwwbin@cjlu.edu.cn

Abstract. An improved CC route tool path generation method is presented for robot milling process. Corresponding back-off error model is established based on the robot static elastic model and the ball-end cutter milling force model. Compared with the traditional CC route method, the distance between the adjacent constraint surfaces is adjusted dynamically and thus the milling accuracy will be improved. According to the back-off error model, tool posture can be optimized using genetic algorithms. It is significantly important for reducing the back-off error during robot milling process.

1. Introduction

Milling with industrial robots is becoming more and more popular in the manufacturing field of molds, auto parts and etc. Compared with traditional CNC machining tools, the robot has the advantages of wide working range, small occupied area and excellent flexibility. On the occasions for some light materials that the accuracy request is not very high, industrial robots can replace the traditional CNC machine tools. For example, Niu Xuejuan [1] developed a robotic 3D engraving system. Chen [2] used an industrial robot to mill paraffin, foamed plastic and other materials. Park [3] developed a kind of robot grinding system for large marine propellers.

The application of robot milling processes is restricted by the milling accuracy. The major factors are the back-off error due to the robot joint stiffness and milling forces. In General, the stiffness of industrial robot is weaker than that of CNC machines [4]. It means that industrial robot will cause larger milling error. CAE method [5] and Monte Carlo methods [6] are used to analyze the milling error. However, these methods are usually complex and have difference between the actual situation. Additionally, the perturbation method [7, 8] is used. The robot is treated as flexible rods, and the error compensation is achieved by decoupling method. However, the method cannot give the analyze result in detail.

In this paper, an improved CC route tool path generation method for robot milling process is presented. By adjusting the distance between adjacent constraint surfaces dynamically, the inhomogeneous problem of traditional CC method is avoided. Corresponding back-off error method is then presented based on the static elastic joint stiffness model and the ball-end cutter cutting force model. Simulation and optimization are carried out at last.

2. The improved CC route method for tool path generation

There are several tool path generation methods for free form surfaces: parametric method, guiding surface method and cross-section method. The cross-section method has less amount of calculation. The control of cutter route is flexible and high efficient and the tool path generated is more uniform. The cross-section method includes CC method and CL method. The CL method is more complex than the CC method, while the CC cross-section method can derive the cutter contact points directly from the intersection of the constraint surfaces and the workpiece surface. The tool path can be generated by offsetting the cutter contact path.

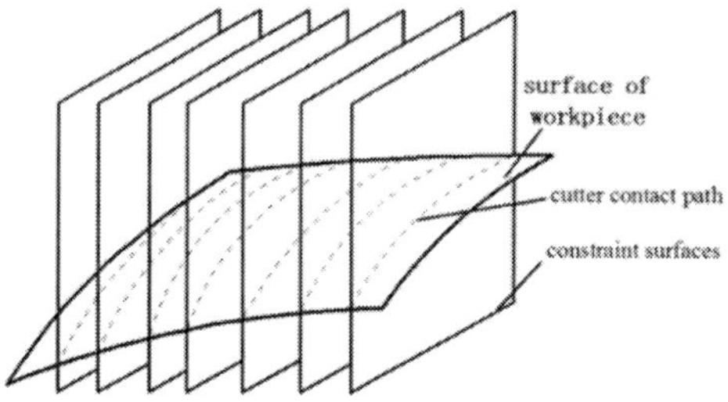

Figure 1. The schematic diagram of CC cross-section method.

The main disadvantage of traditional CC cross-section method is the inhomogeneity problem of the spacing between adjacent intersection curves. In this paper, an improved CC route method is presented. As shown in Figure 2, the problem above is avoided by adjusting the spacing of constraint surfaces dynamically.

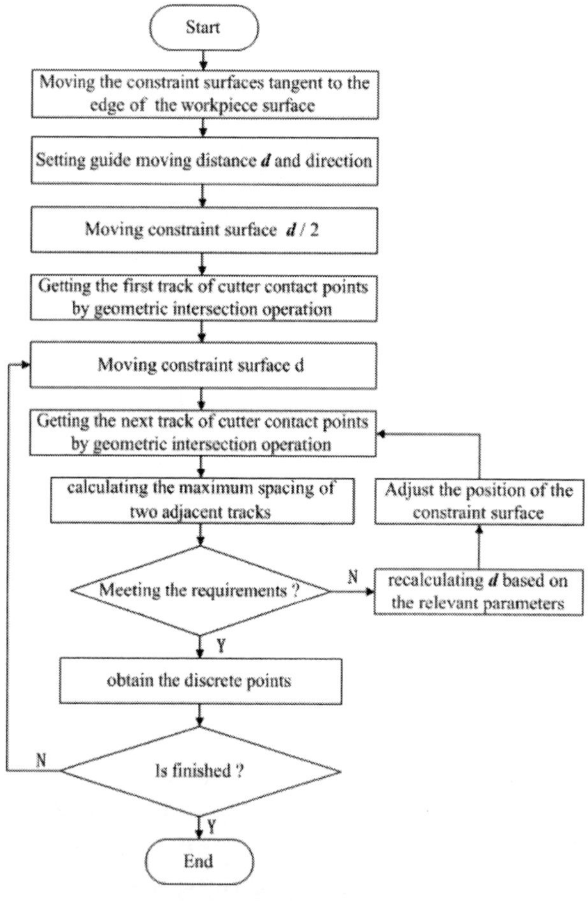

Figure 2. Flowchart of the improved CC route method.

In the algorithm, each constraint surface moves with a determined distance d at the beginning, and the cutter contact points can be obtained from the intersection curves. Then calculate the maximum spacing value of two adjacent paths and adjust d if it is out of range.

3. Modelling and optimization of back-off error

The motion of robot is usually not too fast when milling work-pieces and thus can be regarded as quasi-static moving process. The rods of the robot are treated as rigid bodies as their deformation can be ignored compared with the back-off errors.

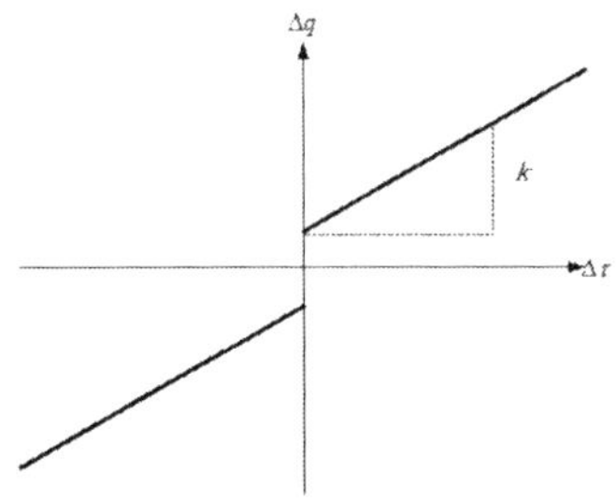

Figure 3. Robotic joint stiffness model.

Figure 3 is the robot joint stiffness model reflecting the linear relationship between the joint torque and the rotation angle. Because of the influence of gap, the curve is not continuous at the original point. However, the gap can be ignored as it is usually not too large and thus the robot joint elastic model can be established as follows:

$$\boldsymbol{\omega} = \mathbf{K}_X \boldsymbol{\delta}_d = \mathbf{J}^{-T} \mathbf{K}_0 \mathbf{J}^{-1} \boldsymbol{\delta}_d \tag{1}$$

Where $\boldsymbol{\omega}$ is the twist which is applied to the end of the robot. $\mathbf{K}_X$ is the robot stiffness matrix in the Cartesian coordinate system. $\boldsymbol{\delta}_d$ is a 6×1 vector containing the displacement and rotation. $\mathbf{K}_0$ is the diagonal stiffness matrix in joint space. $\mathbf{J}$ is the Jacobian matrix of the robot.

In the process of robot milling, the twist $\boldsymbol{\omega}$ in equation (1) is the reacting cutting force $\mathbf{F}_C$ applied to the cutter. Ball-end cutter is considered in this paper and an empirical model is used to calculate $\mathbf{F}_C$.

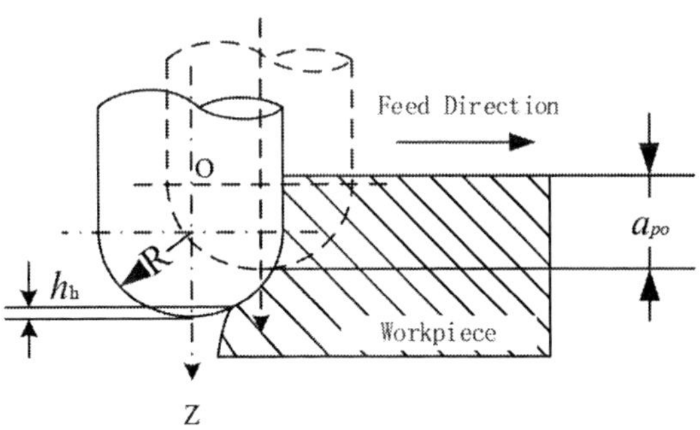

Figure 4. The schematic diagram of ball-end cutter milling.

The ball-end cutter is frequently used for free-form surface milling. As shown in Figure 4, a_{po} is the milling depth. R is the spherical radius of the milling cutter. h is the height of residual areas between milling rows. h_h is the height difference between the cutter tip and the start point of the cutting edge which is involved in cutting. The model of main cutting force is as follows.

When $a_{po} \le R$ and $h_h \le h$

$$F_C = \frac{a_{eo} f_z Z p_2}{2\pi}[\arcsin(1-\frac{h}{R}) - \arcsin(1-\frac{a_{po}}{R})] + \frac{f_z Z p_1 (h-h_h)}{\pi} \tag{2}$$

When $a_{po} \le R$ and $h_h > h$

$$F_C = \frac{a_{eo} f_z Z p_2}{2\pi}[\arcsin(1-\frac{h_h}{R}) - \arcsin(1-\frac{a_{po}}{R})] \tag{3}$$

Where a_{eo} is the milling spacing. f_z is the feed engagement. Z is the number of milling teeth. p_1 and p_2 are unit cutting forces of different cutting layers.

$$p_1 = p_5 \left[\pi / (2 f_z)\right]^{\mu}$$

$$p_2 = p_5 / h_{avp}^{\mu} = p_5 / \left[\int_{R-a_{po}}^{R-h} h_{av} / (a_{po} - h)\, \mathrm{d}z \right].$$

According to the equations (1), (2) and (3), the milling error δ_F caused by the cutting force $\mathbf{F_C}$ can be derived as:

$$\delta_F = \left\| \mathbf{K}_X^{-1} \mathbf{F}_C \right\| \tag{4}$$

The back-off milling error in robot milling process is mainly affected by the cutting force $\mathbf{F_C}$ and the robot stiffness matrix $\mathbf{K_X}$. The cutting force $\mathbf{F_C}$ is mainly depended on the cutting parameters, such as cutting speed v_c, cutting depth a_p and feed rate f, etc. Meanwhile, the robot stiffness matrix $\mathbf{K_X}$ is determined by the robot joint angles θ_1~θ_6. As a result, the back-off milling error of milling process can be reduced and the cutting accuracy will be improved by optimizing the cutting parameters and joint angles.

In this paper, the influence of robot joint angles to the back-off error is studied while the cutting parameters are not considered. Once the cutter contact path is determined, the back-off error is mainly influenced by the posture of the robot which is described by RPY angles (θ_R, θ_P, θ_Y). Genetic algorithm is used here to find the optimal RPY angles. The objective function can be described as:

$$\min \delta_F(\theta_R \quad \theta_P \quad \theta_Y) \tag{5}$$

4. Simulation and analysis

The following simulation is carried out to analyze the influence of robot postures on back-off milling error. The robot is ABB-IRB140 and the material of the work-piece is magnesium alloy. The parameters used in simulation are as follows:

- Unit of cutting force p_5 = 25MPa;
- μ = 0.19;
- Feed engagement f_Z = 0.03mm/Z;
- Milling spacing a_{eo} = 3mm;
- Radius of ball-end cutter R = 5mm;
- Cutter tooth number Z = 2;

- Robot joint stiffness k_i=6×10^{5}N/m, i=1, 2…6.

The work-piece is a hemispherical object and its radius is 0.1m. The center point in the robot coordinate system is (0, 0.7, 0.7). Milling error at P (0, 0.7, 0.8) is simulated. α and β are the angles between the tool axis and XOZ or YOZ planes. Assume that the available range of α and β is [-1, 1]. According to the back-off milling error model, the distribution of back-off error is shown in Figure 5.

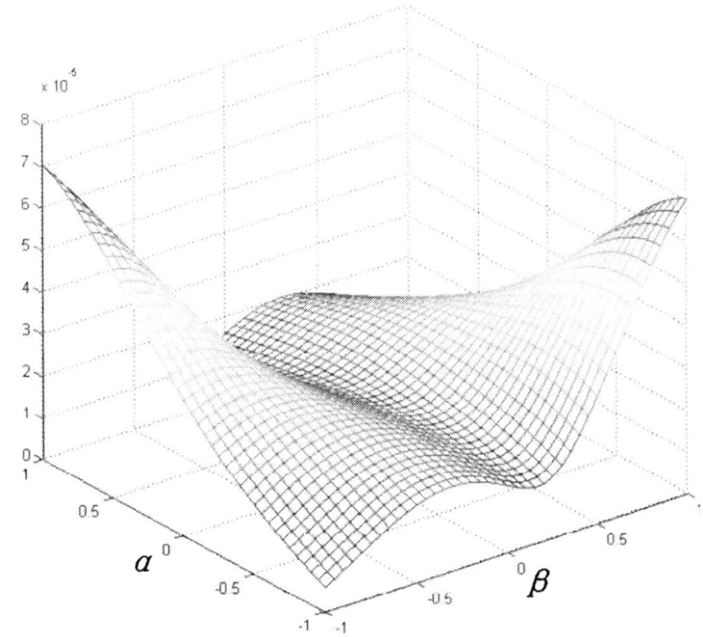

Figure 5. Distribution of milling error.

In Figure 5, the minimum value is 6.045×10^{-6}m and the maximum value is 7.064×10^{-5}m. The RPY angles of the cutter posture optimized by genetic algorithms are [0.814, 0.775, -0.862]. The corresponding milling error is 8.858×10^{-6}m which is close to the minimum value.

5. Conclusion

In this paper, the problems of tool path generation and back-off milling error in robot milling process are discussed. The back-off error model during the milling process is established and the cutter postures optimization method using genetic algorithms is given. Simulation is carried out and shows the feasibility of the optimization to reduce the back-off milling errors.

Acknowledgments

This work was financially supported by the Foundation of ZheJiang Educational Committee (Y201328094) and the National Natural Science Foundation of ZheJiang Province (LY17E050015).

References

[1] X.J. Niu, J.T. Liu, L. Sun, et al.: submitted to IEEE International Conference on Robotics and Biomimetics (2007)

[2] Y. N. Hu, Y. H. Chen: submitted to The International Journal of Advanced Manufacturing Technology (1999)

[3] Park J W, Cho H U, Chung C W, et al.: submitted to Journal of mechanical science and technology (2012)

[4] G. Alici, B.J. Shirinzadeh, Enhanced Stiffness Modeling,: submitted to IEEE Transactions on

Robotics (2005)

[5] C. Dumas, S. Caro, M. Chérif, etc.: submitted to IEEE/RSJ International Conference on Intelligent Robots and Systems (2010)

[6] M. Wan, W.H. Zhang, G. Tan, et al.: submitted to Journal of materials processing technology (2008)

[7] Z.Z. Liu, H.Y. Liu, Z. Luo, et al.: submitted to Journal of agricultural machinery (2013)

[8] Y.G. Zhang J.G. Gao W.Z. Liu, et al.: submitted to Journal of agricultural machinery (2013)

[9] Y.G. Zhang J.G. Gao, et al.: submitted to Journal of agricultural machinery (2014)

A parallel optimization method for product configuration and supplier selection based on interval

Jian Zheng*, **Meng Zhang and Guoxi Li**

College of Mechatronics Engineering and Automation, National University of Defense Technology, Changsha 410073, China

*Corresponding author e-mail: gfkd_zheng@163.com

Abstract. In the process of design and manufacturing, product configuration is an important way of product development, and supplier selection is an essential component of supply chain management. To reduce the risk of procurement and maximize the profits of enterprises, this study proposes to combine the product configuration and supplier selection, and express the multiple uncertainties as interval numbers. An integrated optimization model of interval product configuration and supplier selection was established, and NSGA-II was put forward to locate the Pareto-optimal solutions to the interval multiobjective optimization model.

1. Introduction

Product configuration is an effective way to achieve rapid product development. In the environment of mass customization, the product configuration design based on product family, with high efficiency and low cost of mass production to achieve the design and production of personalized products, so as to enhance the enterprise's rapid response ability and market competition ability. After the product configuration scheme is determined, according to the enterprise business strategy making production plan and outsourcing parts procurement plan, there is a selection of the best suppliers for parts. Rational selection of suppliers can help companies reduce costs and increase management flexibility, improve the core competitiveness of enterprises, and long-term cooperation with high quality suppliers, so as to effectively reduce the risk of procurement and maximize the profits of enterprises.

At present, domestic and foreign scholars have done a lot of research on product configuration and supplier selection, considering the integration of product configuration and supplier selection is still rare. Jiang and other research on inventory driven product configuration optimization problem, the establishment of multi-objective optimization model and genetic algorithm [1]; Ostrosi and others regard the product configuration problem as the mapping process between multiple fuzzy models, and optimize the effective configuration scheme by the self-defined fuzzy configuration syntax to get the optimal solution [2]; considering the uncertain information of the new components, Liu et al. have established a fuzzy multi-objective configuration optimization model, and use mixed integer nonlinear programming algorithm to solve [3]; Yeh and others in the supplier selection criteria framework to add green guidelines, and the establishment of a multi-objective mixed integer nonlinear programming model, the use of multi-objective genetic algorithm for optimal solution [4]; Luo and other people to establish a product family design and supplier selection of a unified optimization model to maximize the total income of the product family optimization goals, and the use of genetic algorithms to solve the model [5].

There is a lot of research on product configuration and supplier selection, but they are usually regarded as two separate issues to carry out analysis and research. Through the above analysis, we know that the product quality, cost and delivery time are the basic elements of modern commodity, they rely on suppliers to provide spare parts, modules or subsystems in a considerable extent, therefore, it is necessary to study the integration of product configuration and supplier selection. However, the research on the process integration and parallel optimization has not been widely concerned. Further, the influence of multiple uncertainties exist widely in the complex market environment for the development of new products cannot be ignored. The existing research focuses on the environment based on deterministic or only considering the fuzzy information, which is inconsistent with the objective facts of actual environment. In order to obtain the precise probability distribution of uncertain or fuzzy membership function often is very difficult. For a practical problem, it is much easier to obtain the possible range of uncertain parameters, and the uncertainty information is greatly reduced.

According to the actual process of product development, this paper proposes to combine the product configuration and supplier selection, and express multiple uncertainties with interval numbers. An integrated optimization model of interval product configuration and supplier selection is established. The objective function and constraint conditions are transformed by interval ranking relation and interval satisfactory degree, then a series of configuration selection schemes are obtained by optimizing the integration model with NSGA-II algorithm. Finally, this method is validated by manufacturing data from typical lathe products.

2. The design of the functional assembly of product family

The integration of product configuration and supplier selection is a complex dynamic process of comprehensive optimization, which is a typical environment of multiple uncertainties. In order to avoid distortion or deviation of results, it is not allowed to ignore the uncertain factors in the actual decision-making. The establishment of the integrated optimization model is the key to solve the problem of new product development. Analyzing the integrated design process shown in Figure 1, abstracting the problem and establish the mathematical model. Product configuration is based on the pre-established product family structure, driven by customer demand for the product family of configurable components to select, combine and optimize, to get the configuration results which meet the design constraints and customer needs. Configurable components include functional components and instance components

Assume that the function component set of the product family is $FM = \{FM_1, FM_2, \cdots, FM_n\}$, A functional component FM_i can be implemented by a number of instance components $IM_i = \{IM_{i1}, IMi2, \cdots, IM_{im(i)}\}$, $m(i)$ is the number of instance components, $IM_{ij}(1 \le i \le n,\ 1 \le j \le m(i))$ represents the jth instance component of the functional component. For a manufacturing enterprise, instance components may include owned, purchased, and new developed components. The product configuration process can be expressed as: the customer's functional requirements are mapped to the functional component set, and the target functional component set of the configured product is $FM^C = \{FM_1^C, FM_2^C, \cdots FM_w^C\}(w \le n)$. One instance component is selected in the corresponding instance set $IM_{ij}(1 \le i \le w,\ 1 \le j \le m(i))$, respectively. And ultimately constitute a customer satisfaction with the configuration of products. When the supplier selection is incorporated, the choice of instance components needs to take into account the differences between the instance component attribute information provided by different suppliers. To do so, the product family configuration model is integrated with the supplier list to establish an integrated model, that is based on the product family configuration model, a supplier list S_{ijk} is added for each instance component, S_{ijk} ($1 \leqslant i \leqslant n$, $1 \leqslant j \leqslant m(i)$, $1 \leqslant k \leqslant s(i, j)$) represents the kth supplier of the jth instance component IM_{ij} of the functional component FM_i.

It can be seen from the above analysis that the model parameters exhibit multiple uncertainties, which may be expressed as a definite number, a random number, a fuzzy number or an interval number form. In order to effectively solve the model, the multiple uncertainties are expressed as interval numbers, and an integrated optimization model of product configuration and supplier selection based on interval is established.

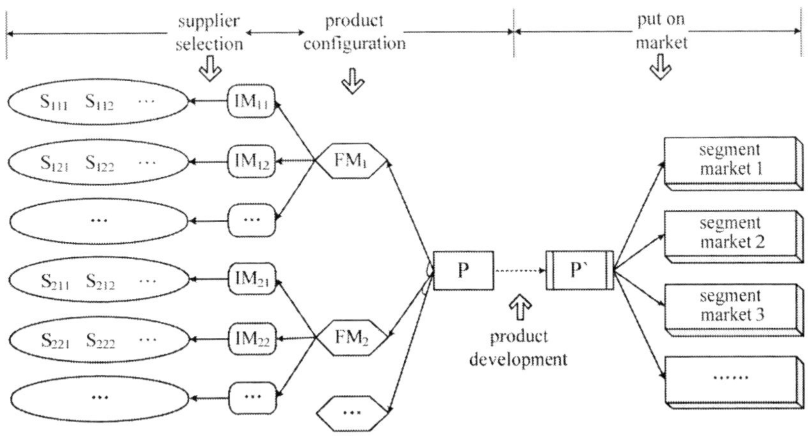

Figure 1. Product structure and supplier combination of parallel configuration process.

3. Product cost driven interval integrated optimization model

3.1. Build the Cost Matrix of the Instance Components

$$C=[\underline{C},\overline{C}]=\Big[[\underline{c}_{111},\overline{c}_{111}]\cdots[\underline{c}_{1m(1)s(1,m(1))},\overline{c}_{1m(1)s(1,m(1))}]\cdots [\underline{c}_{ijk},\overline{c}_{ijk}]\cdots[\underline{c}_{nm(n)s(n,m(n))},\overline{c}_{nm(n)s(n,m(n))}]\Big] \quad (1)$$

Where c_{ijk} denotes the price interval provided by the kth supplier of the jth instance component of the ith functional component set, $c_{ijk}=\left[\underline{c}_{ijk},\overline{c}_{ijk}\right]$.

3.2. Mathematical Optimization Model of Product Cost

The functional components of the product family consist of required components and optional components, and the mathematical model is established as:

$$\min C_z=\sum_{i=1}^{n}\sum_{j=1}^{m(i)}\sum_{k=1}^{s(i,j)}(\varepsilon_{ijk}\cdot c_{ijk})+(\sum_{i=1}^{n}\sum_{j=1}^{m(i)}\sum_{k=1}^{s(i,j)}\varepsilon_{ijk}-1)\cdot C_A, \quad (1+\alpha)\cdot C_z\le C_{\max} \quad (2)$$

Where C_z is the total cost range of the configured product, ε_{ijk} represents the decision variable corresponding to the instance component IM_{ij} and its supplier list S_{ijk} (1≤k≤s (i, j));When $\varepsilon_{ijk}=1$, it means that the instance IM_{ij} is involved in the configuration, and the corresponding supplier is selected as S_{ijk}; when $\varepsilon_{ijk}=0$, it means that the instance component IM_{ij} does not participate in the configuration and the corresponding kth supplier S_{ijk} is not selected; For an instance component IM_{ij}, if $\sum_{k=1}^{s(i,j)}\varepsilon_{ijk}=0$, it means that the instance component is not involved in the configuration. C_A is the average assembly cost between the modules, $C_A = [\underline{C}_A, \overline{C}_A]$; α is the profit margin of the enterprise, C_{max} is the floating range of the highest price that the customer can afford, $C_{max} = (\underline{C}_{max}, \overline{C}_{max})$.

3.3. Establishment of Constraint Conditions

(1) Price constraints

$$(1+\alpha)(\underline{C}_z, \overline{C}_z) \le (\underline{C}_{max}, \overline{C}_{max}) \tag{3}$$

(2) Configuration constraints

$$\begin{aligned} &\sum_{k=1}^{s(i,j)} \varepsilon_{ijk} = 1, i = 1,2,\cdots,n; j = 1,2,\cdots,m(i) \\ &\begin{cases} \sum_{j=1}^{m(i)} \sum_{k=1}^{s(i,j)} \varepsilon_{ijk} = 1 & \text{when } FM_i \text{ is an optional component set;} \\ \sum_{j=1}^{m(i)} \sum_{k=1}^{s(i,j)} \varepsilon_{ijk} \le 1 & \text{when } FM_i \text{ is a required component set.} \end{cases} \\ &i = 1,2,\cdots,n \end{aligned} \tag{4}$$

4. Transformation of interval optimization problems

4.1. Problem Description

When uncertainties exist in the problem and uncertainties are described by interval, the generalized nonlinear interval uncertain multi-objective optimization problem can be expressed as:

$$\begin{aligned} &\min_x f(\boldsymbol{x},\boldsymbol{a}) = \{f_1(\boldsymbol{x},\boldsymbol{a}), f_2(\boldsymbol{x},\boldsymbol{a}), \cdots, f_k(\boldsymbol{x},\boldsymbol{a})\} \\ &\text{s.t.} \quad g_j(\boldsymbol{x},\boldsymbol{a}) \le V_j = [\underline{V}_j, \overline{V}_j], (j = 1,2,\cdots,m) \\ &h_k(\boldsymbol{x}) = B_k = [\underline{B}_k, \overline{B}_k], (k = 1,2,\cdots,l) \\ &\boldsymbol{a} = (a_1, a_2, \cdots, a_q) \subset R^q, a_i = [\underline{a}_i, \overline{a}_i], i = 1,2,\cdots q; \\ &x_{il} \le x_i \le x_{iu}, (i = 1,2,\cdots,n) \end{aligned} \tag{5}$$

Where $\boldsymbol{x}$ is the variable, x_i is the ith variable of the decision vector, x_{il} and x_{iu} are the lower and upper limits of the design variable, n is the number of optimization variables, $f(\boldsymbol{x},\boldsymbol{a})$ is the objective function vector, $f_k(\boldsymbol{x},\boldsymbol{a})$ is the kth interval objective function; the vector $\boldsymbol{a}$ is a q-dimensional uncertainty, a_i is the ith variable of the interval vector, $\overline{a}_i$ and $\underline{a}_i$ are the upper and lower bounds, respectively; $g_j(\boldsymbol{x},\boldsymbol{a})$ is the jth constrained interval inequality, and $h_k(\boldsymbol{x})$ is the kth constraint interval equation.

4.2. The Transformation of Uncertain Optimization

Firstly, the interval objective function is transformed by the interval relation, then the interval constraint is transformed based on the interval possibility degree [6]. According to a definition of the interval ranking relation "$\prec=$", arbitrary two intervals $A = [\underline{A}, \overline{A}]$ and $B = [\underline{B}, \overline{B}]$ interval ranking relation is as follows:

$$\begin{cases} A \prec= B, \text{if } m(A) < m(B) \\ \qquad \text{or } m(A) = m(B) \text{ and } w(A) \geq w(B); \\ A \prec B, \text{ if } A \prec= B \text{ and } A \neq B. \end{cases} \tag{6}$$

Where $m(A)$ and $m(B)$ are the midpoints of the interval A and B, $m(A) = (\overline{A} + \underline{A})/2$, $m(B) = (\overline{B} + \underline{B})/2$. $w(A)$ and $w(B)$ are the widths of the interval A and B, $w(A) = (\overline{A} - \underline{A})/2$, $w(B) = (\overline{B} - \underline{B})/2$.

According to the above-mentioned interval ranking relation, the objective function can be transformed into:

$$\begin{cases} \min\limits_{x}[m(f_i(\boldsymbol{x},\boldsymbol{a})), w(f_i(\boldsymbol{x},\boldsymbol{a}))] \\ m(f_i(\boldsymbol{x},\boldsymbol{a})) = \dfrac{1}{2}\left(\overline{f}_i(\boldsymbol{x}) + \underline{f}_i(\boldsymbol{x})\right) \\ w(f_i(\boldsymbol{x},\boldsymbol{a})) = \dfrac{1}{2}\left(\overline{f}_i(\boldsymbol{x}) - \underline{f}_i(\boldsymbol{x})\right) \\ i = 1, 2, \cdots, k, \end{cases} \tag{7}$$

Where $m(f_i(\boldsymbol{x},\boldsymbol{a}))$ and $w(f_i(\boldsymbol{x},\boldsymbol{a}))$ are the midpoint and width of the ith objective function $f_i(\boldsymbol{x},\boldsymbol{a})$, $\underline{f}_i(\boldsymbol{x})$ and $\overline{f}_i(\boldsymbol{x})$ are its lower and upper limits, $\underline{f}_i(\boldsymbol{x})$ and $\overline{f}_i(\boldsymbol{x})$ values are calculated as follows:

$$\begin{aligned} \underline{f}_i(\boldsymbol{x}) &= \min_{a\in\Gamma} f_i(\boldsymbol{x},\boldsymbol{a}) \\ \overline{f}_i(\boldsymbol{x}) &= \max_{a\in\Gamma} f_i(\boldsymbol{x},\boldsymbol{a}) \end{aligned} \tag{8}$$

Where $\Gamma = \{a \mid \underline{a}_i \leq a_i \leq \overline{a}_i, i = 1, 2, \cdots, q\}$.

5. Verification and analysis

5.1. Design of Function Component Assembly for Lathe Product Family

In this paper, we refer to the data in paper [7] and improve it, a typical lathe product assembly of an enterprise is taken as an example, module unit assembly cost range is $C_A = [372, 376]$ yuan. For details, please refer to the main parts list of the lathe. According to the standardization and modularization technology, the enterprise has now formed four series of 22 different types of modular products .Each supplier provides examples of the shipment interval and product price range as follows (some data have been omitted):

Table 1. The design of functional assembly of lathe product family.

Serial number	Module name	Instance number	Supplier name	Price(yuan)
1	Active Power Module	M_{11}	S_{111}	[15500,16800]
			S_{112}	[15000,16000]
		M_{12}	S_{121}	[14500,15500]
			S_{122}	[17000,17800]
			S_{123}	[14800,15800]
		M_{13}	S_{131}	[18000,19000]
			S_{132}	[18800,19800]
2	The horizontal and vertical feed module	M_{21}	S_{211}	[2600,3000]
			S_{212}	[3000,3200]
			S_{213}	[2450,2600]
		M_{22}	S_{221}	[2500,2800]
			S_{222}	[2300,2550]
		M_{23}	S_{231}	[3100,3400]
			S_{232}	[2800,3000]
……	……	……	……	……
11	Lighting module	M_{111}	S_{1111}	[80,100]
			S_{1112}	[75,90]

5.2. Multi-objective Configuration Optimization and Supplier Selection for Lathe Products

Using the method proposed in Chapter 2, the two-dimensional decision variable ε_{ijk} is taken as the design variable, and the cost C_z as the objective function is minimized. The product configuration optimization and supplier selection are carried out under the condition of formula (3) and (4).Because NSGA-II algorithm first fast and then slow search characteristics, its propagation algebra K is not so high that the number of iterations is $K = 150$, the configuration scheme population size is $N = 100$, the crossover probability is $P_c = 0.7$, and the mutation probability is $P_m = 0.02$ .Using Matlab to optimize and simulate based on NSGA-Ⅱ,the dynamic change curve of the interval objective function into the midpoint objective function and the radius objective function is shown in Fig.2 and Fig.3.

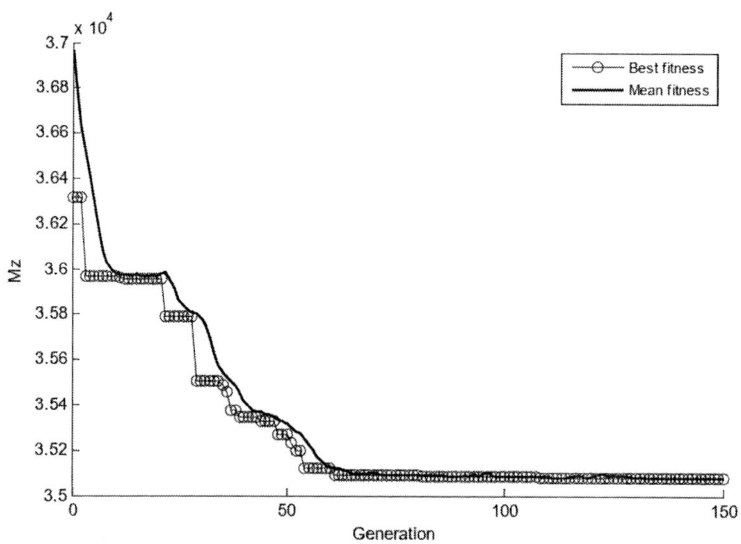

Figure 2. Change curve of midpoint objective function.

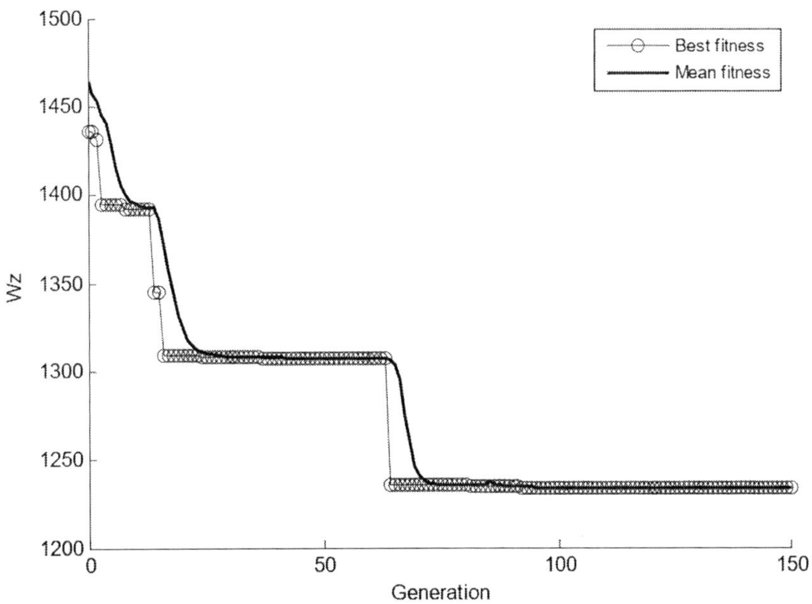

Figure 3. Change curve of midpoint objective function.

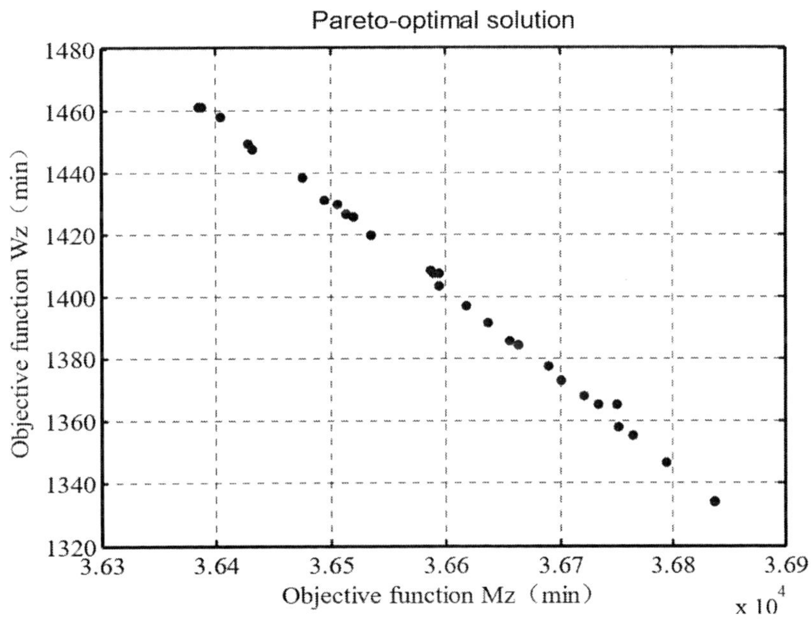

Figure 4. Configuration scheme Pareto set.

Under the influence of uncertainty, the midpoint objective function M_z and the radius objective function W_z have their own meanings: the midpoint objective function represents the average level of the cost, and the radius objective function represents the fluctuation of the cost. If the enterprise focuses on the average product cost is small, from the Pareto set of the upper left side of the election point; if the enterprise bias in the product cost fluctuation range is small, from the Pareto set the lower right side of the point. It can be seen from Fig. 4 that the results of the configuration are ideal in terms of the number and diversity of solutions.

IOP Conf. Series: Materials Science and Engineering **207** (2017) 012038 doi:10.1088/1757-899X/207/1/012038

Therefore, according to different needs, enterprises can decide to focus on the cost of the average level or cost fluctuations, and then decide which point is the best, and from the Pareto set to find the corresponding value of the design variable to get the most satisfactory products configuration and supplier selection. Table 2 shows the comparison of the three configuration schemes selected from the Pareto set according to the different requirements of the average cost level and the cost fluctuation.

Table 2. Optimal configuration scheme.

Num	Cost/yuan	Module number	Instance	Supplier
1	[34620,37510]	9	M12M21M31M41M51M61M73M81M91	$S_{122}S_{211}S_{311}S_{411}S_{511}S_{611}S_{731}S_{811}S_{911}$
2	[34087,36981]	10	M12M23M31M43M51M62M73M81M92M102	$S_{122}S_{231}S_{311}S_{432}S_{511}S_{622}S_{732}S_{811}S_{923}S_{1021}$
3	[35944,38622]	11	M12M21M32M43M51M61M73M82M91M102M111	$S_{122}S_{211}S_{322}S_{431}S_{511}S_{612}S_{731}S_{823}S_{912}S_{1021}S_{1111}$

6. Conclusion

This study will help to solve the problem of parallel optimization of product configuration and supplier selection, and it can effectively deal with the impact of multiple factors of uncertainty .Moreover, in theory, the method can promote the basic research of product and supply chain integration and optimization design; in practice, the method can provide a scientific basis for the parallel decision making of product design and supplier selection in the complex market environment, which has important theoretical significance and practical value.

Acknowledgments

This research is supported by National Natural Science Foundation of China. (No.51505488)

References

[1] Zhaoliang Jiang, Sisi Xuanyuan, Lin Li, et al, Inventory-shortage driven optimisation for product configuration variation, International Journal of Production Research. 49(2011)1045-1060.

[2] W. Ostrosi E, Ferney M, Fuzzy agents for product configuration in collaborative and distributed design process, Applied Soft Computing. 12 (2012) 2091–2105.

[3] Yiliu Liu, Zixian Liu, Multi-objective product configuration involving new components under uncertainty, Journal of Engineering Design. 21 (2010) 473-494.

[4] G. Yeh W C, Chuang M C, Using multi-objective genetic algorithm for partner selection in green supply chain problems, Expert Systems with Applications. 38 (2011) 4244-4253.

[5] Cao Y, Luo X G, Kwong C K, et al, Joint optimization of product family design and supplier selection under multinomial logit consumer choice rule, Concurrent Engineering. 20 (2012) 335-347.

[6] Jin Cheng, Gui-fang Duan, Zhen-yu Liu, Interval multiobjective optimization of structures based on radial basis function, interval analysis, and NSGA-II, Journal of Zhejiang University-SCIENCE A. 15 (2014) 774-788.

[7] Wei W, Tan J R, Feng Y X, et al, Multi-objective configuration optimization of modularized product based on NSGA-II, Computer Integrated Manufacturing Systems. 13 (2007) 2092-2085.

IOP Conf. Series: Materials Science and Engineering **207** (2017) 012039 doi:10.1088/1757-899X/207/1/012039

Effect of scanning path on temperature field in laser cladding

Xinyong Gong*, **Jianguo Luo and Dezhi Hu**

School of Mechanical and Electrical Engineering, North China Institute of Science and Technology, Hebei 065201, China

*Corresponding author e-mail: huagu1984@126.com

Abstract. Temperature field of planar laser cladding with length direction, width direction and helical scanning path were simulated and analysed. Temperature distribution could be effected significantly by the laser scanning path. For short-path scanning, heat was concentrated in the local area, resulting in larger (/smaller) temperature gradient in the length (/width) direction of the substrate. However, such characteristics of long-path scanning were on the contrary. By comparison, the heat affected zone (HAZ) of the substrate with helical scanning mode was the smallest and its temperature fluctuations were more dispersed.

1. Introduction

Laser cladding was an important application of laser processing technology, which could be used to the surface modification [1], manufacture of three-dimensional solid parts [2], and remanufacturing of damaged parts [3]. As fine microstructure, high strength of combined interface, small HAZ, laser cladding had many applications in aerospace, metallurgy, chemical, automobile, coal and other industries. Meanwhile, some basic researches related had been developed widely [4].

The temperature field, stress/strain field, microstructure/properties of the laser cladding process had coupling effects. The theoretical calculations of the latter two required the temperature solution to be introduced into the model firstly [5-6]. In the plane cladding process, warpage of the substrate would be affected by the laser scanning path. Thus, difference of temperature characteristics corresponding to multiple scanning paths, which were calculated by ANSYS software, was discussed in this paper.

2. Model Building

The planar laser cladding could be done with different scanning modes, such as long, short, and spiral path (Fig.1). Long (/short) path referred to grating movement of the laser beam along with length (/width) side of the processing plane. The spiral mode meant straight scanning of approximate helix, then such path from outside to inside was chosen in this paper.

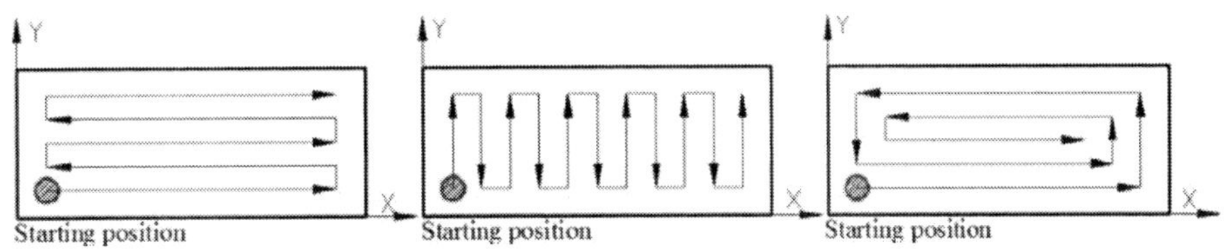

Figure 1. Schematic diagram of laser scanning (a) long-path, (b) short-path, (c) spiral-path.

Laser cladding of 316L stainless on 45steel substrate was set as physical model (Fig.2). To meet the actual working conditions, a large plate (45steel) was arranged under above substrate to transfer its internal heat. Size of the cladding layer, substrate, plate were set as 60 mm×20 mm×0.3 mm, 80 mm×25 mm×10 mm, 160 mm×50 mm×10 mm, respectively. Main process parameters were determined, such as laser power (P=2000 W), scanning speed (V=4 mm/s), powder feeding rate (Vf=3 g/min), overlap rate (n=0). Some other parameters of physical models corresponding to above three scanning paths were listed in Tab.1. The cladding process could be completed by three cycles of helical scanning, in which the single cladding length decreased with the number of scanning cycles (60 mm→52 mm→44 mm、 12 mm→4 mm→0 mm), so the cladding time became shorter (15 s→13 s→11 s、 3 s→1 s→0 s). Above parameters were all introduced into APDL programs of ANSYS software to complete the simulation calculation.

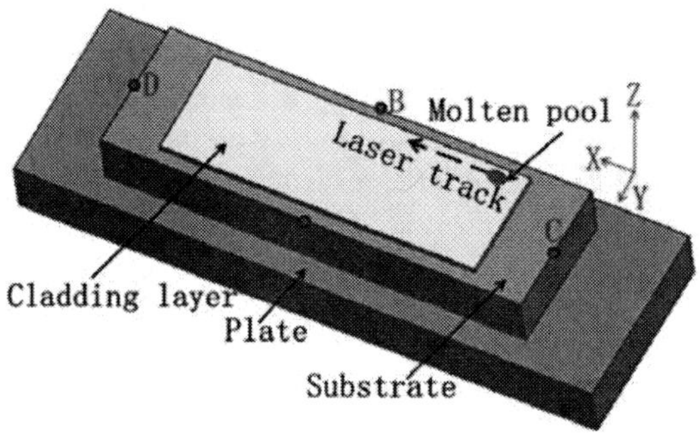

Figure 2. Physical model and distribution of key points.

Table 1. Physical model parameters.

	Long-path	Short-path	Spiral-path		
			1st circle	2nd circle	3rd circle
Cladding area	60 mm × 20 mm				
Spot size	4 mm × 4 mm				
Scanning speed	4 mm/s				
Single track length	60 mm	20 mm	60 mm 12 mm	52 mm 4 mm	44 mm
Single track time	15 s	5 s	15 s 3 s	13 s 1 s	11 s
Cladding number	5	15	2 2	2 2	1
Total time	75 s	75 s	75 s		

In addition, meshing and mathematical model were similar to the literature [7], and unnecessary to go into detail.

3. Results and analysis

Temperature distribution of laser cladding in 20 s (a, d, g), 45 s (b, e, h), 75 s (c, f, i) corresponding to long (a, b, c), short (d, e, f), and spiral (g, h, i) scanning path were shown in Fig.3, respectively. Melting point of 316L stainless steel powder is about 1510℃, so the red zone (>1510℃) could be regarded as molten pool. Shape feature of each molten pool was stable and consistent with experimental phenomena, which presenting a nearly elliptical shape, so process parameters chosen in this paper were suitable. The green cloud diagram represented 600~900℃ physical domain, and its long axis were consistent with the scanning direction, so a significant impact of scanning path on temperature field of cladding parts could be inferred. In addition, for above three models, low temperature region (<100℃) decreased and high temperature region (200~600℃) increased with the cladding process (25 s→50 s→75 s), which indicating that the overall substrate trended to be warming and heat accumulated until the end of cladding.

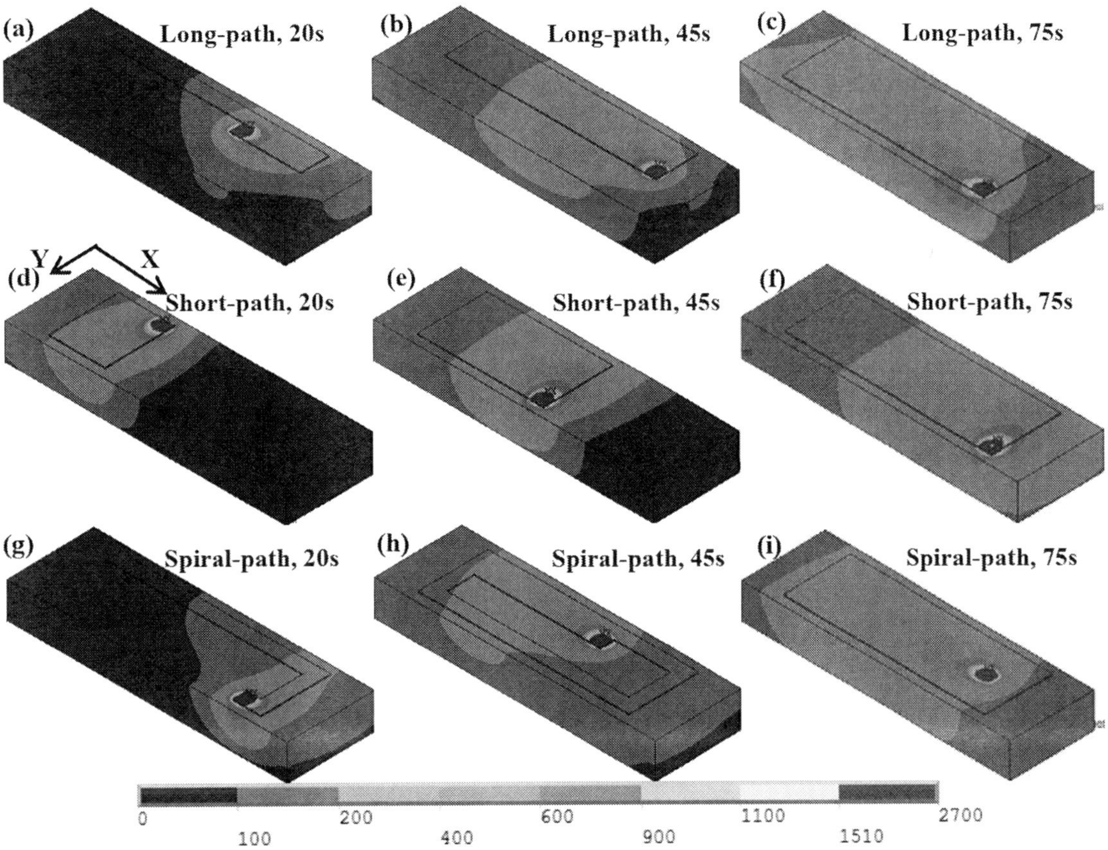

Figure 3. Temperature field distribution in 20s (a, d, g), 45s (b, e, h), 75s (c, f, i).

For short-path scanning, convective heat transfer surface of high temperature and direction of heat conduction were relatively scarce, so the heat was easily accumulated in the local area, which resulting in a large temperature difference in length-side (X-) direction, but even in width-side (Y-) direction. Temperature field characteristic of long-path scanning was just on the opposite. As alternate scanning direction (X→Y→-X→-Y→······), by which temperature contour was affected, temperature field for spiral-path would be a compromise distribution between long- and short-path models.

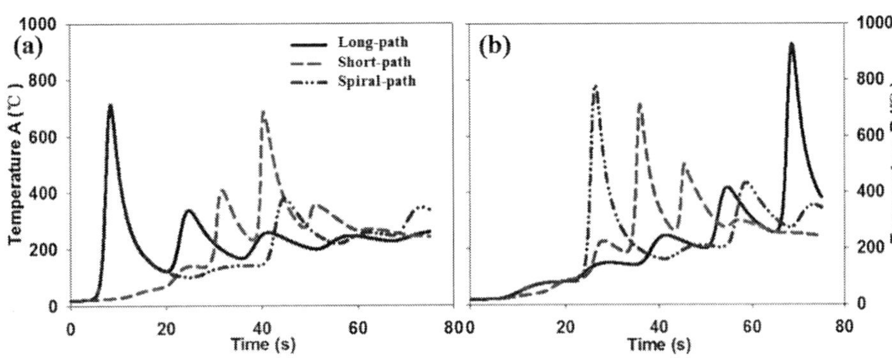

Figure 4. Transient temperature curve of point A, B.

Transient temperature curves of A, B point (Fig.2) during the entire cladding stage (0~75 s) were shown in Fig.4, and significant difference existed due to scanning path. The beginning and terminal in wide-side direction at middle region of the substrate were represented by A, B point respectively, which all going through temperature fluctuation for 3~4 times. As increased temperature of each trough, heat storage process in nearby regions for point A, B could be concluded. However, great difference existed in the time interval (T) of temperature peaks between the three scanning conditions, namely T-short < T-long < T-spiral. Peaks for short-path were concentrated in the time domain, but relatively dispersed for spiral-path, by which heat-affected zone near the substrate edge in width-side direction was the smallest.

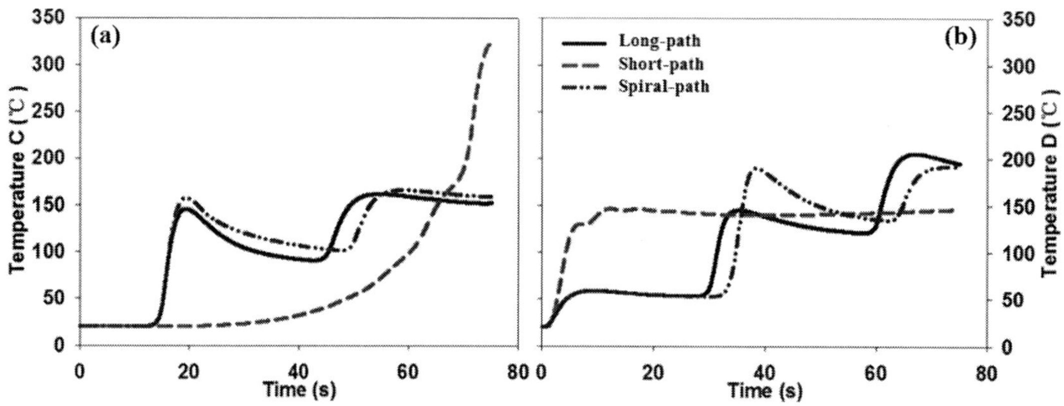

Figure 5. Transient temperature curve of point C, D.

Transient temperature curves of two substrate ends (C, D point) in length-side direction were illustrated in Fig.5, where similar temperature variation history was loaded for long-path and spiral-path models. During the whole cladding process, the laser beam moved to above ends 3 times, and the third peak of point C did not appear in the 0~75 seconds, so point C (/D) had experienced two (/three) thermal cycles. Temperature characteristics of the short-path model in these two regions were obviously different. Curves did not fluctuate with the reciprocating motion of the laser beam, which meant temperature variation reflected overall temperature rise of the substrate, rather than transient heating from the laser beam. In addition, the temperature at the beginning (point D) of laser scanning was constant after a brief temperature rise, where the heat conducted from the adjacent area could be offset by loss of surface convection. But the terminal (point D) experienced a long dormant period (0~40 s), slow (40~60 s) and accelerated (60~75 s) heating successively. According to this, a large temperature gradient existed in the length-side direction of the substrate.

4. Conclusion

The ANSYS software was used to simulate the laser cladding process with long, short, and spiral scanning path, and their temperature field were solved. By comparing and analyzing, the following conclusions were obtained.

(1) The isotherm of laser cladding model was elliptical nearly, and its long axis coincided with the moving direction of laser beam, by which a significant effect of laser scanning path on temperature field characteristics could be inferred.

(2) As the heat for short-path model being prone to accumulate in the local area, a larger (/smaller) temperature gradient existed in the length (/width) direction of the substrate. Characteristic of long-path model was just on the opposite.

(3) Effect on heat affect zone varied depending on the location. For substrate ends in width-side direction, the largest HAZ appeared in the short-path model, then long-path, spiral-path models were

in turn. And in length-side direction, the latter two models showed fluctuation characteristics of temperature rise.

Acknowledgments
Financial support from Basic Scientific Research Service Fee of Central University under the contract number of 3142015012 and 3142015097 are greatly appreciated.

References

[1] F. Weng, H. Yu, J. Liu, Microstructure and wear property of the Ti5Si3/TiC reinforced Co-based coatings fabricated by laser cladding on Ti-6Al-4V, Optics & Laser Technology. 92 (2017) 156-162.
[2] M. Song, X. Lin, F, Liu, The purification of AlSl 420 stainless steel in laser solid forming, Materials & Design. 89 (2016) 1035-1040.
[3] Z. Liu, Q. Jiang, T. Li, Environmental benefits of remanufacturing: A case study of cylinder heads remanufactured through laser cladding, Journal of Cleaner Production. 133 (2016) 1027-1033.
[4] M. Leino, J. Pekkarinen, R. Soukka, The role of laser additive manufacturing methods of metals in repair, refurbishment and remanufacturing-enabling circular economy, Physics Procedia. 83 (2016) 752-760.
[5] F. Weng, H. Yu, C. Chen, Effect of process parameters on the microstructure evolution and wear property of the laser cladding coatings on Ti-6Al-4V alloy, Journal of alloys and compounds. 692 (2017) 989-996.
[6] I. Shishkovsky, Heat transfer and diffusion-controlled kinetics of liquid-solid phase in titanium matrix composite during selective laser melting, Optical and quantum electronics. 30 (2017) 299-303.
[7] X. Gong, Fundamental research on repairing of TC11 titanium alloy blades by laser cladding deposition, Ph D Thesis, General research institute for nonferrous metals, China (2014).

IOP Conf. Series: Materials Science and Engineering **207** (2017) 012040 doi:10.1088/1757-899X/207/1/012040

Development of a conformable electronic skin based on silver nanowires and PDMS

Haopeng Wang*

Information Science Academy, China Electronics Technology Group Corporation, Beijing 100086, China

*Corresponding author e-mail: wanghp_cetc@163.com

Abstract. This paper presented the designed and tested a flexible and stretchable pressure sensor array that could be used to cover 3D surface to measure contact pressure. The sensor array is laminated into a thin film with 1 mm in thickness and can easily be stretched without losing its functionality. The fabricated sensor array contained 8×8 sensing elements, each could measure the pressure up to 180 kPa. An improved sandwich structure is used to build the sensor array. The upper and lower layers were PDMS thin films embedded with conductor strips formed by PDMS-based silver nanowires (AgNWs) networks covered with nano-scale thin metal film. The middle layer was formed a porous PDMS film inserted with circular conductive rubber. The sensor array could detect the contact pressure within 30% stretching rate. In this paper, the performance of the pressure sensor array was systematically studied. With the corresponding scanning power-supply circuit and data acquisition system, it is demonstrated that the system can successfully capture the tactile images induced by objects of different shapes. Such sensor system could be applied on complex surfaces in robots or medical devices for contact pressure detection and feedback.

1. Introduction

E-skin for contact pressure measurement is a key component for next generation humanoid robots and medical devices [1-3]. It is formed through the embedment an array of pressure sensors in a flexible, stretchable, or foldable substrate. Nowadays, the skin-like tactile sensor has been widely applied to the robotics and medical devices [4, 5]. Smart e-skin which acquires pressure information can be used to cover the surfaces of robotics and medical devices to sense the physical contact with human/environment [6, 7]. Flexibility and stretch ability are two of the most important factors for the application of such sensor array to the humanoid robots and medical devices. The flexibility of the senor array is desired to make the sensor conform to the shape of its applied device. The stretch ability of the sensor has advantages that the sensor is nonbreakable and that the sensor can be easily mounted on complex surfaces or deformable parts such as the joints of a device [8, 9].

Several research studies on flexible or stretchable pressure sensor array have been reported, which resulted in various skin-like pressure sensor arrays. Most of these sensor arrays are based on capacitive or piezoresistive force sensing principles. Reference [10] presents a stretchable and multifunctional capacitive sensor. This sensor was made through the embedment of gold-trace stripes into silicone rubber thin films. Although the mechanical compliance of the gold-trace stripes and silicone membranes allows the device to bend, fold, and stretch, the low conductivity of the gold thin films and its poor adhesion with the substrate impede the integration of the sensor array. Bao et al. [11] developed a skin-like pressure and strain sensor array based on transparent elastic films of carbon nanotubes with PDMS as the substrates. The applied pressure was measured through the change of the capacitance among the grid of capacitors formed by the perpendicular strips of nanotubes. Their device can detect pressures above 50kPa. Someya et al. [12] demonstrated an application that used organic field-effect transistors (FETs) with pressure-sensitive rubber to realize a stretchable pressure sensor matrix. The sensor's stretch ability was achieved through its fishnet style. Although the device can detect pressure up to 10kPa, its stretch ability may lose when the working condition requires the device to stick to the host surface. Pan et al. [13] proposed an ultra-sensitive resistive pressure sensor based on hollow-sphere microstructure induced elasticity in conducting polymer. The pressure sensor based on the elastic microstructured thin film enables the detection of pressures of less than 1 Pa and exhibits a short response time, good reproducibility, excellent cycling stability and temperature-stable sensing. However, the low conformability limit its application. Yang et al. [14] designed a highly twistable tactile sensor array. Extendable spiral electrodes were employed as elastic interconnects of sensor array. The tactile sensing elements were formed by dispensing conductive polymer on the spiral electrodes. The fabricated sensor array can be twisted up to 70° without any damage in the structure or functionality. However, the sensor array fabrication method may deteriorate the non-uniformity of the sensing elements which is one of the drawbacks that the sensor arrays usually suffer from. In this paper, we present a novel design of a flexible and stretchable pressure sensor array based on piezoresistive principle for e-skin application. The AgNWs/PDMS based conductors instead of metals are served as interconnects and electrodes for the sensor arrays to improve its flexibility and stretch ability. The sandwiched sensing layer consisting of the porous PDMS film and circular conductive rubber enables the sensor array to be highly flexible and stretchable. The characteristic and performance of the sensor array have been measured and discussed.

2. Structure Design

The schematic diagram of the proposed pressure sensor array is shown in Fig. 1(a). In the sensor array, the AgNWs/PDMS conductor strips are used as stretchable interconnects and sensing electrode. The soft metal layer adhered on the conductors and conductive rubber can improve the contact quality between the electrodes and piezoresistive material. The porous PDMS thin film holds the circular piezoresistive elements together and improves the stretch ability of the middle layer. In addition, its isolation among piezoresistive elements eliminates the crosstalk problem [14, 15]. The middle sensing layer is sandwiched between the two electrode layers, each of which brings rows of parallel AgNWs/PDMS based conductive strips in contact with the circular piezoresistive elements. The conductive strips in the top and bottom layers are placed orthogonal to each other. Together with the sensing elements in the middle layer, a array of sensing elements is built up. Fig. 1(b) describes the sensing principle of the pressure sensor array. When a pressure is applied on the sensor, the resistance between conductive strips and piezoresistive film varies, which allows the measurement of the pressure based on the resistance change.

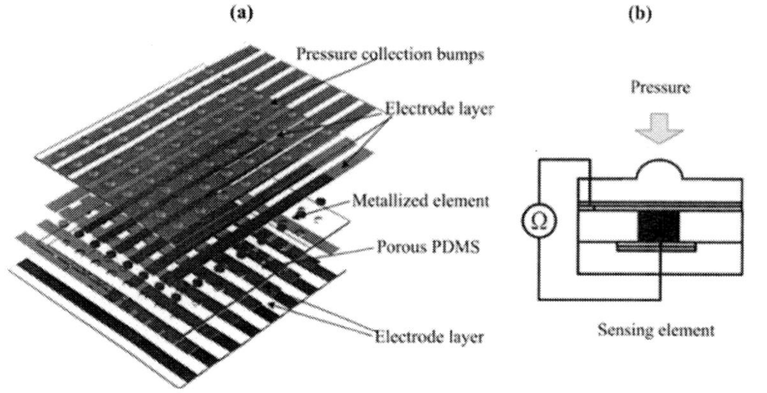

Figure 1. (a) The schematic of the proposed pressure sensor array. (b) The sensing element under pressure.

3. Fabrication Procedure

The pressure sensor array prototype fabricated is shown in Figure 2. The manufacturing cost of each force sensor array is around \$25. The cost will be decreased further if more sensor array are fabricated.

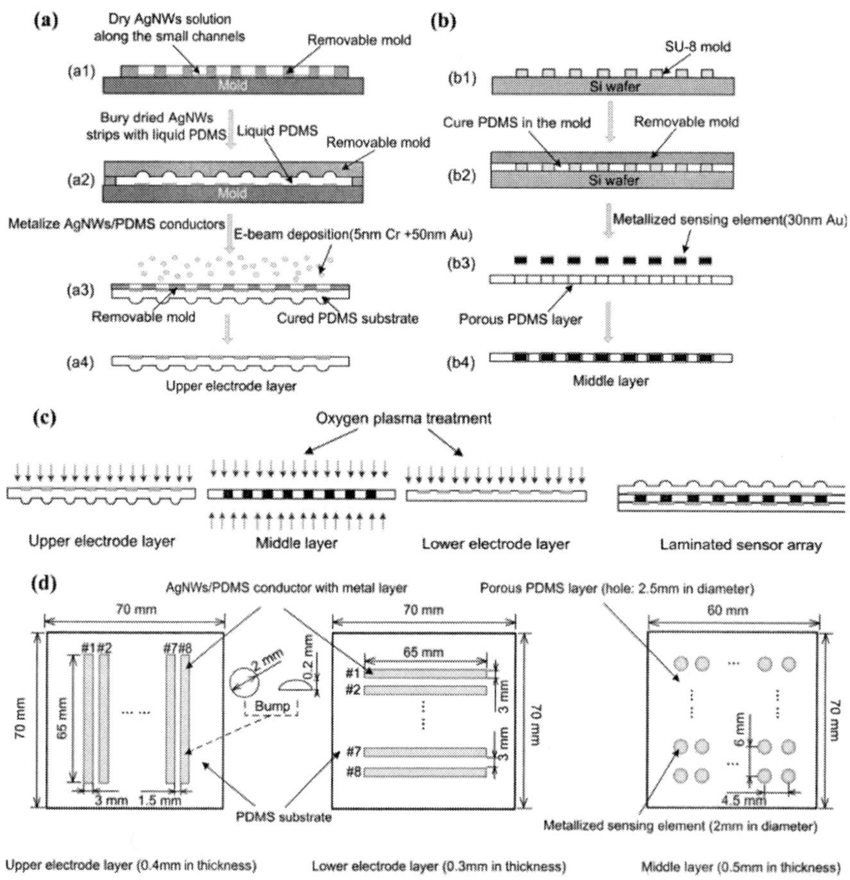

Figure 2. (a) Schematic of the fabrication process of electrode layer; (b) Schematic of the fabrication process of middle layer; (c) Lamination of the e-skin with oxygen plasma; (d) The main dimensions of the components for e-skin.

4. Measurement Results and Discussion

Following the fabrication and integration of the sensor array, some measurements were carried out toward the characteristics and performance of the sensing element. Fig. 3 shows the schematic of the experimental setup for testing the pressure sensor array. An Epson Scara G3 robot holding a load cell (Phidgets, RB-Phi-113) is used to apply the pressure on one sensing element. A voltage division circuit is designed to measure the resistance of the sensing elements. The output of the sensor cell is labeled as R/R0, where R= R0-R; R0 is the original resistance, ~45 kΩ, between the electrodes, R is the resistance under pressed state with minimum value ~400 Ω. During the measurement, with the increase of the pressure, the resistance change of the sensing element was observed. Correspondingly, the value of the resistance was recorded by a PC. The obtained relationship between the R/R0 and the applied pressure on one sensing element during loading and unloading process is shown in Fig. 4.

The performance such as linearity, hysteresis, sensitivity and pressure resolution ratio of the sensing elements are listed in TABLE I. A commercial soft force sensor Flexiforce sensor, a product of the Tekscan, Inc. is used to compare with the designed sensor. The performances of the Flexiforce sensor were tested. The effective range of the designed pressure sensor array is about 180 kPa. The sensitivity decreases with the increasing load applying on the sensing element. When the load exceeds the effective range the output of the sensor changes very little.

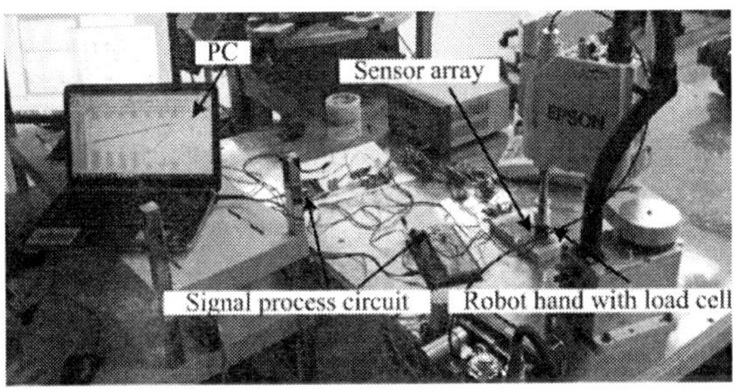

Figure 3. Experimental set-up for testing of the sensor array.

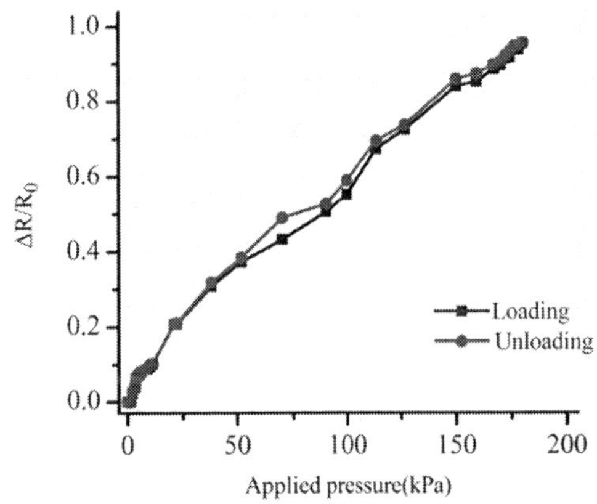

Figure 4. Piezoresistance response of the composites with the applied pressure.

Table 1. Performance comparison between the tactels of the sensor and flexiforce sensors.

	Effective measurement range (kPa)	Linearity (%)	Sensitivity (kPa-1)	Pressure resolution (kPa)	Hysteresis (%)
E-skin	0-180	4.4	0.005	1-5	5.8
Flexiforce sensor	0-80	≤ 10	0.009	> 7	≤ 15

In order to study the capability of the pressure sensor array to map the pressure distribution, a sensor system based on zero potential method was built. Fig. 5 shows the integrated system which includes our e-skin, a scanning power supply circuitry with the scanning frequency at 300Hz, an 8-channel data acquisition system and a PC. A constant voltage (5V) is applied to the row to be selected as power supply, and all the other rows are set to zero volts. The scanned output-voltages of tactile pressure sensor array are transferred to analog-to-digital converter of a micro programmed control unit (MCU). The PC is used to process the obtained signal and displays the pressure distribution as grey-scale images. Fig.6(a) shows that a PVC tube is pressed on the 8×8 sensor array. The Fig.6(b) shows the pressure distribution produced by the contact shape between the PVC tube and sensor array. Obviously, the shape of the PVC tube is clearly resolved by the sensor array.

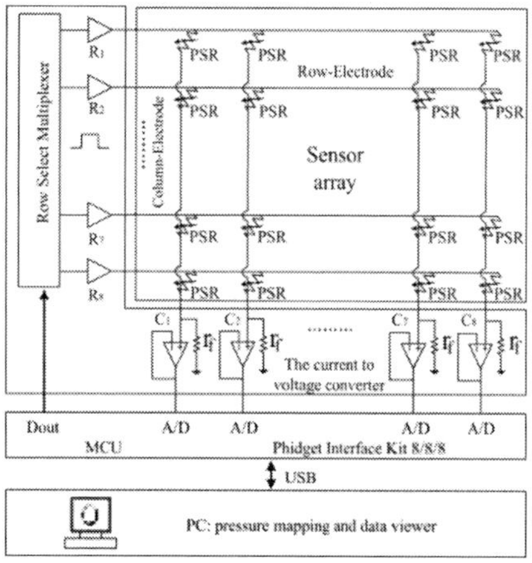

Figure 5. Schematic of the tactile sensor array system.

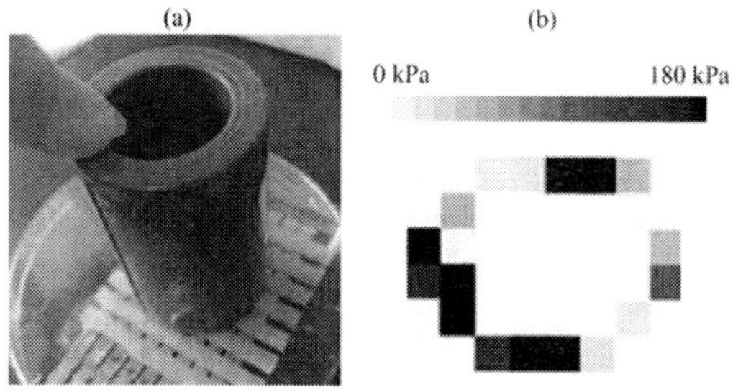

Figure 6. Measured pressure distribution using various objects of different shapes.

5. Conclusion

The development of a novel type of flexible and stretchable 8×8 pressure sensor array is presented in the paper. The conductive rubber is used as the piezoresistive material for pressure sensing. The AgNWs/PDMS conductors covered with nano-scale thin metal film are employed as stretchable interconnects and sensing electrode. The porous PDMS design can effectively increase the flexibility and stretch ability of the sensor array. The discrete distribution of the sensing elements and designed scanning circuit eliminate the crosstalk among the piezoresistive elements. The static characteristic and the dynamic characteristic are investigated in this paper. The sensor has a low hysteresis and good linearity. The sensor array responds fast and stably enough when a variable pressure is applied on it. The pressure distribution on the contact area induced by placing a PVC tube is successfully detected by the sensor array. The skin-like characteristic of the pressure sensor array makes it well suited for the application on the complex surfaces in robots or medical devices for pressure detection and feedback.

References

[1] V.J. Lumelsky, M.S. Shur, S. Wagner, Sensitive skin, IEEE Sensors Journal, 1(2001) 41-51.
[2] M.L. Hammock, A. Chortos, B.C.K. Tee, J.B.H. Tok, Z. Bao, 25th anniversary article: The evolution of electronic skin (e - skin): A brief history, design considerations, and recent progress, Advanced Materials, 25(2013) 5997-6038.
[3] R.S. Dahiya, G. Metta, M. Valle, G. Sandini, Tactile sensing-from humans to humanoids, IEEE Transactions on Robotics, 26(2010) 1-20.
[4] H. Yousef, M. Boukallel, K. Althoefer, Tactile sensing for dexterous in-hand manipulation in robotics-A review, Sensors and Actuators A: Physical, 167(2011) 171-87.
[5] M.I. Tiwana, S.J. Redmond, N.H. Lovell, A review of tactile sensing technologies with applications in biomedical engineering, Sensors and Actuators A: Physical, 179(2012) 17-31.
[6] W. Haopeng, Z. Debao, C. Jianguo, Development of a skin-like tactile sensor array for curved surface, IEEE Sensors Journal, 14(2014) 55-61.
[7] A. Jain, M.D. Killpack, A. Edsinger, C.C. Kemp, Reaching in clutter with whole-arm tactile sensing, The International Journal of Robotics Research, 32(2013) 458-82.
[8] Y. Tada, M. Inoue, T. Kawasaki, Y. Kawahito, H. Ishiguro, K. Suganuma, A flexible and stretchable tactile sensor utilizing static electricity, IEEE/RSJ International Conference on Intelligent Robots and Systems San Diego, CA, 2007, pp. 684-9.

[9] L. Viry, A. Levi, M. Totaro, A. Mondini, V. Mattoli, B. Mazzolai, et al., Flexible three-axial force sensor for soft and highly sensitive artificial touch, Advanced Materials, 26(2014) 2659-64.
[10] D.P.J. Cotton, I.M. Graz, S.P. Lacour, A multifunctional capacitive sensor for stretchable electronic skins, IEEE Sensors Journal, 9(2009) 2008-9.
[11] D.J. Lipomi, M. Vosgueritchian, B.C.K. Tee, S.L. Hellstrom, J.A. Lee, C.H. Fox, et al., Skin-like pressure and strain sensors based on transparent elastic films of carbon nanotubes, Nature Nanotechnology, 6(2011) 788-92.
[12] T. Someya, Y. Kato, T. Sekitani, S. Iba, Y. Noguchi, Y. Murase, et al., Conformable, flexible, large-area networks of pressure and thermal sensors with organic transistor active matrixes, Proceedings of the National Academy of Sciences of the United States of America, 102(2005) 12321-5.
[13] L. Pan, A. Chortos, G. Yu, Y. Wang, S. Isaacson, R. Allen, et al., An ultra-sensitive resistive pressure sensor based on hollow-sphere microstructure induced elasticity in conducting polymer film, Nature Communications, 5(2014) 1-8.
[14] M.Y. Cheng, C.M. Tsao, Y.Z. Lai, Y.J. Yang, The development of a highly twistable tactile sensing array with stretchable helical electrodes, Sensors and Actuators A: Physical, 166(2011) 226-33.
[15] T. D'Alessio, Measurement errors in the scanning of piezoresistive sensors arrays, Sensors and Actuators A: Physical, 72(1999) 71-6.

IOP Conf. Series: Materials Science and Engineering **207** (2017) 012041 doi:10.1088/1757-899X/207/1/012041

Mechanism of piezoelectric micropump with four flexible valves

Changzhi Wei[1, 2, *], Shoushui Wei[3] and Xiaofei Ren[3]

[1]School of Information Science and Engineering, University of Jinan, Jinan 250029, China
[2]Shandong Provincial Key Laboratory of Network based Intelligent Computing, Jinan 250029, China
[3]School of Control Science and Engineering, Shandong University, Jinan 250061, China

[*]Corresponding author e-mail: ise_weicz@ujn.edu.cn

Abstract. A piezoelectric micropump with four flexible valves is proposed and its mechanism is researched. In the micropump chamber four asymmetric flexible valves are used to achieve single direction flow. The instantaneous velocities are analyzed and the peak input and output velocities are 0.14 m/s and 0.2 m/s respectively. The net flow rate is calculated, which is about 2.5×10-11m3/s. Due to the low operating frequency, flexible valves structure, no heat and no restriction on transport fluid type, this kind of micropump can be used for liquid transport and other biological sample containing DNA.

1. Introduction

In recent years, microfluidic systems have been widely used in various fields. Micropumps are the key components of microfluidic systems with applications ranging from biological fluid handling to microelectric cooling [1]. The traditional valve micropump usually utilizes mechanical parts to achieve rectifying [2]. The unidirectional performance is satisfactory, but miniaturization and service life are limited because of moving parts. Especially the cells in the biology sample may be damaged by the valve[3,4]. In this paper, a piezoelectric micropump, with flexible material flaps as valves, is proposed. Due to the low operating frequency, flexible valves structure, no heat and no restriction on transport fluid type, this kind of micropump can be used for liquid transport and other biological sample containing DNA.

This paper researches the mechanism of piezoelectric micropump with four flexible valves. Flow pattern in the micropump chamber is analyzed. The input velocity, output velocity and net volume are calculated. The conclusions can provide theoretical guidance to further analysis.

2. Model and Principle

2.1. Model Geometry

The designed model geometry is shown in Figure 1. The micropump chamber is 600 um in length and 100 um in high. The piezoelectric material, 80 um in length and 10 um in thickness, is put on the middle top of the chamber. The left of the chamber is inlet and the right is outlet. Four tilted flexible flaps are attached to the top and the bottom of the channel to obstruct the counter-flow along the channel length. They are all angled at 45 degrees to the horizontal channel edge.

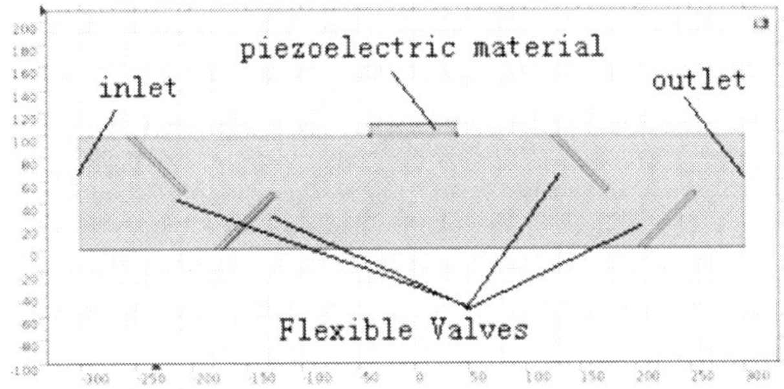

Figure 1. The designed model geometry

2.2. Driving Principle

To achieve sustained flow in a single given direction, four flexible material flaps are used along the chamber length. When the membrane is constricted by the piezoelectric material, the chamber is constricted and the fluid will be push out of the inlet and the outlet. The two flaps of the left will be stand highly to obstruct the flow to inlet and the two of the right will be down to let flow out to the outlet. While the membrane is expanded, the principle is same as above. The flow pulled into the chamber from the inlet is more than the outlet. Although there are both in and out flow in inlet and outlet, in a period the outflow in outlet and inflow in inlet is one direction, it is from inlet to outlet along the chamber length.

2.3. Analysis configuration

The micropump mechanism is simulated by COMSOL. The physics required for the model is configured within the Fluid-Structure Interaction interface. The Linear Elastic Material feature is applied to the four tilted flaps, and then the Fluid-Solid Interface Boundary feature is automatically assigned to the boundaries between the flaps and the fluid in the channel. A sinusoidal driving voltage with a period of 1 s is applied to the piezoelectric electrode via a user input expression. Inlet and Outlet boundary feature are applied to the left and right boundaries of the chamber. The mesh is configured to be tightest around the tilted flaps, in order to resolve the stress within the bending flaps. The mesh is shown in Figure 2. A single time dependent study is performed over a duration of 2 seconds, which corresponds to two full oscillations of the driving voltage.

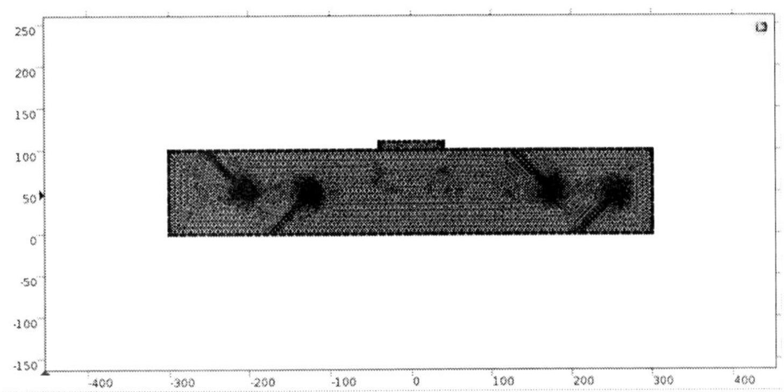

Figure 2. The mesh of the model

3. Results and Analysis

3.1. Flow Pattern in the Micropump Chamber

The mechanism by which the flow direction is regulated can be observed in the Flow and Stress default plot group. During the down stroke, when fluid is pushed from the chamber into the channel, the right-hand flaps are bent down towards the wall of the chamber whilst the left-hand flaps are bent away from the chamber wall. This configuration is shown in Figure 3, which shows the solution at a time of 0.3 s which corresponds to when the vibration amplitude of the piezoelectric material is almost at its maximum. Due to the asymmetric bending of the flaps, fluid can more easily flow out of the right-hand outlet. During the upstroke, when fluid is drawn from the channel into the chamber, the flaps are bent in the opposite directions. This configuration is shown in Figure 4, which shows the solution at a time of 0.8 s. Now the right hand flaps restrict the flow more than the left-hand flaps, and the majority of the fluid that is drawn into the chamber is from the left-hand inlet.

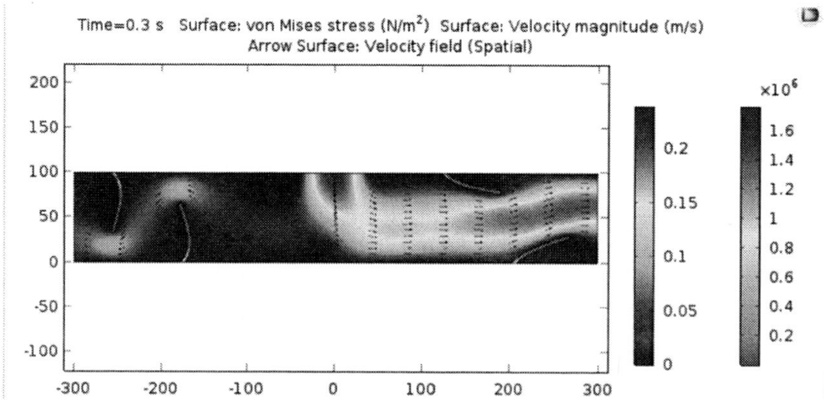

Figure 3. The flow velocity during the down stroke

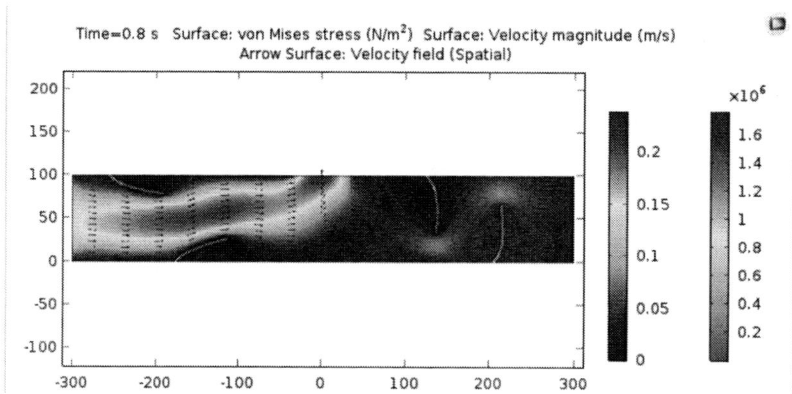

Figure 4. The flow velocity during the upstroke

3.2. The Input and Output Velocity

When the vibration amplitude of the piezoelectric material is 0.96 m/s and the driving frequency is 1Hz, the average velocity at the inlet and the outlet in two periods is calculated as shown in Figure 5. As can be seen, during the first half period, the peak of the input velocity is -0.06 m/s, while the peak of the output velocity is 0.2 m/s. So the fluid pushed out of the micropump chamber is more than the fluid pushed into. During the second half period, the peak of the input velocity is 0.14 m/s, while the peak of the output velocity is -0.02 m/s. So the fluid pushed into the micropump chamber is more than the fluid pushed out.

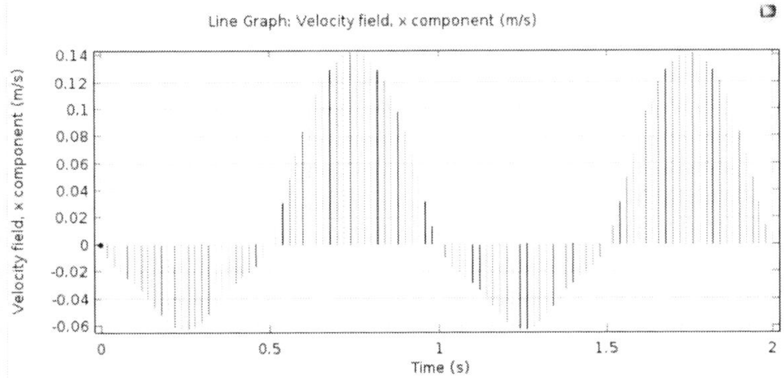

a. The average input velocity

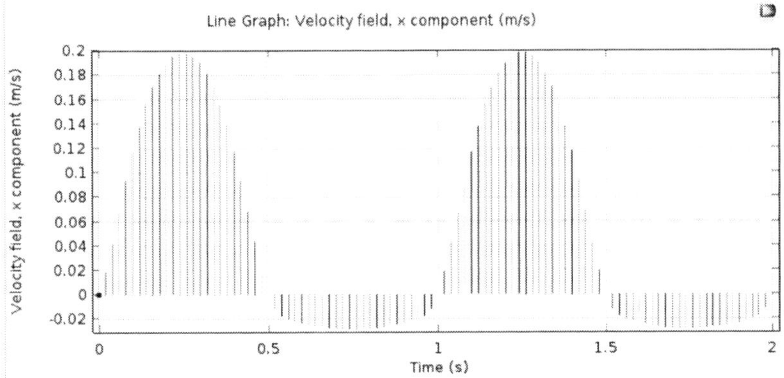

b.The average output velocity

Figure 5. The input and output velocity

3.3. The Net Flow of the Micropump

The net volume of fluid that is pumped from left-to-right is shown in Figure 6. As expected, the gradient of the curve, which is the net flow rate, varies sinusoidally with a period equal to the vibration frequency of the piezoelectric material. The maximum gradients occur at intervals of odd multiples of 0.5 s, which correspond with the peaks in the magnitude of the vibration amplitude. From the curve, the net flow rate is about $2.5\times 10^{-11} m^3/s$, surly that is not large enough in many application fields. To achieve larger flow, larger driving voltage and driving frequency may be considered to utilize.

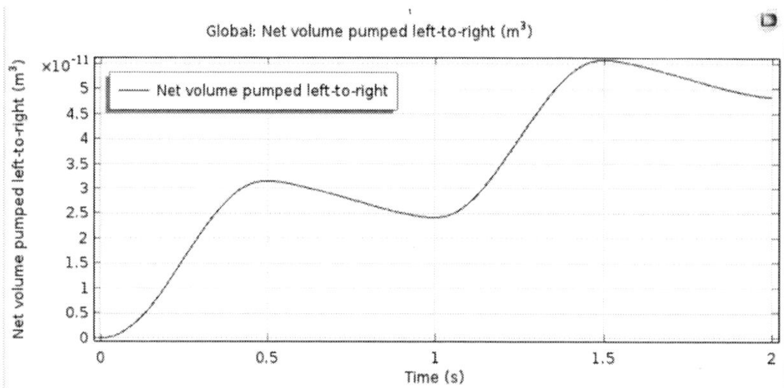

Figure 6. The net volume pumped from left-to-right

4. Conclusion

A piezoelectric micropump with flexible valves is proposed and its mechanism is analyzed. The peak input and output velocities are 0.14 m/s and 0.2 m/s respectively. The net flow rate is about 2.5×10^{-11} m^3/s. Compared with other microfluidic driving methods, this kind of micropump has several advantages and has many possible applications in microfluidic systems. For example, this device could be used to deliver fluid from a droplet reservoir connected to the left-hand outlet into a microfluidic pathway connected to the right-hand outlet. Alternately, this device could be used to create a circulating system where a fluid is pumped around a continuous loop to cool a microelectronic system.

Acknowledgments

This work was financially supported by Dr. Funds of University of Jinan (XBS1535).

References

[1] Huikko K, Kostiainen R, Kotiaho T. Introduction to micro analytical system: bioanalytical and harmaceutical applications. European Journal of Pharmaceutical Sciences. 20(2003)149-171

[2] Lintel H T G V, Pol F C M V D, Bouwstra S. A piezoelectric micropump based on micromachining of silicon. Sensors & Actuators. 15(1988)153-167.

[3] Laser D J. A review of micropumps. Journal of Micromechanics & Microengineering. 14(2004): 35-R64(30).

[4] Chen W F, Wang L X, Fu-Zai L V. A Review of Micropump Based on Piezoelectric Ceramics Actuator. Chinese Hydraulics & Pneumatics, 2014.

The uncertainty research of visual positioning in different mediums

Yanzhu Hu, Song Wang*, Xinbo Ai, Zhen Meng, Wenjia Tian, Jiao Wang, Yu Hu

College of Automation, Beijing University of Posts and Telecommunications, Beijing, 100876, China

*Corresponding author e-mail: wongsangwongsang@163.com

Abstract. In engineering application, there existed deviation when light spreads in different mediums. At the same time miscellaneous medium and random fluctuation may lead to inaccuracy of visual positioning. It explored visual positioning technology in condition of different mediums for binocular camera. Firstly focused on medium material and impurity in real environment, the refractive index formula is derived through binocular visual positioning principle. Then considering deviation when light spreads in different mediums, deductive process of vision positioning is realized according to calculated refractive index. Because cloud model can better describe randomness and fuzziness. It was introduced to deal with problems of random fluctuation in different mediums. Finally simulation was designed to prove the accuracy improvement in different mediums. The result showed that accuracy error had decreased by 66.7% after considering transparent organic mediums. Compared with not knowing refractive index, positioning error had decreased by 49.7%.

1. Introduction

With the development of society, industrial manufacture tends to be mechanized and intelligent. Visual positioning is the key technology to provide position information of target [1-2]. Therefore it has been widely used in a variety of fields like industry, agriculture, tourism. Currently visual positioning was mainly applied in single medium. Some researches concentrated on air medium which can provide position information for industrial and agricultural manipulator. Auto-picking is an example of visual positioning in agricultural production. Especially the application was in agricultural products such as litchi [3], tomato [4] and strawberry [5]. The other researches concentrated on aqueous medium which provide position information for underwater operation. It applies to underwater fishing and submarine reconnaissance [6-7]. But visual positioning often involves the deviation in different mediums. For example, the camera obtains position information from air to water, or remote position happens in special gases. The visual positioning's accuracy is the key to project implementation.

In addition air medium and aqueous medium usually mix with a variety of impurities. So refractive index can vary too. There is still no accurate mechanism formula to describe the situation. At the same time the light may have shape change and random fluctuations from external randomness. It would lead to inaccuracy of position information.

The research mainly includes two parts. The first one focused on refractive index calculation in real environment. The second one focused on random fluctuations from medium. Finally several experiments were designed to prove validity of the research.

2. Visual positioning in condition of different mediums

Derivation process in the paper exemplifies underwater camera. Light spreads through air, transparent organic materials and water. The refractive index of air is 1 by default. The refractive index of transparent organic materials is detected as 1.5 when they are made. The thickness a=3mm. The refractive index of pure water is 1.33. Visual positioning model in condition of different mediums is shown in figure 1.

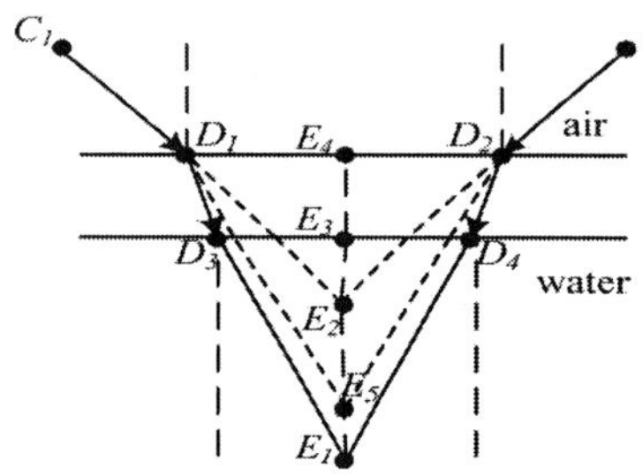

Figure 1. Visual positioning model in condition of different mediums

C_1 and C_2 are two cameras' positions. D_1D_2 and D_3D_4 are planes between different mediums. C_3 and C_4 are projections of C_1 and C_2 in plane D_1D_2. C_5 and C_6 are projections of C_1 and C_2 in plane D_3D_4. E_1 is the real coordinate of target. E_2 is virtual coordinate of target. E_3 and E_4 are projections of target in plane D_1D_2 and D_3D_4. Refractive index is already known as n=1.33.According to visual positioning model in condition of different mediums, refractive index of aqueous medium is inferred in formula (1) and (2).

$$n = \frac{\cos\angle C_1D_1C_3}{a/|D_1D_3|}\cdot\frac{a/|D_1D_3|}{\cos\angle E_1D_3E_4} = \frac{\cos\angle C_1D_1C_3}{\cos\angle E_1D_3E_4} \tag{1}$$

$$n = \frac{\cos\angle C_2D_2C_4}{a/|D_2D_4|}\cdot\frac{a/|D_2D_4|}{\cos\angle E_1D_4E_3} = \frac{\cos\angle C_2D_2C_4}{\cos\angle E_1D_4E_3} \tag{2}$$

In visual positioning of binocular camera, key points coordinates are known as C_1=(0, 0, 0), C_2=(x_1, y_1, z_1), E_2=(x_2, y_2, z_2). Assuming that material plane equation is y=m and y=m+0.3. Therefore C_3=(0, m, 0), C_4=(x_1, y_1+m, z_1), C_5=(0, m+3, 0), C_6=(x_1, y_1+m+3, z_1), E_3=(x_2, m+3, z_2) and E_4=(x_2, m, z_2). C_3,C_4, C_5 and C_6 are the projections of two cameras in planes between different mediums. Because two planes of different mediums are parallel, the position of target will change in the direction of Y axis. It is shown in formula (3).

$$\alpha = 3 - 3\cdot\frac{y_2}{\sqrt{x_2^2+z_2^2}}\tan[\arcsin(\frac{2\sqrt{x_2^2+z_2^2}}{3\sqrt{x_2^2+y_2^2+z_2^2}})] \tag{3}$$

According to equations C_1E_2, C_1E_2, D_1D_2 and D_3D_4, the coordinates of points D_1, D_2, D_3 and D_4 in world coordinate are shown in formulas (4) and (5).

$$D_1 = \begin{bmatrix} x_4 & y_4 & z_4 \end{bmatrix}^{\mathrm{T}} = \begin{bmatrix} m \cdot \dfrac{x_2}{y_2} & m & m \cdot \dfrac{z_2}{y_2} \end{bmatrix}^{\mathrm{T}} \tag{4}$$

$$D_2 = \begin{bmatrix} x_5 & y_5 & z_5 \end{bmatrix}^{\mathrm{T}} = \begin{bmatrix} \dfrac{(m-y_1)(x_2-x_1)}{y_2-y_1} + x_1 & m & \dfrac{(m-y_1)(z_2-z_1)}{y_2-y_1} + z_1 \end{bmatrix}^{\mathrm{T}} \tag{5}$$

According to formulas (4) and (5) position information of "Virtual image" with transparent organic medium is shown in formula (6).

$$E_5 = E_1 + (0, \alpha, 0) \tag{6}$$

The refractive index of water can be obtained through introducing formula (6) to (1) and (2). It is processed in formula (7). It can calculate formula (8) by the formula (7).

$$n = \frac{\cos \angle C_1 D_1 C_3}{\cos \angle E_5 D_1 E_3} = \frac{\cos \angle C_2 D_2 C_4}{\cos \angle E_5 D_2 E_3} \tag{7}$$

$$\frac{|D_1C_3|}{|C_1D_1|} / \frac{|D_1E_4|}{|E_5D_1|} = \frac{|D_2C_4|}{|C_2D_2|} / \frac{|D_2E_3|}{|E_5D_2|} \tag{8}$$

Where, 3D information of points C_1, C_2, C_3, C_4, D_1, D_2, E_4 are already known. There is only one unknown number for Y axis data of E_5. Information above is put in formula (8). The detail is shown in formula (9).

$$\frac{\sqrt{x_2^2+z_2^2}}{\sqrt{x_2^2+z_2^2+1}} / \frac{\sqrt{(x_2-y_2)^2+(z_2-y_2)^2}}{\sqrt{(x_2-y_2)^2+(z_2-y_2)^2+(y_2-m_1y_2/m)^2}} = \frac{\dfrac{\sqrt{[(x_2-x_1)(m-y_1)]^2+y_2^2+[(z_2-z_1)(m-y_1)]^2}}{\sqrt{[(x_2-x_1)(m-y_1)]^2+[(y_2-y_1)(m-y_2)]^2+[(z_2-z_1)(m-y_1)]^2}}}{\dfrac{\sqrt{[(x_2-x_1)(m-y_1)]^2+[(z_2-z_1)(m-y_1)]^2}}{\sqrt{[(x_2-x_1)(m-y_1)]^2+[(y_2-y_1)(m-m_1)]^2+[(z_2-z_1)(m-y_1)]^2}}} \tag{9}$$

Where, x_1, y_1, z_1, x_2, y_2, z_2, m are known variables. So unknown number m_1 can be solved. Then according to 3D information of each points, the unknown refractive index is as shown in formula (10).

$$n = \frac{\sqrt{x_2^2+z_2^2}}{\sqrt{x_2^2+z_2^2+1}} / \frac{\sqrt{(x_2-y_2)^2+(z_2-y_2)^2}}{\sqrt{(x_2-y_2)^2+(z_2-y_2)^2+(y_2-m_1y_2/m)^2}} \tag{10}$$

3. The randomness description of visual positioning based on cloud model

For position information of E_1, wave in medium may effect on light. So position information may have deviation after interference [8]. In order to diminish the interference, cloud model is introduced to describe randomness. Each frame image can provide 3D information. The images come from underwater camera. 3D information is called x_i, y_i, and z_i. Where i=1,…,n. Numerical characteristics of cloud model can reflect the fuzziness and randomness. These are expectation Ex, entropy En and super entropy He respectively [9-10]. The generation algorithm of normal cloud model generator is following.

(1) Expectation of samples is calculated as shown in formula (11).

$$Ex(x_i, y_i, z_i) = \frac{1}{i} \sum (x_i, y_i, z_i) \tag{11}$$

(2) Entropy can be obtained by samples in formula (12).

$$En(x_i, y_i, z_i) = \frac{1}{i}\sum\{[(x_i, y_i, z_i) - E(x_i, y_i, z_i)]^2\} \tag{12}$$

(3) Super entropy of samples is set as formula (13).

$$He(x_i, y_i, z_i) = 0.1 \tag{13}$$

(4) Regarding $Ex(x_i, y_i, z_i)$ and $En(x_i, y_i, z_i)$ as expectation and standard deviation respectively, normal random numbers $(\hat{x}_i, \hat{y}_i, \hat{z}_i)$ are generated.

(5) Regarding $En(x_i, y_i, z_i)$ and $He(x_i, y_i, z_i)$ as expectation and standard deviation respectively, normal random numbers $En'(x_i, y_i, z_i)$ are generated.

(6) Certainty degree is calculated as formula (14).

$$\mu_i = \exp\left[-\sum_{x,y,z}^{x,y,z}[\frac{(x_i - Ex_i)^2}{2(En'_i)^2}]\right] \tag{14}$$

(7) Certainty degree of next frame image is calculated by iteration. Calculation process repeats steps above. The least certainty degree is obtained through comparing certainty degree μ_i of each frame image. The (x_i, y_i, z_i) with least certainty degree are the coordinates of target.

4. Experiment

The two groups of target point's coordinates are [2.332, -2.912, 1] and [1.812, -2.849, 1] respectively. The coordinate of E_2 is [1023.371, -1277.532, 2676. 535]. Transparent organic materials may lead to deviation. So distance from E_5 to E_1 is calculated through formula (1) to (10). Then $\alpha = 2.205$. As known variable m=-800, two points' coordinates are respectively E_1=[1023.371, -1289.737, 2676.535] and E_5=[1023.371, -1287.532, 2676.535]. So the refractive index of water is n=1.35. Considering visual positioning in different mediums, there are four kinds of condition. They are ignoring the shield, including the shield, known refractive index and unknown refractive index.

The real coordinate of target is [1023.5, -1290, 2676.5]. 100 frame images' errors in four conditions are compared as shown in Figure 2.

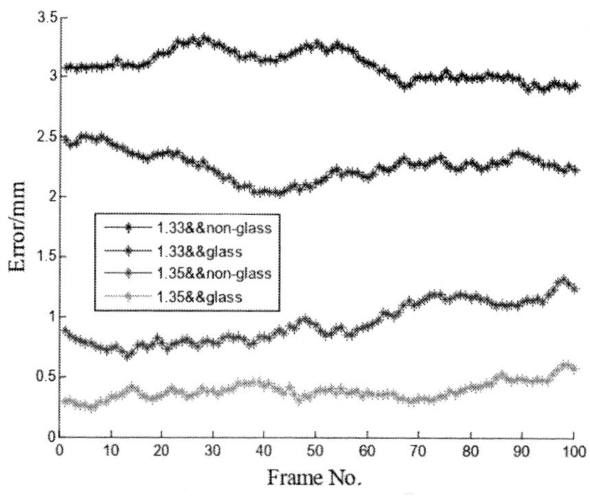

Figure 2. Error Comparison curve of 100 frame image in four conditions.

In the figure above, black line represents condition of ignoring the shield with n=1.33. Blue line represents condition of ignoring the shield with unknown n. Red line represents condition of including the shield with n=1.33. Green line represents condition of including the shield with unknown n. In the condition of unknown n, the average error is 1mm compared with 3mm in condition of n=1.33. The error has reduced by 66.7%. At the same time in condition of including the shield, the error has reduced by 49.7% compared with that in condition of ignoring shield. According to figures above, it is obvious that algorithm in the paper tend to be more accurate. It also fully considers thickness of transparent organic medium and unknown refractive index of water. In order to eliminate effect from random wave, further experiment is required. 100 frame images' data is used to calculate error in the fourth conditions above. According to formula (11) to (14), trend of certainty degree curve is consistent with that of error curve. It is shown in figure 3.

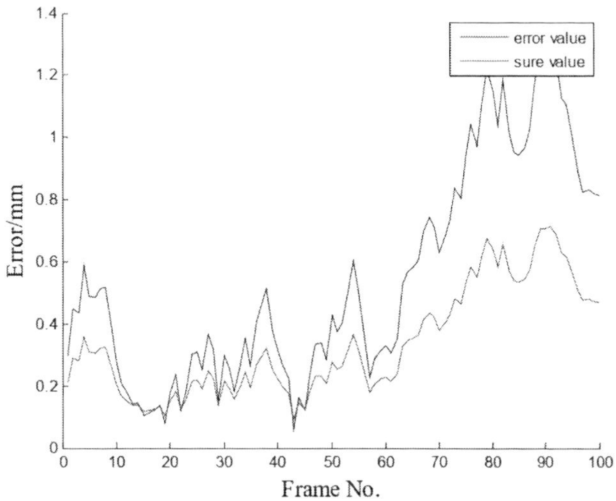

Figure 3. The comparison of certainty degree curve and error curve

The curve trend of error is almost consistent with uncertainty degree. And the point with least uncertainty is most close to the real value.

Repeated experiments above are implemented for a hundred times. Experimental parameters are modified in real environment. They are adjusting the camera position, mixing impurity into water, changing thickness of transparent organic materials and adjusting height of water surface. Coordinates of target are obtained through method in the paper. The conclusion is shown as follows. **Conclusion 1:** The result shows that there is better error in considering organic medium than in un-considering organic medium. **Conclusion 2:** The result shows that there is better error in considering unknown n than un-considering unknown n. **Conclusion 3:** The result shows that there is better error in introducing cloud model than un-introducing cloud model. The experimental method in the paper applies to visual positioning technology in different mediums. It solves the problems of unknown refractive index and medium fluctuations. It provides an effective way for optimizing visual positioning.

5. Conclusion

The effect on visual positioning in different mediums is fully considered. Through deriving positioning formula of bilingual camera, it better solves the light changing problem in different mediums. The refractive index in real environment relative to that of air is worked out. At the same time cloud model can effectively describe randomness and fuzziness. It solves uncertainty problem caused by medium fluctuations in visual positioning. It provides a better way for the development of visual positioning technology in complex environment.

Acknowledgments
This work was partially supported by National Natural Science Foundation of China (No. 61503034&No.61627816), the National Key Technology R&D Program of China (No. 2015BAK40B03& No.2016YFC0701309), Sub project of key project of Beijing, China (No. D161100004916002).

References
[1] Henten E J V, Tuijl B A J V, Hemming J, et al. Field Test of an Autonomous Cucumber Picking Robot [J]. Biosystems Engineering, 2003, 86(3): 305-313.
[2] Zhang Z. A Flexible New Technique for Camera Calibration [J]. IEEE Transactions on Pattern Analysis & Machine Intelligence, 2000, 22(11): 1330-1334.
[3] Xiong J. Design of Visual Position System for Litchi Picking Manipulator [J]. Transactions of the Chinese Society of Agricultural Machinery, 2012, 43: 250-255.
[4] Yin J J, Wu C Y, Yang Simon X, etc. Obstacle-avoidance Path Planning of Robot Arm for Tomato-picking Robot [J]. Transactions of the Chinese Society for Agricultural Machinery, 2012, 43(12): 1296-1297.
[5] Wang L J, Zhang L B, Duan Y H, etc. Fruit localization for strawberry harvesting robot based on visual serving [J]. Transactions of the Chinese Society of Agricultural Engineering, 2015, 31(22).
[6] Wang W J, Xu J J, Lv Z M, etc. Weight estimation of underwater Cynoglossus semilaevis based on machine vision [J]. Transactions of the Chinese Society of Agricultural Engineering, 2012, 28(16): 153-157.
[7] Chen Y J, Zhu K W, Ge Y Z, etc. Binocular vision based locating system for underwater inspection [J]. Journal of Mechanical & Electrical Engineering, 2011, 28(5): 567-573.
[8] Bulian G, Francescutto A, Umeda N, et al. Qualitative and quantitative characteristics of parametric ship rolling in random waves in the light of physical model experiments [J]. Ocean Engineering, 2008, 35(17–18): 1661-1675.
[9] Islam S, Grégoire J C. Giving users an edge: A flexible Cloud model and its application for multimedia [J]. Future Generation Computer Systems, 2012, 28(6): 823-832.
[10] Qin K, Xu K, Liu F, et al. Image segmentation based on histogram analysis utilizing the cloud model [J]. Computers & Mathematics with Applications, 2011, 62(7): 2824-2833.

IOP Conf. Series: Materials Science and Engineering **207** (2017) 012043 doi:10.1088/1757-899X/207/1/012043

A simplified mesoscale model for predicting the mechanical behavior of stitched CFRP laminates

Jun Li[1,*], Bo Wang[2] and Guiqiong Jiao[2]

[1]Department of Structure Design, China Helicopter Research and Development Institute, Jingdezhen 333000, China

[2]School of Mechanics and Civil Construction, Northwestern Polytechnical University, Xi'an 710129, China

[*]Corresponding author e-mail: dasuo@126.com

Abstract. This paper presents a finite element (FE) analysis of the mechanical responses for stitched CFRP laminates under different mechanical loads. Firstly, the through-thickness stitch was simplified to z-pin like reinforcement with a uniform displacement constraint on the upper and lower surfaces of the laminate. Then, a mesoscale 3D representative volume element (RVE) of the stitched composite was proposed and modeled in the FE code ABAQUS, where the reinforcing stitch, composite layers and interfaces were built. A 3D Hashin damage model and built-in cohesive elements were respectively used to predict the mechanical failure of the stitch and the damageable behavior of cohesive interfaces. Simulation results reveal the progressive damage and rupture processes of the RVE under tensile and shear mechanical loads, and macroscopic nonlinear load-displacement responses of the mesoscale model are also captured.

1. Introduction

Carbon fiber reinforced plastic (CFRP) laminates receive wide use in the aerospace industry. However, their usage is affected by their low delamination resistant ability, and consequently the impact damage and the manufacture defect between laminate layers becomes the main safety concern. Adding through-the-thickness stitches or z-pins [1, 2], although with different strengthening mechanisms, is the effective way to improve the delamination fracture toughness of the composite laminates. In the stitched laminates, the stitches are continuous and constantly stretched, and when delamination happens bridging load provided by the stitch will increase with the delamination displacement until the tension stress in the stitch reaches its ultimate strength, then the ruptured stitch can generates little bridging load during the frictional pullout process. In comparison, the frictional bridging load during z-pin pullout process is relatively larger, which is the main mechanism increasing the mode I delamination toughness of the z-pinned laminates.

As the stitching and z-pinning technique become increasingly used in the composite structures, the strength analysis and prediction of these reinforced laminates attracts much attentions along with their application. Many models have been developed in the open literature, which can be roughly categorized into two groups. The first one is to propose theoretical models by analyzing the mechanical process during the tension, bending or shear load of the reinforced structure to obtain closed form solutions [3];

and the other is to build FE models, simplifying the stitches or z-pins into spring elements [4], connecting elements [5], beam elements [2, 6], three-nodes rod elements [7, 8] or cohesive zone elements [9, 10, 11] with material parameters and corresponding load (or traction) versus displacement relationships to predict the mechanical behaviors and load-bearing capacity of the reinforced laminate structures. However, one can't find generally recognized theoretical or FE methods in this research field.

This study aims at proposing a simplified mesoscale FE model for analyzing the stitched laminates, which could be used to predict the strengthening effect of the stitch and its failure modes under different loads, and also capture the macroscopic mechanical responses of the stitched laminate.

2. Mesoscale model

2.1. Model description

In the paper, we take the typical stitched T-shaped sandwich joints as shown in Fig. 1 for analysis. A representative volume element (RVE) was selected from the stitched area which shows periodical geometrical characteristics. Because the stretched stitch restricts the deformation on the upper and lower surfaces of the RVE, we simplify its geometry into a z-pin like stitch, plus a uniform displacement or strain constraint on these surfaces.

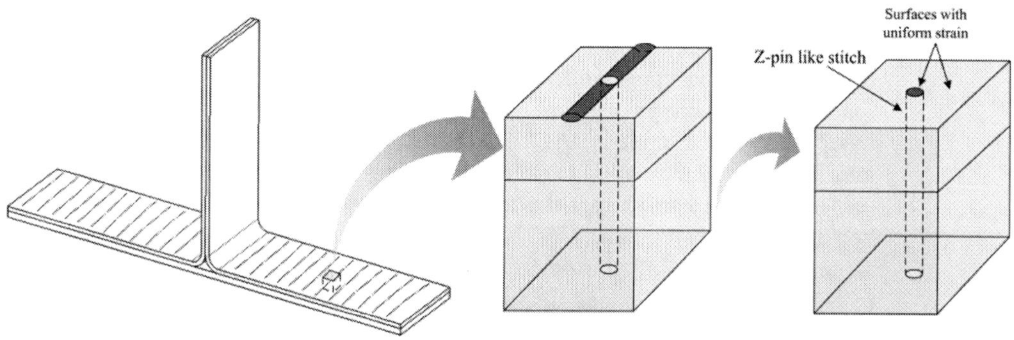

Figure 1. RVE model selection and simplification of the stitched laminates.

Meanwhile, the simplified z-pin like stitch keeps the same cross-section as the original stitch, then the simplified RVE should yield the same through-thickness deformation ΔL under the tensile load F as the original one. Take the cross-section of the RVE for analysis (see Fig. 2), the stitch length of two segments on the upper surface is assumed L_1 and L_2 to respectively, but with only half of the section area of stitch segment in the through-thickness area, which is assumed as $A/2$ and with tensile modulus E. In the simplified RVE, the total length of the through-thickness stitch is comprised by L_3 and L_4, and its modulus E' should satisfy the following equation based on the equivalent displacement assumption under the transverse tension load F (see Fig. 3):

$$\Delta L = \frac{FL_1}{E(A/2)} + \frac{FL_2}{E(A/2)} + \frac{FL_3}{EA} + \frac{FL_4}{EA} = \frac{F(L_3 + L_4)}{E'A} \quad (1)$$

Then, the tensile modulus of the stitch E' in the simplified RVE model can be deduced as:

$$E' = \frac{(L_3 + L_4)E}{(2L_1 + 2L_2 + L_3 + L_4)} \quad (2)$$

In fact, the reinforcing stitch is composed by the Kevlar stitch and resin, and the latter component usually take 60% in the volume, so the longitudinal modulus of the composed stitch should multiply a factor of 0.4 by the modulus provided by the manufactures. In this paper, according to the geometric configuration the stitch length is $L_1 = L_2 = 2.5$mm, and the thickness of the web and flange is $L_3 = 1.6$mm, $L_4 = 8.8$mm, respectively. As a result, one can obtain $E' \approx 0.51E$ in the simplified RVE.

The elastic and strength properties of the reinforcing stitch in the simplified RVE are presented in Table 1and

Table **2**, where the units of modulus and strength parameters are in MPa.

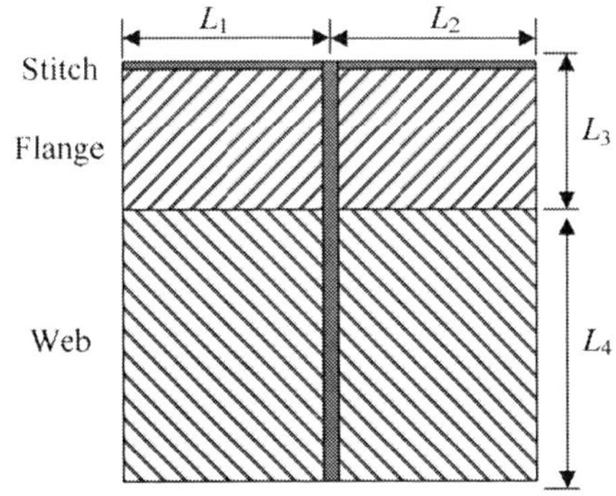

Figure 2. The cross-section of stitched RVE

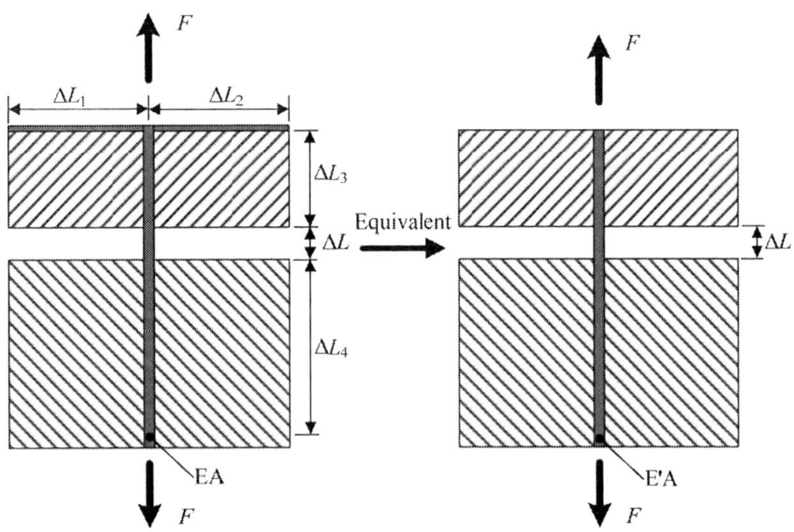

Figure 3. Schematic diagram for equivalent tensile displacement assumption.

Table 1. Elastic properties of the stitch in the simplified RVE model.

E_{11}	E_{22}	E_{33}	v_{12}	v_{13}	v_{23}	G_{12}	G_{13}	G_{23}
25500	4850	4850	0.34	0.34	0.35	1770	1770	1800

Table 2. Strength properties of the stitch.

X_T	X_C	Y_T	Y_C	$S_{XY,XZ}$	S_{YZ}	Z_T
880	186	20	93	40	40	20

Further, the 3D mesoscale RVE model was built in Fig. 4. The model is composed by the flange and web along the thickness direction, and an interface between them was introduced for delamination analysis. In the center part of the model, a cylinder represents the reinforcement stitch with a diameter of 0.72mm, and it was surrounded by a 0.1mm thickness resin rich cylinder area.

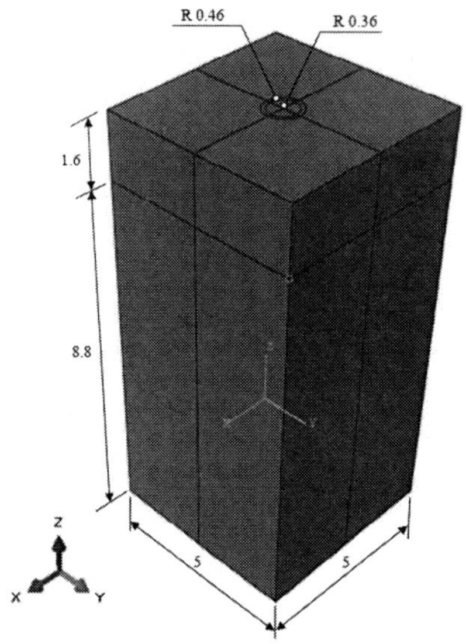

Figure 4. Schematic of the 3D mesoscale RVE model.

2.2. Material model

Under mechanical loads, the stitch firstly sustains elastic stretch and would finally rupture. Hence, the damage and failure behaviors of the stitch should be considered. In this paper, we introduce the well-known Hashin damage criterion to capture these non-linear mechanical behaviors. In the 3D stress state, they have the following expressions considering the characteristics of the stitch:

1) Stitch tensile failure mode (when $\sigma_{11} > 0$)

$$\left(\frac{\sigma_{11}}{X_t}\right)^2 + \left(\frac{\tau_{12}}{S_{XY}}\right)^2 + \left(\frac{\tau_{13}}{S_{XZ}}\right)^2 = 1 \tag{3}$$

2) Matrix tensile failure mode ($\sigma_{22} + \sigma_{33} > 0$)

$$\left(\frac{\sigma_{11}}{X_t}\right)^2 + \left(\frac{\tau_{12}}{S_{XY}}\right)^2 + \left(\frac{\tau_{13}}{S_{XZ}}\right)^2 = 1 \tag{4}$$

3) Matrix compressive failure mode ($\sigma_{22} + \sigma_{33} < 0$)

$$\frac{1}{Y_C}\left[\left(\frac{Y_C}{2S_{YZ}}\right)^2 - 1\right](\sigma_{22} + \sigma_{33}) + \left(\frac{\sigma_{22} + \sigma_{33}}{2S_{YZ}}\right)^2 + \frac{\left(\tau_{23}^2 - \sigma_{22}\sigma_{33}\right)}{S_{YZ}^2} + \frac{\left(\tau_{12}^2 + \tau_{13}^2\right)}{S_{XY}^2} = 1 \tag{5}$$

As the FE analysis code (ABAQUS/Standard) can only provide a 2D Hashin damage material model, a user-defined field subroutine (USDFLD) was programmed to implement the 3D Hashin damage model mentioned above. For simplicity, two state of the material integration point was defined: the elastic state (which is flagged as 0) and failure state (flagged as 1), and three state dependent variables (SDV1, SDV2, and SDV3) were introduced to record the numerical values on the left-hand side of the above three equations, respectively. Besides, two field variables (FV1 and FV2) were introduced with a zero value at the elastic state. When the stitch tensile failure mode is triggered, which means SDV1 reaches and/or exceeds 1.0, FV1 is then set as 1.0. Similarly, when matrix tensile or compressive failure happens (SDV2 or SDV3 $\geq$ 1.0), FV2 is then set as 1.0.

The related stiffness reduction strategy is listed in Table 3. For numerical reasons, the stiffness at the failure material point is set to 1% of the elastic one.

Table 3. The reduction strategy of the stitch's engineering constants.

Failure state	FV1	FV2		E_1	E_2	E_3	ν_{12}	ν_{13}	ν_{23}	G_{12}	G_{13}	G_{23}
No failure	0	0	-->	E_1	E_2	E_3	ν_{12}	ν_{13}	ν_{23}	G_{12}	G_{13}	G_{23}
Stitch tensile failure	1	0	-->	0	E_2	E_3	0	0	ν_{23}	0	0	G_{23}
Matrix faiure	0	1	-->	E_1	0	E_3	0	ν_{13}	0	0	G_{13}	0
Stitch and matrix failure	1	1	-->	0	0	E_3	0	0	0	0	0	0

The ABAQUS/Standard provides cohesive elements capable of simulating the mechanical behaviors of the interface layers between the web and flange, as well as those between the stitch and laminates. Elastic constants of the cohesive elements used in the present analysis are: E_{nn} = 3.0GPa, $G_{ss} = G_{tt}$ = 1.1GPa, and the damage initiation criterion is:

$$\left(\frac{\langle t_n \rangle}{t_n^0}\right)^2 + \left(\frac{t_t}{t_t^0}\right)^2 + \left(\frac{t_s}{t_s^0}\right)^2 = 1 \tag{6}$$

where the symbol <> denotes the Heviside operator, t_n represents normal tensile stress, t_t and t_s represents the shear stresses along two perpendicular direction of the cohesive element, and the tension strength t_n^0 = 65.6MPa, and shear strength $t_s^0 = t_t^0 = 60\text{MPa}$.

Besides, linear damage evolution equation is used to capture the initiation of stiffness change under the mechanical loads. Meanwhile, BK criterion [12] is selected to determine the mixed mode fracture toughness of the cohesive element by setting its power law index η = 1.45, and fracture toughness parameters: G_n^C = 0.25N/mm and $G_s^C = G_t^C$ =0.753N/mm.

Finally, the resin area surrounded the stitch can be treated as a homogonous and isotropic material. It's modulus is 3.5GPa, and Poisson's ratio is 0.35.

2.3. Finite element discretization

Based on the above geometric and material model, the mesoscale RVE model was built in the FE software. In which an 8-node linear brick element with reduced integration (C3D8R) was selected to discrete the flange and web, as well as the resin area, and an 8-node 3D cohesive element (COH3D8) was used for delamination analysis. The geometrical model was appropriately meshed with a relative finer area around the stitch, see Fig. 5. The total number of the solid elements is 245140. Meanwhile, the viscosity factor is set as 5.0×10^{-3} in order to achieve better convergence rate after some simulation tests. As the stitched elements are typically subjected to out-of-plane tension and/or transverse shear loads, two models with different loading conditions are modeled to predict their mechanical responses, respectively.

IOP Conf. Series: Materials Science and Engineering **207** (2017) 012043 doi:10.1088/1757-899X/207/1/012043

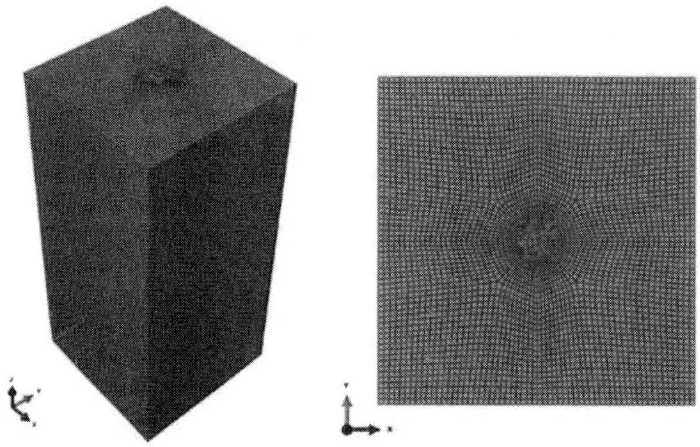

Figure 5. The mesoscale model with a fine mesh.

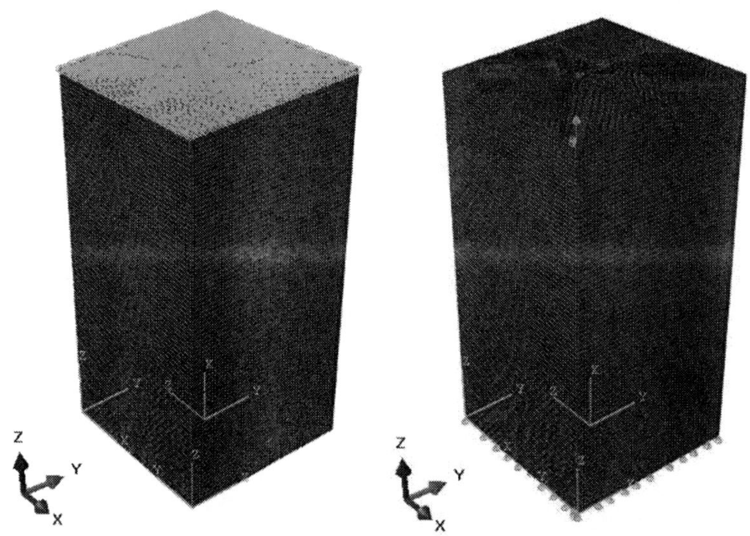

(a) Multi-point constraint on the upper surface (b) Load type

Figure 6. The constraint and load condition on the out-of-plane tension model.

The transverse shear model adopted the same geometrical model and material parameters as the above tension model, and also the same type of multi-points constraints on both the upper and lower surfaces during the shear loading process, see Fig. 7a. In which one more multi-point constraint surrounded the flange laminate was added to conveniently output the reaction force. The transverse shear load is realized by introducing a pair of transverse displacement along and contrary to the Y-axis, see Fig. 7b.

Similarly, the 3D Hashin damage model and cohesive elements were used to simulate the failure behavior of the stitch and delamination of the structure under the shear load.

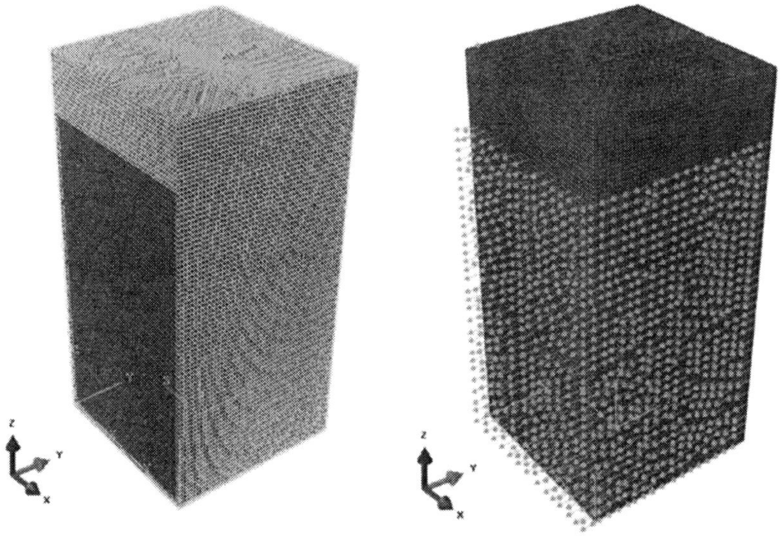

(a) Multi-point constraints (b) Load style

Figure 7. The constraint and load condition on the transverse shear model.

3. Results and discussion

3.1. Out-of-plane tension

From the simulation results, we obtained the relationship between the reaction tension force and the displacement on the loading surface, as shown in Fig. 8. One can observe that the reaction force increases linearly as the out-of-plane tensile displacement increases at the first stage. However, the loading carry capacity of the RVE cell begins to decrease after the reaction force reaches its maximum value. Fig. 9a reveals that at the maximum load the damage factor of almost all the cohesive elements between the web and flange, i. e. the net section reaches 1.0, which means they are closed to failure and delamination begins to occur. The tensile load in this section gradually transfers to the reinforcing stitch, although it has already experienced some minor damages, so the RVE cell still maintains some residual load-carrying capacity.

As the tensile displacement grows, the stitch damages gradually. The state dependent variable SDV1 contour in the section view of the RVE cell (see Fig. 9b) shows that the stitch section near the delaminated surface mainly suffers tensile damage. When the net section of the stitch finally ruptured, the stiffness of RVE cell reduced to 0.

Besides, it can be deduced from Fig. 8 that the nominal tensile strength of the stitched composite is close to 171.04MPa, which is much higher than the tensile strength property of the cohesive element 65.6MPa. This comparison result indicates the introduced stitch can remarkably promote the delamination resistance of the traditional laminates.

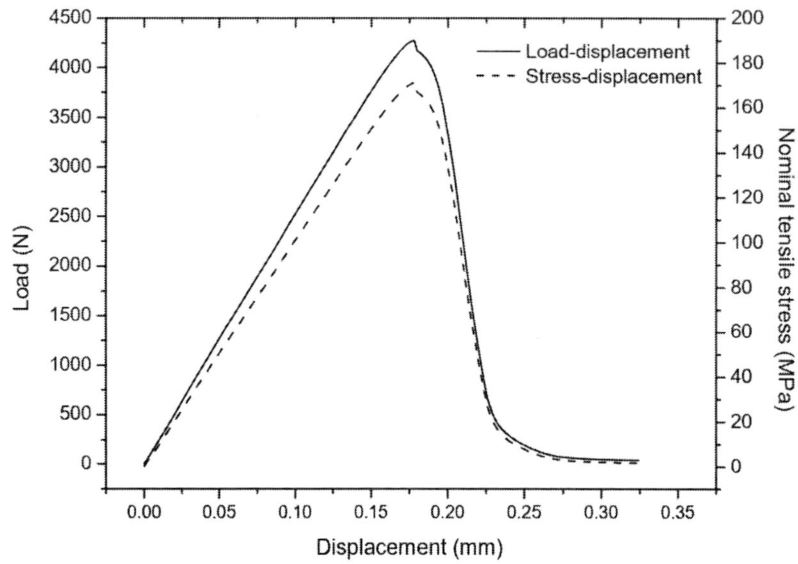

Figure 8. Out-of-plane tensile load/stress versus displacement curve.

3.2. Transverse shear

The Mises stress contour and deformation results of the transversely shear loaded RVE cell at different loading stages are presented in Fig. 10 (given in a symmetric section view for observation purpose). Fig. 10a shows that, at the initial loading stage, the upper and lower plates dislocate under the transverse shear loads, and the stress distribution is not uniform along the loading direction. The cohesive elements near two side-loaded surfaces are subjected to higher stress, and would firstly damage and even failure, then the high stress area gradually transfers to the stitch area (see Fig. 10b). When the shear load reaches its maximum (corresponding to Fig. 10c), the load sheared by the stitch already induces damage and failure in some stitch elements. Finally, the RVE cell ruptures as the effective load area shrinks rapidly.

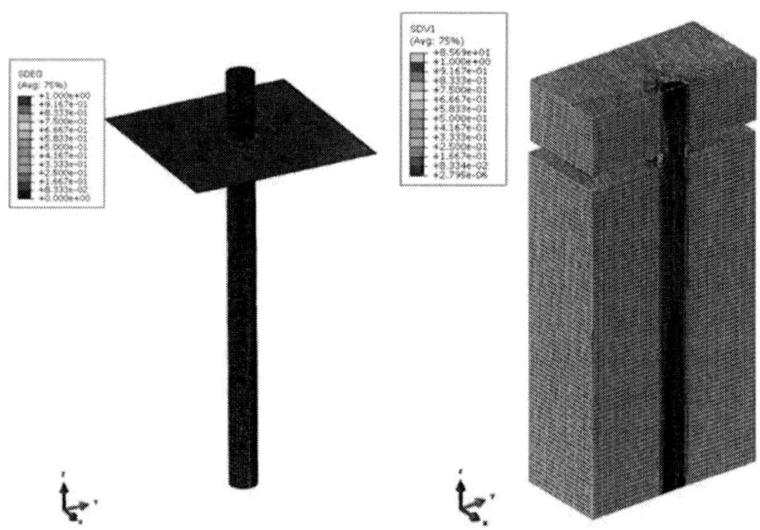

(a) Cohesive damage state at the maximum load (b) Stitch tensile damage at failure load

Figure 9. The damage conditions in the RVE model under tension load.

IOP Conf. Series: Materials Science and Engineering **207** (2017) 012043 doi:10.1088/1757-899X/207/1/012043

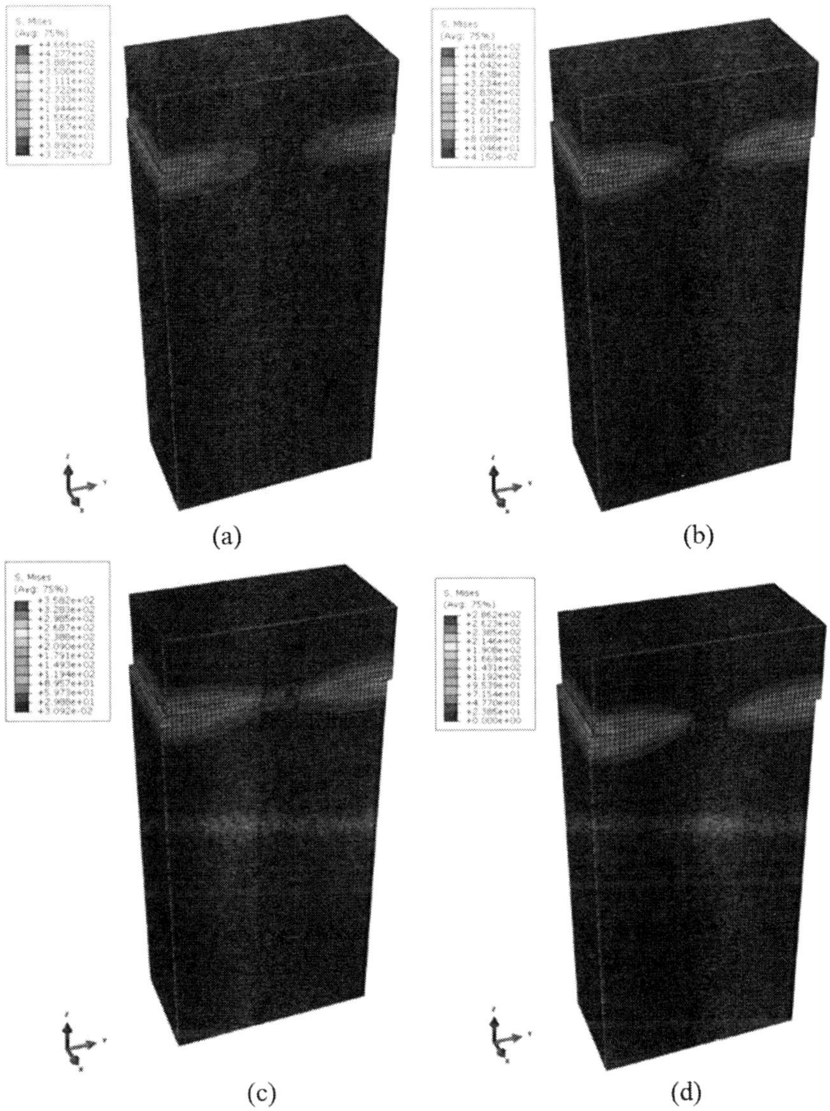

Figure 10. The Mises stress contour and deformation results of the transversely loaded RVE at different loading stages.

Fig. 11 shows the macroscopic shear load .vs. displacement relationship of the stitched RVE. It should be noted that the computation process is aborted due to convergence difficulties, however, one can still clearly see the load-carry tendency of the RVE cell. The relationship between the transverse shear load and the dislocation displacement can be roughly treated as bilinear, which is similar to the tension simulation results. Besides, the maximum averaged shear stress on the loading surface reaches 57.32MPa, it is much higher than the shear strength of the cohesive element (15.50MPa). This result also reveals that the stitch greatly reinforces the shear stress bear capacity of the laminates.

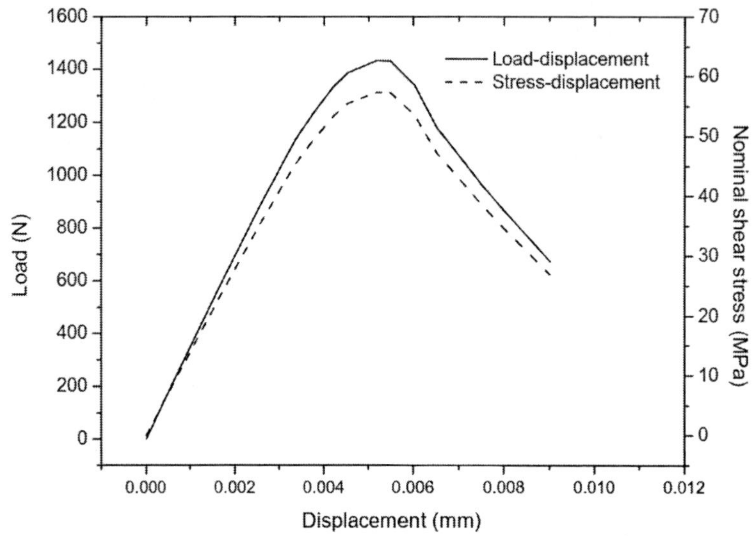

Figure 11. Transverse shear load/stress versus displacement curve.

4. Conclusion

This paper proposed a simplified z-pin like RVE cell for the through-thickness stitch reinforced laminate structure at the mesoscopic level. The model uses a 3D Hashin damage criteria for damage analysis of the stitch and introduces cohesive elements for delamination prediction. The FE simulation reveals the damage and failure process of different components under different mechanical loads.

Non-linear macroscopic load-displacement relationships are captured for the RVE subjected to out-of-plane tension and transverse shear loads. The predicted strengths indicate that the stitch remarkably promote the tension and shear load-carrying capacity compared with the traditional laminates. The proposed mesoscale model is appropriate for microscopic material parameter design, and the predicted strength parameters can be used for structural strength analysis.

Acknowledgments

This work was not financially supported by any fund.

References

[1] A. P. Mouritz, Review of z-pinned composite laminates, Comps. Part. A- Appl. Sci. Manuf. 38 (2007) 2383-2397.

[2] A. Yoshimura, T. Nakao, S. Yashiro, Improvement on out-of-plane impact resistance of CFRP laminates due to through-the-thickness stitching, Comps. Part. A- Appl. Sci. Manuf. 39 (2008) 1370-1379.

[3] L. K. Jain, Y. W. Mai, Analysis of stitched laminated ENF specimens for interlaminar mode II fracture toughness, Int. J. Fracture. 68 (1994) 219-244.

[4] M. D. K. Wood, L. Tong, Q. Luo, Failure of Stitched Composite L-Joints Under Tensile Loading—Experiment and Simulation, J. Reinf. Plast. Composite. 28 (2009) 715-742.

[5] G. Tsai, J. Chen, Effect of stitching on Mode I strain energy release rate, Compos. Struct. 69 (2005) 1-9.

[6] K. I. Tserpes, G. N. Labeas, Mesomechanical analysis of non-crimp fabric composite structural parts, Compos. Struct. 87 (2009) 358-369.

[7] Y. Iwahori, K. Nakane, N. Watanabe, DCB test simulation of stitched CFRP laminates using interlaminar tension test results, Composites Sci. Technol. 69 (2009) 2315-2322.

[8] F. Aymerich, R. Onnis, P. Priolo, Analysis of the fracture behaviour of a stitched single-lap joint,

Comps. Part. A- Appl. Sci. Manuf. 36 (2005): 603-614.
[9] F. Bianchi, X. Zhang, Predicting mode-II delamination suppression in z-pinned laminates, Composites Sci. Technol. 72 (2012) 924-932.
[10] F. Bianchi, X. Zhang, A cohesive zone model for predicting delamination suppression in z-pinned laminates, Composites Sci. Technol. 71 (2011) 1898-1907.
[11] F. Bianchi, T. M. Koh, X. Zhang, Finite element modelling of z-pinned composite T-joints, Composites Sci. Technol. 73 (2012) 48-56.
[12] M. L. Benzeggagh, M. Kenane, Measurement of mixed-mode delamination fracture toughness of unidirectional glass/epoxy composites with mixed-mode bending apparatus, Composites Sci. Technol. 56 (1996) 439-449.

IOP Conf. Series: Materials Science and Engineering **207** (2017) 012044 doi:10.1088/1757-899X/207/1/012044

Calculation method of reliability on combine harvester transmission belt by considering dynamic stress

Zhuohuai Guan, Liang Li and Chongyou Wu*

Nanjing Research Institute for Agricultural Mechanization, Ministry of Agriculture, Nanjing 210014, China

*Corresponding author e-mail: 542681935@qq.com

Abstract. Transmission belt is one of the most likely to fail parts of combine harvester, which affecting the machine reliability seriously. Dynamic strength occurs along with vibration during the operation and must be taken into account when calculating reliability, especially in harsh working environment like harvesting. However, the existing calculation method of reliability on combine harvester transmission belt didn't take the dynamic strength into account. In this research, a reliability calculation method was proposed based on the dynamic analysis of transmission belt. The nonlinear dynamic equation was built using string and beam model. Through the equation, relationship between belt speed and dynamic stress was deduced. Considering dynamic stress and regarding uncertain parameters as random uncertain parameters, reliability calculation model was built. Finally, an example was presented and the above mentioned dynamic reliability calculation method was simulated to verify the theoretical analysis in this paper and tested by the Monte-Carlo method.

1. Introduction

In recent decades, agricultural equipment keep developing rapidly. Reliability has become a new trend of combine harvester, especially in Asia. Combine harvester have a lot of working parts and a complex structure, any problem would cause machine failure and delay the harvesting time [1]. V belt is widely used in combine harvesters, however, the heavy load and strong vibration make it prone to failure. The reliability calculation in use didn't take dynamic stress into account and it is difficult to achieve the desired results. In fact, compared with other parts, transmission belt is most likely to fail during harvesting operations in the field [2]. So it is necessary to consider the influence of dynamic stress when calculating the reliability of belt.

Most of the mechanical models of V-belt are based on the flat-belt mechanics model, including creep theory, shear theory and CMM model [3] .Circumferential and lateral vibration would occurs when V-belt is in operation [4] .Some reliability design methods had been developed based on the basic criteria of V-belt which is having sufficient strength and no slippage occur during power transmission [5]. Qin [6] deduced parameters sensitivity of V-belt reliability and discussed the reliability design method in the case of incomplete probability information. Based on reliability design, Gong [7] and Qing [8] discussed the optimization design algorithm of the belt. Dynamic stress and its effect on failure of belt were analyzed by Wen [9], but the reliability calculation was not involved.

This paper focuses on the reliability calculation method of V-belt by considering dynamic stress. Nonlinear dynamic equation was built and the relationship between belt speed and dynamic stress was deduced. Considering dynamic stress and uncertain parameters, reliability calculation model was built and tested by Monte-Carlo method.

2. Nonlinear dynamic equation of belt

Since the bending stiffness of belt can be neglected during movement process, the dynamic model of belt could be simplified as an axially moving string with viscoelastic properties [10]. Since bending stiffness could be ignored, the belt can be simplified as axial moving string and modelled as an Euler-Bernoulli beam [11] .Selecting a section of belt that does not contact the pulley as subject (shown in Figure.1). Let the horizontal direction be the X axis and the vertical direction be the Y axis to establish absolute coordinate system XOY. The relative coordinate system was established on the belt before deformation occurs. Take the fixed point on the belt as the coordinate origin. Let u and w stand for the displacement of the point (x, y) in the X, Y direction and they were function of position and time.

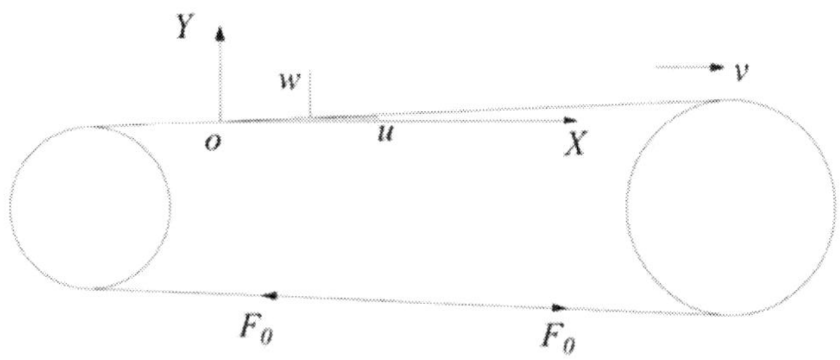

Figure.1 Dynamic model of belt

Relative coordinates can be written as

$$u = u(x, y, t) \quad w = w(x, y, t) \tag{1}$$

The speed relationship between absolute coordinate system and relative coordinate:

$$\frac{dx}{dt} = v + \frac{\partial u}{\partial x}v + \frac{\partial u}{\partial t} \quad \frac{dy}{dt} = \frac{\partial w}{\partial x}v + \frac{\partial w}{\partial t} \tag{2}$$

The acceleration relation:

$$\frac{d^2x}{dt^2} = v^2\frac{\partial^2 u}{\partial x^2} + 2v + \frac{\partial^2 u}{\partial t \partial x} + \frac{\partial^2 u}{\partial t^2} \quad \frac{d^2y}{dt^2} = v^2\frac{\partial^2 w}{\partial x^2} + 2v\frac{\partial^2 w}{\partial t \partial x} + \frac{\partial^2 w}{\partial t^2} \tag{3}$$

The quality of micro element dx was $\rho A dx$ (shown in Figure.2). Ignoring the moment of inertia and shear deformation, dx could be regarded as Euler-Bernoulli beam.

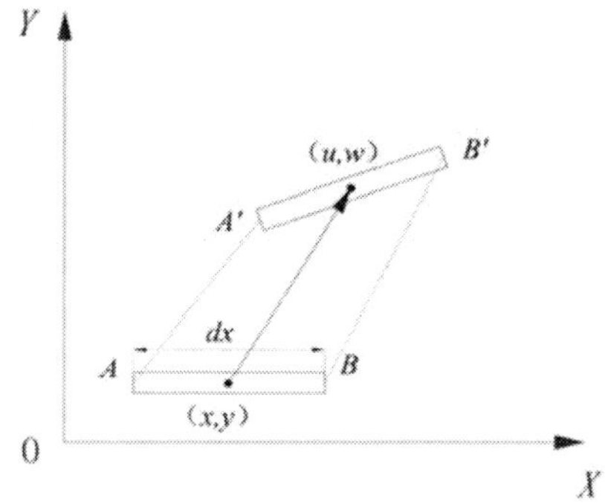

Figure.2 Infinitesimal model of belt

Points AB move to A' B'

$$\overline{AA'} = u\vec{i} + w\vec{j} \quad \overline{BB'} = (u + \frac{\partial u}{\partial x}\vec{i} + w\frac{\partial w}{\partial x}\vec{j}) \tag{4}$$

It can be concluded from the position relation in Figure.2:

$$\overline{AA'} + \overline{A'B'} = \overline{AB} + \overline{BB'} = dx\vec{i} + \overline{BB'} \tag{5}$$

The deformation of AB section is *ds*:

$$ds = \left|\overline{A'B'} - \overline{AB}\right| = \sqrt{\left[(1 + \frac{\partial u}{\partial x})^2 + \left(\frac{\partial w}{\partial x}\right)^2\right]}dx \tag{6}$$

The longitudinal dynamic strain of the belt:

$$\varepsilon_d = \frac{ds - dx}{dx} = \sqrt{\left[(1 + \frac{\partial u}{\partial x})^2 + \left(\frac{\partial w}{\partial x}\right)^2\right]} - 1 \tag{7}$$

Taylor expansion of equation (7) and omit high-order terms, the relationship between dynamic strain and displacement:

$$\varepsilon_d = \partial u/\partial x + \frac{1}{2}(\partial w/\partial x)^2 \tag{8}$$

According to the stress-strain relationship in Kelvin model viscoelastic theory, dynamic stress:

$$\sigma_{1d} = E\varepsilon_d + \eta(\partial\varepsilon_d/\partial t) = E^*\varepsilon_d(x,t) \tag{9}$$

Where E is elasticity modulus, η is dynamic viscous resistance coefficient, E^* is equivalent elastic modulus.

Longitudinal dynamic strain ε_d is only related to $\partial u/\partial x$ and $\partial w/\partial x$, which means the dynamic strain and dynamic stress of are related to the speed of belt. The total tension force of belt:

$$F = F_0 + \sigma_{1d}A = F_0 + E^*\varepsilon_d(x,t)A \tag{10}$$

Where F_0, A are, respectively, initial tension of the belt, cross sectional area of belt.
Linear damping force is proportional to the absolute velocity, the damping force on the *dx*:

$$F_{cx} = cAdx\frac{dx}{dt} \quad F_{cy} = cAdx\frac{dy}{dt} \tag{11}$$

The force of micro element *dx* is shown in Figure.3. Dynamic equation of micro segment *dx*:

$$(F+\frac{\partial F}{\partial s}ds)\cos(\theta+\frac{\partial \theta}{\partial s}ds)-F\cos\theta-cAdx\frac{dx}{dt}=\rho Adx\frac{d^2x}{dt^2} \tag{12}$$

$$(F+\frac{\partial F}{\partial s}ds)\sin(\theta+\frac{\partial \theta}{\partial s}ds)-F\sin\theta-cAdx\frac{dy}{dt}=\rho Adx\frac{d^2y}{dt^2} \tag{13}$$

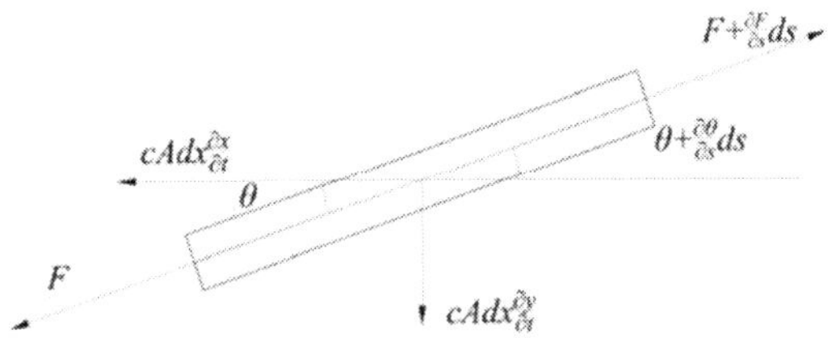

Figure.3 Stress Analysis of the section

When θ is small, the following approximate relation exists: $\sin\theta \approx \theta$ $\cos\theta \approx 1-\frac{1}{2}\theta^2$ $ds = dx/\cos\theta$ $\partial w/\partial x \approx \theta$ .Equation (13) (14) can be changed as:

$$\rho Adx\frac{d^2x}{dt^2}=\frac{\partial \sigma}{\partial t}dx - F\left[\frac{\partial w}{\partial x}\frac{\partial^2 w}{\partial x^2}dx+\frac{1}{2}F\left(\frac{\partial^2 w}{\partial x^2}dx\right)^2\right]-cAdx\frac{dx}{dt} \tag{14}$$

$$\rho Adx\frac{d^2y}{dt^2}=\frac{\partial \sigma}{\partial t}(\frac{\partial w}{\partial x}+\frac{\partial^2 w}{\partial x^2})dx-F\frac{\partial^2 w}{\partial x^2}dx-cAdx\frac{dy}{dt} \tag{15}$$

The first and the second terms of Equation (14) (15) stand for the viscoelasticity of belt; the third term of (14) (15) stand for the linear damping of belt.

The nonlinear dynamical equation of the belt which is only related to displacement can be obtained according to the *Eq*. (2) (3) (10) (14) (15). The dynamic stress σ_{1d} can be calculated by bringing in data.

3. Belt Stress Analysis

The stress of belt mainly include tensile stress σ_1 , bending stress $\sigma_b = Eh/d$, and centrifugal stress $\sigma_c = qv^2/A$. Tensile stress of belt σ_1 includes static tensile stress σ_{1s} and dynamic tensile stress σ_{1d} .

$$\sigma_{1s} = F_1/A \tag{16}$$

$$F_1 = F_{ec}\frac{e^{f\alpha}}{e^{f\alpha}-1} \tag{17}$$

$$F_{ec} = 2F_0 \frac{1-1/e^{f\alpha}}{1+1/e^{f\alpha}} \tag{18}$$

Where F_1, F_{ec}, α, f and F_0 are, respectively, tensile stress of tight side, maximum effective pulling force, wrap angle, friction coefficient and initial tension of the belt .

According to the *Eq.* (17) (18), static tensile stress σ_{1s} can be expressed as

$$\sigma_{1s} = 2F_0 \frac{1-1/e^{f\alpha}}{1+1/e^{f\alpha}} \frac{e^{f\alpha}}{e^{f\alpha}-1} \tag{19}$$

According to the *Eq.* (9) (19), tensile stress of belt σ_1 can be expressed as

$$\sigma_1 = \sigma_{1s} + \sigma_{1d} = 2F_0 \frac{1-1/e^{f\alpha}}{1+1/e^{f\alpha}} \frac{e^{f\alpha}}{e^{f\alpha}-1} + E^{*}\varepsilon_d(x,t) \tag{20}$$

The fatigue strength condition of the belt can be described as

$$\sigma_{max} = \sigma_1 + \sigma_b + \sigma_c \leq [\sigma] \tag{21}$$

Where $[\sigma]$ is allowable stress of belt.

4. Reliability model

Belt speed v, equivalent friction coefficient f, wrap angle α, center distance a and initial tension F_0 are random variables in the belt. Considering these random variables are independent and obeying normal distribution. The set of these random variables are

$$\boldsymbol{x} = [x_1, x_2, \cdots, x_m] \tag{22}$$

Means and variances are

$$\boldsymbol{\mu_x} = [\mu_{x_1}, \mu_{x_2}, \cdots, \mu_{x_m}]^T \quad \boldsymbol{\sigma_x} = [\sigma_{x_1}, \sigma_{x_2}, \cdots, \sigma_{x_m}]^T \tag{23}$$

According to the basic design criteria of belt, the maximum stress must be less than the allowable stress. The dangerous point of the belt is the position that about to leave pulley. We can assume that the belt wouldn't be failure as long as the critical point satisfies the stress condition. The strength reliability function of the drive belt can be expressed as

$$g = [\sigma_1] - \sigma_1 = g(x_1, x_2, \cdots, x_m) \tag{24}$$

The strength failure probability can be expressed as

$$P = P(g \leq 0) \tag{25}$$

The reliability index of strength reliability function is

$$\beta = \frac{\sum_{i=1}^{m} \frac{\partial g}{\partial x_i}(\mu_{x_i} - x_i)}{\sqrt{\sum_{i=1}^{m} (\frac{\partial g}{\partial x_i})^2 \sigma_{x_i}^2}} \tag{26}$$

The failure probability of strength reliability function is

$$P = \Phi(-\beta) \tag{27}$$

The Reliability of belt can be expressed as

$$R = 1 - P \tag{28}$$

Equation (28) is the reliability model of belt by considering dynamic stress. The reliability of the belt can be calculated according to the equations above.

5. Numerical example and discussion

Take some of the belts used in combine harvesters as example. The parameters were shown in Table.1 and the statistical property of random parameters were shown in Table.2.

Table.1 Parameters of belt

parameters	value
diameter of the small pulley	$d_1 = 90mm$
diameter of the large pulley	$d_2 = 315mm$
datum length	$L_d = 1600mm$
depth of section	$h = 0.008m$
mass per unit length	$q = 0.1kg / m$
elasticity modulus of belt	$E = 364MPa$
cross sectional area of belt	$A = 100mm^2$

Tab.2 Statistical property of random parameters

parameters	mean	variance
linear velocity of belt n / r·min^{-1}	1440	20
centre distance a / mm	470	6.2
wrap angle α / rad	2.739	0.03
friction coefficient f	0.47	0.002
transmission power P / kw	4.4	0.4

The reliability of the belt at different speeds were calculated using the proposed method and Monte Carlo calculation results were compared as shown in Figure.4.

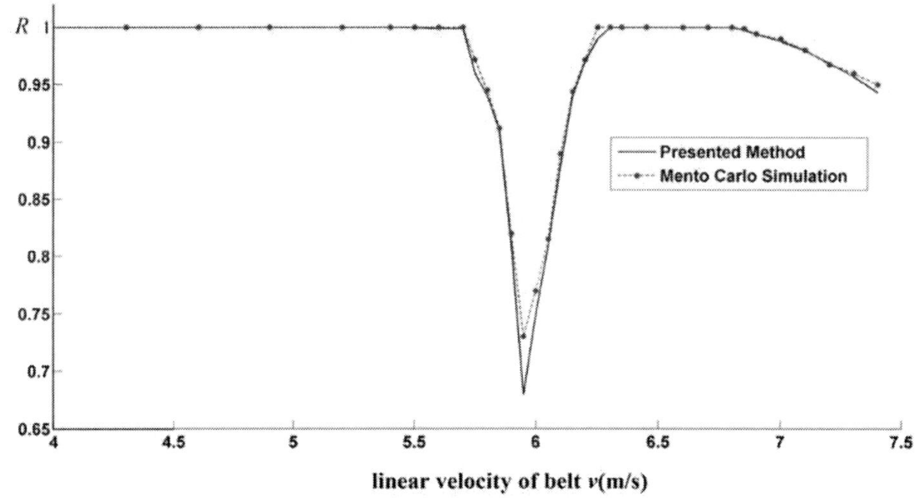

Figure.4 The reliability of the belts at different speeds

As can be seen from Figure.4, when belt speed is less than 5.6m/s, the reliability is close to 1. Because the stress of belt is far less than the allowable stress when belt speed is low.

The reliability of the belt is significantly reduced in the vicinity of 6m/s .It is a dangerous speed range that must be avoid. Dynamic stress increasing significantly and reliability decreasing sharply when the belt speed is close to the critical speed. This dangerous area would not be found if dynamic stress was not taken into account when calculating reliability. Therefore, the reliability calculation method of belt mentioned in this paper could reflect the actual situation more accurately, especially working in vibration environment.

With the increasing of speed and leaving from the dangerous speed range, dynamic stress decreases and reliability improves again. Finally, with the continuously increase of speed, reliability decrease gradually. That is because centrifugal force increase with speed.

The figure shows that the results of the two methods are similar, which means the reliability calculation method proposed in the paper is effective and accurate.

6. Conclusion

The nonlinear dynamic equation was built and the relationship between belt speed and dynamic stress was deduced. Considering dynamic stress and regarding uncertain parameters as random uncertain parameters, reliability calculation model was built. Finally, an example was presented and the dynamic reliability calculation method was simulated to verify the theoretical analysis in this paper and tested by the Monte-Carlo method. Dangerous speed range can be found through reliability calculation. So as to provide some theoretical basis for the design and reliability calculation of transmission belt of combine harvester.

Acknowledgements

We would like to express our appreciation to Technological Innovation Project of Chinese Academy of Agricultural Sciences, Crop harvesting machinery.

References

[1] Cao Yubao. Reliability design of agricultural machinery product [J]. Journal of Agricultural Mechanization Research, 2009, 31(7): 150~152.

[2] Wang ZhongZhang, Wu XinBao. Reliability analysis of semi feed combine harvester made in China [J]. Agricultural Machinery Quality and Supervision, 2004, (6): 21~22..

[3] Ding H, Li D P .Static and Dynamic Behaviors of Belt-Drive Dynamic Systems with a One-

Way Clutch [J]. Nonlinear Dynamics, 2014, 78(2): 1553~1575.

[4] Sun C Q, Ren A H, Sun G X, et al. Calculation of D-section Classical V-belt Power Rating in Different Reliability [J]. Advanced Materials Research, 2011, 338: 446~449.

[5] Qin Daicheng, Wu Junwei. Reliability sensitivity design of V belt drive with incomplete probability information [J]. Machinery, 2013, 51(11): 26~28.

[6] Zhao Hongmei. Research of the belt drive reliability of large travel and heavy duty motion Mechanism [J]. Journal of Mechanical Transmission, 2015, 39(8): 36~39.

[7] Gong Xiaoping, Tong Chonglou. Research on the reliability based on optimum design procedure of V belt transmission [J]. Journal of Air Force Engineering University, 2002, 3(4): 65~68.

[8] Qing Ling, Xie Junsheng. Reliability optimization design of hydraulically automatic lathe V belt transmission [J]. Journal of Mechanical Transmission, 2008, 32(6): 64~67.

[9] Wen Xinrong, Kong Jianyi. Dynamic analysis and life evaluate of belt drive [J]. Journal of Mechanical Transmission, 2013, 27(2): 33~35,50.

[10] Xing M M, Dong S M. A new nonlinear longitudinal vibration model of belt driving system applied in energy saving [J]. Journal of Vibroengineering, 2014, 16(3): 1494~1509.

[11] Li X, Zhao G, Ju X. Dynamics of mass-spring-belt friction self-excited vibration system [J]. Journal of Vibroengineering, 2013, 15(4): 1778.

Price competition and equilibrium analysis in multiple hybrid channel supply chain

Guihua Kuang*, Aihu Wang and Jin Sha

School of Business Administration, South China University of Technology, Guangzhou, Guangdong 510640, China

*Corresponding author e-mail: Guihua_kuang@126.com

Abstract. The amazing boom of Internet and logistics industry prompts more and more enterprises to sell commodity through multiple channels. Such market conditions make the participants of multiple hybrid channel supply chain compete each other in traditional and direct channel at the same time. This paper builds a two-echelon supply chain model with a single manufacturer and a single retailer who both can choose different channel or channel combination for their own sales, then, discusses the price competition and calculates the equilibrium price under different sales channel selection combinations. Our analysis shows that no matter the manufacturer and retailer choose same or different channel price to compete, the equilibrium price does not necessarily exist the equilibrium price in the multiple hybrid channel supply chain and wholesale price change is not always able to coordinate supply chain completely. We also present the sufficient and necessary conditions for the existence of equilibrium price and coordination wholesale price.

1. Introduction

With the rapid development of information technology, human way of life and work has been changed completely. Using Internet, consumers can search, compare and get plenty of information about the goods they want to buy conveniently and economically without restrictions of time, space, weather, etc. On the other hand, the improvement in the efficiency of merchandise delivery provided by the third-party logistics enterprises such as Federal Express and UPS makes merchants have the ability to deliver their products to customers quickly but not costly. The combination of traditional business model with information technology and logistics industry promotes the emergence and development of e-commerce and makes it no longer be a novelty.

The shortening of the length and the reduction of the intermediate links number of supply chain caused by online shopping reduce the price of products and benefit the customers. Meanwhile direct sale online makes enterprises more close to their customers and can obtain more accurate information for production operations management. Sale and purchasing through network became more and more popular and changed both enterprises' sales model and consumers' consumption patterns significantly. When electronic channel has become a critical selling and buying channel, manufacturers who sold only with the help of traditional retailers began to sell products to end customers directly via Internet. Such as IBM, Hewlett-Packard, Nike, Pioneer Electronics, Mattel, Estee Lauder and Cisco Systems, many prominent manufacturers had adopted primarily direct marketing techniques to sell their products [1]-[3]. At the same time of manufacturers started selling online directly to complement their existing retail distribution networks, many traditional retailers also established network channel for obtaining more profits. For instance, in China, Suningyi in 2009 and Wal-Mart in 2010 began to sell products through network in succession.

Although online channel can provides wider coverage with lower operating costs than physical channels and provides easy access to vast numbers of potential customers, it still can't replace the role of physical channel in some ways like setting up the enterprise image and brand, offering after-sales service, etc. For consumers, they need get more detailed information about products through brick-and-mortar stores and take them as the assurance of corporate reputation and product quality. Both manufacturers and retailers will not give up using offline channel for selling, even though they build and operate their online channel. Some outstanding manufacturers even establish exclusive shops to sell their products for example Apple, Louis Vuitton, Haier and so on. If manufacturers or retailers use more than one channel for sales, the hybrid channel supply chain is created, and if manufacturers and retailers both use two or more channels to sell, the multiple hybrid channel supply chain is formed.

In multiple hybrid channel supply chain, manufacturers and retailers sell products with homogeneity through different channels. Consequently, customers have alternatives to choose the channel that is better suited for their needs to purchase [4]. Multiple hybrid channel distribution also enables enterprises to access different customer segments and create a potential for larger demand and market share, but it does not come without complications. While the manufacturers build their exclusive shops and direct channel, their retailer partners are not content with this practice and feel their profits have been deprived. In the market covered by overlapping independent channels, conflicts between manufacturers and retails, direct and traditional channel are almost inevitable. Among all of the conflicts, price competition is undoubtedly the most outstanding. Price is the only way for enterprises to gain profits, so the supply chain participants must come to grip with the problem of pricing strategy. The reasonable price can not only make firms profitable but also alleviate the contradictions in multiple hybrid channel supply chain.

In this paper, first the authors use the different structures of multiple hybrid channel supply chain to analyze pricing strategy of supply chain members and find equilibrium price, and then discuss the effect of wholesale price for supply chain coordination. The rest of this paper is organized as follows: Section 2 provides a review of related literature; Section 3 clarifies the types of multiple hybrid channel supply chain and finishes the preparatory work of modelling; In Section 4, we analyze the equilibrium price of different structures of multiple hybrid channel supply chain; The influence of same channel price competition strategy for the equilibrium price is studied in Section 5; In Section 6, the effect of wholesale price for supply chain coordination is discussed; Finally, conclusion remarks are given in Section 7.

2. Literature Review

With the boom of hybrid channel supply chain, conflicts between channels and members are escalating. In order to coordinate supply chain system effectively, more and more scholars and entrepreneurs have made a lot of effort for it. However, not all effective coordination mechanisms for traditional supply chain are applicable to hybrid channel supply chain. Boyaci observe that simple contracts, such as price-only, buy-back, rebate, revenue sharing, and VMI contracts, cannot coordinate the dual-channel supply chain because of the vertical and horizontal competition in both channels. But he demonstrated that a penalty contract or compensation–commission contract can coordinate the supply chain. Almost all of the literature in this field took it for granted that optimal pricing strategy [6]–is a major issue to attract customers and coordinate hybrid channel supply chain, so a considerable amount of research has been done on price competition and equilibrium of distribution channels.

Balasubramanian [10] build a competition model in a dual-channel environment from a strategic perspective and show that firms can use the level of market coverage as a mechanism to control competition then obtain the price equilibrium. The research of Chiang et al. [11] indicate that manufacturer and retailer who act independently will create a higher retail price, lower sales, and lower profits without a direct channel and the vertically integrated direct channel allows manufacturer to constrain the pricing behavior of retailer. Chun et al. [12] analyze why there are price differences between a conventional retail channel and a direct online channel. Li et al.[13] show that in the markets where retail price consistency across channels is mandatory, an incumbent traditional brick-and-mortar retailer can deter the entry of a pure-play online retailer by strategically refraining from entering online. By contrast, in the markets where price consistency is not a constraint, the incumbent can deter the entry of a pure-play online retailer only if it enters online. In the latter case, the incumbent is willing to cannibalize its own brick-and-mortar business by charging a low online price. Fruchter et al. [14] find an optimal solution when the manufacturer adopts a non-discriminating pricing policy for both direct channel and retail channel. Ahn et al. [15] discuss about the pricing decisions of a dual-channel supply chain under the condition of traditional channel and direct channel operated in spatially separated markets. Cai et al.[16] show that the simple price discount contract can effectively improve both the supplier's and the retailer's performance based on evaluating the impact of a price discount contract and pricing scheme on the dual-channel supply chain competition. Yao et al. [17] build a dual channel supply chain model and discuss how the retailer to use prices and services to attract customers. Chen et al. [18] propose the pricing strategies of manufacturer in a dual-channel supply chain and show that the channel conflict can be resolved by applying two-part tariff or a price sharing agreement. Huang et al. [19] construct a deterministic demand model where the demand on a channel depends on prices, degree of substitution across channels and the overall market potential, then examine several prevalent pricing strategies which differ in the degree of autonomy for the Internet channel. The optimal retail service and prices in centralized and de-centralized dual-channel supply chain are determined in Dan et al.[20]. Xu et al. [21] investigate the impact of establishing a dual-channel supply chain coordinating contract when the supply chain agents are risk aversion under a mean–variance model, and propose the two-way revenue sharing contract to coordinate conflicts between channels. Chen et al. [22] examine how to make the proper decisions of the direct online sales channel together with a retail channel based on the same sales price and the different service levels of dual channel. Ren et al. [23] analyze the optimal decisions in both centralized and decentralized scenarios and design a contract to coordinate supply chain by proposing the dual-channel supply chain models considering consumer returns policies and the price and service competition between channels. Chen et al. [24] consider pricing policies in a dual-channel supply chain and show that improving brand loyalty is profitable for both the manufacturer and retailer and that an increased service value may alleviate the threat of the Internet channel for the retailer and increase the manufacturer's profits. Panda et al. [25] explore pricing and replenishment policies for a dual-channel supply chain with decreasing unit cost of product over its short life cycle.

Although there has been a lot of literature on the price competition of hybrid channel supply chain, the research about multiple hybrid channel supply chain is not common, especially on the supply chain where conflicts between manufacturers and retailers exist in online and offline channel simultaneously. Some of the few articles on this field include: Khouja et al. [26] analyze channel selection and price setting of a manufacturer who has several distribution options including direct channel, manufacturer-owned retail channel, and/or independent retail channel. Li et al. [27] analyze the price competition of the dual-channel supply chain composed of two oligopoly retailers (one sells through network and traditional store, the other sells only through the Internet). Kurata et al. [28] analyze channel pricing in a dual-channel system under competition between a national brand and a store brand and find that wholesale price change does not coordinate the supply chain. Hsiao et al. [29] investigate a channel selection problem between a direct channel and a retail channel for a manufacturer that incurs no production costs in a duopoly and show that the equilibrium where one manufacturer distributes products in both of the channels while the other manufacturer distributes only in the direct channel never arises. Li et al. [30] use the bifurcation theory of dynamical system to perform a numerical simulation on system with dual-channel retailer who makes delay decision considering different conditions and some complex phenomenon such as bifurcation and chaos. But these researches still can't analyze the price competition in multiple hybrid channel supply chain comprehensively. To tackle this problem, this paper constructs two-echelon multiple hybrid channel supply chain models and investigates the price strategies of manufacturer and retailer based on linear demand function. We also examine the effect of wholesale price for supply chain coordination.

3. The model

3.1 Assumptions

Our paper establishes a two-echelon supply chain model with a single manufacturer and a single retailer. There is only one product being sold in the market, but both participants can choose different channel or channel combination for their own sales. For example, manufacturer can sell through direct channel and /or his own retail channel and/or independent retail channel, and at the same time retailer choose network and/or traditional store to sell commodity to customers. This model is based on the following assumptions:

(1) The manufacturer (he) sells his product to independent retailer (she) and customers simultaneously. Firstly, he determines the wholesale price for retailer and the retail prices of his wholly owned channels, and then the retailer determines prices of her direct and/or traditional channel.

(2) The manufacturer and retailer are both risk-neutral and take profit maximization as the goal to determine their channel prices.

(3) To simplify the calculation, we suppose that manufacturer produces and sells and retailer sells a single product at an invariable marginal cost c=0.

3.2 Model construction

Because of the different sales channel combination of manufacturer and retailer choosing, the multiple hybrid channel supply chain has different structures as shown in Figure 1. "M, R, C" represent manufacturer, retailer and customers respectively.

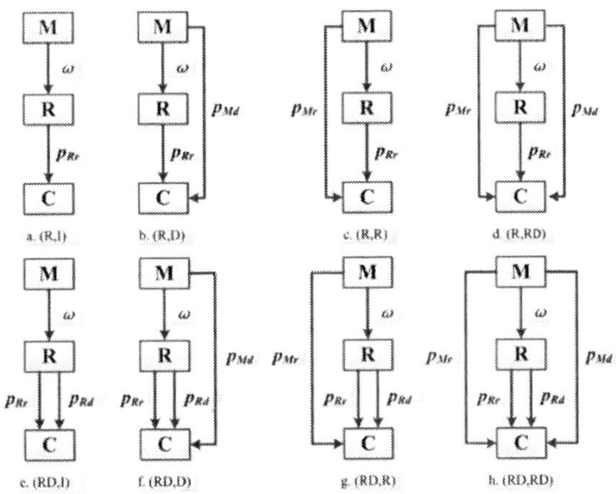

Fig. 1 The structures of multiple hybrid channel supply chain

The manufacturer charges wholesale price ω to retailer for his product and sells them to customers at p_{Mj}, in which M means the parameters of manufacturer and j=r, d represents the channel of his own retail and network. The retailer purchases product from manufacturer and then resell them at p_{Rj}, where R means the retailer's parameters. The retailer has two strategies to choose: "Strategy R" denotes that she resells product only through the traditional channel and "strategy RD" denotes that she distributes in both traditional and direct channel. The manufacturer's "Strategy I" means he sells his product only by independent retailer; "Strategy D" and "Strategy R" mean he uses both independent retailer and his own direct or retail channel for sales; "Strategy RD" means that he sells product through the above three channels simultaneously. In the eight kinds of two-echelon multiple hybrid channel supply chain, the strategy combination (R, D) and (R, R), (RD, D) and (RD, R) are similar both in structure and analytical approach, so we choose strategy (R, I), (R, D), (R, RD), (RD, I), (RD, D), (RD, RD) as study objects to analyze (corresponding to fig. 1-a, b, d, e, f, h).

Considering the effect of different channel prices to channel demand, the demand functions of each channel are given by

$$q_{ij}^{(S_R,S_M)} = Q_{ij} - \alpha_{ij} p_{ij}^{(S_R,S_M)} - \beta_{ij} p_{ij}^{(S_R,S_M)} + \sum_{\substack{i=R,M \\ j=r,d}} \beta_{ij} p_{ij}^{(S_R,S_M)} \tag{1}$$

Where $q_{ij}^{(S_R,S_M)}$ denotes channel demand (subscript i=R, M, j=r, d, and superscript (S_R, S_M) means channel strategy combination of manufacturer and retailer). $\boldsymbol{Q_{ij}}$ denotes the possible largest demand of channel. $\boldsymbol{\alpha_{ij}}$ is the coefficient of price elasticity of channel demand, $\boldsymbol{\beta_{ij}}$ is cross-price elasticity coefficient. It is quite reasonable that $\boldsymbol{\alpha_{ij}} > \boldsymbol{\beta_{ij}} > 0$ i.e., the effect of price sensitivity of a channel is greater than cross-price effect. To facilitate the following discussion, let $\boldsymbol{\alpha_{ij}=\alpha}$, $\boldsymbol{\beta_{ij}=\beta}$. We use $\boldsymbol{\pi}$ represent profits, so the enterprises' profit functions are

$$\pi_R^{(S_R,S_M)} = \sum_{j=r,d} q_{Rj}^{(S_R,S_M)} (p_{Rj}^{(S_R,S_M)} - \omega^{(S_R,S_M)}) \tag{2}$$

$$\pi_M^{(S_R,S_M)} = \sum_{j=r,d} q_{Rj}^{(S_R,S_M)} \omega^{(S_R,S_M)} + \sum_{j=r,d} q_{Mj}^{(S_R,S_M)} p_{Mj}^{(S_R,S_M)} \tag{3}$$

Table 1. Three Scheme comparing.

Numble	Scheme 1	Scheme 2	Scheme 3	Scheme 4	Scheme 5
1	456	456	123	123	123
2	789	213	644	644	644
3	213	654	649	649	649

4. Price competition and Equilibrium analysis

In this section we discuss price competition according to the six kinds of multiple hybrid channel supply chain mentioned in 3.2 and get the equilibrium price combination.

4.1 Strategy combination(R, I)

When manufacturer only through independent retailer and the retailer only through traditional store to sell product to customers, the channel demand is

$$q_{Rr}^{(R,I)} = a - \alpha p_{Rr}^{(R,I)} \tag{4}$$

In which $\boldsymbol{a}$ ($\boldsymbol{a}$>0) denotes the possible largest demand of market. The retailer and manufacturer's profits are respectively

$$\pi_R^{(R,I)} = (p_{Rr}^{(R,I)} - \omega^{(R,I)}) q_{Rr}^{(R,I)} \tag{5}$$

$$\pi_M^{(R,I)} = \omega^{(R,I)} q_{Rr}^{(R,I)} \tag{6}$$

Because the manufacturer can determine the product wholesale price, he is the advantage of the game. According to Stackelberg game rule, we can deduce the optimal decision of supply chain members using backward induction. Based on first-order conditions of the profit function $\partial\pi_R^{(R,I)}/\partial p_{Rr}^{(R,I)} = 0$, we can calculate the optimal reaction function of retailer

$$p_{Rr}^{(R,I)} = \frac{a + \alpha\omega^{(R,I)}}{2\alpha} \tag{7}$$

Substitute formula (7) into (6) and solve $\partial\pi_M^{(R,I)}/\partial\omega^{(R,I)} = 0$, can get

$$\omega^{(R,I)} = \frac{a}{2\alpha} \tag{8}$$

Then

$$p_{Rr}^{(R,I)} = \frac{3a}{4\alpha} \tag{9}$$

4.2 Strategy combination (RD, I)
In this situation, manufacturer only sells through independent retailer but the retailer through both online and offline channel to sell product, the channel demands are

$$q_{Rr}^{(RD,I)} = (1-\rho)a - \alpha p_{Rr}^{(RD,I)} + \beta p_{Rd}^{(RD,I)}$$
$$q_{Rd}^{(RD,I)} = \rho a - \alpha p_{Rd}^{(RD,I)} + \beta p_{Rr}^{(RD,I)} \tag{10}$$

Where ρ $(0<\rho<1)$ is the market share of direct channel. The profit functions of supply chain members are

$$\pi_R^{(RD,I)} = (p_{Rr}^{(RD,I)} - \omega^{(RD,I)})q_{Rr}^{(RD,I)} + (p_{Rd}^{(RD,I)} - \omega^{(RD,I)})q_{Rd}^{(RD,I)}$$
$$\pi_M^{(RD,I)} = \omega^{(RD,I)}(q_{Rr}^{(RD,I)} + q_{Rd}^{(RD,I)}) \tag{11}$$

The retailer's optimal reaction function is

$$p_{Rr}^{(RD,I)} = \frac{\alpha(1-\rho)a + \beta\rho a + (\alpha^2 - \beta^2)\omega^{(RD,I)}}{2(\alpha^2 - \beta^2)}$$
$$p_{Rd}^{(RD,I)} = \frac{\alpha\rho a + \beta(1-\rho)a + (\alpha^2 - \beta^2)\omega^{(RD,I)}}{2(\alpha^2 - \beta^2)} \tag{12}$$

Solving this Stackelberg game, we can get the price equilibrium point

$$p_{Rr}^{(RD,I)} = \frac{\alpha(1-\rho) + \beta\rho + (\alpha+\beta)/4}{2(\alpha^2 - \beta^2)}a$$
$$p_{Rd}^{(RD,I)} = \frac{\alpha\rho + \beta(1-\rho) + (\alpha+\beta)/4}{2(\alpha^2 - \beta^2)}a \tag{13}$$
$$\omega^{(RD,I)} = \frac{a}{4(\alpha - \beta)}$$

4.3 Strategy combination (R, D)
This strategy combination is the dual-channel supply chain that has been widely studied by many scholars. The channel demands and profit functions are respectively

$$
\begin{aligned}
q_{Rr}^{(R,D)} &= (1-\rho)a - \alpha p_{Rr}^{(R,D)} + \beta p_{Md}^{(R,D)} \\
q_{Md}^{(R,D)} &= \rho a - \alpha p_{Md}^{(R,D)} + \beta p_{Rr}^{(R,D)} \\
\pi_{R}^{(R,D)} &= (p_{Rr}^{(R,D)} - \omega^{(R,D)})q_{Rr}^{(R,D)} \\
\pi_{M}^{(R,D)} &= \omega^{(R,D)} q_{Rr}^{(R,D)} + p_{Md}^{(R,D)} q_{Md}^{(R,D)}
\end{aligned} \tag{14}
$$

Because of the advantage of manufacturer, we first calculate the retailer's optimal reaction function

$$
p_{Rr}^{(R,D)} = \frac{(1-\rho)a + \beta p_{Md}^{(R,D)} + \alpha\omega^{(R,D)}}{2\alpha} \tag{15}
$$

Based on first-order conditions of the profit function $\partial\pi_{M}^{(R,D)}/\partial p_{Md}^{(R,D)} = 0$ and $\partial\pi_{M}^{(R,D)}/\partial\omega^{(R,D)} = 0$, we can get the price equilibrium point of strategy combination (R, D)

$$
\begin{aligned}
p_{Rr}^{(R,D)} &= \frac{2\alpha\beta\rho + (3\alpha^2 - \beta^2)(1-\rho)}{4\alpha(\alpha^2 - \beta^2)} a \\
p_{Md}^{(R,D)} &= \frac{\alpha\rho + \beta(1-\rho)}{2(\alpha^2 - \beta^2)} a \\
\omega^{(R,D)} &= \frac{\beta\rho + \alpha(1-\rho)}{2(\alpha^2 - \beta^2)} a
\end{aligned} \tag{16}
$$

4.4 Strategy combination (RD, D)

While the manufacturer uses his own network channel for sales, the retailer also constructs direct channel and uses it with her former traditional channel to sell simultaneously, we call this supply chain structure strategy combination (RD, D). Under this situation, each channel demand function and every enterprise's profit function are as following:

$$
\begin{aligned}
q_{Rr}^{(RD,D)} &= (1-\rho)a - \alpha p_{Rr}^{(RD,D)} + \beta p_{Rd}^{(RD,D)} + \beta p_{Md}^{(RD,D)} \\
q_{Rd}^{(RD,D)} &= (1-\theta)\rho a - \alpha p_{Rd}^{(RD,D)} + \beta p_{Rr}^{(RD,D)} + \beta p_{Md}^{(RD,D)} \\
q_{Md}^{(RD,D)} &= \theta\rho a - \alpha p_{Md}^{(RD,D)} + \beta p_{Rr}^{(RD,D)} + \beta p_{Rd}^{(RD,D)} \\
\pi_{R}^{(RD,D)} &= (p_{Rr}^{(RD,D)} - \omega^{(RD,D)})q_{Rr}^{(RD,D)} + (p_{Rd}^{(RD,D)} - \omega^{(RD,D)})q_{Rd}^{(RD,D)} \\
\pi_{M}^{(RD,D)} &= \omega^{(RD,D)}(q_{Rr}^{(RD,D)} + q_{Rd}^{(RD,D)}) + p_{Md}^{(RD,D)} q_{Md}^{(RD,D)}
\end{aligned} \tag{17}
$$

In which, θ $(0<\theta<1)$ is the market share of manufacturer's direct channel. The retailer's optimal reaction functions are

$$
\begin{aligned}
p_{Rr}^{(RD,D)} &= \frac{\alpha(1-\rho)a + \beta(1-\theta)\rho a + (\alpha+\beta)\beta p_{Md}^{(RD,D)} + (\alpha^2 - \beta^2)\omega^{(RD,D)}}{2(\alpha^2 - \beta^2)} \\
p_{Rd}^{(RD,D)} &= \frac{\alpha(1-\theta)\rho a + \beta(1-\rho)a + (\alpha+\beta)\beta p_{Md}^{(RD,D)} + (\alpha^2 - \beta^2)\omega^{(RD,D)}}{2(\alpha^2 - \beta^2)}
\end{aligned} \tag{18}
$$

Substitute (18) into (17) and calculate the Hessian matrix of $\pi_{M}^{(RD,D)}$ about $\omega^{(RD,D)}$ and $p_{Md}^{(RD,D)}$:

$$
H = \begin{bmatrix} -2(\alpha-\beta) & 2\beta \\ 2\beta & \dfrac{2\beta^2}{\alpha-\beta} - 2\alpha \end{bmatrix} \tag{19}
$$

The necessary and sufficient conditions of (19) being negative definite is $\alpha>2\beta$. When $\alpha>2\beta$, we can get the price equilibrium point as follows. Otherwise, there will be no price equilibrium between manufacturer and retailer.

$$p_{Rr}^{(RD,D)} = \frac{(5\alpha^2 - 9\alpha\beta + 2\beta^2)(1-\rho) + (\alpha^2 + 3\alpha\beta - 6\beta^2)(1-\theta)\rho + 4\beta(\alpha-\beta)\theta\rho}{8(\alpha^2 - \beta^2)(\alpha - 2\beta)} a$$

$$p_{Rd}^{(RD,D)} = \frac{(\alpha^2 + 3\alpha\beta - 6\beta^2)(1-\rho) + (5\alpha^2 - 9\alpha\beta + 2\beta^2)(1-\theta)\rho + 4\beta(\alpha-\beta)\theta\rho}{8(\alpha^2 - \beta^2)(\alpha - 2\beta)} a$$

$$p_{Md}^{(RD,D)} = \frac{\beta(1-\rho) + \beta(1-\theta)\rho + (\alpha-\beta)\theta\rho}{2(\alpha+\beta)(\alpha-2\beta)} a$$

$$\omega^{(RD,D)} = \frac{\alpha(1-\rho) + \alpha(1-\theta)\rho + 2\beta\theta\rho}{4(\alpha+\beta)(\alpha-2\beta)} a \tag{20}$$

4.5 Strategy combination (R, RD)

In strategy combination (R, RD), the manufacturer not only uses direct channel but also builds whole owned retail channel to compete with the independent retailer. But the independent retailer sell product to customers only through her traditional store. The channel demand functions and profit functions are respectively

$$q_{Rr}^{(R,RD)} = (1-\varepsilon)(1-\rho)a - \alpha p_{Rr}^{(R,RD)} + \beta p_{Mr}^{(R,RD)} + \beta p_{Md}^{(R,RD)}$$

$$q_{Mr}^{(R,RD)} = \varepsilon(1-\rho)a - \alpha p_{Mr}^{(R,RD)} + \beta p_{Rr}^{(R,RD)} + \beta p_{Md}^{(R,RD)}$$

$$q_{Md}^{(R,RD)} = \rho a - \alpha p_{Md}^{(R,RD)} + \beta p_{Rr}^{(R,RD)} + \beta p_{Mr}^{(R,RD)}$$

$$\pi_{R}^{(R,RD)} = (p_{Rr}^{(R,RD)} - \omega^{(R,RD)}) q_{Rr}^{(R,RD)}$$

$$\pi_{M}^{(R,RD)} = \omega^{(R,RD)} q_{Rr}^{(R,RD)} + p_{Mr}^{(R,RD)} q_{Mr}^{(R,RD)} + p_{Md}^{(R,RD)} q_{Md}^{(R,RD)} \tag{21}$$

Where ε ($0<\varepsilon<1$) is the market share of manufacturer's offline channel. The retailer's optimal reaction function is

$$p_{Rr}^{(R,RD)} = \frac{(1-\varepsilon)(1-\rho)a + \beta p_{Mr}^{(R,RD)} + \beta p_{Md}^{(R,RD)} + \alpha\omega^{(R,RD)}}{2\alpha} \tag{22}$$

Calculate the Hessian matrix of $\pi_M^{(R,RD)}$ about $\omega^{(R,RD)}$, $p_{Mr}^{(R,RD)}$ and $p_{Md}^{(R,RD)}$:

$$H = \begin{bmatrix} -\alpha & \beta & \beta \\ \beta & \frac{\beta^2 - 2\alpha^2}{\alpha} & \frac{2\alpha\beta + \beta^2}{\alpha} \\ \beta & \frac{2\alpha\beta + \beta^2}{\alpha} & \frac{\beta^2 - 2\alpha^2}{\alpha} \end{bmatrix} \tag{23}$$

The necessary and sufficient conditions of (23) being negative definite is $\alpha<2\beta$. Based on the assumption $\alpha>\beta$, when $\beta<\alpha<2\beta$ we can get the optimal prices of manufacturer as follows:

$$
\begin{aligned}
p_{Mr}^{(R,RD)} &= \frac{(\alpha - \beta)\varepsilon(1 - \rho) + \beta((1 - \varepsilon)(1 - \rho)a + \rho)}{2(\alpha + \beta)(\alpha - 2\beta)} a \\
p_{Md}^{(R,RD)} &= \frac{(\alpha - \beta)\rho + \beta((1 - \varepsilon)(1 - \rho) + \varepsilon(1 - \rho))}{2(\alpha + \beta)(\alpha - 2\beta)} a \\
\omega^{(R,RD)} &= \frac{(\alpha - \beta)(1 - \varepsilon)(1 - \rho) + \beta(\varepsilon(1 - \rho) + \rho)}{2(\alpha + \beta)(\alpha - 2\beta)} a
\end{aligned} \tag{24}
$$

But when $\boldsymbol{\beta<\alpha<2\beta}$, $p_{Mr}^{(R,RD)}$, $p_{Md}^{(R,RD)}$ and $\omega^{(R,RD)}$ are all less than zero, so there is no reasonable price equilibrium in strategy combination (R, RD).

4.6 Strategy combination (RD, RD)

Under this supply chain structure, both of the manufacture and retail sell their production through all kinds of sales channels. In other words, both of the manufacturer and retailer have online and offline channel for sales at the same time. The channel demand functions and profit functions are respectively

$$
\begin{aligned}
q_{Rr}^{(RD,RD)} &= (1 - \varepsilon)(1 - \rho)a - \alpha p_{Rr}^{(RD,RD)} + \beta p_{Rd}^{(RD,RD)} + \beta p_{Mr}^{(RD,RD)} + \beta p_{Md}^{(RD,RD)} \\
q_{Rd}^{(RD,RD)} &= (1 - \theta)\rho a - \alpha p_{Rd}^{(RD,RD)} + \beta p_{Rr}^{(RD,RD)} + \beta p_{Mr}^{(RD,RD)} + \beta p_{Md}^{(RD,RD)} \\
q_{Mr}^{(RD,RD)} &= \varepsilon(1 - \rho)a - \alpha p_{Mr}^{(RD,RD)} + \beta p_{Rr}^{(RD,RD)} + \beta p_{Rd}^{(RD,RD)} + \beta p_{Md}^{(RD,RD)} \\
q_{Md}^{(RD,RD)} &= \theta\rho a - \alpha p_{Md}^{(RD,RD)} + \beta p_{Rr}^{(RD,RD)} + \beta p_{Rd}^{(RD,RD)} + \beta p_{Mr}^{(RD,RD)} \\
\pi_{R}^{(RD,RD)} &= (p_{Rr}^{(RD,RD)} - \omega^{(RD,RD)})q_{Rr}^{(RD,RD)} + (p_{Rd}^{(RD,RD)} - \omega^{(RD,RD)})q_{Rd}^{(RD,RD)} \\
\pi_{M}^{(RD,RD)} &= \omega^{(RD,RD)}(q_{Rr}^{(RD,RD)} + q_{Rd}^{(RD,RD)}) + p_{Mr}^{(RD,RD)}q_{Mr}^{(RD,RD)} + p_{Md}^{(RD,RD)}q_{Md}^{(RD,RD)}
\end{aligned} \tag{25}
$$

Get the retailer's optimal reaction functions as follows:

$$
\begin{aligned}
p_{Rr}^{(RD,RD)} &= \frac{\alpha(1 - \varepsilon)(1 - \rho)a + \beta(1 - \theta)\rho a + \beta(\alpha + \beta)(p_{Mr}^{(DR,DR)} + p_{Md}^{(DR,DR)}) + (\alpha^2 - \beta^2)\omega^{(RD,RD)}}{2(\alpha^2 - \beta^2)} \\
p_{Rd}^{(RD,RD)} &= \frac{\beta(1 - \varepsilon)(1 - \rho)a + \alpha(1 - \theta)\rho a + \beta(\alpha + \beta)(p_{Mr}^{(DR,DR)} + p_{Md}^{(DR,DR)}) + (\alpha^2 - \beta^2)\omega^{(RD,RD)}}{2(\alpha^2 - \beta^2)}
\end{aligned} \tag{26}
$$

We can calculate the Hessian matrix of $\pi_{M}^{(RD,RD)}$ about $\omega^{(RD,RD)}$, $p_{Mr}^{(RD,RD)}$ and $p_{Md}^{(RD,RD)}$:

$$
H = \begin{bmatrix}
-2\dfrac{\alpha^2 - \alpha\beta - \beta^2}{\alpha - \beta} & 2\dfrac{\alpha\beta}{\alpha - \beta} & 2\beta \\
2\dfrac{\alpha\beta}{\alpha - \beta} & -2\dfrac{\alpha^2 - \alpha\beta - \beta^2}{\alpha - \beta} & 2\beta \\
2\beta & 2\beta & -2(\alpha - \beta)
\end{bmatrix} \tag{27}
$$

If and only if $(1+\sqrt{5})\beta/2 < \alpha < 3\beta$, $\pi_{M}^{(RD,RD)}$ is concave. With this constraint we can get the optimal prices of manufacturer as follows:

$$p_{Mr}^{(RD,RD)} = \frac{\beta((1-\varepsilon)(1-\rho)+(1-\theta)\rho+\theta\rho)+(\alpha-2\beta)\varepsilon(1-\rho)}{2(\alpha+\beta)(\alpha-3\beta)}a$$

$$p_{Md}^{(RD,RD)} = \frac{\beta((1-\varepsilon)(1-\rho)+(1-\theta)\rho+\varepsilon(1-\rho))+(\alpha-2\beta)\theta\rho}{2(\alpha+\beta)(\alpha-3\beta)}a$$

$$\omega^{(RD,RD)} = \frac{(\alpha-\beta)((1-\varepsilon)(1-\rho)+(1-\theta)\rho)+2\beta(\varepsilon(1-\rho)+\theta\rho)}{4(\alpha+\beta)(\alpha-3\beta)}a \tag{28}$$

But when $(1+\sqrt{5})\beta/2 < \alpha < 3\beta$, $p_{Mr}^{(RD,RD)}$, $p_{Md}^{(RD,RD)}$ and $\omega^{(RD,RD)}$ are all less than zero, so there is no reasonable price equilibrium in strategy combination (RD, RD).

In the last, the results of the equilibrium price based on six kinds of multiple hybrid channel supply chain are summarized in Table 2.

Table.2 Equilibrium price of multiple hybrid channel supply chain

Strategy		Manufacturer I	Manufacturer D	Manufacturer RD
Retailer	R	$\omega^{(R,I)} = \frac{a}{2\alpha}$ $p_{Rr}^{(R,I)} = \frac{3a}{4\alpha}$	$p_{Rr}^{(R,D)} = \frac{2\alpha\beta\rho+(3\alpha^2-\beta^2)(1-\rho)}{4\alpha(\alpha^2-\beta^2)}a$ $p_{Md}^{(R,D)} = \frac{\alpha\rho+\beta(1-\rho)}{2(\alpha^2-\beta^2)}a$ $\omega^{(R,D)} = \frac{\beta\rho+\alpha(1-\rho)}{2(\alpha^2-\beta^2)}a$	No reasonable price equilibrium
	RD	$p_{Rr}^{(RD,I)} = \frac{\alpha(1-\rho)+\beta\rho+(\alpha+\beta)/4}{2(\alpha^2-\beta^2)}a$ $p_{Rd}^{(RD,I)} = \frac{\alpha\rho+\beta(1-\rho)+(\alpha+\beta)/4}{2(\alpha^2-\beta^2)}a$ $\omega^{(RD,I)} = \frac{a}{4(\alpha-\beta)}$	When α>2β $p_{Rr}^{(RD,D)} = \frac{(5\alpha^2-9\alpha\beta+2\beta^2)(1-\rho)+(\alpha^2+3\alpha\beta-6\beta^2)(1-\theta)\rho+4\beta(\alpha-\beta)\theta\rho}{8(\alpha^2-\beta^2)(\alpha-2\beta)}a$ $p_{Rd}^{(RD,D)} = \frac{(\alpha^2+3\alpha\beta-6\beta^2)(1-\rho)+(5\alpha^2-9\alpha\beta+2\beta^2)(1-\theta)\rho+4\beta(\alpha-\beta)\theta\rho}{8(\alpha^2-\beta^2)(\alpha-2\beta)}a$ $p_{Md}^{(RD,D)} = \frac{\beta(1-\rho)+\beta(1-\theta)\rho+(\alpha-\beta)\theta\rho}{2(\alpha+\beta)(\alpha-2\beta)}a$ $\omega^{(RD,D)} = \frac{\alpha(1-\rho)+\alpha(1-\theta)\rho+2\beta\theta\rho}{4(\alpha+\beta)(\alpha-2\beta)}a$ Otherwise, no price equilibrium.	No reasonable price equilibrium

5. Price competition and Equilibrium analysis

In Section 4, the price competition and equilibrium between the manufacturer and retailer in different structures of multiple hybrid channel supply chain are studied. Sometimes, the same price strategy may be taken by supply chain members for the purpose of competition. The same price means that if the manufacturer and retailer use the same channel to sell simultaneously, they will charge the same channel price of commodity to customers. To distinguish conveniently, we use lowercase superscript to identify parameters of same price strategy. From Fig.1, the same price strategy can be taken in the strategy combination (r, r), (r, rd), (rd, d), (rd, r) and (rd, rd), but (rd, d) and (rd, r) are similar both in structure and analytical approach, so we choose strategy (r, r), (r, rd), (rd, d) and (rd, rd) as study objects. We suppose the demand functions under same price strategy are

$$q_{ij}^{(s_R,s_M)} = Q_{ij} - \alpha p_j^{(s_R,s_M)} + \gamma p_{Cuj}^{(s_R,s_M)} \tag{29}$$

In which, γ is the cross-price elasticity coefficient and $\alpha > \gamma > 0$, "Cuj" means the complementation of set j. By solving the Stackelberg game under different supply chain structures, we can get the equilibrium price of same price strategy shown in Table. 3.

Table.3 The equilibrium price of same price strategy

strategy	Equilibrium Price
(r, r)	$p_r^{(r,r)} = \frac{(3-2\varepsilon)a}{6\alpha}$ $\omega^{(r,r)} = \frac{\varepsilon a}{3\alpha}$
(r, rd)	When $\alpha > \gamma\sqrt{12/13}$ $p_r^{(r,rd)} = \frac{(6\alpha^2-4\gamma^2)(1-\varepsilon)(1-\rho)+2(\alpha^2+\gamma^2)\varepsilon(1-\rho)+5\alpha\gamma\rho}{\alpha(12\alpha^2-13\gamma^2)}a$ $p_d^{(r,rd)} = \frac{\gamma(3(1-\varepsilon)(1-\rho)+5\varepsilon(1-\rho))+6\alpha\rho}{12\alpha^2-13\gamma^2}a$ $\omega^{(r,rd)} = \frac{2\gamma^2(1-\varepsilon)(1-\rho)+(4\alpha^2-\gamma^2)\varepsilon(1-\rho)+4\alpha\gamma\rho}{\alpha(12\alpha^2-13\gamma^2)}a$ Otherwise, no price equilibrium.
(rd, d)	$p_r^{(rd,d)} = \frac{(12\alpha+\gamma)(1-\rho)+11\gamma(1-\theta)\rho+2(\alpha+\gamma)\theta\rho}{20(\alpha^2-\gamma^2)}a$ $p_d^{(rd,d)} = \frac{(2\alpha+11\gamma)(1-\rho)+(10\alpha+\gamma)(1-\theta)\rho+2(\alpha+\gamma)\theta\rho}{20(\alpha^2-\gamma^2)}a$ $\omega^{(rd,d)} = \frac{(2\alpha+\gamma)((1-\rho)+\theta\rho)+\gamma\rho}{10(\alpha^2-\gamma^2)}a$
(rd, rd)	$p_r^{(rd,rd)} = \frac{6(\alpha(1-\varepsilon)(1-\rho)+\gamma(1-\theta)\rho)+(\alpha+\gamma)(\varepsilon(1-\rho)+\theta\rho)}{12(\alpha^2-\gamma^2)}a$ $p_d^{(rd,rd)} = \frac{6(\alpha(1-\theta)\rho+\gamma(1-\varepsilon)(1-\rho))+(\alpha+\gamma)(\varepsilon(1-\rho)+\theta\rho)}{12(\alpha^2-\gamma^2)}a$ $\omega^{(rd,rd)} = \frac{\varepsilon(1-\rho)+\theta\rho}{6(\alpha-\gamma)}a$

6. Effect of wholesale price for supply chain coordination

The overall return of integrated supply chain is optimal in all types, to coordinate multiple hybrid channel supply chain we should calculate the optimal prices of integrated supply chain firstly. Using Π, P and $\boldsymbol{\Omega}$ denote the profits, channel price and wholesale price of centralized decision, the profit function of integrated supply chain can be expressed

$$\Pi^{(S_R,S_M)} = \sum_{\substack{i=R,M \\ j=r,d}} q_{ij}^{(S_R,S_M)} P_{ij}^{(S_R,S_M)} \tag{30}$$

Judge the convexity and concavity of the profit function and maximize $\boldsymbol{\Pi}$, we can get the optimal prices of integrated supply chain under different strategy combinations shown in Table. 4.

Table.4 The optimal price of integrated supply chain

Strategy		Manufacturer I	Manufacturer D	Manufacturer RD
Retailer	R	$P_{Rr}^{(R,I)} = \frac{a}{2\alpha}$	$P_{Rr}^{(R,D)} = \frac{\alpha(1-\rho)+\beta\rho}{2(\alpha^2-\beta^2)}a$ $P_{Md}^{(R,D)} = \frac{\alpha\rho+\beta(1-\rho)}{2(\alpha^2-\beta^2)}a$	No reasonable optimal price. (If you need the proofs, please connect the corresponding author)
	RD	$P_{Rr}^{(RD,I)} = \frac{\alpha(1-\rho)+\beta\rho}{2(\alpha^2-\beta^2)}a$ $P_{Rd}^{(RD,I)} = \frac{\alpha\rho+\beta(1-\rho)}{2(\alpha^2-\beta^2)}a$	No reasonable optimal price. (If you need the proofs, please connect the corresponding author)	When α>3β $P_{Rr}^{(RD,RD)} = \frac{(\alpha-2\beta)(1-\varepsilon)(1-\rho)+\beta((1-\theta)\rho+\varepsilon(1-\rho)+\theta\rho)}{2(\alpha-3\beta)(\alpha+\beta)}a$ $P_{Rd}^{(RD,RD)} = \frac{(\alpha-2\beta)(1-\theta)\rho+\beta((1-\varepsilon)(1-\rho)+\varepsilon(1-\rho)+\theta\rho)}{2(\alpha-3\beta)(\alpha+\beta)}a$ $P_{Mr}^{(RD,RD)} = \frac{(\alpha-2\beta)\varepsilon(1-\rho)+\beta((1-\varepsilon)(1-\rho)+(1-\theta)\rho+\theta\rho)}{2(\alpha-3\beta)(\alpha+\beta)}a$ $P_{Md}^{(RD,RD)} = \frac{(\alpha-2\beta)\theta\rho+\beta((1-\varepsilon)(1-\rho)+(1-\theta)\rho+\varepsilon(1-\rho))}{2(\alpha-3\beta)(\alpha+\beta)}a$ Otherwise, no reasonable optimal price.

If the manufacturer and retailer sell product on the basis of the optimal channel price, they can maximize the overall profits of supply chain. For coordinating multiple hybrid channel supply chain completely, the manufacturer may adjust his channel prices to be equaled to the optimal price. On the other hand, he also can use wholesale price to control retailer's channel price and make them equal to the optimal value. So we let the optimal channel prices in Table.1 equal the optimal reaction functions obtained in Section 4, if the wholesale price Ω can make the equations true, the complete coordination of supply chain can be achieved. The coordination wholesale prices are shown in Table.5.

Table.5 The coordination wholesale prices

Strategy		Manufacturer I	Manufacturer D	Manufacturer RD
Retailer	R	$\Omega^{(R,I)} = 0$	$\Omega^{(R,D)} = \frac{\beta^2(1-\rho)+\alpha\beta\rho}{2\alpha(\alpha^2-\beta^2)}a$	No coordination Ω.
	RD	$\Omega^{(RD,I)} = 0$	No coordination Ω.	When $\alpha>3\beta$ $\Omega^{(RD,RD)} = \frac{2\beta^2((1-\varepsilon)(1-\rho)+(1-\theta)\rho)+\beta(\alpha-\beta)(\varepsilon(1-\rho)+\theta\rho)}{2(\alpha-3\beta)(\alpha^2-\beta^2)}a$ Otherwise, no coordination Ω.

Through the similar calculation, the coordination wholesale prices of same price strategy can be gotten. We show them in Table.6.

Table.6 The coordination wholesale prices of same price strategy

	Equilibrium Price
(r, r)	$\Omega^{(r,r)} = \frac{(2\varepsilon-1)a}{2\alpha}$
(r, rd)	When $\alpha > 3\gamma/2\sqrt{2}$ $\Omega^{(r,rd)} = \frac{(6\alpha^2-4\gamma^2)(1-\varepsilon)(1-\rho)+(4\alpha^2-3\gamma^2)\varepsilon(1-\rho)+2\alpha\gamma\rho}{\alpha(8\alpha^2-9\gamma^2)}a$ Otherwise, no coordination Ω.
(rd, d)	No coordination Ω. (Proofs are provided in Appendix. B)
(rd, rd)	No coordination Ω. (Proofs are provided in Appendix. B)

7. Conclusion

In this paper, the authors divide the multiple hybrid channel supply chain into different types according to the different sales channel combination chosen by the manufacturer and retailer. Then, price competition between supply chains participants are discussed based on Stackelberg game rule. Our results show: (1). When the manufacturer and retailer use different channel price to compete, there is price equilibrium under strategy combination (R, I), (RD, I) and (R, D); If the system parameters satisfy certain conditions, the strategy combination (RD, D) has equilibrium price; There is no price equilibrium under strategy combination (R, RD) and (RD, RD). (2). When supply chain members use same channel price strategy, there is price equilibrium under strategy combination (r, r), (rd, d) and (rd, rd); If the system parameters satisfy certain conditions, the strategy combination (r, rd) has equilibrium price. (3). Using different channel price, the wholesale price change can coordinate supply chain under strategy combination (R, I), (RD, I) and (R, D), can't coordinate (R, RD) and (RD, D), and can coordinate (RD, RD) with certain parameters restrictions. (4). Using same channel price, the wholesale price change can coordinate (r, r), can't coordinate (rd, d) and (rd, rd), and can coordinate (r, rd) if system parameters satisfy certain conditions. This paper discusses the equilibrium prices of different structure multiple hybrid supply chain under the static demand environment and the equilibrium prices of dynamic demand should be the objective of further research.

Acknowledgments
This paper is supported by 2016 Research Project of Guangzhou Key Research Base of Humanities and Social Sciences Guangzhou Financial Service Innovation and Risk Management Research Base (2016GZJFJD02).

References

[1] B. Tedeschi, "Compressed data; Big companies go slowly in devising net strategy," New York Times, March 27, 2000.
[2] A.Tsay, and N. Agrawal, "Channel conflict and coordination in the e-commerce age," Production and Operations Management, vol.13, no.1, pp.93–110, 2004.
[3] G. W. Hua, S. Y. Wang, and T. C. E. Cheng, "Price and lead time decisions in dual-channel supply chains," European Journal of Operational Research, vol. 205, no. 1, pp.113–126, 2010.
[4] K. Takahashi, T. Aoi, D. Hirotani, and K. Morikawa, "Inventory control in a two-echelon dual-channel supply chain with setup of production delivery," International Journal of Production Economics, vol. 133, no. 2, pp. 403–415, 2011.
[5] T. Boyaci, "Competitive stocking and coordination in a multi-channel distribution system," IIE Transactions, vol. 37, no. 5, pp. 407–427, 2005.
[6] L. E. Cardenas-Barron, "Note on: concurrent pricing lot sizing for make-to-order contract production," International Journal of Production Economics, vol. 103, no. 1, pp. 449–450, 2006.
[7] L. Salvietti, N. R. Smith, and L. E. Cardenas-Barron Leopoldo, "A stochastic profit-maximizing economic lot scheduling problem with price optimization, " European Journal of Industrial Engineering, vol. 8, no. 2, pp. 193–221, 2014.
[8] B. Sarkar, S. Saren, and H. M. Wee, "An inventory model with variable demand, component cost and selling price for deteriorating items," Economic Modelling, vol. 30, pp. 306-310, 2013.
[9] S. S. Sana, "Optimal selling price and lotsize with time varying deterioration and partial backlogging," Applied Mathematics and Computation, vol. 217, no. 1, pp. 185–194, 2010.
[10] S. Balasubramanian, "Mail versus mall: a strategic analysis of competition between direct marketers and conventional retailers," Marketing Science, vol. 17, pp. 181–195, 1998.
[11] W. Y. K. Chiang, D. Chhajed, and J. D. Hess, "Direct marketing, indirect profits: a strategic analysis of dual-channel supply chain design," Management Science, vol. 49, no.1, pp. 1-20, 2003.
[12] S. H. Chun, J.C. Kim, "Pricing strategies in B2C electronic commerce: analytical and empirical approaches," Decision Support Systems, vol. 40, pp. 375–388, 2005.
[13] Y. Li, Z. Lin, L. Xu, and A. Swain, "Do the electronic books reinforce the dynamics of book supply chain market?" – A theoretical analysis, European Journal of Operational Research, vol. 245, no. 2, pp. 591–601, 2015.
[14] G. E. Fruchter, C. S. Tapiero, "Dynamic online and offline channel pricing for heterogeneous customers in virtual acceptance," International Game Theory Review, vol. 7, pp. 137–150, 2005.
[15] H. S. Ahn, I. Duenyas, and R. Q. Zhang, "Price competition between retailers manufacturer-owned stores," University of California at Berkeley Working paper 2002.
[16] G. G. Cai, Z. G. Zhang, and Z. Michael, "Game theoretical perspectives on dual-channel supply chain competition with price discounts and pricing schemes," International Journal of Production Economics, vol. 117, no. 1, pp. 80-96, 2009.
[17] D. Yao, and J. Liu, "Competitive pricing of mixed retailed and e-tail distribution channels," Omega, vol.3828, no.1, pp. 235-247, 2005.
[18] J. Chen, H. Zhang, and Y. Sun, "Implementing coordination contracts in a manufacturer Stackelberg dual-channel supply chain," Omega, vol. 40, pp. 571–583, 2012.
[19] W. Huang, and J. M. Swaminathan, "Introduction of a second channel: implications for pricing and profits, " European Journal of Operational Research, vol. no. 1, pp. 258–279, 2009.
[20] B. Dan, G. Xu, and C. Liu, "Pricing policies in a dual-channel supply chain with retail services," International Journal of Production Economics, vol.139, no. 1, pp. 312–320, 2012.
[21] G. Xu., B. Dan, X. Zhang, and C. Liu, "Coordinating a dual-channel supply chain with risk-

averse under a two-way revenue sharing contract," International Journal of Production Economics, vol.147, no. A, pp. 171-179, 2014.
[22] K.-Y. Chen, M. Kaya, and O. Ozalp, "Dual sales channel management with service competition," Manufacturing and Service Operations Management, vol. 10, no. 4, pp. 654–675, 2008.
[23] L. Ren, Y. He, and H. Song, "Price and service competition of dual-channel supply chain with consumer returns," Discrete Dynamics in Nature and Society, Vol.2014, pp.1-10.
[24] Y. -C. Chen, S.-C. Fang, and U.-P. Wen, "Pricing policies for substitutable products in a supply chain with Internet and traditional channels," European Journal of Operational Research, vol. 224, no. 3, pp. 542–551, 2013.
[25] S. Panda, N. M. Modak, S. S. Sana, and M. Basu, "Pricing and replenishment policies in dual-channel supply chain under continuous unit cost decrease," Applied Mathematics and Computation, vol. 256, pp. 913–929, 2015.
[26] M. Khouja, S. Park, and G. G. Cai, "Channel selection and pricing in the presence of retail-captive consumers," International Journal of Production Economics, vol. 125, no. 1, pp. 84–95, 2010.
[27] T. Li, J. Ma, and L. Sun, "Complexity uncertainty analysis of dynamic in a dual-channel energy supply chain model with heterogeneous retailers," Mathematical Problems in Engineering, vol. 2015, pp. 1-10.
[28] H. Kurata, D. Q. Yao, and J. J. Liu, "Pricing policies under direct vs. indirect channel competition and national vs. store brand competition," European Journal of Operational Research, vol. 180, no. 1, pp. 262–281, 2007.
[29] L. Hsiao, Y.-J. Chen, "The perils of selling online: manufacturer competition, channel conflict, and consumer preferences," Marketing Letters, vol. 24, no. 3, pp. 277–292, 2013.
[30] T. Li, and J. Ma, "Complexity analysis of the dual-channel supply chain model with delay decision," Nonlinear Dynamics, vol. 78, no. 4, pp. 2617-2626, 2014.

IOP Conf. Series: Materials Science and Engineering **207** (2017) 012046 doi:10.1088/1757-899X/207/1/012046

Effect of atomization pressure on the flow field distribution of TC4 alloy powder prepared by EIGA

Kuaikuai Guo, Changsheng Liu*, Suiyuan Chen, Jinbao Li and Qian Fu

Northeastern University, Shenyang 110819, China

*Corresponding author e-mail: csliu@mail.neu.edu.cn

Abstract. By means of the electrode induction melt inert gas atomization (EIGA) equipment designed and developed by ourselves, TC4 alloy powder were prepared. In the process of atomization, the pressure plays a decisive role in the morphology and particle size of the powder. In this paper, the effect of atomization pressure on the distribution of fluid structure was studied by computational fluid dynamics (CFD). The results show that with the increase of atomization pressure, the shape of the recirculation zone is expanded and then compressed. When the gas pressure is 2MPa, the suction pressure ΔP is 0.498MPa, thus forming a positive pressure zone. So it is easy to liquid droplets dripping up, resulting in clogging the nozzle, which has been confirmed in the experiment.

1. Introduction

With its significant advantages of high efficiency, high speed and being suitable for processing complex parts, additive manufacture (3D printing) has become the focus of rapid prototyping technology at home and abroad [1,2]. In recent years, the hotspot of 3D printing technology research and development has gradually shifted to the metal material 3D printing technology where the metal material printing equipment has gradually been applied. For the metal additive manufacture, the powder is a dominant material which has a great influence on the final product; meanwhile, powder is the bottleneck of restricting the development of the entire metal 3D printing. The amount of titanium and its alloys can represent a country's aerospace level. The working temperature of TC4 alloy can reach 400°C, widely applied in the field of aerospace, mainly in the preparation of engine fan, compressor disk, blade aircraft structural beam and other important tensional components. However, the local TC4 powder have not reached the standard, China mainly relies on importing 3D printing powder. At present, the preparation of 3D printing titanium alloy powder mainly depends on mechanical polishing, chemical vapor deposition and atomization [3]. As liquid metal doesn't contact with the crucible and directly flows into the atomization zone, the induction Electrode Inert Gas Atomization (EIGA) powder has the following advantages, including high purity, good performance, a high degree of sphere, narrow particle size distribution and high-cooling rate and little environmental pollution. Therefore, EIGA has become the main method of preparing atomized powder of titanium alloy. TC4 titanium alloy powder has been prepared by the self-designed rotating electrode induction melting gas atomization equipment.

In this paper, the software Fluent 6.3 was adopted to simulate the flow field of rotating electrode vacuum atomization in different pressure. The basic characteristics of the gas flow field of the rotating electrode were explored by the simulation of the vacuum atomization flow field of the rotating electrode vacuum atomization.

2. Experiment methods

The atomization nozzle preparation of the TC4 powder experiment is designed of the ring-type HPGA aerosolized nozzle [3-5]. In order to establish an aerosolized hydrodynamic model consistent with the actual situation and higher computational efficiency, it is easier to obtain accurate convergence solution and improve the simulation results and efficiency [6]. The following calculation requirements are designed:

(1) The air flowing inside the nozzle is set to transient flow.

(2) In the process of single-phase simulation of two-dimensional atomized gas, the gas flow field in the atomization model is arranged in the symmetric distribution, which can be optimized and the calculation time is reduced.

(3)The atomization medium interacts with the melt, and the process involves mass transfer, momentum transfer and energy transfer, and the flow type is turbulent flow.

(4) Ignoring atomization medium and droplet gravity.

Since this experiment is an aerosol preparation of 3D printing titanium alloy powder experiments, in addition to the gas flow involved in the basic continuous equations and equations of motion, as well as droplets and high-speed air contact energy exchange energy equation.

Continuous equation:

$$\frac{\partial \rho}{\partial t}+\frac{\partial}{\partial x_i}(\rho u_i)=0 \tag{1}$$

Motion equation (Navier-Stokes equation):

$$\frac{\partial}{\partial t}(\rho u_i)+\frac{\partial}{\partial x_j}(\rho u_i u_j)=-\frac{\partial p}{\partial x_i}+\frac{\partial}{\partial x_j}\left[\mu\frac{\partial u_i}{\partial x_j}-\tau_{ij}\right]+s_i \tag{2}$$

Energy equation:

$$\frac{\partial}{\partial t}(\rho T)+\frac{\partial}{\partial x_j}(\rho u_j T)=\frac{\partial}{\partial X_J}\left[\frac{K}{c}\frac{\partial T}{\partial x_j}-\rho u_i T\right]+s_T \tag{3}$$

Where ρ is the density; u_i and u_j are the velocity components of X_i and X_j directions respectively; μ is the dynamic viscosity; τ_{ij} is the convex tensor; s_i is the generalized source term of the dynamic equation; T is the temperature; K is the thermal conductivity of the gas rate; c is the specific heat capacity of the gas; s_T is the viscous dissipation term.

The flow of atomized gas belongs to turbulent flow, but there are many kinds of models corresponding to turbulent flow, such as one-equation model, two-equation model and Reynolds stress model and large eddy simulation. This test selects the standard k-ε model.

The flow field structure of the annulus HPGA aerosolized nozzle was simulated by "Fluent Version 6.3" hydrodynamic software. First of all, the use of Fluent commonly used pre-processing software Gambit2.4.6 on the nozzle and its fog chamber within the flow field model. When atomization, the distribution of the air flow field is symmetrical distribution of the central line of the nozzle, so taking half of the flow field as a calculation area. The boundary conditions are as follows: the inlet of the nozzle is imported by pressure, the outlet of the atomization chamber is far from the pressure, the center line of the flow field is set as the symmetry surface, and the other sides are all wall type and the wall type is non-slip adiabatic wall. The grid is divided by the division of triangular non-structural network. The calculated area after dividing the grid is shown in Fig. 1.

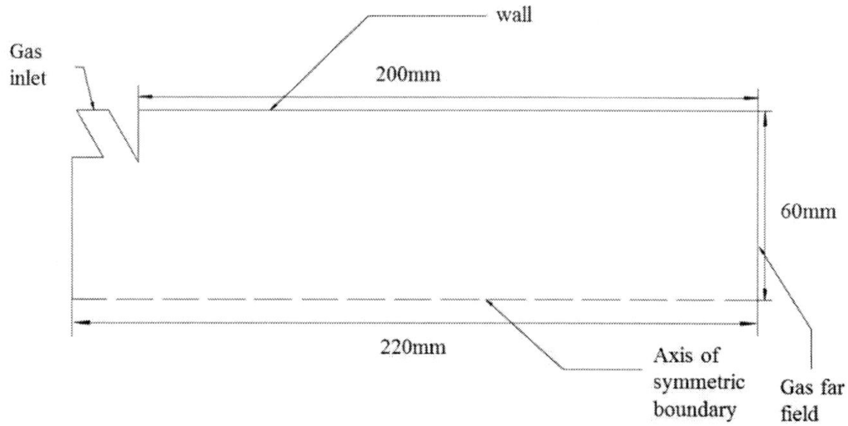

Fig. 1 Grid of gas atomization

The atomization gas is argon and its properties are shown in Table 1.

Table 1 Argon properties in gas atomization

Gas type and properties	Argon
Density($kg \cdot m^{-1}$)	1.6228
Specific heat capacity($J \cdot kg^{-1} \cdot K^{-1}$)	520.64
Thermal conductivity($W \cdot m^{-1} \cdot K^{-1}$)	0.0158
Viscosity($kg \cdot m^{-1} \cdot s^{-1}$)	2.125×10^{-5}
Temperature(K)	273.11

3. Results and discussion

3.1 Velocity distribution of atomized flow field under different pressure conditions

Figure 2 shows the velocity distribution in the atomized flow field of a ring-type HPGA nozzle under different pressure conditions. When gas pressure is 2MPa, the total pressure is relatively low, so the gas pressure inside the nozzle and gas velocity of the gas tank is relatively low, the shock and expansion of the waves are relatively small, and the entire flow field is very small. The recirculation zone shows a thin and short look. And the area formed by the stagnation is almost inverted. When the gas pressure increases to 4MPa, the gas pressure inside the nozzle and the gas pressure in the atomization chamber are increased, and the expansion wave and shock wave at the outlet are all more intense, and the gas velocity at the exit edge in the flow field area also increases significantly, the area of the recirculation zone increased significantly, the stagnation area is no longer concentrated and the state of separation. When the pressure increases to 6MPa, the gas velocity increases, resulting the expansion wave further increasing, while the hysteresis region shape has almost no change. When the air pressure increases to 8MPa, the area of recirculation zone is reduced by the compression, the stagnation area is reduced.

IOP Conf. Series: Materials Science and Engineering **207** (2017) 012046 doi:10.1088/1757-899X/207/1/012046

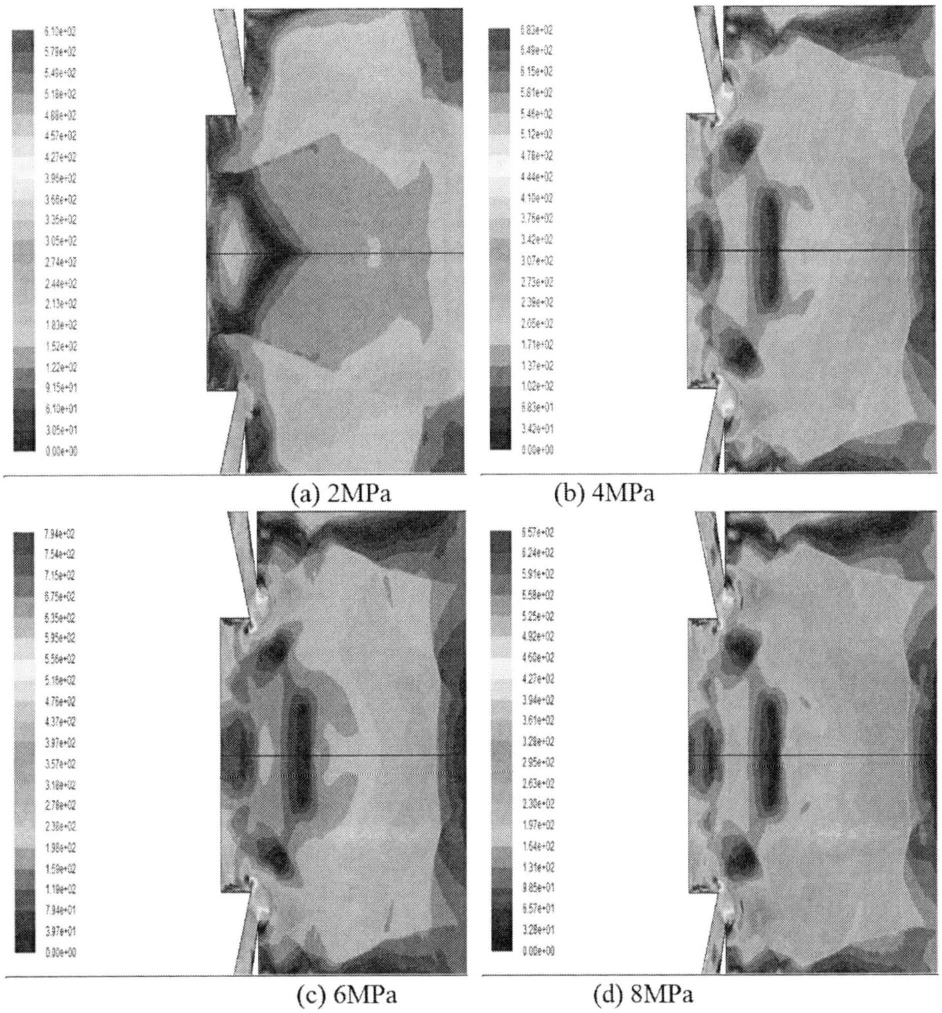

Fig. 2 Velocity distribution graph in different pressure

3.2 Pressure distribution of atomized flow field under different pressure conditions

Figure 3 shows the pressure distribution in the atomized flow field of the ring-type HPGA nozzle under different pressure conditions. When the gas pressure is 2MPa, the overall pressure of the gas flow field is small, the pressure distribution in the recirculation zone is stable at about 1.11MPa, and the pressure around the recirculation zone is mainly 0.612MPa, so the suction pressure ΔP is 0.498MPa. The suction pressure is positive, and the formation of this positive pressure region can be interpreted as: the air flow of two air holes in the nozzle, when the impact is at a certain angle (60 ° in this experiment) for horizontal and vertical components. The two horizontal components of the air flow have the same size and the opposite direction, which are collision at the impact point, because its density and pressure are greater than the surrounding ambient gas, part of the gas moves along the axis of the nozzle. When the molecular packing of air current is larger than that of air jet flow, thus forming a positive pressure zone [7]. So it is easy to liquid droplets dripping up, resulting in the phenomenon of clogging the nozzle, which has also been confirmed in the experiment. While the stagnation pressure remains almost unchanged.

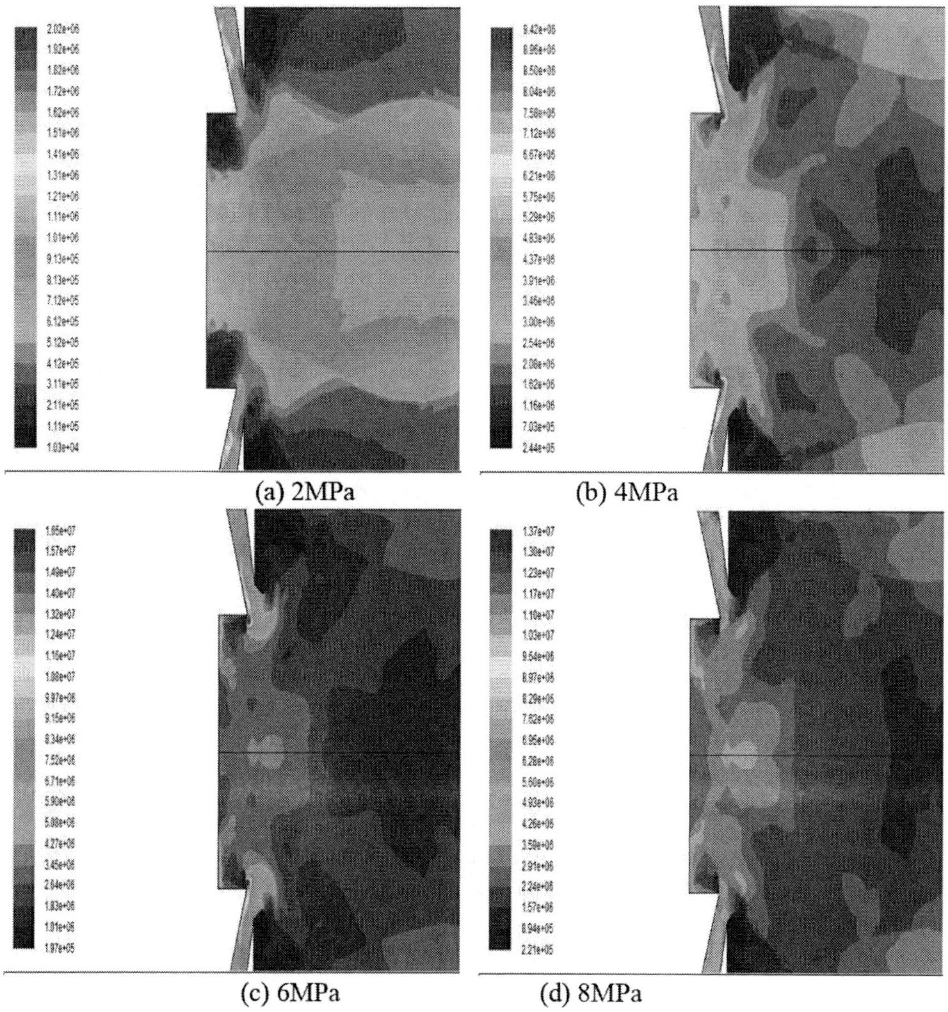

Fig. 3 Pressure distribution in different pressure

When the gas pressure increases to 4MPa, the overall pressure of the gas flow field can be improved, especially in the gas outlet, the pressure has significantly improved and the emergence of a significant expansion area, the pressure of recirculation zone is almost stable, and the pressure within the reflux area has also received a certain degree of increase. The suction pressure corresponding to the recirculation zone can be calculated from the figure. This negative pressure zone is formed by the entrainment of the airflow. When the catheter is near the focal point of the axial jet, the downward flow of the nozzle strong suction action sucks away the gas near the catheter, but it hasn't enough time for gas to be added, thus forming a negative pressure zone [7]. The presence of the negative pressure zone allows the metal droplets to be drawn into the atomization zone to accelerate the flow rate of the metal droplets, which facilitates the atomization process. Compared to 2MPa, the stagnation pressure has slightly increased. When the gas pressure increases to 6MPa, the expansion area is further expanded, the area of the recirculation zone and the pressure in the recirculation zone increases, and the suction pressure in the recirculation zone remains negative. After calculating the pressure difference of the negative pressure zone with range from -2.44 ~ -1.83MPa (6MPa has the maximum absolute of pressure difference), the pressure change on the negative pressure area is not significant, stagnation pressure decreases slightly. When the gas pressure continues to increase to 8 MPa, the expansion zone hardly changes, the area of the recirculation zone is significantly expanded, the suction pressure remains negative and the stagnation pressure decreases.

4. Conclusion

(1) With the increase of atomization pressure, the shape of the recirculation zone first appears short thin state, then compressed and expanded. The area formed by the stagnation is similar to that of the inverted cone after separation and compression, but finally returns to the approximation inverted cone state and the area decreased.

(2) When the gas pressure is 2MPa, the suction pressure ΔP is 0.498MPa, thus forming a positive pressure zone. So it is easy to liquid droplets dripping up, resulting in clogging the nozzle, which has been confirmed in the experiment.

Acknowledgments

This work was financially supported by the Frontier and Key Technical Innovation of Guangdong Province (2015B010122001), the Joint Founds of NSFC-Liaoning (U1508213), and the National Key Research Project (2016YFB1100201).

References

[1]. Visser C W, Pohl R, Sun C, et al. Toward 3D printing of pure metals by laser-induced forward transfer, Advanced Materials, 2015, 27(27): 4087-4092.
[2]. Singh R, Verma M. Investigations for deducing wall thickness of aluminium shell casting using three dimensional printing, Journal of Achievements in Materials & Manufacturing Engineering, 2008, 31(2): 565-569.
[3]. Mullis A M, Bigg T D, Adkins N J. Structure and phase-composition of Ti-doped gas atomized Raney-type Ni catalyst precursor alloys, Intermetallics, 2015, 67:63-68.
[4]. Miller S A. Close-coupled gas atomization of metal alloy, In: Kaysser W A, Huppman J W, eds. Horizons of powder metallurgy. Freiberg, Germany: Verlag Schmidt GMBH, 1986. 29-32.
[5]. Strauss J T. Hotter gas increases atomization efficiency, MPR, 1999, 4(11): 24-28.
[6]. Senecal P K, Schmidt D P, Nouar I, et al. Modeling high-speed viscous liquid sheet atomization, Int. J. Multiphase Flow, 1999, 25: 1073-1097.
[7]. Chen Ping, Liu Fu-ping. Improvement of the atomizer used for producing metal powder, Metal Materials and Metallurgy Engineering, 2009, 37(4): 46-49.

IOP Conf. Series: Materials Science and Engineering **207** (2017) 012047 doi:10.1088/1757-899X/207/1/012047

Research on some factors which influence the aperture of precision micro hole in electronical discharge machining

Min Wang[1], Gangjian Ma[1,*], Jianyong Liu[2,3,4]

[1]Advanced Manufacturing technology of the Key Laboratory of Beijing Municipality, College of Mechanical Engineering and Applied Electronics Technology, Beijing University of Technology, Beijing 100124, China
[2]Beijing Municipal Key Lab of EDM Technology, Beijing 100191, China
[3]Beijing Institute of Electro-Machining, Beijing 100191, China
[4]BeiJing Natural Science Foundation & Beijing Academy of Science And Technology, Beijing 100195, China

[*]Corresponding author e-mail: 1062188911@qq.com

Abstract. The factors which influence the size of the hole during the process of machining precision micro hole are studied in this paper. Through the way that using the same diameter of electrode wire to process large number of micro hole on different materials and different thickness, the processing time, electrode wire loss and pore size was written down. The law of the factors which influence the size of the micro hole is found by analysing the data. It can help to select the appropriate electrical processing parameters to ensure the quality of the aperture during actual operation process. And these are usually ensured by repeated testing and multiple measurements. This research get great help for improving processing efficiency. The study is of great guiding significance to the actual production.

1. Introduction

With the development of science and technology, people pay more and more attention to the precision micro-hole's processing technology in the production. Many countries have invested a lot of manpower and resources to carry out the research [1]. Especially in recent years the technology has been widely used in aerospace, computer technology, instrumentation, precision machinery, automotive, light industry and other industries in some cutting-edge products. At the same time ,as the reason that precision micro-spark hole processing technology has the advantages: non-contact processing, no macro-cutting force ,killing with kindness, precision micro square, realistic and other characteristics, to further determine its broad prospects of development [2,3]. Such as Japan, Switzerland, the United States and other countries have mature products [4]. However the domestic research in this area is mainly to solve the problem of practical application.

The accuracy of the hole is the key to judge whether the processing task is up to or not [5]. It is one of the important criteria to assess the level of precision of precision micro EDM. However, there are many complicated factors such as the length of the processing time, the size of the discharge gap, the loss of the electrode wire and so on. These factors results that the stability of the aperture can not be stabilized during the machining process. In this paper, the test plan that to design different thickness of test pieces, a variety of materials, several groups test program is designed. Record the data of processing time, electrode wire loss, hole's size. And through the data processing to explore the law among these factors.

2. Analysis on the Test Methods of Micro Hole's Size and Real-time Loss of Electrode Wire

In precision micro-hole processing, the accuracy of micro whole's size depends on the discharge state during EDM. While the discharge state is influenced by the pulse width, pulse, working fluid state and many other factors. Therefore, the study that the relationship among aperture and its influencing factors needs large numbers of process tests and test data. In view of this, this paper will use the method of test sample's thickness accumulation: gradually increase the thickness of the work piece to be processed to process larger number of micro holes under the state of liquid and the given parameter. To monitor the electrode wire loss, record the processing time, measure the size of the micro holes.

The method of measuring the size of the micro hole is shown in Figure 1. The sizes which in two vertical directions of the precision micro holes are recorded by the x and y sets of data. Where x_1, x_2 ... x_n denote the dimensions in the direction of the aperture x; y_1, y_2 ... y_n denote the dimensions in the direction of the aperture y.

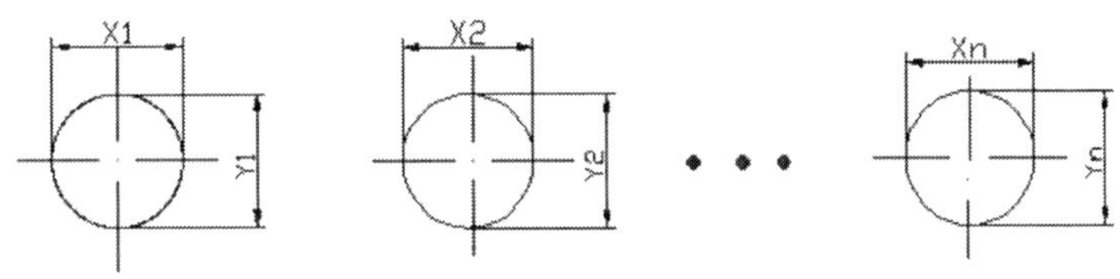

Figure 1. Schematic diagram of the scheme to measure the aperture

The loss of electrode wire is shown in Figure 2, Where l_1, l_2 ... l_n denote the test sample's thickness, Φ_1 Φ_2... Φn denote the test sample's diameter of the micro holes, δ_1, δ_2 ... δn denote the loss of electrode wire. According to the test results, the relationship among electrode wire loss, processing time and pore diameter was analyzed.

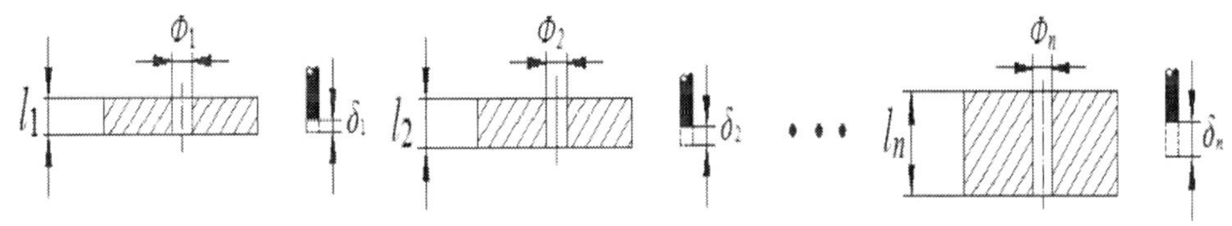

Figure 2 Schematic diagram of electrode wire's loss

3. Test scheme

EDM processing is a special processing ranks [6], influenced by the electrode wire's size, the work piece's material and other factors.

In the case of insufficient number of samples, even under the same processing environment and the same processing parameters to process micro-hole, may also cause instability of micro-aperture, the data of electrode wire loss and the data of processing time .In this paper, TC4 titanium alloy, TC11 titanium alloy and GH4169 spirally with the same thickness of 0.5mm, 0.8mm, 1.0mm, 1.2mm, 1.7mm, 2mm and 3mm were prepared and 40 pieces were processed on each work piece. In the process of processing, the samples with the same material were processed with the same parameters. The processing methods and materials are shown in Table 1.

Table 1 Processing methods and materials

Electrode wire diameter (mm)	Electrode wire material	Cutting fluid	Processing methods
0.19	Tungsten	Deionized water	Rinse processing

Through the statistical analysis of the data, the relationship among the depth of the pores of different depth, electrode wire loss and the working hours were determined. The Processing equipment is the precision micro EDM equipment witch developed independently by Beijing Institute of Electro-Machining, as shown in Figure 3. The machined micro pores were measured by using a high power electronic measuring instrument, as shown in Fig 4.

Figure 3 Precision micro-EDM equipment

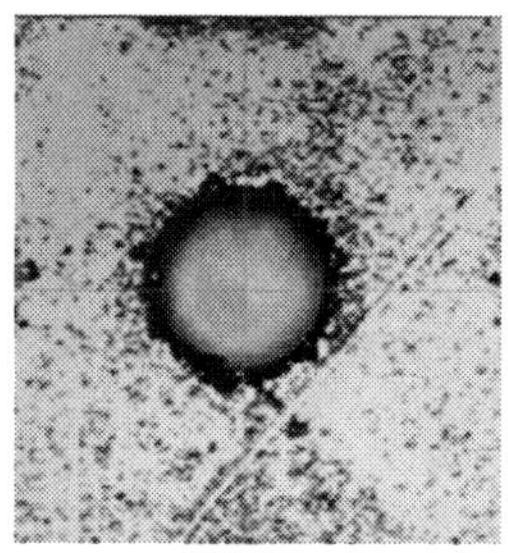

Figure 4 High-power electronic measuring instrument aperture

4. Test data analysis

Since the size of the micro hole is closely related to the discharge state, so the precision micro holes are not absolutely circulars, the front and back sides of the machined holes are also slightly different. The micro holes' size data of TC4 titanium alloy is shown in Table 2.

Table 2 Aperture data of Titanium alloy TC4

Sample thickness/mm	Front aperture/μm	Negative aperture/μm	Hole diameter/μm
0	0	0	0
0.5	224.06375	210.05	217.0569
0.8	228.96	212.913	220.9365
1	231.855	220.22625	226.0406
1.5	237.60375	218.2375	227.9206
2	235.2	223.18	229.19
3	244.77625	216.3638	230.57

If the whole's number is j and its thickness is i, its Processing time is t_{ij}, its Electrode loss is l_{ij}:
So the Average processing time:

$$\bar{t_i} = \frac{\sum_{j=1}^{n} t_{ij}}{n} \tag{1}$$

The Average electrode wire loss:

$$\bar{l_i} = \frac{\sum_{j=1}^{n} l_{ij}}{n} \tag{2}$$

For the pore diameter of the fine micropores just as shown in Figure 1:
The diameter of the j-th hole:

$$\phi_j = \frac{x_j + y_j}{2} \tag{3}$$

The average pore diameter:

$$\overline{\phi_i} = \frac{\sum_{j=1}^{n} \phi_{ij}}{n} \tag{4}$$

According to the above calculation and test data can be obtained in Table 3-Table 5 , Figure 5, 6 shows the processing time and the corresponding relationship between the wire loss.

Table 3 TC4 titanium alloy processing time and electrode wire loss, the corresponding relationship between the apertures

Sample thickness/mm	Processing time/s	Electrode wire loss/μm	Hole diameter/μm
0	0	0	0
0.5	81.4	248.175	217.0596
0.8	108.525	276.725	220.9365
1	114.2	289.4	226.0406
1.5	151.5	343.4	227.9206
2	205.325	422.075	229.19
3	280.225	973.425	230.57

Table 4 TC11 titanium alloy processing time and electrode wire loss, the corresponding relationship between the apertures

Sample thickness/mm	Processing time/s	Electrode wire loss/μm	Hole diameter/μm
0	0	0	0
0.5	86.925	256.25	220.637
0.8	100.85	277.825	228.5285
1	110.275	300.425	225.8435
1.5	142.475	370.125	229.258
2	192.75	465.55	234.8965
3	268.05	951.6	235.5525

Table 5 GH4169 Superalloy processing time and electrode wire loss, the corresponding relationship between the apertures

Sample thickness/mm	Processing time/s	Electrode wire loss/μm	Hole diameter/μm
0	0	0	0
0.5	70.97561	582.8293	250.235
0.7	85.356	850.521	251.3745
1	106.5854	1096.951	254.4525
1.2	132.7561	1481.878	259.097
1.7	208.6829	2341.415	269.3
2	238.4146	2706.268	263.613
3	306.9268	4506.64	296.325

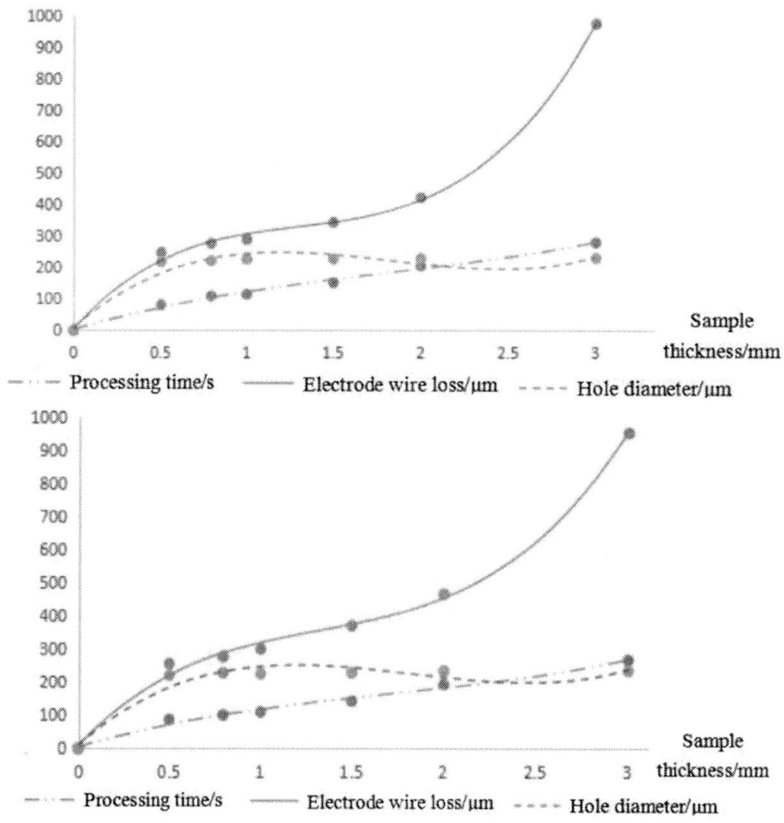

Figure 5 TC4 and TC11 titanium alloy processing time and electrode wire loss, the corresponding relationship between the apertures

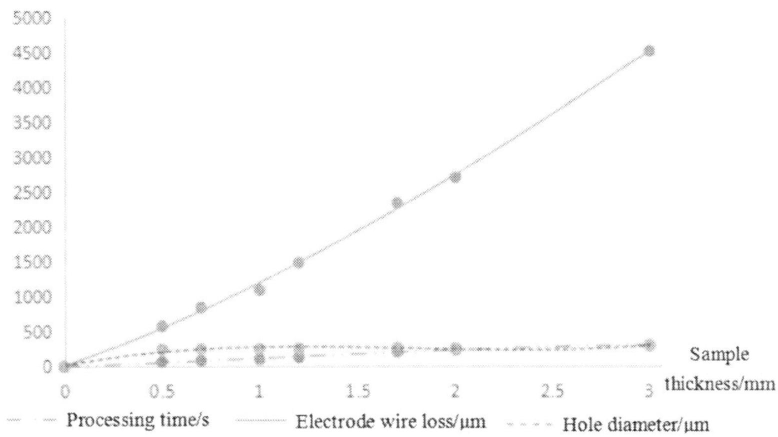

Figure 6 GH4169 Superalloy processing time and electrode wire loss, the corresponding relationship between the apertures

5. Conclusion

From the above analysis, it is obviously to know that the relationship between the processing time in the machining, the real-time loss of the electrode wire and the size of the micro hole can be matched by the different cubic equations of one unknown. The relationship between the working time and the electrode wire loss, the relationship between the wire filament processing time and the working whole diameter and the relationship between the electrode wire loss and the whole diameter were founded to provide a theoretical basis for the identification of abnormalities in the process of machining. And lay the foundation for the identification of abnormal processing in small whole machining.

Acknowledgments

This work was financially supported by National Natural Science Foundation of China (51575014), Beijing Natural Science Foundation & Beijing Academy of Science And Technology (L150006) fund.
Corresponding author: Ma Gangjian

References

[1] LIU Jian-yong,YANG Da-yong,GUO Yan,et al. Research on the Mechanical System of EDM Machine maching for the Precision Reverse Taper Microhole [J]. Electromachining and Mould, 2013, (5): 54-58.
[2] O. Koch W. Ehrfeld F, Michel etal. Recent progress in micro-electro dishcharge machining (Part I): Technology. Proceedings of the 13th International Symposium for Electrod machining ISEM Ⅷ (Vol I.II), 2001: 29-36
[3] JIA Zhi-xin,TENG Xiang-yang,YAN Kai,et al. Novel Technology of Vertical Wire Electrical Discharge Machining with Rotation Wire [J]. Mechanical Engineer, 2006, (5): 19-20.
[4] M.Tsutsumi, Y.Ihara, A.Saito, N.Mishima, S.Ibaraki, M.Yamamoto, M.Kobayashi, T.Yonetani, Standardization of testing methods for kinematic motion of five-axis machining centers—draft proposal for ISO standard, In: Proceedings of the Seventh Manufacturing and Machine Tool Conference, 2008, pp.95–96.
[5] Cao Feng-guo. Special Processing Manual [M]. Beijing:China Machine Press, 2010.
[6] Konig W, Klonche F, Lenzen R. The electrical machining process-what demands will they face in the future [J]. International Journal of Electrical

IOP Conf. Series: Materials Science and Engineering **207** (2017) 012048 doi:10.1088/1757-899X/207/1/012048

Simulation and experimental study of resin flow in fibre fabrics

Fei Yan, Shilin Yan* **and Yongjing Li**

Hubei Key Laboratory of Theory and Application of Advanced Materials Mechanics, Wuhan University of Technology, Hubei 430000, China

*Corresponding author e-mail: yanshl@whut.edu.cn

Abstract. Liquid Composite Moulding (LCM) is gradually becoming the most competitive manufacturing technology for producing large composite parts with complex geometry with high quality and low cost. These parts include those for airplanes, wind turbine blades and automobile components. Fibre fabrics in liquid composite moulding can be considered as dual-scale porous media. In different gap scales, an unsaturated flow is produced during the mould filling process. This particular flow behaviour deviates from the traditional Darcy's law, which is used to calculate the filling pressure and will cause errors. According to sink theory, the unsaturated flow characteristics of this dual-scale porous media were studied in this paper, and a FEM solution program was developed. The results showed that the pressure curves against the position which simulated by sink functions were departure from the position of traditional theory. In addition, the simulation results of partially-saturated region were consistent with the experimental data.

1. Introduction

Fiber fabrics in LCM is often woven or stitched by fibre tow. The gap between fiber tows is at the millimeter level, and the gap inside the fiber tow is at the micron level. Once resin is impregnated into the dual-scale fiber preform, the gap between the fiber tow was impregnated fast at first, and the gap inside the fiber tow was impregnated subsequently. The unsaturated flow would be generated due to the different impregnation rates, thus making existing theory misestimate mold filling pressure and fail to predict mold filling time accurately [1-3].

Several scholars proposed the "sink model" to analyze the unsaturated flow behavior of resin in dual-scale porous fibre preform [4, 5]. This hypothesis would lead to inaccuracy results. For example, the overestimate inlet pressure induced excess in stiffness design of mould, thus increased cost significantly. Failure to simulate the partially-saturated region would ignore the impregnated time inside the fiber tows, so as to underestimate the filling time which cause quality problem of product.

Recently, more and more scholars recognize that the unsaturated flow and its effect on simulated results. Dai Fuhong, Lu Shouzhou and Du Shanyi have established flow control equations that include sink function, and get the distribution of partial saturation by solving the equation. Zhang Yuhua, Chen Xiaojiang and Yan Shilin using Matlab to do study of the pressure curves against the position and the flow front, but its algorithm has some defect and lack of experimental verification [6, 7]. Base on sink theory, the unsaturated flow characteristics of this dual-scale porous media were studied in this paper, and a FEM solution program was developed by our team. In addition, the simulation results of partially-saturated region would be validated.

2. Basic Theory

2.1. Sink Model

To simulate the resin flow behavior, the traditional theory considered the gap between and inside the fibre tows as uniform distribution, when the resin impregnated the preform, the gap would be filled simultaneously. While the sink model considered the different gap between and inside the fibre tows as two independent systems. The resin would impregnate the preform with different rate [8-10]. The volume change rate of resin in fibre tow is not zero and the intra-tow gap is not filled completely. From the mass conservation equation, the following expression can be derived:

$$\nabla \cdot V = S_g \tag{1}$$

Where V is the average velocity and S_g is the sink function. Combined with the Darcy's law, the control equation of unsaturated flow in fibre fabrics can be derived as follow:

$$\nabla \cdot (\frac{K}{\mu} \nabla P) = S_g \tag{2}$$

Where ∇P is the pressure gradient, μ is the liquid viscosity, K is the permeability tensor, S_g is the function of saturation.

2.2. Unsaturated Flow Representation

Due to the characteristics of dual-scale fibre fabrics, the region when resin impregnated was partially saturated, shown in fig. 1(a). The width of partially-saturated region Lps represented the distance for saturated flow front and unsaturated flow front. As the fill process continued, partially-saturated region moved along, this phenomenon can be reviewed during experiment.

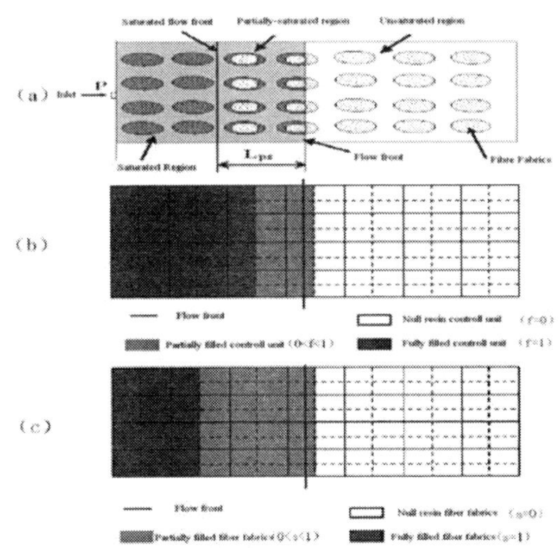

Figure 1. (a) Unsaturated characteristics of resin flow. (b) Distribution of filling factor. (c) Distribution of tow-saturation.

In order to simulate the unsaturated flow, we have to get the flow process during the preform and intra-tow gap. Base on the control volume finite element method (FEM/CV), filling factor f and saturation s is represented the different behavior separately. The control volume is a unit that around the mesh node in FEM. It's a quadrangle which connects by the centroid and each midpoint of the unit. For example, the quadrangles that divide by dash line during in fig. 1 are control volume units. For each control volume unit, it can use filling factor f to represent the filling degree. The relation of filling factor and flow front is shown in fig. 1(b). And the saturation s represents the filling degree at intra-tow gap during partially-saturated region, which is shown in fig. 1(c).

3. Simulation Case

Base on FEM/CV, a solution program was developed by using FORTRAN in order to simulate the unsaturated flow.

In this case, we have study on unsaturated flow at constant injection. The fibre preform is rectangle, which was 90-mm width and 570-mm length. By using triangle unit to discrete, we could get the FEM mesh shown in fig. 2. Where the node number was 204, unit number was 330. The viscosity of resin was 0.059Pa s, and the permeability of fibre fabrics was 7.85×10-10m2, the porosity between fibre tow was 0.3872 and the porosity inside fibre tow was 0.1. The left side was inlet boundary, using line injection. And the flow rate was 1.6×10-2m/s. The right side was out boundary.

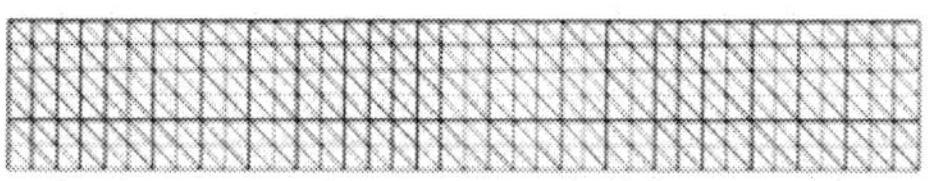

Figure 2. Discrete finite element mesh.

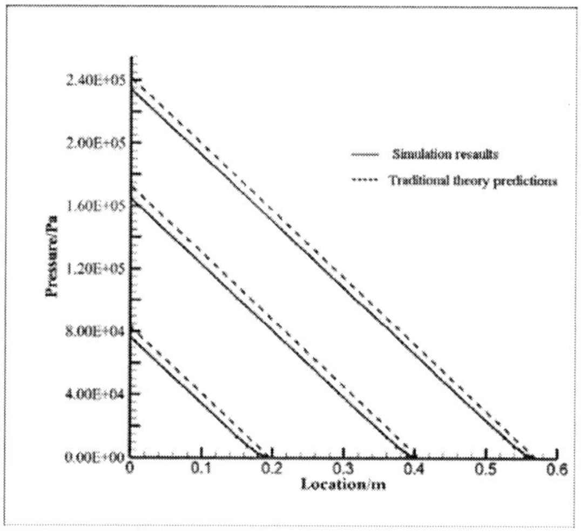

Figure 3. Pressure distribution of different time.

The nonlinear distribution of pressure was indirect exposition of the partially-saturated region. The impregnated degree could be calculated by the saturation of sink model. And the variations of partially-saturated region could get by calculating the saturation of each steady sub step during simulated process. Fig. 4 was the distribution of partially-saturated region when flow front arrived at different locations. For convenient observing, distinct gray scale was represented the difference of saturation. The saturation value between 0 and 1 was represented partially-saturated. From fig. 4 we could find the partially-saturated region moved along the flow front during constant injection. At the first stage, the width of partially-saturated region increased gradually, until the rate of saturated flow and unsaturated flow became sameness. At the second stage, the width of partially-saturated region kept stable and moved along the flow front. At the final stage, the flow front arrived at the end of preform, and the width of partially-saturated region decreased gradually until disappeared.

4. Experimental Research

In order to study further on unsaturated flow characteristics of this dual-scale porous media, an experimental device of constant flow rate was established. This system mainly consisted of a mould, constant flow pump, pressure sensor, data and acquisition terminal. The cavity was designed with 4-mm thickness, 90-mm width and 1000-mm length. It used 7075-T651 aluminium alloy with excellent performance for the mould. To enable observation of the flow process, the upper mould had a 1000 mm × 150 mm × 10 mm transparent organic glass (PMMA) attached to the aluminium alloy frame. The fibre fabrics are both triaxially stitched, and the viscosity of resin was 0.059Pa s. The inlet point was at left side of mould, the flow rate of resin was 1.6×10^{-2}m/s, at the right side of mould there was an exhaust port, shown at fig. 5.

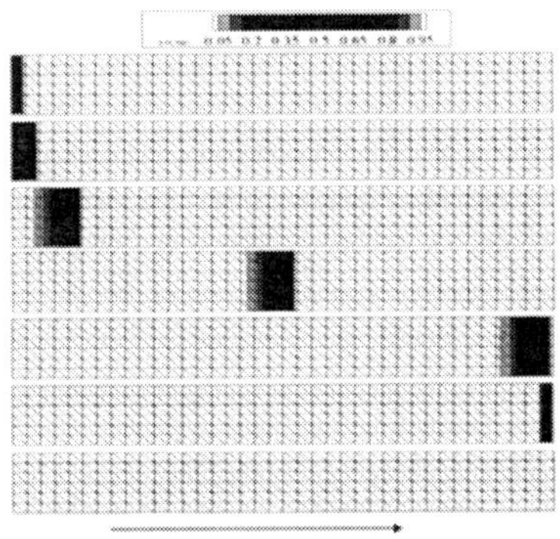

Figure 4. Simulation result of variation tendency of the partially-saturated region.

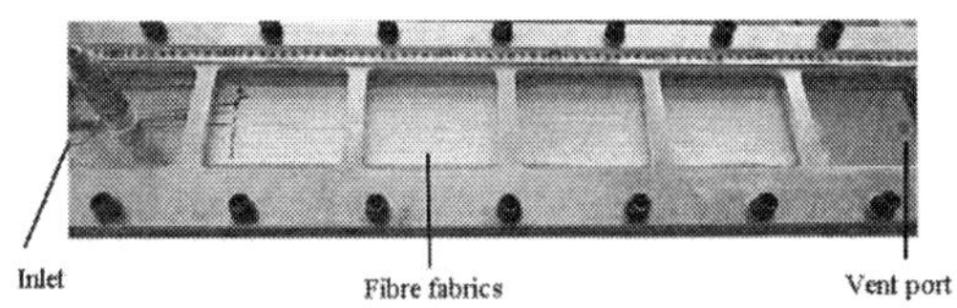

Figure 5. Device of constant injection rate experiment.

The inlet injection was a point, so it hard to achieve the line injection during simulation. The edge effect would lead to the shape of flow front become arc, but after the flow front there was still a uniform arc-shaped partially-saturated region, shown at fig. 6. In order to investigate the variation of partially-saturated region, the distance between arcs was considered as width of partially-saturated region.

Figure 6. Experimental result of variation tendency of the partially-saturated region.

During the filling process, the variation of partially-saturated region against time could be observed, as shown in fig. 6. Fibre tows between the partially-saturated region still showed white color, and the gap inside fibre tows were fully impregnated. The partially-saturated region could get by this feature. The distance between two arcs in fig. 6 represented the width of partially-saturated region. At the early stage of filling process, the width of partially-saturated region increased gradually, and then kept stable. When the flow front arrived at the end of preform, it decreased gradually. These variations were in full accord with simulation results.

5. Conclusion

Aiming at the characteristics of this dual-scale porous media, a method of introducing saturation was developed on response of unsaturated flow during LCM. A FEM solution program was compiled by our team. The results showed that the pressure curves against the position of the saturated region which far from the flow front varying linearly. Before the flow front, pressure curves for partially-saturated region were deviated from linear trend which predicted by traditional theory. According to the simulation results, the partially-saturated region were increased at first and then remained stable. At the end of filling process, it decreased gradually. These results were consistent with the experimental data. It explains the program has good reliability and can provide good references for optimizing the process parameters or controlling the product quality.

Acknowledgments
This work was supported by the National Science Foundation of P.R. China, Grant No. 51073125.

References

[1] M.V. Bruschke, S.G. Advani, A finite element/control volume approach to mold filling in anisotropic porous media, Polym. Composite. 11 (1990) 398-405.
[2] K.M. Pillai, Modeling the Unsaturated Flow in Liquid Composite Molding Processes: A Review and Some Thoughts, J. Compos. Mater. 38 (2004) 2097-2118.
[3] P. Simacek, S.G. Advani, A Numerical Model to Predict Fiber Tow Saturation During Liquid Composite Molding, Compos. Sci. Technol. 63 (2003) 1725-1736.
[4] H. Tan, K.M. Pillai, Multiscale modeling of unsaturated flow in dual-scale fiber preforms of liquid composite molding I: isothermal flows, Compos. Part A. 43 (2012) 1-13.
[5] L. Haichen, W. Biao, Z. Zhengong. Numerical Simulation of Resin Flow in RTM, Acta. Mater. Compos. Sinica. 19 (2002) 18-23.
[6] D. Fuhong, L. Shouzhou , D. Shanyi, Simulations and experiments of unsaturated flow in resin transfer molding process, Acta. Mater. Compos. Sinica, 27 (2002) 84-89.
[7] Z. Yuhua, C. Xiaojiang, Y. Shilin, Flow Simulation of Resin in Dual-scale Porous Medium at Constant Pressure, J. WHUT. 35 (2011) 686-689.
[8] W. A. Walbran, B. Verleye, S. Bickerton, Prediction and experimental verification of normal stress distributions on mould tools during liquid composite moulding, Appl. Sci. Manuf. 43 (2012) 138–149.
[9] C.H. Park, A. Lebel, A. Saouab, J. Breard, Modeling and simulation of voids and saturation in liquid composite molding processes, Compos. Part A. 42 (2011) 658-668.
[10] A. A. Somashekar, S. Bickerton, D. Bhattacharyya, Modelling the viscoelastic stress relaxation of glass fibre reinforcements under constant compaction strain during composites manufacturing, Compos. Part A. 43 (2012) 138-149

Study on the process control of AH36 high strength ship plate

Yanxia Liu, Bing Wang, Yanjun Meng, Ming Chen, Zhongqi Dong*

Matcrial Engineering Department, Hebei College of Industry and Technology, Shijiazhuang 050000, China

*Corresponding author e-mail: dongzhongqi@126.com

Abstract. There are amounts of surplus in the performance of A 36 high strength ship plate at present. The optimal control of smelting process and method of controlled rolling and controlled cooling parameters was used to in order to higher the efficiency on the basis of statistical analysis of production data of A36 high strength ship plate. The influnece of the content of Nb redued and the content of C and Mn controled on the performance of A36 high strength ship plate was investigated. The result show that the content of Nd can reduced by the optinizatation of the composition A36 using the lower plate finishing rolling temperature ang rolling red temperature limited witch is stability controled at process parameters. The cost of A36 high strength ship plate is reducted and the performance can meet the requirements.

1. Introduction

Energy saving, green production has become a trend, in the nine elements of high strength ship plate index meets the ship regulation under the condition of statistical analysis of AH36 plate of high strength ship plate performance, reduce the amount of the substitute alloy, precious alloy, and through smelting and rolling process optimization [1-4], to ensure good mechanical properties of the plate at the same time, but also can to reduce the production cost of ship. Therefore, this study through the control of smelting of AH36 high strength ship plate steel Gao Jie, clarity, shape and size of inclusions, the metallurgical quality of continuous casting billet; optimization of high strength ship plate cold rolling rolling process, obtained the good ship type, performance and surface quality.

2. Comparative analysis of the statistical properties of AH32 steel and the requirements of high strength ship agency

Now AH36 high strength ship steel performance and the requirements of classification performance comparison and analysis are shown in Table 1, the thickness of plate is less than or equal to 25mm AH36 yield strength margin of more than 40MPa accounted for 91.94% of the amount of surplus, the tensile strength above 40MPa accounted for 59.05%, the elongation amount of surplus more than 4% accounted for 91.82% of the impact energy, the proportion of rich margin of more than 40J 98.08%. Plate thickness more than 25-38mm A36 yield strength 40MPa above the amount of surplus accounted for 94.23%, the tensile strength margin above 40MPa accounted for 36.69%, the elongation amount of surplus more than 4% accounted for 95.72% of the impact energy rich margin of more than 40J accounted for 99.81%. As shown in Table 2, statistics on AH36 actual components according to the yield strength margin, yield strength of the surplus is less than 40MPa, 40-80MPa, 80MPa, the composition of control, average [C] = 0.14 - 0.15 %, [Si] = 0.30 %, [Mn] = 1.26 % - 1.30 %, [Als] =

IOP Conf. Series: Materials Science and Engineering **207** (2017) 012049 doi:10.1088/1757-899X/207/1/012049

0.020 % - 0.021 %, [Nb] = 0.023 %- 0.024 %. A large amount of surplus, you can optimize the performance of AH36.

Table 1 The performance analysis of AH36 rolling ship

standard(mm) (Total batch)	Surplus yield (MPa)	Tensile surplus(MPa)	lot number	proportion	Extend surplus (%)	lot number	proportion	Impact surplus(J)	lot number	proportion
8-25 (5211)	≤40	≤40	2134	40.95%	≤4	426	8.18%	≤40	100	1.92%
	>40-80	>40-80	2981	57.21%	>4-8	3011	57.78%	>40-80	341	6.54%
	>80	>80	96	1.84%	>8	1774	34.04%	>80	4770	91.54%
25.5-35 (537)	≤40	≤40	340	63.31%	≤4	23	4.28%	≤40	1	0.19%
	>40-80	>40-80	192	35.76%	>4-8	216	40.23%	>40-80	11	2.05%
	>80	>80	5	0.93%	>8	298	55.49%	>80	525	97.76%

Table 2 The component analysis of AH36 rolling ship

specifications	Surplus yield(MPa)	chemical composition % C			Si			Mn			Als			Nb	
		Min	Max	Av	Min	Max	Av	Min	Max	Av	Min	Max	Av	Min	Max
AH36	≤40	0.12	0.18	0.14	0.2	0.40	0.30	1.18	1.41	1.26	0.009	0.039	0.020	0.020	0.029
8-25mm	>40-80	0.12	0.18	0.14	0.2	0.40	0.30	1.15	1.50	1.27	0.008	0.071	0.020	0.017	0.035
5211	>80	0.12	0.18	0.15	0.2	0.40	0.30	1.15	1.53	1.29	0.008	0.045	0.020	0.017	0.035
AH36 26-	≤40	0.13	0.16	0.14	0.2	0.30	0.30	1.18	1.40	1.26	0.014	0.034	0.021	0.020	0.027
36mm	>40-80	0.12	0.18	0.14	0.2	0.40	0.30	1.15	1.53	1.28	0.009	0.069	0.021	0.020	0.033
537	>80	0.12	0.18	0.14	0.2	0.40	0.30	1.21	1.53	1.30	0.009	0.069	0.021	0.020	0.033

3. Experimental results and analysis

3.1. Control of production process and related technical parameters

Through the performance analysis and optimization of composition, formulate the corresponding optimization test scheme, the specific parameters are as follows: AH36 high strength ship steel smelting, composition, Mn from 1.28% to 1.35%, Nb from 0.025% to 0.015%, the target of smelting cost can be reduced by 29.5 yuan / ton; In the rolling process, the final rolling temperature was reduced by 10 ~30 / C, according to the thickness of the steel plate, and the upper limit of the return temperature was reduced by about 20 DEG C. The first batch of smelting process control strictly according to the Nb component smelting, the average Mn content of 1.28%, the average Nb content of 0.015%, in line with the smelting target, the specific composition of Table 3 shows.

Table 3 The control composition of AH36 down cost performance optimization test smelting

batch number	C%	Si%	Mn%	P%	S%	Als%	Nb%
20907416	0.15	0.24	1.24	0.026	0.013	0.014	0.016
20408887	0.14	0.25	1.27	0.017	0.012	0.02	0.016
20408888	0.14	0.26	1.28	0.007	0.005	0.028	0.015
20907435	0.15	0.26	1.27	0.015	0.008	0.019	0.017
20907436	0.15	0.26	1.35	0.017	0.014	0.029	0.018
20907438	0.13	0.23	1.32	0.021	0.007	0.021	0.013
20408910	0.14	0.24	1.26	0.027	0.014	0.021	0.015
20408911	0.14	0.26	1.27	0.021	0.008	0.025	0.015
20805508	0.13	0.22	1.29	0.022	0.006	0.013	0.015
20703723	0.15	0.24	1.3	0.019	0.014	0.022	0.014
20408981	0.13	0.23	1.27	0.021	0.012	0.024	0.014
20409026	0.16	0.25	1.19	0.018	0.011	0.012	0.016
20907604	0.14	0.27	1.29	0.016	0.006	0.036	0.016
20907605	0.15	0.28	1.27	0.024	0.006	0.024	0.015
20805614	0.13	0.08	1.37	0.044	0.023	0.002	0.013
20409100	0.15	0.28	1.29	0.022	0.007	0.023	0.018
20409101	0.15	0.26	1.28	0.022	0.009	0.023	0.017
20409157	0.14	0.24	1.30	0.021	0.006	0.023	0.015
20907662	0.16	0.26	1.30	0.019	0.01	0.020	0.015

3.2. The performance of AH36 high strength ship after optimization

The first test in rolling plate mill in wire plate, rolling down Nb AH36 ship a total of 9 batches of 73 pieces of 222.487 tons, the rolling thickness is 8mm, 10mm, 20mm, performance inspection of finished products have reached the requirements. The average yield of 8mm steel plate is 57Mpa, the average amount of tensile strength is 58Mpa, the average amount of elongation is 8%, the average amount of impact energy is 48J; 10mm steel plate yield strength average surplus amount of 46.5Mpa, the average tensile strength of surplus 63.5Mpa, the average rate of elongation of 7.5%, the average amount of impact energy surplus 68J; 20mm steel plate yield strength average surplus amount of 50.5Mpa, the average tensile strength of surplus 46Mpa, the average rate of elongation of 6%, the average amount of impact energy surplus 194J. The specific performance is shown in Table 4.

Table 4 The performance of AH36 test plate for reducing cost performance optimization

batch number	thickness (mm)	Yield strength (Mpa)	Tensile strength (Mpa)	Extensibility (%)	0℃ impact of collision (J)			cold bending
2YA03886	8	406	542	29.5	75.9	74.5	72.4	qualified
2YA03887	8	430	551	32	72.8	64.5	66.6	qualified
2YA03888	8	401	551	26	67.6	73.8	67.9	qualified
2YA03889	10	400	552	29	97.6	75.5	104	qualified
2YA03890	10	403	555	28	98.4	95	110	qualified
2LB04437	20	405	536	26.5	205	250	200	qualified
2LB04438	20	405	538	29.5	233	218	229	qualified
2LB04439	20	413	541	25.5	246	232	244	qualified
2LB04440	20	399	530	26.5	206	225	253	qualified

The AH36 plate is optimized by sampling, 4% nital etching after the optical microscope observation of microstructure, as shown in Figure 1 (1/4) by tissue. The matrix microstructure is ferrite and pearlite, microstructure, grain is fine, and the grain size is 10.

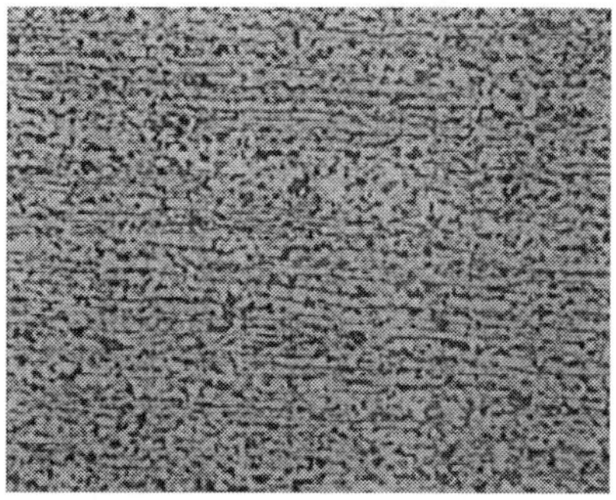

Fig. 1 The microstructure of AH36 drop cost performance optimization test steel plate

After the optimization of AH36 drop Nb, the pilot scheme was converted to the product process specification, and the following adjustment was made on the basis of the original A36 process specification: Reduction of Nb (-0.010%), Mn (0.07%), lower trim plate finishing temperature (-10~30 DEG C) and red temperature limit (-20 C). A36 ship parameters and stability control of subsequent production, the performance of all qualified, the total production of A36 reduced Nb ship a total of 3839 tons, the cumulative drop of 113 thousand and 300 yuan.

4. Conclusion

In the plate by composition in smelting fine C, Mn, down Nb, rolling on the lower plate finishing temperature and red temperature limit, process parameter control stable performance can meet the requirements, to achieve high strength ship plate AH36 performance optimization to reduce costs. According to the degree of improvement of the performance of different alloy plate, to increase the cheap alloy addition amount, reduce the precious alloy, and optimizing smelting and rolling process, which can ensure the good mechanical properties of the plate, but also can reduce the production cost of shipbuilding, improve the level of production technology and resource utilization, has an important significance for the development of circular economy the iron and steel enterprises and sustainable development.

Acknowledgments

Project supported by the science and technology research project of Hebei higher education institutions, the project number: QN2016168 and Hebei College of Industry and Technology ZY2016006

References

[1] G Wensheng, WU Jianzhong, GUO Hanjie, YU Mengxi, DUAN Shengchao, SHI Xiao.Thermodynamic analysis of precipitation of carbon nitride in DH36 steel [J]. Nonferrous Metals Science and Engineering,2016, 8(1):29.

[2] Lei Caihong, Cao Jun. Effect of heat treatment on microstructure and properties of AH36 ship building steel [J].Heat Treatment of Metals,2016,41(7): 130.

[3] WuChunhong, SunLefei, Lai Chaobin, DengZhihao, FuQingxia, YuanChuanquan. Effect of RareEarthTreatment onInclusionof E36SlabforShipbuilding [J]. Iron Steel Vanadium Titanium, 2016, 37(5): 139.

[4] D.M. Sekban, S.M. Aktarer, P.Xue, Z.Y. Ma, G. Purcek. Impact toughness of friction stir processed low carbon steel used in shipbuilding [J]. Materials Science & Engineering A 672 (2016) 40.

IOP Conf. Series: Materials Science and Engineering **207** (2017) 012050 doi:10.1088/1757-899X/207/1/012050

Researching on Rolling Technology of Q460E Plate

Yanjun Meng, Bing Wang, Yanxia Liu, Ming Chen, Zhongqi Dong*

Material Engineering Department, Hebei College of Industry and Technology, Shijiazhuang 050000, China

*Corresponding author e-mail: dongzhongqi@126.com

Abstract. The production process of 3 kinds of Q460E medium and heavy plate produced was studied by controlled cooling rolling process. Results show: that were used in two stage preheating and two stage controlled rolling. The first stage is rolling in the austenite recrystallization zone, and the temperature of the billet is 1050 to 1100 ℃, and the reduction rate is more than 10%; The second stage is rolling in austenite non recrystallization, the rolling temperature is less than or equal to 950 ℃, the finishing temperature is 860 to 790 ℃, until the temperature after the cumulative reduction ratio is more than 50%, more than 12% times the rate of deformation; The laminar cooling is used, so that the steel has good toughness

1. Introduction

Engineering machinery steel with Nb, V, Ti and other micro alloying, using micro alloy elements to form carbon nitride in the steel plate on the dissolution and precipitation, on grain refinement and precipitation strengthening effect of [1~4] Q460E, which has high strength, good toughness and weldability, because of its good performance, widely used in various engineering machinery structure parts manufacturing [5~8]. Micro alloying and controlled rolling and controlled cooling based on the production of Q460E board, become the main way of steel plate during the rolling process of engineering machinery, take controlled rolling and controlled cooling and normalizing process, ensure the steel strength and toughness.

To this end, this paper combined with the characteristics of the equipment and process of the 3500mm rolling mill production line of a certain steel plant, through the reduction of carbon equivalent, using Nb, V microalloying and controlled rolling and controlled cooling technology, to explore the preparation of medium plate Q460E steel.

2. Experimental materials and methods

The thickness of Q460E continuous casting slab is 30~80 mm, the chemical composition is shown in Table 1, and the mechanical properties are shown in Table 2. The rolling is 20, 25 and the 32 mm of the steel strip of the three specifications, the heating system design is shown in Table 3, the rolling process parameters are shown in table 2 . The samples were cut at the strip 1/4, and the mechanical properties of the samples were tested by using BDCL 20mm 80 mm test machine.

IOP Conf. Series: Materials Science and Engineering **207** (2017) 012050 doi:10.1088/1757-899X/207/1/012050

Table 1 chemical composition of test steel (%)

Steel	C	Si	Mn	P	S	Nb	V	Als	CEV
Q460E	0.14-0.18	0.20-0.40	1.45-1.65	≤0.020	≤0.010	0.020-0.035	0.060-0.080	0.015-0.035	0.39-0.45

Table 2 mechanical properties of Q460E

Thickness /mm	σ_s/MPa	σ_b/MPa	δ_5%	V		180℃ bend test mm	
				℃	J	≤16	>16
≤16	≥460	550～720	≥17	-20	34	d=2a	d=3a
16～40	≥440	550～720	≥17	-20	34	d=2a	d=3a
40～63	≥420	550～720	≥16	-40	34	d=2a	d=3a
63～80	≥400	550～720	≥16	-40	34	d=2a	d=3a

Table 3 heating system

preform body	preheating ℃	Heating section ℃	Two heating section ℃	holding zone℃	heating time/ h
cold material	≤850	1040-1100	1230-1290	1180-1240	≥2.5h
Hot material	≤850	1000-1100	1230-1290	1180-1240	≥2h

Table 4 rolling parameters

Finished thickness / mm	Temperature thickness mm	End of temperature ℃	finishing temperature℃	controlled cooling	Return temperature ℃
20	42	870 - 840	840 - 825	y	680 - 700
25	46	870 - 845	830 - 820	y	680 - 700
32	68	860 - 825	840 - 820	y	680 - 700

3. Experimental results and discussion

The mechanical properties of Q460E hot rolled specimens are shown in Table 5, the grain size is shown in Table 6 table 5 and 6 using visible, two stage and two stage preheating control rolling, rolling plate can improve the effective grain size, reach the 10 level, which can significantly improve the steel yield strength and tensile strength; toughness steel plate has no obvious worse, and has a great amount of surplus.

Table 5 mechanical properties of Q460E test

mark	Standard/ mm	Yield strength / MPa	Tensile strength / MPa	Extensibility %	-40℃V Type impact			cold bending
1#	20	485	595	20.5	141	145	140	qualified
2#	20	480	580	27	204	179	217	qualified
3#	20	485	590	28	215	205	210	qualified
4#	20	495	610	23	71.2	63.9	33.1	qualified
5#	25	475	595	26	165	164	170	qualified
6#	25	485	595	25.5	165	166	169	qualified
7#	25	480	610	23.5	184	73.4	150	qualified
8#	25	495	605	27	200	197	203	qualified
9#	25	475	605	28	175	177	180	qualified
10#	32	440	580	25	179	190	195	qualified

The trial plate at 1/4 thickness microstructure as shown in Figure 1. The thickness of steel plate 1/4 tissues of ferrite and pearlite and the grain size is 10, respectively. The volume fraction of ferrite pearlite 38%. 62%, pearlite and fine dispersed, so the strength of the steel plate is obviously improved.

Table 6 Q460E grain size

Steel grade	Inclusion rating	Metallurgical structure	Grain size
Q460D	$A_{0.5}C_{0.5}D_{0.5}$	F+P	10

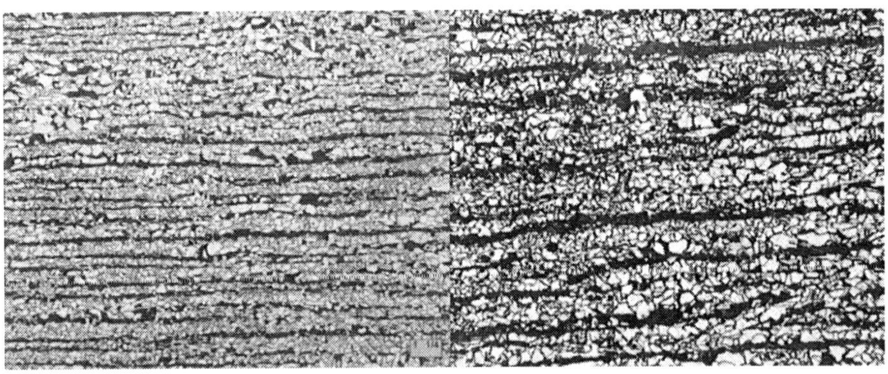

Fig. 1 Q460E metallographic structure

3.1. Temperature schedule

The yield strength, tensile strength and toughness of Q460E plate are improved, which is closely related to the system of slab heating . The slab of Nb, V mainly to large particles of carbon, nitrogen form, when heating at low temperature, which is below 900 DEG C, Nb, V carbon, nitrogen is insoluble, dissolved NB carbon, nitrogen retention in coarse particles after rolling steel plate on the tissue, almost no strengthening effect; began to dissolve, exceed 990 DEG C, therefore, 1000 degrees (or 1040 DEG C) heating temperature to ensure there is enough Nb, V solid solution in the austenite.

At the same time, the heating temperature can not be too high, which is due to the size of austenite grain size after heating will directly affect the grain size after rolling .Usually the austenite grain at 1150 DEG C size uniform, more than 1180 degrees, due to the grain boundary of carbon, nitrogen dissolution, on grain growth inhibition effect disappeared, the austenite grain began to rapidly grow, so the heating temperature should not exceed 1180 degrees. Due to lower heating temperature, is conducive to the precipitation of niobium with carbon, nitrogen, to inhibit the grain growth, and reduce the heating temperature can shorten the residence time in the high temperature zone after rolling, avoid continuous recrystallization of austenite grain in the high temperature zone of the growth, the grain size increases.

3.2. Controlled rolling

The two stage is the rolling control, some special functions of micro alloying elements can only be achieved under certain conditions based on the mechanism of the influence of body performance of material is more sensitive to the process parameters, the two phase control rolling can better play the role of alloying elements such as recrystallization. According to the rolling of niobium in austenite and niobium delay the carbide, nitride dissolution precipitation characteristics.

The austenite recrystallization rolling, rolling billet temperature is 1050 to 1100 DEG C, in the first stage using features of rolling deformation resistance of high temperature steel is low, as far as possible the use of large amount of pressure, to refine the austenite grain, in control of more than 10% pass reduction rate.

To be warm, with decreasing the rolling temperature, into the recrystallized austenite part, in this interval rolling, only part of the austenite recrystallization, resulting in austenite grain size is not uniform, and may form part of the large grain, causing serious mischcrystal. Mischcrystal structure appears, it will be difficult to eliminate, and reduce the steel toughness. To this end, the use of intermediate temperature system, when rolled to the finished steel plate thickness of 2 to 3 times when the temperature began to avoid the austenite recrystallization zone rolling.

To this end, the use of intermediate temperature system, when rolled to the finished steel plate thickness of 2 to 3 times when the temperature began to avoid the austenite recrystallization zone rolling. The type Nb bearing steel rolling, when the temperature is below 950 DEG C that has no dynamic recrystallization can. So the Q460E second stage for the rolling temperature is less than or equal to 950 DEG C, the finishing temperature is 860 ~ 790 degrees. When 12 ~ 20mm thick plate rolling, finish rolling temperature from ceiling to prevent the smooth pass too fast temperature drop, the temperature is too low, have an adverse effect on the roll surface; when the thickness was more than 20mm, the final rolling temperature in the lower limit. Until the temperature after cumulative reduction ratio is more than 50%, the deformation rate greater than 12% times; the last one according to the shape and thickness can be decreased properly, shorten dearance time to grain refinement to improve strength and toughness of steel.

3.3. Controlled cooling process

Control in the process of rolling, recrystallization through repeated deformation, can refine grain and improve the strength and toughness of the steel, but the deformation induced transformation, the Ar3 temperature increased, resulting in ferrite in high temperature precipitation, such as slow cooling after rolling is easy to make the grain growth, so to control the cooling speed. According to the appropriate system the thickness of different laminar cooling after rolling with different cooling pipe, by adjusting the number of groups, Q460E red temperature controlled at 670-700 DEG C.

Control in the process of rolling, recrystallization through repeated deformation, can refine grain and improve the strength and toughness of the steel, but the deformation induced transformation, the Ar3 temperature increased, resulting in ferrite in high temperature precipitation, such as slow cooling after rolling is easy to make the grain growth, so to control the cooling speed. According to the appropriate system the thickness of different laminar cooling after rolling with different cooling pipe, by adjusting the number of groups, Q460E red temperature controlled at 670-700 DEG C.

4. Conclusion

(1) The system of rolling process of Q460E plate for the rolling control two stage preheating and two stages, the first stage in the austenite recrystallization zone rolling billet rolling temperature is 1050 to 1100 DEG C, in control of more than 10% pass reduction rate; the second stage in austenite non

recrystallization zone rolling, rolling the temperature is less than or equal to 950 DEG C, the finishing temperature is 860 to 790 DEG C, until the temperature after the cumulative reduction ratio is more than 50%, the deformation rate greater than 12% times; the laminar cooling patterns.

(2) Two stage control of rolling and rapid cooling after rolling, the grain refinement is the fundamental reason for Q460E to obtain excellent mechanical properties

Acknowledgments
Project supported by the science and technology research project of Hebei higher education institutions, the project number: QN2016168 and Hebei College of Industry and Technology ZY2016006, ZY2016004

References

[1] Wenchao Li,Fangfang Liao,Tianhua Zhou,Harm Askes. Ductile fracture of Q460 steel: Effects of stress triaxiality and Lode angle [J]. Journal of Constructional Steel Research, 2016, 123: 1–17.
[2] Wenchao Li,Fangfang Liao,Tianhua Zhou,Harm Askes. Ductile fracture of Q460 steel: Effects of stress triaxiality and Lode angle [J]. Journal of Constructional Steel Research, 2016 : 917-921.
[3] Yansheng Du,Zhihua Chen,Ming-Xiang Xiong. Experimental behavior and design method of rectangular concrete-filled tubular columns using Q460 high-strength steel [J]. Construction and Building Materials, 2016, 125: 856–872.
[4] Wenchao Li,Luca Susmel,Harm Askes,Fangfang Liao,Tianhua Zhou. Assessing the integrity of steel structural components with stress raisers using the Theory of Critical Distances [J]. Engineering Failure Analysis,2016, 70: 73–89.
[5] CUI Zhanbin, WANG Qian, CUI Zhanhui, FAN Lei, WANG Qingfeng. Effect of self-tempering of temperature on microstructure and mechanical properties of a Q460E class high rise building [J]. Journal of Iron and Steel Research, 2014,(05): 61-63.
[6] ZHANG Qiang. Study on controlled cooling technology for Q460E heavy plate [J]. Steel Rolling, 2014,(05): 61-63.
[7] LIU Haitao, XU LIyong, LIU Donghua, WU Jianding. Welding process research of high strength steel Q460E rolled by TMCP [J]. Electric Locomotives & Mass Transit Vehicles, 2016,(01):64-67.
[8] Mao Guohua. Research and Development of Q460E Micro Alloyed Steel [D]. Xi`an University of Architecture and Technology, 2014.

IOP Conf. Series: Materials Science and Engineering **207** (2017) 012051 doi:10.1088/1757-899X/207/1/012051

Effect of isothermal annealing on Spheroidization of carbides in GCr15 steel

Xiaocai Yang [1,*], Haiqiang Xu [2], Yang Wang [2], Jinlan Ma [1]

[1]Matcrial Engineering Department, Hebei College of Industry and Technology, Shijiazhuang 050000, China
[2]School of Materials Science and Engineering, Hebei University of Science and Technology, Shijiazhuang, Hebei, China 050000, China

*Corresponding author e-mail: dongzhongqi@126.com

Abstract. The spheroidizing effect of GCr15 steel was studied and the microstructure was analyzed by SEM under different isothermal annealing. The results show that: the carbide particles become smaller and smaller which is easy to form and grow into lamellar carbide with the increase of the temperature, and when the heating temperature is increased to 800°C; The carbide particles can grow into spherical shape independently, and even some of the particles begin to grow up with the increasing of the heating time; The optimum process of GCr15 steel is heating temperature 780 °C, heating time 5h, isothermal temperature of 700 °C, isothermal time 3h.

1. Introduction

GCr15 alloy steel has less content of high hardness, good abrasion resistance, excellent mechanical properties and other characteristics have been widely used in bearing manufacturing, die production and defense industry and other fields of [1], but in use, often due to lack of toughness of steel and reduce the service life of the parts, to get high performance needs this heat treatment [2]. In order to make the bearing steel have high strength and toughness [3], it is necessary to obtain the tempered martensite and its microstructure with fine carbide particles. Therefore, the number of control, carbide size, morphology and distribution of bearing steel is the key to high quality, high performance, multi variety and direction for the austenitizing temperature based on different isothermal annealing process in the holding time, the process parameters of isothermal temperature and time on the number of spheroidal carbide particles and to find the optimal distribution, for the industrial production of spheroidization process and provide a theoretical basis.

2. Test materials and methods

2.1. Experimental material

The materials used in the test were taken from the GCr15 steel of a factory, and the main chemical composition was shown in Table 1.

Table 1 chemical composition of GCr15 steel

alloying element	C	Si	Mn	Cr	S	P
content %	0.9	0.17	0.36	1.45	0.01	0.011

2.2. Experimental method

The experiment was carried out in a box type resistance furnace, the sample was processed into a small cylinder of 25x10, and the experiments were carried out at different heating temperatures. The process parameters of sample 1-9, are shown in Table 2, of which 100°C /h heating rate rose to austenitizing temperature, cooling speed to 50 °C /h to isothermal temperature, and then to 50 °C /h to 650 °C out of furnace cooling air cooling. By using TESAN scanning electron microscope and energy spectrum with microstructure analysis.

Table 2 heat treatment process parameters

Sample number	heating temperature /℃	warming-up time /h	Isothermal temperature /℃	isothermal /h
1#	760	4	700	3
2#	780	4	700	3
3#	800	4	700	3
4#	780	5	700	3
5#	780	6	700	3
6#	780	5	680	3
7#	780	5	720	3
8#	780	5	700	2
9#	780	5	700	4

3. Test results and analysis

3.1. Effect of different heating temperature on Microstructure of carbide

Figure 1 is observed with scanning electron microscope in the No. 1, No. 2, the microstructure of three different austenitizing temperature 3 GCr15 samples.

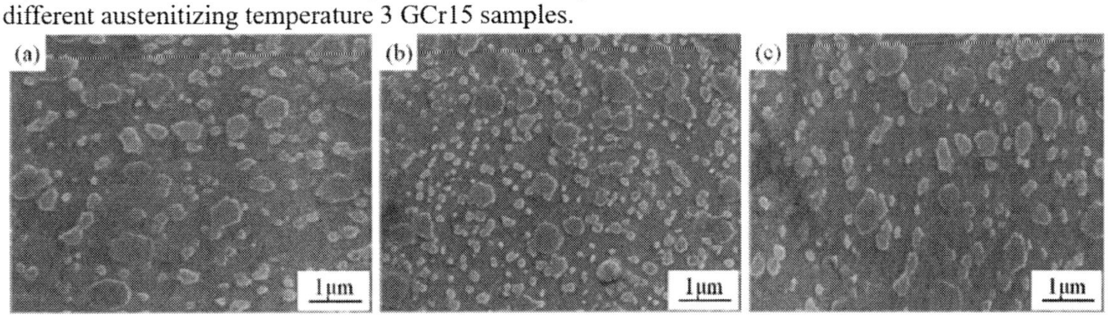

Fig. 1 the microstructure of sample (a)1# (b)2# (c)3#

It can be seen from the figure there are different spheroidizing effect at different temperatures, different morphologies of carbides for three in proportion.(a) and (c) carbide particles in large, there is a small part of a small particle diameter, but mainly to large diameter particles. On the contrary (b) diagram mainly to small granular carbides, only a small part of the large diameter carbide. In the three picture, the carbides have different degrees of spheroidization, some of the large particles can reach 1 m in diameter, the smallest particle diameter of up to 0.2 M. (c) phase (b), compared to some of the growth of carbide particles, the effect is not good. It can be seen from the (b) diagram that a large number of carbide particles are small and dispersed, and the spheroidization effect is obviously better than (a) and (c).

3.2. Effect of different isothermal time on Microstructure of carbide

Figure 2 is observed with scanning electron microscope in the No. 2, No. 4, the microstructure of three different heating time 5 GCr15 samples.

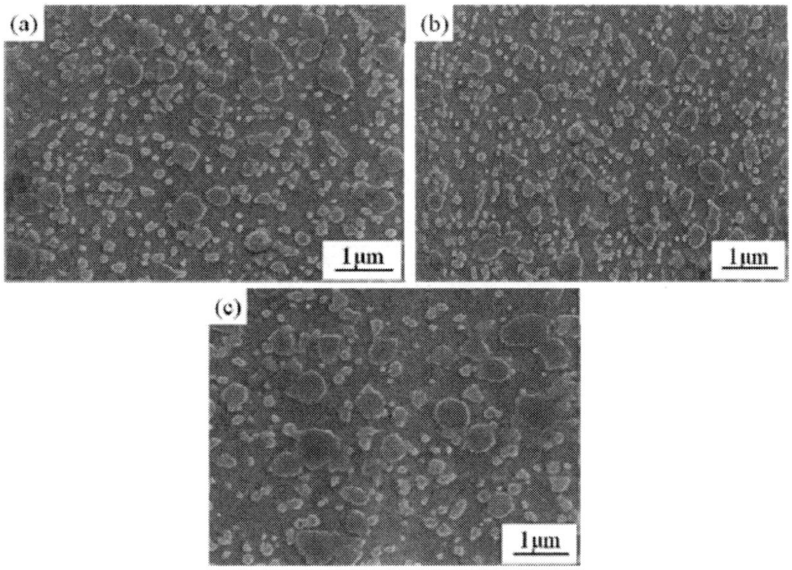

Fig. 2 the microstructure of samples (a)2# (b)4# (c)5#

It can be seen from the figure with the heating time. The carbide morphology, size and number are significantly changed. (a) the austenitizing time is short, there are some undissolved carbide particles in the matrix to form a carbon rich region, and in the subsequent isothermal process as a nucleation center to form a uniform spherical structure. (b) in the austenitizing time, carbide particles dissolve without excessive nucleation centers in the isothermal process followed, only through the carbon depleted zone of ferrite and cementite separate spherical growth, such nucleation speed naturally ratio (a) formation of finer particles in the figure. With the extension of the heating time the carbide is dissolved and some small particles even disappear. When time is too long, the carbon concentration in austenite becomes uniform, and then it becomes a flaky pearlite .

3.3. Effect of different isothermal temperature on Microstructure of carbide

Figure 3 is observed with scanning electron microscope in the No. 6, No. 4, three different isothermal temperature on Microstructure of GCr15 specimen No. 7.

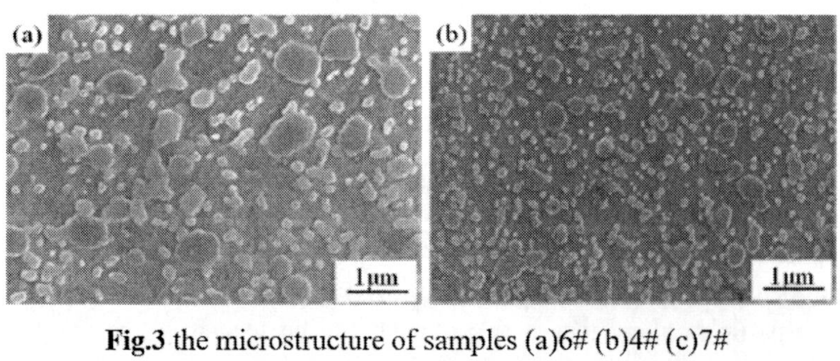

Fig.3 the microstructure of samples (a)6# (b)4# (c)7#

Three figure can be seen in a large number of granular carbides, which is due to the carbide particles in this stage is a spontaneous process, the flake pearlite transformation high surface can the pelletized to reduce the energy. (b) and (a) than the isothermal temperature, carbon atom diffusion and nucleation driven, showing a more uniform dispersion of carbide particles. With the increase of temperature, the carbide particles dissolved residue, non -spontaneous nucleation decreases, in rich and poor carbon zones at the same time from nucleation. You can see a graph of some particles began to grow, the formation of lamellar pearlite.

3.4. Effect of different isothermal temperature and time on Microstructure of carbide

Figure 4 is the microstructure of the samples with different isothermal temperatures observed by SEM in No. 4, No. 9, GCr15. Through the experiment of heating temperature, heating time and isothermal temperature, it can be seen from the (b) that most of the large granular carbides have been dissolved and the size is small. This is because the activation energy and the driving force are suitable, the precipitated carbon atom can be more spontaneous nucleation, the non spontaneous nucleation of carbide particles is also small, so the effect is good. As time went on, some small particles of carbide began to approach the large particles and were swallowed up by large particles. This was because the activation energy and the driving force were too high, and the carbides began to melt and merge.

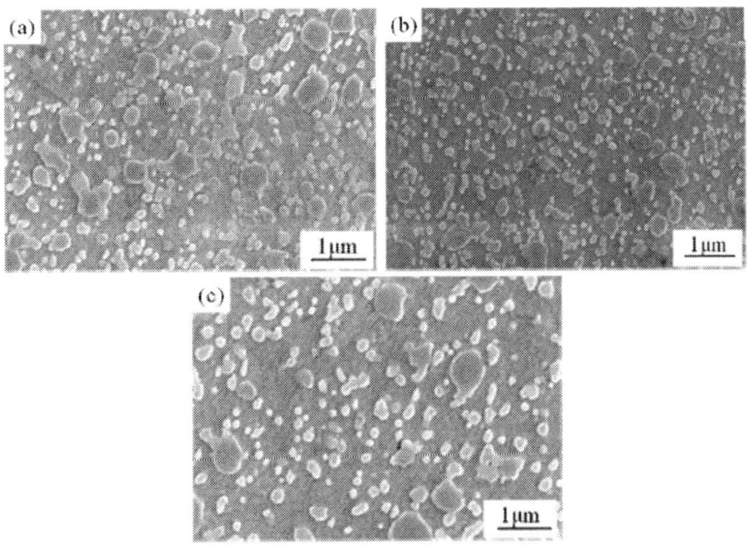

Fig.4 the microstructure of samples (a)8# (b)4# (c)9#

3.5. Effect of heating temperature and heating time on the particle size of carbides

As can be seen from Figure 5, the particle size of the large granular carbides in GCr15 decreases gradually with the increase of annealing time. When the 5h is heated, some of the carbide is dissolved, the original large irregular carbides are dissolved, and the remaining particles become smooth. When heating 6h after some carbide particles began to melt back up, the ball is not the best effect. The heating temperature is too high, the spread of nuclear carbon spacing is too large it is not good to make the ball effect is not good, the excess carbides aggregate to grow at the grain boundaries. When the heating temperature is too low, the diffusion coefficient of carbon is small, and the spheroidizing effect is not good, so the heating temperature is 720 °C, and the heating time is 5h.

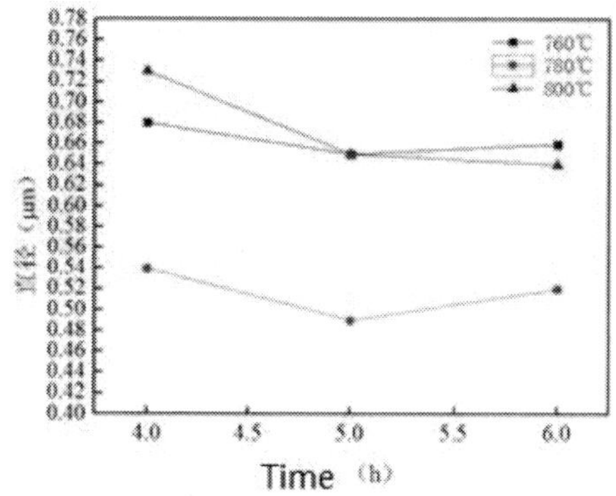

Fig.5 Effect of annealing time on the particle size of carbides

3.6. Effects of isothermal temperature and isothermal time on the particle size of carbides

As can be seen from Figure 6 in the temperature is 720°C with the holding time, the grain diameter GCr15 carbides decreases gradually, this is because with the increase of isothermal time, carbon precipitation gradually increased, the activation energy and the nucleation driving force increases, the carbide grain diameter decreases. And 700 °C, 680 °C degrees of carbide grains is gradually decreased, then the grain began to grow up, this is because some carbide particles began to melt and merge and grow up. The isothermal time is longer, the stored energy is higher, the size will increase. So the best temperature is 700 °C, the isothermal time was 3h.

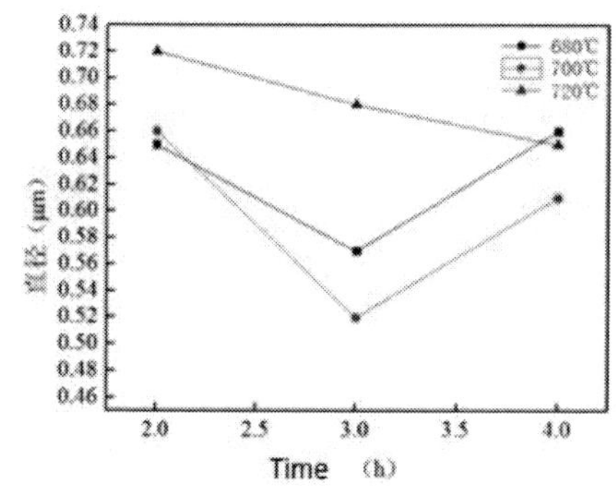

Fig.6 Effect of holding temperature and holding time on the particle size of carbides

4. Conclusion

(1) The isothermal temperature at 760 °C -800 °C change, with the increase of temperature of carbide particles become fine and dispersed when the heating temperature rises to 800 °C when the uniform degree of austenite is too high, easy to nucleate and grow into flake carbide spheroidizing effect is not good.

(2) When the isothermal heating time is extended from 4H to 5h, the carbide particles can grow into spherical shape independently. When the time increases to 6h, the particle size of the carbide is not changed, and even some of the particles begin to melt

(3) The best process of GCr15 steel is heating temperature of 780 °C, heating time 5h, isothermal temperature of 700 °C, isothermal time 3h.

Acknowledgments

Found project: Project supported by the Hebei provincial key research and development program (16211032), Hebei Provincial Higher Education Fund for science and Technology (QN2017138) and the project of Hebei Industrial University (ZY2016006)

References

[1] Dan Zheng, Zhen-bing Cai, Ming-xue Shen, Zheng-yang Li, Min-hao Zhu. Investigation of the tribology behaviour of the graphene nanosheets as oil additives on textured alloy cast iron surface [J]. Applied Surface Science, 2016.

[2] Yun-bo ZHONG, LI Qiang, Yi-peng FANG, Huai WANG, Ming-hu PENG, Li-cheng DONG, Tian-xiang ZHENG, Zuo-sheng LEI, Wei-li REN, Zhong-ming REN. Effect of Transverse Static Magnetic Field on Microstructure and Properties of GCr15 Bearing Steel in Electroslag Continuous Casting process [J]. Materials Science & Engineering A, 2016.

[3] Xiaohua Sha, Wen Yue, Yihui Zhao, Fang Lin, Chengbiao Wang. Effect of sliding mating materials on vacuum tribological behaviors of sintered polycrystalline diamond [J]. International Journal of Refractory Metals and Hard Materials, 2016, 54.

Effect of auxiliary friction on nickel electroforming deposits

Yong Zhang[1, 2, *] and Jian Zhang[1]

[1]Department of Mechanical Engineering, Shazhou Institute of Technology, Zhangjiagang, China
[2]College of Mechanical and Electrical Engineering, Nanjing University of Aeronautics and Astronautics, Nanjing, China

[*]Corresponding author e-mail: zy6982@163.com

Abstract. The nickel deposits were obtained with friction assisted, and the porosity, microstructure, texture and tensile properties of nickel deposits under different parameters were studied. The experiment shows that when the cathode velocity reaches a certain value, non-porous nickel deposits can be obtained under the friction of the microbeads. In addition, the microstructure characteristics of nickel deposits under different parameters were tested and analyzed, and the mechanical properties of nickel deposits before and after annealing were tested.

1. Introduction

Now nickel electroforming has been employed in fabricating precision dies, rocket engine thrust chamber, shaped charge liner, and other special structure parts[1-4].The microstructure and mechanical properties of electroformed nickel are related to electroforming process. In the nickel electroforming, sulfamate electroforming solution is usually used because a higher current density can be used, and the internal stress of the nickel electroforming layer is smaller. However, there are some drawbacks in the traditional electroforming process, such as pinholes, pits and nodules. As deposits grow progressively thicker, the defects on the surface will become worse. The researchers used additives in the solution, pulse power, vibration, ultrasound and other ways to improve the electrodeposition process. But these methods can not completely solve the problem [5-7].

In this paper, electroforming with abrasion-assisted have been applied to the nickel electroforming process to overcome the drawbacks. The nickel deposits were electroformed with attrition assisted and their structures and mechanical properties were examined.

2. Experimental

Figure 1 illustrates the schematic diagram of the experimental apparatus. The system consists of power supply, cathode unit, anode basket, heater with temperature control equipment and an electrolyte circulating system. The electrolyte contained $Ni(NH_2SO_3)_2 \cdot 6H_2O$ 400g/l, H_3BO_3 30g/l, $NiCl_2$ 15g/l. Nickel deposition was controlled at pH=4 in the electrolyte.

All solutions were prepared by freshly distilled water. No sulfur-containing Nickel beads were used as the anode. A stainless cylinder mandrel was used as the cathode, and the deposit area was Ø48 mm×80 mm. Ceramic beads with diameter of Ø1-2mm were chosen as the hard particles.

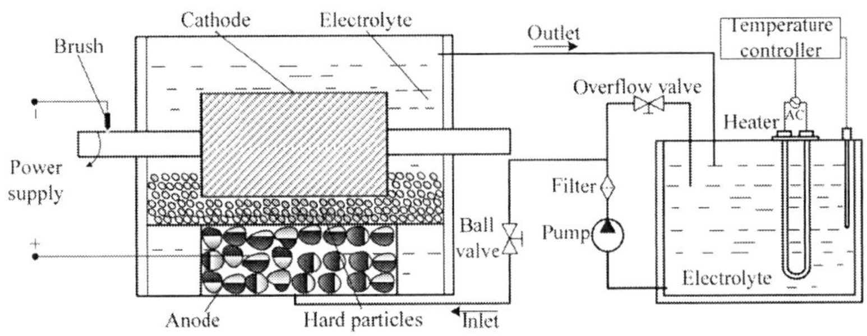

Figure 1. Schematic diagram of the experimental apparatus.

The cathode mandrel was connected to a motor, the speed of the cathode was varied from 0.0025 to 0.08m/s at a fixed current density of 4 A/dm^2 and a fixed temperature of 43℃.

Porosity: the porosity of electroformed nickel is measured by filter paper. Formula: potassium ferricyanide 1g/l, sodium chloride 10g/l, agar 10g/l, T=82~94℃, time 10min. The wetting test solution against filter paper on the sample surface, and after the required time, observe the number of colored spots. In about 30μm thick electroformed layer on the detection, each sample three times, respectively, according to the following formula to calculate the porosity, whichever is the average.

$$\text{Porosity} = \text{pore number (n) / measured area (s)} \tag{1}$$

When the spot diameter is less than 1mm, a spot is calculated by one; when the spot diameter is 1 ~ 3mm, a spot is calculated by three, when the pore diameter is 3~5mm, a spot is calculated by ten.

The structure of the samples was measured by a BRUKER D8 ADVANCE X-ray diffraction (XRD) using Cu Kα radiation. Preferential orientation of the samples was characterized by the orientation index (M) for each plane (hkl), and the formula is shown in Formula (2).

$$TC_{(hkl)} = \frac{I_{(hkl)}/I_{0(hkl)}}{\sum_{i=1}^{n} I_{(hkl)}/I_{0(hkl)}} \times 100\% \tag{2}$$

I (hkl) represents the X-ray diffraction relative intensity of the crystal plane of the deposited, I0(hkl) represents the X-ray diffraction relative intensity of the standard powder(hkl) crystal plane, and n is the number of diffraction peaks.

3. Results and discussions

3.1. Porosity

In the traditional electroforming technology of electroforming, the surface of nickel deposit is dark gray and covered with pitting. The friction assisted electroforming can solve the problem of pinhole formation in nickel deposit, and can adopt high current density in electroforming process. Figure 2 is the graph of the number of porosities at different velocities of cathode, while the current density of cathode is 4A/dm^2. The surface of the nickel deposits is smooth. It can be seen from the curve that the porosities decrease with the increase of velocity of cathode. When the velocity is above 0.0025m/s, the porosity of the nickel deposits is zero.

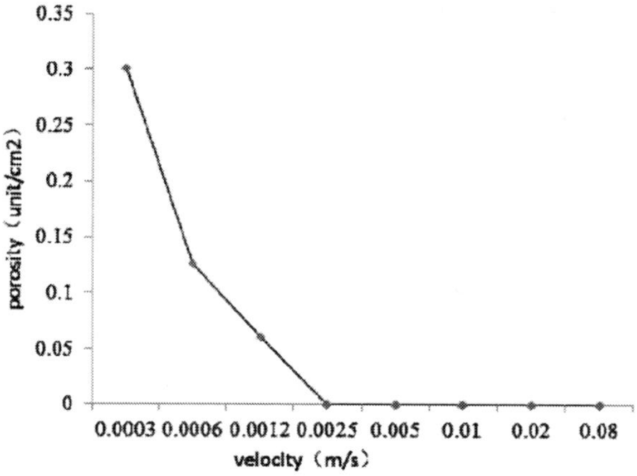

Figure 2. Porosity quantity of nickel deposits under different velocities

3.2. TEM

Figure 3 shows the TEM micrograph of four nickel deposits. As can be seen from the graph, the grain size of nickel deposits is gradually decreased with the increasing of linear velocity of cathode. As shown in Figure 3, when the linear velocity of cathode is 0.0025m/s, the grain size of nickel deposit is coarse, and when the linear velocity of cathode increased to 0.08m/s, the grain size of nickel deposit is significantly decreased. The above results also indicate that hard particles have a role in grain refinement. When speeding up the velocity of cathode, the polishing effect of hard particles can be strengthened.

Figure 3. TEM morphology of nickel deposits under different velocities
(a)0.0025m/s; (b)0.01m/s; (c)0.02m/s; (d)0.08m/s

3.3. Texture coefficient

According to the formula (2), the texture coefficient of the nickel deposit prepared by the traditional electroforming technique is shown in Table 1.

Table 1. The TC (hkl) of nickel deposit prepared by traditional electroforming technology

crystal surface	(111)	(200)	(220)
$TC_{(hkl)}$ (%)	4.79	89.21	6

According to the formula (2), the change trend of texture coefficient of the nickel deposits is obtained at different cathode velocities by using attrition assisted electroforming is shown in Figure 4.

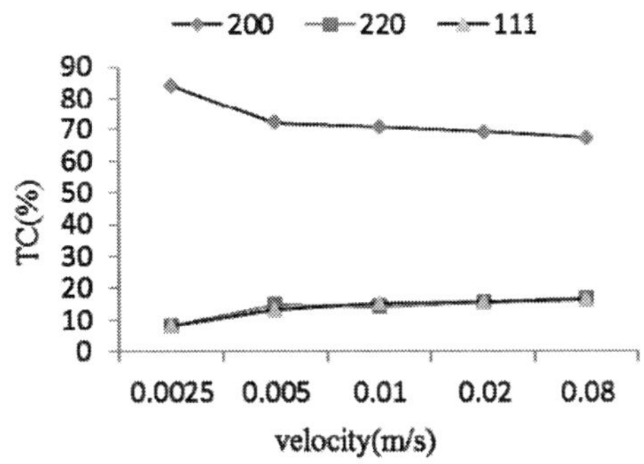

Figure 4. Texture coefficient of nickel deposits with cathode velocities

When using attrition assisted electroforming technology, the change of cathode velocity will affect the texture coefficient of nickel deposit. As shown in Figure 4, with the increase of the cathode velocity, the texture coefficient of (200) crystal surface decreases, while the texture coefficients of (111) and (220) crystal surface are increasing. The above results show that the attrition of the surface layer during the electroforming process can affect the growth of the crystal, and the relative growth rate of each surface is reduced.

3.4. Mechanical property

Electroforming products in some occasions often need to use after annealing, so it is necessary to test and analyze the performance of nickel deposit after annealing. The annealing process used in this paper is as follows: annealing in high temperature vacuum furnace, the temperature rises to 280℃ in 10 minutes, then keep 2 hours at constant temperature, and finally removed after the furnace cooling.

It can be seen from Figure 5 and Figure 6 that the tensile strength of the electroformed nickel increases with the increase of the cathode velocity and the elongation decreases gradually. The tensile strength and the elongation of nickel before and after annealing are the same, but the tensile strength of nickel after annealing has a certain degree of reduction, while the elongation is increased. When the velocity is 0.0025m/s, the tensile strength of nickel decreases from 525MPa before annealing to 452MPa after annealing, and the elongation increases from 24% to 36%, and When the velocity is 0.08m/s, the tensile strength of nickel decreases from 1020MPa before annealing to 841MPa after annealing, and the elongation increases from 4% to 10%.

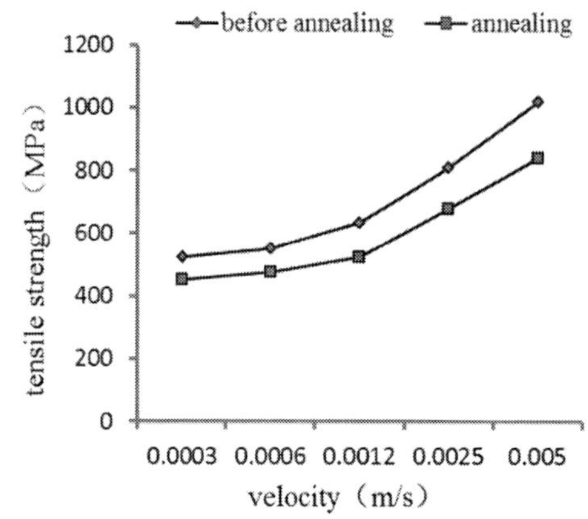

Figure 5. Effect of velocity on tensile strength of electroformed nickel

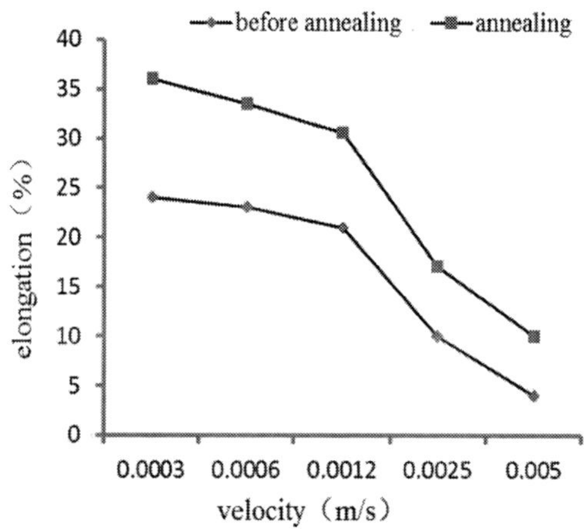

Figure 6. Effect of velocity on elongation of electroformed nickel

4. Conclusion

The nickel deposits were electroformed with attrition assisted and their microstructure were examined. The experimental results show that the attrition of particles could obviously have a role in grain refinement, and when speeding up the velocity of cathode, the grain refinement is enhanced.

It was found that the attrition of hard particles affect the growth of crystallites and their relative speed of growth, which leads to the change of surface diffraction density and orientation index. And it was also found that when the velocity of the cathode was increased, the tensile strength of the electroformed nickel was increased and the elongation was decreased.

Acknowledgments

This work was financially supported by the Jiangsu Planned Projects for Postdoctoral Research Funds, China (Grant No. 1402032C), and the Natural Science Foundation of the Higher Education Institutions of Jiangsu Province, China (Grant No. 16KJB460031).

References

[1] S Semboshi, Y Sakamoto, H Inoue, Electroforming of Oxide-nanoparticle-reinforced Copper-matrix Composite, Journal of Materials Research, 2015, 30(4): 521-527.

[2] M Srivastava, V E Selvi, V K W Grips, Corrosion resistance and mircostrure of electrodeposited nickel-cobalt alloy coatings, Surface & Coatings Technology, 2006, 201(6):

3051-3060
[3] J A Mcgeough, M C Leu, K P Rajurkar, AKMD Silva, Q Liu. Electroforming process and application to micro/macro manufacturing, Annals of the CIRP, 2001, 50(2): 499-514
[4] Xuelei Li, Zengwei Zhu, Yong Zhang, Dong Zhu, Di Zhu, Experimental Research on Electroforming of Complex Parts with Thin Wall, Chinese Journal of Aeronautics, 2010, 31(10): 2068-2074
[5] Mingyong Zhou, Bingyan Jiang, lu Zhang, Injection molding nanostructures by template method and their hydrophobic properties, Optics and Precision Engineering, 2015.23(1): 165-173
[6] M Srivastava, J N Balaraju, B Ravisankar, Development and High Temperature Property Evaluation of Ni-Co-Cr-Al Composite Electroforms, Journal of Materials Engineering and Performance,2015,24(5): 1937-1944
[7] Hezhong Pei, Pan Huang, Qingnan Shi, Influences of Additive and Current Density on the Microstructure of the Electroforming Ni-Co Alloy, Journal of Materials Engineering, 2013, 6:18-24.

IOP Conf. Series: Materials Science and Engineering **207** (2017) 012053 doi:10.1088/1757-899X/207/1/012053

Microstructure and mechanical properties of 980MPa grade Fe-Mn-Al-C lightweight steel

Yinghua Jiang [1, 2, *], **Chunqian Xie** [1, 2]

[1]Shougang Research Institute of Technology, Beijing 100043, China
[2]Beijing key Laboratory of Green Recyclable Process for Iron & steel Production Technology, Beijing 100043, China

*Corresponding author e-mail: yinghuajiang@163.com

Abstract. Fe-Mn-Al-C lightweight high strength steel, adding more Mn, Al and C elements into conventional AHSS, shows amazing mechanical properties, corrosion resistant and weight reduction than conventional AHSS. The mechanical properties and microstructure of Fe-15Mn-6.8Al-0.9C-0.2Ti steel after annealing process were investigated. The results show that the microstructures consisted of secondary phases TiC precipitate and ferrite in the austenite matrix. The tensile strength and elongation of the steel are 985MPa and 36%, respectively. The density is 6.86g/cm3. Continuous strain hardening behavior provides Fe-Mn-Al-C lightweight steel with perfect combination of strength and ductility.

1. Introduction

Recently, much research on the reduction in greenhouse gases and the effective control of global warming has been conducted. In particular, the reduction in vehicle's weight has been devoted to improve fuel efficiency and to reduce vehicle's exhaust emission, which can result in advantages such as eco-friendliness and economic feasibility in automotive steel industries [1-2]. In addition to lightweight needs, automotive steels require a combination of strength and ductility for forming complex shapes as well as sustaining automotive structures [3-4].

Currently, the method of increasing the strength and elongation of steels is to lower the density by adding light element Al to conventional AHSS is mentioned [5-6]. Concepts of Fe-Al-Mn-C-based lightweight steels are fairly simple, but primary metallurgical issues are complicated [7-8]. This is because the lightweight steels can have ferritic, austenitic, or even multiphase structure, depending on contents of primary alloying elements of C, Mn, or Al, which complicatedly changes deformation mechanisms as well [9-10].

In the current study, the microstructural and mechanical properties of Fe–Mn–Al–C steels were investigated. Deformation mechanisms were also investigated in relation with annealed microstructures, and mechanical properties were compared with those of conventional AHSS.

2. Experimental procedure

Alloying elements used for lightweight steels mainly include Mn, Al and C because they decrease the specific weight, due to substitutional or interstitial atoms work to reduce the weight. Both Mn and C are well-known austenite formers. The high C-high Mn grades needed to expand the austenite phase region to improve the ductility. Although Fe–Al-Mn-C steels have attracted interest as lightweight steels, the κ-carbide with perovskite structures occurs with increasing Al content, which causes brittleness of a Fe–Al-Mn-C steel with a high Al content [11-12]. So the formation of k-carbide was inhibited by adding precipitation hardening element Ti. The lightweight steel used in this study was fabricated by a vacuum induction melting method, and its nominal composition is Fe-0.9C-15Mn-6.8Al-0.2Ti (wt. %). After thick plates of 60 mm in thickness were homogenized at 1150 °C for 2 hour, they were hot-rolled 950 °C. They were then cooled in a furnace from 550 °C after holding at this temperature for 1 hour in order to simulate a coiling procedure. The hot-rolled steel sheets of 3 mm in thickness were rolled at room temperature to make 1-mm-thick steel sheets. The sheets were annealed at 830 °C for 500s in a continuous annealing simulator. The density of the studied steel was calculated using JMatPro and then compared with the measurements obtained by Sartoius BSA2245 electronic balance. The mechanical properties of the steels were examined using tensile test machine. Microstructure observations were performed by light optical (LOM) and electron back-scatter diffraction (EBSD). Structural characterization was carried out by X-ray diffraction (XRD) with Cu K radiation and transmission electron microscopy (TEM). The chemical composition was determined by energy-dispersive X-ray (TEM-EDS).

3. Results and discussions

The prediction value obtained using JMatPro was 6.93g/cm3 (Figure 1). It is worth mentioning that the calculated and experimental values show a very good correlation. The density of the lightweight steel was measured to be 6.86g/cm3 by an electronic balance which shows an apparent reduction of 12% in comparison to conventional steels.

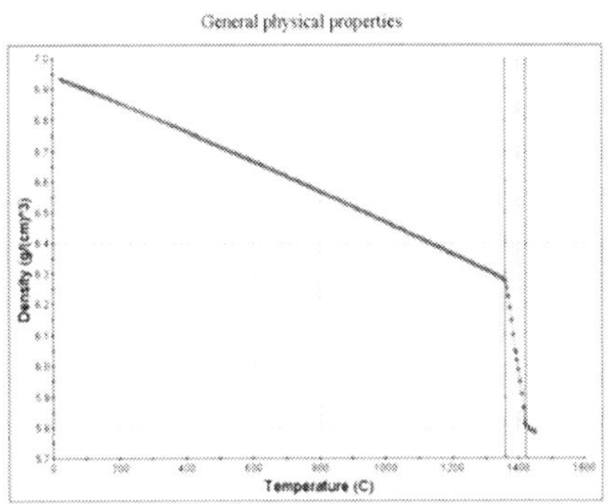

Figure 1. Density vs. temperature predicted for Fe-15Mn-6.8Al-0.9C-0.2Ti lightweight steel by JMatPro software.

Figure 2a shows optical micrograph of the annealed lightweight steel. This microstructure (Figure 2a) contains a small percentage of the ferrite (white-phase) in the austenite matrix (gray-phase). X-ray diffraction pattern of the annealed lightweight steel is provided in Figure 2b. Peaks of austenite and ferrite are observed, there is no peak of κ-carbide, which may be related to Ti addition.

Figure 3 shows the stress-strain curve and strain hardening index (n)-strain curve, which are indicated by black, blue and red line, respectively, for the DP, TRIP and lightweight steels. In the Figure 3a the lightweight steel displays much higher total elongation (36%) and yield strength (700MPa) than DP and TRIP steels. Figure 3b shows the n values of DP and TRIP steels increase at small strains, reach a maximum value. The lightweight steel exhibits much larger strain hardening capability over the whole uniform plastic deformation in comparison to DP and TRIP steels. The n of lightweight steel persists at higher strains where the strain hardening of DP and TRIP steel begins to diminish, which is closely related to the strain-induced twin deformation during tensile deformation.

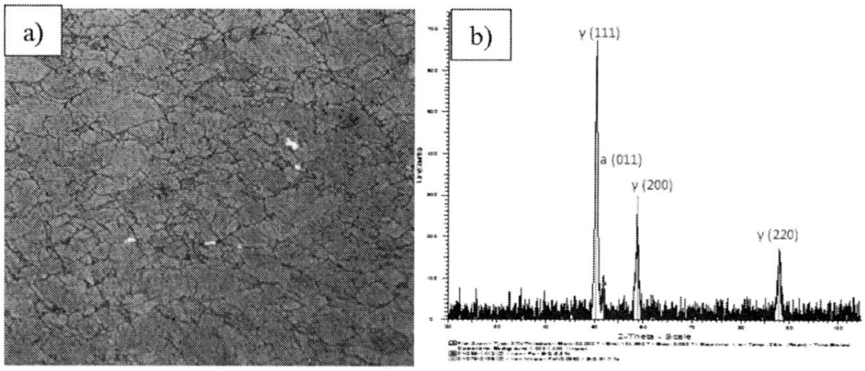

Figure 2. a) Optical micrograph and b) XRD profile of the annealed lightweight steel.

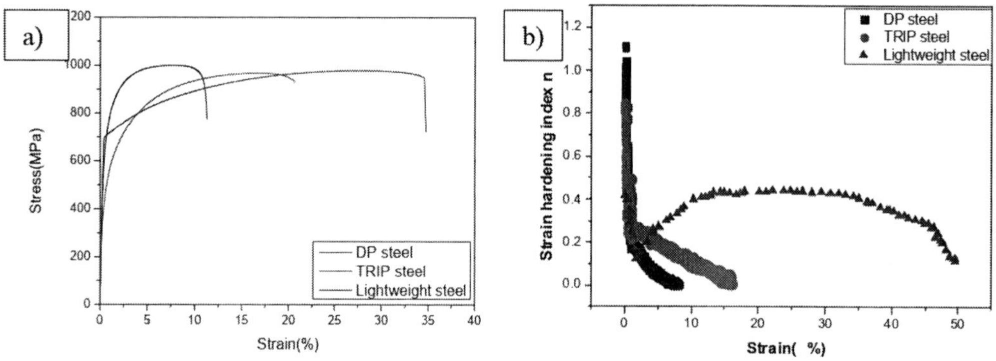

Figure 3. (a) stress–strain curves and (b) strain-hardening index (n)-strain curves, which are indicated by black, blue and red line, respectively, for DP, TRIP and lightweight steel.

TEM micrograph of the lightweight steel after deformation (Figure 4) shows the thin and straight deformation twins with a thickness of a few ten nanometers, which are introduced by tensile deformation. Figure 5 shows EBSD IQ map of lightweight steels as a function of strain. The volume fraction of deformation twins increases with strain progressively. Therefore, the continuously increased volume fraction of deformation twins is the dominant continuous strain hardening mechanism of lightweight steel at higher strain levels.

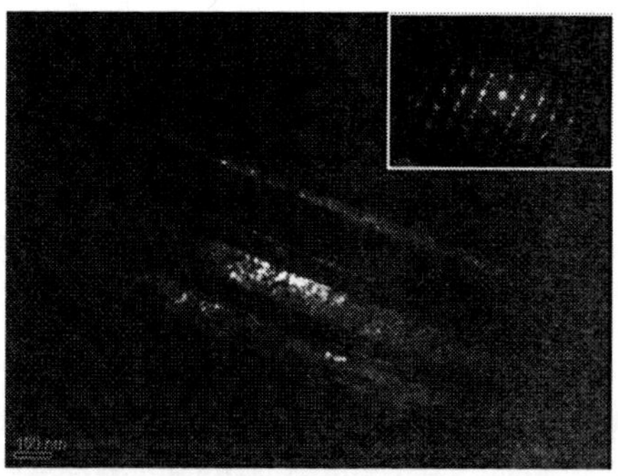

Figure 4. TEM micrograph of tensile deformed lightweight steel.

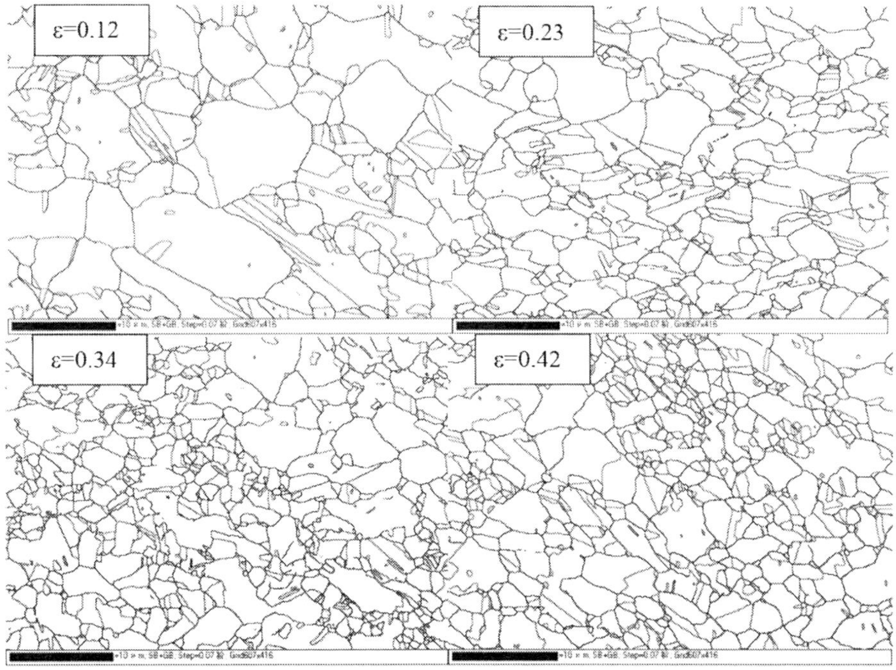

Figure 5. EBSD IQ maps of the lightweight steel as a function of strain

The TEM micrograph of lightweight steel is shown in Figure 6. Figure 6 shows well dispersed particles of a few ten nanometres, which was identified as TiC precipitates due to their chemical composition. This result shows a good agreement with the previous observations that the addition of Ti caused the TiC precipitates to prevent κ-carbide precipitates. And the high YS of lightweight steel is probably owing to precipitation hardening of TiC particles.

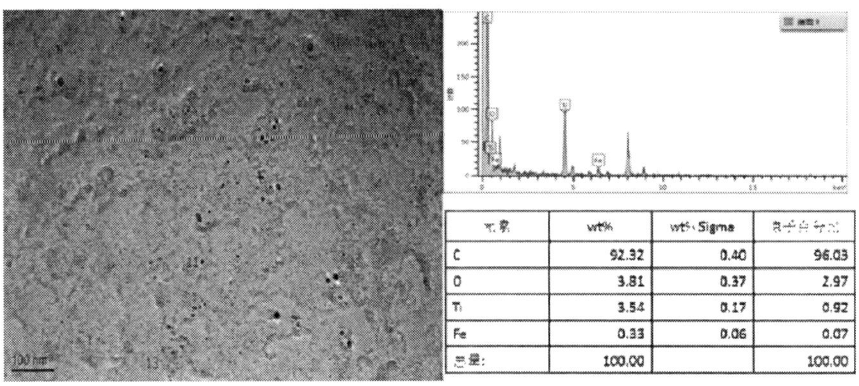

[illegible]	wt%	wt% Sigma	[illegible]
C	92.32	0.40	96.03
O	3.81	0.37	2.97
Ti	3.54	0.17	0.92
Fe	0.33	0.06	0.07
[illegible]	100.00		100.00

Figure 6. TEM micrograph of precipitates in lightweight steel.

4. Conclusion

Microstructure and mechanical properties of Fe-0.9C-15Mn-6.8Al-0.2Ti lightweight steel were investigated in the present study. The lightweight steel showed a small percentage of the ferrite in the austenite matrix with dispersed TiC particles. The elongation and yield strength of the lightweight steel were higher than those of the DP and TRIP steels. And the lightweight steel displays the continuous strain hardening behavior, which was mainly due to the strain induced deformation twins. The suppressed κ-carbide in the lightweight steel was probably attributed to TiC particles.

IOP Conf. Series: Materials Science and Engineering **207** (2017) 012053 doi:10.1088/1757-899X/207/1/012053

References

[1] E. R. H. Fuchs, F. R. Field, R. Roth, and R. E. Kirchain, Strategic materials selection in the automobile body: Economic opportunities for polymer composite design, Composites Science and Technology, 68(2008) 1989-2002.

[2] A. T. Mayyas, A. Qattawi, A. R. Mayyas, and M. Omar, Quantifiable measures of sustainability: a case study of materials selection for ecolightweight auto-bodies, Journal of Cleaner Production, 40(2013) 177-189.

[3] D.K. Matlock, J.G. Speer, E. Moor, and P.J. Gibbs, Recent developments in advanced high strength sheet steels for automotive applications: an overview. JESTECH, 15(2012) 1-12.

[4] R. Kuziak, R. Kawalla, and S. Waengler, Advanced high strength steels for automotive industry. Archives of Civil and Mechanical Engineering, 8(2008)103-117.

[5] Sang-Heon Kim, Hansoo Kim, Nack J. Kim, Brittle intermetallic compound makes ultrastrong low-density steel with large ductility, Nature, 518(2015) 77-82.

[6] Suh, D. W. & Kim, N. J, Low-density steels,Scr. Mater,68 (2013) 337–338.

[7] Kim H, Suh D W, Kim N J, Fe–Al–Mn–C lightweight structural alloys: a review on the microstructures and mechanical properties, Science and Technology of Advanced Materials, 14 (2013).

[8] Sutou, Y., Kamiya, N., Umino, R., Ohnuma, L. & Ishida, K. High-strength Fe-20MnAl-C-based alloys with low density, ISIJ Int.,50(2010) 893–899.

[9] Frommeyer, G. & Bru ¨x, U. Microstructures and mechanical properties of highstrength Fe-Mn-Al-C light-weight TRIPLEX steels, Steel Res. Int., 77 (2006) 627–633.

[10] Jae-Bok Seol, M.J. Yao, P.-P. Choi, D. Raabe and Chan-Gyung Park, Nano-sized carbides in Low-Density Fe-Mn-Al-C alloys: Low-Mn Ferritic and High-Mn Austenitic Steels, The 7th Korea-China Joint Symposium on Advanced Steel Technology, 2015.

[11] RADHAKANTA RANA, CHRIS LAHAYE, and RANJIT KUMAR RAY, Overview of Lightweight Ferrous Materials: Strategies and Promises, The Minerals, Metals & Materials Society, 66(2014) 1734-1746.

[12] M. WITKOWSKA, A. ZIELIŃSKA-LIPIEC, J. KOWALSKA and W. RATUSZEK, Microstructural changes in a high manganese austenitic Fe-Mn-Al-C steel, Archives of Metallurgy and Materials, 59(2014) 971-975.

Effect of die's annual width on the extrudate swell of plastic tube

Zhong Ren[1, 2*] and Xingyuan Huang[2]

[1]Key Laboratory of Optic-Electronic and Communication, Jiangxi Science and Technology Normal University, Nanchang, 330038, China

[2]School of Mechanical and Electrical Engineering, Nanchang University, Nanchang, 330031, China

*Corresponding author e-mail: renzhong0921@163.com

Abstract. To ascertain the effect of annual width on the extradite swell of plastic tube, the 3D extrusion forming of plastic tube was numerically studied. The geometric models and finite element meshes with different annual widths were established. Under the same material properties, boundary conditions, and numerical methods, the extradite swell situations of plastic tube (i.e., inner radius, outer radius, and wall thickness) were obtained. Numerical results show that the extradite swell ratio of plastic tube decreases with the increasing of the annual width.

1. Introduction

Extrusion forming is one of most important methods of manufacturing the plastic tube. The plastic tube can be widely used in some fields, such as, architecture engineering, medical diagnosis, optical communication, etc. The extrusion forming of plastic tube is achieved by using the extruder, cooling device, and traction device. Moreover, the extrusion die of plastic tube is also an important component. The structure of the extrusion die directly impacts the quality of the extruded plastic tube. As we know, the polymer melt has the strong viscoelastic characteristics, which will generate the serious extrusion swell problem of the extruded plastic tube [1, 2], especially for the higher flow velocity. The extradite swell of plastic tube will bring the more difficulty to the adjusting of the extrusion process. The extradite swell of plastic tube is dependent to many factors, such as, the physical properties of polymer melt, screw speed of extruder, temperature, structure of the extrusion die, etc. [3]. In this paper, the effect of the structure of extrusion die on the extradite swell of plastic tube was studied. For the extrusion die of plastic tube, there is the mandrel in the channel of die, which lead to form the annual channel. To ascertain the effect of die's annual width on the extrusion swell of plastic tube, the numerical simulation of polymer melt flow in the annual channel of die was investigated. The velocity distributions, pressure distribution, and stress distribution were obtained. The mechanism of extradite swell of polymer melt induced by the annual width was analyzed.

2. Numerical simulation

2.1. Geometric model.
The geometric model of plastic tube is shown in Figure 1.

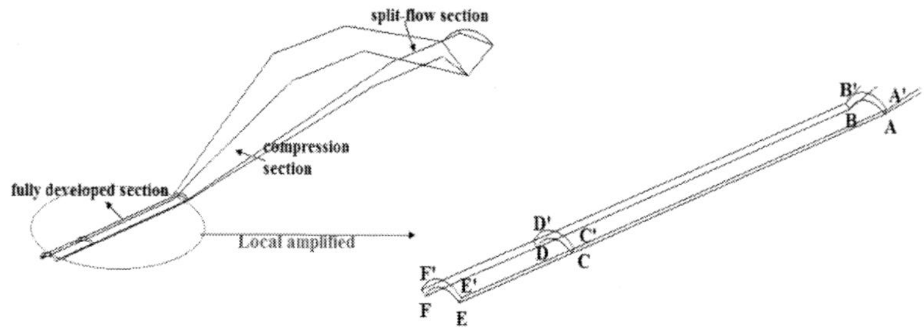

Figure 1. Geometric model of plastic tube extrusion forming

From Figure 1, it can be know that the structure of plastic tube consists of the split flow section, compression section, and fully developed section. The local amplified figure of the fully developed section is shown in the right side of the Figure 1. That is, AA'BB'CC'DD' is the fully developed section of plastic tube. The length of fully developed section is about 30mm. The split-flow angle is about 51°, the length of compression section is about 58mm. The diameter of the mandrel is 5mm.

2.2. Governing equations
The governing equations are given as follows:

$$\textit{Continuity equation:}\ \nabla \cdot v = 0 \tag{1}$$

$$\textit{Momentum equation:}\ \nabla p - \nabla \cdot \tau = 0 \tag{2}$$

Where ∇ is Hamilton operator, v is the velocity of melt, p is the pressure, τ is the extra stress tension.

In this paper, PTT model [4] was used as the constitutive equation, which is given as follows:

$$\exp\left[\frac{\varepsilon\lambda}{(1-\eta_r)\eta} tr(\tau_1)\right]\tau_1 + \lambda\left[\left(1-\frac{\xi}{2}\right)\overset{\nabla}{\tau_1} + \frac{\xi}{2}\overset{\Delta}{\tau_1}\right] = 2(1-\eta_r)\eta D \tag{3}$$

where $\eta_r = \eta_2 / \eta$ is the viscosity ratio, η is the total viscosity of the melt, η_2 is the Newtonian viscosity component of the melt, λ is the relaxation time, ε and ξ are the parameters of the melt correlated with the material tensile and the shear characteristics, respectively. $\overset{\nabla}{\tau_1}$ Is the upper convicted derivative of the extra stress tensor τ_1, η_1 is the Non-Newtonian component viscosity of the melt, D and is the strain-rate of the tensor.

In this paper, the viscosity (η) of PTT construction model is 6000 Pa.s, relaxation time (λ) is 0.2 s, the parameter of ε and ξ are 0.18, and 0.23, respectively. The viscosity ratio is 0.12.

2.3. Boundary conditions
(1) Inlet: Supposed that the flow of melt is looked as the full-developed, steady and laminar flow:

$$\partial v_z / \partial z = 0, v_x = v_y = 0 \tag{4}$$

Where v_x , v_y , and v_z are the flow velocities of melt at the direction of x, y, and z , respectively.

(2) Wall: The no-slip boundary condition is imposed on the wall of die's channel:

$$v_n = v_s = 0 \tag{5}$$

Where n, and s are the normal and tangential direction, respectively.

(3) Free boundary: the zero normal stress and entangle velocity are imposed, i.e.

$$f_n = 0 , f_s = 0 \text{ and } v_n = 0 \tag{6}$$

At the same time, the surface tension was ignored. There is no any pressure in the lumen of plastic tube.

(4) Symmetric boundary: The following conditions should be satisfied,

$$v_n = 0, f_n = 0, f_s = 0 \tag{7}$$

(5) Outlet: The zero trance force and entangle velocity are imposed, i.e.

$$f_n = 0 , v_s = 0 \tag{8}$$

3. Numerical results and analysis

3.1. Extradite swell ratio

In this paper, three extradite swell ratios, i.e., inner radius, outer radius, and wall thickness were computed. The extradite swell ratios of plastic tube with different annual widths are obtained, which are shown in Figure 2.

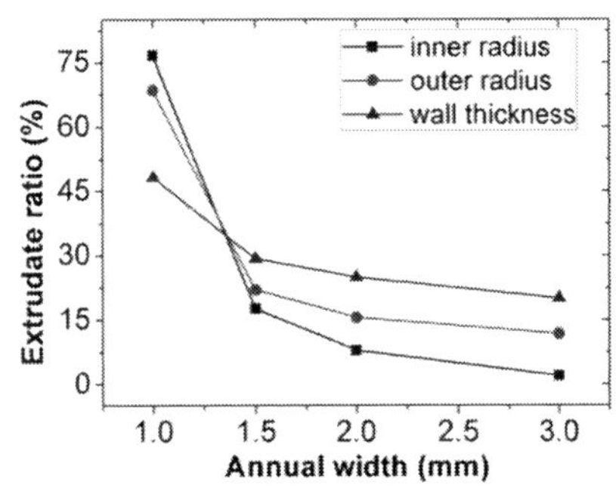

Figure 2. Extradite swell ratio of plastic tube with different annual widths

From Figure 2, it can be seen that, when the annual width is equal to 1mm, the extradite swell ratio of inner radius is larger than others, the extradite swell ratio of wall thickness is smallest. However, when the annual width is larger than 1.5mm, the extradite swell ratio of inner radius is smallest, but the extradite swell of wall thickness is largest. Moreover, from Figure 2, it can be seen that with the increase of the annual width, the extradite swell ratio of plastic tube decreases. In order to ascertain the mechanism of the effect of annual width on the extradite swell of plastic tube, the physical field distributions should be analyzed.

3.2. X velocity distribution

In this paper, four different annual widths of plastic tube were used, i.e., 1mm, 1.5mm, 2mm, and 3mm. On the inlet face, the same inlet volume flow rate of melt was imposed, i.e., Q=25mm^3/s. The X velocity distributions of plastic tube at the outlet of the die are shown in Figure 3.

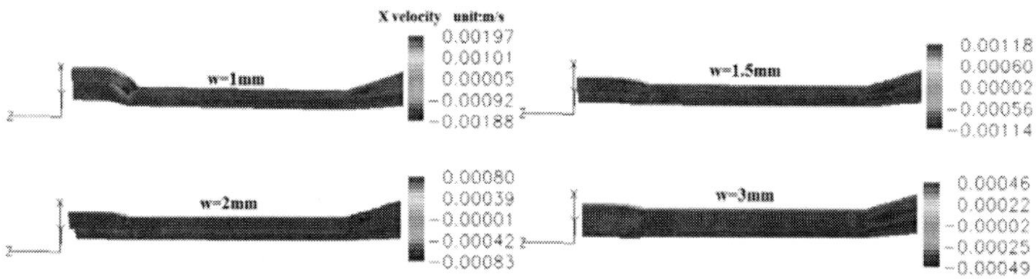

Figure 3. X velocity distributions of plastic tube with different annual widths

From Figure 3, it can be seen that with the increase of the annual width (i.e., w), the X velocity of polymer melt at the outlet of the die decreases.

3.3. Pressure distribution

The pressure distributions of plastic tube with different annual widths are shown in Figure 4.

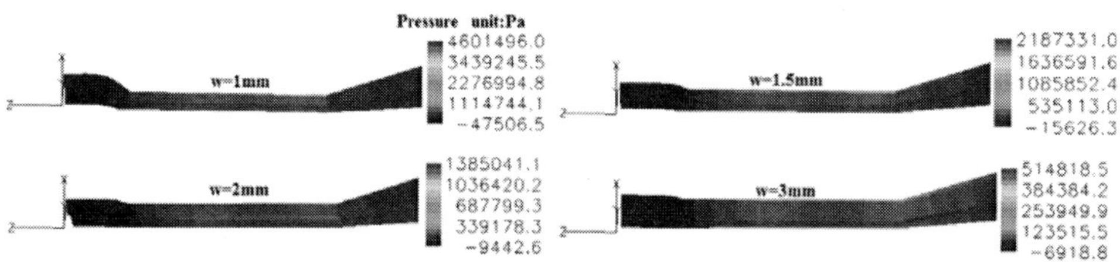

Figure 4. Pressure distributions of plastic tube with different annual widths

From Figure 4, it can be seen that, when the melt with same volume flow volume was imposed on the inlet surface of the die, with the increasing of the annual width of die, the pressure of the plastic tube's melt decreased.

3.4. First normal stress difference distribution

The first normal stress difference (i.e., N1) distributions of plastic tube with different annual widths are shown in Figure 5.

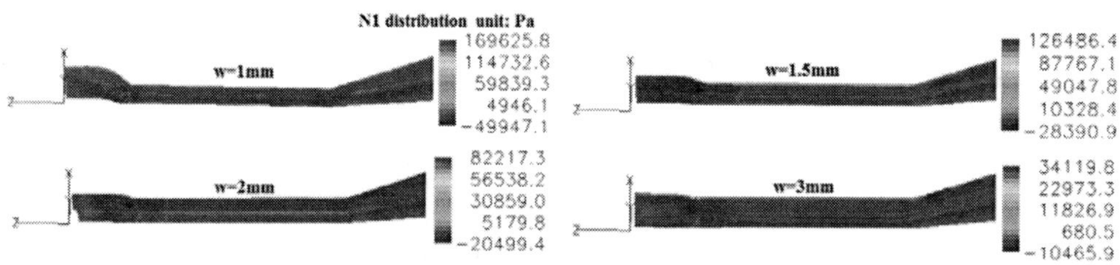

Figure 5. N1 distributions of plastic tube with different annual widths

From Figure 5, it can be seen that, when the melt with same volume flow volume was imposed on the inlet surface of the die, with the increasing of the annual width of die, the first normal stress difference of the plastic tube's melt decreased.

According to Figure 3, 4, and 5, the reason of decreasing the extradite swell for the plastic tube can be expressed as follows: under the same inlet volume flow rate, with the increasing of the annual width, the X velocity at the outlet of die, pressure, and first normal stress difference of plastic tube all decrease. Therefore, the extradite swell effect should be fully considered when the plastic tubes with thin wall thickness are extruded.

4. Conclusion

In this paper, the effect of annual width on the extradite swell of plastic tube was numerically studied by using the computed fluid dynamics software POLYFLOW. The geometric model and finite element meshes of plastic tube with different annual widths are established. The same material properties, boundary conditions, and numerical methods are imposed. Numerical results show that with the increase of the annual width, the extradite swell of plastic tube decrease.

Acknowledgments

This work was supported by the National Natural Science Foundation of China (61650402), the Scientific Research Foundation of Jiangxi Provincial Education Bureau (No. GJJ150814), Top-notch talent foundation project of JXSTNU (2014QNBJRC004), and 2011 Cooperative Innovation Center Project of Jiangxi Province (2014XTCX005).

References

[1] J. M. Dealy, Effect of filler content and additives on the extrudate swell of polyethylene pipe resin, Polymer Engineering & Science 31 (1991) 1176-1181.
[2] Y. C. Ahn, M. E. Ryan, Analysis of extrudate swell from an annular die, Computers & Fluids 21 (1992) 267-289.
[3] S. Tanoue, Y. Iemoto, Effect of die gap width on annular extrudates by the annular extrudate swell simulation in steady-states, Polymer Engineering & Science 39 (1999) 2172-2180.
[4] N. P. Thien, R. I. Tanner, A new constitutive equation derived from network theory, J. Non-Newtonian Fluid Mech. 2 (1997) 353 -365.

IOP Conf. Series: Materials Science and Engineering **207** (2017) 012055 doi:10.1088/1757-899X/207/1/012055

The application of a new type of hydraulic classification equipment: swirl continuous centrifugal separator

Haiyun Xie[1, 2], Rongxin Liu[1, *], Yuanhong Li[1], Pengfei Zhang[2], Chao Ding[1], Luzheng Chen[1, 2] and Xiong Tong[1, 2]

[1]Faculty of Land Resource Engineering, Kunming University of Science and Technology, Kunming, China
[2]State Key Laboratory of Complex Nonferrous Metal Resources Clean Utilization, Kunming, China

*Corresponding author e-mail: liu_rongxin@163.com

Abstract. Swirl Continuous centrifugal concentrator was based on principle of hydro cyclone and fluidized bed separation, which is new hydraulic grading equipment. The effect of new equipment on the classification of Panzhihua ilmenite was studied under different operating conditions. The results show that the underflow pressure, the overflow rate, the feed pressure and the recoil pressure have a significant effect on the grading effect of the cyclone centrifugal continuous concentrator. At the same time, the classification effect on the Titanium magnetite of this concentrator is significant. The new equipment is hopeful to be generalized and gradually used in practice of hydraulic classification.

1. Introduction

Centrifugal separation is an effective gravity concentration process for selective separation of mineral particles in the centrifugal force field. As the centrifugal dressing operation is simple, low cost, environmental protection and separation efficiency, the fluidized bed centrifugal separation technology represented by Knelson and Falcon centrifuges has been widely used in the mining field [1-4], but the problems existing in the centrifugal separation such as intermittent ore discharge and low processing capacity [5-7]. Based on the centrifugal separation technology and fluidized bed separation principle, the centrifugal continuous separator was developed in Kunming University of Science and Technology [8], and in the foundation of material classification by use of centrifugal force, while the introduction of fluidization principle to improve the classification efficiency of backwash water. In this paper, the experimental type cyclone centrifugal continuous sorting machine as a test equipment, for the -2mm titanium magnetite grading experiments were carried out to analyze the impact of different factors on the classification, which provides a reference for industrial application of cyclone centrifugal continuous separator.

2. Equipment and Method

2.1. Equipment

The principle of cyclone centrifugal separator is the continuous flow constructed by rotary centrifugal force field, and the coupling fluidization effect of backwash water resistance process of mixed material classification.

The main parts of the CHC-100 rotary swirl centrifuge contains: tangential feed pipe, vortex column, separation chamber, overflow pipe and so on. The ring groove structure in the sorting cavity, a fluidized pore structure on the separation chamber wall, it can be seen from figure 1.

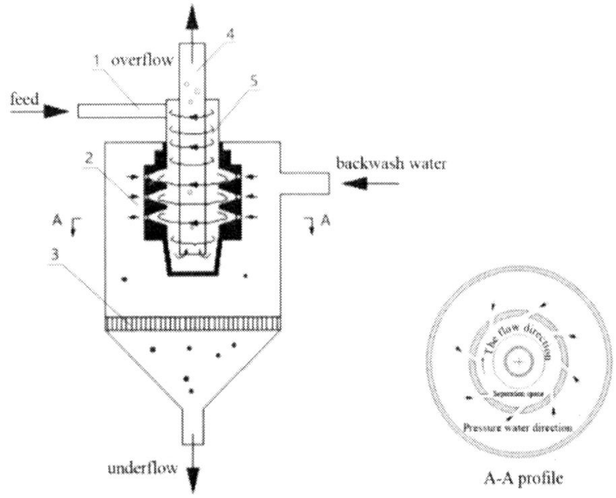

Fig.1 Schematic diagram of continuous centrifugal concentrator

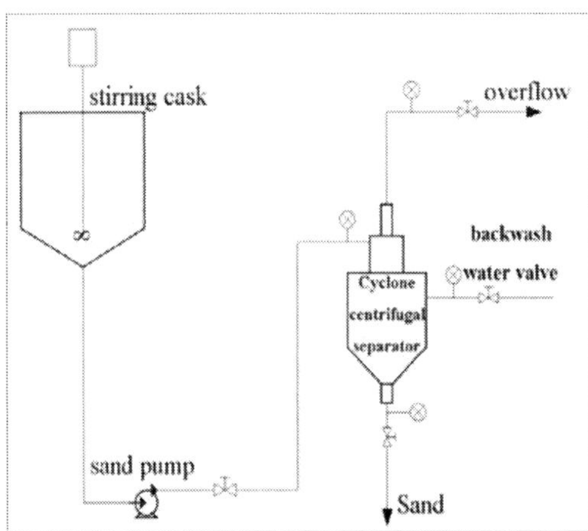

Fig.2 Experimental setup of continuous centrifugal concentration1—tangential feed pipe: 2—sorting cavity: 3—separation chamber: 4—overflow pipe: 5—vortex column

2.2. Particle size characteristics of ore samples

Test ore samples for the titanium magnetite, ore from Panzhihua Iron and Steel Group concentrator. The particle size distribution of the ore samples is shown in table 1.

IOP Conf. Series: Materials Science and Engineering **207** (2017) 012055 doi:10.1088/1757-899X/207/1/012055

Table1 particle size distribution

Size fraction/mm	**Yield/%**	**Negative accumulation /%**
+0.15	37.60	100.00
-0.15~0.0.74	40.90	62.40
-0.074	21.50	21.50
Total	100.00	-

Table 1 shows that the ore is mainly coarse and middle grain size, and low -0.074mm fine grain content.

2.3. Equipment and method

Titanium magnetite test classification test device shown in figure 2. After the end of each classification test, sand and overflow products were filtered, dried and weighed. In this study, we studied the sedimentation rate of + 0.15mm, -0.15 ~ +0.074mm and -0.074mm for sedimentation, and calculated the classification efficiency. The classification efficiency formula (1) is adopted.

$$E = \frac{\theta(\beta-\alpha)}{\alpha(\beta-\theta)} \times 100\% \quad (1)$$

In thc formula (1): E - classification efficiency percentage, α - calculation of content in feed percentage, β- calculation of content in the overflow percentage, θ - calculation of content in the sand percentage.

3. Test results and discussion

3.1. The effect of ore pressure

It is important influence for the classification efficiency. The minimum feed pressure is 64 kPa and the maximum feed pressure is 72 kPa. Other test conditions: feed concentration is 10%, underflow pressure and overflow pressure is 50kPa, backwash water pressure is 19kPa. The test results are shown in Fig.3.

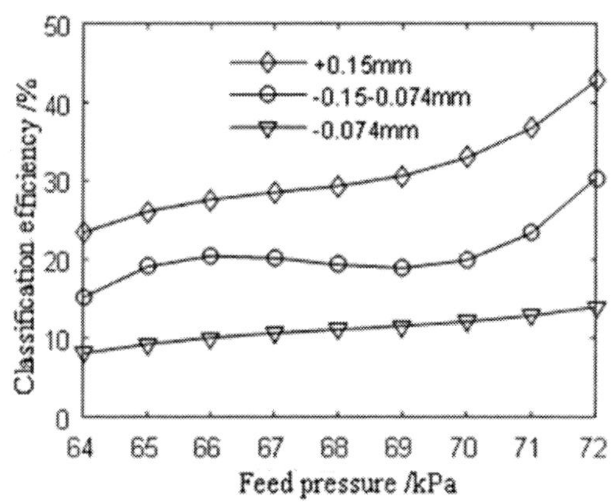

Fig.3 Relationship of Spigot pressure and grading effect

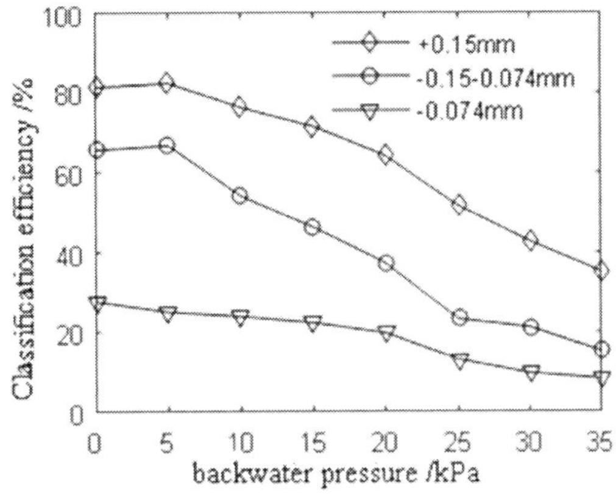

Fig.4 Relationship of backwater pressure and grading effect

In figure 3, with the increase of feed pressure, the classification efficiency of each particle materials such as +0.15mm, -0.15~0.074mm, -0.074mm are increased, the classification efficiency of coarse and intermediate grade material can be improved significantly if increasing feed pressure. The recommended feed pressure range is 70 to 72 kPa.

3.2. Effect of backwash water pressure

The effect of the backwash water pressure on the classification of the material is reflected in the selective barrier mechanism for the different particle size materials. During the test, the backwash water pressure varies from 0 kPa to 35 kPa, the other conditions: the ore concentration is 10%, the feed pressure is 72 kPa, the underflow pressure was 55 kPa, and the overflow pressure was 50 kPa. The test results for different backwash water pressures are shown in Fig.4.

As can be seen from the results in Fig. 4, when the backwash water pressure increases, the grading efficiency of each particle size tends to decrease. When the recoil water pressure is greater than 5kPa, the grading efficiency of each grain grade is rapidly reduced, which is not conducive to grading. The results showed that the appropriate backwash water pressure is beneficial to improve the efficiency of coarse particle classification. The best pressure in the test is 5kPa.

3.3. Effect of underflow pressure

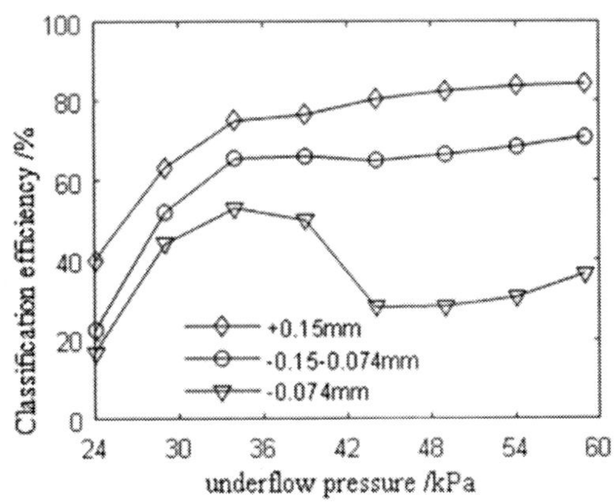

Fig.5 Relationship of underflow pressure and grading effect

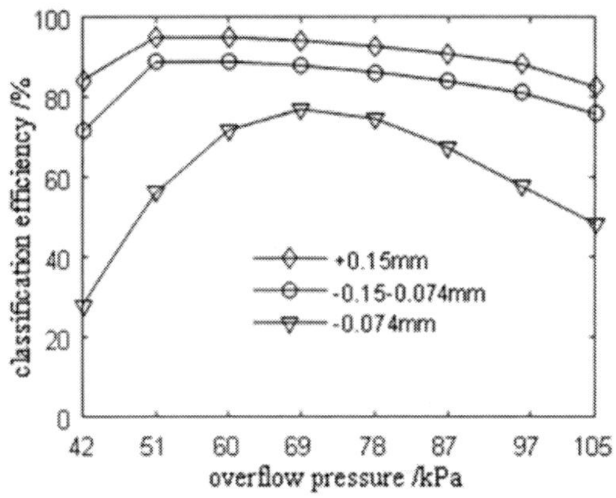

Fig.6 Relationship of overflow pressure and grading effect

During the test, the pressure change of underflow is 24 ~ 60kPa, the ore concentration is 10%, the feed pressure is 72kPa, the overflow pressure is 50kPa, the recoil water pressure is 5kPa, the test results of different underflow pressure are shown in figure 5.

According to the results of figure 5, the classification efficiency of coarse and medium grain size is 80%~84% and 65%~70%. Respectively, the classification efficiency is 22%~26%, and the classification effect is better when the spigot pressure is 45~54kPa. It is determined that the pressure range of the spigot pressure is 45~54kPa.

3.4. Effect of overflow pressure

In this experiment, the influence of the overflow pressure on 42 ~ 105kPa was investigated. The ore concentration is 10%, the feed pressure is 72kPa, the underflow pressure is 50kPa, the recoil water pressure is 5kPa, and the test results are shown in figure 6.

In figure 6, with the overflow pressure from 42kPa to 60kPa, + 0.15mm, -0.15mm ~ +0.074mm, -0.074mm material classification efficiency has improved significantly. The results show that the appropriate increase in overflow pressure is conducive to the realization of coarse grading and intermediate grain grade grading. The optimum overflow pressure range is 42~49kPa.

4. Conclusion

(1)Swirl Centrifugal Continuous Separation Machine is a new type of centrifugal separation equipment based on the principle of fluidized bed centrifugal sorting. The main factors that influencing the classification efficiency of the new equipment on the titanium magnetite include the ore pressure, the recoil water pressure, the underflow pressure, the overflow pressure and the feed concentration.

(2) The optimum range of the influencing factors is obtained. The feed pressure is 70kPa ~ 72kPa, the recoil water pressure is 0kPa ~ 5kPa, the underflow pressure is 42kPa ~ 48kPa, the overflow pressure range is 42kPa ~ 49kPa and the best feeding concentration is 2% ~ 12%.

(3) The classification effect of the hydrothermal grading equipment on the titanium magnetite is obvious, which provides a basis for the further application of the new equipment to the classification or separating other minerals.

Acknowledgments

This work was financially supported by the Natural Science Foundation of China (Grant No. 51464030).

References

[1] Juncheng Li, Zhancheng Guo, Jintao Gao. Enriching Perovskite Phase from CaO-TiO2-SiO2-

Al2O3-MgO Melt by Super Gravity during Cooling Process [J]. CHINESE JOURNAL OF RARE METALS, 2014, 38(1): 93-101.
[2] Chen, Xu. Study on the Application of Centrifugal Separation Technology in Separation of Clinker Slurry [J]. Light metal, 2004, (6): 15-18.
[3] Liansuo An, Yang Yang, Chunyang Liu, Guoqing Shen. Influence on performance of gypsum cyclone with central rod [J]. CIESC Journal, 2014, 65(2): 468-473.
[4] Yucheng Liu, Junzhong Xu, Jianmei Yuan, Yinlong Zhang, Tian Feng. Experiment on thermochemical demulsification and centrifugal separation of abandoned oil-based drilling fluid [J]. Chinese Journal of Environmental Engineering, 2013, 7(6): 2333-2339.
[5] Jialin Wang, Chunmao Chen, Shuangchun Yang, Guangxu Yan, Shaohui Guo. Experimental study on purifying oily wastewater by air-sparged hydrocyclone and its application [J]. Modern Chemical Industry, 2008, 38(3): 62-64.
[6] Wenbing Chen, Bingnan Lu, Hongyu Lu. Study on the Development of Oil - field Wastewater Treatment by Water - tight Cyclone [J]. CHINA WATER &WASTEWATER, 2003, 19(13): 44-46
[7] Qiang Y, Wang H L, Yi L, et al. Solid/liquid separation performance of hydro cyclones with different cone combinations[J]. Separation & Purification Technology, 2010, 74(3):271-279.
[8] Luzheng, Chen. Rotary continuous centrifugal separator [P]. China: CN101890393A [P], 2010.

IOP Conf. Series: Materials Science and Engineering **207** (2017) 012056 doi:10.1088/1757-899X/207/1/012056

Research on Evaluation of resource allocation efficiency of transportation system based on DEA

Zhehui Zhang, Linan Du*

China Waterborne Transport Research Institute, P.R.China

*Corresponding author e-mail: dulinan@vip.163.com

Abstract. In this paper, we select the time series data onto 1985-2015 years, construct the land (shoreline) resources, capital and labor as inputs. The index system of the output is freight volume and passenger volume, we use Quantitative analysis based on DEA method evaluated the resource allocation efficiency of railway, highway, water transport and civil aviation in China. Research shows that the resource allocation efficiency of various modes of transport has obvious difference, and the impact on scale efficiency is more significant. The most important two ways to optimize the allocation of resources to improve the efficiency of the combination of various modes of transport is promoting the co-ordination of various modes of transport and constructing integrated transportation system.

1. Instruction

Transportation is an important basic, guiding and strategic industry of the national economy. In the new era, the internal and external environment and the demand of the development of the reform and development of transportation have changed deeply. The transportation industry has started entering the key stage of the main feature in structural adjustment, the quality and efficiency of development and integration. Construction of modern integrated transport system, is to adapt to grasp the new normal economic development, promote supply-side reform and support the objective requirement of building a moderately prosperous society in an all-round way. Study and evaluation of the efficiency of resources allocation in the integrated transport system is an important basis to improve the effectiveness of various modes of transport and to promote the coordinated development of various modes of transport, which is a great significance to play the role of comprehensive transportation supported to lead the development of the social economy and to improve the core competitiveness of the country.

2. DEA methods and models

Data envelopment analysis, DEA, is an evaluation method of efficiency for multi input and multi output decision-making units. DEA method is the intersection of mathematics, operations research, mathematical economics and management science, which use Mathematical programming models to evaluate the relative effectiveness of multiple input, multiple output units, or decision making units (DMU). The data envelopment analysis model is based on the relative efficiency evaluation. According to Multi input and multi output, this model evaluated the relative effectiveness or benefit from the same type of department or unit. Among them, the model can be used to evaluate the overall effectiveness of the scale of the decision making unit. Therefore, this paper evaluated resource allocation efficiency of integrated transportation system in China with Archimedean infinite ε in DEA model. In practical application, the dual programming model is used to make the calculation more convenient. And the formula for the model is as follows:

$$(DC^2R)\begin{cases}\min[\theta-\varepsilon(S^-+S^+)]\\ \sum_{j=1}^{n}X_j\lambda_j+S^-=\theta X_0\\ \sum_{j=1}^{n}Y_j\lambda_j-S^+=Y_0\\ \lambda_j\geq 0,\ j=1,\ldots,n\\ S^-\geq 0,\ S^+\geq 0\end{cases} \tag{1}$$

In which: θ is effective value for DMU_{j0}, The effective utilization of DMU_{j0} of inputs relative to output, is technical efficiency which means that In the case of a given set of inputs, the actual output of a department is the ratio of the maximum output under the same input.

εis an Archimedean infinite small number which is less than and real number and always larger than 0.

S^- And S^+ are Slack variables which represent Excessive investment and Output deficiency.

λ_j Related to DMU_{j0} is Re construct an effective DMU in the ratio of j times to combine the decision making units.

When $\theta = 1$, $S^- = S^+ = 0$ that means DEA is effective which means that the output has been achieved an optima with Original input. When $\theta = 1$ and $S^- \neq 0$ or $S^+ \neq 0$,that means Weak DEA is effective which means the original output does not change when input is reduced or increase the output while the original input does not change. When $\theta < 1$, that means that the DEA is inefficiency which can be combined to reduce the proportion of the original input and maintain that the original output does not change.

3. Evaluation index selection

This paper studies the comprehensive transportation mode of highway, railway, water transport, civil aviation and other transportation mode. Time range is defined to evaluate the efficiency of the resource allocation of various modes of transport which is the number and value of the products and services that is obtained by the unit resource input of various transportation modes. In this sense, the resource allocation of transportation system is divided into two parts, which are resource input and output. The input factors include various transport land (coastlines), capital and labor. Output is reflected on the transport services - freight and passenger volume.

Table 1. The Input and Output Index System of Transportation Industry

Items		Indicators	Representative symbol	unit
Input	Land (shoreline) resources	Transit mileage (roads, railways, inland waterways)	X_1	Million kilometres
		Terminal shoreline length	$X_{1'}$	Meter
	funds	Investment in Fixed Assets	X_2	Billion
	labor force	Number of Employees	X_3	Million people
Output	shipment service	Freight volume	Y_1	Million tons
		Passenger traffic	Y_2	Million people

(1) Input indicators: X1 is for railway, highway, and the land with Water transport (including inland and coastal transportation), shoreline, waterways and other resources. X2 is investment in fixed assets for railways, highways, water transport and civil aviation. X3 is number of employees for various modes of transport.

(2) Output indicators: Y1 and Y2, respectively, represent the volume of freight and passenger traffic completed in the year of various modes of transport.

(3) Data collection: The time range of this study was 1985-2015 years. The land resources, the coastal port quay, number of employees, freight and passenger traffic volume data are from the China Statistical Yearbook. Fix asset investment data is from Statistical yearbook of Chinese fixed assets investment, Chinese Transportation Industry Development Statistics Bulletin and national railway administration and the State Civil Aviation Administration.

4. Model calculation results and analysis

This paper uses DEAP2.1, the software, to evaluate the Dynamic efficiency of resource allocation for Railway, highway, waterway and civil aviation in 1985-2015, and analyzes both horizontal and vertical ways based on evaluation results deeply. The first, from the perspective of time series, to analyze the dynamic evolution process of resource allocation efficiency of various modes of transport; the second, from a horizontal contrast, to compare and research differences in resource allocation efficiency of various modes of transport.

4.1. Dynamic efficiency analysis

4.1.1 Railway transportation. It can be seen from the following table that in 1985-2015, the comprehensive efficiency of railway transportation in China was 1 in 1985/1990/2005/2011/2013-2015, which means that DEA is effective, and scale income is unchanged and also reach the optima. And in the remaining years, the performance of the DEA shows inefficiency. All of these years, return to scale increase gradually. In the case of other unchanged conditions, the ratio of output to increase is greater than the proportion of inputs of various factors of production, which means that namely, the increase in the input factors of railway transportation brings the economic benefit.

In 1995 and 2000, the pure technical efficiency of railway transport is 1 which mean that input and output do not need increase. And scale effective is less than 1. Return to scale is increasing. That indicates its size, input and the output does not match. The scale need to be increased.

Table 2. DEA Evaluation of Railway Transportation

firm	1985	1990	1995	2000	2005	2010	2011	2012	2013	2014	2015
crste	1	1	0.985	0.972	1	0.947	1	0.977	1	1	1
vrste	1	1	1	1	1	0.99	1	0.985	1	1	1
scale	1	1	0.985	0.972	1	0.956	1	0.992	1	1	1
	-	-	irs	irs	-	irs	-	irs	-	-	-

4.1.2 Road transport. It can be seen from the following table that form 1985 to 2015, the comprehensive efficiency of highway transportation in 1985/1990/2000/2011/2012 was 1, which means that the DEA is effective. The scale income does not change which reach the optima. And the remaining years, the DEA is ineffective. Only in 2010, return to scale increase gradually. In others, return to scale is decreasing, which means that Production factor inputs of road transport to fail to bring about the same proportion of output, showing diseconomy of scale.

In which, the pure technical efficiency of highway transportation was 1 in 2010, which means that the input and output do not need to be increased. The scale efficiency is less than 1, and the return toscale is increasing, which means that its size does not match input and output, so the scale should be increased. On the contrary, the pure technical efficiency of the road transport in 2014 was 1, but the return to scale is diminishing, so the scale should be narrowed.

Table 3. DEA Evaluation of Road Transportation

firm	1985	1990	1995	2000	2005	2010	2011	2012	2013	2014	2015
crste	1	1	1	1	0.941	0.994	1	1	0.925	0.979	0.903
vrste	1	1	1	1	0.959	1	1	1	0.938	1	0.914
scale	-	1	1	1	0.981	0.994	1	1	0.987	0.979	0.988
		-	-	-	drs	irs	-	-	drs	drs	drs

In 2015, our country road transport performance for DEA is no efficiency. And the input redundancy and output problems exist at the same time. Three input indicators can decline 39.252, 4010.53 and 111.191 input to reach effective DEA.

Table 4. Road transport DEA input, output effectiveness

variable		original value	radial movement	slack movement	projected value
output	1	3150000	0	0	3150000
output	2	1619100	0	1894846	3513946
input	1	457.73	-39.252	0	418.478
input	2	16513.3	-1416.08	-2594.45	12502.77
input	3	387.966	-33.27	-77.921	276.775

4.1.3 Water transport. Road transport. It can be seen from the following table that form 1985 to 2015, the comprehensive efficiency of water transportation in 1985/1995/2000/2015 was 1, which means that the DEA is effective. The scale income does not change which reach the optima. And the remaining years, the DEA is ineffective. All of these years, return to scale increase gradually. Among them, in 2012 /2014, water transport pure technical efficiency was 1, while the scale effective is less than 1, return of scale is increasing, which means that Its size does not match input and output, so the scale should be increased.

Table 5. DEA Evaluation of Water Transportation

firm	1985	1990	1995	2000	2005	2010	2011	2012	2013	2014	2015
crste	1	0.982	1	1	0.765	0.887	0.935	0.999	0.967	0.989	1
vrste	1	1	1	1	0.957	0.989	0.99	1	0.995	1	1
scale	1	0.982	1	1	0.799	0.896	0.945	0.999	0.972	0.989	1
	-	irs	-	-	irs	irs	irs	irs	irs	irs	-

4.1.4 Airport transport. It can be seen from the following table that form 1985 to 2015, The comprehensive efficiency of airport transportation in 1990/2000/2005/2010 was 1, which means that the DEA is effective. The scale income does not change which reach the optima. And the remaining years, the DEA is ineffective. Only in 1985 and 1995, return to scale increase gradually. Form 2011, the return to scales increase gradually.

In 2015, our country airport transport performance for DEA is 1. And the input redundancy and output problems do not exist. On the other hand, the scale efficiency is less than 1, showing that the income scales decreases gradually, which means that only when the scale is narrowed, can the input matches the output.

Table 6. DEA Evaluation of Airport Transportation

firm	1985	1990	1995	2000	2005	2010	2011	2012	2013	2014	2015
crste	0.487	1	0.627	1	1	1	0.871	0.735	0.587	0.612	0.601
vrste	1	1	0.805	1	1	1	1	0.944	0.836	0.942	1
scale	0.487	1	0.778	1	1	1	0.871	0.779	0.703	0.65	0.601
	irs	-	irs	-	-	-	drs	drs	drs	drs	drs

4.2. Horizontal comparative analysis
Comparing the comprehensive efficiency of four modes of transport, railways, highways, water transport and civil aviation (graph 1), we can see the first, comprehensive efficiency of DEA of railway transport is higher. More than half of these years achieve DEA effective, and the efficiency fluctuation is not obvious, and is relatively stable. The second, comprehensive efficiency of DEA of road transport is related high. But in recent years, the performance of returns to scale diminishes, and the relative efficiency tend to reduce. The third, comprehensive efficiency of DEA of water transport is form decrease to increase. The lowest point is at 2005 and from 2005 to 2015, the DEA reaches valid. The forth, the fluctuation of comprehensive efficiency of DEA of civil aviation transportation is the most obvious. Except 1990/2000/2005/2010 years, DEA achieved effective. Other years of comprehensive efficiency value are almost significantly lower than other modes of transport.

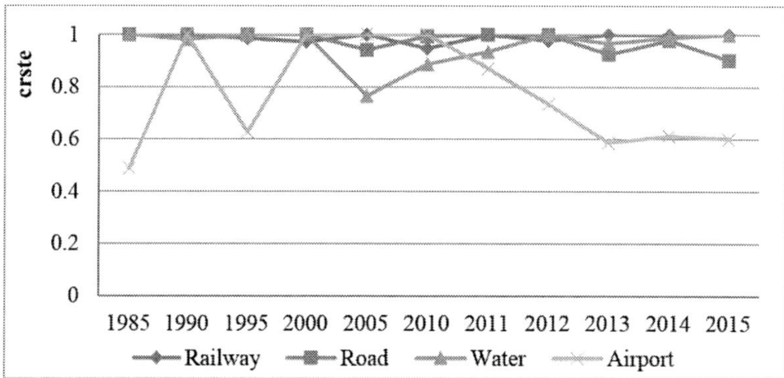

Figure 1. Comparison of Comprehensive Efficiency of Various Transportation Modes

5. Conclusions and implications
For a long time, our country various transportation modes are independent of each other, and follow the respective technical and economic characteristics and development needs, which is less based on the integrated transport system integration development requirements to design the total network size, scale structure, layout structure, which lead to uneven development of various transport modes uncoordinated and unbalanced. Allocation of resources is unreasonable. This paper uses data envelopment analysis to evaluate resources allocation efficiency of railway, highway, water transport and civil aviation and other transport modes in 1985-2015. The result is that the resource allocation efficiency of various modes of transport is not balanced. The scale efficiency of water transport and railway transport increased significantly. The return to scale of Road transport and civil aviation is diminishing. Reasonable adjustments according to the advantages and characteristics of various modes of transport push Co-ordination of various transport modes, which make the formation of the integrated transport network of the Division of labor, organic combination, reasonable layout and association. Construction of integrated transportation system is the important way to optimize the allocation of resources to improve the efficiency of the combination of various transport modes.

References

[1] Wang Dan, Yang Zan. Evaluation on the Resource Allocation of the Transportation System Based on DEA. Journal of Wuhan University of Technology, 2009,33 (2): 275~278.

[2] Zhou Qing-feng, Ma Si, Luo Yuan-yuan. Harmonious Relations of Transportation Energy Consumption in Integrated Transport Development. Journal of Transportation Systems Engineering and Information Technology, 2012,12(4): 11~16.

[3] GENG Yanbin, LI Ke, CHEN Jing, XIE Dian. Study on the Realization Routes of Comprehensive Transportation System With Chinese Characteristics. China Transportation Outlook, 2016,38 (5): 1~9.

[4] TANG Wen, KONG Hui-zhen, HUI Hong-qi. Analysis on Scientific Efficiency Based and on Technical Resource Allocation Data Envelopment Model. Science and Technology Management Research, 2011(13): 187~191.

IOP Conf. Series: Materials Science and Engineering **207** (2017) 012057 doi:10.1088/1757-899X/207/1/012057

An automatic detection method for the boiler pipe header based on real-time image acquisition

Yi Long[1,2], YunLong Liu[1,2,*], Yongliang Qin[3], XiangWei Yang[1,2], DengKe Li[1,2] and DingJie Shen[1,2]

[1]State Grid Hunan Electric Power Company Research Institute, Changsha, China
[2]Hunan Xiangdian Test Research Institute Co.Ltd. Changsha, China
[3]Yuanyang Hu Power Plant, Shenhua Guoneng Energy Group Corporation Limited, Yinchuan, China

*Corresponding author e-mail: 13874809964@163.com

Abstract. Generally, an endoscope is used to test the inner part of the thermal power plants boiler pipe header. However, since the endoscope hose manual operation, the length and angle of the inserted probe cannot be controlled. Additionally, it has a big blind spot observation subject to the length of the endoscope wire. To solve these problems, an automatic detection method for the boiler pipe header based on real-time image acquisition and simulation comparison techniques was proposed. The magnetic crawler with permanent magnet wheel could carry the real-time image acquisition device to complete the crawling work and collect the real-time scene image. According to the obtained location by using the positioning auxiliary device, the position of the real-time detection image in a virtual 3-D model was calibrated. Through comparing of the real-time detection images and the computer simulation images, the defects or foreign matter fall into could be accurately positioning, so as to repair and clean up conveniently.

1. Introduction

The boiler pipe header is one of the most important of the thermal power plants [1]. It is necessary to detect the inner part of the boiler pipe header during A-type repair (usually every 5 to 8 years). Usually, an endoscope is used to test the inner part of the thermal power plants boiler pipe header [2-3]. However, since the endoscope hose manual operation, the length and angle of the inserted probe cannot be controlled. Besides, it has a big blind spot observation subject to the length of the endoscope wire. Additionally, it is difficult to determine the specific location of the endoscope probe [4-5]. Therefore, it is impossible to determine the position of the defects or foreign matter accurately [6].

In this paper, an automatic detection method for the boiler pipe header based on real-time image acquisition and simulation comparison techniques was proposed. The magnetic crawler with permanent magnet wheel could carry the real-time image acquisition device to complete the crawling work and collect the real-time scene image. According to the obtained location by using the positioning auxiliary device, the position of the real-time detection image in a virtual 3-D model was calibrated. Through comparing of the real-time detection images and the computer simulation images, the defects or foreign matter fall into could be accurately positioning, so as to repair and clean up conveniently.

2. Principle of automatic detection system

The principle block diagram of the automatic detection system is shown in Fig. 1. The magnetic crawler with permanent magnet wheel could carry the real-time image acquisition device to complete the crawling work on the inner surface of the enclosed chamber. Then the image acquisition device is used to complete the real-time collection of the scene image of the enclosed chamber, and the positioning and distance measuring device is used to achieve the positioning of the crawler. According to the obtained location by using the positioning auxiliary device, the position of the real-time detection image in a virtual 3-D model was calibrated. Through comparing of the real-time detection images and the computer simulation images, the defects or foreign matter fall into could be accurately positioning, in order to clean up and eliminate system error conveniently without opening the cover of the hydropower turbine runner chamber.

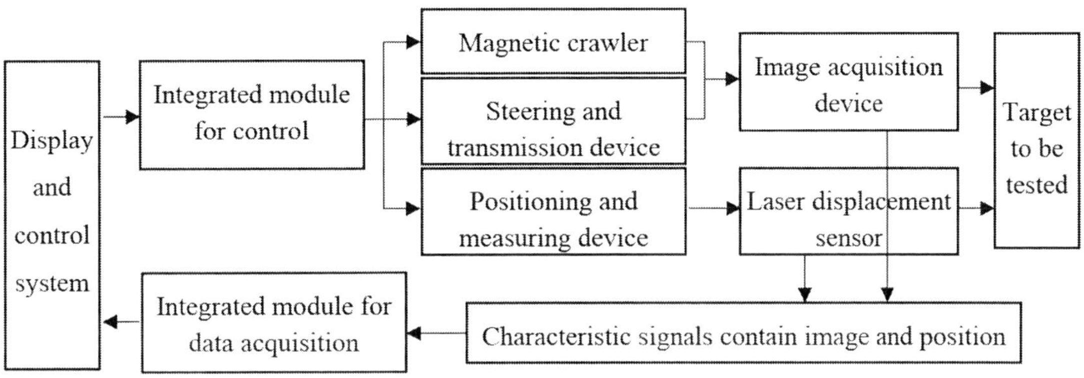

Figure 1. Principle block diagram of the automatic detection system

3. Principle and detection steps

The flow chat of the detection control system is shown in Fig. 2.The detection control system contains the motor control system and the data acquisition and processing system, which could realize the coordinated control of the motors and complete interactive communication with data acquisition and processing module. The detection control system contains 4 motors, and a half of them are used to control the motion of the magnetic crawler, and the others are used to achieve the perspective control of the image acquisition and illumination device.

Firstly, the magnetic crawler which carries the real-time image acquisition device and the positioning measuring device was put into the boiler pipe header. According to the collected positional parameters by using the positioning measuring device, the magnetic crawler could be positioned; according to the acquired angle parameters between the machine vision centerline and x-axis, y-axis and z-axis, the real-time acquisition feature image could be positioned.

Secondly, the virtual 3-D model of the boiler pipe header was established and the position of the corresponding feature image in the virtual model could be calibrated, according to the obtained positional parameters, the angle parameters and the structure parameters. Then, the real-time acquisition feature image and the corresponding feature image in the virtual model could be split-screen display in the computer control interface. At this time, once a foreign matter was detected by the image acquisition device, it would be displayed on the real-time image display screen. Through comparing of the real-time images and the virtual model, the corresponding virtual image with the same size virtual foreign matter would be displayed on another screen at the same time.

Finally, the position and size of the defects or foreign matters fall into the boiler pipe header could be obtained and be accurately calibrated, so as to repair and clean up conveniently.

IOP Conf. Series: Materials Science and Engineering **207** (2017) 012057 doi:10.1088/1757-899X/207/1/012057

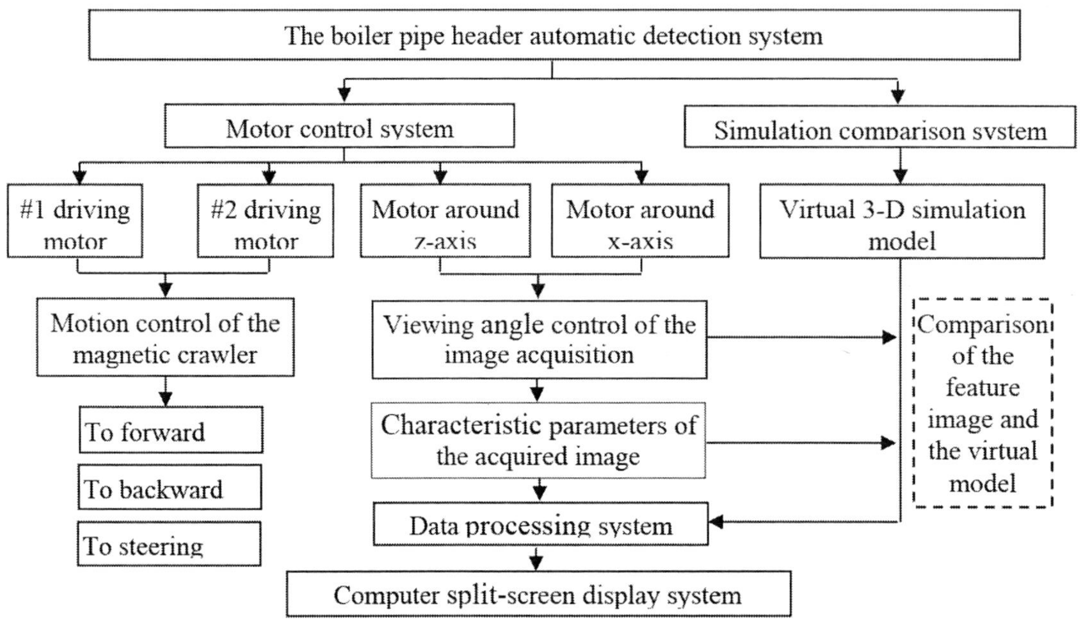

Figure 2. Flow chat of the detection control system

4. Application test

4.1. Processing magnetic crawler

The CAD design diagram and the processing physical photo of the magnetic crawler are shown in Fig. 3 and Fig. 4. A group of rechargeable batteries are used to supply energy, and the wireless wifi technology is used to achieve communication connection with computer control system, which has 40m effective signal transmission and apply to the detection of the structure of the inner flat and simple closed structure.The magnetic crawler has compact and airtight structure and be able to meet the testing requirements of the thermal power plants boiler pipe header.

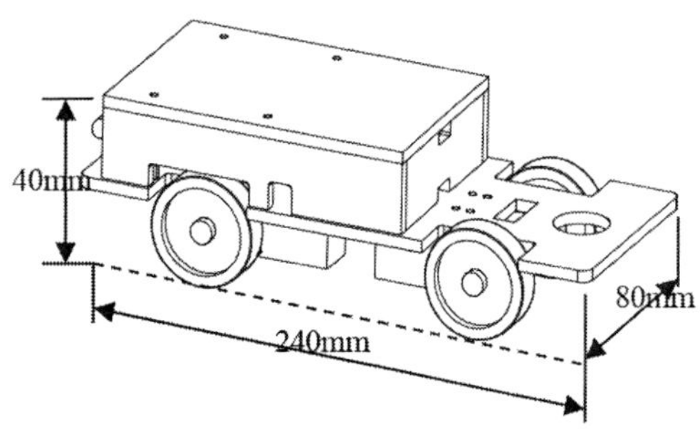

Figure 3. CAD diagram of magnetic crawler

Figure 4. Photo of wireless magnetic crawler

4.2. Detection of thermal power plants boiler pipe header

Generally, an endoscope is used to test the inner part of the thermal power plants boiler pipe header. However, since the endoscope hose manual operation, the length and angle of the inserted probe cannot be controlled. Additionally, it has a big blind spot observation subject to the length of the endoscope wire. Taking a high-temperature reheater outlet header of a thermal power plant as a testing example, its size parameters are Φ 914 mm× 90 mm× 18314 mm, as shown in Fig. 5.

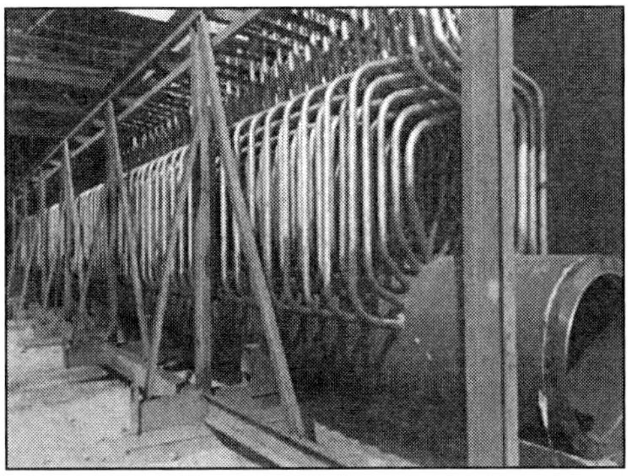

Figure 5. Photo of boiler pipe header

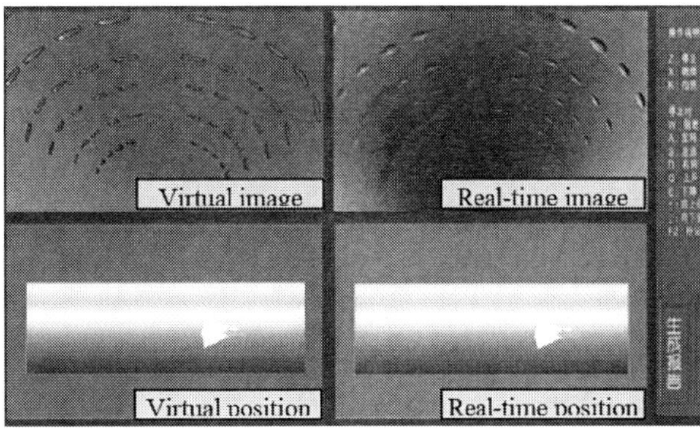

Figure 6. Split screen display interface of system

First of all, the virtual 3-D model of the high-temperature reheater outlet header must be established due to the actual structure and size parameters. And then, putting the wireless magnetic crawler into the entrance of the outlet header and connecting with the computer control system, the magnetic crawler was controlled to walk foreword. Meanwhile, the machine perspective in virtual model followed up according to the acquired real-time position parameters of the magnetic crawler. As a result, it was realized that the real-time acquisition of images and the virtual extract images split displayed comparably, as shown in Fig. 6.

In the high-temperature reheater outlet header, the magnetic crawler could complete the turn, U-turn and other actions. From the test results, it shows that the magnetic crawler has stable, reliable and well real-time operation, and detection system has clear image, stable signal and well synchronization transmission, to be suit for internal inspection of large-scale pipeline and boiler pipe header.

5. Conclusion

By proposing the automatic detection method, the problems of the existing testing methods for the thermal power plants boiler pipe header was improved significantly. The application test showed that the designed detection device could calibrate accurately and clean up the defects or foreign matter conveniently. According to the existing testing technical problems, such as the poor reliability, the larger detection blind spots and the short detection distance, it showed that the proposed detection method is effect and reliable.

Acknowledgments

This research was financially supported by the State Grid Hunan Electric Power Company (NO. 5216A514004A) and the Hunan Xiangdian Test Research Institute Co.Ltd. (NO.18).

References

[1] HU Jiansheng and LIANG Leibo. The ultrasonic test of the narrow gap arc welding seam in thick-walled header of a 1000 MW power plant, NDT. 9 (2013) 66-68
[2] SONG Shaohe and HAN Zhenhua. Ultrasonic testing of thick-wall header and pipe, NDT. 1 (2015) 79-80,95
[3] ZHANG Liang. Analysis on Fault Handling Methods in Turbine Operation, China High Technology Enterprises. 32 (2015) 61-62
[4] QU Zhao, ZHU Jianlin and LAI Xu. Study on hydroelectric generating unit on-line state monitoring system based on virtual instruments, Electric Power Construction. 2 (2005) 61-64.
[5] DONG Baoguo. Detection of Transformer Oil Leakage Based on Image Processing, Electric Power Construction. 11 (2013) 121-124.
[6] GAO Qiang and WEN Yao. Sag Calculation of Transmission Lines Based on Segmentation Image Analysis, Electric Power Construction. 11 (2010) 38-42.

Laser lever method in the application of young's modulus measurement

Yingyuan Fan*, **Zhaoyun Qiu and Renhong Zhao**

Weifang Medical University, Weifang 261053, China

*Corresponding author e-mail: yyfan@126.com

Abstract. Young's modulus of solid material is usually measured by tensile test. We adopt a new laser lever to instead of the traditional optical lever in this article. Contrast experiments with two different methods are designed to measure the steel-wire's Young's modulus. The results show that the new method is more accurate and efficient than the old one.

1. Introduction

Young's modulus is an important physical quantity in describing the ability of solid material to resist elastic deflection. It is necessary to measure the Young's Modulus if you want to get insight into some kind of material, such as metal, semiconductor, optical fiber, ceramic rubber and rubber material.

2. The traditional optical lever method of Young's modulus measurement

Young's modulus is the modulus of longitudinal elasticity. The traditional measuring device is shown in figure1: One plane mirror with which deflection surface parallel to the axis of the telescope and connected with the steel wire is installed on three pivot. If the length of the steel wire changed by different hook weights, the plane mirror will revolve. And so the light reflected by the plane mirror will revolve. Of course, the scale value in telescopic views will also change.

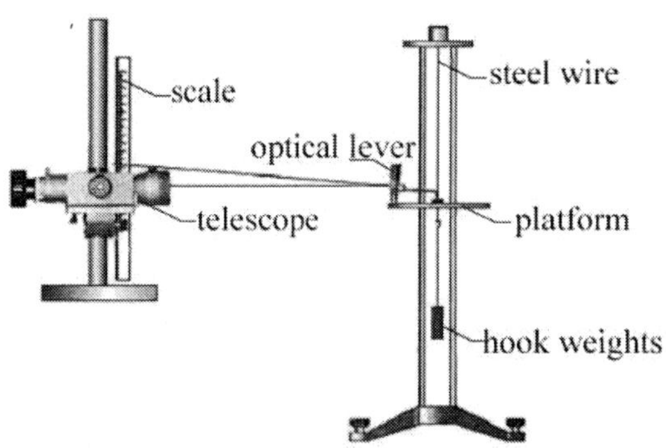

Figure 1. The traditional device to measure Young's modulus by optical lever

When the steel wire stretched by hook weights, the forces (F) is caused by the interaction of the adjacent material. According to Hooke's Law, the force of unit cross-sectional area ($\frac{F}{S}$) is proportional to the elongation of unit length ($\frac{\Delta L}{L}$). The relationship between them is shown as formula 1.

$$\frac{F}{S} = E\frac{\Delta L}{L} \tag{1}$$

In formula 1, E is Young's modulus and S is the cross-sectional area of the steel wire. Assuming the diameter of the steel wire is d , then the cross-sectional area $S = \frac{1}{4}\pi d^2$, and formula 2 was deduced from formula 1 as flowing.

$$E = \frac{4FL}{\pi d^2 \Delta L} \tag{2}$$

F L and S are easy to measure, yet the small variation of the steel-wire's length(ΔL) is relatively difficult. Optical lever method shown in figure2 is usually used to measure ΔL .

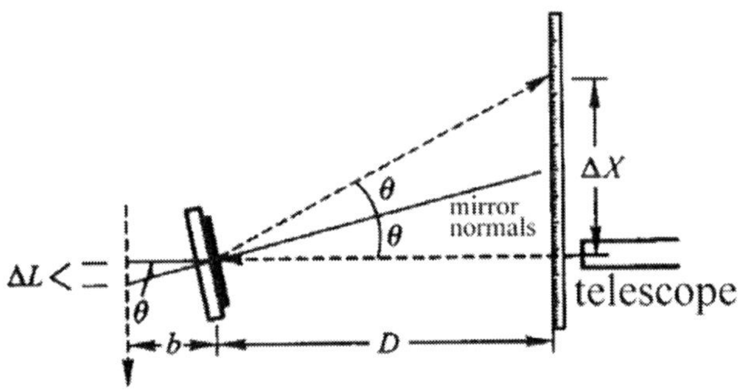

Figure 2. Magnification of the optical lever method

In figure2, the deflection angle of the plane mirror normal is θ, and then the angle between the incident and reflected ray is 2θ . Usually numerical value of θ is tiny enough that $\tan\theta \approx \theta$ and $\tan 2\theta \approx 2\theta$. From the geometrical relationship in figure 2, we can easily point out the equation as formula 3:

$$\Delta L = \frac{b}{2D}\Delta X \tag{3}$$

It can be seen from formula 3: The small variation of the steel-wire's length (ΔL) correspond to the length of two intersections (ΔX)in figure2. So the magnifying multiple of ΔL is $\frac{2D}{B}$ and formula 3 can be rewritten as formula 4:

$$E = \frac{8FLD}{\pi d^2 b \Delta X} \tag{4}$$

In formula 4, F , D , L , d , b respectively represent the tensile force, the distance between the plane mirror and scale, the steel-wire's length, the steel-wire's diameter and the optical lever's length. Naturally, E represent the Young's modulus of the steel wire.

3. Laser lever method of measuring Young's modulus by the improved device

The light of the semiconductor laser is better in monochromatic and directionality. In figure1: The telescope is replaced by a small cross laser connected to the scale. The laser beam produced by the cross laser vertically shots out through a tiny hole. And then the laser beam will be reflected by the plane mirror on the scale. It is easy to judge the angle of the laser reflection by the improved measurement device in figure3. Because the principle of measurement between the instrument in figure1 and in figure3 is selfsame, the way to calculate E by laser lever is identical to formula 4.

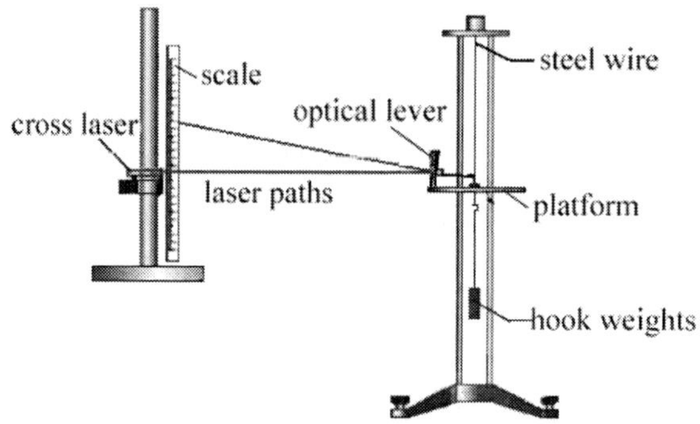

Figure 3. The improved device to measure Young's modulus by laser lever

4. Contrast experiment with two different measuring devices

A contrast experiment is designed to bear out the instrument precision of the improved measuring device to measure Young's modulus by laser lever. A piece of steel wire is measured separately by optical lever method and laser lever method. In the process of measurement, the mount of hook weights is firstly increased and secondly decreased to adapt to steel-wire's delayed effect.

Table 1. Measured result by optical lever method

Number of weights	0	1	2	3	4	5	4	3	2	1	0
X /cm	9.51	9.81	10.10	10.40	10.70	10.98	10.69	10.40	10.12	9.81	9.52
$\overline{X}$ /cm	9.515	9.81	10.11	10.40	10.695	10.98					

Note: X is the reading of the telescope. $\overline{X}$ is the average of X with same number of weights

$L = 43.10\text{cm}$, $d = 2.90\text{mm}$, $b = 9.50\text{cm}$, $D = 112.80\text{cm}$

Gravitational acceleration: $g = 9.80\text{m/s}^2$

Each weight quality: $m = 360\text{g}$

The gravity of the weight caused the tensile force (F) during the measurement, so $F = mg$. And so the formula 4 means:

$$E = \frac{8mgLD}{\pi d^2 b\Delta X} \tag{5}$$

The length of two intersections (ΔX) is obtained by gradual deduction method on the basis of above data: $\Delta X = 0.293\text{cm}$.

Result was obtained through the calculation according to formula5:

$$E = 1.866 \times 10^{11}\,\text{N/m}^2$$

Table 2. Measured result by laser lever method

Number of weights	0	1	2	3	4	5	4	3	2	1	0
X /cm ($D = 105.11\text{cm}$)	7.18	7.45	7.72	8.01	8.29	8.60	8.32	8.02	7.73	7.46	7.20
$\overline{X}$ /cm ($D = 105.11\text{cm}$)	7.19	7.455	7.725	8.015	8.305	8.60					
X /cm ($D = 174.38\text{cm}$)	4.13	4.61	5.06	5.53	6.02	6.47	5.99	5.53	5.08	4.65	4.16
$\overline{X}$ /cm ($D = 174.38\text{cm}$)	4.145	4.63	5.07	5.53	6.005	6.47					
X /cm ($D = 297.22\text{cm}$)	6.09	6.89	7.71	8.50	9.25	10.06	9.26	8.49	7.72	6.91	6.12
$\overline{X}$ /cm ($D = 297.22\text{cm}$)	6.105	6.90	7.715	8.495	9.255	10.06					

Note: The parameters in this research are the same as the first experiment except the parameter D.

X is the reading of the laser reflection point on the scale. $\overline{X}$ is the average of X with same number of weights.

When $D = 105.11\text{cm}$, Then ΔX =0.282cm, and then $E = 1.806 \times 10^{11}\,\text{N/m}^2$

When $D = 174.38\text{cm}$, Then ΔX =0.465cm, and then $E = 1.817 \times 10^{11}\,\text{N/m}^2$

When $D = 297.22\text{cm}$, Then ΔX =0.791cm, and then $E = 1.821 \times 10^{11}\,\text{N/m}^2$

5. Conclusion

The theoretical value of the steel-wire's Young's modulus is given that $E = 1.820 \times 10^{11}\,\text{N/m}^2$. The results obtained in two different methods are in close proximity to the theoretical value. Furthermore the result obtained by laser lever method is more accurately than the one by optical lever method. Another conclusion is that the result obtained by laser lever method is even more accurately with a longer D.

Conclusions could be drawn from this paper: Firstly, the improved measurement device is a precise, simple and easily operated method to measure Young's modulus. Secondly, the method by laser lever is more accurately and predominantly than by optical lever. Lastly, a longer distance between the plane mirror and scale should be designed during the measuring process.

Acknowledgments

This work was financially supported by Educational Reform Project Fund (2015Y018 and 2015Y008) of Weifang Medical University and Scientific Planning Project of Higher Medical Education in Shandong Province (YJKT201511).

IOP Conf. Series: Materials Science and Engineering **207** (2017) 012058 doi:10.1088/1757-899X/207/1/012058

References

[1] Shuwu Yang. General Physics and Experiment (Mechanics and Thermology fascicle) [M], Higher Education Press, Beijing, 2007, pp.152-154 (In Chinese).

[2] Wenbin Li. College Physical Experiment [M], Hunan Science and Technology Press, Changsha, 2003, pp.60-64. (In Chinese).

[3] Ying Liu, Jiayu Zhang, Ying Wang, Rong Shang, Laser bending light lever to measure Young's modulus, Physical Experiment of College.2015(12), pp.28-30. (In Chinese).

[4] Xie Hui, Optical lever calibration in atomic force microscope with a mechanical lever, Review of Scientific Instruments. 2008 (9).

[5] Panduputra Yohannes, Ng Tuck Wah, Neild Adrian, Ling William Yeong Liang, Adhesion force studies using a dangling optical lever with variable sensitivity, J. Optics Letters. 2011, pp.175-177.

IOP Conf. Series: Materials Science and Engineering **207** (2017) 012059 doi:10.1088/1757-899X/207/1/012059

The judgment of the All-melted-moment during using electron beam melting equipment to purify silicon

Xiaojie Han[1], Jianxiong Meng[1], Shuaiye Wang[2], Tonghao Jiang[2], Feng Wang[2], Yi Tan[1, 2] and Dachuan Jiang[1, 2, *]

[1]School of Materials Science and Engineering, Dalian University of Technology, Dalian 116023, China
[2]DUT (Qingdao) Research Institute of New Energy Materials Technology Co., Ltd, Qingdao 266234, China

[*]Corresponding author e-mail: jiangdachuan@dlutlg.com

Abstract. Experiment has proved that the rate of impurity removal depends on the pressure and the temperature of the vacuum chamber during using electron beam to smelt silicon, and the amount of removed-impurity depends on time when other conditions are the same. In the actual production process, smelting time is a decisive factor of impurity removal amount while pressure and temperature of the vacuum chamber is certain due to a certain melting power. To avoiding the influence of human control and improving the quality of production, thinking of using cooling water temperature to estimate the state of material during metal smelting is considered. We try to use the change of cooling water temperature to judge that when silicon is all melted and to evaluate the effectiveness of this method.

1. Introduction

Electron beam melting, referred to as EBM, is a method by changing the kinetic energy of high-speed electron beam to heat energy to smelt metal [1, 2] in a vacuum environment (10^{-3} Pa). This method is usually used to smelt metal and alloy of high melting point, heat-resisting material and high-quality alloy for special purpose [3, 4]. It also can be used to smelting silicon to remove P and metal impurity, such as Al, Fe and so on [5].

Shi S et al. researched the mechanism of volatile impurities transfer during using electron beam to smelt silicon, and they have proved that the rate of impurity removing depending on pressure and temperature of vacuum chamber during using electron beam to smelt silicon. At this stage, the removal of impurity is sensitive to smelting time when pressure and temperature is certain. This research is mainly aimed at Fe, Al and Ca and so on and it has been proved that electron beam melting equipment has purification effect generally [6-8]. But the research did not judged when silicon is all melted definitely. It is hard to detect temperature of molten pool since the chamber of electron beam melting equipment has an environment of high temperature and high vacuum. In actual production process, we usually judge the moment of all-melted by manually watching, but this method is volatility especially disturb the removal of impurity at initial stage of smelting [9] when the impurity is volatilize quickly. In addition, electron beam melting equipment has high energy consumption. So for the stability of product and reducing energy consumption, it is important to judge when the silicon is all melted.

Considering cooling water can reflect the temperature of molten pool, in our experiment, we try to use the change of cooling water temperature to judge the time when silicon is all melted and evaluate the effectiveness of this method.

2. Experiment

During production, the main equipment is EBS-500 electron beam furnace and accessories are vacuum system include mechanical, diffusion and roots pump, electronic control system and cooling system. We use the silicon on the bottom of crucible after the preparation of mono-crystalline silicon using Czochralski method as raw material, which is n-type semiconductor.

Table 1. Main steps and parameters.

Step	Load material	Increasing vacuum	Preheat electron gun	Electron beam emission	Smelting	Solidification
Parameter	50-60kg each ingot	0.5Pa for Electron gun 0.2Pa for smelting	Vacuum:0.5Pa Up to 1000mA In 30min	Vacuum:0.2Pa Up to 250kW	20min	Electron beam attenuation

During the process smelting, vacuum degree of chamber will be influenced by gas from silicon. And electron gun will be shut down when the pressure of chamber is higher than security line. This phenomenon often occurs when purifying silicon whose purity is poor. We can know from the change in temperature of cooling water that the temperature of chamber reduce quickly. And on the bottom of molten pool, there can be liquid silicon curdling because of a lot of heat being taken away by cooling water. After turn on electron gun, silicon will be all welt at a time, but we can not make sure when it is. In our experiment, we collect temperature value of cooling water displayed in electronic control system each 10 second in the process of melting and smelting.

3. Result

3.1. Judge the moment of all-melted

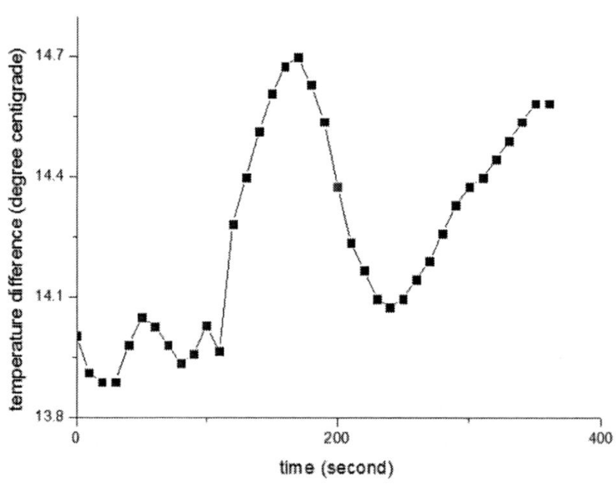

Figure 1. The change of cooling water temperature difference nearby the moment of all-melted.

IOP Conf. Series: Materials Science and Engineering **207** (2017) 012059 doi:10.1088/1757-899X/207/1/012059

We draw figure 3.1.1 through the cooling water temperature nearby the moment of all-melted, which is one of a series, the red point means all-melted observed manually. Through the similarity we can know that the change of cooling water temperature difference is fluctuant before the moment of all-melted. The reason is that solid-liquid transition occurs repeatedly at the bottom of molten pool because a lot of heat is absorbed by cooling water when the silicon is almost melted, on the contrary, the heat absorption and release of this process influence the temperature of cooling water. We can also know that the cooling water temperature difference is always increase after silicon is all melted because the temperature of melt is increase fast and continually, so if we can find the similarity of this process, the similarity can be the gist of all-melted.

Figure 3.1.2 is drawn through the temperature difference of whole smelting progress which is one of a series, the red point means all-melted moment watched manually, the green point means electron gun being shut down and yellow point means the power of electron gun reaching up to 250kW again.

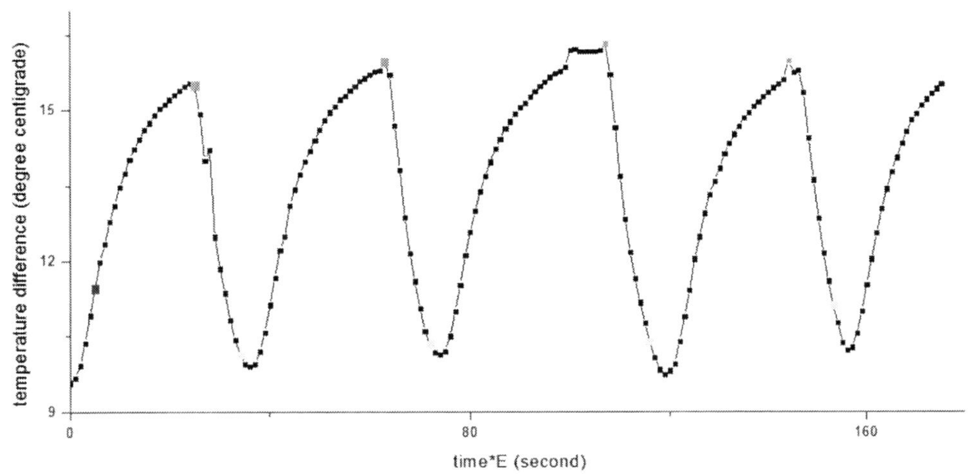

Figure 2. The change of temperature difference of whole smelting progress.

We know that the impact of electron beam make the melt temperature and the cooling water temperature difference increase. Since the melt temperature will not increase infinitely, it will remain stable after a while, which is the same for the cooling water temperature difference. This phenomenon shows in figure 3.1.2 before the third time electron gun is shut. When electron gun is shut down, the melt temperature and the cooling water temperature difference fall quickly, and the decrease lasts for a short time after the power of electron gun being up to 250kW again. So we can speculate that there is a small amount of melt solidifying again during the electron gun is shut. After turn on electron gun, the heat offered by electron beam is lower than the heat absorbed by cooling water for a short time which result in the decrease of cooling water temperature difference. As the power of electron gun increasing, the former counteract the latter and the speed of temperature decrease is getting slow. Finally the heat offered by electron beam is higher than the heat absorbed by cooling water and the solid silicon melt again. So the change of cooling water temperature difference is decreasing first and then increasing.

Now we fit the curve from the red point to the first green point and the result is showed in figure 3.1.3.

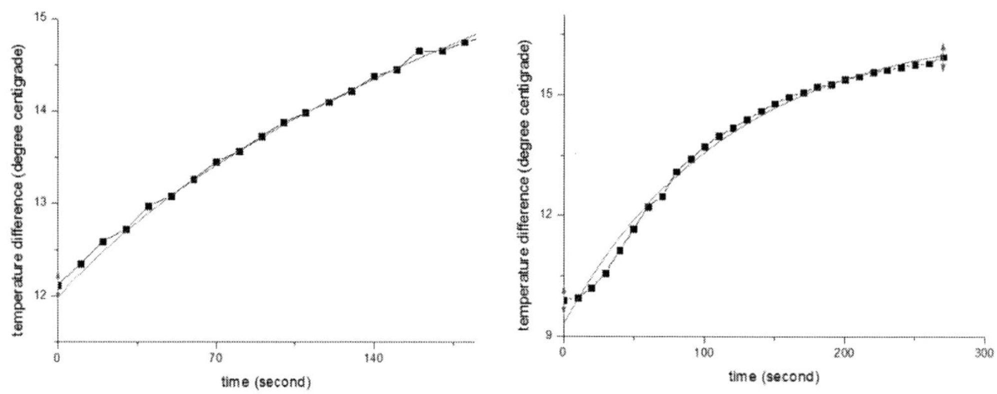

3.1.3-a 3.1.3-b

Figure 3. The fitted curve of the change in cooling water temperature difference

Through the fitted curve we can know that from one point which is after the red point, the curve is upward as exponential function, and the function expression is:

$$y = A_1 \exp(-x / B_1) + A_2 \exp(-x / B_2)$$

The value of A_1, B_1, A_2, B_2 depends on the environment of production, such as the in-flowing cooling water temperature, the melt quality and so on. We can draw the same conclusion if we fit any other curves which is similar to this process

In the situation of the electron gun being shut down, we fit the curve from the moment when the cooling water temperature difference begin to increase to the moment when the electron gun is shut again, and the result is showed in figure 3.1.3-b. We can know from the fitted curve that from one moment the curve is upward as exponential function, and the function expression is:

$$y = A_1 \exp(-x / B_1) + A_2 \exp(-x / B_2)$$

The value of A_1, B_1, A_2, B_2 depends on the environment of production. We can draw the same conclusion if we fit any other curves which is similar to this process.

In conclusion, when the cooling water temperature difference begin to rise as exponential function no matter it is in the process of melting or smelting, the moment of beginning is the all-melted-moment.

3.2. The evaluation of effectiveness

After smelting, the content of B is still the same and the p-n type of product is unstable. The reason is that the purification make donor impurity and acceptor impurity balance each other out and the small difference of smelting time will lead to the change of p-n type. Combing with figure 4.1, the logical relation curve of electrical resistivity and smelting time, we can know the degree of P removal [10, 11].

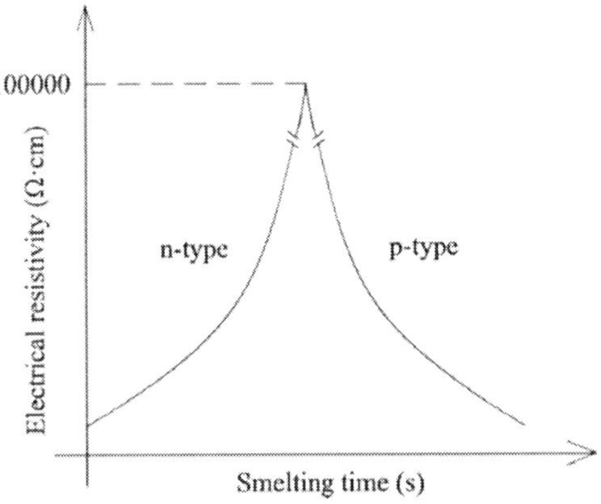

Figure 4. The logical relation curve of electrical resistivity and smelting time.

In the actual production process, the time of electron gun being shut down can be 1-6 times even more and the influence of total time errors can not be ignored. If we use the method above to calculate the smelting time and combine with the scatter-gram of product's electrical resistivity, we can evaluate the effectiveness of this method. We fit all the curves which is rising, calculate the smelting time, draw the scatter-gram and gather them to table 4.1.

Table 2. Smelting time and scatter-gram of product's electrical resistivity.

Smelting time (s)	1030	930	840
product' s electrical resistivity(the shadow means n-type)	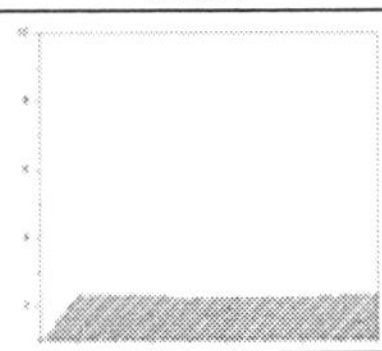		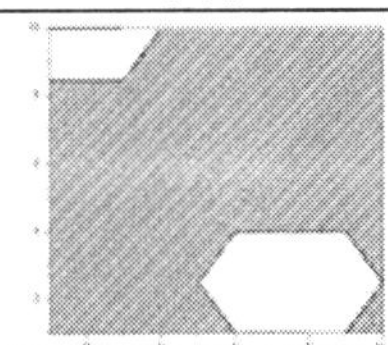

We can know from figure 4.1 that around the balance of donor impurity and acceptor impurity, the product will be p-type if the smelting time is longer slightly and will be n-type if the smelting time is shorter slightly. Combine figure 4.1 and table 4.1 we know that, the product is mainly n-type whose melting time is shorter relatively and is mainly p-type whose smelting time is longer relatively, which means this method of judging the all-melted-moment is efficient.

4. Conclusion

Through the experiment and analysis we draw a conclusion that during the process of using electron beam melting equipment to purify silicon, if the change of cooling water temperature difference is upward as exponential function at a moment, and the function expression is:

$$y = A_1 \exp(-x / B_1) + A_2 \exp(-x / B_2)$$

The silicon is all melted at this moment and the value of A_1, B_1, A_2, B_2 depends on the environment of production, such as the in-flowing cooling water temperature、 the melt quality and so on. Using this method to calculate the smelting time and combing smelting time with figure 4.1 and scatter-gram of product's p-n type, we can deem that the method by using the change rule of cooling water temperature difference to judge the all-melted-moment is efficient.

Acknowledgments
The authors gratefully acknowledge financial support from the National Natural Science Foundation of China (Grant No. 51304033 and No. 51404053), the Specialized Research Fund for the Doctoral Program of Higher Education (Grant No. 20130041110004) and the Fundamental Research Funds for the Central Universities (Grant No. DUT15LAB06).

References
[1] Tan Y, Shi S, Jiang D C. Progress in Research and Development of Solar-grade Silicon Preparation by Electron Beam Melting [J]. Journal of Inorganic Materials, 2015, 30(8): 785.
[2] Bakish R. The substance of a technology: Electron-beam melting and refining [J]. JOM, 1998, 50(11): 28-30
[3] Analysis of the Electron Beam Smelting Technology and Developing Trend [J]. Journal of Hebei Institute of Architecture & Civil Engineering, 2008.
[4] Li-Pu M A, Liu W C. Electron Beam Melting Technology and Its Application [J]. Nonferrous Metals Processing, 2008.
[5] Dong W, Wang Q, Peng X, et al. Aluminum evaporation from metallurgical silicon in electron beam melting process [J]. Chinese Journal of Materials Research, 2010, 24(6): 592-596.
[6] Peng X, Dong W, Tan Y, et al. Removal of aluminum from metallurgical grade silicon using electron beam melting [J]. Materials Science Forum, 2011, 675-677(4): 45-48.
[7] Wang Q, Dong W, Tan Y, et al. Impurities evaporation from metallurgical-grade silicon in electron beam melting process [J]. Raremetals(English), 2011, 30(3): 274-277.
[8] Shi S, Dong W, Xu P, et al. Evaporation and removal mechanism of phosphorus from the surface of silicon melt during electron beam melting [J]. Applied Surface Science, 2013, 266(2): 344 - 349.
[9] Jiang D, Shi S, Tan Y, et al. Segregation and evaporation behaviors of aluminum and calcium in silicon during solidification process induced by electron beam [J]. Semiconductor Science & Technology, 2015, 30(3): 035013
[10] Perichaud I, Trassy C, Martinuzzi S. Multicrystalline silicon wafers prepared from upgraded metallurgical feedstock [J]. Solar Energy Materials & Solar Cells, 2008, 92(10): 1269-1273.
[11] Binetti S, Libal J, Acciarri M, et al. Study of defects and impurities in multicrystalline silicon grown from metallurgical silicon feedstock [J]. Materials Science & Engineering B, 2009, s 159 - 160(11): 274-277

IOP Conf. Series: Materials Science and Engineering **207** (2017) 012060 doi:10.1088/1757-899X/207/1/012060

Effects of progressive solution treatment on microstructure and property of the aluminium alloy electrical round rod

Xiejun Hu[1], Renguo Zhang[2], Xiaoyan Zhang[1,*], Tingyi Lin[1], Junjie Zhang[1], Jun Yan[1]

[1]College of Material and Metallurgy, Guizhou University, Guiyang 550003, China
[2]Guizhou Herun new Aluminium Materials Technology Co., Ltd, Guiyang Guizhou 550000, China

*Corresponding author e-mail:ivzhangxiaoyan@163.com

Abstract. This paper aims to research the effects of progressive solution treatment on property of aluminium alloy electrical round rod, we mainly research the influences of different temperature on tensile strength and electrical conductivity of the aluminium alloy electrician round rod. The component and phase constitution of the aluminium alloy was analysed by means of OLYMPUSPMG-3 horizontal metallurgical microscope, SUPRA40 scanning electron microscope, APOLLO-10Xenergy spectrometer, tensile strength and conductivity of aluminium alloy electrical round rod under various progressive solution treatment. The results show that: comprehensive performance of specimen is optimal by means of scheme(350℃×1h+510℃×6h),the conductivity is 56.6%IACS and the tensile strength is 103MPa.

1. Introduction

Transmit the electricity from western areas to East China, north-south power exchange, intelligent networking have become the focus of the national power grid construction strategy. Although the annual output of China's cable ranks in the forefront of the world, technology content is not high, especially with the power cable distribution [1].The use of annealed copper wire is more than 90% and the proportion of aluminium or aluminium alloy wire is very small. After the Second World War, the world's copper resources are generally in short supply, and the price is expensive, but the aluminium resources are abundant[2-4].Therefore, it has important economic significance to replace copper wire with aluminium wire as soon as possible[5-6].This paper is to study the effect of the progressive solution and the single-stage solid solution on the microstructure and properties of the aluminium alloy round rod so that we can improve the mechanical properties and electrical conductivity. Laying the foundation for researching and developing the processing technology of high performance electrician round aluminium alloy rod.

2. Experimental

2.1. Experimental materials

The composition of raw materials in the experiment is listed in Table 1.In order to guarantee the experimental accuracy, there are three samples of each state, and the experimental data was obtained by average when testing the performance of the sample.

Table 1. The chemical composition of the experimental alloy (wt. / %)

Element	Al	B	Si	Fe	Mg	Cr	Ni	Ti	others
Content	99.685	0.020	0.067	0.111	0.005	0.001	0.007	0.033	0.071

2.2. Experimental processing technology

The experiments were carried out to compare the progressive solution with the single-stage solid solution to determine the best experimental processing technology. Concrete parameters of the progressive solution are 250℃×2h+530℃×8h, 300℃×2h+530℃×8h, 350℃×2h+530℃×8h, 400℃×2h+530℃×8h, 450℃×2h+530℃×8h, Concrete parameters of the single-stage solid solution is 530℃×8h.

2.3. Analysis and test methods

Rolling process was conducted by the F50-150 type rolling round machine at room temperature; Intron - 8501 materials tensile tester with the stretching rate of 2 mm/min was used to test the tensile strength and elongation of the specimens. QJ57 Arms Bridge was performed to analyse the electrical conductivity of specimens; Microstructure and element distribution of the specimens were observed by Olympus-BH2 optical microscope (OM), SUPRA40 field emission scanning electron microscope (SEM) and APOLLO-10Xspectrometer; Phases were analysed by X' Pert PRO X ray polycrystalline diffraction (XRD).

3. Results and discussion

3.1. Effect of progressive solution on electrical conductivity and tensile strength

As shown in Fig.1,with the increase of the primary solution temperature, the conductivity first decreases and then increases while the tensile strength first increases and then decreases when the secondary solution temperature parameters is 530℃×8h.When the parameters of the progressive solution is 250℃×2h+530℃×8h,the electrical conductivity is 56%IACS and the tensile strength is 95MPa;Comprehensive performance of specimen is optimal by means of scheme(350℃×2h+530℃×8h),the conductivity is 55.7%IACS and the tensile strength is 103MPa.The conductivity of the progressive solution decreased slowly and the tensile strength improved obviously compared with the single-stage solid solution (the conductivity is 55.8%IACS and the tensile strength is 99MPa),indicating that the progressive solution is better.

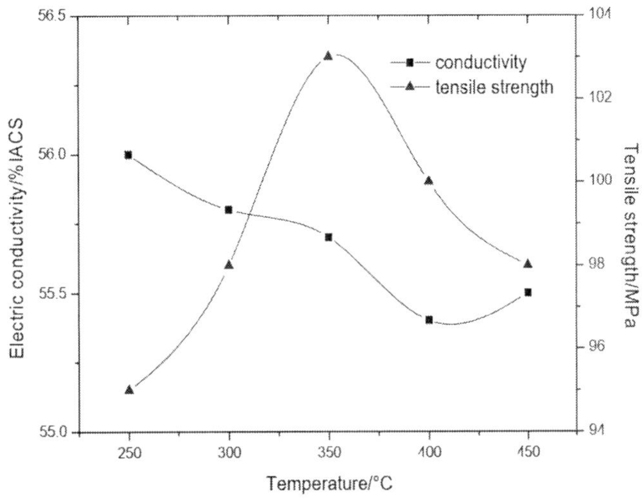

Figure 1. Impact of the primary solution temperature on the conductivity and tensile strength of specimens

3.2. Microstructure of specimens

Figure 3 shows the metallographic structure of specimens with different of the progressive solution treatment. With the increase of the primary solution temperature, the grains become refining and then become coarse. The grain boundary of specimens was coarse (Fig.2a).The grain boundary of specimens was slender and distributed in grain boundary evenly (Fig.2b).The alloy elements are solid solution adequately, the grain is slender, the microstructure is uniform and the grain boundaries are clean because the temperature of the primary solution is low, there is a part of Sub-grain and smaller grain boundary, the grain boundary migration rate is lower, so that the specimen can obtain smaller grain boundary microstructure in high temperature stage(Fig.2c).The grain is refining and have a small amount of second phase(Fig.2d).The grain is coarse relatively (Fig.2e).

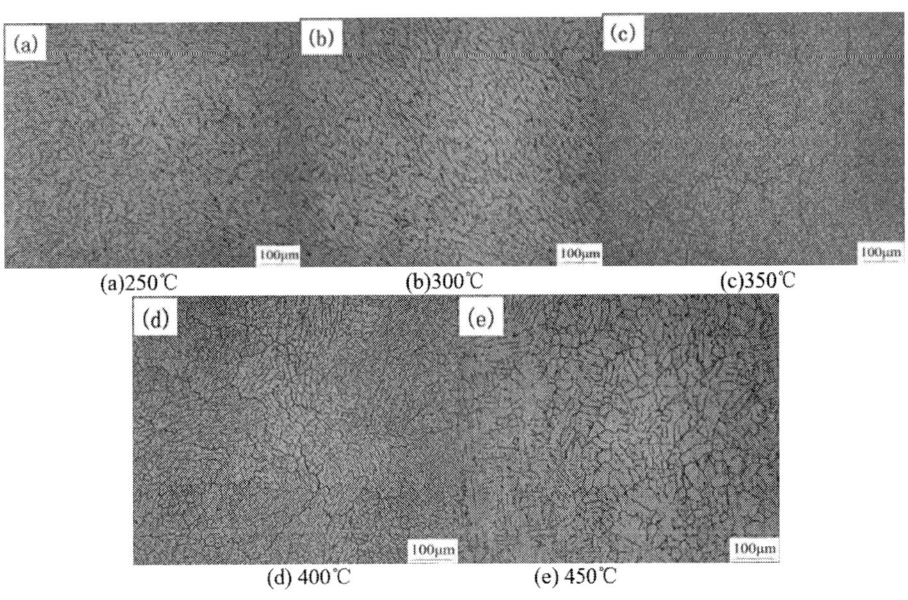

Figure2. Microstructure of specimens on different the primary solution temperature

3.3. SEM and EDS analysis of specimens

Figure 3 demonstrates the SEM and EDS analysis of specimens by means of scheme (350℃ ×2h+530℃×8h).As shown in Fig. 3, the grain boundary and the crystal are relatively clean, almost no white particles in the crystal, only a very small amount of white particles precipitation on the grain boundaries. It is pointed out that a large number of the alloy element is too late to precipitate in the temperature, and the precipitate phase is too late to grow up, the size is small and the solid solution is sufficient [7-10].

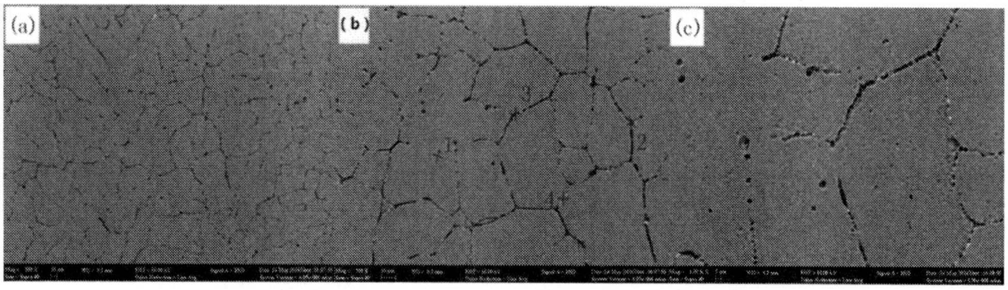

Figure3.SEM of sample under different multiples

Component analysis was performed at positions 1, 2, 3 and 4 in Figure 3 (b). As can be seen from the results in Table 2: there are a mass of Al and few Cu by analysing the precipitates of point 1. The copper content of point 2 is much more than point 1, and the content of Y element is relatively small, which may be the result of the formation of Cu elements in the solution process. Point 3 at the grain boundaries, the main components are Al, Cu, La, Ce, Y. Point 4 is white particles, and the content of alloying elements are higher, it may be mixed precipitation of Al, Cu, La, Ce, and Y elements. It can be seen that the distribution of Cu, Ce, La, Y elements is more uniform after progressive solution treatment, and these elements exist in the grain and grain boundaries, which shows that the alloying elements are completely solid solution [11-13].

Table 2.EDS analysis result of specimens /%

Elements		Al	Cu	La	Ce	Y
Wt.%	1	99.12	0.88	-	-	-
	2	93.37	5.84	-	-	0.79
	3	92.82	2.13	2.78	1.69	0.59
	4	85.26	5.29	4.82	4.25	0.38

3.4. XRD analysis of specimens

Figure 4 shows the XRD analysis of the specimens by means of scheme(350℃×2h+530℃×8h).From the diffraction peak in Fig.4, the main phases of the specimens are Al, Al9Si, Cu3Si, Al3Y, Al0.3Fe3Si0.7 and so on. The main phases of the strongest diffraction peaks are Al, Al9Si. There is no Ce phase in each peak, the reason may be that the Ce phase has been fully melted into the matrix, indicating that Ce phase are completely solid solution after the progressive solution treatment [14-15].

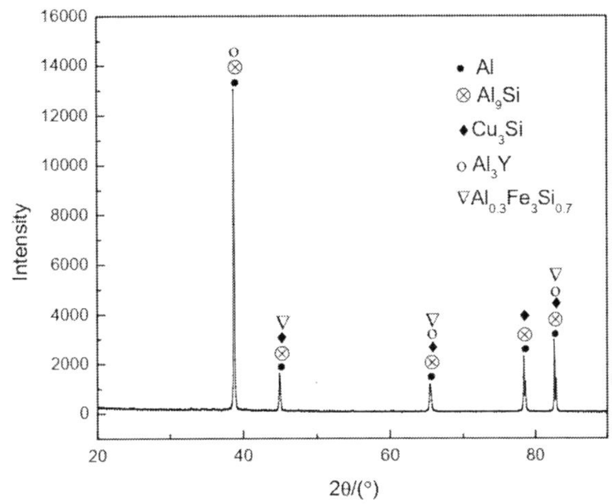

Figure.4 XRD analysis of the sample

4. Conclusion

(1) The grain boundary and the crystal of the specimens are relatively clean, almost no white particles in the crystal, only a very small amount of white particles precipitation on the grain boundaries, metallurgical structure is uniform and the alloy elements are solid solution adequately by means of scheme (350℃×1h+510℃×6h).

(2) Comprehensive performance of specimen is optimal by means of scheme(350℃×1h+510℃×6h),the conductivity is 56.6%IACS and the tensile strength is 103MPa. Compared with the single-stage solution treatment (530℃×8h), the tensile strength of the progressive solution treatment is improved obviously and the conductivity of the sample is reduced very little.

Acknowledgments
Industrial research project (20153004-2) Supported by the Guizhou Province
Corresponding author: Xiaoyan Zhang (1960-), woman, professor, engaged in High-strength cast aluminium wires, non-ferrous metal surface modification Research. Tel: 0851-81158902, e-mail:ivzhangxiaoyan@163.com.

References

[1] Zwichau E C, Freiberg U T. Possibilities for the of industrial Cal culation heat treatment diagrams of industrial Al Zn Mg(Cu) alloys[J].Aluminum,1999,75:90-96.
[2] Kamran D, Atiye N. Interactive effects of aging parameters of AA6056 [J]. Metals and Materials International, 2011, 18(5): 757-767.
[3] Li J, Song R G, Chen Xiao M, Ma X C. Study on Strengthening Solution Treatment for 7050 High Strength Aluminum Alloy[J]. Hot Working Technology, 2009,38(6):125-128.
[4] Wang X X, M L. Effect of Rare Earth Elements in Aluminum Alloy[J]. Foundry Technology,2010,31(12):1617-1620.
[5] Zhang X M, Mao W M, Zhu W Z. Influence of Zn, Mg and Cu Contents on Hot Cracking Behavior and Microstructure of 7075 Aluminum Alloys[J]. Special Casting & Nonferrous Alloys, 2014, 34(12): 1336-1339.
[6] Zou L, Pan Q L, He Y B, et al. Effect of minor Sc and Zr addition on microstructures and mechanical properties of Al-Zn-Mg-Cu alloys[J]. Transactions of Nonferrous Metals Society of China, 2007,17(2):340-345.
[7] Chao R Z, Guan X H, Guan R G, et al. Effect of Zr and Sc on mechanical properties and

electrical conductivities of Al wires [J]. Transactions of Nonferrous Metals Society of China, 2014,24(10):3164-3169.

[8] You C Y. Mechanism of the heat-resistant aluminum alloy conductor and its application in power transmission lines. Electric wire & cable. 2004,04:4-10.

[9] Wu L, Dong W P, Research and Progress of Thermomechanical Treatment of Al Alloy. Hot Working Technology. 2014,43(04):27-31.

[10] Gao H X, Zhang X Y. The effect of Zr, Cu elements and rare earth Ce on microstructure and property of the electrical rod. Functional materials. 2015,3(46): 3073-3081.

[11] Han Y, Liu D Y. Influence of RE on Conductivity and Heat-resistant Properties of Aluminum Alloys. Hot working technology. 2014, 43(14):28–31.

[12] Chen W . Study on Rare Earth Aluminum Alloy Conductor Material of High-performance. Hefei: Hefei university of technology.2013.

[13] Kristian B P,David B,Peter K Kristensen,et al.Interface structure and strength of ultrasonically wedge bonded heavy aluminium wires in Si-based power modules[J].J Mater Sci: Mater Electron ,2014,25:2863–2871.

[14] Liu D Y, Li W J. Alloying design of a thermal-resistant aluminum alloy conductor material with high conductivity. Transactions of materials and heat treatment. 2014,35(06):17-21.

[15] Mcallen P,Phelan P.Ductile fracture by central bursts in drawn 2011 aluminium wire[J].International Journal of Fracture,2005, 135:19–33.

IOP Conf. Series: Materials Science and Engineering **207** (2017) 012061 doi:10.1088/1757-899X/207/1/012061

Research on Finite Element Model Generating Method of General Gear Based on Parametric Modelling

Yulong Lei, Bo Yan, Yao Fu*, Wei Chen& Liguo Hou

Jilin University, State Key Laboratory of Automobile Simulation and Control, Changchun 130022;

*Corresponding author e-mail: fu_yao@jlu.edu.cn

Abstract. Aiming at the problems of low efficiency and poor quality of gear meshing in the current mainstream finite element software, through the establishment of universal gear three-dimensional model, and explore the rules of unit and node arrangement. In this paper, a finite element model generation method of universal gear based on parameterization is proposed. Visual Basic program is used to realize the finite element meshing, give the material properties, and set the boundary / load conditions and other pre-processing work. The dynamic meshing analysis of the gears is carried out with the method proposed in this pape, and compared with the calculated values to verify the correctness of the method. The method greatly shortens the workload of gear finite element pre-processing, improves the quality of gear mesh, and provides a new idea for the FEM pre-processing.

1. Introduction

At present, in view of the problems such as NVH, lightweight, structural strength checking and so on, the traditional tedious and complicated analytical calculation method cannot meet the requirement of Engineering problems. With the development of computer simulation technology, the design and check of the finite element method has been widely used in automotive transmission, but the process in finite element calculations, with extremely tedious become the main problem in engineering and technical personnel .

In the finite element software, the finite element modeling of the gear and the application of the pretreatment conditions also depend on the manual operation. ANSYS, Hyper mesh, Abacus and other mainstream commercial finite element analysis software for the finite element analysis of the gear, and there is no better solution.

This paper first establish general gear tooth profile curve equation of the unified coordinate system, and then follow the INP file rule generating gear unit node coordinates and the corresponding arrangement of law, impose conditions with material properties and the boundary are given at the same time, finally compile executable program in the Visual Basic 6 environment .

2. Establishment of gear tooth profile equation

Taking the circle center of the gear as the coordinate origin, the gear tooth profile curve coordinate system is set up with the single tooth face as the left and right symmetry, and the gear tooth profile coordinate system is shown in figure 1. In the tooth profile curve coordinate system, the tooth profile

IOP Conf. Series: Materials Science and Engineering **207** (2017) 012061 doi:10.1088/1757-899X/207/1/012061

involute (2-3) equation and the tooth root transition curve (3-4) equation of a single gear are deduced theoretically.

2.1 gear tooth profile involute equation

The polar coordinate equation of involute gear tooth profile is shown in figure 2.

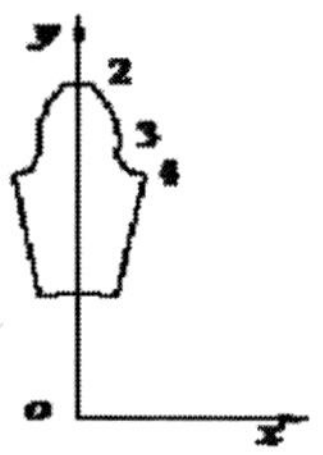

Fig.1 Coordinate of gear tooth profile curve

From the graphical parameters:

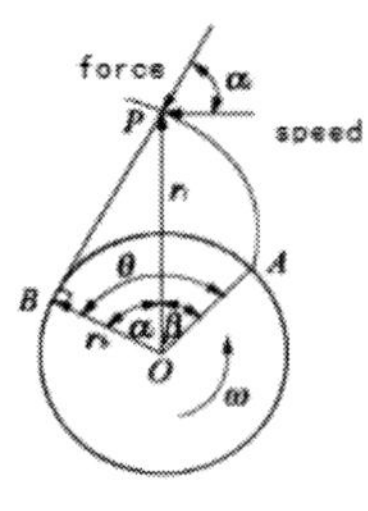

Fig.2 Polar coordinates of gear involute curve

$$\tan\alpha = \frac{\overline{BP}}{\overline{OB}} = \frac{\overset{\frown}{AB}}{r_b} = \frac{r_b(\alpha+\beta)}{r_b} = \theta \tag{1}$$

In the formula: ri is the radius of any point; rb is the radius of the base circle; α is the pressure angle at any point; θ is AB segment angle.α+β=θ.

The gear involute polar coordinate system is converted into a rectangular coordinate system, as shown in figure 3.

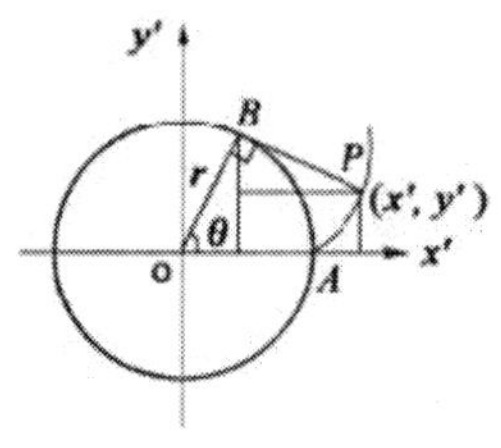

Fig.3 Rectangular coordinate system

In the xoy coordinate system, the coordinates of any point P (x, y) can be expressed as:

$$\begin{cases} x = x'\cos\delta + y'\sin\delta \\ y = -x'\sin\delta + y'\cos\delta \end{cases} \tag{2}$$

δ is the rotation angle.

Thus, the expression of P at any point on the involute gear tooth profile in xoy coordinate system can be obtained:

$$\begin{cases} x = (r_b \sin\theta - \tan\alpha_k \cdot r_b \cdot \cos\theta)\cos\delta - \\ (r_b \cos\theta + \tan\alpha_k \cdot r_b \cdot \sin\theta)\sin\delta \\ y = (r_b \sin\theta - \tan\alpha_k \cdot r_b \cdot \cos\theta)\sin\delta + \\ (r_b \cos\theta + \tan\alpha_k \cdot r_b \cdot \sin\theta)\cos\delta \end{cases} \quad (3)$$

2.2 dedendum transition curve equation

The gear machining process, gear cutter fillet processing gear root transition curve should be established, as shown in Figure 1 of the gear tooth profile curve coordinate system, get the right gear tooth root transition curve of the equation (3-4).

The schematic diagram of the tooth profile of the cutter is shown in figure 4.

The following parameters are shown in the diagram:

$$\begin{cases} a = h_a^* m + c^* m - r_\rho \\ b = \dfrac{\pi m}{4} + h_a^* m \times \tan\alpha + r_\rho \cos\alpha \\ r_\rho = \dfrac{c^* m}{1 - \sin\alpha} \end{cases} \quad (4)$$

Using the rack type cutter to process the gear, the gear tooth profile is the cutting tool processing line and the gear processing pitch circle (graduation circle) is tangent to the pure rolling to form, as shown in figure 5. P is a node, NN is the normal line of the contact point between the fillet of the tool and the transition curve, and the angle between the NN and the cutting line of the cutter. If the coordinate system is shown in Figure 5, the involute equation can be:

$$\begin{cases} x_2 = r\sin\varphi - \left(\dfrac{a}{\sin a'} + r_\rho\right) \times \cos(a' - \varphi) \\ y_2 = r\cos\varphi - \left(\dfrac{a}{\sin a'} + r_\rho\right) \times \sin(a' - \varphi) \end{cases} \quad (5)$$

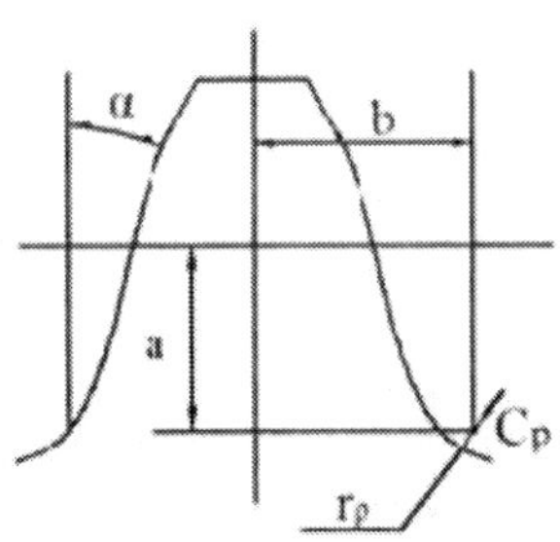

Fig.4 Schematic diagram of tool profile parameters

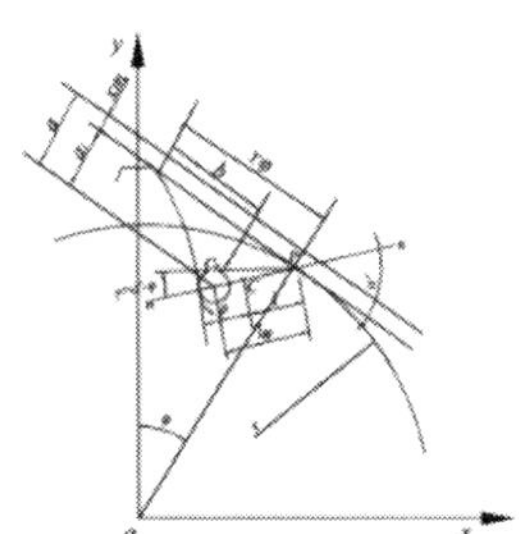

Fig.5 Schematic diagram

3. INP file unit and node writing rules

INP file is a text file, which contains a complete description of the model, in the preprocessor (such as ABAQUS/CAE) and Solver (ABAQUS/Standard or ABAQUS/Explicit) establishes a data transfer bridge [9]. At present, most of the former processors (such as MSC.PATRAN, FEMAP, and Hyper Mesh) are mostly supported in INP file format to output the model.

The INP file consists of a series of data blocks, each of which represents a particular part of the model. A data block is always started with a keyword with a "*", followed by a corresponding parameter, and one or more data lines.

For the preparation of the pretreatment conditions INP file rules, refer to the fourth pretreatment conditions.

4. Programming

4.1 establish the form of gear mesh

This paper adopts the gear mesh arrangement form as shown in Figure 6, the right end face of single tooth mesh arrangement form as shown in Figure 7, the gear meshing gear form can satisfy the calculating accuracy in stress and bending stress, and can make the gear mesh reasonably, improve computational efficiency, but also for different requirements of precision, but also through the position adjusting control parameters in Figure 8 and 9 point 10 points, the number of grid is more reasonable.

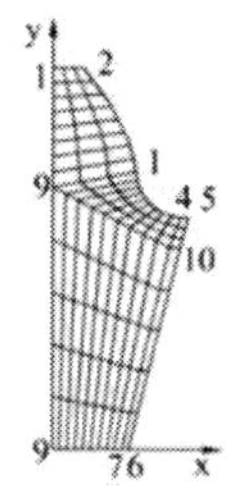

Fig.6 Overall arrangement form of gear mesh **Fig.7** Right end of the single tooth mesh

4.2 gear unit node coordinate generation method

Firstly, on the basis of the element node coordinates of the second section is the gear tooth profile curve equation to generate a single tooth right end, because the tooth on the symmetry properties of Y axis, then the element node coordinates can generate a single tooth left end, through the acquisition of spiral angle and coordinate rotation transformation formula of element node coordinates a tooth is obtained, and then the rotation transformation can be obtained by all the nodes of the whole gear coordinate values, and finally explores node numbering rules, and give the corresponding coordinate value of node number .

4.3 node numbering and distribution

According to the INP rules, to the gear unit node coordinates generated number, in addition to as far as possible to meet each node is connected, continuous changes, but also try to make the node number rationalization, provide the corresponding rules for subsequent node key arrangement unit. If the node number is not regular, it is almost impossible to arrange the information model of a gear model which contains thousands of nodes and node numbers.

5. Pretreatment conditions (material, assembly, contact, load, etc.)

5.1 give gear material properties

Define the gear material properties, the default material name is: elastic modulus of 210000 MPa, Poisson's ratio of 0.3, density of 7.8 x t/, and finally define the section properties.

5.2 analysis step definition

In the analysis step, we used 0.1s to display the dynamic analysis, and the time of the preset analysis is chosen. An explicit algorithm is used to determine the solution without any iteration, and the computational resources are less implicit than the previous incremental step.

5.3 field variables and historical variables

The contact pressure, contact force, stress, displacement and so on are the output variables of the field variables.

5.4 define contact and contact properties

(1) Create gear on contact properties.

(2) The selected contact involute tooth surface and fillet surface, tooth top surface node driven gear, which were placed into the unit set M_SURF-1, S_SURF-1.

(3) The creation of face-to-face contact and choose the calculation method of contact method, the finite slip formula.

5.5 set reference point

The main axis of the driven gear is taken as the reference point, and the coupling rigid constraint is established between the reference point and the inner ring and the profile.

6. Example

The preparation of gear finite element mesh generation based on parametric modeling program, GUI interface as shown in Figure 8, a calculation of dynamic stress of gear meshing gear meshing, the basic parameters of the large number of gear teeth was 24, the number of gear teeth was 20, 3mm module and pressure angle of 20 degrees addendum, 3mm, tooth width is 40mm, the large gear resistance moment is 100Nm.

The compiler generated INP file, the mesh as shown in Figure 9, the grid quality inspection shows that the finite element model of a total of 145920 units, of which the minimum Jacobian is 0.79, the mesh quality is good.

The INP file is imported into ABAQUS submitted by operation, the finite element calculation of stress nephogram and gear contact pressure contours are shown in Figure 10 and Figure 11, get the maximum contact stress is 591MPa, the maximum bending stress is 128.6MPa.

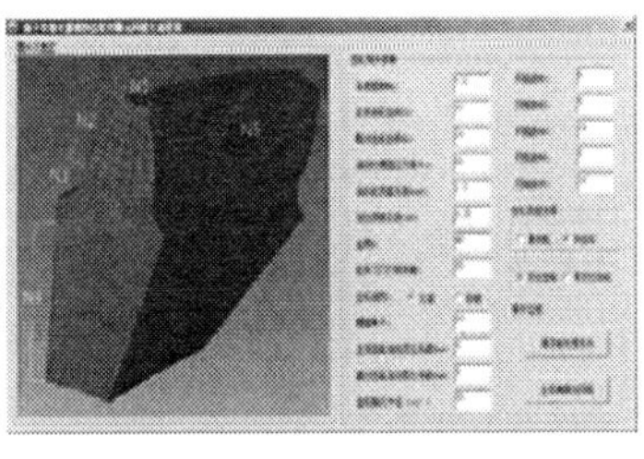

Fig.8 gear parameter input GUI

Fig.9 numerical example

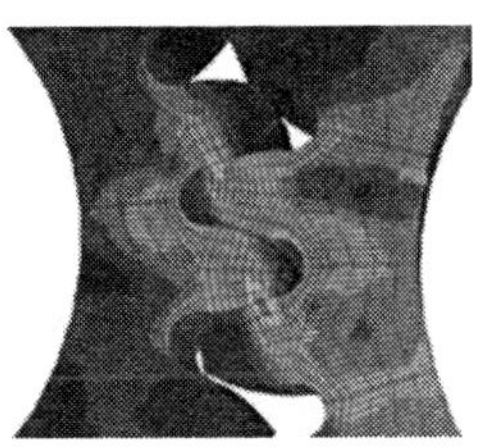

Fig.10 finite element calculation results

Fig.11 gear contact

Contact stress:

$$\sigma_{Hi} = Z_H Z_E Z_\varepsilon Z_\beta Z_{BD} \sqrt{\frac{2000T_i}{d_1^2 b}\frac{u+1}{u} K_{Vi} K_{H\beta i} K_{H\alpha i}} \tag{6}$$

Bending stress:

$$\sigma_{Fi} = \frac{2000T_i}{d_1^2 b} Y_F Y_s Y_\beta K_{Vi} K_{F\beta i} K_{F\alpha ii} \tag{7}$$

By theoretical calculation, the contact stress Hi is 573.1MPa, the bending stress Fi is 120.5MPa.

According to the calculation results, the tooth surface contact stress, finite element calculation results and theoretical calculation results by 3.29%, bending stress, finite element calculation results and theoretical calculation results is 6.72%, due to the application of the penalty function in the finite element model, the differences in the reasonable range, which verified the validity of the established model, and verified the correctness of the proposed modeling based on parameterized finite element model of gear generating method.

7. Conclusion

(1)The gear tooth surface equation of general unified coordinates are deduced according to INP file rule generation gear mesh node coordinates, grid node distribution inquiry unit, thus obtained gear finite element mesh node information.

(2)We propose a method to generate value coordinate numerical iterative linear equation based on the method of obtaining the gear generation node finite element mesh information. The method of arranging the nodes of the mesh based on the meshing parameters is given.

(3)The INP file to add gear contact stress and bending stress for pretreatment conditions, such as material properties, boundary conditions and loads.

(4)Finally through a pair of gear case is verified by the program to calculate the gear contact stress and bending stress is consistent with the ISO, to verify the feasibility of the method.

(5)For the involute tooth root transition curve fitting is good, can be used for precise gear calculation.

Acknowledgment

This work was supported by International S&T Cooperation Program of China（Grant No. 2014DFA71790）,the Key Scientific and Technological Project of Jilin Province（Grant No. 20160519008JH）， Chinese National Natural Science Foundation(Grant No. 51575220) and Graduate Innovation Fund of Jilin University（Grant No. 2016170）".

References

[1] YANG Shenghua. Finite element analysis of gear contact [J]. Chinese Journal of Computational Mechanics, 2003, 02: 189-194.

[2] HAO Dongsheng, WANG Delun. An approach to building a FEM of helical gears with accuracy contact analysis [J]. Journal of Dalian University of Technology, 2011, 06: 825-831.

[3] LI Chang, HAN Xing. Digital Finite Element Simulation of Gear Box Based on Explicit Dynamics [J]. Acta Armamentaria, 2009, 07: 978-983.

[4] TANG Jinyuan, PU Taiping. Spiral Bevel Gear Meshing Stiffness Calculations Based on the Finite Element Method [J]. Journal of Mechanical Engineering, 2011, 11: 23-29.

[5] Sun Huan, Chen Zuomo. Mechanical principle (Seventh Edition) [M]. Beijing: Higher Education Press, 1997: page.

[6] Wu Jize, Wang Tong. Root transition curve and root stress [M]. Beijing: National Defence Industry Press, 1989: page.
[7] Wu Xutang. Gear meshing principle [M]. Beijing: Mechanical Industry Press, 1984: page.
[8] Fu. Differential geometry and gear meshing principle [M]. Beijing: Petroleum University Press, 1999: page.
[9] Shi Yiping, ABAQUS finite element analysis examples [M]. Beijing: Machinery Industry Press, published: page number:page.
[10] Zhuang Zhuo, by Ogawa, Liao Jianhui, Shen Xinpu, Liang Minggang, cen song, based on ABAQUS finite element analysis and application [M]. Beijing: Tsinghua University press, page 2009: page.
[11] Cao Jinfeng, Shi Xu. Finite element analysis of the ABAQUS FAQ [M]. Beijing: Mechanical Industry Press, 2009: page.
[12] HOU Xiangying, FANG Zongde, CAI Xiangwei, NING Chengfeng. Pretreat ment and Post Treatment of Tooth Contact Analysis Based on ABAQUS [J]. Mechanical Science and Technology for Aerospace Engineering, 2015, 07: 993-996.
[13] TANG Jinyuan, ZHOU Wei, and CHEN Siyu. Contact-impact Analysis of Gear Transmission System [J]. Journal of Mechanical Engineering, 2011, 07: 22- 30.
[14] TANG Jinyuan, LIU Xin, And DAI Jin. Study on corner contact shock of gear transmission by ANSYS/LS-DYNA software [J]. Journal of Vibration and Shock, 2007, 09: 40- 44+ 50+ 169.

IOP Conf. Series: Materials Science and Engineering **207** (2017) 012062 doi:10.1088/1757-899X/207/1/012062

Mechanical Deformation Mechanism of Fiber Reinforced Composites T-joints

Wei Su[1, 3 *], Ke-qiang Cheng[1], Ren-huai Liu[1, 2, 3], Shi-qing Huang[2, 3]

[1]Scienceand Technology on Reliability Physics and Application Technology of Electronic Component Laboratory, China Electronic Product Reliability and

Environmental Testing Research Institute, Guangzhou, China
[2]Department of Mechanics and Civil Engineering, Jinan University, Guangzhou, China
[3]MOE Key Lab of Disaster Forecast and Control in Engineering, Jinan University, Guangzhou, China

*Corresponding author e-mail: suwei5@126.com

Abstract. In this paper, we focus on the need for ideal theory analytical solution of mechanical deformation mechanism for T-joints. Through removing unnecessary constraints, stress equivalence system of T-joints can be established by force method. Solution equations of unknown nodal forces and the total vertical displacement can also be obtained, and this result can be used to evaluate strength of FRC T-joints.

1. Introduction

Fiber reinforced composite materials (FRC) have advantages of strength and modulus for the super ratio, and should be widely used in modern aerospace structures. As a kind of typical structural unit, T-joints is made up of three pieces of laminated plates and filler. Photograph of fiber reinforced composites T-joints is shown in Fig.1. This structure has assembly advantage over other connection of engineering. T-joints can save a lot of mechanical connection, further reduce the weight and improve the fatigue life of structure.

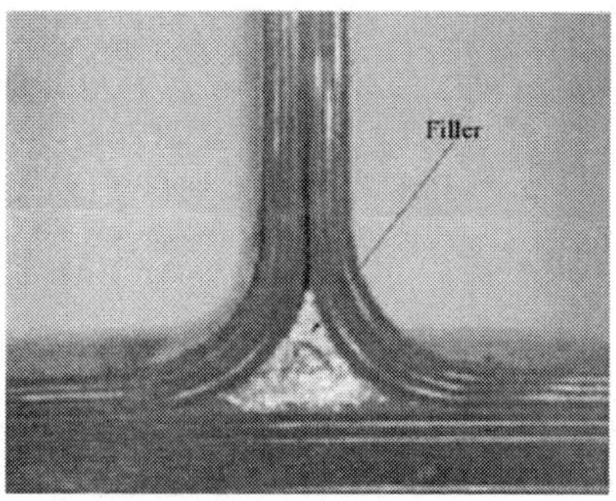

Figure 1. Photograph of fiber reinforced composites T-joints

On the other hand, the industry is slow to accept this joint structure, because the main reason for hindering the joint structure to get more extensive application is lack of information on the reliability and understanding of its mechanical deformation mechanism [1-3]. Composite joint development is a complex and comprehensive strong technical problem which is considered bearing capacity as the main index of the design requirements, the varieties and properties of structure, fatigue life, experimental and testing, manufacturing process and a series of problems. Some researchers [4-7] analyze finite element method of T-joints structure strength. Because of the complexity of the problem, it is difficult to get ideal theory analytical solution.

2. Deformation theory model of T-joints

T-joints is composed of three co-curing laminated plates in this paper, with film bonding interface and solidification. Arc transitional area is in the Root of joint, and the joints interior is full of filler commonly used foam or one-way belt. Padding around is connected with the laminated plates by film. T-joints measures 200 x 50 x 120mms. The sub-layer number of layer 1, 2, laminated plates is 13 layer, Angle of the layer and the order is (-45/0/45/0/-45/0/0/0/45/0/-45/0/45) with thickness of 1.5 mm. layer 3 laminated plates is 32 layer, and angle of the layer and the order is (45/0/-45/0/0/45/0/-45/0/0/45/0/-45/0/45/-45)with thickness of 4 mm.

Considering the structure symmetry, half of the T-joints is taken to analyze. The length of the AB is l_1, and the height is h_1. The length of the BC is l_2, and the height is h_2. BD Is the circular arc with radius R, The stiffness of AB, BC, BD is EI_1, EI_2, EI_3. Pull force F act on area of D, First, the constraint reaction everywhere can be calculated, then bending moments can also be obtained.

The strength and stiffness value of filler material are often far less than the fiber reinforced composite material, so influence on deformation and damage of T-joints by filler can be ignored. The stress drawing of T-joints is shown in Fig.2.

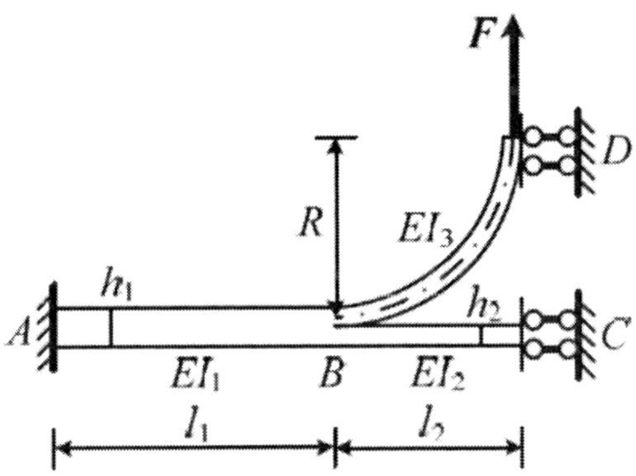

Figure 2. Stress drawing of T-joints

Assuming the width unit of component is 1, section inertia of each part can be obtained as follows.

$$I_1 = \frac{h_1^3}{12}, \quad I_2 = \frac{h_2^3}{12}, \quad I_3 = \frac{(h_1 - h_2)^3}{12} \tag{1}$$

3. Mechanical deformation mechanism

Support reaction of the fourth statically indeterminate structures can be solved by force method, and stress equivalence system of T-joints can be obtained through removing unnecessary constraints of C, D. Combined equations have to be solved in the analysis of the super static determined structure,

And unknown nodal forces can be replaced by X_1, X_2, X_3, X_4. The stress equivalence system of T-joints is shown in Fig.3.

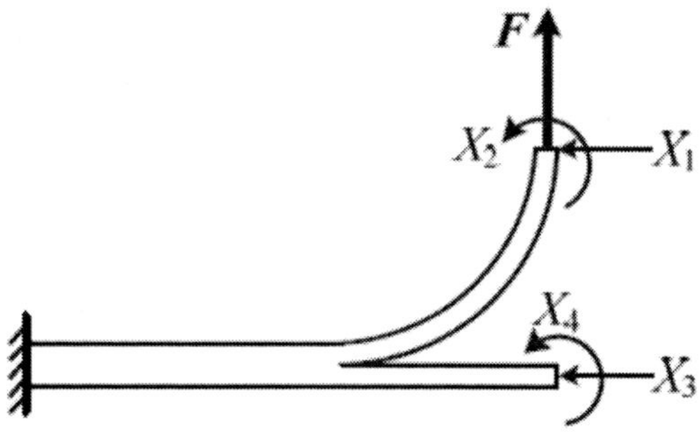

Figure 3. The stress equivalence system of T-joints

Solution equations of unknown nodal forces can be expressed as follows:.

$$\begin{aligned}
&\delta_{11}X_1 + \delta_{12}X_2 + \delta_{13}X_3 + \delta_{14}X_4 + \Delta_{1P} = 0 \\
&\delta_{21}X_1 + \delta_{22}X_2 + \delta_{23}X_3 + \delta_{24}X_4 + \Delta_{2P} = 0 \\
&\delta_{31}X_1 + \delta_{32}X_2 + \delta_{33}X_3 + \delta_{34}X_4 + \Delta_{3P} = 0 \\
&\delta_{41}X_1 + \delta_{42}X_2 + \delta_{43}X_3 + \delta_{44}X_4 + \Delta_{4P} = 0
\end{aligned} \tag{2}$$

In the equation, each coefficient and free items can be obtained by each mechanical subsystem. The coefficient substituted into the force method equation can solve the unknown reactions as follows:

$$\begin{aligned}
X_1 &= \frac{2F[-I_3 l_1^2(\pi - 2) + I_1(\pi - 4)R^2]}{I_1(\pi^2 - 8)R^2} \\
X_2 &= -\frac{F[I_3 l_1^2(\pi - 4) + I_1(\pi^2 - 2\pi - 4)R^2]}{I_1(\pi^2 - 8)R} \\
X_3 &= \frac{F\{2I_3 l_1^2 l_2(8 - 3\pi) + R[(-I_2 l_1^2(\pi^2 - 8) + 8I_1 l_2(\pi - 3)R]\}}{(h_1 - h_2)I_1 l_2(\pi^2 - 8)R} \\
X_4 &= -\frac{FI_2 l_1^2}{2I_1 l_2}
\end{aligned} \tag{3}$$

Vertical displacement under the action of X_1, X_2, X_3, X_4, F can be expressed as follows:.

$$\begin{aligned}
\Delta_{X_1} &= \frac{1}{EI_1}[\int_0^{l_1} X_1 \times R(l_2 + x)dx] \\
\Delta_{X_2} &= \frac{1}{EI_1}[\int_0^{l_1} X_2 \times (l_2 + x)dx] \\
\Delta_{X_3} &= \frac{1}{EI_1}[\int_0^{l_1} -(l_2 + x) \times X_3 \times \frac{h_1 - h_2}{2} dx] \\
\Delta_{X_4} &= \frac{1}{EI_2}[\int_0^{l_2} X_4 \times x dx] + \frac{1}{EI_1}[\int_0^{l_1} X_4 \times (l_2 + x)dx] \\
\Delta_F &= \frac{1}{EI_1}[\int_0^{l_1} F \times (R + x) \times (l_2 + x)dx]
\end{aligned} \tag{4}$$

The total vertical displacement of C can be expressed by $\Delta_C = \Delta_{X_1} + \Delta_{X_2} + \Delta_{X_3} + \Delta_{X_4} + \Delta_F$, then putting equation (4) substituting into equation (3), the equation (5) can be obtained as follows:

$$\begin{aligned}\Delta_C = Fl_1\{&-6h_1(h_1-h_2)^3 l_1^2 l_2(l_1+2l_2)(3\pi-8) - l_1[3h_1h_2^3 l_1(l_1+2l_2)+(h_1^3h_2l_2-h_1^4l_2)\\ &(4l_1+3l_2)](\pi^2-8)R + 24h_1^4 l_2(l_1+2l_2)(\pi-3)R^2\} / Eh_1^6(h_1-h_2)l_2(\pi^2-8)R\end{aligned} \tag{5}$$

4. Conclusion

The result of research shows that T-joints support reaction of statically indeterminate structures can be solved by force method and stress equivalence system of T-joints can be obtained through removing unnecessary constraints. Solution equations of unknown nodal forces and the total vertical displacement can be obtained. Ideal theory analytical solution of mechanical deformation for T-joints is established at last and this result can be used to strength evaluation of FRC T- joints.

Acknowledgments

This work was financially supported by Key Lab Fund (JAB1728150) and the Distinguished Young Scientist Program of Guangdong Province (2015A030306002), National Natural Science Foundation of China (51505089), Natural Science Foundation of Guangdong Province (2016A030313672) and Pearl River S&T Nova Program of Guangzhou (2014J2200086), Science and Technology Research Project of Guangdong (2015B090912002, 2015B090901048).

References

[1] Wong C M S,L Matt hews F.A finite element analysis of single and two2hole bolted joint s in fibre reinforced plastic[J]. J ournal of Composite Materials 1981, 15: 481 - 491.
[2] Lie S T, G Yu,Z Zhao. Analysis of mechanically fastened composite joint s by boundary element methods[J].Composites Part B,2000, 31: 693 - 705.
[3] Tsujimoto Y,D. Wilson.Elasto plastic failure analysis ofcomposite bolted joints[J]. J ournal of Composite Materials,1986, 20: 236 - 252.
[4] Chang F K,Y Chang K.Post failure analysis of bolted composite joint s in tension or shear out mode failure[J]. Journal of Composite Materials,1987, 21: 809 - 833.
[5] Lessard L B, Shokrieh M M.Two dimensional modeling of composite pinned joint failure [J]. Journal of Composite Materials Science and Engineering A,1995, 29: 671 - 697.
[6] Hung C L, Chang F K.Strengt h envelope of bolted composite joint s under bypass loads [J]. Journal of Composite Materials 1996, 30: 1402 - 1435.
[7] Hung C L,K. Chang F.Bearing failure of bolted composite joints: Model and verification [J]. Journal of Composite Materials Science and Engineering A,1996, 30: 1359 - 1400.

IOP Conf. Series: Materials Science and Engineering **207** (2017) 012063 doi:10.1088/1757-899X/207/1/012063

Fracture failure analysis of Fiber Reinforced Composites T-joints

Wei Su[1, 3 *], De-zhi Kong[1], Ren-huai Liu[1, 2, 3], Shi-qing Huang[2, 3]

[1]Scienceand Technology on Reliability Physics and Application Technology of Electronic Component Laboratory, China Electronic Product Reliability and Environmental Testing Research Institute ,Guangzhou, China

[2]Department of Mechanics and Civil Engineering, Jinan University, Guangzhou, China
[3]MOE Key Lab of Disaster Forecast and Control in Engineering, Jinan University, Guangzhou, China

*Corresponding author e-mail: suwei5@126.com

Abstract. This paper discusses staple FRP T-joints on the project based on the classical beam theory of Euler-Bernoulli and a theoretical model of energy release rate is set up to discuss fracture propagation of composite T-joints. The paper is further analysis on fracture failure affected by composite pane depth and joints chamfer dimension parameters and so on. The conclusion benefit for extensive use of projects is also summarized.

1. Introduction

In recent years, Fiber reinforced composite materials (FRC) T-joints get more and more extensive Application in the actual projects. The strength is affected severely because stress concentration is around the structure cracks. As a kind of typical structural unit, T-joint is made up of three pieces of laminated plates and filler. Fiber reinforced composites T-joints is shown in Fig.1. T-joints can save a lot of mechanical connection, further reduce the weight and improve the fatigue life of structure.

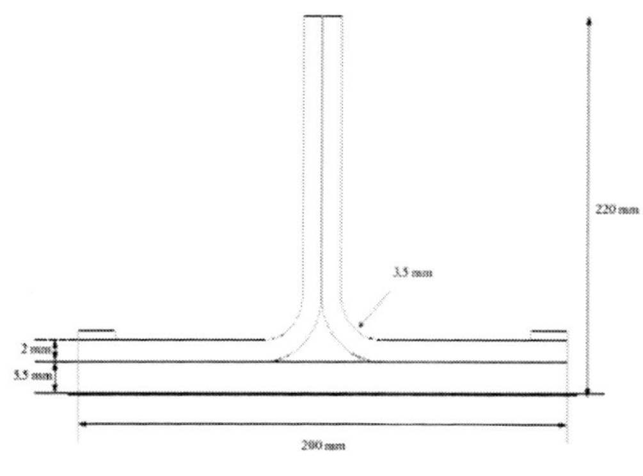

Figure 1. Fiber reinforced composites T-joints

Composite joint development is a complex and comprehensive strong technical problem which is considered bearing capacity as the main index of the design requirements [1-4], the varieties and properties of structure, fatigue life, experimental and testing, manufacturing process and a series of problems. Pradhan [5] uses the method of finite-element to research the influencing factors of strength in mixed joints composites. Kayapo and Dzenis [6] set up the crack of nonlinear finite element model of single lap joint. Strain energy release rate is preferred parameters of fracture mechanics method, and in this paper FRP T-joints on the project based on the classical beam theory of Euler-Bernoulli and a theoretical model of energy release rate is set up to discuss fracture propagation of composite T-joints.

The theoretical model of energy release rate for T-joints

According to the theory of brittle fracture of Griffith, T-type cracks need to absorb energy, at the same time release part when they propagate and form new surface. If the energy absorption rate is greater than the strain energy release rate, crack will be stable, otherwise unstable. Considering the structure symmetry, half of the T-joints is taken to analyze. The length of the AB is l_1, and the height is h_1. The length of BC is l_2, and the height is h_2. BD Is the circular arc with radius R, The stiffness of AB, BC, BD is EI_1, EI_2, EI_3. Pull force F act on area of D, First, the constraint reaction everywhere can be calculated, and bending moments can also be obtained. The strength and stiffness value of filler material are often far less than the fiber reinforced composite material, so influence on deformation and damage of T-joints by filler can be ignored. Simplified free-body diagram of T-joints is shown in Fig.2.

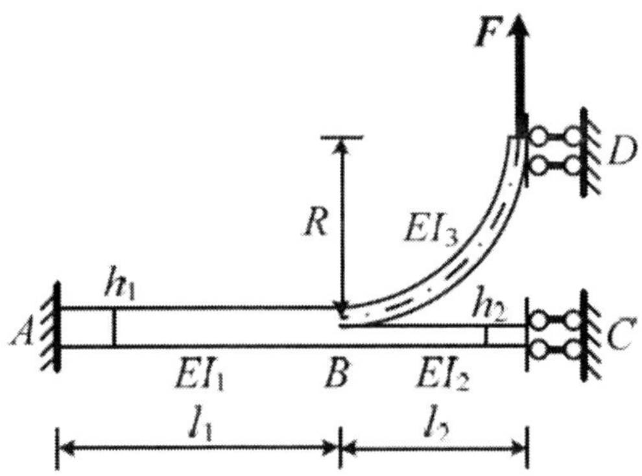

Figure 2. Simplified free-body diagram of T-joints

On the basis of calculation results, the total vertical displacement of C can be expressed by equation (1).

$$\Delta_C = Fl_1\{-6h_1(h_1-h_2)^3 l_1^2 l_2(l_1+2l_2)(3\pi-8)-l_1[3h_1h_2^3l_1(l_1+2l_2)+(h_1^3h_2l_2-h_1^4l_2)(4l_1+3l_2)](\pi^2-8)R+24h_1^4l_2(l_1+2l_2)(\pi-3)R^2\}/Eh_1^6(h_1-h_2)l_2(\pi^2-8)R \tag{1}$$

Elastic strain energy in linear elastic system can be expressed as follows:

$$U=\frac{1}{2}F\Delta_C \tag{2}$$

According to the theory of brittle fracture of Griffith, when the external load is constant, can be expressed as follows:

$$G = \frac{1}{\pi}\frac{\partial U}{\partial R} = \frac{F}{2\pi}\frac{\partial \Delta_C}{\partial R} \tag{3}$$

Then putting equation (1) substituting into equation (3), the equation (4) can be obtained as follows:

$$G = \frac{6F^2 l_1 (l_1 + 2l_2)[(h_1 - h_2)^3 l_1^2 (3\pi - 8) + 4h_1^3 (\pi - 3)R^2]}{2\pi E h_1^5 (h_1 - h_2)(\pi^2 - 8)R^2} \tag{4}$$

Numerical experiments

T-jionts is composed of three co-curing laminated plates in this paper, with film bonding interface and solidification. Arc transitional area is in the Root of joint, and the joints interior is full of filler commonly used foam or one-way belt. Set F=20kN, E=135GPa, π=3.14, R=10mm, l_1=90mm, l_2=10mm, h_1=5.5mm, h_2=4mm, the thickness of the skin h_2 and the flange h_1- h_2, chamfering radius R, the length of the flange l_1+ l_2 can be gained. This paper discusses the relationship between the change flange and skin thickness and energy release rate in the case of constant force F, also discusses the relationship between the change flange length and the chamfering radius and energy release rate. The Relational graph of G and h2 when h1=5, 6, 7, 8mm can be shown in Fig.3

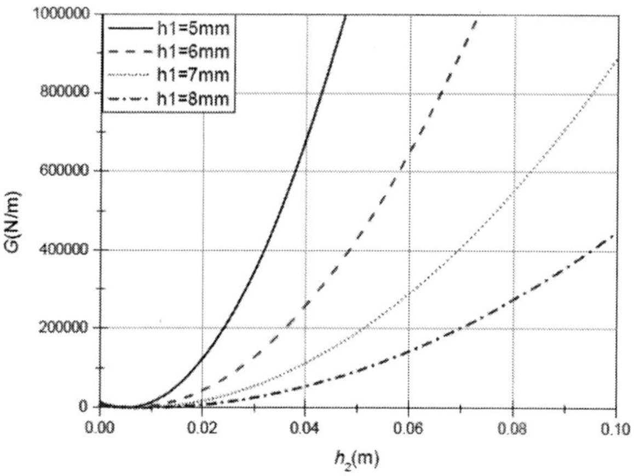

Figure 3. Relational graph of G and h2 when h1=5, 6, 7, 8mm

Also, given the height of the flange and the skin, we also can analyze the influence of the flange length and the change of the chamfering radius on energy release rate, as shown in Fig.4 and Fig.5.

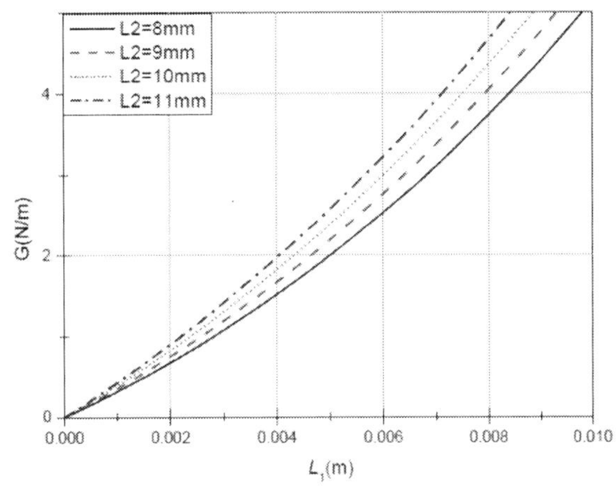

Figure 4. Relational graph of G and l1 when l2=8, 9, 10, 11mm

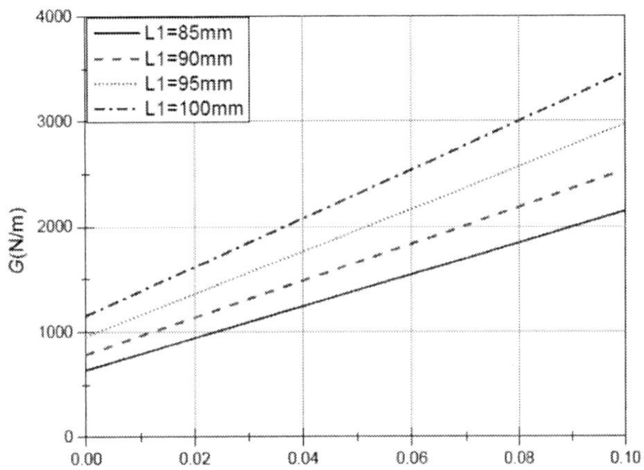

Figure 5. Relational graph of G and l2 when l1=85, 90, 95, 100mm

Conclusion

The result of research shows that based on the classical beam theory of Euler-Bernoulli, fracture failure is affected by composite pane depth and joints chamfer dimension parameters and so on. From Fig.3, energy release rate G increases with the sum of the flange thickness and the thickness of skin.As shown in Fig.4 and Fig.5, energy release rate G increases with the length of the flange. The conclusion Provides a valuable reference to further study the T-joints mechanical performance under complex conditions.

Acknowledgments

This work was financially supported by Key Lab Fund (JAB1728150) and the Distinguished Young Scientist Program of Guangdong Province (2015A030306002), National Natural Science Foundation of China (51505089), Natural Science Foundation of Guangdong Province (2016A030313672) and Pearl River S&T Nova Program of Guangzhou (2014J2200086), Science and Technology Research Project of Guangdong (2015B090912002, 2015B090901048).

References

[1] Pradhan S C,Y Lam K.A finite element approach, International Journal Adhesion and Adhesives[J]. TayTE.Deternination of fracture parameters of laminated thermoplastic

composite materials,2000, 20: 395-401.
[2] Kayupov S,Dzenis.In-plane biaxial crushing of honeycombs- Part II: analysis[J].1999, 36(29):4397-4423.
[3] Xie D, Chung J, M Waas A,al et.Failure analysis of adhesively bonded structures:from coupon level data to structure level predictions and verification[J]. INTERNATIONAL J ournal of fracture,2005, 134:231-250.
[4] Wong C M S,L Matt hews F.A finite element analysis of single and two2hole bolted joint s in fibre reinforced plastic[J]. J ournal of Composite Materials 1981, 15: 481 - 491.
[5] Lie S T, G Yu,Z Zhao. Analysis of mechanically fastened composite joint s by boundary element methods[J].Composites Part B,2000, 31: 693 - 705.
[6] Tsujimoto Y,D. Wilson.Elasto plastic failure analysis ofcomposite bolted joints[J]. J ournal of Composite Materials,1986, 20: 236 - 252.

IOP Conf. Series: Materials Science and Engineering **207** (2017) 012064 doi:10.1088/1757-899X/207/1/012064

Effectiveness Evaluation Method of Anti-Radiation Missile against Active Decoy

Junyao Tang[a], Fei Cao[b] and Sijia Li[c]

Xi'an Research Institute of High Technology, Xi'an, China

[a]514492278@qq.com, [b]caofeimm@163.com, [c]13259798735@163.com

Abstract. In the problem of anti-radiation missile against active decoy, whether the ARM can effectively kill the target radiation source and bait is an important index for evaluating the operational effectiveness of the missile. Aiming at this problem, this paper proposes a method to evaluate the effect of ARM against active decoy. Based on the calculation of ARM's ability to resist the decoy, the paper proposes a method to evaluate the decoy resistance based on the key components of the hitting radar. The method has the advantages of scientific and reliability.

1. Introduction

Anti-Radar Missile (Anti-Radar Missile), also known as anti-radar missiles, is a special attack radar and other radiation source of fast, accurate tactical missile. It guides the enemy radar as a guide, tracks and destroys the radiation source. In the modern war, such as the Gulf War, the Kosovo war, the Vietnam War in the hands and feet, has become a veritable "radar killer". In order to resist the threat of ARM to the entire information war, a series of defensive measures have appeared. Active decoy technology has become one of the most effective anti-ARM methods, which greatly affects the missile hit accuracy. In order to improve the development of ARM technology, it is necessary to study the method of evaluating the effect of decoy.

2. Effect evaluation method

The effectiveness of anti-radiation missile anti-active decoy, anti-decoy effect, not simply to ARM hit the target and the miss two logic to judge. In order to obtain the operational effectiveness, the damage effect of the radar and bait should be evaluated. The western countries on the typical target damage specific classification, such as the flight is divided into: Class KK (the first time the complete destruction of the target (five minutes), Class Class A (five minutes to eliminate the target),Class B (two hours to eliminate targets), C (the main task that the enemy wants to complete is blocked, it can only be used for secondary purposes), Class D (resulting in additional repair and repair of the enemy), Class E (resulting in the enemy is difficult to complete the task) and other damage levels. Armored vehicles are divided into: Class K (completely destroyed), Class F (fire loss), Class M (caused by the inability to maneuver) and other damage levels. Therefore, the evaluation of anti-decoy effect can refer to the above damage classification level, and estimate the damage probability of ARM warhead to the target. In the case of non-confrontation environment and system failure, the damage probability of single ARM to a single radiation source is called ARM warhead damage probability [1].

3. The calculation of ARM anti decoy damage probability

The damage probability is one of the most important indexes to evaluate the performance of ARM against active decoy. The damage probability includes the damage probability of the protected radar and the damage probability of the i active decoy.The traditional evaluation method is based on the distance from the point source of the ARM to the source of the ARM, which can be used to deduce the probability of the radar seeker [2].Let ARM fight into the category of tribal decoy array, if the target and the radar active decoy is effective damage, the distance between the requirements (1):

$$|O_iO''| \le R_d + R_r \tag{1}$$

R_d Is the ARM warhead damage radius, R_r is a safe point for radiation point source. $|O_iO''|$ is the distance between the final placement of the ARM warhead and the i-th point source. Assuming the placement of O'' coordinates is (x'', y'', z''), the coordinates of the i-th point sources is (x_i, y_i, z_i).So then

$$|O_iO''| = \sqrt{(x_i - x'')^2 + (y_i - y'')^2 + (z_i - z'')^2} \tag{2}$$

There are N ARM drop points on the location of the source-induced system, If the ARM placement position and the i-th point source do not meet the above inequality have n_i points, then the i-th point source damage probability is:

$$P_i = {}^{n_i}\!/_{N} \tag{3}$$

According to the probability of knowledge available, the probability of damage to the entire active induced system is:

$$P_{总} = \sum_{i=0}^{n} P_i \tag{4}$$

Based on the above description of the probability of damage, this article will ARM hit the i-th radiation source probability is defined as:

$$P_i = \begin{cases} 0 & |O_iO''| > R_d + R_r \\ \dfrac{(R_d + R_r)^2}{|O_iO''|^2} & |O_iO''| \le R_d + R_r \end{cases} \tag{5}$$

There are N ARM drop points on the location of the source-induced system, and the average of N points is:

$$\overline{P_i} = \frac{1}{N}\sum_{j=1}^{N} P_i(j) \tag{6}$$

$P_i(j)$ Represents the probability of damage to the i-th point source. The probability that the i-th radiation source and the entire active ejection system are destroyed are:

$$P_i = \frac{1}{N}\sum_{j=1}^{N} P_i(j) \quad P = \frac{1}{N}\sum_{i=0}^{n}\sum_{j=1}^{N} P_i(j) \tag{7}$$

4. Evaluation method of anti - seizure based on hit down of radar key components

Damage probability depends on the vulnerability of the radiation source, ARM warhead fuze parameters, seeker guidance parameters, ARM with the radiation source of the encounter conditions. In the same coordinate system, the coordinate distribution density $\varphi(X,Y,Z)$ of the drop point (X,Y,Z) can be called the error rule. If the ARM warhead at the point of explosion, there will be a corresponding with the corresponding radiation source damage probability, called the coordinate damage law [3].Coordinate damage laws are closely related to ARM warhead efficiency and radiation source vulnerability. Because the position of ARM drop point has a certain randomness, the probability of anti-perturbation is a reasonable probability of damage, and the conditional distribution density is related to ARM guidance error and fuze start law. According to the probability formula, the probability of successful single-shot is:

$$P = \int\int_{-\infty}^{\infty}\int G(X,Y,Z)\varphi(X,Y,Z)dXdYdZ \tag{8}$$

ARM warheads are usually fragmented. The warhead uses the high-speed collision of the fragments to effectively dampen the radiation source such as radar. Thus, the probability of damage is related to the number of fragments and the motion parameters of the fragment at the point of impact of the radar radiation source. Fragmentation Warhead Coordinate Damage Probability Description Method is characterized by the probability of the damage of the coordinates written as a single fragment damage probability function. The damage probability of the fragment is related to the motion parameters of the radar radiation source, the vulnerability of the radiation source of the radar and the distance from the radiation source to the radar radiation source. The number of fragments of the radar radiation source is related to the coordinates of the landing point and the critical size and distribution. According to the characteristics of radar radiation source and the power of warhead, the relationship between damage probability and drop point coordinates is established by using geometric relation and vector synthesis.

The application of analytical calculation method to determine the ARM attack efficiency index is difficult, but can be simplified, the analytical solution obtained, the biggest advantage of this solution is more intuitive, has great reference value for understanding ARM's attack efficiency.

The fragment of ARM warhead is composed of metal shell. In order to meet the requirements of a certain shape, size and weight of fragments, the prefabricated or semi prefabricated shell structure is adopted in the design of the shell, so as to obtain the fragments with relatively regular shape, weight and quantity. For this kind of shell structure, some parts of the shell metal explosion after being crushed into a small powder particles, the radar radiation source is not effective, the weight loss of fragments in general about 10%~15%.

The warhead is detonated by a laser proximity fuze or a trigger fuze, and the velocity of ARM is relative to the vector of the ARM velocity and the initial velocity of the fragment. Even if all the fragments of the same growth rate, because the moving direction of the fragment, the synthesized value of initial velocity of fragments is also different. The initial velocity of the fragment relative to the ARM projectile at the tip of the warhead is one of the research aspects [4].

Calculate the initial velocity of the fragments when abroad often use Gurney equation, the premise is: product expansion speed along the direction of the linear distance distribution; ARM warhead fuze signal instantaneous detonation; loaded warhead in explosive energy in internal energy conversion for all products and product shell energy. The law of conservation of energy:

$$V_0 = D_e\sqrt{\frac{\beta}{5(\beta+2)}} \tag{9}$$

Among them, D_e is the instantaneous detonation velocity of explosive; β is the mass ratio of explosive loaded to the shell.

Warhead detonated after a damage zone in space, with the effective damage radius, direction angle and scattering angle to describe the damage zone size and shape. Radius of damage with vulnerability of radar radiation source is not a different direction angle and scattering angle of flight varies with ARM. Therefore, the research on the size and shape of the damage area should be combined with the characteristics of radar emitter, warhead and the state of the encounter. The first analysis of the static scattering feature of the fragments, and then study the dynamic scattering characteristics.

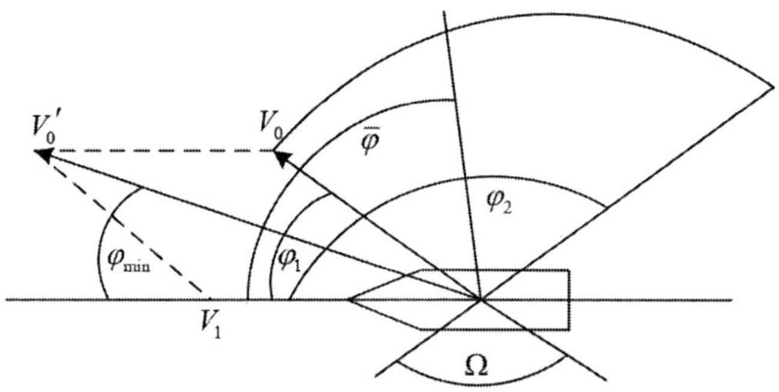

Fig.1 Schematic diagram of the fragment

Warhead shape, structure and loading of explosives and detonating manner decided static scattering characteristic fragment. And the scattering characteristics are defined by the parameters such as the scattering angle Ω, the mean direction angle $\bar{\varphi}$ and the variance σ_φ of the distribution density of the fragments. Flying angle Ω in the axial section, the cone angle centered on the center of gravity of the charge, including 90% effective fragments:

$$\Omega = \frac{40\sqrt{Q_V}}{D_e\sqrt{\frac{1}{K_a}-\frac{1}{2}}}(\cos\varphi_2 - \cos\varphi_1) \tag{10}$$

Q_V Is the explosion to produce heat; K_a = explosive weight / (explosive weight + shell metal weight); φ_1, φ_2 is the angle between the shell surface and the wave direction when the detonation wave reaches the front and back limits. The average direction angle $\bar{\varphi}$ refers to the angle between the axis of the axle and the center of the distribution of the density of the core. For the case where the fragments are completely symmetrical along the distribution center, the geometric bisector of the flying angle is both the centerline. Fragment distribution function in the form of a normal distribution, from the above we can see that 90% of the number of fragments in the scattering angle is valid, then the distribution of variance σ_φ and the relationship between the scattering angle:

$$\sigma_\varphi = \frac{1}{3.3}\Omega \tag{11}$$

The dynamic scattering velocity V_{0g} is the vector sum of the velocity V_m and the static scattering velocity X, and the mathematical expectation of the scattering direction and the dynamic scattering velocity is:

$$\begin{aligned} \overline{\varphi'} &= \arctan\frac{\overline{V_0}\sin\overline{\varphi_c}}{\overline{V_0}\sin\overline{\varphi_c} + V_m} \\ \overline{V_{0g}} &= \sqrt{\overline{V_0}^2 + V_m^{\ 2} + 2\overline{V_0}V_m\cos\varphi_c} \end{aligned} \tag{12}$$

Through the above analysis, and then calculate the probability of damage. Probabilistic calculations are based on the assumption that the damage caused by personal injury and breakage is not taken into account; the damage caused by the secondary effect is not considered; in the case of combat mode and condition known, the position of the missile hit the radar radiation source is randomly distributed.

When the ARM Warhead explodes, it is assumed that the damage to any of the two critical parts of the radiation source is an independent event that can be considered as a non-destructive accumulation condition. According to the principle of probability, it can be seen that damage to any critical component under the condition of no damage accumulation will damage the whole radar radiation source. The radiation damage pattern of radar radiation source is as follows:

$$G(X,Y,Z) = 1 - \prod_{i=1}^{K}[1 - p_i(X,Y,Z)] \tag{13}$$

$p_i(X,Y,Z)$ Is the probability that the warhead will damage the i-th piece when it explodes at point (X,Y,Z); K is the number of vital parts. Assuming that $p(m)$ is the probability of hitting the block fragments on the critical components of the radar radiation source, A is the probability of damage to the m pieces broken on the part. By the full probability formula available:

$$p_i(X,Y,Z) = \sum_{m=0}^{\infty} p(m)\cdot G(m) \tag{14}$$

Law of the fragment by normal distribution description:

$$\Psi_R(\varphi_R) = \frac{1}{\sqrt{2\pi}\sigma_{\varphi_R}}\exp\left[-\frac{(\varphi_R - \overline{\varphi_R})^2}{2\sigma_{\varphi_R}^2}\right] \tag{15}$$

Among them, $\Psi_R(\varphi_R)$ is the relative scattering average distribution density and φ_R is the mathematical expectation of relative motion scattering.

$$\sigma(\varphi_R)=\frac{1}{(2\pi)^{3/2}\sigma_{\varphi_R}\sin\varphi_R}\exp\left[-\frac{(\varphi_R-\overline{\varphi_R})^2}{2\sigma_{\varphi_R}^2}\right] \quad (16)$$

The number of fragments between the critical parts of the radar radiation source:

$$\overline{m}=\frac{NS_{OT}\sin^2(\overline{\varphi_p}-\gamma\cos\beta)}{(2\pi)^{3/2}\rho^2\sigma_{\varphi_R}\sin\varphi_p}\exp\left[-\frac{(\varphi_p-\overline{\varphi_R})^2}{2\sigma_{\varphi_R}^2}\right] \quad (17)$$

Among them, S_{OT} is the key parts of the area, ρ,β,φ_p is the key parts of the center in the bombing point as the origin of the spherical coordinates of the ball, γ for the radar radiation source - fragments of the relative coordinate system and the rotation angle between the body coordinate system [5].

Assume that a ARM fragment warhead, the number is 2000, the static flying speed is 1020m/s, the relative motion of flying fragments is expected to 60 degrees, when the warhead explosion, if a crucial part of the projection area of 0.5 square meters, as shown in Figure 2 for the number of the fragments and hit fall curve distance.

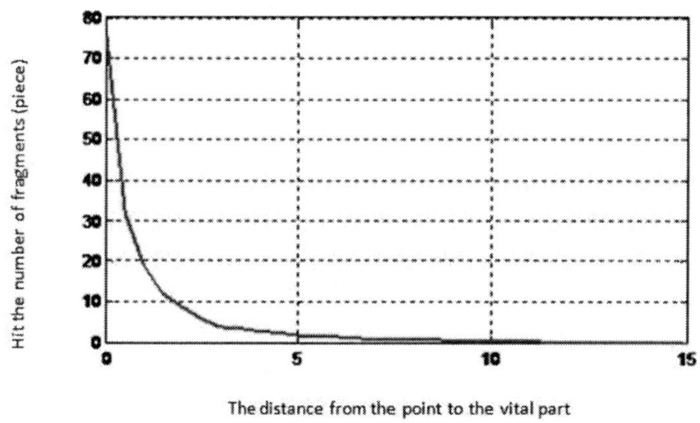

Fig.2 the relationship between the number of hits and the landing point distance

For the fragmented warhead, the total number of fragments N is much larger than the number m of fragments that hit a critical component of the radar radiation source, which is in accordance with the Poisson distribution conditions:

$$p_N(k=m)=C_N^m p^m q^{N-m}\approx\frac{\lambda^m}{m!}e^{-\lambda} \quad (18)$$

Where $\lambda=np=\overline{m}$, $\overline{m}$ is expected to hit the vital parts of the fragment count. For the condition of no damage accumulation, the damage probability of the coordinates is subject to the law of exponential damage:

$$G(m)=1-(1-p_1)^m \quad (19)$$

Among them, the probability that the key component is damaged for a fragment. Integrated formula (13) to formula (19), we can get:

$$G(X,Y,Z)=1-\exp\left[-\sum_{i=1}^{K}\overline{m_i}(X,Y,Z)p_{1i}\right] \tag{20}$$

Among them, p_{1i} is the i-th critical components of a single fragment damage probability [6].

5. Conclusion

It is very important to improve the hit accuracy of ARM by studying the effective damage probability of anti-radiation missile. By calculating the damage probability, we can judge the performance of ARM against active decoy. In this paper, the effect evaluation model is deeply studied, and the probability of damage and the probability of anti-decoy are given, which makes the evaluation more accurate, reliable and scientific.

References

[1] Mingshun Ai: Research on signal processing technology of ARM against active decoy, ph.D., China, 2011.
[2] Fuzhong Bai. Active decoy system against ARM scheme. Winged missile. 2011.
[3] Yuanjun Ma. Radar active decoy design. Radar and countermeasure. Vol.2002 (2), p. 32-35.
[4] Yankui Wang. Analysis of ARM seeker technology development. Winged missile.Vol.2009 (3), p. 39-44.
[5] Hongsheng Li. Analysis of ARM induced polarization performance. Fire Control Radar Technology. Vol. 2006, 35(3), p. 42-44.
[6] Mingshun Ai, Hongguang Ma. An evaluation index of ARM anti active decoy performance. Vol. 2010, 32(10), p. 13-16.

IOP Conf. Series: Materials Science and Engineering **207** (2017) 012065 doi:10.1088/1757-899X/207/1/012065

The relationship between the super plasticity of laser welding joint of titanium alloy and hydrogen treatment

Zean Cao, Donghai Cheng*, XunyanJiang, Dean Hu, Yiping Chen

Nanchang Hang Kong University, Nanchang, China

*Corresponding author e-mail: 1250985008@qq.com

Abstract. The superplastic deformation uniformity of laser welded joint of TC4 titanium alloy is improved by hydrogen treatment. The non-uniform deformation coefficient K was introduced to quantification ally characterize the non-uniform deformation. The results show when the content of hydrogen exceeds 0.29%, the super plasticity of the titanium alloy welded plate decreases with the increase of the hydrogen content. The decrease of the shrinkage of the base material is larger than that of the weld section with the increase of hydrogen content. The K can be used to describe the non-uniform deformation of the weld and the base material during the superplastic deformation of laser welded joint of the TC4. The K value increases with increaseing hydrogen content, increaseing deformation temperature and decreaseing strain rate.The K value reaches the maximum of 0.84 with hydrogen content of 1.299%, deformation temperature of 920 ℃, strain rate of $10^{-4}S^{-1}$.

1. Preface

The welding / super plasticizing technique is a processing method that combines welding with superplastic forming. The welding / superplastic forming technology of titanium alloy can not only overcome the shortcomings of poor processing performance of titanium alloy, but also produce multi-layer hollow lightweight and complex structure, which has unique technical prospects in the aviation and aerospace. However, laser welding technology is very suitable for combination with superplastic forming process because of its characteristics of fast processing speed, easy automatic welding, and small welding deformation.

The domestic and foreign scholars carry out a series of researches on the laser welding / superplastic forming technology of titanium alloy [1-6]. However, the research shows that there are obvious deformation inhomogeneity in the laser welded plate of titanium alloy after superplastic deformation. Up to now, the study on the improvement of superplastic deformation uniformity of titanium alloy laser welded joint is not reported.

Hydrogen treatment is a commonly used method for titanium alloy processing. When the amount of hydrogen in the titanium alloy welded plate is low [7-8], hydrogen can reduce the $\alpha+\beta/\beta$ transition temperature, improve the amount of βpromote superplastic deformation and improve the superplastic deformation capacity. When the titanium alloy welding plate in the high hydrogen, the microstructure of the joint can be adjusted by the hydrogen element to improve the uniformity of the TC4 laser welding joint. The machining hardening leads to the differences in microstructure and mechanical properties between the weld and the base metal during laser welding. It make the welded joint has uneven deformation in the superplastic deformation process, which directly affects the mechanical properties, reliability and the service life after forming. Adding excessive hydrogen to welding parts can improve the superplastic deformation uniformity of laser welding joints TC4 to meet the requirements. Therefore, the method of hydrogen treatment is adopted to improve the uniformity of superplastic deformation of titanium alloy laser welded joints.

2. Experimental materials and methods

The material used for the test was 0.8 mm thick TC4 titanium alloy sheet, and the welded joint was in the form of a butt joint. Before welding, the oil of plate is washed by 5% to 10% NaOH ethanol solution. Oxides of surface is removed by 5% HF + 30% HNO3 aqueous solution pickling. And then the plate is rinsed by water, dried and stored in a clean dish.

The welding test uses a CO2 axial-flow laser with a power range of 100-4000W and TEM01 mode. Test parameters is off-focus △ f of -0.5 mm, laser power P of 900 W, welding speed V of 3.0 m/min.

The hydrogen test is carried out in a tubular hydrogen furnace with a temperature of 700℃. The hydrogen content of the sample after hydrogen deposition is measured by weighing method with high precision physical balance. The mass fraction is 0,092%, 0.869%, 1.073%, 1.299%

The high temperature tensile test is carried out on a CMT4104 tensile tester. Before drawing, the Ti-5 glass protective coating is coated on the surface of the sample. The coating has good protection effect at 800 ~ 1000 ℃. The test range is initial strain rate of $10^{-3}S^{-1}$, 10^{-4} S^{-1}, deformation temperature of 900 ℃, 920℃. The plate is cooled in furnace. The tensile specimen size is shown in Figure 1.

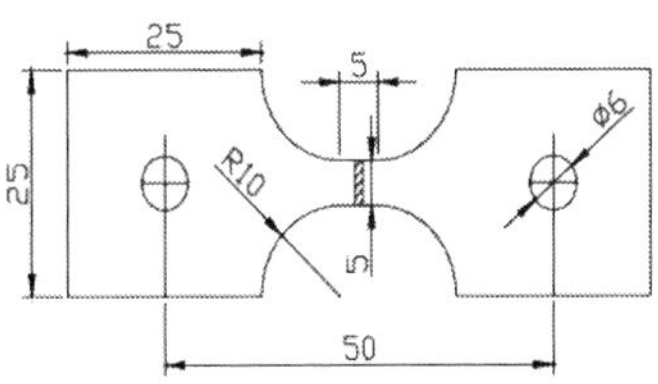

Fig.1 Dimension of hot tensile specimens

3. Experimental results and discussion

3.1 Effect of hydrogen on the mechanical properties of superplastic deformation

Fig.2 shows the effect of hydrogen production on the peak flow stress of superplastic deformation of joints. It can be seen from the figure that the peak flow stress of the non-hydrogen welded joint is less than 10MPa, and the peak flow stress of the joint after hydrogen treatment is more than 10MPA. The hydrogen treatment improves the peak stress of the joint deformation and reduces the superplastic deformation .This is due to the solid solution strengthening of hydrogen. With the increase of hydrogen content, hydrogen-induced solid solution strengthening will become very strong. And after the content of hydrogen exceeding the saturated solid solubility, the formation of the sheet hydride δ will strengthen the alloy and further increase the flow stress of the material. When the deformation temperature and the initial strain rate are constant, the peak flow stress of the welded plate increases with increasing hydrogen content. As shown in Fig.2, the hydrogen content increases from 0.291% to 1.299% and the peak flow stress increases from 35.3MPA to 48.1MPA at the initial strain rate of 10-3S-1, the deformation temperature of 900℃.

It can be seen from Fig.2 that at the same hydrogen content and the same initial strain rate, the peak flow stress of the sample decreases with increasing deformation temperature, because enhanced thermal activation effect can reduce the critical shear stress, improve the free energy of atoms and promote grain boundary slip with increasing deformation temperature. As shown in Fig.2, the peak flow stress decreases from 30.7MPA to 21.3MPA at hydrogen content of 0.291% and strain rate of $10^{-4}S^{-1}$ when the deformation temperature rises from 900℃ to 920℃. At the same hydrogen content and deformation temperature, the peak flow stress increases with the initial strain rate. When the strain rate is high, the dislocation density increases rapidly, which causes dislocation plug. Diffusion creep and dislocation slip cannot effectively coordinate its grain boundary slip. Grain size is not exactly the same. Its shape is not absolutely equated, so the sliding and rotation of the grains must be blocked in some places. It results in stress concentration. The internal deformation of the material cannot go on and strain hardening cannot be fully eliminated, which results in increasing flow stress.

3.2 Effect of Hydrogen Content on Superplastic Deformation and Uniformity of Joint

Titanium alloy base metal and welded joints are very different in the phase composition, microstructure and distribution, so there is uneven problem between deformation of the base metal and that of weld deformation. In order to quantitatively measure the unevenness of the deformation of the base material and the welded joint during the superplastic forming of the welded plate, the superplastic deformation unevenness coefficient K of the welded joint is introduced.

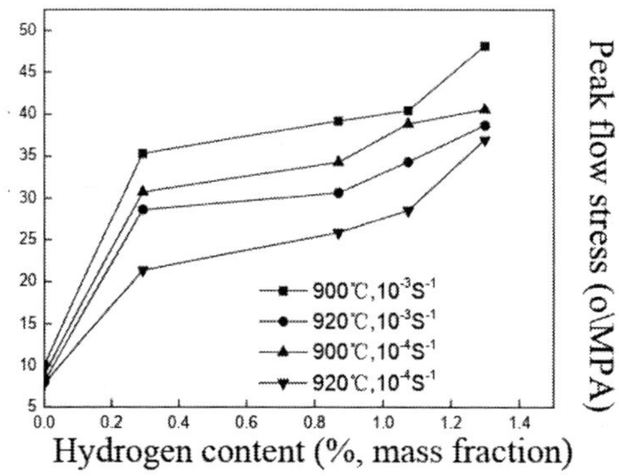

Fig.2 Relationship between peak flow stress and hydrogen content

Define the non-uniformity coefficient K of the superplastic deformation of welded joints as the ratio of the shrinkage of the weld section and the shrinkage of the base metal in the welded joint:

$$K = \Psi h/\Psi m$$

In the formula, Ψh is the cross section shrinkage of weld seam, Ψm is the cross section shrinkage of base material of weld.

it can be seen from the above analysis, the K value is larger, the welded joint make greater contribution to the deformation of the welded plate, the unevenness of the joint deformation is smaller. But because the deformation rate of the welded joint is smaller than that of the base metal, the large K value also reflects the poor ability of superplastic deformation of base material to a certain extent.

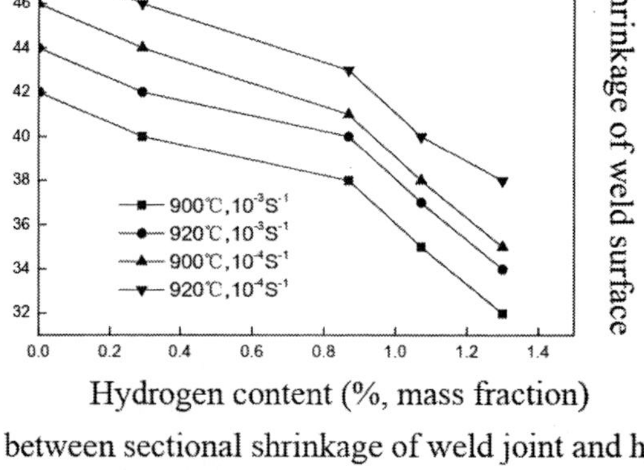

Fig.3 Relationship between sectional shrinkage of weld joint and hydrogen content

Figure 3 shows the effect of hydrogen on the shrinkage of the weld section. It can be seen from the figure that the hydrogen treatment reduces the cross section shrinkage rate of the whole weld. But, the decrease in the shrinkage of the weld section is relatively large with increasing hydrogen content, because the weld is needle-type marten site. It have small contact surface with hydrogen at high temperature and absorb less hydrogen. On the other hand, hydrogen atoms in the weld is not easy to spread among the atoms because of self-processing hardening and high energy among atoms.

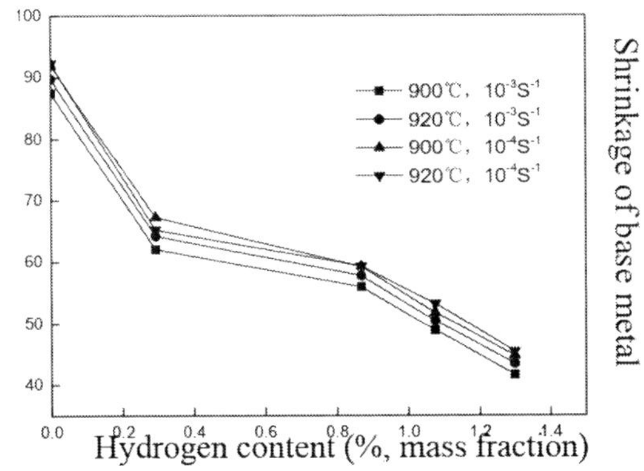

Fig.4 Relationship between elongation of weld plate base metal of weld plate and hydrogen content

Figure 4 shows the effect of hydrogen content on the shrinkage of the base metal. The shrinkage of the base material decreases greatly due to the solid solution strengthening of hydrogen. The solid solution strengthening caused by hydrogen becomes very strong with the increasing hydrogen content. When the hydrogen content increases to a certain extent, the strengthening effect dominates. Hydrogen fixed in the titanium alloy base material produce brittle phase, which leads to the rapid decline of the superplastic deformation of the base metal. When the deformation temperature and the strain rate are constant, the shrinkage rate of the base metal decreases with increasing the hydrogen content.

Comparing with Fig.3 and Fig.4, it can be seen that the hydrogen content is the main factors that affect the deformation ability of the weld and the base metal. It has an effect on the final section shrinkage by influencing the growth of the grain inside the sample and the degree of recrystallization. Section shrinkage of welded plate rate decreases with increasing hydrogen content. Hydrogen dissolves in the titanium alloy laser welding plate to form brittle phase so that the strength and hardness of titanium alloy laser welding plate increase and the cross section shrinkage of the welding plate declines. When the deformation temperature and strain rate are constant, the decrease of shrinkage of the base metal is larger than that of the weld with increasing hydrogen content. There is mainly $\alpha+\beta$ phase in the base metal. The proper amount of hydrogen can stabilize the β phase and promote the superplastic deformation. But when the excess hydrogen is applied, it can form a gap solid solution in the base metal to hinder the superplastic deformation. On the other hand, the tensile stress is easy to gather at the stress concentration due to the strong diffusion of hydrogen at high temperature. It results in high hydrogen concentration at the defects such as grain boundary, which the pins dislocation.

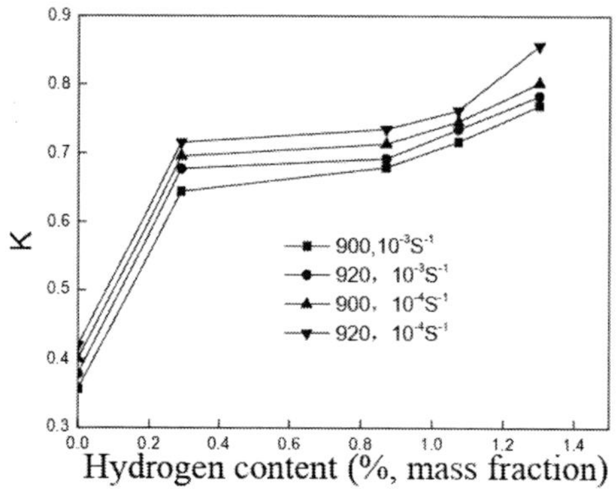

Fig.5 Relationship between K value and hydrogen content

Fig. 5 shows the relationship between the non-uniform deformation coefficient K and the hydrogen content. It can be seen that the K value increases with increasing hydrogen content, increasing deformation temperature and decreasing strain rate. In the case of no hydrogen, the maximal non-uniform deformation coefficient is 0.5 at strain rate of $10^{-4}S^{-1}$, deformation temperature of 920 ℃, which shows the difference between the shrinkage of the weld and the base metal during the superplastic deformation is obvious. The minimal K is 0.62 and the maximal K is 0.84 at hydrogen content of 0.291% -1.299%, deformation temperature of 920℃and 900℃ and strain rate of $10^{-3}S^{-1}$ and $10^{-4}S^{-1}$. The K value is more than 0.5. It suggested that hydrogen can adjust the phase composition of welded joints to improve the superplastic deformation uniformity of the titanium alloy welding joints. Hydrogen treatment decreases the cross section shrinkage of the weld and base metal. When the deformation temperature and strain rate are constant, the shrinkage of the base metal decreases faster with increasing hydrogen content. The higher the K value is, the better the coordination of deformation of the weld and the base metal is. The K value can characterize the deformation heterogeneity during the superplastic deformation of TC4 laser welded joint under certain conditions.

4. Conclusion

1) When the hydrogen content exceeds 0.29%, the super plasticity of the titanium alloy laser welding plate decreases with increasing the hydrogen content. It also shows the peak flow stress increases and the cross section shrinkage decreases.

2) The decrease of the shrinkage of the base metal is larger than that of the weld section with increasing hydrogen content. Hydrogen treatment improves the uniformity of superplastic deformation of titanium alloy laser welding plate.

3) The non-uniform deformation coefficient K can be used to describe the heterogeneity of the weld and the base metal during superplastic deformation. The K value increases with increasing hydrogen content, increasing deformation temperature and decreasing strain rate. The K value reaches the maximum of 0.84 at hydrogen content of 1.299%, deformation temperature of 920 ℃ and strain rate of $10^{-4}S^{-1}$.

Acknowledgments
This work was financially supported by Natural Science Foundation of China (51465042) fund.

References

[1] Cheng Donghai, Huang Jihua, Lin Haifan, Zhang Hua. Microstructure and mechanical analysis of Ti-6Al-4V laser butt weld joint [J]. Transactions of The China Welding Institution, 2009, 30 (2):103-106.
[2] Cheng Donghai, Huang Jihua, Yang Jing, Zhang Hua, Guo Hepin. Superplastic Deformation Mechanical Behavior of Laser Welded Joint of TC4 Titanium Alloys [J]. Rare Matel Meterials and Engineering, 2010, 39 (2):277-280.
[3] Cheng Donghai, Chen Yiping, Hu De'an, Wei Qiang. Microstrucutre analysis of superplastic

deformation on laser butt weld Ti-6Al-4V joint [J]. Transactions of the China Welding Institution, 2011, 32 (9):81-84.

[4] Cheng Donghai, Huang Jihua, Chen Yiping, Hu De'an, Microstructure evolution characterization of superplastic deformation of titanium alloy [J].Transactions of The China Welding Institution, 2012,33(7):89-92.

[5] J Mazumder, W M Steen. Microstructure and Mechanical Properties of Laser Welded Ti-6Al-4V [J]. Metallurgical Transactions A. 1982, 13A (5):865-871.

[6] Will Jeffrey, D Kistner, G Matthew. Fabrication of Laser Beam Welded Superplastically Formed Multi-Sheet Structure Using Advanced Titanium Alloys [J]. International SAMPE Technical Conference. 1996, 28(4-7):651-663.

[7] Guo long, Bai Bingzhe, Hou Hongliang. Superplasticity of Ti-6Al-4V by processed hydrogenation[J]. Chinese Journal of rare metals,2009,33(4):467-471.

[8] Hou Hongliang, Huang Chongguo, Wang Yaoqi. Microstructure evolution and superplasticity of hydrogenated titanium alloys[J]. Journal of University of Science and Technology Beijing,2008,30(11):1270-1274.

IOP Conf. Series: Materials Science and Engineering **207** (2017) 012066 doi:10.1088/1757-899X/207/1/012066

Comparative experiments on the technological performance of disc filters

Guangrong Liu[1,2], Hua Jiang[1,2*], Dongmei Liao[2], Yingchun Deng[2]

[1]Hubei Key Laboratory of Waterjet Theory and New Technology, Wuhan University
[2]College of Power and Mechanical Engineering, Wuhan University, Wuhan, China

*Corresponding author e-mail: jianghuawhu@126.com

Abstract. The filtration performance of JY disc filters was compared with that of Arkal disc filters when a parallel filtration system was designed under lab-scale devices. The comparisons of working and backwash performance were made for different filtration level. The experimental results showed that the two kinds of filters had similar process performance and effluent water quality under the same experimental condition. JY filter has higher filter performance index (FPI) than Arkal disc filters with the micron rating from 20 to 55μm discs, on the other hand, FPI value of JY filter has lower FPI than Arkal disc filters with the micron rating from 100 to 200μm discs. The removal rates of suspended solid(SS) for both filters vary in the range from 64% to 91.5%, moreover, the mean removal rate of SS for JY and Arkal filter was 78.2% and 79% respectively. The removal rates of turbidity were lower than 13% and the mean removal rate of turbidity was 7% for both filters. The critical backwash time and the critical backwash water loss for both filters were 5 seconds and 15 litres respectively.

1. Introduction

Disc filters have been applied on many industries. Compared with other filtration methods, the disc filters have the features of high filtration accuracy, adjustable, low energy consumption vast suitable range and less covered area etc. The automatic switching and quickly self-cleaning process to a completely new state are realized in disc filers. Currently, disc filters are the advanced technology in water treatment field [1, 2, 3].

In this paper, the filtration performance of JY disc filters made in China) was compared with that of Arkal disc filters made in Israel). By adding various amounts of reagents into discs with different filtration level, we compared the difference of both filters in parameters such as turbidity, suspended solids, periodic quantity of water produce, FPI, head loss, flux, and backwashing and so on. By sampling discs of the filters, the mechanism of groove's clogging was initially discussed. Moreover, backwash performance of disc filters was studied. The experimental results can be taken as a reference in practice.

2. Experimental

2.1 Analytical Methods

The turbidity of water sample was determined using a portable turbid meter (Model 2100P, Hach Company, and Loveland, CO, USA). The results were reported in cephalometric turbidity units (Ntu). Standard methods [4] were used to measure the value of suspended solids (SS). Supernatant was filtered with 0.45μm filter after sample was taken from the water tube. The head loss was measured by piezometer (0.5-2.5Mpa, Lei-da Instrument Factory, Zhejiang, and China) and the filtering velocity was measured by glass rotameter (Model LZB-50, Yuyao, China).

2.2 The evaluation parameters of filters performance

In general, filter performances are considered to be in better condition when the time of filtration is longer and head loss is less. Existing filter evaluation indexes are mainly FPI formula, FI formula method, F formula, SC formula, and JP formula method. The filtration performance can be partially evaluated by above indexes, but they are not suitable for evaluating the filtration performance of disc filters. Water quality is judged by turbidity in the traditional evaluation method, however, the disc filter is a kind of filter with mechanic sieving as a main role, which has a low removal rate of turbidity. In this paper, FPI is adopted as the evaluation parameter of filters performance [5].

$$FPI = \frac{VT}{H_t} \tag{1}$$

Where: V—— Filtration velocity, m/h; T—— Filtration time, h; H_t—— Head loss in the end of filtration, m.

The index of FPI has no dimension and the filtration velocity necessary for computing FPI is the average filtration velocity for a whole filtration cycle. The index of FPI is the ratio of cyclic water production and head loss in the end of filtration, which is used to evaluate the filtration performance. The bigger the FPI value, the better the filter performance. It means the quantity of water production will be obtained at best when the energy is consumed at least. However, the index of FPI cannot involve filter evaluation with water quality evaluation in the influent and effluent. To some extent, FPI can reflect the performance of the filter.

2.3 The design of experimental apparatus

Since disc filter has a larger flow of work, laboratory cannot meet such conditions of work. The methods of decreasing disc number were adopted in order to reduce flow. The original disc filter load height is 22.95 cm and the current disc filters load height in the test is 6.00 cm. Consequently, the flow rate of current filter in test is one fourth that of the original filter. In this way, the basic operating conditions, such as pressure, filtration rate, were consistent with actual operating conditions. The diameter of both filter disc is 11.6 cm and the working area is 0.02185 m^2. The surface area ratio of current filtration and original filtration is 0.26. The comparisons of operating parameters between the current filter and the original filter are shown in table 1. The comparisons of disc's number of both filters in fieldwork are listed in table 2. It is found from table 2, each slice of Arkal disc is a little bit thinner and flatter than that of JY disc.

Table 1 The comparisons of operating parameters of filters

Operating parameters	Original filter	Current filter
Working pressure	80~800KPa	80~280KPa
Backwash pressure	280~750KPa	150KPa/280KPa
Backwash time	5~20s	1~10s(modification)
Backwash water volume	3~30L	1~20L(modification)
Backwash flow	8m^3/h	1.7 m^3/h
head loss	1~80KPa	1~80 KPa

Table 2 The comparisons of disk's number of both filters in fieldwork

Different disc level(μm)	20	55	100	200
JY disc filters slices)	101	120	60	38
Arkal disc filters slices)	104	120	65	40

Diatomite has the properties of being easily dispersed into small size, stable solution, small pilot error, whereas activated carbon has relatively larger size and is suitable to simulate poor water quality. Therefore, diatomite is used in the filter operation with the micron rating from 20μm to 55μm, meanwhile, granular activated carbon was added in the filter operation with the micron rating from 100μm to 200μm. The parallel system of filters was designed under the condition of lab owing to the comparative test.

3. Comparison of filters performance with different precision discs

In this paper, comparisons between JY disc filters and Arkal disc filters are made in these areas of filtration effects, head loss-time chart, flow-time chart, effluent quality, filter discs performance and backwash property[6].

3.1 Comparison of filtration effects

Table 3 Influent/effluent water quality parameters during operation

Precision	Filter	Influent turbidity (Ntu)	Effluent turbidity (Ntu)	Influent SS_0 (mg/L)	Effluent SS_1 (mg/L)	Cyclic water production (m^3)	FPI	Removal rate of turbidity (%)	Removal rate of SS (%)
20μ	Arkal	0.50	0.48	4.5	0.6	0.66	15.68	4.0	86.6
	JY	0.51	0.49	4.5	0.5	0.56	18.60	3.9	88.9
	Arkal	1.55	1.45	8.8	0.8	0.2	4.83	6.5	90.0
	JY	1.53	1.43	8.8	0.8	0.16	5.31	6.5	90.0
	Arkal	1.6	1.46	10.4	1.0	0.18	3.77	8.1	90.4
	JY	1.55	1.40	10.5	0.9	0.12	3.98	8.8	91.5
55μ	Arkal	5.80	5.40	49.4	14.1	0.7	9.14	6.9	71.5
	JY	5.95	5.50	49.6	14.6	1.4	8.57	7.6	70.6
	Arkal	6.40	6.05	76.4	23.2	0.4	5.67	5.5	69.6
	JY	6.50	6.11	77.1	25.2	0.56	6.65	6.0	67.4
	Arkal	10.70	10.20	99.1	34.6	0.36	4.20	4.6	65.1
	JY	10.60	10.26	99.5	35.9	0.56	6.52	3.2	63.9
100μ	Arkal	7.43	6.50	105.5	46.2	6.4	138	12.5	56.2
	JY	7.40	6.64	106.2	45.1	4.5	90	10.3	57.5
200μ	Arkal	8.27	7.60	130.2	60.7	11	110.86	8.1	53.3
	JY	8.63	7.86	129.9	59.1	10	94.78	8.9	54.5

Under different influent quality, filtration properties with different precision discs (20, 55, 100, 200μm) were contrasted. The results are listed in table 3.

In common, the removal rate of turbidity is low for both filters with the micron rating from 20 to 55μm discs. On the whole, the operation effectiveness of 55μm filter is quite opposite to that of 20μm filter. For JY filter (55μm), the volume of cyclic water production and the value of FPI is larger (48 mg/L exceptions), whereas it's suspended solid (SS means suspended solid) removal rate is lower than that of Arkal filters. The results show the connection between the volume of filtered water per cycle and effluent effect was inverse for disc filters.

The operation effectiveness of 100μm discs is similar to that of 200μm discs. For Arkal filters, the volume of filtered water per cycle and the value of FPI is larger, whereas its suspended solid removal rate is lower than that of JY filters. For JY filters, the cycle of water production is shorter, the volume of filtered water per cycle and the value of FPI is smaller, whereas its suspended solid removal rate is higher than that of JY filters. Similarly, the removal rates of turbidity are very low for both disc filters. The removal of turbidity is related to the absorption of van der Waals force and Coulomb force [7], so that the role of contact flocculation is very weak for disc filter.

3.2 Comparison of head loss-time chart and flow-time chart

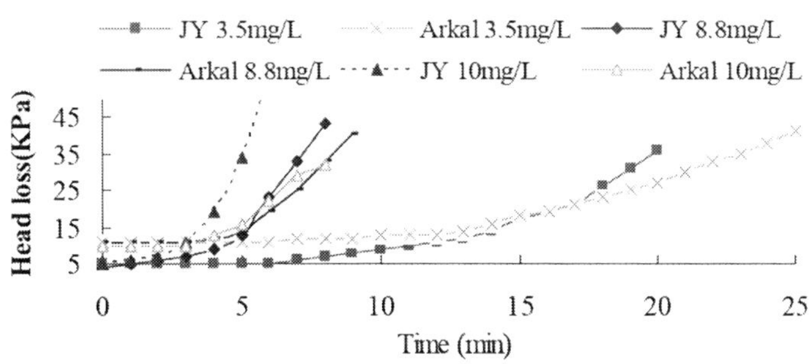

Figure1 Head loss chart with 20μm disc

The operations of four kinds of filtration rating were done in the experiment, however, the tendency were similar for head loss-time charts and flow-time charts. In order to save space, we just take the disc of one kind of filtration rating as an example. Head loss-time chart and flow-time chart of JY and Arkal filters were measured (as shown in figure 1 and figure 2) when dosage of suspended solids in the influent was 3.5 mg/L, 8.8 mg/L, 10 mg / L, respectively.

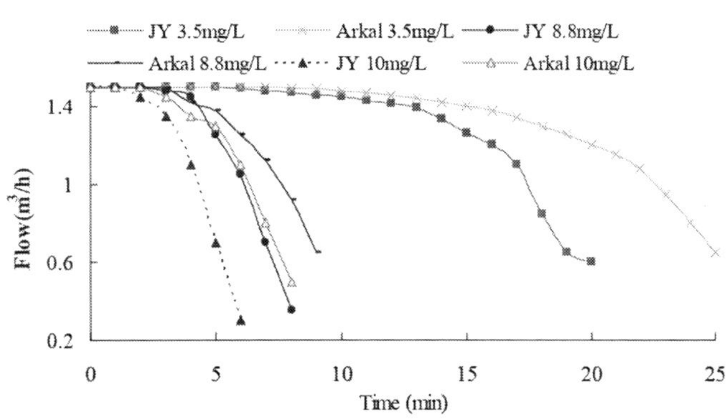

Figure 2 Flow-time chart with 20μm disc

Obvious conclusion from figure 3 and figure 4 are as follows. For the same kind of filter, the lower the suspended solids in the influent, the longer the water making cycle. Under the same conditions in the influent, the cycle of Arkal filter is longer than that of JY filters. There exists inflection point in the head loss-time and flow-time graphics, and inflection point is the formation of relatively stable conditions, namely the pressure is about 0.015 MPa, flow is about 1.3m^3/h. The head loss-time and flow-time curve of 55, 100, 200μm disc filter are basically similar to those of 20μm disc filter. Only the value of head loss and flow is different when a turning point is come into being.

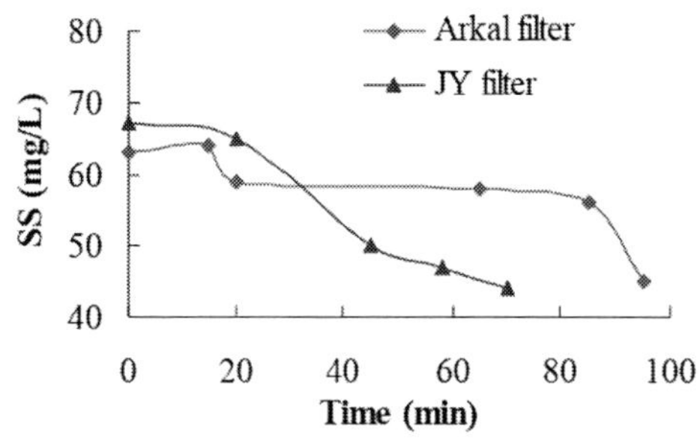

Figure 3 SS change with 100μm disc

3.3 Comparison of changes in effluent quality

In order to compare the changes in effluent quality of JY and Arkal filters, the experiments of both filters have been made with the micron rating 100μm discs. The reason to select 100μm filter as an example is that its longer water cycle can facilitate the sample analysis. The suspended solids in the influent are controlled as the scope of 111.0±2.0mg/L, and the changes of effluent SS are shown in figure 4. From figure 4, the quality of effluent is gradually improved in both filters, and a faster decline of SS value is found in the effluent of JY filter. At the same time, the formation of sieve layer can be observed through transparent cylinder. It is obvious that in-surface filtration exists in the filtration mechanism of discs filter, namely filter cake factors.

These phenomena can be explained as follows: in the early stages of filtration, a bigger gap existed in relatively clean discs, therefore only relatively larger particles were blocked; in the middle stages of filtration, the surface aperture of disc was monished, some middle-size particles were also closured; by the end of filtration, since internal closure of filter has not yet fully played the role, the surface will continue to play a role in filtration, causing the surface to the lower deck over the closure of impurities, resulting in seriously intercepting of filtration channel. At that time, the aperture of filtration was further reduced, thus product flow was reduced and effluent quality was improved.

For disc filter, the frequency of backwash will be reduced when the cycle of water production is long in the whole operation process, which means energy consumption will be reduced. From figure 4 and results in 3.2, we can observe that the energy consumption of Arkal filter is low and the effluent quality is poor. Contrarily, the energy consumption of JY filter is high and the effluent quality is fine.

4. The determination of filter discs performance

4.1 The cross test

Discs with the micron rating 55μm were selected as an example to make cross-test. Namely the discs of JY filter are put into Arkal filter, contrarily, the discs of Arkal filter are put into JY filter in the operation. The tendency of pressure-time curve and flow-time curve in the experiments are similar to those of figure 2 and figure 3. The corresponding data is listed in table 4, obvious difference doesn't exist after cross tests are done. The results show that disc is the main factors to affect the filtration performance.

Table 4 Results of cross test

Operation	Filter	Influent turbidity(Ntu)	Effluent turbidity(Ntu)	Influent SS (mg /L)	Effluent SS (mg /L)	Periodic product flow (m^3)
Normal test	Arkal	17.7	16.0	106.7	53.6	0.37
	JY	17.6	16.1	107.8	57.8	0.74
Cross test	Arkal	17.6	16.0	106.5	53.5	0.39
	JY	17.5	16.0	107.2	58.0	0.75

4.2 Analysis of disc surface observation

Before the backwash performance of disc filters was produced, from the observation of disc surface, it is found that the contaminated site was focused on the outside disc of Arkal filter, while the contaminated site was extended to the internal disc of JY filter and was not penetrated yet. Combined cyclic water production and water quality data with disc surface observation, it can be concluded that the filtration mechanism of Arkal filter is mainly in-surface filtration, while the filtration mechanism of JY filter is not only in-surface filtration but also in-depth filtration [8, 9].

5. Comparison of backwash property

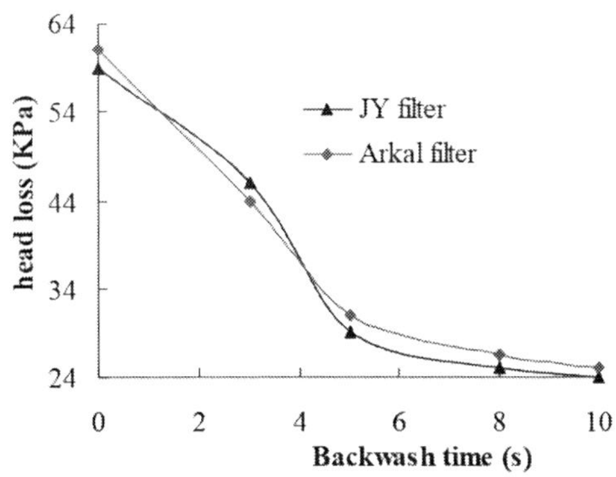

Figure 4 backwash head loss-time with 100μm disc

When the spring's pressure is increased to some extent, it means the discs are failed and backwash starts. The parameters, such as time, water consumption, and the recovery of the initial pressure are recorded in backwash experiments. After backwash started, critical backwash time could be measured when the initial pressure has reached 100% recovery [10]. Backwash head loss charts of Arkal filter and JY filter with 100μm disc are shown in figure 4. The figure shows that the shapes of backwash head loss charts are almost same for both filters. The head loss threshold of JY and Arkal filter is 59kPa, 61kPa when filter backwashing started. Collected at different moments, the backwash effluent quality is measured and the disc cleaning effects are observed. The results are as follows:

① for filter discs with the same micron rating, the shape of backwash curve is alike as the initial backwash flow is same. ② Minimum pressure of start-up backwash: 0.25 MPa for Arkal filter and 0.15 MPa for JY filter. ③ the critical backwash time for both filters are 5 seconds, then head loss is resumed to initial value. After backwash is completed, both filter discs are clean and bright. ④ the critical backwash water loss for both filters are 15 liters. When backwash is finished, the effluent water quality is the same as influent water quality.

6. Conclusion

(1) Both filters have general similar craft performance. A reliable backwash time is greater than 5 s, and water loss is about 15 L. The recovery rate on initial pressure is 100%.

(2) Turbidity removal rate is relatively low, and around 10% to 20%. The removal of turbidity is related to the absorption of van der Waals force and Coulomb force, so that the role of contact flocculation is very weak for disc filter.

(3) Suspended solids removal rate is relatively stable for both disc filters, such as 20μm filter about 90%, 55μm filter about 70%;

(4) It can be concluded that the filtration mechanism of Arkal filter is mainly in-surface filtration, while the filtration mechanism of JY filter is not only in-surface filtration but also in-depth filtration.

Acknowledgments
This work was financially supported by "Twelfth Five-Year" national science and technology major projects (2014ZX07104-005) and the experimental teaching center opening experiment project of Wuhan University in 2017.

References
[1] H.K. Lin, H. Zhou, the Application of Discal Filter on Industrial Water Treatment Field, Petro-Chemical Equipment Technology [J], 23 (2) (2002)16-18(Ch).
[2] Anon. Disc filter approved by California Water Recycling Criteria for wastewater reuse, Filtration & Separation. 45(9) (2008) 15-15.
[3] C.I. Ratiu, Modular disc filter with integrated and automated self-flushing operator. Bulletin of the University of Agricultural Sciences and Veterinary Medicine, 61(2004)464-464.
[4] APHA. Standard Methods for the Examination of Water and Wastewater 20th ed. Washington DC: American Public Health Association, 1995.
[5] Y.H. Jing , Y.H. Jin, J.C. Fan, Indicators to Estimate the Homogeneous Filtrating Material, Water & Wastewater Engineering, 26(3)(2000)13-16(Ch).
[6] T.A. Ribeiro, J.E. Patterning, R.P. Aroldis, et a1.Comparison between disc and non-woven synthetic fabric filter media to prevent emitter clogging. Transactions of the Asabe, 51(2) (2008)441-453.
[7] D.Z. Qian Water treatment engineering in power plant, China Electric Power Press, 1998 (Ch).
[8] J.L. Guo, J. Meng, G.P. Li, Z.K. Luan, H.X. Tang, Physicochemical interaction and its influence on deep bed filtration process, Journal of Environmental Sciences-China, 16(2)(2004) 297-301.
[9] B. Eikebrok, Coagulation-direct filtration of soft, low alkalinity humic waters, Water Science and Technology, 6 (1999)55-62.
[10] W.C. Dong, A study on the filtration and back flushing performance of the filters with crushed quatz sand for drip-irrigation, Journal of Hydraulic Engineering, 12(1997)72-77(Ch).

IOP Conf. Series: Materials Science and Engineering **207** (2017) 012067 doi:10.1088/1757-899X/207/1/012067

Photocatalytic decomposition of methylene orange using nitrogen-doped titania prepared by ultrasonic technique

Huifang Zhang[1, 2, *], Ling Sun[1], Wenping Cao[1] and Hanhu Liu[2]

[1]College of Environmental Engineering, Xuzhou Institute of Technology, Xuzhou Jiangsu 221000, China

[2]School of Environmental Science and Spatial Informatics, China University of Mining and Technology, Xuzhou Jiangsu 221000, China

[*]Corresponding author e-mail: kdzh82@163.com

Abstract. Nitrogen-doped titania (TiO_2) photocatalysts were prepared by sol-gel process coupled with ultrasonic irradiation at room temperature. The photocatalytic decomposition of methyl orange (MO) solution under ultraviolet and visible light irradiation was used as a probe reaction to estimate their photocatalytic activities. The physico-chemical properties of the samples were investigated by powder X-ray diffraction (XRD), transmission electron microscopy (TEM), diffuse reflectance spectroscopy (DRS), and X-ray photoelectron spectroscopy (XPS). A more uniform nanoparticle dispersion, marginally higher surface area, smaller grain size, and higher phase purity are some advantages of preparing TiO_2 by the ultrasonic technique. The nitrogen-doped titania prepared by ultrasonic technique were more effective than the samples prepared by conventional stirring method. Its UV–Vis diffuse reflectance spectra was red shifted compared to the un-doped sample and it showed high decomposition of MO under ultraviolet as well as visible light irradiation.

1. Introduction

Titania (TiO_2) has been extensively used in air purification , wastewater treatment ,and hazardous waste control because of its desirable properties, such as outstanding chemical stability ,high photocatalytic activity, relatively low cost, and non-toxicity [1,2]. However, it can only be excited under ultraviolet light of less than 387 nm due to its very wide band gap (3.2 eV). Electron-hole recombination can also inhibit the photocatalytic reaction process. These issues restrict its practical application.It is therefore of great interest and importance to extend the absorption wavelength range of TiO_2 to the visible light region.Previous studies have found that nitrogen-doped TiO_2 (N-TiO_2) has a higher visible photocatalytic activity since it can absorb visible light [3-5]. Different physical and chemical methods have been studied for the synthesis of N-TiO_2, such as hydrothermal synthesis [6], ion exchange [7], ion implantation [8], sol-gel methods, magnetron sputtering, high-speed milling machine-chemical method, high-temperature calcination and oxidation of TiN, etc. [9,10]. However, most are very time-consuming and require specific pH control and high temperature or through complex multistep reaction processes. Ultrasound reactions are generally carried out at ambient temperature and pressure, and are currently under focus for the preparation of optical, magnetic, and catalytic nanomaterials. Powerful ultrasound is used to stimulate chemical processes in liquids since it causes novel chemical reactions and physical changes which would not otherwise occur. These special physical and chemical conditions accelerate the hydrolysis reaction, decrease the grain and particle size of TiO_2, improve its dispersion, and increase its surface area [11, 12]. The size and morphology of TiO_2 nanocomposites are closely related to the ultrasonic frequency; the minimum size is obtained at low frequency (20 kHz) [13]. Moreover, acoustic cavitation can promote incorporation of the doping component into TiO_2 [14].

Herein, we report a novel ultrasonic technique (at frequency 28 kHz) coupled with a simple sol-gel method for the synthesis of nitrogen-doped titania. On one hand, nitrogen doping enlarges the light absorption range of TiO_2 to enhance its photocatalytic efficiency under visible light. On the other hand, ultrasonication can control the particle size and distribution of TiO_2.

2. Methods

2.1. Synthesis

Nitrogen-doped nanosized TiO_2 was synthesized using tetra butyl titanate ($Ti(OC_4H_9)_4$, TBOT) in the presence of water and ethyl alcohol absolute under ultrasonic irradiation. 17 ml TBOT and 40 ml ethanol were mixed and then added drop wise to a mixture containing acetic acid, deionized water, and ethyl alcohol. This mixture was sonicated under vigorous mechanical stirring and high intensity ultra sonication of 28 kHz. Then, 0.7 ml triethylamine was added drop wise as the nitrogen source. The mixture was then irradiated in an ultrasonic cleaning tank (28 kHz) for 1 h, followed by aging in a closed beaker at room temperature for 4 d. The wet gel was then dried at 80 °C (353 K) for approximately 10 h, collected, and ground into a fine powder. The dried powder was finally calcined in a muffle furnace at 500 °C (773 K) for 2 h and labeled as UN-TiO_2. The other nitrogen-doped TiO_2 sample was prepared under conventional stirring method without ultrasonic irradiation, labeled as N-TiO_2. As a reference, pure TiO_2 was prepared using the same procedure except that without triethylamine doping or ultrasonic irradiation, labeled as O-TiO_2.

2.2. Characterization

TiO_2 samples were characterized by XRD, TEM, XPS, and UV-vis DRS. The chemical composition of TiO_2 dried powders were obtained using a D/Max-III diffract meter with an applied accelerating voltage of 40 kV and current of 30 mA. The morphology of aggregated particles and crystallite size were obtained via TEM imaging with applied scan accelerating voltage and magnification of 100 kV and 135,000 times, respectively. XPS was used to identify the atom level doping of N in TiO_2 samples with AlKαX radiation. All binding energies were referenced to the C1s peak at 284.6 eV of surface adventitious carbon. The diffuse reflectance spectra (200 to 700 nm) of TiO_2 dry powders were obtained on an UV-vis spectrophotometer.

2.3. Photocatalytic decomposition of methyl orange (MO)

Photocatalytic experiments with MO solution were performed in a glass photocatalytic reactor. Irradiation was carried out using 365 nm UV and 250 W xenon lamps. MO solution (250 mL) was magnetically stirred and ultrasonically irradiated. Oxygen was continuously pumped into the reaction solution before and during illumination. Ten mL sample was taken every five or ten min and centrifuged (5000 rpm). Absorption at 464 nm was obtained on a UV-vis spectrophotometer and the changes in MO concentration were tracked.

3. Results and discussion

3.1. XRD analysis

Fig. 1 shows the XRD patterns of the dry TiO_2 powders. Identification of the TiO_2 phase, confirmed by comparison with accepted standard peaks from JCPDS, revealed that the three kinds of powders contained only the anatase phase; no peaks corresponding to the rutile phase were observed. The full-width at half-maximum of the UN-TiO_2 samples are broadened, which might be attributed to ultra sonication. The average particle size of UN-TiO_2, N-TiO_2 and O-TiO_2 obtained using the Scherrer formula was 19.85, 22.85 and 22.18 nm, respectively. We note that samples prepared with ultra sonication have narrow particle size range compared to those prepared by conventional stirring methods. The apparent width and intensity of the (101) peak of N-TiO_2 are large. This can be explained by the ultrasonic cavitation process. The implosive collapse of bubbles in a liquid solution can produce extremely high temperatures during hydrolysis, which accelerated the hydrolysis reaction and promoted formation of the anatase phases [12].

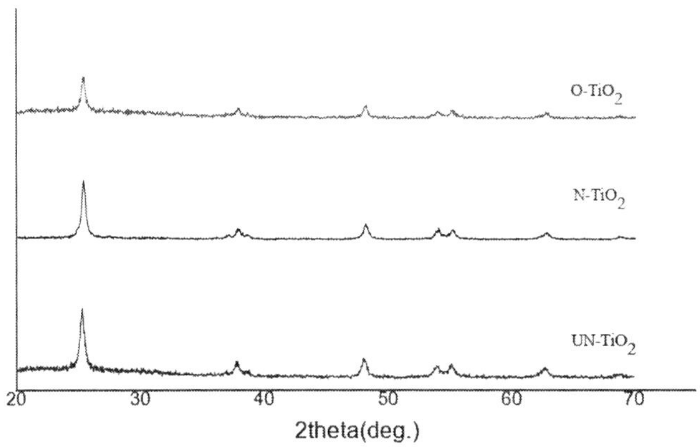

Figure 1. XRD patterns of samples.

3.2. *TEM images*

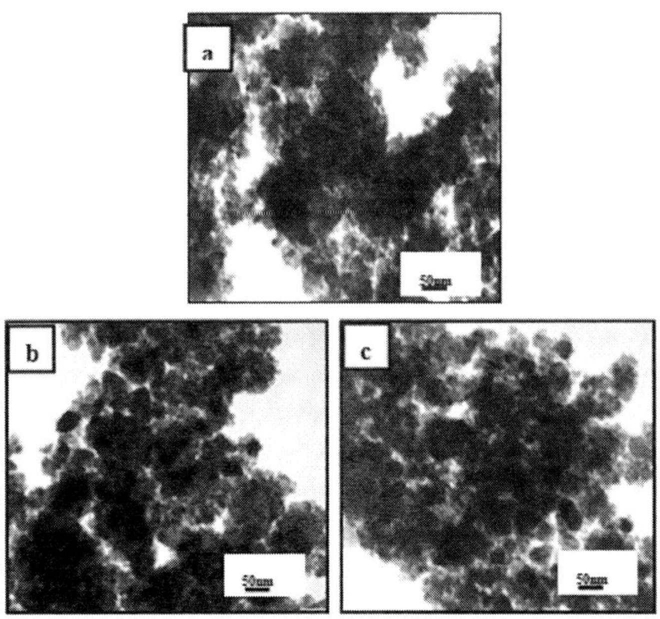

Figure 2.TEM photographs of samples.

In order to further elucidate the size and crystal structure of the nan composites, TEM micrographs of O-TiO_2, N-TiO_2 and UN-TiO_2 were obtained, as shown in Fig.2.These three catalysts are all nanosized.

The size of the primary particle is 20 ± 2 nm, which agrees with the values determined from XRD analyses. As shown in Fig. 2a and Fig. 2b, both O-TiO_2 and N-TiO_2 powder show a dense, aggregated structure, resulting in a decreased surface area. Fig. 2c shows that UN-TiO_2 contains more uniformly sized and less agglomerated particles compared to N-TiO_2. The TEM and XRD data indicate that particles are likely to be nan agglomerates of 2-3 crystallites in the UN-TiO_2 sample and 5-6 crystallites in the N-TiO_2 sample, in which agglomeration is more pronounced. This clearly reveals the role of ultra sonication, which results in a less agglomerated and more homogeneous sample. It is well known that collapsing bubbles create extreme temperatures and pressures during aqueous sonochemical processes. Cavitation bubbles undergo oscillations and increase in size before finally collapsing, resulting in higher temperature and more turbulent flow conditions. Cavitation and shockwaves can accelerate solid particles to high velocities, and antiparticle collisions can disrupt surface adsorption between particles and reduce agglomeration [15].

3.3. UV-vis DRS

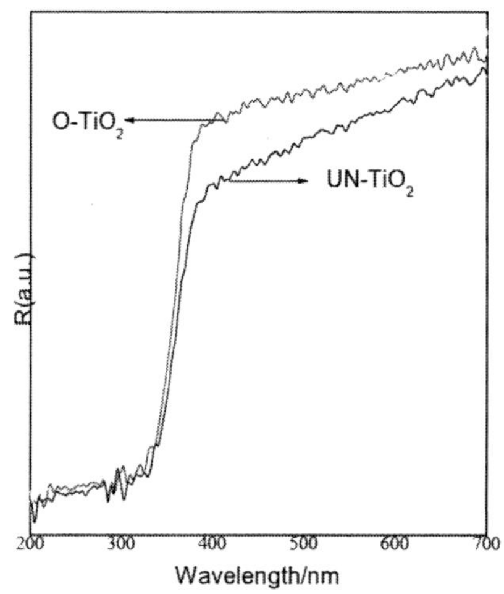

Figure 3. UV-Vis diffuse reflectance spectra of O-TiO_2 and UN-TiO_2.

The ability of TiO_2 to absorb light usually changes after nonmetal doping [16]. Fig. 3 shows the diffuse reflectance spectra over the wavelength range of 300–700 nm for O-TiO_2 and UN-TiO_2. It is clear that both samples have strong absorption in the ultraviolet region. UN-TiO_2 exhibits stronger absorbance intensity in the visible light region compared to O-TiO_2, however, it is still weak compared to that in the ultraviolet. Compared to O-TiO_2, UN-TiO_2 shows an obvious red shift at the absorption edges around 400 nm and a shoulder peak at longer wavelengths. Although the band gap energy of the samples was not ascertained, it was identified that the band gap of UN-TiO_2 could be greatly narrowed as a result of the red shift displayed by the UN-TiO_2 catalyst. These features further confirm mixing of the 2p states of the doped N anion with the O2p states, which is responsible for narrowing of the band gap energy of the titanate and the red shift of the valence band edge [10]. As mentioned earlier, ultrasonic cavitation provides a highly unusual and reactive chemical environment, i.e. extreme pressure and high temperature. The sol-gel method enhanced by ultra sonication therefore facilitates incorporation of N atoms into the TiO_2 lattice during preparation.

3.4. XPS

To identify the atom level doping of N in UN-TiO_2, XPS analyses were carried out. Fig. 4 shows the XPS survey spectrum and the high-resolution spectra of the N1s region of the UN-TiO_2 sample. As shown in Fig. 4a, UN-TiO_2 contains a small amount of N atoms (binding energies between 397 and

401 eV). Two XPS peaks at binding energies of 397.3 and 400.1 eV are observed in Fig. 4b. Different N1s peak positions represent various forms of UN-TiO_2. Assignment of N1s XPS peaks has been under dispute. As reported by Irie and Saha, N1s peaks can be decomposed into two contributions corresponding to different chemical states. One is attributed to β-N1s (binding energy at approximately 397eV), which belongs to N1s of TiN, the other to molecularly chemisorbed γ-N_{1s} (binding energy at approximately 400 eV) [17]. The peak at around 400.00 eV has been frequently observed for N-TiO_2 materials [18]. At least one kind of nitrogen doping in the N-TiO_2 sample was therefore believed to be responsible for the visible light photoactivity.

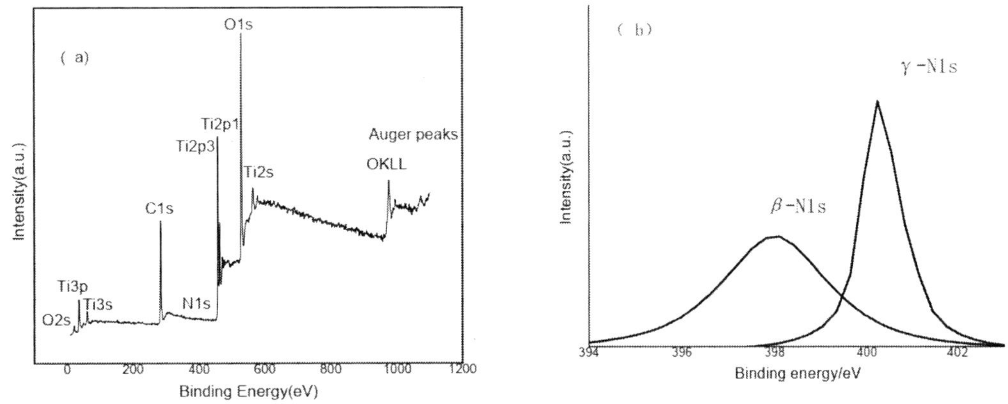

Figure 4. XPS survey spectrum (a) and N1 score level (b)of sample.

3.5. *Photocatalytic activity*

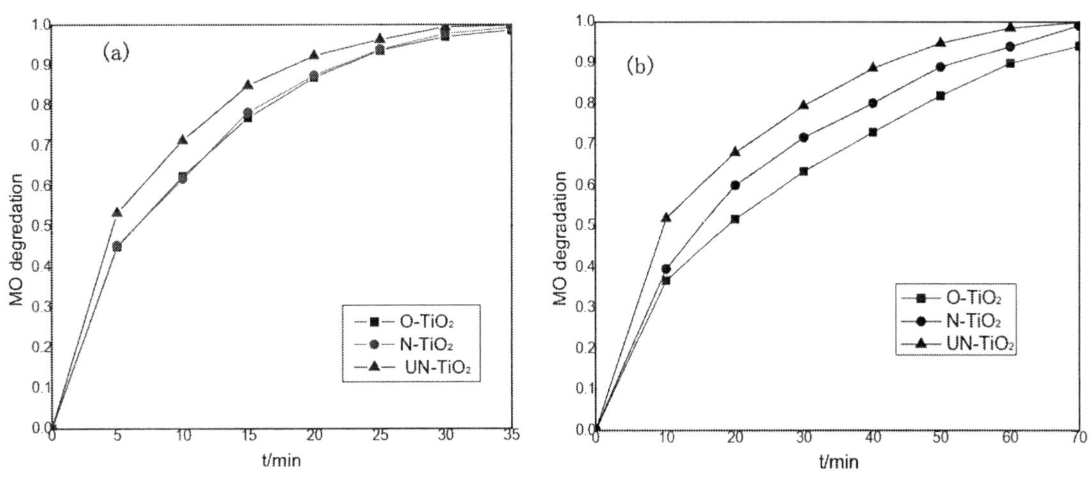

Figure 5. Photocatalytic decomposition of MO under UV light (a) and visble light (b)

To explore the photocatalytic activity of samples, their degradation of MO under visible and UV light was investigated (Fig. 5). Under UV light irradiation, UN-TiO_2 could degrade 53 and 99% MO in 5 and 30 min, respectively. The photocatalytic activity of N-TiO_2 under UV light exceeded that of O-TiO_2 and N-TiO_2. As shown in Fig. 5b, the decomposition rate of UN-TiO_2 towards MO under visible light irradiation is significantly enhanced compared to O-TiO_2 and N-TiO_2. In particular, UN-TiO_2 could degrade 99% MO in 60 min, whereas O-TiO_2 could decompose only 88%. The enhanced UV photocatalytic performance of UN-TiO_2 could be attributed to the fact that ultra sonication creates a unique environment for the hydrolysis of titanium lakeside, decreases the grain and particle size of TiO_2, and increases surface area, as indicated by the TEM results shown in Fig. 2. The lifetime of photogenerated electrons and holes can therefore be increased, which led to higher photocatalytic activity. The high visible photocatalytic activity of UN-TiO_2 is due to nitrogen doping, as indicated by the XPS results shown in Fig. 4, which is responsible for the red-shifted absorption edge of TiO_2 [19](as indicated by the DRS results shown in Fig. 3).

4. Conclusion

Anatase form Nitrogen-doped TiO_2 was synthesized by a sol-gel process under ultrasonic irradiation. Ultrasonication reduces the crystallite size, and reduce agglomeration between particles. UN-TiO_2 prepared under ultrasonic irradiation showed higher UV and visible photocatalytic activity, which was mainly attributed to its small particle size, high surface area, and extended light absorption range. In particular, MO decomposition by UN-TiO_2 under visible light reached 99% in 60 min, while that of pristine TiO_2 prepared without doping or ultrasonic irradiation was only 88%.

Acknowledgments

The work was supported by grants from the Science and Technology Plan Projects of Ministry of Housing and Urban-Rural Development of the People's Republic of China under Grant Nos. R22016061; and the Young Project of Xuzhou Institute of Technology under Grant No. XKY2016231.

References

[1] X.Chen, J.Zhang, Y. Huang,Catalytic reduction of nitric oxide with carbon monoxide on copper-cobaltoxides supported on nano-titanium dioxide. Environ Sci. 21(2009) 1296-1301.
[2] L.Gao,S.Zheng,Q.Zhang,Photocatalysis material of nano-titania and its application Beijing: Chemical Industry Press. 2002: 13-15.
[3] S.Sato,Photocatalytic activity of NOx-doped TiO2 in the visible light region. Chem. Phys. Lett. 123(1986)126-128.
[4] R.Asahi, T.orikawa, T.Ohwaki,Visible-light photocatalysis in nitrogen-doped titanium oxides. Science 293 (2001)269-271.
[5] L.Yu,X.Yang,J.He,Y.He,D.Wang,A fluorine free method to synthesize nitrogen and lanthanum co-doped TiO2 nanocrystals with exposed {0 0 1} facets for enhancing visible-light photocatalytic activity. Journal of Molecular Catalysis A: Chemical 399 (2015) 42–47.
[6] C.C.Hu, T.C.Hsu, S.-Y.Lu,Effect of nitrogen doping on the microstructure and visible light photocatalysis of titanate nanotubes by a facile cohydrothermal synthesis via urea treatmen. Applied Surface Science 280 (2013) 171– 178.
[7] S.S.Juliana,K.Klaus,V.B. Pinheiro Mauricio,Visible-light photocatalytic activity of NH4NO3 ion-exchanged nitrogen-doped titanate and TiO2 nanotubes. Journal of Molecular Catalysis A: Chemical. 394 (2014) 48–56.
[8] C.H.Yong,U.B.Chan,H.S.Dong,S.U.Han,Band gap narrowing of TiO2 by nitrogen doping in atmospheric microwave plasma. Chem.Phys. Lett. 413 (2005) 454–457.
[9] S.Mozia,M.Tomaszewska,B.Kosowska,Decompositon of nonionic surfacant on a nitrigen-doped photocatalyst under visible light irradiation. Appl. Catal.B. 56(2004)285-290.
[10] G.Liu, Y.Zhao, C. Sun, F. Li, L. Max, H.Cheng,Synergistic effects of B/N doping on the visible-light photocatalytic activity of mesoporous TiO2. Angew. Chem. Int. Ed. 47(2008)4516–4520.
[11] P.S. Awati,S.V. Awate,P.P.Shah,V.Ramaswamy,Photocatalytic decomposition of methylene blue using nanocrystalline anatase titania prepared by ultrasonic technique. Catal. Commun.4 (2003) 393–400.
[12] S. Senthilkumaar, K. Porkodi, R.Vidyalakshmi,Photodegratation of a textile dye catalyzed by sol-gel derived nanocrystalline TiO2 via ultrasonic irradiation. J. Photochem. Photobiol.A 170(2005)225-232.

IOP Conf. Series: Materials Science and Engineering **207** (2017) 012067 doi:10.1088/1757-899X/207/1/012067

[13] D.Lin, J.Wang, X.Li, H. Liu, Effect of ultrasonic frequency on the structure and sonophoto-catalytic property of CdS/TiO2 Nanocomposite. Mat.Sci.Semicon. Proc.5(2012)152-156.

[14] J.Ouyang, M.Chang, Y. Zhang, X. Li, CdSe-sensitized TiO2 nanotube array film fabricated by ultrasonic-assisted electrochemical deposition and subsequently wrapped with TiO2 thin layer for the visible light photoelectrocatalysis. Thin Solid Films. 520(2012)2994-2999.

[15] K.S.Suslick,M.M.Fang,T.Hyeon,M.M.Mdleleni,Applications of sonochemistry to materials synthesis. Sonochemistry and Sonoluminescence 524(1999)291-320.

[16] J.Yu, G.Dai, Q.Xiang, Fabrication and enhanced visible-light photocatalytic activity of carbon self-doped TiO_2 sheets with exposed 001 facets. *J Mater Chem*.1 (2011) 1049–1057.

[17] H.Irie, Y.atanabe, K.Hashimoto,Nitrogen-concentration dependence on photo- catalytic activity of $TiO_{2-x}N_x$ powders. J Phys Chem.B 23 (2003)5483-5486.

[18] X.Li, P. Liu, Y.Mao, M.Xing, J. Zhang, Preparation of homogeneous nitrogen-doped mesoporous TiO2 spheres with enhanced visible-light photocatalysis. Appl. Catal. B-Environ.164 (2015) 352–359.

[19] Y.Dong, Y.Wang, T.Cai, Long Kou, G.Yang, Z.Yan,Preparation and nitrogen-doping of three-dimensionally ordered macroporous TiO2 with enhanced photocatalytic activity. Ceram. Int. 40 (2014) 11213–11219.

Impact of long-term application fertilizer on soil total sulphur and valid sulphur

Mengyu Gao[1, 2], Xiaoling Lu[1, 2], Yuqian Huang[1, 2], Ning Liu[1, 2], Jinfeng Yang[1, 2, *]

[1]College of Land and Environment, Shenyang Agricultural University, Shenyang, 110866 China
[2]National Engineering Laboratory of High Efficient Use on Soil and Fertilizer Resources P. R. China

[*]Corresponding author e-mail: yangjinfeng7672@163.com

Abstract. The object of this study was to investigate the effect of the long-term application fertilizer on soil total sulphur and valid sulphur. The results showed that applying fertilizer can improve total sulphur and valid sulphur. In comparison with the low level of nitrogen fertilization treatment, the high one total sulphur and valid sulphur were obviously increased by 29.41% and 19.0%, respectively. Compared with in application of different levels nitrogen and the low level of organic fertilizer, the high level treatment total sulphur and valid sulphur contents were significantly increased by 10.73% and 23.47% than the low one. In application of organic fertilization can also.improve total sulphur and valid sulphur The total sulphur and valid sulphur content were higher than organic fertilization only treatment 34.14% and 455.89% in comparison with high levels of organic fertilization mix with nitrogen, phosphorous and potassium fertilization treatment.

1. Introduction

Sulphur is of the essential element for plant growth and development. The organic carbon content, the type of fertilizer and soil pH are the main factors to influence the distribution of soil sulphur element forms. Especially the application of organic fertilizer plays an important role which can improve soil properties, increase soil organic sulphur content and potential sulphur [1]. In recent years, the sulphur content in the soil declined year by year with crops multiple cropping index enhanced, yield increased and atmospheric sulphur emissions reduced [2,3].The soil sulphur has become limit factor to improve agricultural crops yield, quality and the utilization ratio of fertilizer[4]. It is very important to know soil sulphur condition so that we can ensure the sustainable development of agriculture and improve soil sulphur biological effectiveness. There were many researches about it such as Zhong [5] found that the contents of soil total sulphur was increased by 65. 42 mg/kg in NPK plus organic manures treatment compared with the origin samples, while the ratio of available sulphur in NPK plus organic manures treatment was lower than that in NPK. The content of soil adsorbed sulphur were improved by 177. 4% and 93. 3% in NPK plus organic manures treatment compared with the origin samples. Gao's [6] research showed that the contents of soil total sulphur was improved in the treatment which application of inorganic nitrogen fertilizer, organic fertilizer and combined application of organic fertilizer and inorganic nitrogen fertilizer and combined application of nitrogen and phosphorus and potassium fertilizer. But the more organic fertilizer and inorganic nitrogen fertilizer was used, the more contents of soil total sulphur was increased.

As a result, this study was based on a thirty seven-year long-term fertilization experiment and employed the impact on soil total sulphur and valid sulphur.

2. Materials and methods

The long-term experiment was located in Liaoning Province at the Shenyang Agricultural University Peanut Scientific Research Center (40°48'N, 123°33' E). The soil at the experimental site was a brown soil classified as an alfisol, with hydromica as a dominant clay mineral. This experiment began in 1977 and the basic physical and chemical properties are detailed in Table 1.

Table 1 Basic properties of soil at the beginning of the experiment

Soil layer (cm)	Organic matter (g/kg)	Total N (g/kg)	Total P (g/kg)	Total K (g/kg)	Available N (mg/kg)	Available P (mg/kg)	Available K (mg/kg)	pH
0–20	13.1	0.53	0.67	18.8	56.2	12.5	89.6	6.81

There three split-plot designs, chemical plot, low levels of organic fertilizer, high levels of organic fertilizer. Fifteen treatments were established in 1979 and end in 2014: nitrogen and phosphorous fertilization (N_1P), nitrogen, phosphorous and potassium fertilization (N1PK), low levels of nitrogen fertilization (N1), high levels of high levels of nitrogen fertilization (N2), no fertilization (CK), low levels of organic fertilization mix with nitrogen and phosphorous fertilization (M1N1P), low levels of organic fertilization mix with nitrogen, phosphorous and potassium fertilization (M1N1PK), low levels of organic fertilization and nitrogen fertilization (M1N1) low levels of organic fertilization and high levels of nitrogen fertilization (M1N2), low levels of organic fertilization (M1), high levels of organic fertilization mix with nitrogen, phosphorous and potassium fertilization (M2N1PK), high levels of organic fertilization and nitrogen fertilization (M2N1) , high levels of organic fertilization and high levels of nitrogen fertilization (M2N2), high levels of organic fertilization (M2). The experimental field cropping system consisted of maize- maize- peanut rotation cropping. Soil samples at depths of 0–20 cm and 20–40 cm were collected randomly in 2014.

The total and valid sulphur were determined by ICP-AES method. Microsoft Excel 2010 and SPSS 19.0 were used for the statistical analysis of the experimental data.

3. Results and discussion

3.1. Effect of different fertilization on soil total sulphur

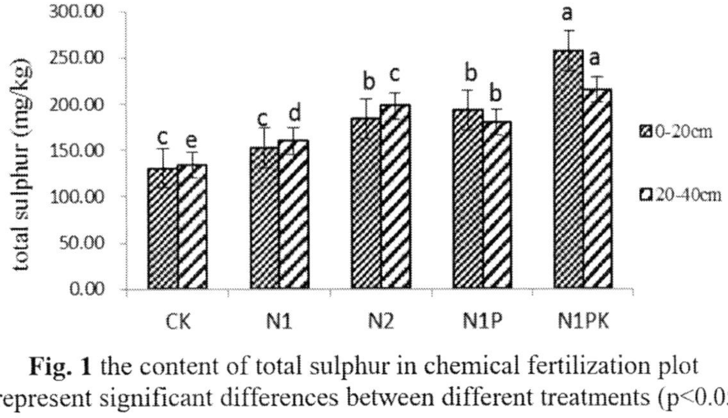

Fig. 1 the content of total sulphur in chemical fertilization plot
Different letters represent significant differences between different treatments (p<0.05; Duncan's test)

In the depth of 0-20cm, the contents of soil total sulphur with chemical fertilizations treatments were all higher than no fertilization treatment. The application of low high level of nitrogen significantly increased 29.41% in comparison with the high one. Compared between N1PK, N1P and N1treatments, the N1PK treatment had the highest total sulphur content that was 499.73mg/kg and significantly higher than N1 treatment 1.91 times. It showed that applied in more abundant chemical fertilizations can improve the content of total sulphur. The result was as same as Xu's [7]. The total sulphur content of every treatment in 20-40cm depth was lower than that in 0-20cm depth. Ren's result [8] had the same result. In comparison with no fertilization applied, the total sulphur content was improved when chemical fertilizations were used. The content of sulphur in low levels of nitrogen treatment was increased significantly 28.71% compared to the high level treatment. It was different with the result in 0-20cm depth. The N1PK treatment total sulphur was the highest in comparison with N1P and N1 treatment and that significantly higher than nitrogen fertilization treatment 0.55 times.

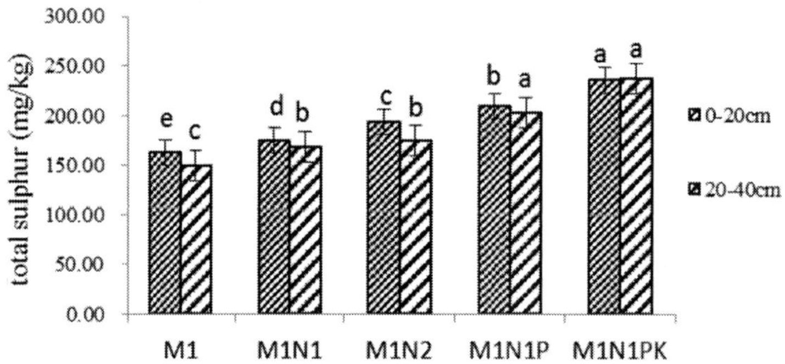

Fig. 2 the content of total sulphur in low levels of organic fertilization plot

In the low levels of organic fertilization plot, the contents of total sulphur were obviously higher in application of chemical fertilizations than organic fertilization only in the depth of 0-20cm.Obviously, the effect was better when organic and chemical fertilizations applied together. In comparison with different levels of nitrogen fertilizers, the high level treatment total sulphur content was significantly increased to 10.73% than the low one. The content of total sulphur with low levels of organic fertilization and nitrogen and phosphorous fertilization applied was the highest 235.33mg/kg and it obviously higher than M1N1P treatment 12.36% and M1treatment 21.82%.respectively. It showed that the more kinds of fertilizations were used the higher total sulphur content was. In the depth of 20-40cm, the regulars were similar as 0-20cm depth but the total sulphur contents were lower than that. The content of total sulphur of M1N1PK treatment was significantly increased to 58.36% compared with no chemical fertilizers were used.

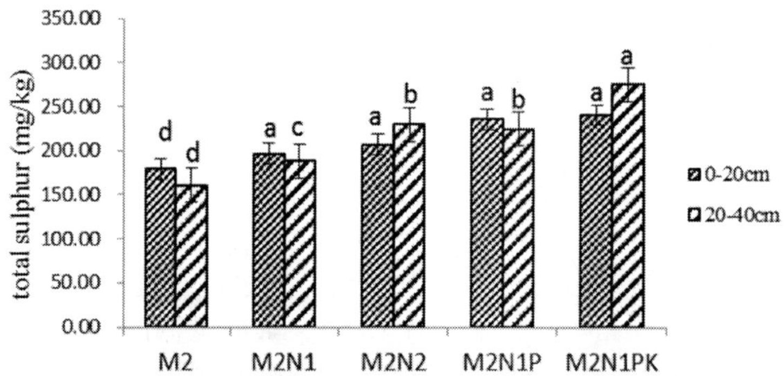

Fig. 3 the content of total sulphur in high level of organic fertilization plot

In the high levels of organic fertilization plot, the total sulphur content of chemical fertilization mixed with organic fertilization treatments were higher than organic fertilization only treatment. It same as Gao's [6] result. In comparison with high level of organic fertilizer only treatment, total sulphur content was significantly increased by 34.14% in M1N1PK treatment. The total sulphur content of high level of nitrogen treatment was higher than the low one 5.29%. However, there was no significant difference. There were similar regulars in 20-40cm depth. The total sulphur content of high level of nitrogen treatment was significantly higher than the low one 22.00%.The nitrogen fertilization was helpful to increased total sulphur content in deeper soil depth.

3.2. *Effect of different fertilization on soil valid sulphur*

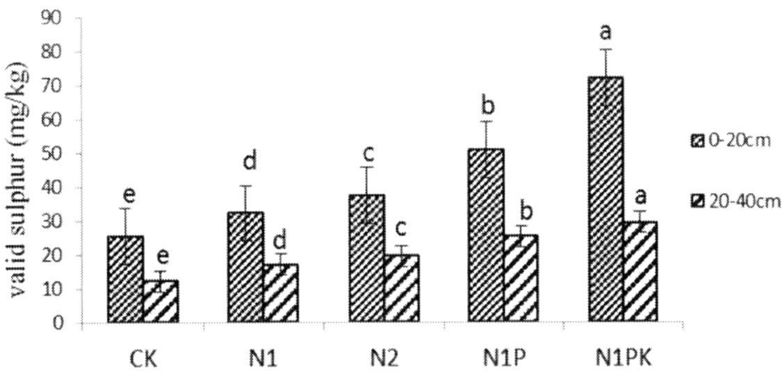

Fig. 4 the content of valid sulphur in chemical fertilization plot

In chemical fertilization plot, no matter what kinds of fertilizations were applied, the contents of valid sulphur were improved compared with in application of no fertilizer. N1PK treatment had the highest valid sulphur and it higher than CK treatment 1.38 times. In comparison with different levels nitrogen fertilization treatments, the valid sulphur in N2 treatment was significantly increased by 16.05%. The result confirmed that more chemical fertilization can improve valid sulphur content. The chemical fertilizers were helpful to improve soil valid sulphur. The reason was that soil pH significantly reduced when chemical fertilizer applied in soil. There were more positive charges at soil surface increased so that it can absorb more SO_4^{2-} [9].The regulars in the depth of 20-40cm as same as in 0-20cm depth.

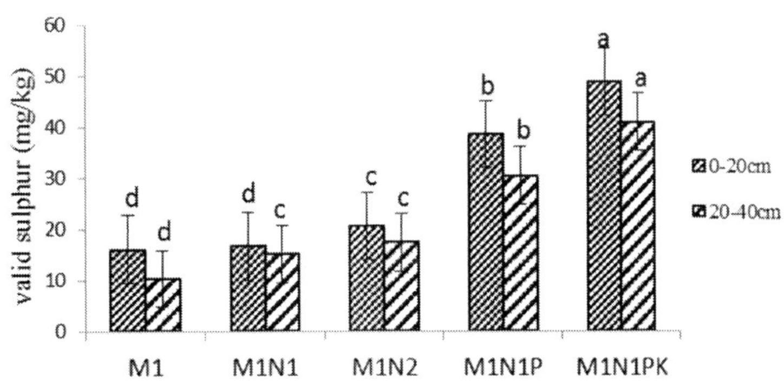

Fig. 5 the content of valid sulphur in low levels of organic fertilization plot

In 0-20cm depth of low level organic fertilization plot, the valid sulphur content was the highest in M1N1PK treatment and it significantly higher than M1N1 treatment 26.49% ,M1N1% treatment 192.61%, respectively. In comparison with low level of nitrogen treatment, valid sulphur content was significantly increased by 23.47% in M1N2 treatment. However, there was no significant difference between M1N1 and M1 treatment. The content of valid sulphur was higher than only organic fertilization treatment 0.60mg/kg compared with chemical fertilizer mixed with organic fertilization treatment. There were similar regulars in 20-40cm depth but no obvious difference between different levels of nitrogen treatments. Organic fertilizer application brought a lot of anion groups and inhibited SO_4^{2-} by competing adsorption sites in the soil adsorption cumulative [10].

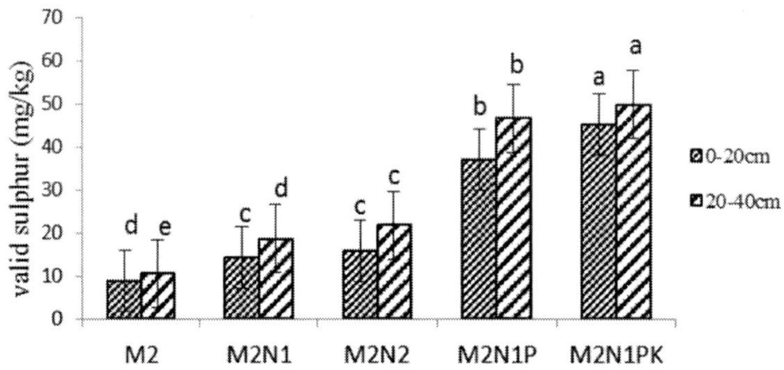

Fig. 6 the content of valid sulphur in high levels of organic fertilization plot

In high levels of organic fertilization plot, every chemical mixed organic fertilization treatments had more valid sulphur in comparison with organic fertilization only treatment in 0-20 cm depth of soil. It in M2N1PK treatment was higher than in M2 treatment 4.14 times. There was no significant difference between different levels nitrogen treatments and valid sulphur content in M2N2 treatment was higher than that in M2N1 treatment 10.28%. However, in 20-40cm depth, there was significant difference. Except that, the regulars in different depths were same. In application of organic fertilizer can improve soil carbon. Carbon bonds sulphur is the main source of organic sulphur mineralization. This part of the organic sulphur can be converted into valid sulphur effectively when drying or microorganisms acting [11].

4. Results

1) After long-term fertilization for 37 years, in comparison with in application of no fertilizer, after other only chemical fertilization treatments the soil total sulphur and valid sulphur content were increased.

2) Compared with organic only fertilization treatment, in application of organic fertilization mixed with chemical fertilization can improve total sulphur and valid sulphur better. The effect of high levels of organic fertilization mix with nitrogen, phosphorous and potassium fertilization was significantly greater.

Acknowledgements
This study was carried out with National Engineering Laboratory of High Efficient Use on Soil and Fertilizer Resources/Monitoring & Experimental Station of Corn Nutrition and Fertilization in Northeast Region funded by National Natural Science Foundation of China (41371287) and Special Fund for Agro-scientific Research in the Public Interest (201203030-04-03).

References

[1] Li S T, Lin B. The status and evaluation of plant available sulphurin soils, J. Plant Nutrition and Fertilizer Science (1998) 4 (1) 75-83.
[2] Eriksen J. Soil sulphur cycling in temperate agricultural systems, J. Advances in Agronomy (2009)102: 55-89.
[3] Zhao F J, Lehmann J, Solomon D, Fox M A, Mcgrath S P. Sulphur speciation and turnover in soils: Evidence from sulphur K-edge XANES spectroscopy and isotope dilution studies, J.

Soil Biology and Biochemistry(2006)38: 1000 -1007.

[4] Frster S, Welp G, Scherer H W. Sulphur specification in bulk soil as influenced by long-term application of mineral and organic fertilizers, J. Plant, Soil and Environment (2012)58: 316-321

[5] Zhong H F, Zhang S L, Guo T,Shi X J. Variation of inorganic sulphur forms under long-term fertilization in purple soil, J. Plant Nutr Fert Sci(2015)(06): 1652-1656.

[6] Gao S T, Li W X, Zhong X J, Liu S T, Yin Y C. Effects of long- term located fertilization on dynamic changes of total sulphur in the non- calcareous fluro- aquic soil, J. Chin Agric Bull (2008) (10): 306-309.

[7] Xu C, Wang S S, Li J M, MaY B, Sun W T, Luo L, Zhang S Z. Effects of long-term fertilization on evolution of S forms in a red soil and a black soil, J. Chin J Appl Ecol (2014) (04): 1069-1075.

[8] Ren J, Zhun P, Bian X Z, Peng C, Guo J R, Gao H J, Yan X G, Li Q, Liu J Z,Wu J G. Effect of Different Fertilization Systems on Content and Space Variance of P and S in the Black Soi, J. Journal of Soil and Water Conservation (2010)(03): 105-108.

[9] Scherer H W. Sulphur in crop production: Invited paper, J. European Journal of Agronomy (2001)14: 81-111

[10] Frster S, Welp G, Scherer H W. Sulphur specification in bulk soil as influenced by long-term application of mineral and organic fertilizers, J. Plant, Soil and Environment (2012) 58: 316-321.

[11] Solomon D, Lehmann , de Zarruk K K, Julia D, James K, Liang B Q, Stephen M. Speciation and long- and short-term molecular-level dynamicsof soil organic sulphur studied by X-ray absorption nearedge structure spectroscopy, J.Journal of Environmental Quality (2011) 40: 704-718.

IOP Conf. Series: Materials Science and Engineering **207** (2017) 012069 doi:10.1088/1757-899X/207/1/012069

Effects of different tillage systems and amendments on root properties

Mengyu Gao[1, 2], Yang Yan[1, 2], Na Li[1, 2], Peiyu Luo[1, 2], Jinfeng Yang[1, 2, *]

[1]College of Land and Environment, Shenyang Agricultural University, Shenyang 110866, China

[2]National Engineering Laboratory of High Efficient Use on Soil and Fertilizer Resources, P. R. China

[*]Corresponding author e-mail: yangjinfeng7672@163.com

Abstract. The object of this study was to investigate the effect of different tillage systems and amendments on root properties. There were five treatments: maize continuous cropping, maize and peanuts rotation, peanuts continuous cropping, peanuts continuous cropping with low level of amendment and peanuts continuous cropping with high level of amendment. The results showed that maize continuous cropping increased total root length by 118.95%, projected area by 204.86%, projected area by 150.70%, total root volume by 20.66%, and average root diameter by184.53%. The amendments also improved root properties and the high level of amendment had much more better effect.

1. Introduction

The root is main organ to absorb nutrients and moisture for plants. The ability of crops absorbing mineral nutrient and moisture is about the development of roots [1]. The quality of root directly influences the growth of aerial part. The continuous cropping inhibits the crops growth and yield improvement [2]. Scientific and reasonable rotation is an effective measure to solve continuous cropping obstacle. Reasonable rotation with other crops can coordinate the relationship between the crop and soil, reduce the harmful microbes in the soil [3], improve soil ecological environment and promote high yield. Application of soil amendment is one of the important measures of restoration of degraded soil. Soil conditioner can effectively improve soil physical and chemical properties, soil nutrient status. It has a positive impact on soil microorganisms and productivity of degraded soil [4]. There were many researches about the effect of different tillage systems and amendments on root, such as Wang [5] set up five years of continuous monoculture sorghum as the control was compared to alfalfa and scallion planted in the fourth year. The results showed that incorporation of alfalfa and scallion into the rotation significantly improved sorghum shoot growth. Specifically, sorghum grain yield increased by 16.5% in the alfalfa rotation plots compared to the CK. The rotations also increased sorghum root system growth, with alfalfa or scallion rotation increasing sorghum total root length by 0.3 and 0.4 times, total root surface area by 0.6 and 0.5 times, root volume by 1.2 and 0.6 times. Dong's research results showed that treatments with amendments, except for lime, significantly increased the intensity of bleeding sap, root shoot ratio, biomass in bleeding sap.

As a result, this study was based on a six-year long-term experiment in order to investigate the effect of different tillage systems and amendments on root development and providing theoretical basis for solving peanut continuous cropping obstacle

2. Materials and methods

The long-term experiment was located in Liaoning Province at the Shenyang Agricultural University Peanut Scientific Research Center (40°48'N, 123°33' E). The soil at the experimental site was a brown soil classified as an alfisol, with hydromica as a dominant clay mineral. This experiment began in 2011 and the basic physical and chemical properties are detailed in Table 1.

Table 1 Basic properties of soil at the beginning of the experiment

Soil layer (cm)	Organic matter (g/kg)	Total N (g/kg)	Total P (g/kg)	Total K (g/kg)	Available N (mg/kg)	Available P (mg/kg)	Available K (mg/kg)	pH
0–20	13.1	0.53	0.67	18.8	56.2	12.5	89.6	6.81

The cultivars were Huayu 33 peanuts and Zhendan958 maize. There were five treatments: MC: maize continuous cropping, applied 487.4g carbamide +4441g $(NH_4)H_2PO_4$+405g K_2SO_4 every 27 m^2 plot. MPR: maize and peanuts rotation cropping and maize was cropped in 2016, applied 487.4g carbamide +4441g $NH_4H_2PO_4$+405g K_2SO_4 in every 27m^2. PC: peanuts continuous cropping, applied 162g carbamide +486g $NH_4H_2PO_4$+607.6g K_2SO_4 in every 27m^2 plot. PCCA: peanuts continuous cropping, applied 2024g amendment1+162g carbamide +486g$(NH_4)H_2PO_4$+607.6g K_2SO_4 in every 27 m^2 plot. PCBA: peanuts continuous cropping, applied 1619g amendment 2+162g carbamide +486g $NH_4H_2PO_4$+607.6g K_2SO_4 in every 27m^2 plot.

Microsoft Excel 2010 was used for the statistical analysis of the experimental data.

3. Results and discussion

3.1. The effect of different tillage systems on root

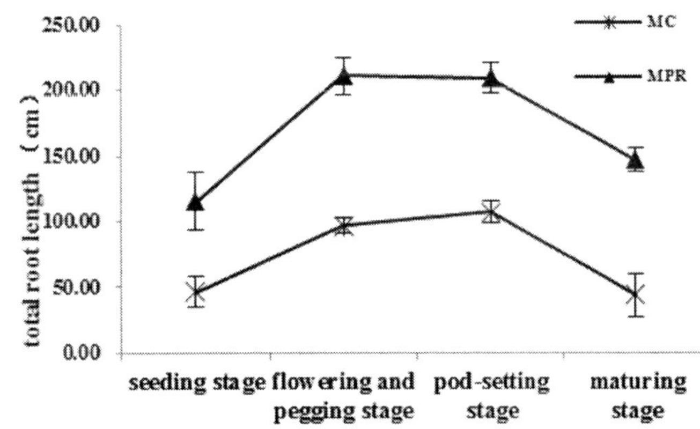

Fig.1 total root length in different stage

IOP Conf. Series: Materials Science and Engineering **207** (2017) 012069 doi:10.1088/1757-899X/207/1/012069

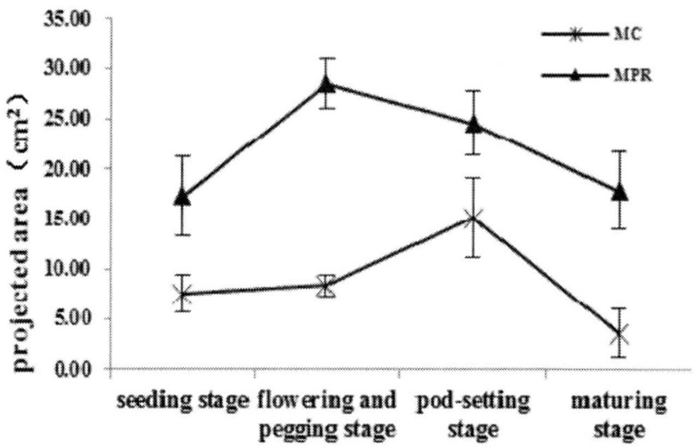

Fig.2 projected area in different stages

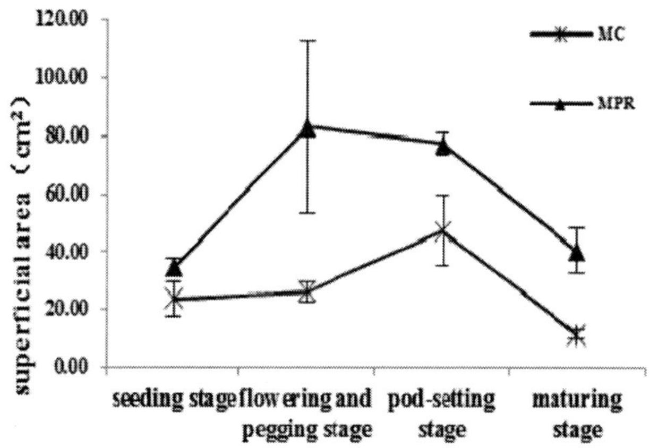

Fig 3. superficial in different stages

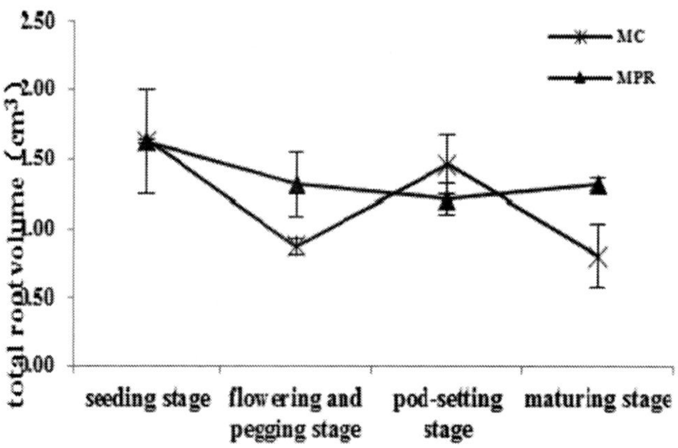

Fig.4 total root volume in different stages

IOP Conf. Series: Materials Science and Engineering **207** (2017) 012069 doi:10.1088/1757-899X/207/1/012069

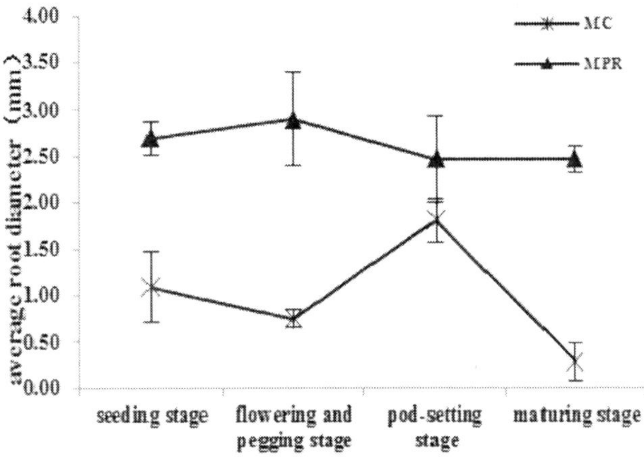

Fig.5 average root diameters in different stages

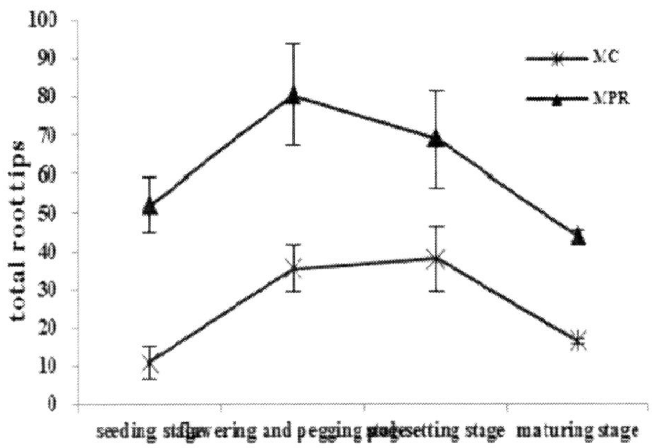

Fig.6 total root tips in different stages

The figure 1 showed that total root length in each growth period with continuous cropping was higher than that with rotational cropping and it was 43.36 ~106.73 cm and 115.49~210.96 cm, respectively. The total root length reached a maximum of 210.96 cm in flowering and pegging stage and it increased by 118.95% compared with the same stage with continuous cropping

The projected area of root was 3.67-15.11cm^2 and 17.34-28.53 cm^2 when continuous cropping and rotational cropping were used, respectively. In comparison with maize continuous cropping treatment, the projected area of root was higher than rotational cropping treatment in every stage and that averagely increased by 204.86%.

The superficial areas in each stage were 11.53-47.47cm^2 with continuous cropping and 34.51-77.40cm^2 with rotational cropping. The MPR treatment's results were averagely higher 1.5times than MC treatment.

No matter continuous or rotational cropping, the total root volume was decreased.MC and MPR treatment decreased to 50.92% and 19.02%, respectively. Expect in pod-setting stage, the volume with rotational cropping decreased by 20.66% in comparison with continuous cropping. It was much higher than in any else stage.

From figure 5, in continuous cropping treatment, the average diameters were on the decline in all stages and decreased by 74.54%, except it increased from flowering and pegging stage to pod-setting stage and up to a maximum of 1.81 mm. In rotational cropping treatment, the average diameter increased from seeding stage to flowering and pegging stage and reached to 2.90mm. After that, it decreased to 2.47mm in pod-setting stage and kept stable. Overall, the root average diameter with rotation cropping was greater than with continuous cropping and the average growth rate was 312.46%.

The root tips were 11-38 and 44-81 when continuous cropping and rotational cropping were used. In comparison with maize continuous cropping treatment, it was higher than rotational cropping treatment in every stage and that averagely increased by 184.53%.

In general, total length, projected area, superficial area and tips of root had the same change regulars. With continuous cropping, every index slowly growing from seeding stage to flowering and pegging stage, and reached the maximum in pod-setting stage. After that it declined gradually. In rotation cropping treatment, every index rapidly increased and up to the maximum in flowering and pegging stage. It gradually declined and had the minimum in mature stage.

The total volume and average diameter had similar regular. In continuous cropping treatments, the index declined from seeding stage. They increased rapidly from flowering and pegging stage to pod-setting stage. After that, it declined to the minimum in maturing stage. All index decreased when rotation cropping was used.

After long-term experiment for 6 years, rotation cropping treatments increased total length, project area, superficial area, average diameter and total tips of root in comparison with continuous cropping. Is same as the results of WANG's [6]. Rotational cropping is a kind of effective measures to relieve the continuous cropping obstacle [7], promote crop growth and increase yield [8]. It will balance the relationship between crop and soil and lead to the changes of agricultural ecological environment by using reasonable rotation cropping. The reason of rotational cropping can increase root properties may be related to it can improved soil biological properties. Yue's [9] study found that AWCD value of soybean and tobacco rotation cropping was obviously higher than that of tobacco continuous cropping. Rotational cropping increased the soil microbial carbon source and utilization ability. Roots are the main parts of the plants absorb the moisture and nutrient. Good root growth can increase crop yield, therefore, rotational cropping is conducive to solve obstacles of continuous cropping and promote the sustainable development of agriculture.

3.2. The effect of different amendment on root properties

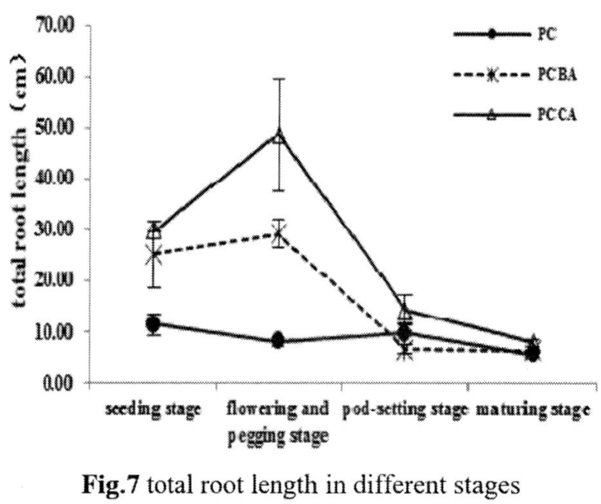

Fig.7 total root length in different stages

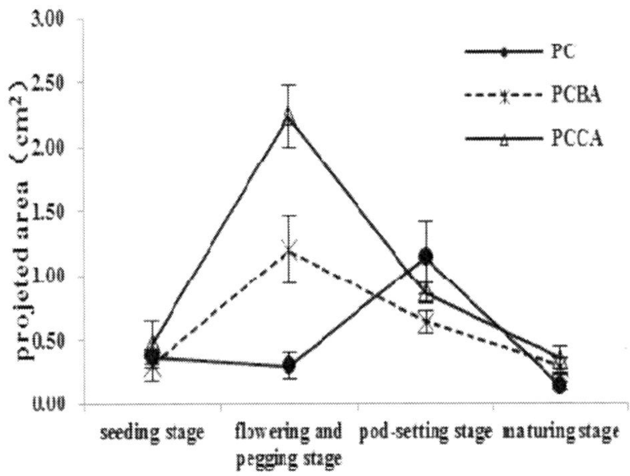

Fig.8 projected area in different stage

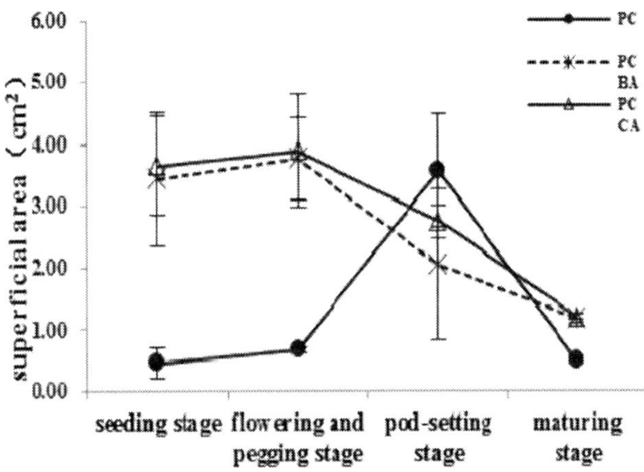

Fig.9 superficial areas in different stages

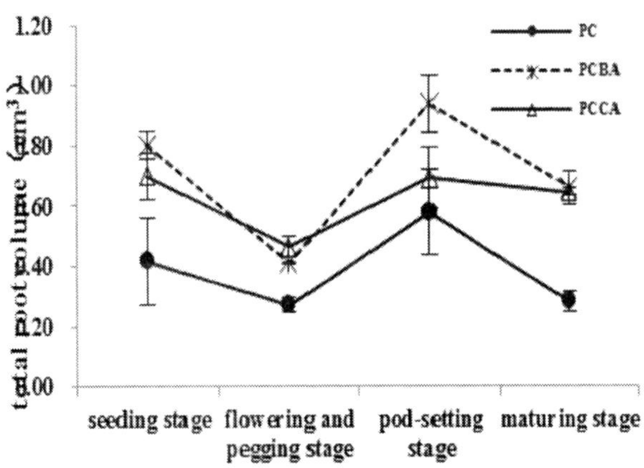

Fig10. total root volume in different stages

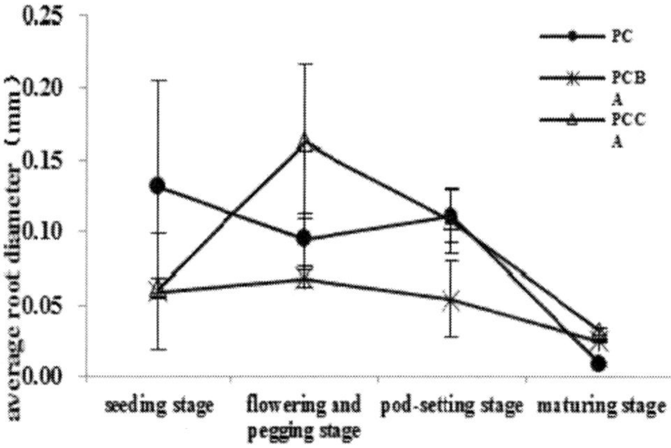

Fig.11 average root diameters in different stages

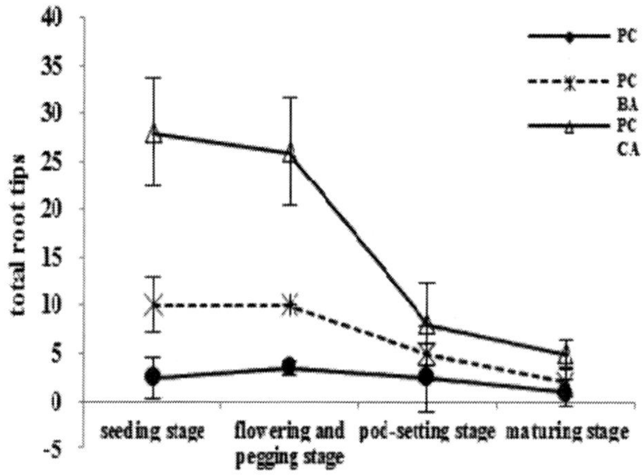

Fig.12 total root tips in different stages

From figure 7, the different treatments of total length of root in every stage was 8.06-48.71cm, 6.51-29.32cm, 5.86-11.55cm, respectively and the sequence was PCCA>PABA>PC. In flowering and pegging stage, it reached the longest with PCBA and PCCA treatments and in comparison with PCBA treatment, PCCA treatment was longer 66.13%.

The projected area of root with amendment 2 applied was bigger than in application of amendment 1 in all stages and they were 0.35-2.24cm^2 and 0.3-1.2cm^2. In flowering and pegging stage, the total length reached the maximum and PCCA treatment increased by 86.67% compared with PCBA treatment. Expect in pod- setting stage, the projected area with amendment applied was bigger than without it. The projected area was the biggest in pod-setting stage and it was 1.14cm^2. In comparison with PCBA and PCCA treatment, it increased by 78.13% and 31.03%, respectively.

In figure 9, in application of amendment 2, the superficial areas in all stages were bigger than amendment1 applied. They were reached the biggest in flowering and pegging stage, 3.90cm^2 and 3.77cm^2, respectively. The superficial area of PCCA treatment was increased by 3.45% compared with PCBA treatment. Except in pod-setting stage, superficial area in all stage obviously smaller than application of amendment in comparison with there was no amendment applied. In pod-setting stage, it bigger than PCBA and PCCA treatment 74.15% and 28.36% compared with PC treatment.

Overall, the total volume was the biggest with amendment 2 applied treatment and no amendment used treatment was the smallest. The volume reached the maximum in pod- setting stage whatever which treatments and they were 0.94 cm^2, 0.69cm^2.0.58cm^2, respectively. It increased to 36.23% and 62.07% when amendment 2 used in comparison with amendment1 applied and no amendment used treatment.

In figure 11, the average diameter was the biggest in flowering and pegging stage in PCCA and PCBA treatment and they were 0.16mm and 0.07mm. In comparison with application of amendment 1, the diameter with amendment 2 applied was increased by 128.57%. In other stage, it in PC treatment was bigger than in PCBA treatment.

The total tips of root when amendment 2, amendment1 and no amendment applied were 5-28, 2-10, 1-4 respectively. The tips were the most in seeding stage in PCCA and PCBA treatment and it in PCCA treatment was more than 180% compared with PCBA.

Relative to no amendment applied, amendment can obviously improve the total root length, the projection area, surficial area, total volume, average diameter and tips of root directly. The more content of amendment used, the better the effect was. Reason may be that the peanut continuous cropping decreased the soil quality and single crop continuous cropping consumed soil nutrient which cannot even replenished. The amendment added nutrients for the growth of peanuts [10].

4. Results

1) The rotation cropping can improve total length, projected area, projected area, total volume, average diameter and tips of root in comparison with continuous cropping.

2) The amendment was helpful to increase total length, projected area, projected area, total volume, average diameter and tips of root. The more amendment was applied, the effect was better

Acknowledgements

This study was carried out with National Engineering Laboratory of High Efficient Use on Soil and Fertilizer Resources/Monitoring & Experimental Station of Corn Nutrition and Fertilization in Northeast Region funded by National Natural Science Foundation of China (41371287) and China Agriculture Research System (CARS-14).

References

[1] Yamaguchi J, Tanaka A. Quantitative observation on the root system of various crops growing in the field, J. Soil Science and Plant Nutrition (1990) 36: 483-493.

[2] Qin S H, Cao L, Zhang J L et al. Effect of rotation of leguminous plants on soil available nutrients and physical and chemical properties in continuous cropping potato field. Acta Agronomica Sinica (2014) 40(8): 1452-1458.

[3] Kurle J E, Grau C R, Oplinger ES, Mengistu A. Tillage, cropsequence and cultivar effects on sclerotinia stem rot incidence and yield in soybean, J. Agronomy Journal(2001)3: 973－982.

[4] Wu Z F. Soil Amendments [M]. Beijing: Science Press (1976) 24-34.

[5] Wang J S, Fan F F, Guo J, Wu A L , Dong E W, Bai W B, Jiao X Y,Effects of biochar on stalk characters and yield of maize, J. Chin J Appl Ecol (2016) 07: 2283-2291.

[6] Dong W J, Zheng H P, Huang X, Xie K Z, Xu P Z, Zhang R Z, Effect of different soil amendments on root activity and yield of rice in cold spring paddy soils, J. Chin Agric Bull (2013) 36: 182-187.

[7] Salgado E, Cautin R. Avocado root distribution in fineand coarse-textured soils under drip and microsprinkler irrigation, J. Agricultural Water Management (2008) 95: 817-824

[8] Liu X, Zhang S L, Liu GF , et al. Control of continuous potato monoculture barrier via biological soil disinfestation method in Yellow River irrigation areas of central Gansu Province, North-west China. Chinese Journal of Applied Ecology (2015) 26 (4): 1205-1214.

[9] Yue B B. The rhizosphere soil microbial diversity is changed by flue-cured tobacco continuous cropping. Ph D Thesis. Harbin: Northeast Forestry University, 2012.

[10] Wang F, Li S G, Xu F H , Gu J G, The research progress of mechanism of continuous cropping obstacles, J. Soil and Fertilizer Sciences (2013)(5): 6-13.

IOP Conf. Series: Materials Science and Engineering **207** (2017) 012070 doi:10.1088/1757-899X/207/1/012070

Model updating approach for icing forecast of transmission lines by using particle swarm optimization

Haigang Tian[1,*], Naifeng Liang[2], Puzhi Zhao[2], Xuan Wang[3], Chuangwei Zhu[3], Sanchun Zhang[1], Wei Wang[1]

[1]State Grid Xin Jiang Economic Research Institute, Urumqi 830000, China
[2]State Grid Xin Jiang Electric Power Company, Urumqi 830000, China
[3]State Grid Xin Jiang Yi Li Electric Power Company, Urumqi 835000, China

[*]Corresponding author e-mail: axl6533503@163.com

Abstract. With the purpose of predicting icing of transmission lines, a model updating approach is presented in this study. The changes of structural dynamic response of transmission lines that is caused by icing is studied firstly by using finite element method. Then, model updating method and particle swarm optimization is implemented to indentify the thickness of icing according to the alternation of natural frequencies. The results show that the proposed methodology is meaningful to monitor line icing.

1. Introduction

Icing of transmission lines would cause line galloping, insulator flashover and even tower collapse which is a great threat for the operation of power grid [1-2]. Thus, effective monitoring and condition assessment of line icing is the key issue which needs to be solved, and the establishment of icing forecast plays an important role in the operation of power grid.

It's very difficult to evaluate the condition of lines manually considering that transmission lines are located sparsely. Hence, the on-line monitoring technique has attracted increasing attentions in engineering application. The current monitoring strategy are mainly based on measuring the weight, angle of insulator, wind speed, temperature and humidity. The equivalent thickness of line icing can be estimated according to measured data [3-5], based on that, related staff will be alerted as long as the thickness evaluated beyond a predefined value. However, the actual icing condition is usually distributed non-uniformly along the length of line, thus, the monitoring technique based on the evaluation of equivalent thickness is unable to get more information of icing distribution in details. In this work, a methodology for icing forecasting based on model updating method is presented. The modal frequencies are introduced to identify the icing condition of transmission lines with the help of artificial intelligence technique, and the results show that the proposed approach is able to identify and predict the distribution of icing.

2. Identification of icing based on model updating process

2.1. Numerical model of transmission lines

Transmission line is one kind of cable-suspended structure which cannot bear compressive force and flexural moment, and a large deformation would be generated when load is applied. In this study,

ANSYS, a commercial FE (finite element) method software is implemented to establish numerical model, and the transmission line can be modeled appropriately by using LINK10 element according to its mechanical characteristics.

The FE analysis of transmission line is a typical geometrical non-linearity problem. It is very important to obtain the geometric shape under load (including gravity), and the solution of each load step highly depends on the results in the previous one. Therefore, to complete the process of shape finding, the initial FE model should be established based on catenary theory. Then, the FE model needs to be updated as long as gravity is introduced, and the converged results is the real FE model of transmission line.

2.2. Model updating method

Due to the uncertainties in simplifying assumptions of structural geometry, material properties and boundary conditions, there is always a difference between FE analysis and experimental test. The propose of model updating method is to adjust numerical model with experimental results which are considered to be accurate, thus, a more realistic FE model can be achieved and the unknown structural parameters and conditions can be identified according to the process which behaves as an inverse problem [6]. Dynamic responses, such as natural frequencies and modal shapes, are able to reflect the structural properties and are widely used for model updating method [7-8], hence, it will be introduced with the purpose of identifying icing in this study. A flowchart of model updating procedure is illustrated in Fig 1.

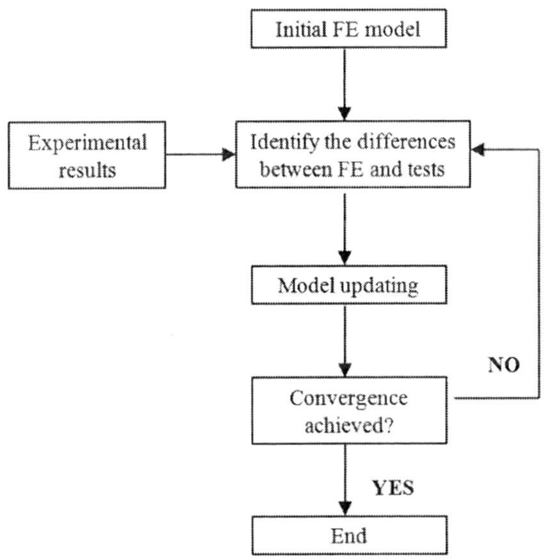

Figure 1. Flowchart of model updating method

2.3. Icing forecasting based on model updating method

The selection of parameter for updating is a key problem for model updating process which needs to be sensitive enough to structural response and changing of structural parameters. For transmission line, icing would increase structural mass leading to the alternation of dynamic behaviors, and conversely structural dynamic measurements are able to reflect the icing conditions. Therefore, the thickness of icing is chosen as the updating parameter in this study, and it will be transferred to the mass of each element in order to carry out the FE analysis and model updating procedure.

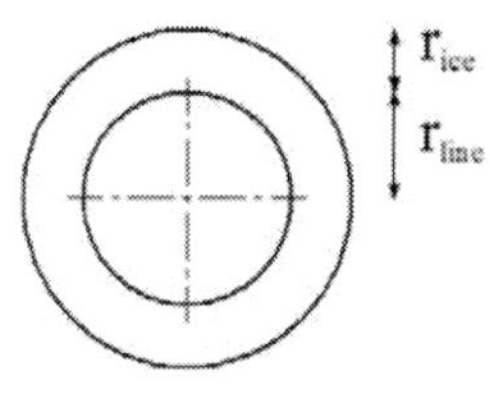

Figure 2. The area of transmission line with icing

As shown in Fig 2, the radii of line is r_{line}, the equivalent thickness of icing is r_{ice}, with the purpose of simplifying model updating process, the equivalent area of line A_{eq} is introduced herein as the updating parameter as shown in Eq. 1

$$A_{eq} = A_{line} + \frac{\rho_{ice}}{\rho_{line}} \times A_{ice} \tag{1}$$

where A_{ice}, A_{line} and ρ_{ice}, ρ_{line} represent the area and density of icing and transmission line, where the icing area is

$$A_{ice} = \pi\left[(r_{line} + r_{ice})^2 - r_{line}^2\right] \tag{2}$$

Until now, the changing of structural mass caused by icing is transferred to equivalent area of transmission line by using Eqs. (1)-(2), and the model updating process can be completed by updating the area of each element in FE model.

Furthermore, single objective optimization is the widely used approach in model updating procedure, and the formation of objective function predominates the final results. Dynamic response (such as natural frequencies, modal shapes, frequency response function e.g.) are usually selected to construct objective functions [8], hence, the objective function F applied in this study is shown as follows

$$F = 1 - MAC = 1 - \frac{\left|f_a^T f_t\right|^2}{(f_a^T f_a)(f_t^T f_t)} \tag{3}$$

where f_a and f_t denote the vector constructed by natural frequencies which are obtained by numerical analysis and experimental test, respectively. MAC represents modal assurance criteria, whose value equals 1 if f_a and f_t correspond perfectly and it equals 0 conversely.

2.4. Optimization algorithm for model updating

As presented previously, the identification of line icing has been transferred to an optimization problem which needs to minimize the objective function in Eq. (1), besides, the optimization algorithm plays an important role in model updating procedure. With the development of artificial intelligence, some evolutionary algorithms are implemented in this field, such as particle swarm optimization (PSO). PSO performs quite well in global searching in comparison of conventional optimization methodology, besides, the advantages of fast convergence and easy programming makes it become promising in optimization and model updating [9-10].

As an evolutionary optimization approach, PSO starts searching optimal solution in the entire space in the following way

$$v_i^{k+1} = w \cdot v_i^k + c_1 \cdot r_1 \cdot (pbest_i - x_i^k) + c_2 \cdot r_2 \cdot (gbest - x_i^k) \quad (4)$$

$$x_i^{k+1} = x_i^k + v_i^{k+1} \quad (5)$$

where x_i^k , v_i^k denote the location and velocity of i-th particle in k-th iteration, c_1 and c_2 are the constant related to convergence of swarm which usually equals 2. r_1 and r_2 are random number between 0 and 1 in order to avoid falling into local optimum, and w denotes the inertial item that controls the velocity of swarm.

With the purpose of identify icing condition and distribution along transmission line in this work, the equivalent thickness of icing in each element r_{ice}^j are used to construct updated vector $\overline{r} = (r_{ice}^1, r_{ice}^2, \ldots r_{ice}^n,)$, and PSO is implemented based on experimental test data. The procedure is illustrated in Fig 3 as follows.

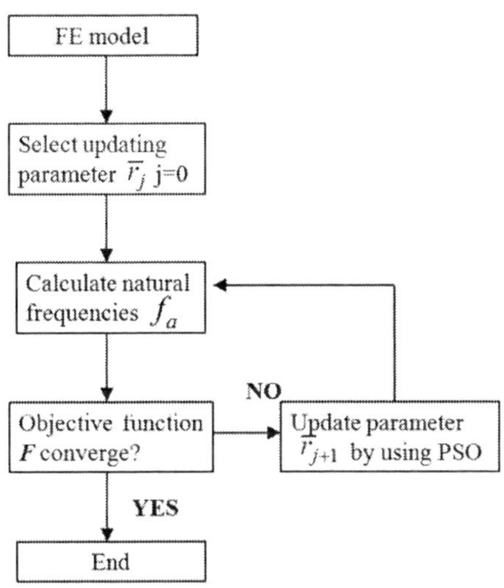

Figure 3. The flowchart of identification of line icing

3. Case study

To evaluate the accuracy and efficiency of proposed approach, a case study is carried out for a transmission line with geometrical and material properties as shown in Tab 1.

Table 1. Geometrical and material properties of transmission line

Span (m)	435
Area (mm^2)	275.96
Elastic modulus (MPa)	73000
Mass (kg/km)	922.2
Fracture force (N)	75620

First of all, FE model is constructed based on ANSYS, and initial strain and gravity are applied to employ the shape finding process by updating structural geometrical location. Then, the natural frequencies can be obtained as shown in Tab 2.

Table 2. The first six natural frequencies

Natural frequencies	(Hz)
f_1 (out-of-plane symmetrical)	0.3283
f_2 (in-plane symmetrical)	0.3381
f_3 (in-plane anti-symmetrical)	0.6575
f_4 (out-of-plane anti-symmetrical)	0.6577
f_5 (out-of-plane symmetrical)	0.9887
f_6 (in-plane symmetrical)	0.9890

The structural changing caused by icing would lead to alternation of dynamic behaviors. In order to evaluate the influences of different thickness of icing on natural frequencies, three cases are assumed: the line is divided into six parts uniformly, three icing conditions with thickness of 5,8,10 mm are applied on the third part, respectively. Then, the thickness of icing is transferred to equivalent area of line by using Eqs. (1)-(2), and the first six natural frequencies are calculated under the three cases. The effect of icing thickness on changes of modal frequencies is illustrated in Fig. 4.

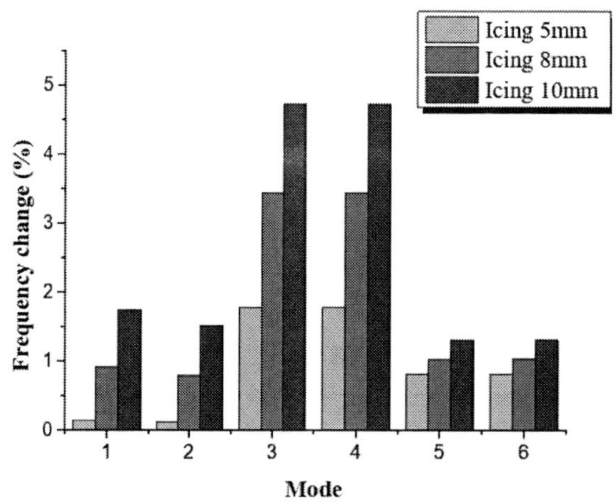

Figure 4. Influences of icing thickness on changes of natural frequencies

It is clear in Fig. 4 that the changing of natural frequencies increases with the increasing of icing thickness, and natural frequencies has the capability to reflect the alternation of icing which demonstrates that frequencies are sensitive enough to form objective function for model updating process. To verify proposed approach for the identification of icing, the transmission line with 10mm thickness of icing is chosen as a benchmark, and the natural frequencies are calculated as pseudo experimental data f_t to construct objective function as shown in Eq. (3). The thickness of icing r_{ice} is transferred to equivalent area of line A_{eq}, and the vector $\overline{r}$ which is formed by the thickness of icing in each element r_{ice} will be updated by employing PSO algorithm to minimize objective function in Eq (3). The results of identification is shown in Fig 5.

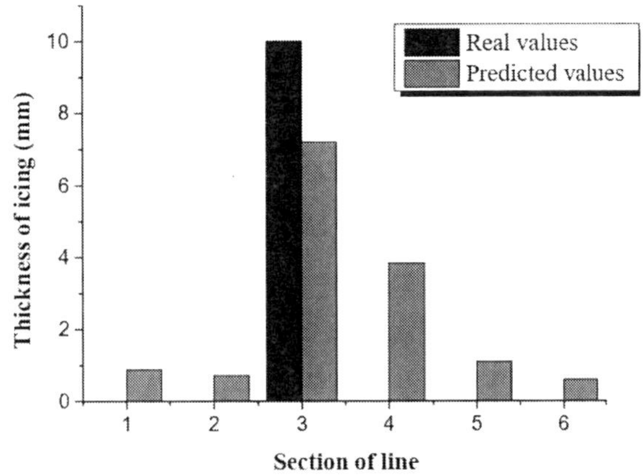

Figure 5. Identification of thickness of icing

From the results it can be seem that icing is located at the middle span of line which correspond quite well with predicted values. Thus, based on the changing of natural frequencies it is able to identify line icing by using proposed model updating approach and PSO algorithm. Compared with conventional strategy of icing detection based on static measuring, the presented method focuses more on the distribution of icing along the length of transmission line.

4. Conclusion

With the purpose of identify icing of transmission line, the thickness of icing is transferred into equivalent area of line in this study, and modal frequencies are selected to form objective function in order to carry out model updating process. The results show that modal frequencies are sensitive enough to reflect structural alternation. Besides, as an evolutionary algorithm, PSO is effective in searching optimum solution for the identification of icing. In this work, more attentions were paid to the distribution of icing along the length of power line rather than global equivalent thickness of icing. Furthermore, it should be noticed that symmetrical changing in symmetrical structure would cause same variations for structural responses, thus, more dynamic information (modal shapes e.g.) should be introduced to increase accuracy.

References

[1] M. Farzaneh, Atmospheric icing of power networks, Springer, Berlin, 2008.
[2] M. Farzaneh, T. Baker, A. Bernstorf, Insulator icing test methods and procedures a position paper prepared by the IEEE Task Force on insulator icing test methods, IEEE Transactions on Power Delivery, 18, 4(2003)1503-1515.
[3] E.J. Goodwin, J.D. Mozer, A.M. Di Gioia, Predicting ice and snow loads for transmission line. Proceeding of the First IWAIS, 1983.
[4] P. Mc Comber, J. Lafontaine, J. Druez, A comparison of neural network and multiple regression transmission line icing models, Proceedings of 7th International Workshop on Atmospheric Icing of Structures, Reykjavik, Iceland, (1998)175-180.
[5] M. Huneault, C. Langheit, J. Caron. Combined models for glaze ice accretion and de-icing of current-carrying electrical conductors, IEEE Trans on Power Delivery, 20, 2(2005)1611-1616.
[6] A. Teughels, G. De Roeck, Damage detection and parameter identification by FE model updating, Archives of Computational Method in Engineering,12, 2(2005) 123-164.
[7] J.E. Mottershead, M.I. Friswell, Model updating in structural dynamics: a survey, Journal of Sound and Vibration, 167, 2(1993)347-375.
[8] M.I. Friswell, J.E. Mottershead, Finite element model updating in structural dynamics, Kluwer Academic Publishers, Dordrecht, 1995.
[9] R.C. Eberthart, Y.Shi, Particle swarm optimization: developments, applications and resources. In:

Proceedings of the 2001 congress on evolutionary computation, 1, 81-86 (2001).
[10] T. Marwala, Finite-element-model Updating Using Computational Intelligence Techniques, Springer, London, 2010.

IOP Conf. Series: Materials Science and Engineering **207** (2017) 012071 doi:10.1088/1757-899X/207/1/012071

A weight modification sequential method for VSC-MTDC power system state estimation

Xiaonan Yang[1, *], Hao Zhang[2], Qiang Li[1], Ziming Guo[2], Kun Zhao[1], Xinpeng Li[2], Feng Han[1]

[1]China Electric Power Research Institute, Beijing 100192, China
[2]State Grid Jibei Electric Power Co., Ltd, Beijing 100032, China

[*]Corresponding author e-mail: yangxiaonan@epri.sgcc.com.cn

Abstract. This paper presents an effective sequential approach based on weight modification for VSC-MTDC power system state estimation, called weight modification sequential method. The proposed approach simplifies the AC/DC system state estimation algorithm through modifying the weight of state quantity to keep the matrix dimension constant. The weight modification sequential method can also make the VSC-MTDC system state estimation calculation results more ccurate and increase the speed of calculation. The effectiveness of the proposed weight modification sequential method is demonstrated and validated in modified IEEE 14 bus system.

1. Introduction

As a result, the negative impact of the highly developed industry is not enough to prevent the adverse impact, leading to the global three major crises: shortage of resources, environmental pollution and ecological damage. Flexible DC transmission technology can effectively use new energy to help improve the ecological environment. To support flexible DC Power transmission, voltage source converter plays an important role in power systems. It is now considered to be a feasible solution for connecting offshore wind farms due to its significant advantages over both current source converter (CSC) HVDC and traditional AC system [1] and the first project which is 400 MW, ±150kV two terminal VSC HVDC system for wind farms connection has been commissioned in Germany in 2010 [2]. The concept of "European Offshore Super grid" [3] has been proposed to build a VSC based subsea meshed DC grid which is used to connect the renewable offshore wind farms to the AC networks. Apart from line commutated converter (LCC), the VSC technology makes the extension to multi-terminal HVDC connections relatively easy. Quite a few papers have been published on VSC-MTDC steady state modeling for load flow [4, 5].

State estimation plays important role in Energy Management System (EMS). With rapid development of HVDC technologies, development of practical methods for AC/DC hybrid power system state estimation is an emergent demand. However, most papers have been published on AC systems, and only a few studies have so far been carried out on AC/DC systems state estimation which includes a generalized VSC MTDC model. In [6, 7], the state estimation was proposed for the two-terminal VSC-HVDC system and the MTDC system was not considered. Glover and Sheikoleslami [8] emphasized on the DC network state estimation without including the AC/DC interconnection system. Leita da Silva et al. [9] separated an AC/MTDC interconnected system into three subsystems, namely

AC system, AC/DC interface and DC system. Jagatheesan and Duraiswamy [10] extended the fast decoupled state estimation technique originally applied on the AC system to the AC/DC system and proposed a sequential solution scheme to solve P-Q-DC iteratively. The sequential method is simple and can be easily appended to existing AC fast decoupled state estimators (FDSE). Zhang boming [11] presents a new state estimation algorithm, called improved sequential method, which is based on the polar formulation.

Since VSC-MTDC has significant advantages in power systems and state estimation has drown more attentions in EMS, this paper presents a new state estimation algorithm, called weight modification sequential method. This method decouples the AC system and the DC system by mathematic method according to the information matrices formed by the alternating solution method. And the residual error information of AC/DC system interface is fully utilized to realize accurate and rapid convergence of AC/DC system by weight modification. Results obtained on IEEE14 test systems have been presented.

2. AC state estimation algorithm

Nowadays, state estimation has become an indispensable part of energy management system. The goal of the state estimator in the power system is to give the best estimate of the state of the power system to minimize the effect of the measurement error. The mathematical model of state estimation can be expressed as:

$$z = h(x) + v \tag{1}$$

Usually assume that the network parameters and wiring status is accurate to each node voltage amplitude and phase angle for the state estimation vector X, to obtain the measured value Z, for the measurement error v.

The weighted least square (WLS) state estimation is designed to find the system state x by solving the following optimization problem:

$$J(x) = [z - h(x)]^T R^{-1}[z - h(x)] \tag{2}$$

The solution of the nonlinear equation (2) can be obtained by an iterative procedure as shown below:

$$\Delta\hat{x}^k = [H^T(\hat{x}^k)R^{-1}H(\hat{x}^k)]^{-1}H^T(\hat{x}^k)R^{-1}[z - h(\hat{x}^k)] \tag{3}$$

$$\hat{x}^{k+1} = \hat{x}^k + \Delta\hat{x}^k = \hat{x}^k + [H^T(\hat{x}^k)R^{-1}H(\hat{x}^k)]^{-1}H^T(\hat{x}^k)R^{-1}[z - h(\hat{x}^k)] \tag{4}$$

According to the above iterative correction, until convergence, this time close to the minimum value of the objective function.

Since the basic weighted least squares estimation method needs a large amount of memory and a long computation time, a fast decoupled state estimation (FDSE) method is usually adopted. This is a well-recognized model of utility performance on high voltage networks. The quantity measurement, the state quantity and the measurement equation are decomposed into active and reactive power, and the hypothesis is introduced in the Jacobian matrix H:

$$\frac{\partial h_a}{\partial V} \approx 0 \quad \frac{\partial h_r}{\partial \theta} \approx 0 \tag{5}$$

IOP Conf. Series: Materials Science and Engineering **207** (2017) 012071 doi:10.1088/1757-899X/207/1/012071

The modified equation of the state estimation of the fast decoupled method can be obtained:

$$\begin{aligned} A\Delta\theta^{(l)} &= a^{(l)} \\ B\Delta V^{(l)} &= b^{(l)} \end{aligned} \tag{6}$$

Where A-the symmetry matrix of active constants
B-the symmetrical matrix of reactive power
a-the active free vector, b-the reactive free vector.

The above state estimation approach is only suit for the AC system. When the MTDC system is added to the AC power system, the system state cannot be calculated by the WLS state estimation.

3. DC state estimation algorithm

3.1. VSC-MTDC Model

The VSC-MTDC steady-state model is shown as follows:

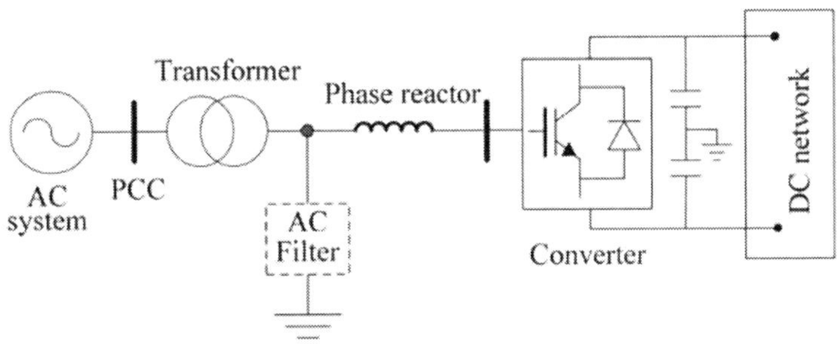

Figure 1. Schematic diagram of VSC –MTDC.

From figure 1, where $\dot{U}_{ci} = U_{ci}\angle\delta_{ci}$ is the converter input voltage. $\dot{U}_{ti} = U_{ti}\angle\delta_{ti}$ is from the AC side the node voltage which is connected to inverter. $R_i + X_{li}$ is the equivalent impedance for the rheological and commutation bridges. $P_{si} + Q_{si}$ is the inverter input power.

Since $R_i << X_{li}$, the converter input power is:

$$P_{si} = \frac{U_{ti}U_{ci}}{X_{li}} sin\delta_i \tag{7}$$

$$Q_{si} = \frac{U_{ti}}{X_{li}}\left(U_{ti} - U_{ci}cos\delta_i\right) \tag{8}$$

The converter input voltage U_{ci} is:

$$U_{ci} = \mu_d M_i U_{di} / \sqrt{2} \tag{9}$$

Where: $\delta_i = \delta_{si} - \delta_{ci}$, M_i is the modulation degree, μ_d is the DC voltage utilization.

As the impedance of the converter bridge forward, so the active power P_{ci} into the converter bridge is equal to the active power output P_{di}, Q_{ci} is the reactive power into the converter bridge, I_{di} is the DC current.

$$P_{di} = U_{di} I_{di} \tag{10}$$

$$P_{di} = P_{ci} = \frac{U_{ti} U_{ci}}{X_{li}} sin\delta_i \tag{11}$$

$$Q_{ci} = \frac{-U_{ci}}{X_{li}}(U_{ci} - U_{ti} cos\delta_i) \tag{12}$$

DC system network equation is:

$$I_{di} = U_{di} G \tag{13}$$

Where: G is network conductance.

3.2. *VSC-MTDC measurement equations*

The state estimation of flexible DC transmission usually adopts the WLS method. There are a total of 11 variables appeared: $U_{di}, U_{ci}, U_{ti}, M_i, \delta_i, I_{di}, P_{si}, Q_{si}, P_{ci}, Q_{ci}, P_{di}$.

The DC nodes' power flow equations in the VSC-MTDC system are:

$$\Delta P_{ti} = P_{ti} - \sum f_{P_{aci}}(U_{aci}, \delta_{aci}, U_{ti}, \delta_{ti}) - f_{P_{si}}(U_{ti}, \delta_{ti}, U_{di}, \delta_{ci}, \mathrm{M}) = 0 \tag{14}$$

$$\Delta Q_{ti} = Q_{ti} - \sum f_{Q_{aci}}(U_{aci}, \delta_{aci} U_{ti}, \delta_{ti}) - f_{Q_{si}}(U_{ti}, \delta_{ti}, U_{di}, \delta_{ci}, \mathrm{M}) = 0 \tag{15}$$

The AC nodes' power flow equations in the VSC-MTDC system are:

$$\Delta P_{ai} = P_{ai} - P_{ai}(U_{ai}, \delta_{ai}) \tag{16}$$

$$\Delta Q_{ai} = Q_{ai} - Q_{ai}(U_{ai}, \delta_{ai}) \tag{17}$$

4. AC/DC state estimation

The actual measurement equations in AC/DC system are shown as follows:

$$\Delta P_{si} = P_{si} - f_{Psi}(U_{ti}, \delta_{ti}, U_{di}, \delta_{ci}, \mathrm{M}) = 0 \tag{18}$$

$$\Delta Q_{si} = Q_{si} - f_{Qsi}(U_{ti}, \delta_{ti}, U_{di}, \delta_{ci}, \mathrm{M}) = 0 \tag{19}$$

$$\Delta P_{di} = U_{di} I_{di} - f_{Pci}(U_{ti}, \delta_{ti}, U_{dI}, \delta_{ci}, M_i) = 0 \tag{20}$$

$$\Delta I_{di} = I_{di} - U_{di} G = 0 \tag{21}$$

$$U_{ti}^{m} - U_{ti}^{t} = 0 \tag{22}$$

$$U_{di}^{m} - U_{di}^{t} = 0 \tag{23}$$

$$\delta_i^{m} - \delta_i^{t} = 0 \tag{24}$$

$$M_i^{m} - M_i^{t} = 0 \tag{25}$$

$$I_{di}^{m} - I_{di}^{t} = 0 \tag{26}$$

Where U_{ti}^{m} is measurement value, U_{ti}^{t} is true value.

The VSC HVDC converter has four control modes when operating:

1) U_{dc}, Q control: The DC bus voltage and the reactive power keep constant. The U_{dc}, Q control constraint is

$$\begin{aligned} U_{di} - U_{di}^{ord} &= 0 \\ Q_{si} - Q_{si}^{ord} &= 0 \end{aligned} \tag{27}$$

2) U_{dc}, U_{ac} control: The DC bus voltage and AC bus voltage keeps constant and the U_{dc}, U_{ac} control constraint is

$$\begin{aligned} U_{di} - U_{di}^{ord} &= 0 \\ U_{ti} - U_{ti}^{ord} &= 0 \end{aligned} \tag{28}$$

3) *P, Q* control: The active power and the reactive power is constant and the *P, Q* control constraint is

$$\begin{aligned} P_{si} - P_{si}^{ord} &= 0 \\ Q_{si} - Q_{si}^{ord} &= 0 \end{aligned} \tag{29}$$

4) *P*, U_{ac} control: The active power and AC bus voltage is constant and the *P*, U_{ac} control constraint is

$$\begin{aligned} P_{si} - P_{si}^{ord} &= 0 \\ U_{ti} - U_{ti}^{ord} &= 0 \end{aligned} \tag{30}$$

Since the different control mode has different constant state measurement. Under different control modes, the Jacobian matrix and corresponding factor table will change in dimension, and because of the switch of control mode, the number of Jacobian matrix dimension is changed through the different states' adding and exiting.

In this paper, in order to keep the dimension unchanged, we give a large weight $R_{\max}$ to the known quantity measurement under different control modes in the weighted least squares estimation. Put large weight $R_{\max}$ to the states as follows:

1) U_{dc}, *Q* control: Put large weight $R_{\max}$ to U_{di}, Q_{si}
2) U_{dc}, U_{ac} control: Put large weight $R_{\max}$ to U_{di}, U_{ti}
3) *P, Q* control: Put large weight $R_{\max}$ to P_{si}, Q_{si}
4) *P*, U_{ac} control: Put large weight $R_{\max}$ to P_{si}, U_{ti}

This approach achieves the goal of making the state estimation algorithm of high precision, high speed and low memory. The estimate of known state measurements is obtained using WLS by minimizing the objective function:

$$J(x) = [\Delta z - H(x_0)\Delta x]^T R_{\max}^{-1} [\Delta z - H(x_0)\Delta x] \tag{31}$$

Where, the H dimension is not changed.

By using sequential method, there are three unknown variables between AC system and DC system they are P_{si} ,Q_{si}, U_t. The AC and DC decoupling calculations are implemented. Each iteration process involves AC iterations and DC iterations until they both converge independently and the AC/DC system convergence overall.

Under different control mode, the AC/DC system has different state estimation convergence sign. The convergence sign of the state estimation analyzed as follows:

1) U_{dc}, Q control:

The difference $\Delta P_{si}, \Delta U_t$ are the convergence signs of the state estimation between results of the AC and DC systems.

2) U_{dc}, U_{ac} control:

$\Delta P_{si}, \Delta Q_{si}$ are the convergence signs of the state estimation

3) P, Q control:

ΔU_t is the convergence sign of the state estimation

4) P, U_{ac} control:

ΔQ_{si} is the convergence sign of the state estimation

A weight modification sequential method is proposed so as to improve accuracy and fast convergence of AC/DC system. The algorithm gives an initial weight $R_{ac}^{(0)}$ to the unknown state quantity $x_0^{(0)}$ from AC side and gives an initial weight $R_{dc}^{(0)}$ to the unknown state quantity $x_0^{(0)}$ from DC side in the first state estimation calculation. After the convergence of the AC system iterative calculation and the DC system iteration calculation, the state quantity x_{ac} is calculated by the AC system and the state quantity x_{dc} is calculated by the DC system.

The absolute value of the difference between the AC and DC systems is calculated:

$$\Delta c_{ad} = |x_{ac} - x_{dc}| \tag{32}$$

If the absolute value is more than the system adaptive comparison threshold Δc_{ε}:

$$\Delta c_{\varepsilon} < \Delta c_{ad} \tag{33}$$

The next state $x_0^{(l+1)}$ after calculating is different in each case:

$$x_0^{(l+1)} = \frac{R_{dc} x_0^{(0)} + R_{ac} x_0^{(0)}}{R_{dc} + R_{ac}} \tag{34}$$

Where $x_0^{(l+1)}$ is measurement during the next iteration, $l \geq 0$, $x_0^{(0)}$ is initial measurement.

Next, put this average measurement $x_0^{(l+1)}$ into iterative calculation in AC system and DC system separately. And add bigger weight $R_{\max}$ to this average measurement $x_0^{(l+1)}$:

$$J(x) = [\Delta z - H(x_0^{(l+1)})\Delta x]^T (R_{ac} / R_{dc} + R_{\max})^{-1} [\Delta z - H(x_0^{(l+1)})\Delta x] \tag{35}$$

IOP Conf. Series: Materials Science and Engineering **207** (2017) 012071 doi:10.1088/1757-899X/207/1/012071

Keep iterative calculate until satisfied the equation:

$$\Delta c_{ad} < \Delta c_{\varepsilon} \tag{36}$$

Based on the effective method of weight modification, it provides a good data section for AC / DC power network, which provides technical support for power grid dispatching control. The modification of the weight of the AC / DC system can effectively accelerate the convergence of the state estimation. While minimizing the interference of undesirable data, the algorithm satisfies the AC / DC grid state estimation with VSC-MTDC.

The program flow chart is as follows:

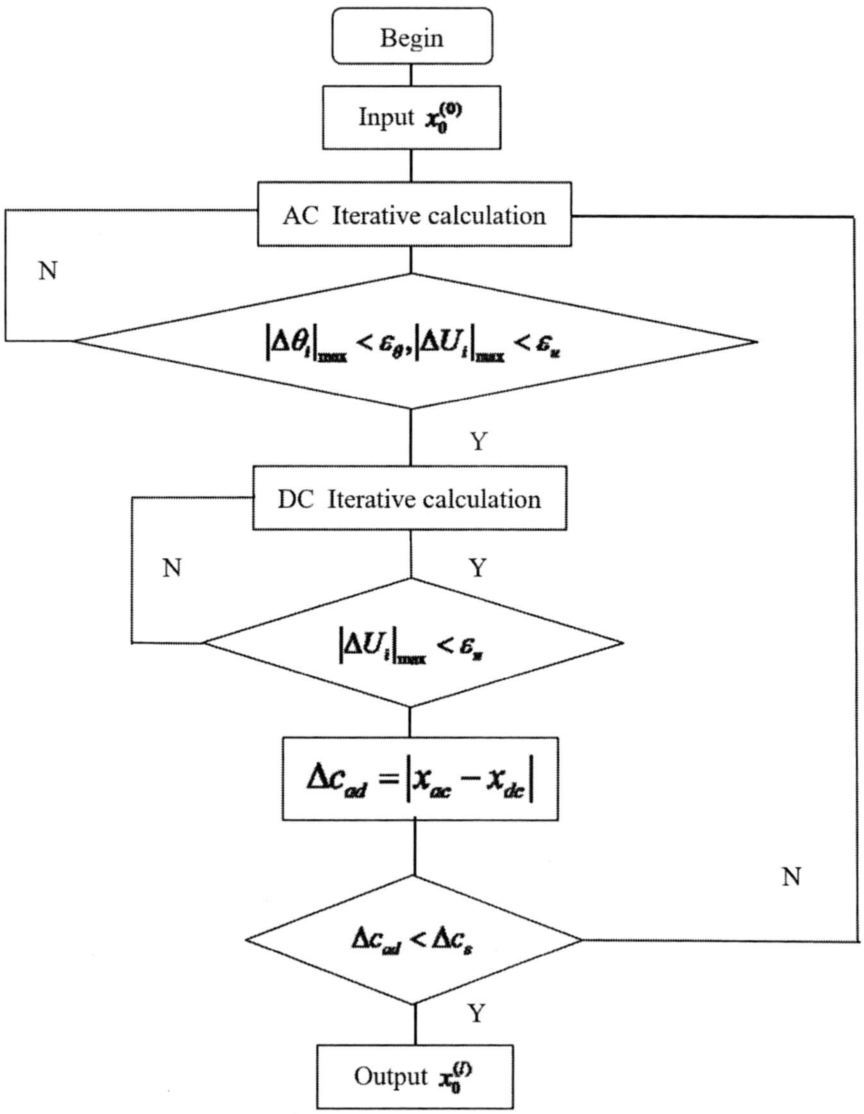

Figure 2. Program flow chart

5. Case Study

IEEE 14 bus system is used for simulation to test the approach proposed in this paper. Figure 3 is the modified IEEE14 node AC-DC hybrid system, where VSC1, VSC2 and VSC3 are assumed to operate at the node 2, 5, 4 orderly. Improved adaptive weighting function method and sequential method are applied to solve the system. All the Equivalent reactance is set to XL= 0.15, R = 0.006 and the resistance of DC circuit is Rd= 0.03, VSC1 is in U_{dc}, Q control mode (U_{d1}, ref = 2.0000, Q_{s1}, ref = 0.2099), VSC2 is in P, Q control mode (P_{s2}, ref = -0.3601, Q_{s2}, ref = 0.0187); VSC3 is in P, Q control mode (P_{s3}, ref = - 0.8663, Q_{s3}, ref =- 0.0864).

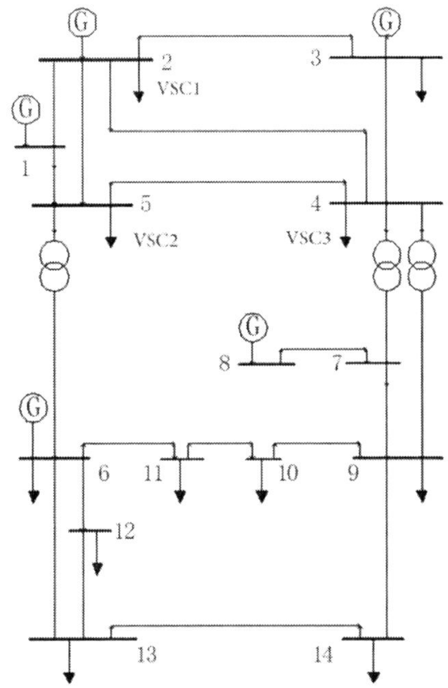

Figure 3. Modified IEEE -14 -bus AC -DC system

Table 1. State estimation results of AC system.

Method	Proposed method		Sequential method	
Node No.	U_{ti}	δ_{ti}	U_{ti}	δ_{ti}
1	1.0600	0.000	1.0600	0.000
2	1.0289	-6.049	1.0289	-6.049
3	1.0119	-15.129	1.0120	-15.129
4	1.0406	-14.123	1.0406	-14.122
5	1.0113	-4.034	1.0113	-4.034
6	1.0702	-12.159	1.0702	-12.159
7	1.0666	-15.698	1.0666	-15.696
8	1.0920	-15.693	1.0919	-15.691
9	1.0541	-16.540	1.0541	-16.539
10	1.0477	-16.038	1.0476	-16.037
11	1.0538	-14.240	1.0538	-14.240
12	1.0556	-13.261	1.0556	-13.262
13	1.0491	-13.596	1.0491	-13.597
14	1.0323	-16.237	1.0323	-16.236

Table 2. State estimation results of DC system

DC node No.	method	$P_s(MW)$	$Q_s(MVAr)$	$\delta(°)$	M	U_d	I_d
VSC1	Proposed	124.6502	20.9900	10.3325	0.8220	2.0000	0.6195
	Sequential	124.6502	20.9900	10.3323	0.8220	2.0000	0.6197
VSC2	Proposed	-36.0099	1.8700	-3.0340	0.8293	1.9920	-0.1805
	Sequential	-36.0100	1.8700	-3.0340	0.8293	1.9920	-0.1808
VSC3	Proposed	-86.6303	-8.6400	-6.7073	0.8743	1.9893	-0.4389
	Sequential	-86.6300	-8.6400	-6.7073	0.8743	1.9893	-0.4389

Table 1 and Table 2 show the state estimation results of AC/DC system by using different state estimators. Obviously the results of two algorithms are very close.

Table 3. Comparison of performances between improved and sequential algorithms

Method	Improved method	Sequential method
No. of iterations	9	11
Iteration time/s	0.0908	0.1186

From table 3, it is noted that the sequential estimator needed more iterations than the weight modification sequential estimator. Because the weights of the same state quantities in different systems are modified.

6. Conclusion

This paper presents a weight modification sequential method which is suitable for the state estimation in the VSC-MTDC system. The new state estimation method keeps the matrix dimension unchanged through changing the weight of matrix. The weight modification sequential method has been described in detail above. It is only through extending an existing AC sequential method can realize the proposed method. Simulation results in modified IEEE 14 bus system demonstrate the effectiveness of the weight modification sequential method.

Acknowledgments

This work was financially supported by Beijing Municipal Science & Technology Commission Project (Z161100004816025) and State Grid Corporation of science and technology project fund.

References

[1] ABB Group, It's time to connect – Technical description of HVDC Light technology, [Online].
[2] ABB Grid connection of offshore wind farms – BorWin1, [Online].
[3] D. Van Hertem and M. Ghandhari, "Multi-terminal VSC HVDC for the European Supergrid: Obstacles," Renew. Sustain. Energy Rev., vol. 14,no. 9, pp. 3256-3163, Dec. 2010.
[4] J. Beerten, S. Cole, and R. Belmans, "Generalized Steady-State VSCMTDC Model for Sequential AC/DC Power Flow Algorithms," IEEETans. Power Syst., vol. 27, no. 2, pp. 821-829, May 2012.
[5] Xiao-Ping Zhang, "Multiterminal Voltage-Sourced Converter-BasedHVDC Models for Power Flow Analysis," IEEE Tans. Power Syst., vol.19, no. 4, pp. 1877-1884, Nov., 2004.
[6] Antonio de la Villa Jaén, Eonrique Acha and Antonio Gómez Expósito,"Voltage Source Converter Modeling for Power System StateEstimation: STATCOM and VSC-HVDC," IEEE Tans. Power Syst., vol. 23, no. 4, pp. 1552-1559, Nov., 2008.
[7] Guoqiang Sun, Yuyan Li, Zhinong Wei and Fang Ye, "State estimationof power system with VSC-HVDC," Electric Power AutomationEquipment, vol. 30, No. 9, pp.6-12, Sep., 2010.

[8] J. D. Glover and M. Sheikoleslami, "State estimation of interconnectedHVDC/AC systems," IEEE Trans. Power Apparatus and Systems, vol.102, pp. 1805–1810, June 1983.
[9] A. M. Leita da Silva, R. B. Prada, and D. M. Falcao, "State estimationfor integrated multi-terminal DC/AC systems," IEEE Trans. Power Apparatus and Systems, vol. 104, pp. 2349–2355, Sept. 1985.
[10] R. Jegatheesan and K. Duraiswamy, "AC/multi-terminal DC powersystem state estimation—A sequential approach," Electric Machinesand Power Systems, vol. 12, pp. 27–42, 1987.
[11] Qifeng Ding, T. S. Chung and Boming Zhang, "An improved sequential method for AC/MTDC power system state estimation," in IEEE Transactions on Power Systems, vol. 16, no. 3, pp. 506-512, Aug 2001.

IOP Conf. Series: Materials Science and Engineering **207** (2017) 012072 doi:10.1088/1757-899X/207/1/012072

Numerical study of convection in phase change material based on Lattice-Boltzmann method

Tianyu Zhang[1,*], Ying Feng[2] and Zhening Zhao[1]

[1]North China Electric Power Research Institute, Beijing 100032, China
[2]Institute of Economics and Technology of State Grid Chengdu power supply company, Chengdu 610000, China

[*]Corresponding author e-mail: zzztttyyy456@126.com

Abstract. In this paper, the lattice Boltzmann method was studied for the phase change process with convective heat transfer in phase change energy storage materials. Firstly, the macroscopic heat transfer equations for the phase change process with convective heat transfer was given, by which we built the lattice Boltzmann equations for solving the problems. In the model, the speed model of D2Q9 was selected, and the boundary conditions including of non-equilibrium extrapolation and bounce back scheme were selected. Then, the effects of different Rayleigh number on the temperature field and velocity field were analyzed. Further research in a square cavity heat transfer processes with high temperature object and low temperature object were studied, in order to observe the effects of different temperature objects in the phase change process using the changes of phase field.

1. Introduction

Phase change energy storage [1] is a key technology of energy conservation and pollution reduction, with many applications in various fields. In the building energy conservation, phase change material is used to build walls and floor heating to regulate room temperature [2]. In the power peak-loading, the air conditioning and heating system based on PCM have contributed to cut the load [3]. Besides, PCM have been extensively used in solar energy, greenhouses, textiles and thermostat in spacecraft.

Present paper studied the convection in PCM based on Lattice-Boltzmann method, taking relatively stable Double distribution function method and common meshing of D2Q9. The interface of solid and liquid is solved with bouncing format.

2. Governing equations

The convection in phase change divide enthalpy into sensible and latent item, namely. H means total enthalpy, means sensible heat and means latent heat, where. Thus, for the convection equations in phase change [4], under the premise of Boussinesq approximation of incompressible liquid, can be simplified as following:

$$\nabla \bullet \mathbf{u}^{*} = 0 \tag{1}$$

$$\frac{\partial \mathbf{u}^*}{\partial t^*} + \mathbf{u}^* \bullet (\nabla \mathbf{u}^*) = -\nabla P^* + \Pr \nabla^2 \mathbf{u}^* - \Pr Ra\beta T^* \tag{2}$$

$$\frac{\partial T^*}{\partial t^*} + \mathbf{u}^* \bullet \nabla T^* = \nabla^2 T^* - \frac{1}{St}\frac{\partial f_l}{\partial t^*} \tag{3}$$

Where

L	—	length
T_1	—	superheat temperature
T_0	—	initial temperature
f_l	—	liquid fraction
C_p	—	specific heat
ν	—	kinematic viscosity
β	—	thermal expansion
g	—	gravity,

The introduction of double distribution and D2Q9 meshing is elaborated in [5, 6], and the velocity $\mathbf{e}_i$ is:

$$\mathbf{e}_i = \begin{cases} (0,0), i=0 \\ (\cos((i-1)\pi/2), \sin((i-1)\pi/2), i=1,2,3,4 \\ \sqrt{2}(\cos((2(i-5)+1)\pi/4, \sin((2(i-5)+1)\pi/4), i=5,6,7,8 \end{cases} \tag{4}$$

Collision equation:

$$f_i(\mathbf{x}+\mathbf{e}_i\Delta t, t+\Delta t) = f_i(\mathbf{x},t) - \frac{\Delta t}{\tau}(f_i(\mathbf{x},t) - f_i^{eq}(\mathbf{x},t)) + \Delta t \times 3w_i g\beta \mathbf{e}_y T(\mathbf{x},t)\rho(\mathbf{x},t) \tag{5}$$

$$g_i(\mathbf{x}+\mathbf{e}_i\Delta t, t+\Delta t) = g_i(\mathbf{x},t) - \frac{\Delta t}{\tau_h}(g_i(\mathbf{x},t) - g_i^{eq}(\mathbf{x},t)) \tag{6}$$

Where,

$\omega_0 = 16/36$, $\omega_i = 4/36, (i=1,2,3,4)$, $\omega_i = 1/36, (i=5,6,7,8)$, sonic velocity is $c_s^2 = c^2/3$, pressure p is $P = \rho c_s^2$, kinetic viscosity $\nu = c_s^2 dt(\tau - 0.5)$,thermal expansion coefficient $\kappa = \mathbf{c}_s^2 dt(\tau_h - 0.5)$.

Static distribution function for f_i:

$$f_i^{eq} = \rho\omega_i\left[1 + 3(\mathbf{e}_i \bullet \mathbf{u}) + \frac{9}{2}(\mathbf{e}_i \bullet \mathbf{u})^2 - \frac{3}{2}\mathbf{u}\bullet\mathbf{u}\right] \tag{7}$$

Static distribution function for g_i :

$$g_i^{eq} = \mathrm{T}\omega_i\left[1+3(\mathbf{e}_i \bullet \mathbf{u})+\frac{9}{2}(\mathbf{e}_i \bullet \mathbf{u})^2-\frac{3}{2}\mathbf{u}\bullet\mathbf{u}\right] \tag{8}$$

Where, $\mathbf{e}_i$ and w_i is coefficient, $\mathbf{u}$ is velocity in macroscope.
Density ρ and momentum $\rho\mathbf{u}$ are defined as:

$$\rho=\sum_{i=0}^{8} f_i ;\ \rho\mathbf{u}=\sum_{i=0}^{8} f_i\mathbf{e}_i \tag{9}$$

Macro temperature is:

$$\mathrm{T}=\sum_{i=0}^{8} g_i \tag{10}$$

Based on non-equilibrium extrapolation, according boundary condition is set up as shown in Fig. 1. Point 3-0-1 lie on the boundary, while 6, 2 and 5 lie in the flow field. Before collision, the distribution function $f_a(O,t)$ needs to be figured out and separated into states of equilibrium and non-equilibrium.

$$f_\alpha(O,t)=f_\alpha^{eq}(O,t)+f_\alpha^{neq}(O,t) \tag{11}$$

The part of equilibrium is obtained through macroscopic quantity of boundary nodes. If node O has unknown macroscopic quantity, then this quantity is replaced by neighbor node's quantity. The distribution function of boundary node O is approximated as:

$$f_\alpha(O,t)=f_\alpha^{eq}(O,t)+[f_\alpha(B,t)-f_\alpha^{eq}(B,t)] \tag{12}$$

Taking collision effect into consideration, the distribution function can be expressed:

$$f_\alpha^{+}(O,t)=f_\alpha^{eq}(O,t)+(1-\frac{1}{\tau})[f_\alpha(B,t)-f_\alpha^{eq}(B,t)] \tag{13}$$

Where, $f_\alpha^{+}(O,t)$ is the distribution function after collision, and τ relaxation time:

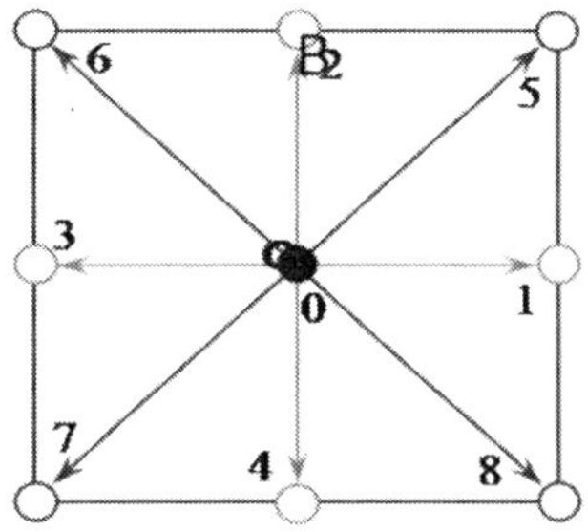

Fig. 1 Non-equilibrium extrapolation scheme

3. Numerical simulation and setup

2-D square is taken as calculation area, investigating different cases of convection in phase change process.

3.1. Convection of right side heated in square

The square's initial temperature is set 200K. Then the left wall is heated, keeping a constant temperature of 400K. The top and bottom walls are adiabatic, taking a set meshing of 64×64. Change Rayleigh number to see the change in melting and convection. The calculation is determined to reach convergence if the error reduces to 10^{-6}. Set Pr=0.02 and St=0.1, see the convection in phase change at Ra=1000, 10000 and 50000.

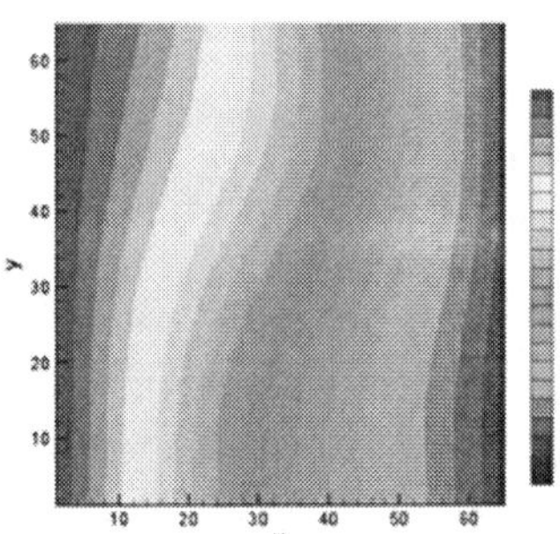

Fig. 2a Temperature field (left) at Ra=1000

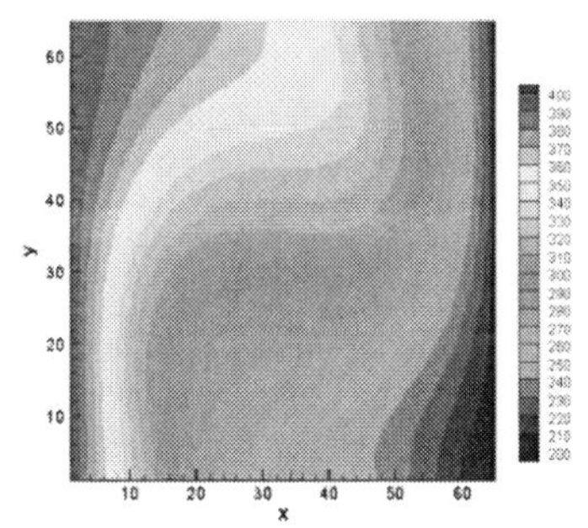

Fig. 2b Temperature field (left) at Ra=10000

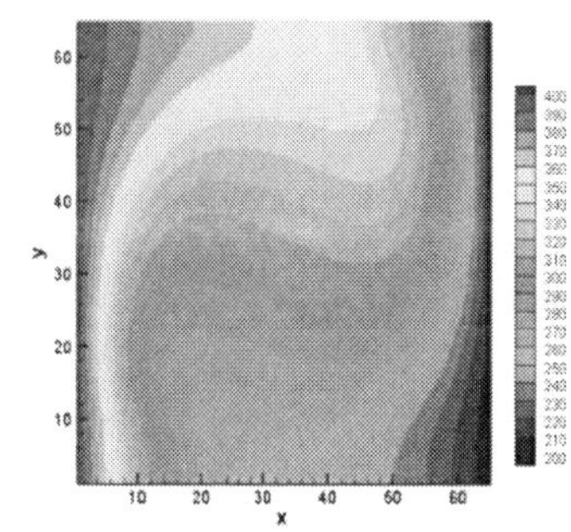

Fig. 2c Temperature field (left) at Ra=50000

When Ra increases, the convection process enhances. And the temperature distribution gets more irregular. The hot liquid moves upwards, transferring some heat to solid to make it melt and leading to collision transfer [7].

Under the same condition, the movement of phase interface with time is shown below (Pr=0.02, St=0.01 and Ra=10000).

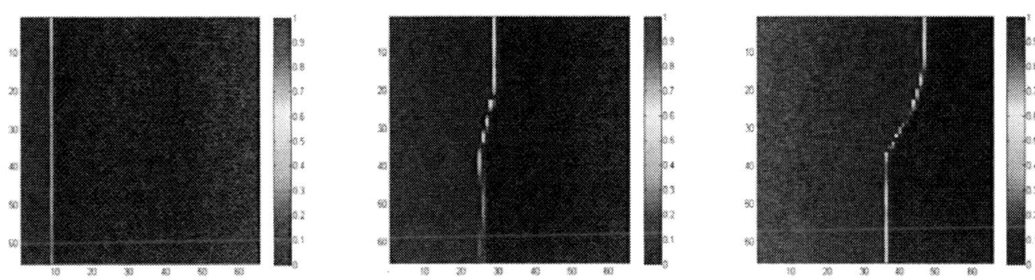

Fig. 3 Trend diagram of phase interface with time

In the initial period, the conduction is dominated compared with convection, leading to a parallel interface to wall. As time goes, the heated liquid moves upwards due to the difference in density. And convection begins to take lead, fasten top part melting.

Keep the parameters constant, and add a constant object of 200K and 400K. Take time point of 23min and 43min.

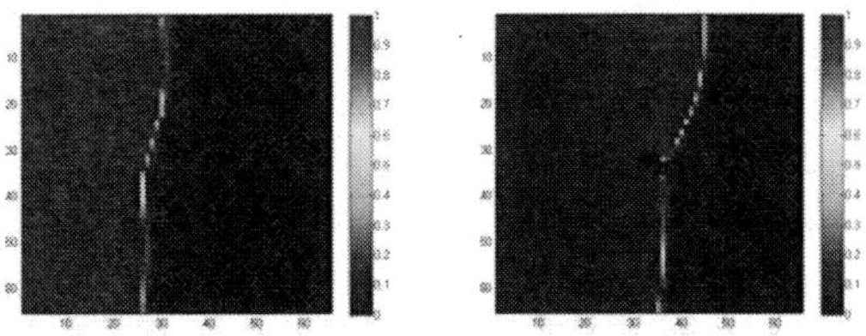

Fig. 4 The phase interface changes with time with low temperature object

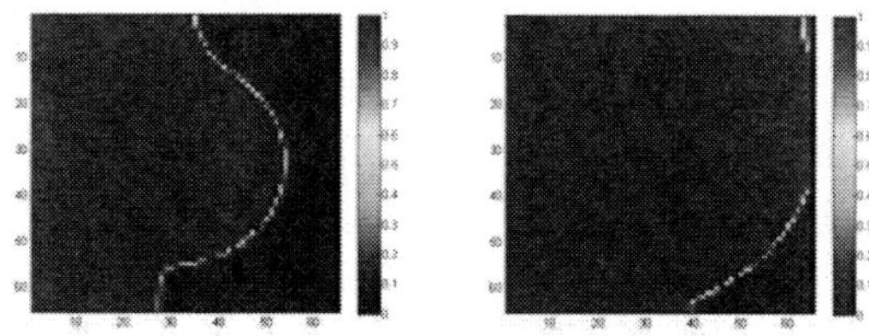

Fig. 5 The phase interface changes with time with high temperature object

As shown in Fig. 4 and 5, the cool object in the square makes little difference, while hot object can make contribute to enhancement of melting to a great extent.

3.2. Convection of bottom side heated in square

All the parameters are set as same as in 3.1. Only change is that the heating boundary is bottom wall [8, 9].

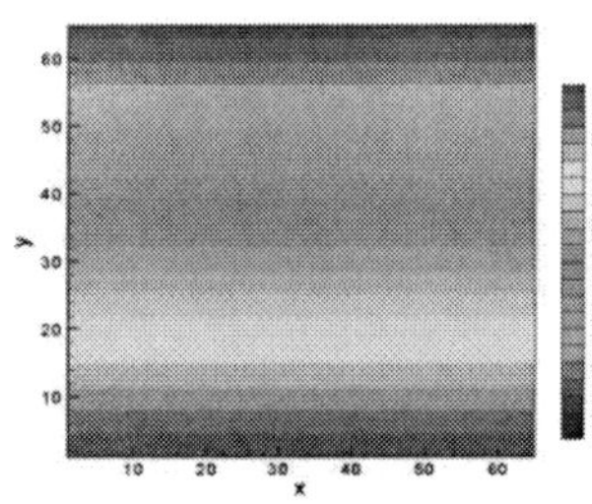

Fig. 6a Temperature field (left) at Ra=1000

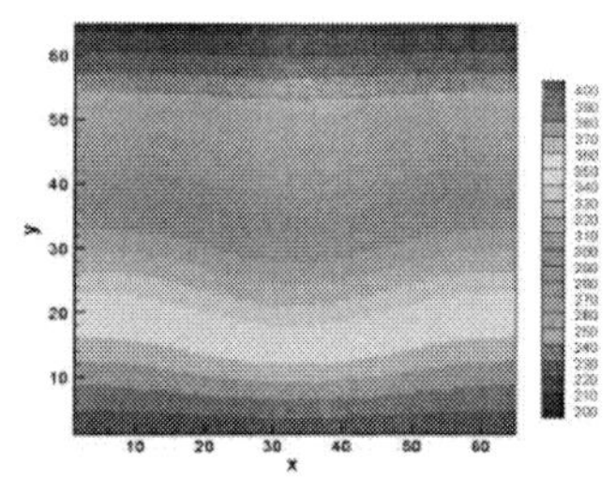

Fig. 6b Temperature field (left) at Ra=10000

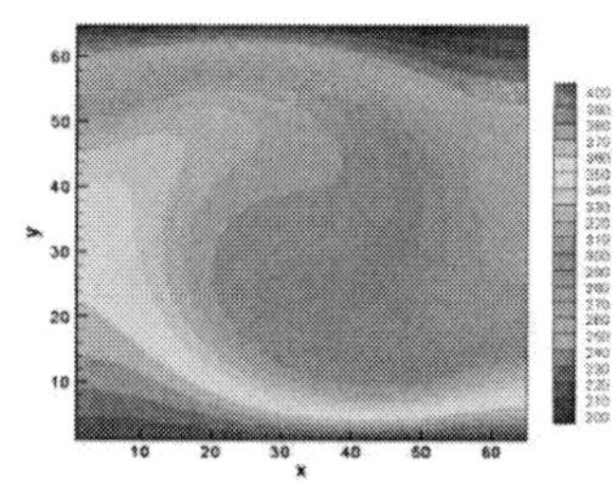

Fig. 6c Temperature field (left) at Ra=50000

With the increase of Ra, convection effect enhances and temperature field changes greatly. Compared to left wall heated, the convection effect is more obvious. It is mainly because the heated liquid upward movement is significant, leading to significant convection effect.

Under the same condition, the movement of phase interface with time is shown below (Pr=0.02, St=0.01 and Ra=10000).

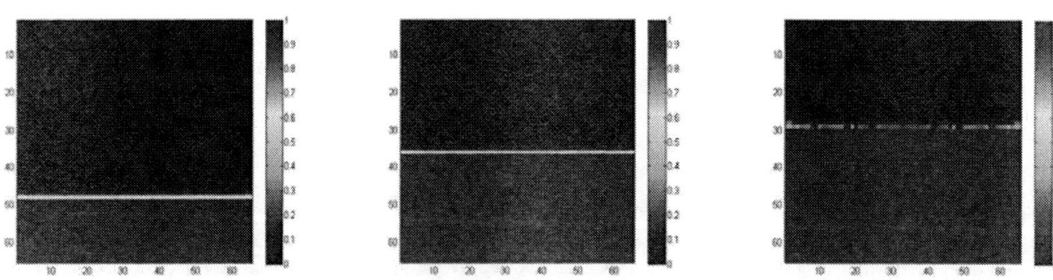

Fig. 7 Trend diagram of phase interface with time

As shown in the figure, in the initial period the conduction is dominated and the interface is parallel to wall. As time goes, the convection begins to take lead. However, due to bottom wall heated, bottom heated liquid all moves upwards, so the interface is still parallel to bottom wall.

Keep other parameters same as previous, insert a cool (200K) and hot (400K) object in the middle. Calculate the movement of phase interface with time, and take time point of 13min and 33min (Pr=0.02, St=0.01, Ra=10000).

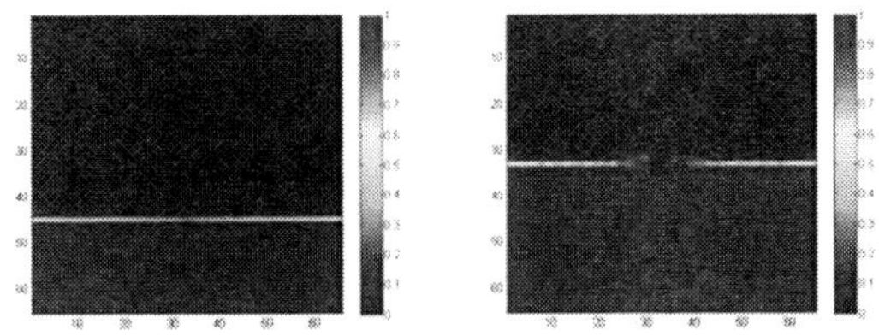

Fig. 8 The phase interface changes with time with low temperature object

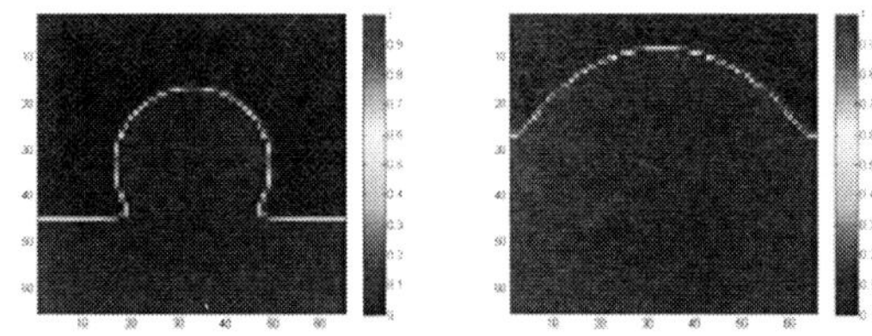

Fig. 9 The phase interface changes with time with high temperature object

As shown above, in the case of cool object in middle, when the interface has not reached the object, the movement of interface is the same as no object. When the interface has reached the object, the other part of interface is not influenced. In the case of hot object, the melting process is fastened obviously.

4. Conclusion

This paper analyzes the melting process of PCM in square based on Lattice-Boltzmann method. Firstly, the convection equations are introduced, and are solved by Lattice-Boltzmann method. Different boundary conditions are studied and according movement of interface is investigated. The thermostatic object is inserted in the middle of square. From the results obtained, we can conclude that when the interface has not reached the cold object, the interface is not influenced. If interface reaches the cold object, the other phase interface is still not influenced. While for the hot object, after reaching the object, the whole melting process is greatly fastened.

References

[1] Manish K, Rathod, Jyotirmay B. Thermal stability of phase change materials used in latent heat energy storage systems: A review [J]. Renewable and Sustainable Energy Reviews, 2013, 18: 246–258.

[2] Qi Shen. Analysis of application of PCM in building conservation [J]. Conversation and Environmental Protection, 2016, 1: 14-15.

[3] Wei Dong, Zengchan Zhou, Bin Guo. Applications of PCM in agricultural apparatus [J]. Crafts in Greenhouses, 2015, 12: 48-52.

[4] Christian H, Andrea P, Bastien C, et al. Lattice Boltzmann model for melting with natural convection [J]. International Journal of Heat and Fluid Flow, 2008, 29: 1469–1480.

[5] Gong S, Cheng P. A lattice Boltzmann method for simulation of liquid–vapor phase-change heat transfer [J]. International Journal of Heat and Mass Transfer, 2012, 55: 4923–4927.

[6] Wu J, Shu C. A solution-adaptive lattice Boltzmann method for two-dimensional incompressible viscous flows [J]. Journal of Computational Physics, 2011, 230: 2246–2269.

[7] Niranjan N G, Kannan N I. Modified enthalpy method for the simulation of melting and solidification [J]. Indian Academy of Sciences, 2013, 38: 1259–1285.

[8] Guo Z L, Zheng C G, Shi B C. Non-equilibrium extrapolation method for velocity and pressure boundary conditions in the lattice Boltzmann method [J]. Chinese Physics, 2002, 11: 366-374.

[9] Gallivan M A, Noble D R, Georgiadis I G, et al. An evaluation of the bounce-back boundary condition for lattice Boltzmann simulations [J]. Int. J. Numer. Methods Fluids, 1997, 25: 249-263.

The effect of Y^{3+} and Eu^{3+} on properties of $Zn_3V_2O_8$ phosphor

Jiaolian Luo[1, 2, *], Ruxue Chen[1, 2], Zhenglin Zou[1, 2]

[1]Special and Key Laboratory of Guizhou Provincial Higher Education for Green Energy-Saving Materials. Guizhou Minzu University, Guiyang 550025, P.R. China
[2]School of materials science and engineering, Guizhou Minzu University, Guiyang 550025, P.R.China

*Corresponding author e-mail: jiaolianluo@qq.com

Abstract. In this paper, the $(Zn_{1-x}Y_x)_3V_2O_8$ and $(Zn_{1-x}Eu_x)_3V_2O_8$ phosphors were successfully prepared by high temperature solid state method. The effect of the doping amount of Y^{3+} and Eu^{3+} on the luminescent properties of phosphors was discussed. The phase, surface morphology and luminescent properties of the phosphors were characterized by X ray diffraction (XRD), scanning electron microscopy (SEM) and photoluminescence (PL). The results show that the phosphor is the target product , the sample of the composite particles is bulky, the excitation spectral range of $(Zn_{1-x}Y_x)_3V_2O_8$ phosphors is 300-400nm, the excitation peak is at 361nm, and $(Zn_{1-x}Y_x)_3V_2O_8$ phosphor emission spectra in the range of 420-690nm, the emission peak in 559nm. The results show that the crystal structure of $Zn_3V_2O_8$ and $(Zn_{1-x}Eu_x)_3V_2O_8$ were the same, and the particle uniformity of $(Zn_{1-x}Eu_x)_3V_2O_8$ was better than $Zn_3V_2O_8$. The emission band $Zn_3V_2O_8$ and $(Zn_{1-x}Eu_x)_3V_2O_8$ were present at the wave length of 420-690 nm, the luminescence property of $(Zn_{1-x}Eu_x)_3V_2O_8$ was enhance by Eu-doped.

1. Introduction

It is a hot spot for researchers to search for luminescent materials with high performance and high luminous efficiency. Choi[1] et al synthesized successfully Ca_3Sr_3 - $(VO_4)_4$: Eu^{3+} phosphors, the photoluminescence peak at 618 nm; Junjiao Kou[2] spin crossover successfully synthesized met vanadic acid zinc ($Zn_2(OH)_3VO_3$:Eu^{3+}) and zinc pyro vanadate ($Zn_3(OH)_2V_2O_7$.$2H_2O$: Eu^{3+}) and study its optical properties; Xiao Fen[3] et al, with sol gel method, the Eu3+ activated $Ca_3(VO_4)_2$ red phosphors were successfully synthesized and their luminescent properties were studied; Shumei Lin[4] et al were prepared by hydrothermal method with good crystallinity ($Zn_3(OH)_2V_2O_7$. nH_2O nanobelts, were obtained ($Zn_3(OH)_2V_2O_7$. nH_2O nanotubes by changing the reaction temperature; Donghua Chen, et al[5] $Ca_3(VO_4)_4$: Eu^{3+}, Sm^{3+} phosphors was synthesized successfully with high temperature solid phase method, and studied the related performance; Shibin Ni[6] ,the ($Z_{n3}(OH)_2V_2O_7$ .nH_2O nanobelts and nanosheets were prepared by hydrothermal method; Liu Hao et al[7].The green phosphor $Ca_5Mg_4(VO_4)_6$ was synthesized by solid state method, the effects of sintering temperature and heat treatment temperature on the luminescent properties of phosphors were also discussed; Shi[8] also researched the preparation and property of nano materials of $Zn_3(OH)_2V_2O_7.2H_2O$ deeply; The flower-type structure $Zn_3V_2O_8$ was obtained using the chemical bath deposition and the high-heat treatment withnot add any template agent and surfactant by Shi Rui[9],Zhou[10] et al $Ca_3(VO_4)_2$:Eu^{3+} nanocrystals were synthesized at low temperature and their luminescence properties were analyzed ; the $Zn_3V_2O_8$

nanosphere was obtained using the chemical bath deposition and the high-heat treatment with the base of amino acetic acid by Wang Miao[11],Jiaolian Luo[12] etal synthesized $Zn_3V_2O_8$ and $(Zn_{1-x}Eux)_3V_2O_8$ phosphor successfully by using solid phase method. In this paper, V_2O_5, Y_2O_3, ZnO as raw materials, using high temperature solid state method, the $Zn_3V_2O_8$:Y^{3+} phosphor was prepared successfully, the structure, surface morphology and luminescent properties of the synthesized phosphors were characterized by XRD, SEM and PL. And the effects of Y^{3+} concentration on the properties of $Zn_3V_2O_8$ phosphors were analyzed.

2. Experiment

The raw materials are V_2O_5, Y_2O_3, Eu_2O_3, ZnO, according to the stoichiometric ratio of the reactants, weighing in a mortar, and fully grinding 50min, the reaction to the crucible in the ground, and then transferred to the muffle furnace; to muffle furnace in 100min to set the calcination temperature of 600℃. At preset temperatures for 4h, and then set the muffle furnace is reduced from calcination temperature to room temperature in the range of 120min, last, take out of sample. The experimental procedure of $(Zn_{1-x}Eu_x)_3V_2O_8$ is same to it, Set the calcination temperature as shown in table 1:

Table 1 The temperature of the muffle furnace process

Reaction	Temperature	Time
Warming up stage	RT -600℃	80 min
calcination stage	600℃	240 min
cooling stage	600℃-RT	120 min

3. The experimental result and discussion

3.1. The XRD analysis result of sample $Zn_3V_2O_8$ and $(Zn_{1-x}Y_x)_3V_2O_8$

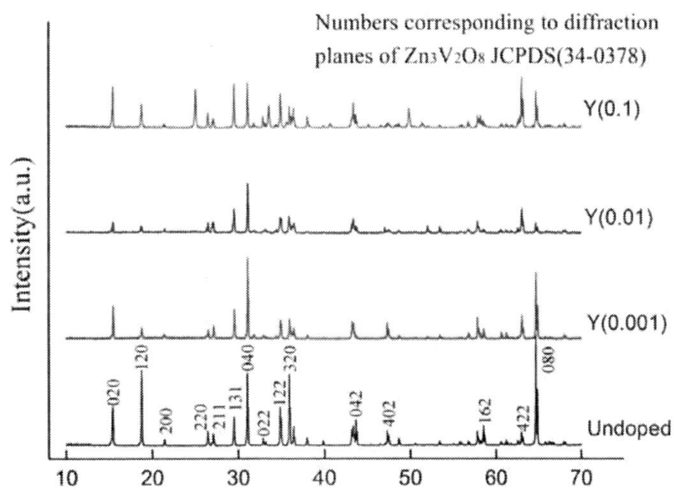

Figure 1. The XRD of the $Zn_3V_2O_8$ and $(Zn_{1-x}Y_x)_3V_2O_8$ phosphor under the resultant temperature of 600℃

Figure 1 is the XRD (DY1632) of the samples of $Zn_3V_2O_8$ and $(Zn_{1-x}Y_x)_3V_2O_8$ phosphors calcined at a temperature of 600℃ for 4h. It can be seen from the spectrogram that all the diffraction peaks are matched with the results of $Zn_3V_2O_8$ (JCPDS34-0378) standard card, which belongs to the orthorhombic system, in which the lattice constants a=8.2290, b=11.5284, c=6.1116, Z=4. Y^{3+} doped with a certain amount of $Zn_3V_2O_8$ fluorescent powder, controlling the molar amount, when the doping amount of Y^{3+} is x=0.001, compared with the undoped, the characteristic peak of the product is no obvious change, no impurity peaks appeared in the crystal structure, and complete. The positions of the main peaks of the XRD spectra are in good agreement with each other, and the main diffraction peaks of the sample $(Zn_{1-x}Y_x)_3V_2O_8$ are basically matched with the results of the $Zn_3V_2O_8$ standard card.

3.2. The XRD analysis result of sample $Zn_3V_2O_8$ and $(Zn_{1-x}Eu_x)_3V_2O_8$.

Figure 2 is the XRD of the $Zn_3V_2O_8$ and $(Zn_{1-x}Eu_x)_3V_2O_8$ phosphor under the resultant temperature of 600℃.The lattice parameter, phase structure and the status of Eu-doping of the sample $Zn_3V_2O_8$ and $(Zn_{1-x}Eu_x)_3V_2O_8$ were analyzed by X-Ray diffraction (XRD) (DY1632). The operating voltage is 45 KV, the operating current is 40 mA, radiation source is Cu target, the scope of the scanning angle in the range of 10°～80°, the scanning time is 3 min. Figure 2 is the XRD of the $Zn_3V_2O_8$ and $(Zn_{1-x}Eu_x)_3V_2O_8$ phosphor under the resultant temperature of 600℃. The XRD of $Zn_3V_2O_8$ shows that the largest emission peak was appeared at the diffraction angle of 35°, and the largest emission peak was 5000. The XRD of $(Zn_{1-x}Eu_x)_3V_2O_8$ shows that the crystal structure of $Zn_3V_2O_8$ wasn't changed, the intensity of diffraction, fluorescence properties and color rendering index of $Zn_3V_2O_8$ was increased by Eu-doped.

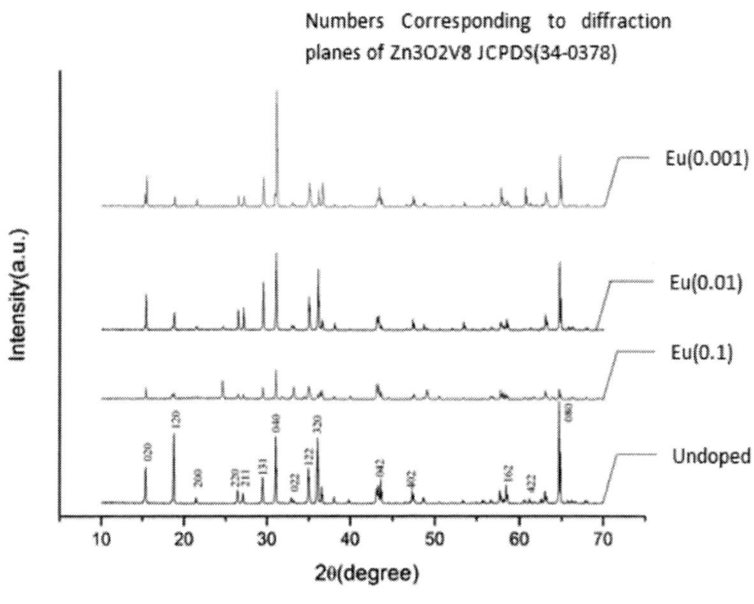

Figure 2. The XRD of the $Zn_3V_2O_8$ and $(Zn_{1-x}Eu_x)_3V_2O_8$ phosphor under the resultant temperature of 600

3.3. The SEM analysis result of sample $Zn_3V_2O_8$ and $(Zn_{1-x}Y_x)_3V_2O_8$

Figure 3 is the SEM of $(Zn_{1-x}Y_x)_3V_2O_8$ phosphor powder calcined at a temperature of 600 ℃ and the $Zn_3V_2O_8$ (EVO18) spectrum of the samples after isothermal 4h. It can be concluded from the figure that the sample size is large, and there is no agglomeration. The SEM diagram of $Zn_3V_2O_8$ phosphor is magnified 300 times, and the particle size is about 30μm. And the scan of $(Zn_{1-x}Y_x)_3V_2O_8$ phosphor powder is magnified 300 times, 600 times, 1000 times, the particle size of 32-35μm, which is consistent with the results of the spectrogram and XRD spectra of the same. It can be seen from the SEM spectra of rare earth doped fluorescent powder surface is rough, the two phosphor particles of $Zn_3V_2O_8$ and $(Zn_{1-x}Y_x)_3V_2O_8$ are same, they are suitable for practical application. However, the rare earth doped fluorescent powder particles is greater.

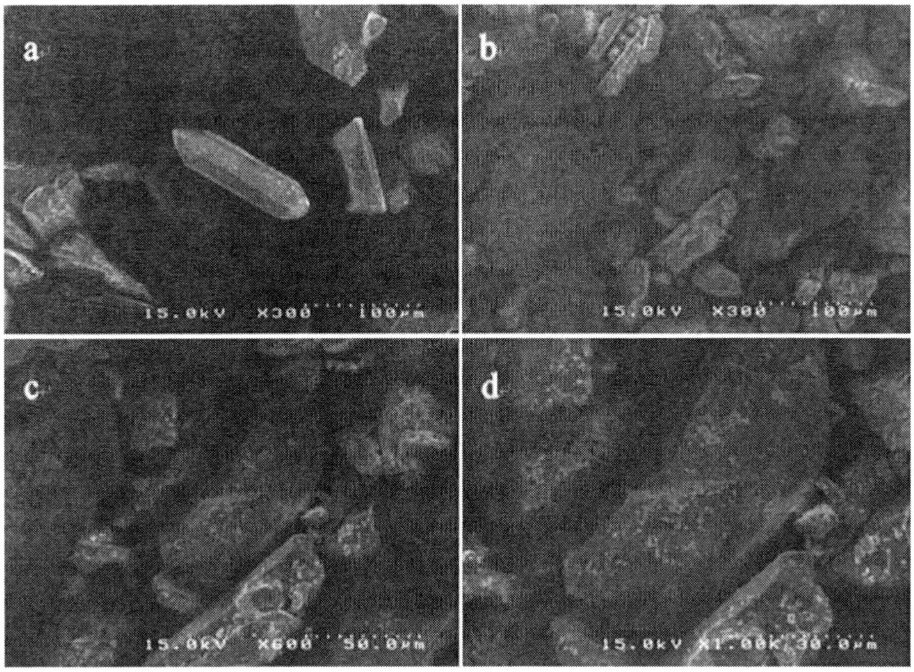

Figure 3. The SEM analysis result of sample $Zn_3V_2O_8$ (a) and $(Zn_{1-x}Y_x)_3V_2O_8$ (b, c, d)

3.4. The SEM analysis results of the sample $Zn_3V_2O_8$ and $(Zn_{1-x}Eu_x)_3V_2O_8$.

Figure 4 is the SEM analysis results of the sample $Zn_3V_2O_8$ and $(Zn_{1-x}Eu_x)_3V_2O_8$. The figure 4 shows that the $Zn_3V_2O_8$ and $(Zn_{1-x}Eu_x)_3V_2O_8$ phosphor have the rules morphology structure, uniform particles rod structure, and haven't reunion phenomenon. The SEM of $Zn_3V_2O_8$ phosphor was amplified with 200 times, the SEM of $Zn_{1-x}Eu_x)_3V_2O_8$ phosphor was amplified with 500 times, this results were correspond with XRD standard drawing. At the same time, the SEM of $(Zn_{1-x}Eu_x)_3V_2O_8$ show that the phosphor surface was rough, the uniformity of particles is better, so the $(Zn_{1-x}Eu_x)_3V_2O_8$ phosphor is more suitable for the requirements of the powder in practical application.

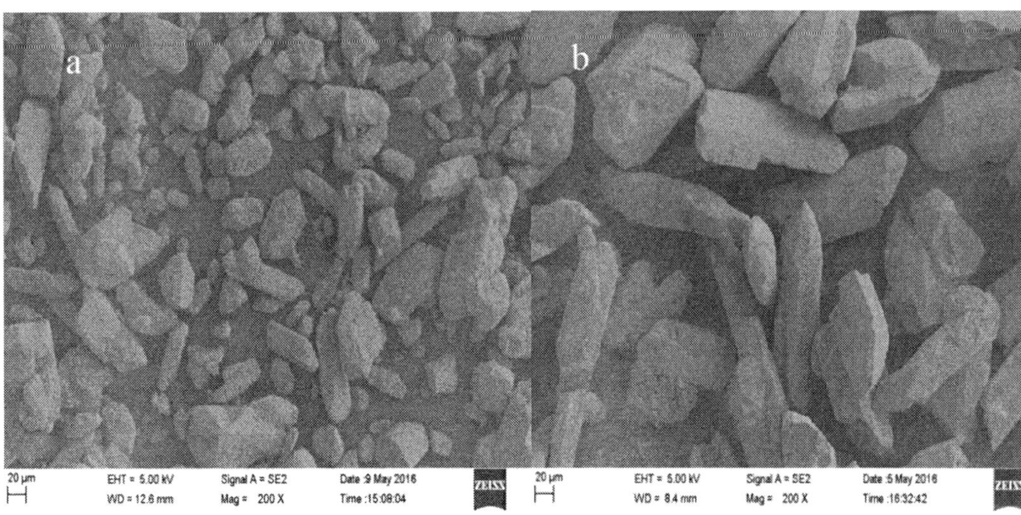

Figure 4. The SEM analysis results of the sample $Zn_3V_2O_8$ (a), (Zn1-xEux)$_3V_2O_8$ (b)

3.5. The exaction spectrum and emission spectrum analysis of the sample $(Zn_{1-x}Y_x)_3V_2O_8$ and $(Zn_{1-x}Eu_x)_3V_2O_8$

Figure 5 is the excitation spectra and emission spectra of $Zn_3V_2O_8$ phosphor and $(Zn_{1-x}Y_x)_3V_2O_8$ phosphor powder at 600℃ with 4h constant temperature, can be drawn from the figure, when the doping amount of rare earth Y^{3+} x=0.001, the luminescent properties of $(Zn_{1-x}Y_x)_3V_2O_8$ phosphor is the best, when the doping amount $Y3^+$ is x=0.01 and x=0.1 respectively, luminescence properties of $(Zn_{1-x}Y_x)_3V_2O_8$ phosphors were low compared to undoped Y^{3+}, $Zn_3V_2O_8$ and $(Zn_{1-x}Y_x)_3V_2O_8$ phosphors have wide excitation spectrum between 300-400nm, $Zn_3V_2O_8$ phosphors excitation peak at 361nm, which belongs to the near ultraviolet range, the excitation peak of $(Zn_{1-x}Y_x)_3V_2O_8$ phosphor is in 361nm too. The results indicate that the two kinds of phosphor can be effectively excited by near UV, so the two kinds of fluorescent powder can be used as a near ultraviolet excitation chip LED phosphors. At the same time, the emission spectrum range of $(Zn_{1-x}Y_x)_3V_2O_8$ phosphor is between 420-690nm, and the peak value of the emission spectrum is 559nm, which belongs to the range of visible yellow and green light. Compared with the two phosphors, the luminescent properties of rare earth doped Y^{3+} (x=0.001) phosphors were better. It is of great significance to the application of white LED.

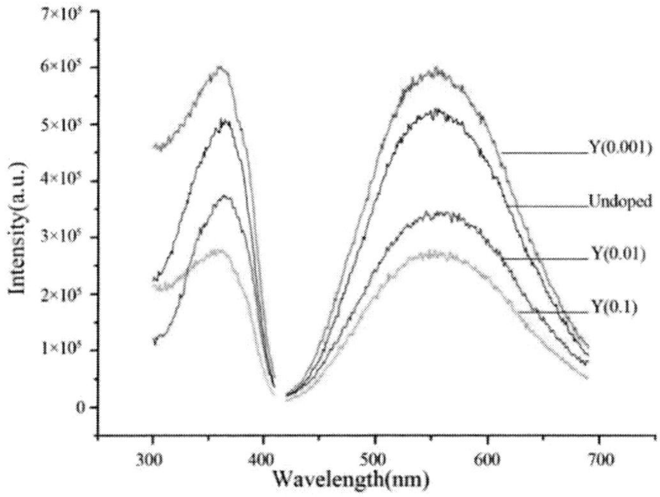

Figure 5. The luminescence property of the $Zn_3V_2O_8$ and $(Zn_{1-x}Y_x)_3V_2O_8$ phosphor

Figure 6 were the excitation spectrum and emission spectrum of the $Zn_3V_2O_8$ and $(Zn_{1-x}Eu_x)_3V_2O_8$ phosphor under the resultant temperature of 600℃. The figure 4 shows that the excitation spectrum of $Zn_3V_2O_8$ and $(Zn_{1-x}Eu_x)_3V_2O_8$ phosphor have the wavelength range of 300-400nm, the excitation spectrum of $Zn_3V_2O_8$ appeared at the wave length of 361 nm, was belong to the scope of ultraviolet light. The excitation spectrum of $(Zn_{1-x}Eu_x)_3V_2O_8$ was moving to the direction of the red shift, appeared at 380 nm, and was present between ultraviolet and near ultraviolet. The experimental results shows that the excitation spectrum was not affected by Eu-doping, and existed the white LED that using the near ultraviolet excitation by rare earth elements. At the same time, the emission spectrum of $Zn_3V_2O_8$ and $(Zn_{1-x}Eu_x)_3V_2O_8$ phosphor have the wavelength range of 420-690 nm, the emission spectrum of $Zn_3V_2O_8$ appeared at the wave length of 560 nm, belong to the yellow green range of visible light. The emission spectrum of $(Zn_{1-x}Eu_x)_3V_2O_8$ was moving to the direction of the red shift, appeared at 570 nm, fall with the scope of visible light yellow green-yellow light. Comparing the experimental results we find the luminescence property was improved by Eu-doped. More important, the intensity of $(Zn_{1-x}Eu_x)_3V_2O_8$ is more stonger than $(Zn_{1-x}Y_x)_3V_2O_8$ when x=0.001.

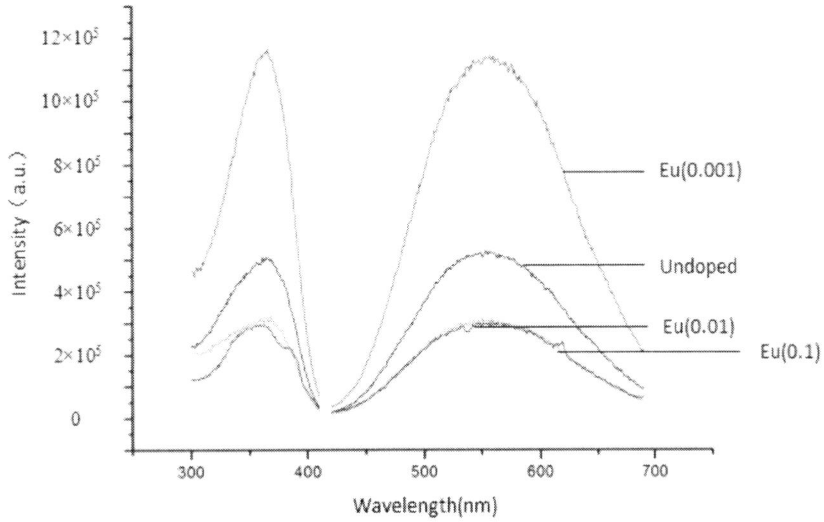

Figure 6. The luminescence property of the $Zn_3V_2O_8$ and $(Zn_{1-x}Eu_x)_3V_2O_8$ phosphor

3.6. The effect of Y^{3+} and Eu^{3+} doping on Luminescence Properties

the Y^{3+} doped $Zn_3V_2O_8$ phosphors with different concentrations of light intensity variation, rare earth ion as the luminescent center, the doping concentration has an important influence on the luminescence properties of rare earth ions by fluorescent powder preparation requirements to different doping concentration to obtain the best luminescent properties not, under normal circumstances, the higher the concentration was, the better the luminescent intensity. From Figure 1 and Figure 3 shows: with the increase of Y^{3+} concentration, due to the concentration quenching phenomenon, the diffraction peak intensity of the product gradually weakened, while the emission intensity between 420-690nm, $(Zn_{1-x}Y_x)_3V_2O_8$ fluorescent powder is also gradually reduced. The effect of Eu^{3+} doping on luminescence properties is same as Y^{3+}.but when x=0.001, the the intensity of $(Zn_{1-x}Eu_x)_3V_2O_8$ is more stonger than $(Zn_{1-x}Y_x)_3V_2O_8$.

4. Conclusion

In this paper, $Zn_3V_2O_8$: Y^{3+} phosphors were synthesized by high temperature solid phase method. The results show that when the amount of Y^{3+} doping is x = 0.001, the luminous intensity of the obtained phosphor is stronger than that of the undoped $Zn_3V_2O_8$ phosphor, and the luminescence intensity of the phosphor is stronger than that of the undoped $Zn_3V_2O_8$, But the Y^{3+} concentration had no obvious effect on the lattice structure of the zinc vanadate phosphor. However, with the increase of Y^{3+}, As a result of the concentration quenching phenomenon, the product of the diffraction peak intensity gradually weakened, $(Zn_{1-x}Y_x)_3V_2O_8$ phosphor at 420-690nm, the emission peak intensity of the product is gradually decreasing. The effect of doped Eu^{3+} is same as Y^{3+}.And when x=0.001, the the intensity of $(Zn_{1-x}Eu_x)_3V_2O_8$ is more stonger than $(Zn_{1-x}Y_x)_3V_2O_8$.

Acknowledgments

Project supported by the Special and Key Laboratory of Guizhou Provincial Higher Education for Green Energy-Saving Materials; The national college students' innovative entrepreneurship training program of the Chinese Education Department, in 2015,China(201510672001) ;The Science and Technology Foundation of Guizhou Province, China(The contract LH of Guizhou No.[2014]7384);The Natural science Foundation of Guizhou Province education department, China(The provincial construction projects of undergraduate course teaching, in 2015); The introduction of talent research foundation of Guizhou Minzu University, China (NO.2014(12)).

References

[1] Choi S, Moon Y M, Kim K, et al. Luminescent properties of a novel red-emitting phosphor: Eu^{3+} -activated $Ca_3Sr_3(VO_4)_4$ [J]. Journal of Luminescence, 2009, 129(9): 988-990.

[2] Kou J Solvothermal Synthesis and Luminescence property Study of Rare Earth Europium Ions Doped $Zn_2(OH)_3VO_3(Zn_3(OH)_2V_2O_7{\cdot}2H_2O)$ Nanocrystals. [D]. Bohai University, 2016.1-10.

[3] Xiao F, Xue Y, Zhang Q. Preparation and luminescence properties of $Ca_3(VO_4)_2$: Eu^{3+}red phosphor for white light-emitting diodes [C]//The optical and optoelectronic technology exchange of China ordnance Societyin 2009. 2009.

[4] Lin S, Hydrothermal prepared $Zn_3(OH)_2V_2O_7.nH_2O$ nanocrystals [D]. Lanzhou, Lanzhou University, 2009.41-44.

[5] Huang J, Li Q, Chen D. Preparation and luminescence properties of Ca_3 $(VO_4)_2$: Eu^{3+}, Sm^{3+}, phosphor for light-emitting diodes [J]. Materials Science & Engineering B, 2010, 172(2): 108-113.

[6] Ni S, The preparation and application of Fe_3O_4 and $Zn_3(VO_4)_2$ metal oxide the transition research, [D]. Lanzhou University, 2010, 86-98.

[7] Liu H, Ruan J, Han J.et al. Preparation and Photoluminescence Properties of Green-emitting Vanadate Phosphor $Ca_5Mg_4(VO_4)_6$ [J]. Joural of silicate, 2013, 41(3): 334-341.

[8] Shi R., Wang Y, Zhou F.et al.$Zn_3(OH)_2V_2O_7.2H_2O$ and $Zn_3V_2O_8$ nanostructures controlled fabrication and photocatalytic performance. [J]. J. Mater. Chem, 2011,21: 6313-6319.

[9] Sheng L J, Li Bo, Wang Z Z etal. Rare earth luminescent material vanadate system is reviewed [J]. Rare earth, 36 (2015) 129-130.

[10] Zhou X C, Wang X J, Kuang R Y, et al. Low Temperature Synthesis and Photoluminescence of $Ca_3(VO)_2$:Eu^{3+} Nanocrystals with Li^+ Addition [J]. Inorg Mater, 2010, 46(11): 1244-1247.

[11] Pan Yuexiao, Wu Mingmei, Su Qiang. Synthesis of Eu^{3+} doped calcium and strontium carbonate phosphors at room temperature [J]. Materials Research Bulletin, 38 (2013) 1537-1544.

[12] Luo J, Cen W, Zhang X, et al. The effect in the production and Luminescence property of $Zn_3V_2O_8$ with Eu-doping[C]// International Conference on Advanced Materials and Information Technology Processing. 2016.

IOP Conf. Series: Materials Science and Engineering **207** (2017) 012074 doi:10.1088/1757-899X/207/1/012074

Research and Improvement on Characteristics of Emergency Diesel Generating Set Mechanical Support System in Nuclear Power Plant

Yang Zhe

Suzhou Nuclear Power Research Institute Co., Ltd., Suzhou, China

Abstract. There are often mechanical problems of emergency power generation units in nuclear power plant, which bring a great threat to nuclear safety. Through analyzing the influence factors caused by mechanical failure, the existing defects of the design of mechanical support system are determined, and the design idea has caused the direction misleading in the field of maintenance and transformation. In this paper, research analysis is made on basic support design of diesel generator set, main pipe support design and important components of supercharger support design. And this paper points out the specific design flaws and shortcomings, and proposes targeted improvement program. Through the implementation of improvement programs, vibration level of unit and mechanical failure rate are reduced effectively. At the same time, it also provides guidance for design, maintenance and renovation of diesel generator mechanical support system of nuclear power plants in the future.

1. Introduction

Nuclear power plant has a high reliability of power system and each unit is equipped with two emergency diesel alternating current power supply as the backup. When the plant unit bus line causes a power failure due to the main power supply outside power plant or failure of bus line, the emergency diesel generator set will provide power supply for safety system and equipment of power plant within 10 seconds so that the power can be provided to safety system and equipment of power plant in time and guarantee emergency safety shutdown of reactor to prevent the damages of important components caused by the power failure of power system, especially it can prevent the core melting accident from occurring.

Nuclear power plant emergency diesel generator set often has mechanical failures, which seriously affects safety and reliability of emergency power supply in power plant; and as reciprocating power machinery, the complexity, diversity and non-linearity of this system determines the fault diagnosis and processing difficulties. The power plant hopes to further study the vibration performance of diesel engine and reduce the mechanical failure rate of unit, so that the relevant important equipment and accessories generally achieve a good running state to eliminate the adverse factors caused by failure factors.

2. Analysis of Main Mechanical Faults of Diesel Engine and Overview of Support System

After various mechanical faults of diesels in nuclear plant are sorted and analyzed, it is found that a large number of mechanical faults are related to the severe vibration of its components. And the factors that cause severe vibration include two factors, unit body and design of support system.

Through the overall dynamic test of unit, according to GB/T10397-1989 *Small and Medium-power Diesel Vibration Rating Standards* and GB/T6075.6-2002 (or ISO10816-6: 1995) *mechanical vibration —evaluating of machine vibration by measurements on non-rotating parts —part 6 : reciprocating machines with power ratings above 100kW "standard*. Through the analysis of the data, the final conclusion shows that the vibration level of the diesel engine body should be in the A / B area specified by national standard, which belong a good level, thus the diesel engine would have the capacity to run for a long time.

Considering the nuclear safety design, nuclear power plant emergency diesel generator set must be isolated from the outside world, to avoid external factors as much as possible such as earthquake. Therefore, it is required that there should be isolation design which is independent of surrounding environment for main support systems including foundation support, pipe support, equipment accessories support. The support system design and installation failure are the main factors in causing mechanical failures.

Study is made on the design and performance of the support system of the emergency diesel generator set in a nuclear power plant of China General Nuclear Power Group with analysis on three factors including base, typical equipment support and the pipeline support.

3. Research on Base Characteristics of Diesel

The vibration isolation should be taken into consideration of the base design of heavy duty diesel generator. On the one hand, it should be avoided transmitting dynamic loads to external surroundings during the operation, which could cause the base sink:On the other, decoupling from the external interference should be designed to protect the unit from the damage caused by the earthquake and so on. The simple mechanical model of the diesel generator set is shown as Figure 1.

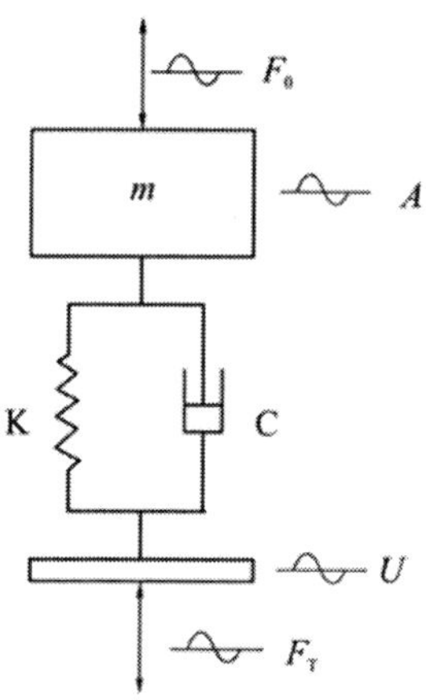

Figure 1. Diesel generator mechanics model

Assuming that the exciting force is a sine function, its absolute transfer rate (or transfer coefficient) is expressed as follows:

$$\eta_A = \frac{U}{A} = \frac{F_T}{F_0} = \sqrt{\frac{1+4\xi^2\lambda^2}{(1-\lambda^2)^2+4\xi^2\lambda^2}} \tag{1}$$

A is the amplitude of the device m:U is the base amplitude:F_T is the dynamic load transmitted to the foundation base after vibration isolation:F_0 is the exciting force of the device:K and C represent the elastic and damping elements respectively: ξ is damping factor of system: λ is the frequency ratio, that is, the avoidance degree of equipment operating speed ω and the critical speed and other natural frequencies (frequency ratio).

The mechanical support design of diesel engine unit in nuclear power plant is based on the above mechanical model. The support system made of springs and dampers has functions such as filtering the vibration of the equipment and isolating the safe shutdown earthquake (SSE).

Table 1. Parameters Comparison of Nuclear Power Plant Diesel Generating Set Base Support System

	Diesel generator weight / kN	**Base weight / kN**	**Spring box performance /kN·mm^{-1}**	**Damper performance / kN·m^{-1}**
Old unit	502	1 465	K_v=5.8	C_v=250
			K_h=6.8	C_h=250
			Spring box number:16 pieces	Damper number:4 pieces
New unit			K_v=5.8	C_v=62.5
			K_h=6.8	C_h=62.5
			Spring box number:16 pieces	Damper number:16pieces

Although the base mechanical support design of different sets are slightly different, their performances are basically the same with specific parameters shown in table 1. First, both of them meet the design criteria under the earthquake SSE and reduce the impact of external vibration such as earthquakes; secondly, it can effectively isolate the vibration and reduce the impact of equipment operation on the foundation and surrounding equipment. According to the calculation based on design parameters, the efficiency of vibration isolation can reach 98%, which means that during the process of the diesel generating set operation, the support system almost filters all dynamic loads, then the basic column only bear static load basically. Therefore, through low-frequency damping suppression and high-frequency dynamic decoupling, the base design satisfies requirements of SSE anti-seismic performance and has a high vibration isolation efficiency, which could meet the needs of long-term operation.

Compared with the vibration change before and after the foundation transformation, the vibration value kept below 2.0 mm/s. It can be seen that the changes of the anti-seismic support system have little effect on the vibration of diesel engine.

4. Research on Supporting Characteristics of Typical Equipment Accessories

The analysis is made on the case of the influence of turbo charger.

4.1 Overview of problems on Turbo Charger

The horizontal connection bolt between the old diesel engine supercharger and intercooler fractured frequently. In order to solve this problem, the support mode of supercharger bracket has been improved by using the U-shaped bracket and cancelling horizontal connection with the intercooler to avoid the interaction between the intercooler and the supercharger. Comparison of supercharger support before and after the transformation is shown in Figure 2.

In the re-identification of function, it was found that the vibration amplitude exceeded the control standard 65 mm / s, and there were still a number of partial weld crack leakage and other failures in the subsequent operation. In order to further find out vibration reasons and completely solve the problem of supercharger vibration, as well as, eliminate the risk of vibration deterioration, a series of investigation, measurement and analysis have been carried out.

After making comparison of old and new diesel supercharger-related connection structure, it is found that there are different supporting forms for exit of supercharger. The exit of old diesel supercharger is connected with cooling water pipe through the H-shape angle steel support, which can provide assistant support to cooling water pipeline. However, there are no supports between exit of new supercharger and cooling pipe, which is separated from each other, as shown in Figure 3 and Figure 4.

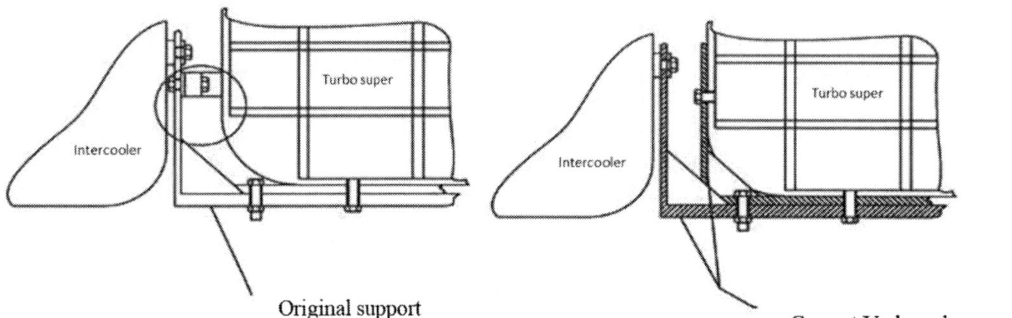

a) **Before** the transformation b) after the transformation

Figure 2. Form of U support

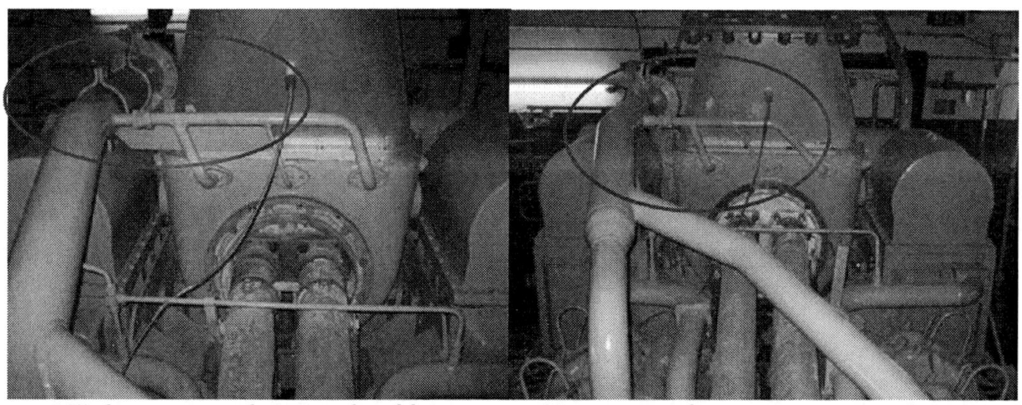

a)H-shape support between the old supercharger and cooling water pipe **b)**no support between new supercharger and cooling water pipe

Figure 3. Platform Comparison of the top structure of the supercharger between new and old diesel

Figure 4. Platform of the H-support between of old supercharger and cooling water pipe

4.2 Characteristics Analysis of Turbocharger Support

4.2.1 Analysis on Structural stress

Source of structural stress:

① The structural stress generates during the process of U-shaped support field installation and ribbed slab welding for reinforcement. The Survey shows that the way to install U-shape support and weld ribbed slab for reinforcement both in old and new unit cannot meet standard fully. There is no torque requirement for the bolt fastening between supercharger and U-shape support, but there are also two types of welding, one is full welding and the other is partial welding. The main differences between the two types welding of ribbed slab are the effects on deformation and structural stresses of supports. As the supercharger bracket material is made of 16Mn steel, its strength is high. If the ribbed slab is adopted with full welding, the welding heat is higher than partial welding, resulting in different deformation and structural stress of the support's side plate and floor plate. If U-shaped bracket and supercharger which had large deformation or structural stress could not install correctly, there would be larger possibility to produce vibration in the course of the operation.

② The structural stress generates during the process of H-shape support on-site installation. There is no torque requirement for the H shape support fastening slot which connects the supercharger with cooling water pipes. In the experiment of adjusting H shape bracket fastening bolts, it is found that the degree of tightening bolts is not the same, so the vibration and stress between supercharger and cooling water pipe are not the same.

③ The stress generated by thermal deformation of equipment. During the operation, the equipment has the tendency of expansion and deformation, and the expansion will be obstructed when the bolts are completely fastened, and the stress doesn't release well, which presents as vibration level rise. After releasing the connecting bolts, the vibration level significantly reduces due to the release of the structural stress.

4.2.2 Analysis on support characteristics

(1)The impact of two type's brackets. The change of the natural frequency before releasing the U-shaped brackets and the H-brackets is compared with that after releasing the U-shaped brackets and H-shaped brackets respectively, and the natural frequency of the supercharger changes within 1Hz. The H-shaped bracket has a significant effect on the natural frequency of the structure. Compare to the U-shaped bracket, H-shaped bracket acts on the force and torque from the top of the supercharger at the same time. It is not difficult to understand the structural characteristics and the impact of vibration.

(2) Support factor analysis. The support of supercharger of old unit include the support and fastening of the U-shaped bracket and the bottom of the connecting bolts, the confinement on both sides of the U-bracket and the fastening of the bolts, and the fixing of the top of the H-bracket. U-brackets and bottom bolts are essential for support and fastening. After releasing all the U-shaped bracket and the H-shaped bracket, the vibration of the supercharger body is close to the vibration level of the U-shaped bracket and equivalent with the vibration level of the diesel engine body, which indicates that there is no obvious mechanical looseness; and from the perspective of the elimination of structural stress, the degree of tightening the supercharger of the two brakes should not be too high.

In summary, although the vibration source of the supercharger vibration is the vibration of the diesel engine body, the reason for causing the high level vibration is the comprehensive influence of the local factors.

By loosening the U-shaped bracket horizontally connecting bolts and H-brackets, and sacrificing some mechanical looseness, it avoids structural resonance and eliminates the effect of structural stress to a certain extent, so as to achieve the effect of effectively reducing the vibration level.

4.2.3 Improvement scheme and effects. By eliminating the H-shaped bracket of the diesel engine turbocharger of old unit and transforming the bottom U-shaped bracket, it effectively reduces the diesel engine supercharger vibration level, which makes it close to the diesel engine body vibration level. At the same time, based on the stress evaluation, removing the H-shaped bracket does not affect the fatigue life and seismic requirements of connected cooling water pipe. The mechanical failure and high vibration problems of diesel engine supercharger of old unit can be solved.

5. Researches on Characteristics of Main Pipeline Support System

5.1 Analysis of Main Problems of Main Pipeline

Different from the design concept of modern diesel engine which need compact structure, decreasing vibration and noise reduction, supercharger and cooler and connected pipe locating at both ends of the unit, is a typical cantilever structure (as shown in Figure 5), and it forms an absolute rigid connection between the body and surrounding space. Some of the hoses used for vibration reduction are far away from the diesel engine body, and the effect of vibration reduction is rather limited. The main pipe span is too long, and has too many elbows. The piping itself, pipe fittings and stents constitute piping system is regarded as a vibration system of mechanical structure. When the pressure pulsation acts on the bend of pipeline or pipe cross-section whose size changes, it will produce unbalanced force. This force will cause mechanical vibration of the pipe. In particular, when the frequency of the interference force is proportional to the natural frequency of the structural system, the vibration will increase significantly. As the diesel engine's vibration is 12.5 Hz and harmonic-based broadband vibration, it is easy to be similar to the frequency of attachment structure, which forms resonance phenomenon.

Because the seismic design of the diesel engine requires it must be isolated from the outside surroundings, with the large span of the pipeline floating direction, the original design of the pipe support was limited by spatial structure that the cooling water inlet pipe and the backwater pipe and the body are fixed mutually by bundled, as Figure 6 shows. This type of support design will lead the previous pipeline vibration and bracket fracture problem into misunderstanding, which could improve the strength of the bracket by inertia as a control mean. In fact, the rigid support plays a role of vibration transmission between the vibrating source and the pipe, which can increase the mutual interference between the pipes. Increasing the bracket to a sufficient strength can prevent it from breaking, but the maximum stress position of the pipe is transferred, which often causes failure on other stress sensitive locations on the pipe.

In addition, there is no strict specification of the bracket on-site installation, such as the difference between the bolts tightening torque, welding process and quality differences, which often determines the difference between the pipeline vibration and the probability of failure. A large number of test results show that improper bracket fastening is likely to cause the phenomenon of pipeline resonance.

Figure 5. Typical partial cantilever beam structure

Figure 6. Typical Bundled Fixed Form

5.2 Diesel engine pipeline vibration reduction test

In order to test the effect of the fixed bracket between the inlet pipe and the return pipe in the main pipeline system of the diesel engine, the pipe support adjustment test was carried out. By adjusting the support rigidity of each bracket in the main pipeline system, the overall level of pipeline vibration shows a decreasing trend of consistency, and the vibration phenomenon is improved comprehensively. The maximum vibration of the pipe of one of the units drops from 122 mm / s to 28 mm / s, and the frequency of vibration (RMS) decrease by 77%. The peak frequency is 37.5 Hz before that decrease by 87%, which significantly eliminates the influence of pipeline resonance.

5.3 Improvement test of main pipeline support system

According to the overall seismic design requirements of the diesel engine and the characteristics of the pipeline space structure, combining with the modeling analysis, and introducing the damping support produced by professional agencies, a set of pipeline support program is finally designed, as shown in Figure 7:

(1) Installing soft pad in the pipe clamp between No. 1 support and No. 3 support.

(2) The No. 2 support is changed from ▽shaped to V-shaped and rubber resistors are added to each support.

(3) Adding No.4 support (which is between cylinder A5 and A6 cylinder head). The rubber vibration isolator is assembled in the middle of the No.4 support.

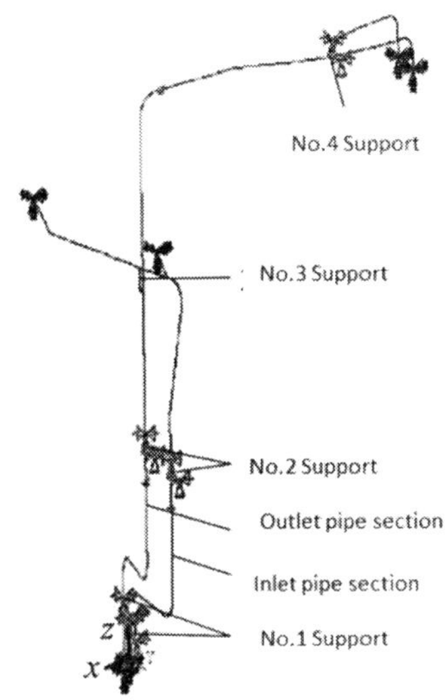

Figure 7. Finite element calculation model for cooling water pipes

5.4 Renovation and Effect of Main Pipeline Support System

5.4.1 The purpose of pipeline support system transformation

(1) Improve the design defects of the pipe bracket: Due to the severe vibration and mechanical failure caused by the unreasonable design of the support, the design of the support is improved. Reducing the transmission rate of the interference force between the pipes, this can effectively eliminate the resonance effect and reduce the stress and other time-related potential effect, and unify the formation of contradictory pipes and brackets.

(2) Improve installing defects of the pipe bracket: Weakening impacts of human factors produced by the torque-free tightening bracket, in order to make the pipeline vibration stay in a more stable state.

5.4.2 The effect of pipeline support system transformation. According to the test plan, all the main pipeline support systems of the diesel engine unit have been transformed. The vibration, stress and seismic capacity of the pipeline after the improvement are compared with that before the improvement. The effect is as follows: ① overall vibration level of the pipeline is improved on the basis of the adjustment of the bracket, and several points with high vibration level were improved obviously. ② meeting the seismic requirements; ③ the stress range of components with poor fatigue resistance is improved to varying degrees.

Through years of running practice, the parameters are stable and mechanical failure rate is greatly reduced.

5.4.3 Evaluation of modified bracket effect. The improving effect of the pipeline support determines the availability of the elastic support. The rigid support which is designed originally is compared with the elastic bracket after transformed, as shown in Table 2.

Table 2 Advantages and disadvantages comparison of diesel engine pipe bracket

Considerations	**Original design rigid support**	**Modified elastic support**
Vibration	Easily leading to pipeline resonance. In fact, it transfers the diesel engine body vibration source to the pipeline, with an increase of the transmission rate of interference	Adjusting the natural frequency of the pipeline, which could effectively suppress the resonance amplitude and significantly reduce the resonance effect; Effectively isolating the vibration source and reducing the transmission rate of interference, in order to offset and consume vibration energy
Stress	Improper bracket fastening will make the pipeline accumulate stress in thermal expansion, resulting in deterioration of vibration or weld fatigue fracture and other phenomena; under the reaction force, the fracture of bracket is easy to occurred at the weld position	After the installing the vibration damping brackets, the alternating stress of the components with poor fatigue resistance of the pipe is improved to varying degrees; releasing stress, which can decrease the interaction force between the pipeline and the support and have a potential effect on the long-term safe operation of the unit
Anti-seismic requirements	Meeting the anti-seismic requirements	The impact on the seismic capacity is not obvious; meeting the anti-seismic requirements
Maintenance	Installing the bracket without the torque requirements that leads more human factors which could cause high vibration; obtaining raw material locally and low maintenance cost	The damping block requires a certain amount of compression; installation should be standardized; spare parts need to be prepared; higher maintenance costs

6. Conclusion

(1) A large number of mechanical faults in the emergency power generation unit of the nuclear power plant are related to the design defect of the supporting system. Due to the isolation requirements of the diesel engine from the outside world, there is a need to design an independent support system. While with the mistakes of manufacturers design ideas, the daily maintenance of power plant diesel engine and the transformation work is led to errors.

(2) It is necessary to carry out a comprehensive analysis and evaluation of the overall status and causes of the diesel engine unit. The blind work should be avoided. The diesel engine foundation design is ideal and there is no need for reformation.

(3) For the main pipe support system and supercharger and other important equipment components support design, it needs to effectively reduce the transmission rate of interference between each other. In practices, it proves that the use of flexible support is a practical support improving measure.

References

[1] Gb/t 6075.6 - 2002 Measurement and evaluation of mechanical vibration of machines on nonrotating parts: Reciprocating machines with power greater than 100 [s]. State Administration of quality supervision, inspection and quarantine, kW (2002.12.)

[2] Zhu Shijian. Vibration theory and vibration isolation technology. Beijing: National Defence Industry Press, 2006.06

IOP Conf. Series: Materials Science and Engineering **207** (2017) 012074 doi:10.1088/1757-899X/207/1/012074

[3] Zhao Mei. Mechanical vibration and noise. Science Press .2005.04
[4] Suzhou Thermal Power Research Institute. The evaluation report of the vibration state of the cooling water return pipe of nuclear power plant emergency diesel engine, 200703
[5] Suzhou Thermal Power Research Institute. Nuclear power plant emergency diesel generator set vibration test. 2007.12
[6] FRAMTTOME.PGX42EOM001FSAM45SS.1992.06
[7] FRAMTTOME.PGXDXG00003EM5142GN.1993.11

IOP Conf. Series: Materials Science and Engineering **207** (2017) 012075 doi:10.1088/1757-899X/207/1/012075

Facile preparation of α-Fe_2O_3 nanorods and the promotion in Cr(VI) adsorption

Jie Liu[1], Ying Li[1] Huining Xiao[1] and Jun Ke[2,*]

[1]Department of Environmental Science & Engineering, North China Electric Power University, Baoding 071003, China
[2]School of Chemistry and Environmental Engineering, Wuhan Institute of Technology, Wuhan 430073, China

*Corresponding author e-mail: kejunmars@hotmail.com

Abstract. We present a simple and effective method for the fibracation of α-Fe_2O_3 nanorods by a hydrothermal process. Field-emission scanning electron microscopy (SEM), X-ray diffraction (XRD) and N_2 adsorption-desorption are used to analyze the morphology, phase structure and surface area for the as-prepared α-Fe_2O_3 nanorods, implying the fine dispersion of nanorods as well as the much increased surface area. The adsorption capacity of the material was evaluated by the adsorption of Cr(VI) in waste water, and the α-Fe_2O_3 nanorods exhibts the enhanced performance compared with the normalα-Fe_2O_3 nanoparticles. The adsorption of Cr(VI) on α-Fe_2O_3 nanorods fits the Langmuir adsorption model, and was controlled by the chemical process according to the high correlation to the pseudo-second-order model.

1. Introduction

Among heavy metals in the natural environment, chromium is one of the widely distributed pollutants in water and soil, existing in two common oxidation states, Cr(III) and Cr(VI) [1]. In general, Cr(III) occurs naturally, while the emitted Cr(VI) is commonly derived from industrial processes, such as electroplating, metal finishing, pigment manufacture, tannery facilities, and chromium mining. Cr(VI), existing in most aquatic environments as water soluble oxy-anions, HCrO4- or CrO42-, is highly toxic and environmentally harmful once introduced into the environment [2]. Therefore, the development of technologies to prevent further chromium discharge and remediate Cr(VI) contamination are of great importance.

Up to date, various multifunctional materials, such as biochar [3], inorganic nanoparticles [4], metals [5], and polymer [6] have been employed as absorbents for the rapid removal of metal ions. Among these materials, hematite (α-Fe_2O_3) being one of the most stable, non-toxic, environment-friendly and corrosion-resistant metal oxide has been paid more and more attention in wastewater treatment [7]. Moreover, α-Fe_2O_3 has many applications such as catalysis, gas sensors, magnetic materials, drug delivery, and light-induced water splitting [8-12]. In the past, some groups have prepared different morphologies and crystal structures of Fe_2O_3 through hydrothermal methods, where it is shown that the Cr(VI) adsorption capacity of α-Fe_2O_3 is highly related with morphologies and exposed crystal facet [13]. On the basis of the surface adsorption theory, adsorption capacity to substances to a large degree depends on surface area and surface chemical properties of adsorbent, which is closely correlated with preparation parameters. Furthermore, the adsorption behavior of Cr(VI) mainly occurs on the surface, and this means the surface micro-structure of α-Fe_2O_3 may significantly affect the adsorption capacity. However, it is still a challenge to investigate relationship between the adsorption properties, different morphology, micro-structures and preparation parameters of α-Fe_2O_3.

In the present work, uniform nanorod and microsphere α-Fe_2O_3 are successfully synthesized by a facile hydrothermal method. The adsorption ability of as-synthesized α-Fe_2O_3 toward Cr(VI) is examined, and possible adsorption mechanism is deduced.

2. Experimental

2.1. *Synthesis of α-Fe_2O_3 nanorods*

In a typical preparation procedure, hexahydrate ferric chloride ($FeCl_3 \cdot 6H_2O$, 1.0836 g), ascorbic acid ($C_6H_8O_6$, 0.7045 g) and urea ($CO(NH_2)_2$, 0.6006 g) were dissolved in 70 ml DI water. The mixed solution was transferred to a Teflon autoclave (100 ml) and reacted at 160 °C for 4 h. After cooling to the room temperature, the obtained slurry was washed by DI water and ethanol, dried at 70 °C for 12 h, and calcined at 400 °C for 4 h. The obtained Fe_2O_3 was denoted as α-Fe_2O_3-NR. To investigate the effect of morphology on the adsorption capacity of Fe_2O_3 materials, another Fe_2O_3 material was also prepared, in which ascorbic acid was removed from the precursors, while the others step was same with the synthesis of α-Fe_2O_3-ND. The obtained Fe_2O_3 material was denoted as Fe_2O_3-NP.

2.2. *Characterization*

The surface morphology of the as-prepared materials was observed by filed-emission scanning electron microscope (FE-SEM, Neon40esb Zeiss). The Brunauer-Emmett-Teller (BET) surface area of the sample was determined by nitrogen adsorption at 77 K, carried out on Micromeritics ASAP 2020. X-ray diffraction (XRD) was carried out on a Bruker D8-Advanced X-ray instrument using Cu Kα radiation (λ = 1.5418 Å).

2.3. *Adsorption of Cr(VI) ions*

In the present study, $K_2Cr_2O_7$ was used as the source of Cr(VI) ions. Different concentrations of Cr(VI) ions were prepared and the pH value was adjusted to 3 by HCl. In a typical adsorption capacity measurement, the concentration of adsorbents was 1 $g \cdot L^{-1}$, and the concentration of Cr(VI) ions was determined by a UV-vis spectrophotometer (JASCOV-670) at the characteristic absorption wavelength of Cr(VI) of 325 nm. The amount of metal ions adsorbed at adsorption time t (q_t, $mg \cdot g^{-1}$), the amount of metal ions adsorbed at equilibrium (q_e, $mg \cdot g^{-1}$), and the removal efficiency (E%) of Cr(VI) was calculated by the following equations:

$$q_t = \frac{(C_0 - C_t)V}{m} \tag{1}$$

$$q_e = \frac{(C_0 - C_e)V}{m} \tag{2}$$

$$E(\%) = \frac{(C_0 - C_t)}{C_0} \times 100\% \tag{3}$$

where C_t ($mg \cdot L^{-1}$) is the concentration of Cr(VI) at time t, V is the volume of the solution (L), q_t and q_e ($mg \cdot g^{-1}$) are the amounts of adsorbed Cr(VI) at time t and at equilibrium time, respectively, m is the mass of adsorbents (g), and C_0 and C_e ($mg \cdot L^{-1}$) are the initial and final concentrations of Cr(VI), respectively.

3. Results and Discussions

3.1. Characterizations of α-Fe_2O_3 materials

The morphology of as-prepared Fe_2O_3 materials is observed by SEM measurement. From Figure 1a and b, the typical morphology of Fe_2O_3-ND is uniform nanorods, with the length and width of 485 and 77 nm, respectively, while the morphology of Fe_2O_3-NP is sphere with rough surface, the diameter of which is 10~30 μm. The surface area of Fe_2O_3 materials were investigated by N_2 adsorption-desorption isotherms, which are 50.3 and 12.9 m^2/g for Fe_2O_3-ND and Fe_2O_3-NP, respectively. The SEM images, combining with N_2 adsorption-desorption results, imply that the addition of ascorbic acid greatly benefits the formation of uniform morphology of Fe_2O_3 nanoparticles and favors the enlargement of surface area. Figure 1c displays the XRD patterns of Fe_2O_3-ND and Fe_2O_3-NP. Both the Fe_2O_3 materials are assigned to α-Fe_2O_3 (PDF#33-0664), and the diffraction peaks at 24.2°, 33.2°, 35.7°, 40.9°, 49.6°, 54.2°, 62.5° and 64.2° are assigned to (012), (104), (110), (113), (024), (116), (214) and (310) planes of α-Fe_2O_3, respectively. By comparison, the Fe_2O_3-ND exhibits the much stronger intensity, suggesting that the introduction of ascorbic acid greatly facilitates the crystallization of α-Fe_2O_3.

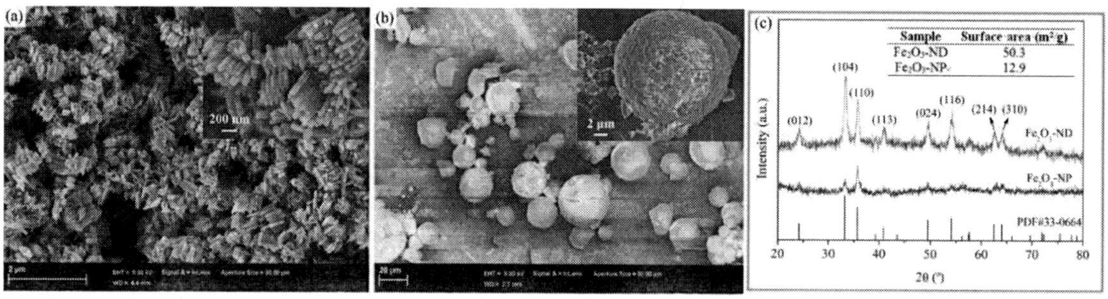

Figure 1. SEM images of Fe_2O_3-ND (a) and Fe_2O_3-NP (b), and XRD patterns of Fe_2O_3 samples (c) (insert: surface areas of Fe_2O_3 samples).

3.2. Cr(VI) adsorption properties of α-Fe_2O_3 samples

Owning the large surface area and the uniform morphology, the Fe_2O_3-ND sample is expected to exhibit excellent adsorption capacity, which were evaluated using the toxic Cr(VI) as the aimed pollutants. Figure 2a shows the adsorption isotherms of Cr(VI) on Fe_2O_3-ND and Fe_2O_3-NP samples, obtained with different initial concentrations of Cr(VI) ranging from 10 to 110 $mg \cdot L^{-1}$, which indicates that Fe_2O_3-ND exhibits the better adsorption capacity, with a maximum adsorption capacity of 28.3 $mg \cdot g^{-1}$, while the maximum adsorption capacity of Fe_2O_3-NP is 14.6 $mg \cdot g^{-1}$. Both the Langmuir and Freundlich isotherm models are used to analyze the adsorption data presented in Figure 2a, which are described by Equations (4) and (5), respectively:

$$\frac{C_e}{q_e} = \frac{1}{K_L q_m} + \frac{C_e}{q_m} \tag{4}$$

$$\log q_e = \log K_f + \frac{1}{n}\log C_e \tag{5}$$

From Figure 2b and c, the Langmuir adsorption model is adopted to fit the equilibrium adsorption data better for both the Fe_2O_3-ND and Fe_2O_3-NP samples according to the higher correlation coefficients (R^2), which are 0.9729 and 0.9822 for Fe_2O_3-ND and Fe_2O_3-NP, respectively, while the value of R^2 of Freundlich models for Fe_2O_3-ND and Fe_2O_3-NP are 0.9684 and 0.8899, respectively.

Figure 3a displays the adsorption rate curves of Fe_2O_3-ND and Fe_2O_3-NP samples for Cr(VI) ion with the concentrations of 10 and 30 $mg \cdot L^{-1}$. Fe_2O_3-ND exhibits the much more rapid adsorption to Cr(VI), and the excellent adsorption capacity is attributed to the increased surface area caused by the uniform nanoscale morphology. Figure 3b shows a simple kinetics analysis of adsorption using the pseudo-second-order model, which is expressed as equation (6):

$$\frac{t}{q_t} = \frac{t}{q_e} + \frac{1}{k_2 q_e^2} \tag{6}$$

where k_2 is the pseudo-second-order rate constant ($g \cdot mg^{-1} \cdot min^{-1}$). The correlation coefficients R^2 show high correlation for the two initial Cr(VI) concentrations for the two as-prepared Fe_2O_3 materials, thus the Cr(VI) adsorption on the two samples are both controlled by the chemical process [14-16].

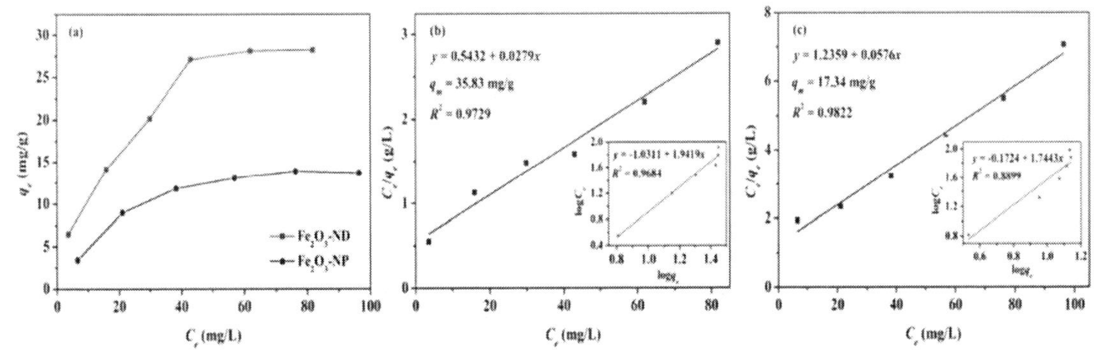

Figure 2. (a) Cr(VI) ion adsorption curves of Fe_2O_3-ND and Fe_2O_3-NP samples. (b) Linear Langmuir adsorption isotherm of Fe_2O_3-ND for Cr(VI) ion, insert: linear Freundlich adsorption isotherm of Fe2O3-ND for Cr(VI) ion. (c) Linear Langmuir adsorption isotherm of Fe_2O_3-NP for Cr(VI) ion, insert: linear Freundlich adsorption isotherm of Fe_2O_3-NP for Cr(VI) ion.

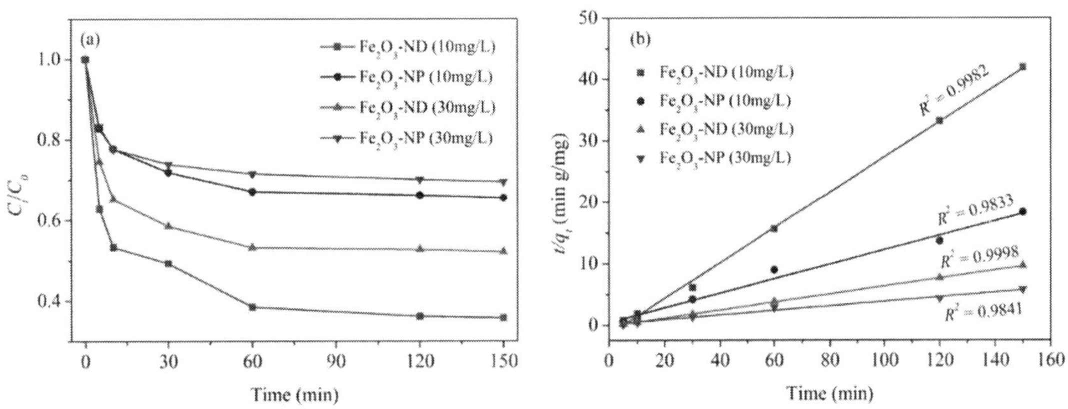

Figure 3. (a) Adsorption rate curves of Fe_2O_3-ND and Fe_2O_3-NP samples for Cr(VI) ion with the concentrations of 10 and 30 mg·L-1. (b) The pseudo-second-order kinetic model of Fe_2O_3-ND and Fe_2O_3-NP samples for Cr(VI) ion adsorption with the concentrations of 10 and 30 mg·L-1.

4. Conclusion

In the present study, a novel α-Fe_2O_3 nanorods was fabricated by a one-step hydrothermal method, and the uniform morphology could be attributed to the introduction of ascorbic acid in the precursors during the preparation. The α-Fe_2O_3 nanorods exhibited a rapid adsorption to Cr(VI), and the excellent performance was mainly attributed to the fine dispersion and the consequently increased surface area. The adsorption of Cr(VI) on α-Fe_2O_3 nanorods fit the Langmuir adsorption model, and the kinetic analysis confirmed that the adsorption could be described by pseudo-second-order model, meaning that the adsorption was controlled by the chemical process.

Acknowledgments
This work was supported financially by the National Nature Science Foundation of China (21507029, 21501138), Nature Science Foundation of Hebei Province (B2016502063), the Natural Science Foundation of Hubei Province (2015CFB177), Open Foundation of Key Laboratory of Industrial Ecology and Environmental Engineering (KLIEEE-15-02) and China Ministry of Education and the Fundamental Research Funds for the Central Universities (2016MS109).

References

[1] J. Fan, T.O. Okyay, D.F. Rodrigues, The synergism of temperature, pH and growth phases on heavy metal biosorption by two environmental isolates, J. Hazard. Mater. 279 (2014) 236-243.
[2] W. Jiang, Q. Cai, W. Xu, M. Yang, Y. Cai, D.D. Dionysiou, K.E. O'Shea, Cr(VI) adsorption and reduction by humic acid coated on magnetite, Environ. Sci. Technol. 48 (2014) 8078-8085
[3] Z.H. Ruan, J.H. Wu, J.F. Huang, Z.T. Lin, Y.F. Li, Y.L. Liu, P.Y. Cao, Y.P. Fang, J. Xie, G.B. Jiang, Facile preparation of rosin-based biochar coated bentonite for supporting α-Fe_2O_3 nanoparticles and its application for Cr(VI) adsorption, J. Mater. Chem. A 3 (2015) 4595-4603.
[4] M. Baikousi, A.B. Bourlinos, A. Douvalis, T. Bakas, D.F. Anagnostopoulos, J. Tuček, K. Šafářová, R. Zboril, M.A. Karakassides, Synthesis and characterization of γ-Fe_2O_3/carbon hybrids and their application in removal of hexavalent chromium ions from aqueous solutions, Langmuir 28 (2012) 3918-3930.
[5] M. Kumaria, C.U.P. Jr, D.Mohan, Heavy metals [chromium (VI) and lead (II)] removal from water using mesoporousmagnetite (Fe_3O_4) nanospheres, J. Colloid Interface Sci. 442 (2015) 120-132.
[6] Y.P. Huang, H. Ma, S.G. Wang, M.W. Shen, R. Guo, X.Y. Cao, M.F. Zhu, X.Y. Shi, Efficient catalytic reduction of hexavalent chromium using palladium nanoparticle immobilized electrospun polymer nanofibers, ACS Appl. Mater. Interfaces 4 (2012) 3054-3061.
[7] Z. Mai, Y. Hua, P. Huang, X. Zhang, X. Dong, Y. Fang, C. Wu, J. Cheng, W. Zhou, Outside-in stepwise bi-functionalization of magnetic mesoporous silica incorporated with Pt nanoparticles for effective removal of hexavalent chromium, Powder Technol. 312 (2017) 48-57.
[8] T. Zhou, C. Li, H. Jin, Y. Lian, W. Han, Effective adsorption/reduction of Cr(VI) oxyanion by halloysite@polyaniline hybrid nanotubes, ACS Appl. Mater. Interfaces 9 (2017) 6030-6043.
[9] D. Zitoun, N. Pinna, N. Frolet, C. Belin, Single crystal manganese oxide multipods by oriented attachment, J. Am. Chem. Soc. 127 (2005) 15034-15035.
[10] C.C. Yu, X.P. Dong, L.M. Guo, J.T. Li, F. Qin, L.X. Zhang, J.L. Shi, D.S. Yan, Template-free preparation of mesoporous Fe_2O_3 and its application as absorbents, J. Phys. Chem. C 112 (2008) 13378-13382.
[11] L. Liu, H.Z. Kou, W. Mo, H. Liu, Y. Wang, Surfactant-assisted synthesis of α-Fe_2O_3 nanotubes and nanorods with shape-dependent magnetic properties, J. Phys. Chem. B 110 (2006) 15218-15223.
[12] E. Thimsen, F.L. Formal, M. Grätzel, S.C. Warren, Influence of plasmonic Au nanoparticles on the photoactivity of Fe_2O_3 electrodes for water splitting, Nano Lett. 11 (2011) 35-43.
[13] Z. Liu, R. Yu, Y. Dong, W. Li, B. Lv, The adsorption behavior and mechanism of Cr(VI) on 3D hierarchical α-Fe_2O_3 structures exposed by (001) and non-(001) planes, Chemical Engineering Journal 309 (2017) 815-823.
[14] G. Liu, Q. Deng, H. Wang, S. Kang, Y. Yang, D.H.L. Ng, W. Cai, G. Wang, Synthesis and characterization of nanostructured Fe_3O_4 micron-spheres and their application in removing toxic Cr ions from polluted water, Chem. Eur. J. 18 (2012) 13418-13426.
[15] G. Chen, C. Qiao, Y. Wang, J. Yao, Synthesis of magnetic gelatin and its adsorption property for

Cr(VI), Ind. Eng. Chem. Res. 53 (2014) 15576-15581.
[16] C. Wan, J. Li, Facile synthesis of well-dispersed superparamagnetic γ-Fe_2O_3 nanoparticles encapsulated in three-dimensional architectures of cellulose aerogels and their applications for Cr(VI) removal from contaminated water, ACS Sustainable Chem. Eng. 3 (2015) 2142-2152.

IOP Conf. Series: Materials Science and Engineering **207** (2017) 012076 doi:10.1088/1757-899X/207/1/012076

Conductivity and electrochemical performance of $LiFePO_4$ slurry in the lithium slurry battery

Caimei Feng[1, 2], Yongchong Chen[1, 2, *] Dandan Liu[1] and Ping Zhang[3]

[1]Energy Storage Technology Research Group, Institute of Electrical Engineering, Chinese Academy of Science, Beijing 100190, China
[2]University of Chinese Academy of Sciences, Beijing 100049, China
[3]Beijing Hawaga Power Storage Technology Company Ltd., Beijing 100085, China

[*]Corresponding author e-mail: ycchen@mail.iee.ac.cn

Abstract. Lithium slurry battery is a new type of energy storage technique which uses the slurry of solid active materials, conductive additions and liquid electrolyte as the electrode. The proportion of conductive addition and the active material has significant influence on the conductivity and electrochemical performance of the slurry electrode. In the present work, slurries with different volume ratios of LiFePO4 (LFP) and Ketjenblack (KB) were investigated by the electrochemical workstation and charge-discharge testing system (vs. Li/Li+). Results show that the conductivity of the slurry increases linearly with the addition of KB, and the measured specific capacity of the slurry reaches its theoretical value when the volume ratio of KB to LFP is around 0.2. Based on this ratio, a slurry battery with higher loading of LFP (19.1 wt.% in the slurry) was tested, and a specific capacity of 165 mAh/g at 0.2 mA/cm2 and 102 mAh/g at 5 mA/cm2 was obtained for LFP.

1. Introduction

The concept of lithium slurry battery was proposed based on the semi-solid lithium flow battery, which is characterized by flowable semi-solid suspension and independent design of power and energy density [1, 2]. There has been much research work on the semi-solid lithium flow battery [3-9]. However, the low electrical conductivity, high viscosity of the electrode suspension, and high pumping loss caused by the high viscosity are the critical challenges for the application of the semi-solid lithium flow battery [10-12]. The lithium slurry battery can avoid such limitations since it cancels the external circulating system, and the electrode slurry can be viscous and have high loadings of active materials and conductive additions, which is favorable to improve the energy density and decrease the internal resistance. In the lithium slurry battery, the Li-intercalation active materials are not coated on the current collector but dispersed in the electrolyte, then the problems of electrode materials pulverizing and peeling off from the current collector after long-term cycling can be avoided. Theoretically, the lithium slurry battery will implement the charge and discharge function as long as the crystal structure of the active material is not damaged. Therefore, the lithium slurry battery has the potential advantages of high energy density, good cycling performance and also low manufacture cost in large-scale applications.

In the present work, LiFePO4 (LFP) was chosen as the positive active material since it has a long, high and stable voltage plateau, which is convenient for the analysis of polarization [13, 14]. Slurries with different ratios of LFP and Ketjenblack (KB) were assembled as half cells (vs. Li/Li+) to investigate the conductivity and electrochemial performance. Based on the analysis of the experimental results, optimized LFP slurry was prepared and can be used in the lithium slurry battery.

2. Experimental procedures

2.1. Preparation of the electrode slurry

Ball-milling method was used to prepare the electrode slurry. Firstly, powders of LFP (SLFP-ES09, Tianjin STL Energy Technology Co., LTD.) and KB (ECP-600JD, Shanghai Cuikehuagong Technology Co., LTD.) were weighed and mixed in the ball-milling tank after being dried at 100 ℃ for 12 hours. Zirconia balls were added with the mass ratio of 10:1 to the solid particles. After that, electrolyte (1 M $LiPF_6$ dissolved in EC: DMC: EMC of 1: 1: 1 in volume) was added to the tank in the argon-filled glove box with oxygen and water levels maintained below 5 and 0.1 ppm, respectively. Then, the ball-milling tank was sealed with the lid, moved out of the glove box and fixed on the planetary ball-milling tight. The milling time was 8 hours and the rotate speed was 500 rpm. After milling, the electrode slurry was poured into reagent bottle in the glove box and sealed up for the following steps. To analyze the influence of the ratio of KB to LFP on the conductivity and electrochemical performance, multiple component contents (volume percentage) of the electrode slurry were designed, as shown in Table 1.

Table 1. Contents of LFP and KB in the positive slurry

Number	Content of LFP (vol%)	Content of KB (vol%)	$R_a(\Omega)$	σ_a(mS/cm)
1	10.0	0.00	45.7	2.4
2	5.3	0.16	28.4	3.8
3	5.3	0.24	23.4	4.6
4	5.3	0.32	22.0	4.9
5	2.9	0.44	17.1	6.3
6	2.4	0.44	14.2	7.6
7	2.4	0.55	12.1	8.9

2.2. Measurement of the conductivity

The method to evaluate the conductivity of the slurry in semi-solid lithium battery has been discussed in our previous work [15], hence the proposed method of alternating current resistance test at fixed frequency was directly used in the present work. A modified mould was used here with a cylindrical test geometry sandwiched by two stainless steel plates. The electrode slurry was encapsulated between the stainless steel plates which were separated by insulating layer to form two electrode poles. The measurement was conducted on the mould by the ZENNIUM electrochemical workstation (ZAHNER-elektrik GmbH & Co. KG, Germany) at a fixed frequency of 1×10^5 Hz and an alternating current amplitude of 0.01 mA. The apparent conductivity of the slurry, σ_a, can be calculated by the formula below:

$$\sigma_a = L/R_s S \tag{1}$$

Where L is the thickness of the electrode slurry, S is the cross-sectional area of the electrode slurry and R_a is the measured apparent resistance. In the experiments, it was designed that L is 0.25 cm and S is 2.32 cm^2.

IOP Conf. Series: Materials Science and Engineering **207** (2017) 012076 doi:10.1088/1757-899X/207/1/012076

2.3. Measurement of the electrochemical performance

The batteries were assembled and tested in a half cell using a 0.5 mm thick Li metal foil (99.9%, Sigma-Aldrich) as the counter electrode and Celgard 2325 as separator. A plastic insulating frame with thickness of 1 mm was used to form the cavity of the positive LFP slurry. Stainless steel enclosures were used not only to form lugs but also seal the battery. All the procedures were conducted in the glove box under argon atmosphere.

The charge-discharge tests were carried out with a LAND-CT2001A battery test system. Galvanostatic charge-discharge tests were performed between 2.0 V and 4.5 V versus Li/Li^+. The lowest current density was set as 0.2 mA/cm^2 and the highest current density was set as 5 mA/cm^2.

3. Results and discussion

3.1 Influence of the ratio of KB to LFP on the conductivity

The measured apparent resistance and conductivity for the slurry with different contents of LFP and CB are shown in Table 1. The relationship between the ratios of KB to LFP and the conductivity was plotted in Figure 1. It can be seen that the apparent conductivity of the slurry is almost increases linearly with the addition of KB, and when the ratio of KB to LFP is 0.23, the apparent conductivity reaches 8.9 mS/cm.

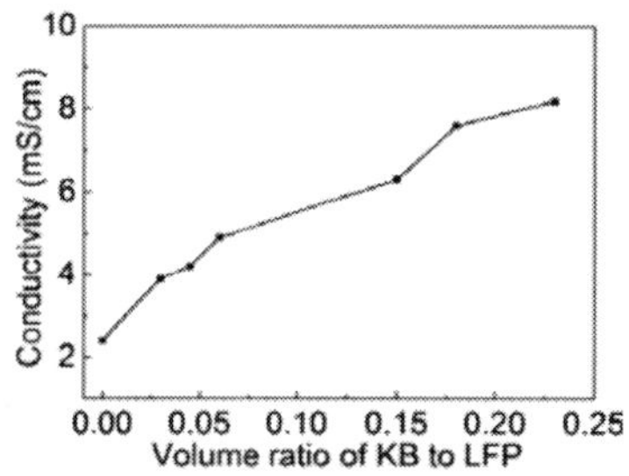

Figure 1. Relationship between the conductivity and the volume ratios of KB to LFP in the slurry.

3.2 Influence of the ratio of KB to LFP on the electrochemical performance

Figure 2 shows the charge-discharge curves of batteries using different LFP slurries at 0.2 mA/cm2. In Figure 2(a), the slurry without KB illustrates the largest polarization, and no charge/discharge voltage plateau was seen. As the content of KB increases, the polarization moderates slightly. The discharge specific capacity of LFP increases to 92 mAh/g when the KB content is 0.32 vol% along with 5.3 vol% of LFP, which is still far lower that the theoretical value. In Figure 5(b), a long and stable discharge voltage plateau was seen for the 5# and 6# slurry, especially for the 6# slurry, which exhibits the smallest polarization and highest specific capacity of 146 mAh/g for LFP.

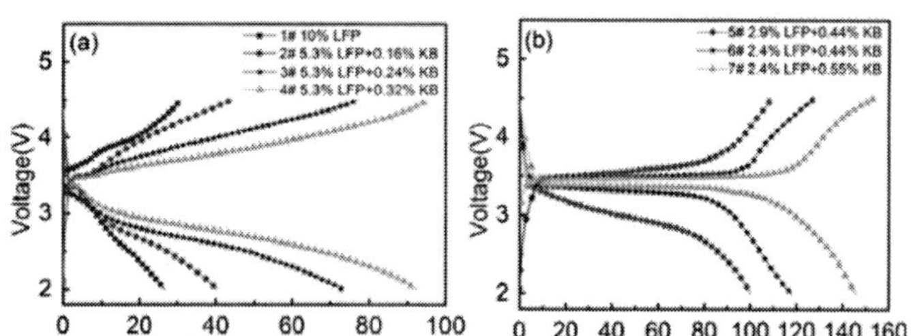

Figure 2. Charge-discharge curves of different electrode slurries at 0.2 mA/cm2.

Figure 3 presents the LFP specific capacity and discharge overpotential for the battery with the slurry of different volume ratios of KB to LFP. The electrode slurry with higher relative ratio of KB to LFP performs smaller overpotential and better charge-discharge property, which is consistent with the changing tendency of the conductivity in Figure 1. When the volume ratio of KB to LFP is above 0.2, the overpotential is less than 0.1 V.

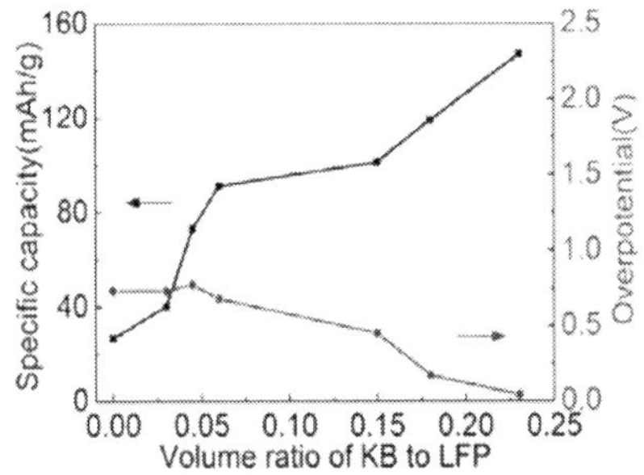

Figure 3. The LFP specific capacity and overpotential of the battery with the slurry of different volume ratios of KB to LFP.

Except for the specific capacity of active material LFP, the energy density of the electrode slurry is also important for the battery technology. The theoretical specific capacity of the electrode slurry, Cc, can be calculated by the following formula:

$$C_c = m_1 \times C_t / (m_1 + m_2 + m_3) \quad (2)$$

Where Ct is the theoretical capacity of the active material, m1, m2 and m3 are the mass of active material, conductive agent and electrolyte, respectively. Figure 4(a) compares the theoretical specific capacity of the slurry at different volume percentages of LFP as the volume content of KB is changing from 1% to 10%. It can be seen that the influence of the volume content of KB on the theoretical capacity is almost negligible, while the content of LFP is the dominant factor that determines the theoretical capacity of the slurry. In Figure 4(b), the gap between the theoretical and measured specific capacity shows that when the volume ratio of KB to PVDF is low, only small part of the active materials take part in the electrochemical reaction, while when the proportion of KB to LFP is increased to around 0.2, the actual measured capacity is nearly equal to the theoretical value. Therefore, a relatively high KB ratio can not only maintain a high theoretical capacity, but also form a better conductive network and promote the fully reaction of LFP.

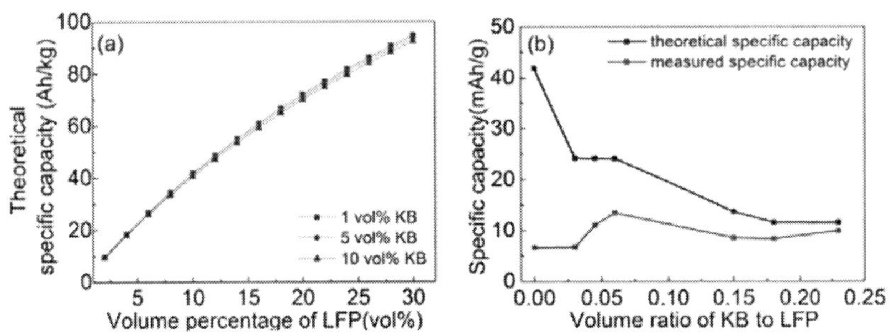

Figure 4. (a) Relationship between the theoretical specific capacity of the slurry and the volume percentage of LFP when the volume content of KB is 1, 5 and 10 vol%. (b) Comparison of the theoretical and measured specific capacity of the slurry at different volume ratios of KB to LFP.

3.3 Optimization of the performance of the LFP slurry

When the content of KB particles in the electrode slurry is small, the particles cannot form continuous conductive network, and result in high internal resistance. Along with the increase of the KB content, some independent clusters gradually appear but are still not connected in the slurry, which cannot improve the conductivity remarkably. Long-range associativity between particles appears until the content of KB increases to a certain critical value and then consecutive conductive network forms in the electrode slurry. The electrical conductivity of the slurry improves and the velocity of electrochemical reaction increases accordingly. The above experiments shows that there can be an effective conductive structure when the KB content is 0.55 vol%.

When LFP and KB particles are dispersed in the electrolyte uniformly, the electronic transfer path exists among the whole solid particles. Small amount of LFP does not affect the electrical conductivity of electrode slurry in the existed conductive network, but the continuing addition of LFP particles may destroy the consecutive structure. Therefore, the volume fraction of conductive particles in the all solid particles is an important factor. Kirkpatrick [16] calculated the critical volume fraction of spherical particles at different filling ways of such as simple cubic, face-centered cubic, body centered cubic and random accumulation with the classic percolation theory. It was pointed out that the conductive network under different filling ways has the same critical volume fraction. Considering the present experimental results that when the volume ratio of KB to LFP is around 0.2, an effective conductive network can be formed, we designed an optimized slurry with 7.5 vol% of LFP and 1.5 vol% of KB (ratio of KB to LFP is 0.2) which has a higher loading of active material (19.1 wt.% of LFP in the slurry with designed specific capacity of 32.5 Ah/kg) and tested its electrochemical property in the following.

The viscosity of the electrode slurry increases with the increasing content of KB due to its nano-structure and high specific surface area. Thus, a ball milling process with less milling time and slow rotate speed (1 hour and 100 rpm) was applied to this designed slurry. Figure 5 shows the measured charge/discharge curves at different current densities. The measured LFP specific capacity is 165 mAh/g at current density of 0.2 mA/cm2, corresponding to 31.4 Ah/kg of the slurry, which means nearly full reaction of LFP. The rate property of the battery is also good, which has a specific capacity of 102 mAh/g at current density of 5 mA/cm2, corresponding to 19.4 Ah/kg of the slurry.

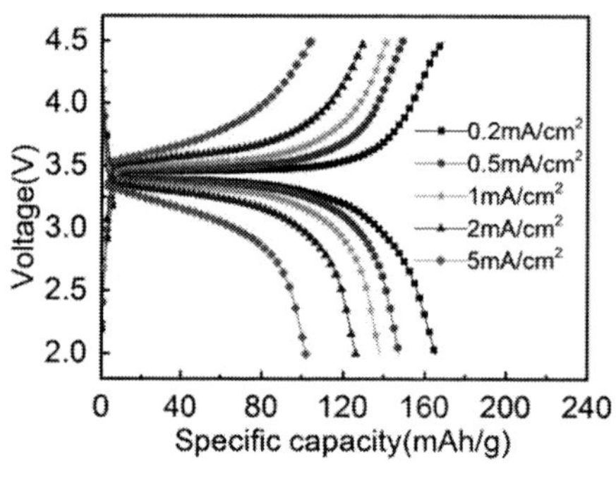

Figure 5. The charge-discharge curves of LFP slurry with optimized ratio of KB and LFP.

4. Conclusion

The influence of the ratios of KB to LFP in the electrode slurry on the conductivity and electrochemical performance of the LFP lithium slurry battery was investigated in the present work. Ball milling method was used to prepare the LFP electrode slurry. The results from electrochemical workstation and charge/discharge test show that the performance of the slurry increases with continuing addition of conductive KB, and a 0.2 volume ratio of KB to LFP was found to be a critical value for the slurry to form an effective conductive network and exhibits good specific capacity. Finally, electrode slurry with 7.5 vol% LFP and 1.5 vol% KB was tested and exhibits excellent performance with specific capacity of 165 mAh/g at 0.2 mA/cm2 and 102 mAh/g at 5 mA/cm2 for LFP.

Acknowledgments
This work was financially supported by the National Natural Science Foundation of China (Grant No. 51477170 and 51607175) and the Science and Technology Planning Project of Beijing (Grant No. Z161100000416001).

References

[1] Y.M. Chiang, M. Duduta, R. Holman, P. Limthongkul, T. Tan, U.S. Patent 8,993,159 B2. (2015)
[2] Y.C. Chen, C.M. Feng, Y.P. Zhang, P. Zhang, CN. Patent 201510164222.1 (2015)
[3] Q.Z. Huang, H. Li, M. Gratzelc, Q. Wang, Reversible chemical delithiation/lithiation of LiFePO4: towards a redox flow lithium-ion battery. Phys. Chem. Chem. Phys. 15 (2013) 1793-1797.
[4] H.N. Chen, Y.C Lu, A high-energy-density multiple redox semi-solid-liquid flow battery. Adv. Energy Mater. 6 (2016) 1502183.
[5] F.R. Brushett, J.T. Vaughey, A.N. Jansen, An all-organic non-aqueous lithium-ion redox flow battery, Adv. Energy Mater. 2 (2012) 1390-1396.
[6] Y.C. Chen, M.X. Wu, Y.K. Ren, L.B. Kang, Y. J. Li, L. Han, D.Y. Lin, Q.P. Wang, Research progress in lithium ion flow battery, Adv. Technol. Electr. Eng. Energy 31 (2012) 81-85.
[7] S. Hamelet, T. Tzedakis, J.B. Leriche, S. Sailler, D. Larcher, P.L. Taberna, P. Simon, J.M. Tarascon, Non-aqueous Li-based redox flow batteries, ECS 159 (2012) A1360-A1367.
[8] Y.K. Ren, Y.C. Chen, C.M. Feng, G.R. Fang, L. Han, Q.P. Wang, Electronic conductivity model and simulation of electrode suspension of lithium-ion flow battery, Mod. Sci. Instrum. 24 (2014) 84-89.
[9] L. Madec, M. Youssry, M. Cerbelaud, P. Soudan, D. Guyomard, B. Lestrieza, Electronic vs ionic limitations to electrochemical performance in $Li_4Ti_5O_{12}$-based organic suspensions for lithium-redox flow batteries, J. Electrochem. Soc. 161 (2014) A693-A699.
[10] J.J. Biendicho, C. Flox, L. Sanz, J.R. Morante, Static and dynamic studies on $LiNi_{1/3}Co_{1/3}O_2$-based suspensions for semi-solid flow batteries, ChemSusChem 9 (2016) 1938-1944.
[11] M. Duduta, B. Ho, V.C. Wood, P. Limthongkul, V.E. Brunini, W.C. Carter, Y.M. Chiang, Semi-solid lithium rechargeable flow battery, Adv. Energy Mater. 1 (2011) 511-516.
[12] K.C. Smith, Y.M. Chiang, C.W. Craig, Maximizing energetic efficiency in flow batteries utilizing non-newtonian fluids, J. Electrochem. Soc. 161 (2014) A486-496.
[13] M.S. Whittingham, Lithium batteries and cathode materials, Chem. Rev. 104 (2004) 4271-4301.
[14] A. Milev, L. George, S. Khan, P. Selvam, G.S.K. Kannangara, Li-ion kinetics in LiFePO4/Carbon nanocomposite prepared by a two-step process: the role of phase composition, Electrochim. Acta 209 (2016) 565-573.
[15] C.M. Feng, D.D. Liu, Y.C. Chen, J.J. Du, P. Zhang, Evaluation method of the conductive property of the electrode slurry in semi-solid lithium battery, J. Func. Mater., (2017) (accepted).
[16] S. Kirkpatrick, Percolation and conduction, Rev. Mod. Phys. 45 (1973) 574-588.

IOP Conf. Series: Materials Science and Engineering **207** (2017) 012077 doi:10.1088/1757-899X/207/1/012077

First-principles study of the photovoltaic properties of orthorhombic $Ca_2Si_{1-x}P_x$ (0<x<0.5)

Weifu Cen[2,a], Yinye Yang[1,2,*], Menghui Fan[1,2]

[1]Key Laboratory of Pattern Recognition and Intelligent Systems of Guizhou Province, Guizhou Minzu University, Guiyang, 550025, China
[2]College of materials science and engineering, Guizhou Minzu University, Guiyang, 550025, China

[*]Corresponding author e-mail: Ythin1969@sohu.com, [a]cenwefu1988@sina.cn

Abstract. Electronic structure and optical properties of simple orthorhombic $Ca_2Si_{1-x}P_x$ have been calcu-lated by the first-principles methods. The result shown that the band gap of the simple ortho-rhombic $Ca_2Si_{1-x}P_x$ was still a direct semiconductor and the band gap was increased with the concentration of P impurity increased. The dielectric constant and refractive index were in-creased with the concentration of P impurity increased. The extinction coefficient, the reflectivi-ty, the absorption and the conductivity have significant changed in the energy range of 0eV~1eV. Therefore, Transmission property of energy band structure and optical properties of simple or-thorhombic Ca2Si1-xPx were changed by changing the impurity concentration. It is a useful method for modulating photoelectric transmission property of the simple orthorhombic $Ca_2Si_{1-x}P_x$.

1. Introduction

Metal silicides have recently the subject of extensive studies to improve the under-standing of their structural, electronic, and optical properties due to their possible tech-nological applications. Among this wide family of compounds, the so-called “Kankyo semiconductors” [1] play an increasing role in a device’s fabrication because they are composed of naturally abundant and less toxic elements. In particular, Ca2Si is considered to be a promising candidate for applications. Ca2Si has a simple orthorhombic structure and a cubic structure, respectively. The orthorhombic structure, which is a stable phase [2] with lattice constants of a=0.7667nm, b=0.4799nm, c=0.9002nm. It is a semiconductor with an energy gap of 1.9eV [3] (theoretics volume=0.36eV [4]), and its electronic band structure is investigated[5-7]. The cubic phase Ca2Si which the lattice constants a=0.7016nm with an energy gap of 0.56eV [8] (theoretics volume). Recently, it is been reported that Ca2Si film in Ca-Si system was grown directly on Si(100) substrates by sputtering and annealing [9-12]. However, Ca2Si can be applied to the industry still needs further im-provement. New technique should be used for investigating in the further, In addition, the electronic structure and optical properties of orthorhombic Ca2Si bulk influenced by doping obtains excellent electronic structure and optical properties.

2. Calculation methods

The calculations were performed by using the CASTEP (Cambridge serial total energy package) package of the Material Studios of Accelrys [18]. All the possible structures are optimized by the BFGS algorithm (proposed by Broyden, Fletcher, Goldfarb and Shan-non), the optimization is performed until the forces on the atoms are less than 0.01eV/A and all the stress components are less than 0.02GPa, the tolerance in the self-consistent field (SCF) cal culation is 1×10-6meV/atom. Ultrasoft pseudo-potentials are expanded within a plane wave basis set with 330eV cutoff energy in the process of optimization. Exchange and correlation effects are shown in scheme of Perdew-Burke-Eruzerhof (PBE) generalized gradient approximation (GGA), the k-point sampling are 3×4×4 mesh in the Brillouin zone (BZ).

3. Result and discussion

3.1 Electronic structure

Fig.1 (a) show that the simple orthorhombic Ca_2Si is a direct semiconductor with the energy band gap of 0.29868eV at Γ, this result value is smaller than the value of 0.35eV at Γ reported by Migas etal[6], the result compare well with the results in the literature [4]. We have extended our studies to the $Ca_2Si_{1-x}P_x$ system so as to test the band structure of the possible P-doped concentration. Fig.1 (b)~(i) shown the band structure of different simple orthorhombic $Ca_2Si_{1-x}P_x$ ($0<x<0.5$) compositions. The energy band gap is 0.31459eV, 0.36668eV, 0.46026eV, 0.52698eV, 0.59817, 0.68751eV, 0.70274eV and 0.77489eV for the corresponding concentration of P impurity 6.25 at %, 12.5 at %, 18.75 at %, 25 at %, 31.25 at %, 37.5 at %, 43.25 at %, 50 at %, respectively.

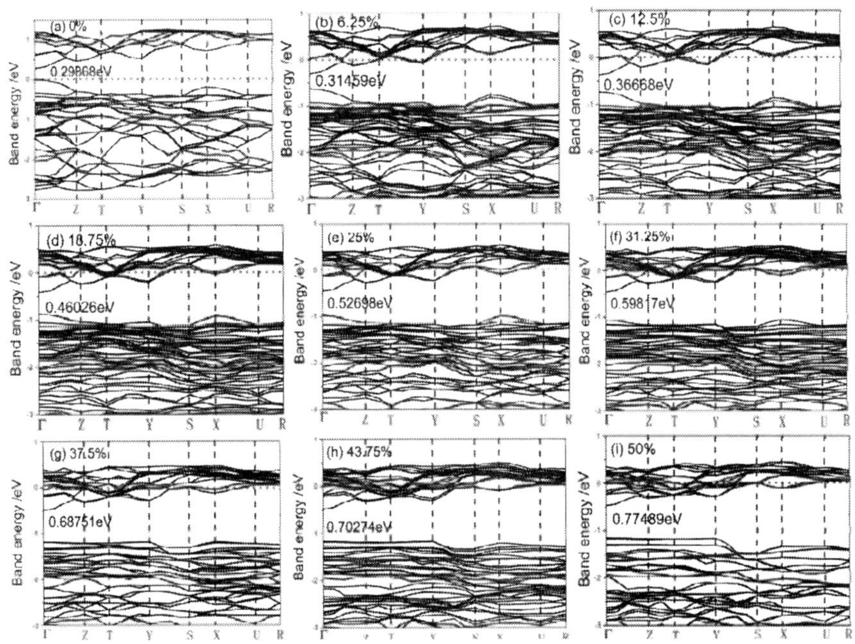

Fig.1 The band structure of $Ca_2Si_{1-x}P_x$

3.2 Optical properties

Fig.2 show the dielectric function of simple orthorhombic $Ca_2Si_{1-x}P_x$。The real component of dielectric function for this figure shows that the static dielectric constant is 8.7073, 8.6104, 15.8948, 19.2846, 26.4672, 43.6826, 51.8764, 117.8906 and 97.1167 for the corresponding concentration of P impurity 0 at %, 6.25 at %, 12.5 at %, 18.75 at %, 25 at %, 31.25 at %, 37.5 at %, 43.25 at %, 50 at %, respectively. In the range of 6.25 at % ~ 43.25 at %, the static dielectric constant is increased with the corresponding concentration of P impurity increased. The imaginary component of dielectric function figure show that the prominent is increased with the corresponding concentration of P impurity is increased. The peak is produced by electron transition, it exhibit's the information of band structure and spectral, the intensity of electron transition is modulated by the changing of the corresponding concentration of P impurity, but it can't modulate the electronic transition probability.

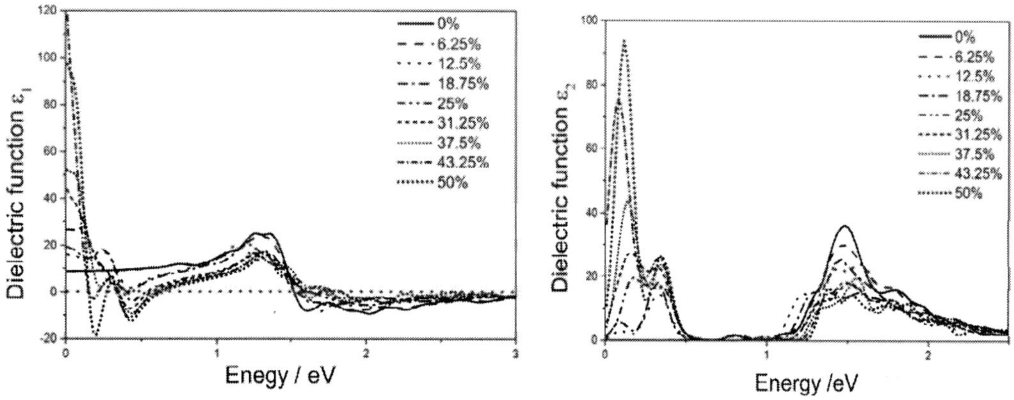

Fig.2 The dielectric function of $Ca_2Si_{1-x}P_x$

Fig.3 shows the refractive index *n* and the extinction coefficient *k* of simple orthorhombic $Ca_2Si_{1-x}P_x$. The picture of refractive index *n* show that the refractive index is 2.9508, 2.9344, 3.9877, 4.3939, 5.1469, 6.6239, 7.2081, 10.9847 and 9.8705 for the corresponding concentration of P impurity 0 at %, 6.25 at %, 12.5 at %, 18.75 at %, 25 at %, 31.25 at %, 37.5 at %, 43.25 at %, 50 at %, respectively. The changing of refractive index *n* for simple orthorhombic $Ca_2Si_{1-x}P_x$ is mainly appeared in the energy range of 0eV~1.5eV, the refractive index *n* is significant increasingly with the corresponding concentration of P impurity is increased, so the refractive index *n* is increased. The extinction coefficient picture shows that the extinction coefficient curve edge of Ca_2Si is appear at the energy of 1.01eV, and the extinction coefficient *k* is increased rapidly in the energy range of 1.01eV~1.57eV.

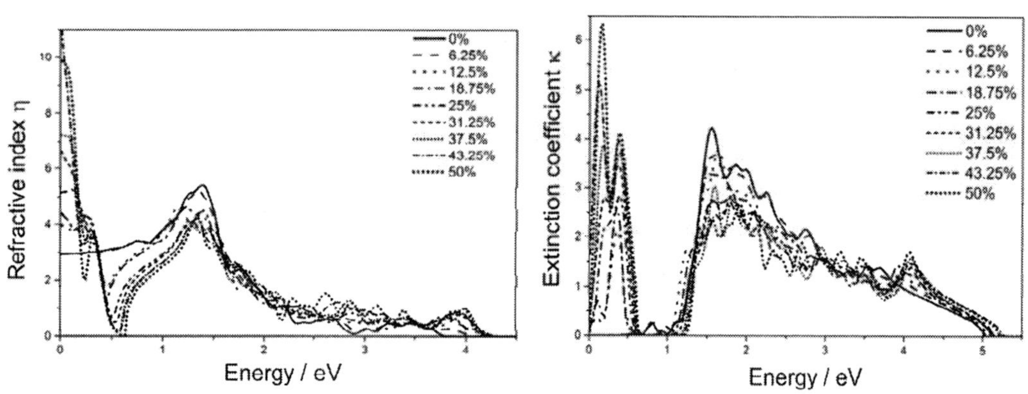

Fig.3 The refractive index and extinction coefficient of $Ca_2Si_{1-x}P_x$

Figure.3 shows the reflectivity of simple orthorhombic $Ca_2Si_{1-x}P_x$. The tendency of reflectivity is similar to the refractive index in the energy range of 0eV~1.5eV, the reflectivity is changed flat but it has a little oscillation in the energy range of 1.5eV~3.7eV, and the reflectivity significant increasingly in the energy range of 3.7eV~5.2eV. The reflectivity is attained 90% or more, even occur total reflection phenomenon and presents metal emission characteristic in the energy range of 3.7eV~5.2eV for the corresponding concentration of P impurity 0 at %, 6.25 at %, 12.5 at %, 18.75 at %, 25 at %, 31.25 at %, 37.5 at %, 43.25 at %, 50 at %, respectively. The total reflection is moved to the direction of high energy with the corresponding concentration of P impurity is increased. However, the most significant changed is appear in the energy range of 0eV~0.7eV, the reflectivity value of $Ca_2Si_{1-x}P_x$ is greater than the reflectivity value of Ca_2Si and it is increased with the corresponding concentration of P impurity is increased. The cause of this result is that P 3p orbital electronic is stimulated emission, and lead to the reflectivity occurred change.

4. Conclusion

In this work, we have reported the electronic structure and optical properties of simple orthorhombic $Ca_2Si_{1-x}P_x$ with the changing of corresponding concentration for P impurity by the first-principle method, it was shown that the fundamental band gap of the simple orthorhombic $Ca_2Si_{1-x}P_x$, is stayed a direct semiconductor and the band gap is increased with the corresponding concentration of P impurity in increased. The optical properties have a significant change with the corresponding concentration of P impurity is changed. The dielectric constant and refractive index are increased with the corresponding concentration of P impurity is increased. The extinction coefficient, the reflectivity, the absorption, the conductivity are have a significant changed in the energy range of 0eV~1eV. It is a useful method for modulating the photoelectric transmission property of the simple orthorhombic $Ca_2Si_{1-x}P_x$.

Acknowledgements

Project supported by the science and technology foundation of Guizhou Province, China (NO.LH[2017]7077 and NO.LH[2015]7218), the youth science and technology talents growth fund program of GuiZhou province education department, China (NO.KY[2016]166,) and This research was supported by Innovation Group Major Program of Guizhou Province (No.KY[2016]029).

References

[1] Yang Y Y, Xie Q.A Single Phase Semiconducting Ca-silicide Film Growth by Sputtering Conditions, Annealing Temperature and Annealing Tim[J].Journal of Materials Science,2009,44(14):3877-3882.

[2] Lebègue S, Arnaud B, Alouani M. Calculated quasiparticle and optical properties of orthorhombic and Ca2Si [J]. Physical Review B, 2005, 72(8):1-8.

[3] Matsui H, Kuramoto M, Ono T, et al. Growth of Ca_2Si layers on Mg_2Si/Si(111) substrates [J] .Cryst Growth,2002,237-239:2121-2124.

[4] Imai Y, Watanabe A. Energetics of alkaline-earth metal silicides calculated using a first-principle pseudo potential method [J]. Intermetallics, 2002, 10(4):333-341.

[5] Imai Y, Watanabe A, Mukaida M. Electronic structures of semiconducting alkaline-earth metal silicides [J].Journal of Alloys and Compounds, 2003, 358(1-2):257-263.

[6] Migas D B, Miglio L, Shaposhnikov V L et al. Comparative Study of Structural, Electronic and Optical Properties of Ca_2Si,Ca_2Ge,Ca_2Sn and Ca_2Pb [J]. Physical Review. B, 2003, 67(20): 205203,1-205203,7.

[7] Takagi N, Sato Y, Matsuyama T et al. Growth and Structural Properties of Mg_2Si and Ca_2Si Bulk Crystals [J].Applied Surface Science.2005,244(1-4):330-333.

[8] Cheng J, Yang Y Y, Liang Y, Selective Growth of Ca2Si Film in Ca-Si System by R.F MS by Annealing, 3M-NANO, 2012, 9:75-79.

[9] Yang YY, Xie Q. The Selective Growth of a Single Phase Ca-silicide Film from Ca-Si System by R.F. Magnetron Sputtering and by Annealing [J]. International Journal of Information and Electronics, 2008, VI, N2-4: 6-9.

[10] Ma J L, Zhang H M, Song J J et al. Energy band structure of uniaxial-strained silicon material on the (001) surface arbitrary orientation [J]. Acta Phys. Sin, 2011, 60(02):027101.

[11] Wang G Y, Ma J L, Zhang H M etal. Model of intrinsic carrier concentration of [110]/(001)-uniaxial strained Si [J]. Acta Phys. Sin, 2011, 60(07): 077105.

[12] Cui D M, Jia R, Xie q etal. First-priciples Calculations of Stressed Ru_2Si_3 [J]. Chinese joural of Luminescence. 2011, 32(09): 907-912.

[13] Segall M D, Lindan P J D, Probert M J, et al. First-principles simulation: ideas, illustrations and the CASTEP code [J].Journal of Physics,2002,14(11):2717-2744.

IOP Conf. Series: Materials Science and Engineering **207** (2017) 012078 doi:10.1088/1757-899X/207/1/012078

Formation mechanism and prevention analysis in plateau damp environment

Yongjun Meng[1, a], Fengxia Chi[2, 3, b]

[1]College of Civil Engineering and Architecture, Guangxi University, Key Laboratory of Disaster Prevention and Structural Safety of Ministry of Education, Nanning 530004, China
[2]Zhejiang Institute of Communications, Hangzhou 311112, China
[3]Zhejiang Scientific Research Institute of Transport, Hangzhou 311305, China

[a]hitmengyj@163.com, [b]chifengxia@126.com

Abstract. This paper analyses and researches road condensation ice formation mechanism, so as to find out the effective and economical method to prevent road frozen ice and govern it. The method of prevention can be divided into passive inhibition technology and proactive inhibition technology. Passive inhibition technology cannot inhibit the formation of condensation. In the process of research of prevention condensation, early warning system for snow and ice research also should not ignore. To establish early warning systems and take it into implementation will greatly reduce the extent of the ice and snow disasters.

1. Introduction

Most of the Southwest China belongs to the plateau damp mountains areas. In the winter, roads are susceptible to condensation ice, which is characterized by hardness, thinness and wideness. After the road surface condensation ice formed, it not only has a serious influence on road traffic safety, but also poses a serious threat to the people's lives and property safety. Having recognized the threat the road surface condensation ice may bring about, it is necessary to have a systematic study on condensation ice formation mechanism in damp environment on the plateau. On the basis of this analysis, the road condensation ice prevention was conducted in order to solve or reduce the danges road condensation.

2. The formation mechanism of condensation ice

2.1 Condensation ice

Condensation ice is a state of water condensing on the road surface. In a broad sense, condensation ice can be divided into condensation ice of snow melting, condensation ice on road surface and frozen condensation ice. Condensation ice of snow melting often occurs in the northern area, which are mainly caused by temperature difference during the day melting and at night producing condensation ice with lower temperature and traffic influences. It is characterized by lasting for a long time but easily getting removed. Condensation ice on road surface is formed when the water falling on the road during the day encountered the below zero temperature at night. It is characterized by lasting for a short time, being thin and hard removing. Frozen condensation ice which is different from the former two condensation ice is a kind of natural disaster in the country's plateau damp mountains areas in the southwest.

To find out the formation mechanism of the country's plateau damp mountains areas in the southwest. The formation mechanism of the freezing rain must first be made clear. By studying the spatial and temporal distribution and structure characteristics of temperature and humidity of freezing rain in our country, and according to the two kinds of freezing rain formation mechanism, it can be concluded that winter freezing rain formation mechanism in our country is mainly divided into two kinds-- the north (> 30° N) is mainly ice phase mechanism of freezing rain, and the south (< 30° N) is given priority to warm rain mechanism of freezing rain [1].

According to the temperature, the clouds can be divided into warm and cold cloud, warm cloud temperature being above 0℃ and the cold cloud below 0℃. When the cloud droplets grow to a certain size, it can't stay in the sky, which finally landed and produce precipitation. According to the different altitude of the clouds temperature and the thickness of the cloud, precipitation pattern can be divided into snow, sleet, freezing rain and rain, etc. Usually in winter, the surface above 3000m is the snow condensation layer with the temperature below 0 ℃. Ranging 1500m to 3000m, this layer usually is warmer layer which the temperature is slightly higher than the snow condensation layer. Close to the ground within 1500m is the air layer. The water, ice and snow in the slightly below 0℃ snow condensation layer transformed into liquid droplets when entering the 1-4 ℃ atmosphere, a comparatively warmer layer. When liquid droplets enter the surface air layer which temperature is slightly lower than 0℃,it becomes super-cooled water droplets. Super-cooled water droplets come into contact with the ground which the temperature is lower than 0℃ to form the so-called freezing rain. When landing on the surface of road, thus, it forms the condensation ice. Figure 1 is a schematic diagram of the formation of condensation ice.

Therefore, we can draw that the ice formation of southwest plateau damp area should have three necessary conditions: ①the formation of precipitation conditions; ②special structure of "cold-warm-cold" layer in the atmosphere; ③the ground temperature below 0℃.

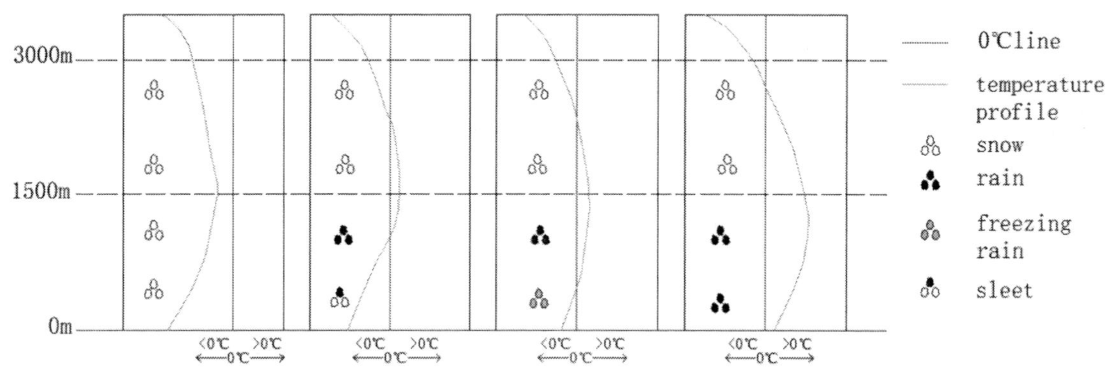

Figure 1. Schematic diagram of the formation of condensation ice.

3. The hazard of road condensate ice and prevention analysis

3.1 The hazard of road condensate ice

It is easy for the plateau damp mountains areas in southwest to form condensation ice in winter. Thus, it reduces the pavement skid resistance ability, which lowers road traffic capacity seriously. According to relevant data analysis [2-3], the existence of road condensation ice greatly reduces the adhesion coefficient of road surface, thus worsening the adhesion to the vehicle's braking stability and operation stability. It will finally cause the brake failure, loss of direction control, vehicle easily skidding and wandering and braking distance being prolonged. Road condensation ice not only poses a serious threat to people's lives and property, but also hindered the local economic development and social construction. Therefore, it is very necessary for pavement condensation prevention and governance.

3.2 Prevention analysis

3.2.1 Passive inhibition condensation ice technology. Passive inhibition condensation technology is a kind of passive way to remove snow, when snow is coming or already came. As shown in table 1, it is the classification of the advantages and disadvantages of passive inhibition condensation ice technology.

Table 1. Classification of passive inhibition condensation ice technology

Snow-melting agent	Sprinkle sand and gravel	Manual removal	Mechanical removal	Heart removal
convenient, deicing effect good before freezing, but easily corrode road and, pollute atmosphere.	convenient and fast for the early ice and now, economic and environmental-friendly.	low efficiency, cost highly, suitable for small amount of snow and the difficult sections of ice and snow.	fast, suitable for the large amount of snow and ice, but affected by the ice thickness.	need for corresponding equipment, cost highly, limited application and promotion.

Sprinkling snow-melting agent mainly sprinkles to the pavement a certain amount of anti-freezing chemical materials. It can make snow and ice on the pavement melting or reduce pavement freezing degree. Commonly used snow-melting agents are in solid and liquid forms, mainly concluding the edible salt, calcium chloride, potassium slag, acetate etc.

Manual removal can remove ice and snow more thoroughly, but it is low in efficiency, time-consuming and labor-consuming. It will affect traffic and pose a threat to the personal safety of workers at the same time when being conducted. It cannot be applied to a wide range of operation for a long period of time, but for difficult small sections of ice and snow.

Adopting mechanical equipment to remove snow has been applied for dozens of years. This method can be divided into two categories-- mechanical shoveling snow and ice and mechanical blowing snow. Mechanical shoveling snow and ice is suitable for large area clearing with heavy snow and before ice freezing. Mechanical blowing snow is suitable for the thin pavement snow without crushed, and it usually only applies to the airport and other small scopes where snow removal is manageable.

Heat removal is the use of high temperature medium to be in contact with the snow. By the using of heat transmission generated by temperature difference, it can increase the temperature of snow to go above the melting point and transform into flowing water to drain off. This method requires a lot of heat and is not affordable. Heat removal can be also subdivided into the following three ways. Spraying water melting method is suitable for the area that is small and has a good drainage system so that the melting ice water can flow away quickly, otherwise, it is easy to freeze again, leading to more serious traffic accidents. Heat pipe method is that the pipe geothermal is transferred to the ground surface to melt snow and ice, but construction and installation of the heating pipe are complicated and take a long time. Fluid heating method is to use natural hot water or stored solar energy for heating. Taking the method of solar heating method for example, stores the solar energy in the summer, then heat energy storage is used to melt ice and snow in winter. Construction and installation price of it is high, and it is not applicable.

The methods above belong to the traditional methods of dealing with road condensation ice which can be divided into two major categories of removing and melting methods. The former is mainly by means of artificial or mechanical equipment crushing snow and condensation ice and moving from the road to the other place to ensure road traffic safety. The latter is mainly through the adoption of a certain snow-melting agent or heating method to melt snow and ice to flowing water and discharge it.

3.2.2 P roactive inhibition condensation ice technology. Proactive inhibition condensation ice technology is the use of pavement mechanical characteristics or structural characteristics, through the vehicle repeatedly rolling, spontaneously to clear condensation ice or change the state of ice and snow so as to improve the traffic. Proactive inhibition condensation ice technology mainly has the following three ways.

(1) The road of storage salt

Germany first used a substance that is called "verglimi" and is a kind of a salt to deice [4]. This salt is covered by a hydrophobic substance. "Verglimit" has good temperature stability and it can be directly added to the asphalt mixture in the process of mixing. In order to make the pavement work, porosity of the mixture in engineering applications is about 3-4% (volume), the porosity of Marshall Specimen in the experiment is about 2-3% (volume). In order to improve the durability of the mixture, the mixture of the binder content should be improved.

Besides "verglimit"," mafilon" is the same type of material. Mafilon is a pale yellow powder, and the powder is made of sodium chloride which is a hydrophobic material on the surface. As a result of the existence of this kind of hydrophobic material, it can ensure that the salt can exist in the water for a long time and gradually release under the effect of wheel friction. And this kind of hydrophobic material has good high temperature performance, making the road surface to release less amount of salt in the heat than in a freezing point temperature. And it won't last a damp phenomenon on the road surface during the summer.

The road of storage salt technology isn't applied widely. Japan mainly carried out some research about latest installation technology of inhibiting freeze. Its purpose is essentially to slow the progress of the pavement freezing by adding the sodium chloride and calcium chloride in the asphalt mixture to reduce the freezing point, so as to achieve anti-freezing effect.

(2) Elastic rubber asphalt pavement

The research for the technology of rubber particles in asphalt mixture began in the late 1980s. Rubber used in the research of elastic pavement is mainly rubber particle and rubber pieces.

Based on the rubber particles related study [5], it shows: (1) As the needle flake content in the rubberized concrete decreasing, the mixture's compact ability, stability and durability is improved. The characteristic differences of rubber particle surface affect the performance of the mixture slightly. (2) The rubberized concrete concrete's mixing molding temperature is 10 to 20 degrees higher than ordinary asphalt concrete. Mixing process should be controlled strictly. The step is to dry mix rubber particles and stone around 30s, and then adding asphalt to mix about 90s. When molding it should be twice compaction molding, and pavement construction also should use twice compaction technology. (3) The deicing effect of rubber particles asphalt pavement is affected by rubber particles content, distribution location, pavement thickness, ice thickness and ambient temperature. When the temperature is below -12℃ or the ice thickness is more than 9mm, deicing effect almost disappear.

As to application studies of rubber block, a Japanese company sprinkled 2cm diameter rubber block and pressed it into the just finished asphalt road's surface with roller. It showed that the method improves elasticity of the pavement and has superior performance in terms of deicing and anti-slipping etc, but durability and flatness is not ideal.

According to the research on frost-resisting pavement in recent years, it was found there are several deicing technology with rubber materials as following: (a) Using rubber granular mixtures, that is to add certain specifications of the waste rubber particles to the mixture to increase the flexibility of surface to curb freezing; (b) Cutting a diameter of 50mm, 25mm deep small hole in the surface of asphalt concrete according to a certain spacing and then pressuring into the same size of the rubber cylinder, and filling the gap with fiber at the same time. Depending on the rubber's stress which generates by the effect of vehicle, ice will automatically break; (c) Soaking with polyurethane in the groove of drainage pavement to pave, it can avoid ice adhesion through its elasticity and smoothness, thus to improve the ability to resist freeze; (d) filling the groove of the pavement surface with rubber particles to reduce the freeze.

(3) Rough pavement of anti-sliding

Rough pavement of anti-sliding is to make up the loss of structural depth which is brought about by condensation ice, provide enough road anti-sliding ability, and reduce the effect of condensation ice on road traffic safety by rebuilding pavement wearing layer of great structural depth, so as to reduce the influence of snow and ice on the road traffic capacity. Larger surface and the interstices of internal aggregate can smoothly provide drainage channel, and pavement anti-sliding function can be effectively maintained when producing condensation ice. Rough pavement has the application value in damp regions where the snow is not deep, and condensation ice is thinner.

As to rough pavement, the most often used thing is Open Graded Friction Course (OGFC) [6]. Because of the pore connectivity of the material, it has less condensation ice when compared to the conventional pavement and can improve anti-sliding performance. But with the increase of ice and snow, the internal space is filled and the pavement anti-sliding performance will gradually decline. At this time, using salt to deice costs about 20% more than conventional road. So it is not economical, and the technology is used too narrowly to apply widely.

From the above analysis, we can know that proactive deicing technology is the direction of future research of deicing. As to the research of proactive deicing technology, foreign countries studied earlier than china, but china began to research the technology widely in recently years. If the technology has a major breakthrough, I believe that it will be applied widely in the near future. It will greatly reduce the probability of traffic accidents, and ensure the safety of vehicles and pedestrians.

3.2.3 The early warning of pavement condensation ice. From the study of material of paved roads, it is an important approach to solve pavement condensation ice. And this method is also the main solution of solving condensation ice, but before producing condensation ice, informing the drivers in time of the time of producing condensation ice and severity forecast will reduce traffic accidents to a certain extent. Condensation ice forecast has a lot relevance with the establishment of disaster warning system of snow and ice. The so-called disaster warning system of snow and ice is to establish Highway Meteorological Service System [7] or Road Weather Information System (RWIS) [8]. It can improve the intelligent level of traffic management and traffic safety of the public service ability through the accurate and timely meteorological Information.

Based on the traffic accidents caused by the complex weather and road conditions, developed countries are successively setting up different forms of highway meteorological service system. Traffic management institutions in some countries integrated the RWIS and Intelligent Transport System (ITS) and successively developed a series of Pavement Meteorological Monitoring System. While on the research of road traffic safety early warning, some foreign related research institutions are studying.

The research about this aspect started relatively late at home. In recent years, we have accumulated preliminary experience about analysis theory methods of climate effects on road traffic safety, but we have not formed systematic research results in the weather forecast service system and also failed to carry out large-scale highway meteorological service. The National Climate Center has analyzed and reviewed the effects of climate and severe weather on the traffic system and researched the severe weather and highway traffic, but not reach to the level of the public service yet.

As the increasing of the highway traffic meteorological service demand, some meteorological observatories in our country are beginning to build severe weather monitoring system along the highway. However, due to funding constraints, such progress is slow and focuses on monitoring a single meteorological phenomena. The early warning measures about disaster from snow and ice has not been studied systematically, it remains to be studied further.

4. Conclusion

According to the above analysis, it has to meet some conditions to form condensation ice. Condensation ice will greatly reduce the friction coefficient of pavement, seriously influence the traffic, and even lead to frequent occurrence of traffic accidents. In order to ensure safety, it is necessary to prevent and govern the pavement condensation ice. Some traditional methods mainly include the removal method and melting method. In view of the rapid development of our countries transport network system, these methods could not have a wide range of applications, it only applies to the city and highway in some key areas.

At present, china and other countries focus on proactive inhibition condensation ice technology, which has a variety of techniques to select. But each technology is not perfect. One technology can solve the problem of condensation ice, but it brings the durability of pavement and the stability of driving at the same time. One does not affect the durability of pavement, but deicing effect is not ideal. So this technology remains to be studied further. It not only can carry out from the perspective of road materials to prevent and govern çondensation ice, but also the research of disaster warning system of snow and ice should not be neglect. Only by combining the two kinds of direction organically can we ensure the safety of driving to the maximum extent.

Acknowledgments

This research was supported by Guangxi Natural Science Foundation (Number:2014jjBA60003) and Zhejiang Provincial Natural Science Foundation of China under Grant No. LY13E080006.

References

[1] Jian-Jun, O. U., et al. "Analyses on Spatial-Temporal Distributions and Temperature-Moisture Structure of Freezing Rain in China." Plateau Meteorology 30.3(2011): 692-699.
[2] Zhu, Yunsheng, et al. "Analysis on Dynamic Response Between Asphalt Pavement and Tire Under Glazing-ice Condition." Journal of Wuhan University of Technology (2014).
[3] Higgins, Daniel D., et al. "Morphology of ice wear from rubber–ice friction tests and its dependence on temperature and sliding velocity." Wear 265.5–6(2008): 634-644.
[4] Li Yongju, Gao Yiping, Tian Baoxia, Paving method of urban pavement of anti-freezing in the Sapporo City of Japan. Highways & Transportation in Inner Mongolia, 2001,4(4): 16~17.
[5] Zhou Chunxiu. Research on application technology of asphalt mixture of rubber-particles in areas with snow and ice. Doctoral dissertation, Harbin Institute of Technology(2006).
[6] Li Hongping, Wu Dejun, Yang Chenguang. Research on pavement performance of different modified asphalt pavement drainage(OGFC-13). Highway, 6(2009): 151-154.
[7] Zhang Wenhong, Wang Wenjin. Meteorological service system design overview. Energy Research and Management, 2 (2012): 9-12.
[8] Zhou Hong. Expressway meteorological support service system. China ITS Journal, 5(2011): 117-118.

Research on Experiment of Islanding Protection Device of Grid-connected Photovoltaic System Based on RTDS

Ning Zhou[1], Jia Yang[2], Zheng Cheng[3], Bo Chen[1], Yong Chun Su[1], Zhan Shu[1], and Jin Zou[1]

[1]State Grid Jiangxi Electric Power Research Institute Nanchang 330096, China
[2]State Grid Electric Power Research Institute Nanjing 210061 China
[3]State Grid Jiangxi Electric Power Company 330077, China

Abstract. Solar photovoltaic power generation is the power generation using solar cell module converting sunlight into DC electric energy. In the paper an equivalent model of solar photovoltaic power generation system is built in RTDS. The main circuit structure of the two-stage PV grid-connected system consists of the DC-DC, DC-AC circuit. The MPPT (Maximum Power Point Tracking) control of the PV array is controlled by adjusting the duty ratio of the DC-DC circuit. The proposed control strategy of constant voltage/constant reactive power (V/Q) control is successfully implemented grid-connected control of the inverter when grid-connected operation. The closed-loop experiment of islanding protection device of photovoltaic power plant on RTDS, verifies the correctness of the simulation model, and the experimental verification can be applied to this type of device.

1. Introduction

With the continuous growing of power demand, the supply of global energy sources becomes increasingly tight, and the use in solar energy is drawing more and more attentions currently. TO widely use the solar energy in distributed generation can not only cut down the consumption of traditional energy, but also reduce environmental pollution [1-5].

This paper proposes to use V/Q control mode to realize the grid-connected control of inverter, which will not need to use the complicated phase-locked loop control technology, and under the circumstance of constant electric current at the DC side of inverter, just adjusting DC busbar voltage and controlling the steady state reactive value at the side of grid to be 0 can guarantee the grid-connected operation requirement that the power factor of inverter is 1. The MPPT control of photovoltaic array is also one part of the control system in this paper, which adopts the constant voltage tracking method to track the MPP—Maximum Power Point quickly, and through the connection between great inductance and DC side of inverter, it can guarantee that the DC input by inverter is smooth, the simulation results indicate reasonable system structure, and the control system has high performance. Through the experiment of islanding protection device in photovoltaic power station, the photovoltaic grid-connected simulation system is verified to be able to use in the research of grid-connected simulation in large-scale photovoltaic power station and the closed-loop experiments of all kinds of devices.

IOP Conf. Series: Materials Science and Engineering **207** (2017) 012079 doi:10.1088/1757-899X/207/1/012079

2. Main topology of photovoltaic power generation system

The photovoltaic grid-connected system usually consists of three parts: photovoltaic array, photovoltaic inverter and power grid. The topological structure of grid-connected is shown in Fig 1.

Among them, the photovoltaic array consists of several photovoltaic modules, and the photovoltaic array output power is equivalent to the power in series-parallel connection of the component photovoltaic modules. The photovoltaic grid-connected inverter is to convert the DC output from solar energy cells into AC meeting power grid requirements and input it into the devices of power grid, which is the core of the energy transformation and control of grid-connected photovoltaic system.

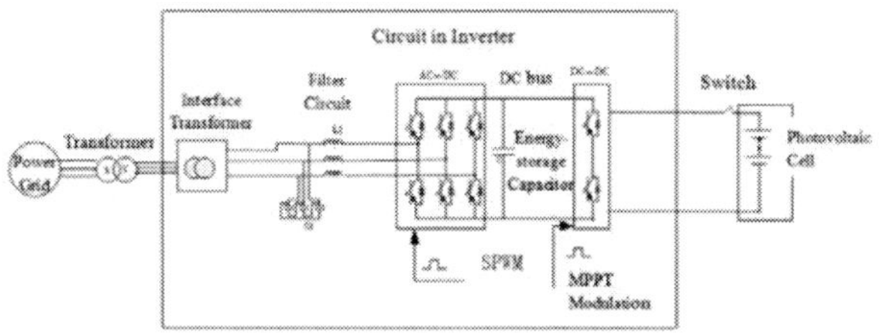

Fig 1. The topological graph of photovoltaic grid-connected system

This figure of grid-connected topology structure is the photovoltaic grid-connected inverter system for two-stage conversion. A DC energy-storage capacitor with enough capacity is usually installed between the pre-DC-DC convertor and post-DC-AC convertor, between 10 and 100e3 μF generally. This DC energy-storage capacitor can also have decoupling function in the control of pre-convertor and post-convertor, at the same time of energy change in the pre-convertor and post-convertor when buffering. Therefore, the control strategy on the pre-convertor and post-convertor can generally be researched independently [3].

The photovoltaic array at DC side and DC-DC converter circuit in this paper adopt the interfacing technology with double controlled source in connection, while at three phase AC side, it is connected between interface transformer and load. In this paper, under the V/Q control mode of inverter, it is supposed that: firstly, the photovoltaic array shall be working under the STC (Standard Test Conditions).STC refers to the input condition with the external temperature 25℃ and the illumination intensity 1000 W/m^2 [1-2].

3. Design of Control Circuit

The energy-storage capacitor C in Figure 1. Divides DC-DC converter circuit and DC-AC converter circuit into two parts, and their control designs can be carried out separately. The photovoltaic system controlled circuit in this paper is divided into two parts, the first part DC-DC convertor mainly realizes the MPPT control in photovoltaic cell array; the second part DC-AC convertor adopts V/Q control mode, and realizes the output voltage and frequency stabilization of inverter. There are two basic control goals in this part: firstly, to keep DC voltage stabilization in the front end; secondly, to realize grid-connected current control, or even need to adjust the reactive power of power grid according to instructions [7-8].

3.1. Controller design in DC-DC converter circuit (Photovoltaic cell array MPPT control)

In order to increase the utilization efficiency of solar energy, we shall try our best to make the solar energy cell array run at MPP under STC, which is the basic requirement for solar energy cell to run. Thus, we shall carry out MPPT control on photovoltaic array [5][9]. The control block diagram is shown as Fig 2.

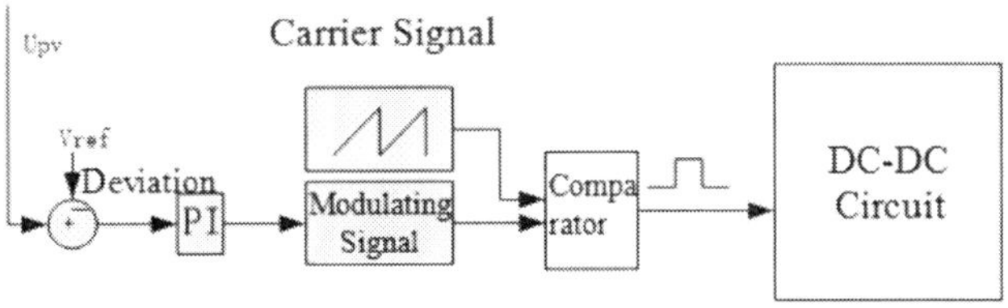

Fig 2. Photovoltaic array MPPT control block diagram

Among them: U_{PV} is the voltage value at the end of photovoltaic array; V_{ref} is the set value of the voltage value at the end of photovoltaic array.

The adoption of DC-DC circuit to realize MPPT control in solar energy cell array is a kind of the most common control modes. It uses PI regulator to control the duty ratio of switch tube D, and set reasonable PI control parameters. By comparison between the sampled signal of voltage at the end of photovoltaic array and the reference value of voltage, the modulating signal will be obtained after the error is adjusted by PI regulator. The signal is just the duty cycle of DC-DC circuit switching tube, and the carrier signal amplitude is set as 0-1. It is stipulated that, when the modulating signal is greater than or equal to the carrier signal, the switch tube will be connected, and otherwise it will be closed. In this way, it guarantees that the modulating signal amplitude is just the duty ratio D of switching tube. If there is any requirement on the abrupt change in photovoltaic array power, the reference value voltage can be changed to realize power adjustment. This is a easy way in adjustment, and the constant voltage tracking method is easy to use under STC.

3.2. The controller design in DC-AC converter circuit

The single-phase circuit principle in the operation of connected grid for photovoltaic power generation is shown in Fig. 3 as below. The power grid is considered as the voltage source with infinite capacity, and U_G, i_G, Z_G are respectively the voltage, current and resistance of the power grid, the inverter is equivalent to current supply, the output current is i_o, and the voltage, current and resistance in the equivalent load are U_L, i_L and Z_L respectively.

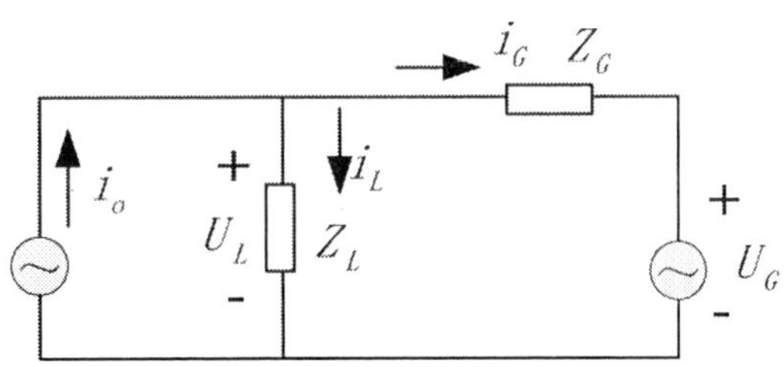

Fig 3. Equivalent circuit of single-phase grid-connected

Because of the great inductance and smooth current connected at the side of DC busbar, it is set to be I_0. If DC busbar voltage V is controlled, it is equivalent to control the active power P input into the inverter.

$$P=U*I \tag{1}$$

It can be obtained from the above Fig.3 that

$$,U_G = U_L + Z_G * i_G \tag{2}$$

Because the line impedance Z_G of power grid is little, it can be obtained that:

$$U_G = U_L \tag{3}$$

As can be seen in Formula (3), the output voltage of inverter is decided by the voltage of power grid. The active power P output from inverter and reactive power Q can be also expressed as:

$$P = U_G * I_0 * \cos(\angle \dot{U}_G - \angle \dot{I}_0) \tag{4}$$

$$Q = U_G * I_0 * \sin(\angle \dot{U}_G - \angle \dot{I}_0) \tag{5}$$

In order to input the high-quality electric energy to the power grid, the reactive power shall be reduced to the greatest extent in the power output from the inverter, and make the power factor be close to 1, namely require Q=0. Because P input into power grid is positive, it can be obtained according to Formula (4) and Formula (5) that, $\angle \dot{U}_G = \angle \dot{I}_0$, which can achieve the goal of the same phase.

According to the above analysis, control the reactive power Q=0, which can make the power factor at the side of grid be 1.

The V/Q control mode adopted in this paper uses the principle of SPWM (Sinusoidal Pulse Width Modulation), which can realize the control of power factor at the side of grid well, and guarantee the grid-connected waveform quality. The V/Q control block diagram is shown in the below Fig. 4.

Among them: V is DC busbar voltage; V_{ref} is DC the set value of busbar voltage; Q is the reactive power transmitted into the power grid; Q_{ref} is the set value of reactive power; $Angle$ is the phase position of sinusoidal modulated wave (rad); Mag is the amplitude of sinusoidal modulated wave; the carrier frequency is 3kHz, with amplitude range of 1,double polar modulation. SPWM modulated wave frequency adopts the grid frequency f, and guarantees the requirement of the same frequency in grid-connected waveform and power grid.

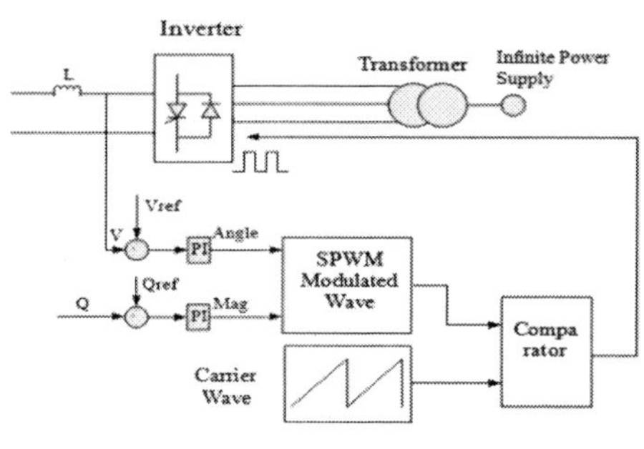

Fig.4 Circuit diagram of V/Q control

4. The experimental verification of islanding protection device in photovoltaic power station

According to the above discussion, build the islanding protection simulation model in the photovoltaic power station on RTDS/RSCAD, as shown in below Fig.6.

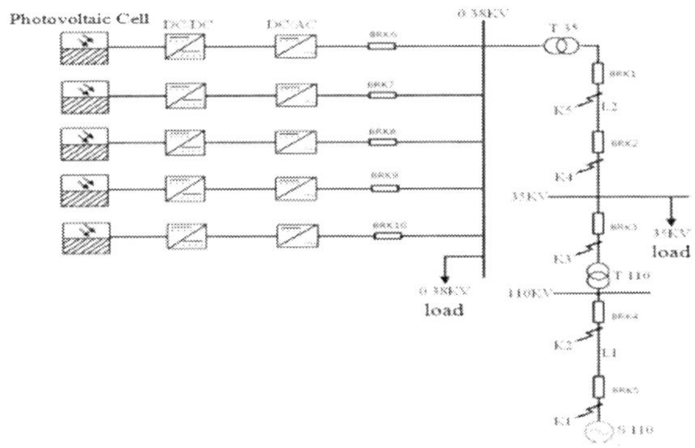

Fig.5 The experiment of photovoltaic power plant islanding protection device

Fig.5 is the system chart of the experiment of islanding protection device in photovoltaic power station. DC/DC and DC/AC are the two-stage conversion of DC-DC,DC-AC,L1 is 110kV circuit,L2 is 35kV circuit,T35 and T110 are both transformers,S110 is 110kV large power grid system. K1 to K5 in the figure are fault points, BRK1-BRK10 are breakers, and BRK1 is the grid-connected switch of photovoltaic system. This photovoltaic power station can be equivalent to 5 photovoltaic power generation models with the same capacity. The number of modules in series of every photovoltaic array N_s=200, the number of modules in parallel N_p=100, the number of cells in every module N_c=36. The open-circuit voltage value of single module under STC is V_{oc}=21.7V, the DC output voltage at MPPT calculated according to the theory of photovoltaic power generation is about 3.4kV, which value is the reference voltage of PI regulator V_{ref} . The maximum power of every equivalent photovoltaic array under STC is about 1MW. The power grid connects with LC filter circuit by interface transformer, and the filter capacitor among them usesΔ connection way, the model of interface transformer is Yd11.The major parameters in the system are shown in the below Table 1.

Tab.1 The main parameter of photovoltaic system

Parameters	Numerical Valuea
DC capacitance C	80e3 μF
Inductance L in DC-DC circuit	0.01H
AC filter inductance L2	0.0005H
Filter capacitor Cf	10uF
Changes in small-step and large-step interface transformers	380V/380V

All of the photovoltaic modules shown in the above Fig.5 adopt the model building proposed in this paper. In this control model, the PI regulator parameters in DC-DC circuit are set as below: proportionality coefficient K_p=0.5; integral coefficient K_i=10. In DC-AC circuit, proportionality coefficient K_p=0.2; integral coefficient K_i=10. The simulation waveform of single photovoltaic modules is shown in Fig. 6 as below.

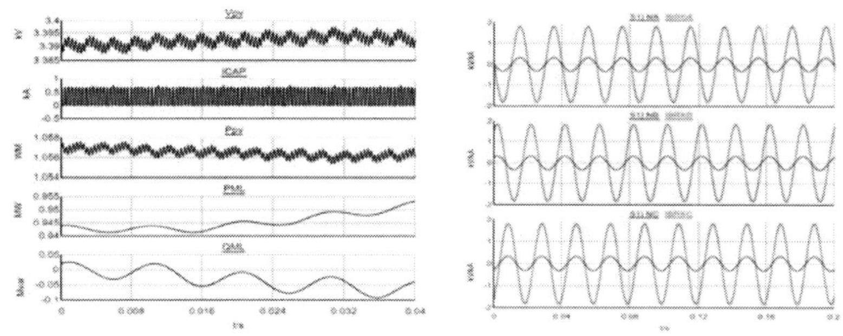

Fig.6 The waveform of single PV array voltage, current and power simulation

In the above Fig.6, V_{pv} is the actual value of voltage at the end of photovoltaic array; I_{CAP} is the actual value of current at the end of photovoltaic array; P_{pv} is the actual value of the output power of photovoltaic array. PML is the actual value of the active power of input grid, and QML is the actual value of the reactive power of input grid. NA, NB and NC are 380V busbar grid-connected point voltage of the photovoltaic modules, and IBRKA, IBRKB, IBRKC are the electric current of every photovoltaic modules at the grid-connected location. The total harmonic contents of three-phase current (in consideration of the first seven times of harmonic waves) are respectively: 0.54%, 0.49%, 0.53%, so the harmonic contents are very little. Therefore, in the above-mentioned photovoltaic modeling approach, the output power is stable, which can be used in the application of the simulated engineering of large-scale photovoltaic power station.

This experiment connects the islanding protection device in photovoltaic power station into the three-phase voltages at the side of grid (380V of busbar voltage) and the grid-connected point (the exit of 35kV boosting transformer circuit) of the inverter in the photovoltaic new energy field and the position signals of Switch BRK1 to Switch BRK5 in the system. The actuating signals of switch BRK1 for cutting down or connecting with the grid in the islanding protection device will be connected with oscillograph and simulation system respectively.

According to the judging logic of the islanding protection device in photovoltaic power station, when there is any fault-free switch trip at K1 to K5, interphase in two-phase, two-phase grounding and three-phase fault, the grid-connected switch will all be cut down. The below Fig.7 is the experimental recorded waveform of ABG at K2 in the near-end of 110kV circuit.

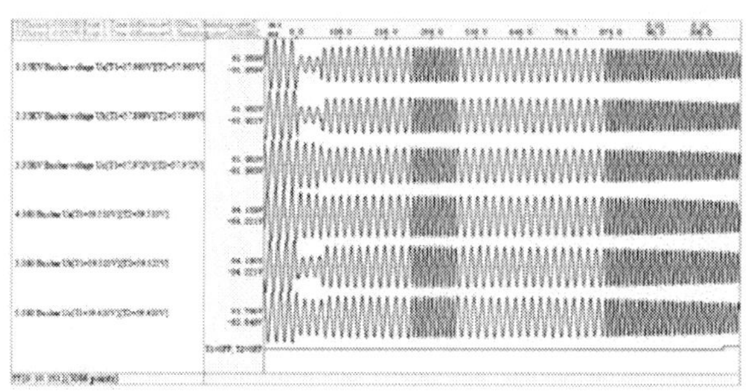

Fig.7 The Oscillography of 110kV line near K2 point ABG fault

In Fig.7, 1-T-BRK1 is the actuating signal of islanding detection device BRK1.When there is any fault in K2, the islanding detection device can detect that the corresponding voltages in 35kV busbar and 380V busbar reduce, and send out the grid-connected switch trip order after the delayed period of 300ms. The simulated grid-connected voltage in photovoltaic power station can output ideal voltage waveform under different operation conditions. The tested device can run reliably under the environment of photovoltaic grid connected, and the action of device is in conformity with the theoretical judgement. In the meantime, it also verifies the simulation model of photovoltaic grid connected built on RTDS can apply to the experiment of photovoltaic grid-connected device.

5. Conclusion

The photovoltaic grid-connected system uses RTDS/RSCAD in modeling and simulation, adopts V/Q control mode to realize the control in photovoltaic grid connected, simplifies the design in control system, and can realize the adjustable power factor well with good control effect. Through the verifications of all experiments in the closed loop of islanding protection device, the photovoltaic grid-connected model built by RTDS can apply to engineering practice.

Acknowledgments

NING Zhou (1982-), Female, Senior Engineer, is occupied in the safety and stability of large power grid and simulation analysis at present.
JIA Yang (1980-), Female, Senior Engineer, is occupied in the simulation research on electrical power system at present.
ZHENG Cheng (1979-), Male, Senior Engineer, is occupied in the safety and stability of large power grid and simulation analysis at present.
BO Chen (1986-), Male, Senior Engineer, is occupied in the simulation research on electrical power system at present.
YONG Chun Su (1973-), Male, Senior Engineer, is occupied in the safety and stability of large power grid and simulation analysis at present.
ZHAN Shu (1977-), Male, Senior Engineer, is occupied in the simulation research on electrical power system at present.
JIN Zou (1978-), Male, Senior Engineer, is occupied in the simulation research on electrical power system at present.

References

[1] ZHAO Zheng-ming, LIU Jian-zheng, SUN Xiao-ying, et al. Photovoltaic Power Generation of Solar Energy and its Application [M]. Beijing: China Machine Press, 2005.
[2] ZHANG Xing, CAO Yun-xian, et al. Grid-connected Photovoltaic Power Generation of Solar Energy and its Inverting Control [M]. Beijing: China Machine Press, 2010.
[3] ZHAO Zheng-ming, LEI Yi, HE Fan-bo, et al. Overview of Large-scale Grid-connected Photovoltaic Plants [J].Automation of Electric Power System 2011, 35(12): 101-107
[4] LIU Dong-ran, CHEN Shu-yong, MA Min, et al. A Review on Models for Photovoltaic Generation System[J]. Power System Technology, 2011, 35(8): 47-52.
[5] ZHI Teng-fei, WANG Jin-mei, ZHOU Chen, et al. Research on Conrtol Strategy of Photovoltaic Power Plant Inverter Based on Optimized Way of Photovoltaic Module Operation. [J] Electrical Measurement & Instrumentation. 2013(10):59-63.
[6] YE Lin,YANG Ren-gang,YANG Ming-hao et al. Real time digital simulator and its applications [J]. Electrotechnical Journal 2004(5): 49-52.
[7] REN Li-ping, DONG Xing-lv, YU Yue-hai et al. Based on RTDS Solar Cell Modeling and Maximum Power Point Tracking Control[J] Jiangsu Electrical Engineering. 2011, 30(5): 17-19.
[8] SU Li-ping, CHEN Kan, Li Guo-jie et al. Real-time simulation study of photovoltaic grid-connected system by RTDS[J]. Power System Protection and Control 40(15)110-115.
[9] YANG Fan, PENG Hong-wei, HU Weibing, et al.The study of solar cell maximum power point tracking technology. [J] Electrical & Electronics 2008, 31(4): 1081-1084.
[10] FAN Xiao-cong, FANG Jiang-an, Design of Grid-conected photovoltaic system based on fuzzy-PID control[J] Journal of Mechanical &Electrical Engineering. 2010, 27(9): 104-107.

IOP Conf. Series: Materials Science and Engineering **207** (2017) 012080 doi:10.1088/1757-899X/207/1/012080

Research on Electromagnetic Force Distribution and Vibration Performance of A Novel 10/4 Switched Reluctance Motor

Ziyu Fu[1, a], Xinyu Wang[2, b] Cheng Cao[3, c] Meng Liu[3, d] and Kangxi Wang[3, e]

[1]School of Electrical Engineering & New Energy, Three Gorges University, Yichang 443000, China;

[2] School of Electrical Engineering & New Energy, Three Gorges University, Yichang 443000, China.

[3] School of Science and Technology, Three Gorges University, Yichang 443000, China.

[a]839294952@qq.com, [b]727040645@qq.com, [c]843787885@qq.com,

[d]517038505@qq.com [e]2059998639@qq.com

Abstract. Radial electromagnetic force is one of the main reasons causing the vibration and noise of the switched reluctance motor. Based on this, the novel structure of 10/4 pole switched reluctance motor is proposed, which increases the air gap flux and electromagnetic torque by increasing the number of stator poles. In addition, the excitation current of the stator winding is reduced by early turn-off angle. Through the finite element modelling analysis, the results show the superiority of the new type of switched reluctance motor. In the end, the vibration characteristics of the conventional motor and the new motor are compared and analysed, and the effect of the structure of this new type of switched reluctance motor is verified.

1. Introduction

Switched Reluctance Motor (SRM or SR Motor for short) is a structure of doubly salient .It is a new kind of variable speed drive system. When it is compared with the traditional motor, SRM has the advantages of simple structure, low cost and good regulating performance [1-6]. Since the 1990 s, it has been increasingly applied in electric vehicle, mine, oilfield, textile machinery and other industrial areas [7-9]. However, it has been found in the study of SRM that the serious problem of the noise and vibration has limited the popularization and application of it in some occasions [10-11]. The study shows that the effect on the stator radial force is the main cause of noise and vibration [12-15]. It has designed a kind of semi-anechoic chamber to isolate noise in literature [16].However, the use of the semi-anechoic chamber have limitations which cannot fundamentally solve the problem of vibration. In literature [17], In order to reduce the noise and vibration, the gear of switched reluctance motor is designed as inclined structure to achieve the effect of flatting radial electromagnetic force. But it has higher requirements for the mechanical design .In literature [18], Based on the theory of magnetic permeability which can decrease the amplitude of radial force, the rotor has been set no filler ring holes to reduce the vibration and noise. But the round hole would make the stress concentration problem obviously which has bad effect to the rotor structure. In literature [19], to achieve noise reduction and vibration damping by using the method of adding auxiliary coil, however, because of the existence of strong winding coupling, the reliability of motor is reduced, fault tolerant ability is decreased. In literature [20], the axial type switched reluctance motor is proposed. It can improve the electromagnetic torque. But it has no further ways to reduce the radial force.

This paper has designed a new type of five phase 10/4 structure of switched reluctance motor. By increasing the number of stator pole to increase the air gap flux and electromagnetic torque.

2. Basic principles and innovative approaches

2.1. Basic principles

The static torque of a switched reluctance motor (SRM) can be obtained by the partial derivative of its magnetic field energy or magnetic co-energy to the rotor position angle.

$$n = \left.\frac{\partial W_m'}{\partial \theta}\right|_{i = const} \tag{1}$$

In formula 1, n is the static torque; W_m' is magnetic co-energy; θ is the position angle of the rotor; and i is the electric current of exciting winding.

When magnetic circuit of motor is not saturated, which means the magnetic circuit is assumed to be linear, then

$$W_m = W_m' = \frac{1}{2} i\psi = \frac{1}{2} i^2 L \tag{2}$$

In formula 2, W_m is the magnetic energy of the whole magnetic field; ψ is the flux linkage of conduction phase windings; L is the inductance of conduction phase windings.

And then the calculation formula of torque T is obtained:

$$T = \frac{\partial W_m'}{\partial \theta} = \frac{1}{2} i^2 \frac{\partial L}{\partial \theta} \tag{3}$$

In formula 3, i, ψ, and L are the electric current, flux linkage and inductance which conduct phase windings.

The above equation shows that the torque of SRM is obtained because terminal inductance changes with the different spatial positions and the torque is always trying to maximize the inductance. When $\frac{\partial L}{\partial \theta} > 0$, a positive torque is generated. When $\frac{\partial L}{\partial \theta} < 0$, a braking torque is generated.

Accordingly, the torque characteristic $T = f(\theta)$ can be obtained from the curve $L = f(\theta)$ which shows that the inductance L of SRM varies with rotor position angle θ, as shown in Fig 1.

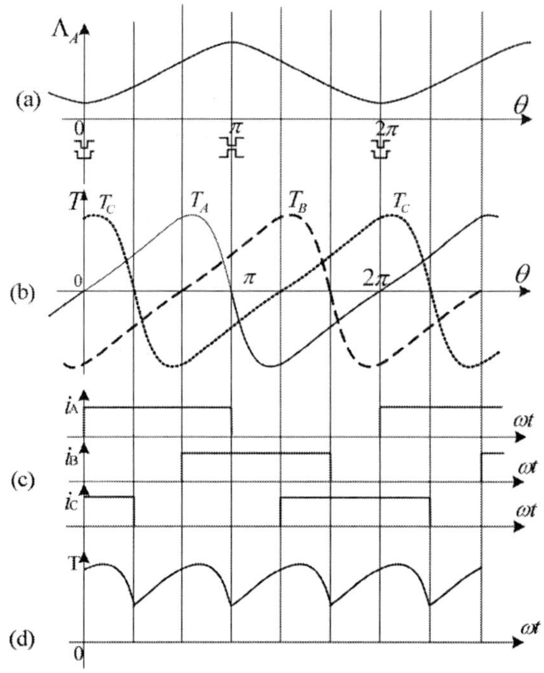

(a)A-phase permeability curves; (b) Torque characteristic;
(c) Winding conduction time; (d) Resultant torque

Fig.1. Operating principle of SRM

Fig.1. (a) is the inductance curve $L_A(\theta)$ of A-phase winding, and the inductance is the smallest when the stator polar axis is aligned with the rotor slot axis, now let $\theta = 0^\circ$;the inductance is the largest when stator polar axis is aligned with the rotor polar axis, and now set the mechanical angle $\theta = 45^\circ$ (the electrical angle is 180 degrees). On this basis, the combined torque of the final SRM (Fig. 1 (d)) is obtained according to different winding lead-on time (Fig. 1 (c)). This paper studies the distribution rule of electromagnetic force of a new type of SRM with this basic idea.

2.2. Innovative approaches

In conventional SRM, commutation often occurs when the stator polar axis and the rotor polar axis coincides with each other. And since the radial electromagnetic force reaches the maximum value at this moment, large noise and vibration of the SRM are generated when running. Given that, this paper proposes a new method for designing the structure of the 5-phase 10/4 SRM by increasing the number of stator poles to increase the air-gap flux and the circumferential electromagnetic force and by turning on and off in advance the stator windings current to reduce the radial electromagnetic force. In this way, not only electromagnet torque of SRM can be increased, but the vibration and noise of it can be reduced.

Fig. 2 shows the structural diagram and operating principle of the new five-phase 10/4 SRM in which the pole pairs of the stator poles is 5 and that of rotor poles is 2. When αA =20° is detected (Fig. 3-1 (a)), A-phase winding is turned on and the rotor is rotated by the electromagnetic force. When the rotor rotates to αD=20°(Fig.3-1(b)), A-phase winding is turned off, D-phase winding is turned on and generates electromagnetic force to drive the rotor to continue its rotation. When the rotor rotates to αB=20° (Fig.3-1(c)), D-phase winding is turned off, B-phase winding turns on and generates electromagnetic force to drive the rotor to continue its rotation. When the rotor rotates to αE =20° (Fig.3-1(d)), B-phase winding is turned off, E-phase winding is turned on and generates electromagnetic force to drive the rotor to continue its rotation. When the rotor rotates to αC=20°

IOP Conf. Series: Materials Science and Engineering **207** (2017) 012080 doi:10.1088/1757-899X/207/1/012080

(Fig.3-1(e)), E-phase winding is turned off, C-phase winding is turned on and generates electromagnetic force to drive the rotor to continue its rotation. When the rotor rotates to αA=20° (Fig.3-1(f)), C-phase winding is turned off, A-phase winding is turned on and generates electromagnetic force to drive the rotor to continue its rotation. The rotor will continue to rotate with the repetition of the process showed above.

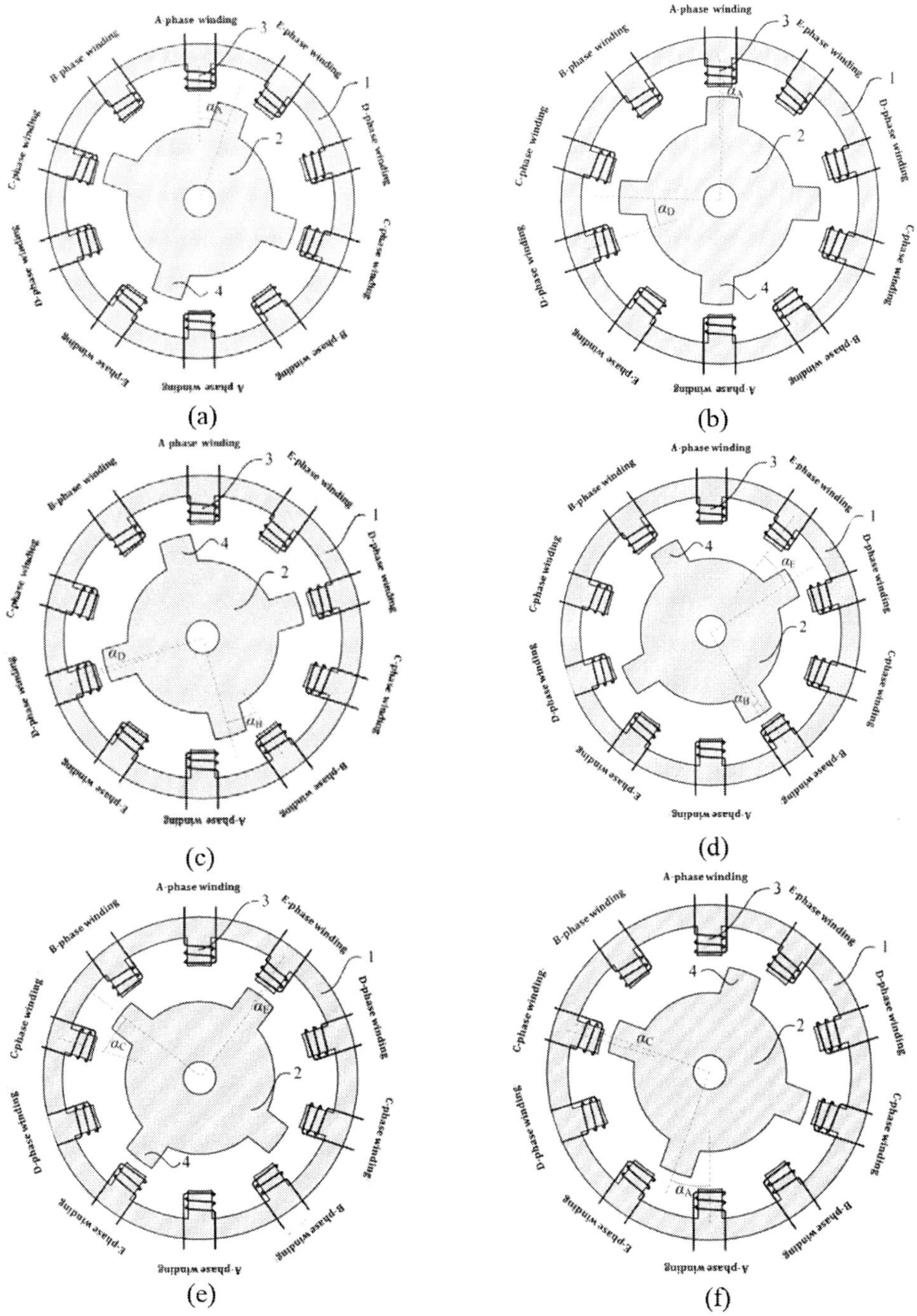

Fig.2. Operating principle of the new five-phase 10/4 SRM

3. Distribution Character of Electromagnetic Force

Whether this new type of switched reluctance motor can achieve the effect of vibration and noise reduction depends on whether it can effectively reduce the radial electromagnetic force. In this paper, the finite element model of the traditional switched reluctance motor and the new type of switched reluctance motor is established, the distributions of inductance, torque and radial electromagnetic force are analyzed respectively, and the rationality of the new switched reluctance motor is verified.

1/2 axisymmetric model is established by ANSYS simulation software; B-H curve is used for stator and rotor material, and relative permeability is 1 in other areas. The field winding applies a DC load that is applied to the different field windings depending on the rotor position. In this study, the number of pairs of stator poles of conventional switched reluctance motor is 3 and the number of pole pairs of rotor is 2. The number of pairs of stator poles of the new switched reluctance motor is 5 and the number of rotor poles is also 2. As shown in Fig.3.

The inductance of the SRM is obtained by the electromagnetic energy method, that is, the magnetic energy in the whole region is obtained by the finite element analysis, and the inductance is finally determined according to the loaded current value. The electromagnetic torque is obtained by the ANSYS macro command TORQSUM, and the radial electromagnetic force is obtained by getting the coordinate transformation after acquiring the relevant force data.

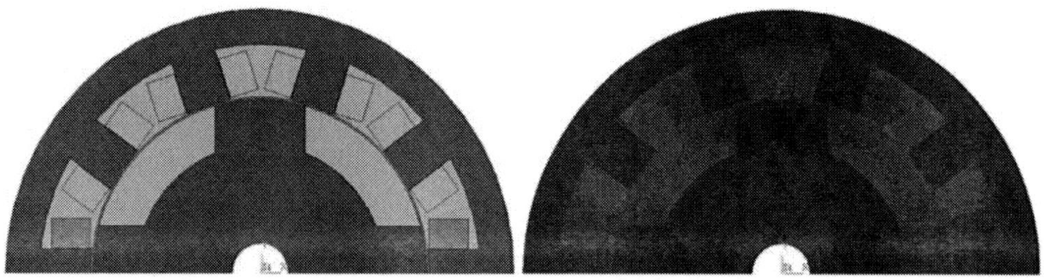

(a) (b)

Fig.3. new switched reluctance motor geometric model and grid division

According to the above-mentioned basic principle, the change law of the inductance of the switched reluctance motor in an electric cycle is analyzed, as shown in Fig.4. The maximum value of the inductance of the switched reluctance motor appears near the electrical angle of 0 and the minimum occurs at an electrical angle of approximately 180 degrees; the rapidly decreasing inductor area occurs within an electrical angle range of 30 to 90 and the rapid increase in inductance occurs at an electrical angle of 270 To 330, The above-mentioned principle shows that the electrical angle and the mechanical angle meet the condition of Nr times (the number of pole pairs of the stator); for the number of stator poles of new switched reluctance motor is 5, and the mechanical angle of one electrical cycle is 72 degrees; , The relative distance between the stator and rotor is shorter, so the inductance rising or falling area appears more steep change.

The distribution of inductance directly affects the electromagnetic torque of the switched reluctance motor. The electromagnetic torque of the switched reluctance motor is mainly determined by the spatial change of the motor inductance. The specific distribution is shown in Fig.5. The electromagnetic torque of the switched reluctance motor is almost zero in the region where the rate of change of inductance is small (electrical angle between 120 to 240 degrees); the electromagnetic torque of the switched reluctance motor is negative in the region where the inductance decreases fastest; In the region where the inductance rises fastest, the electromagnetic torque of the switched reluctance motor is positive, and the positive torque region is the normal working range of the switched reluctance motor. At the same time, the new switched reluctance motor inductance change rate is greater, so in the same excitation conditions it can obtain greater electromagnetic torque, an increase of 51.7%.

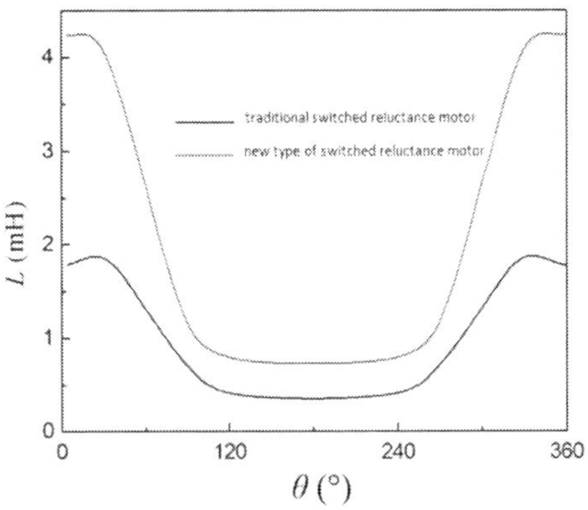

Fig.4. Inductance variation law of SRM in an electrical cycle

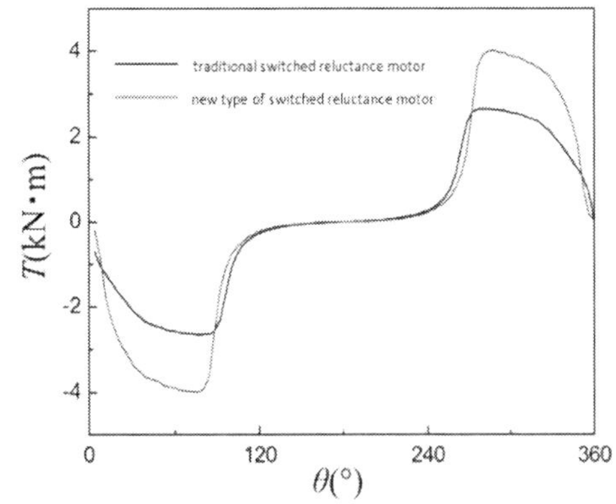

Fig. 5. Torque variation law of SRM in an electrical cycle

In order to verify the distribution of the radial force of the new SRM, the distribution of the radial electromagnetic force of SRM is analyzed in an electric cycle, as shown in Fig.6. The results show that, compared with the conventional switched reluctance motor, the amplitude of the radial electromagnetic force decreases by 45.3%, while the new SRM achieves 51.7% increase of the electromagnetic torque. The result shows the superiority of the new SRM.

Based on the analysis of the electromagnetic torque of single-phase loading, single-cycle SRM and distribution law of radial electromagnetic force, combined with the characteristics of SRM excitation, we further study the electromagnetic torque of SRM in multiphase loading and the distribution character of radial electromagnetic force. In order to more clearly reflect the difference between the electromagnetic torque and the radial electromagnetic force distribution between the traditional motor and the new motor, Fig. 7 shows the comparison between these two. Red lines represent the new motor, black lines represent the traditional motor; it can be seen that the new SRM has greater electromagnetic torque capacity and smaller radial electromagnetic force, and its overall performance is much better.

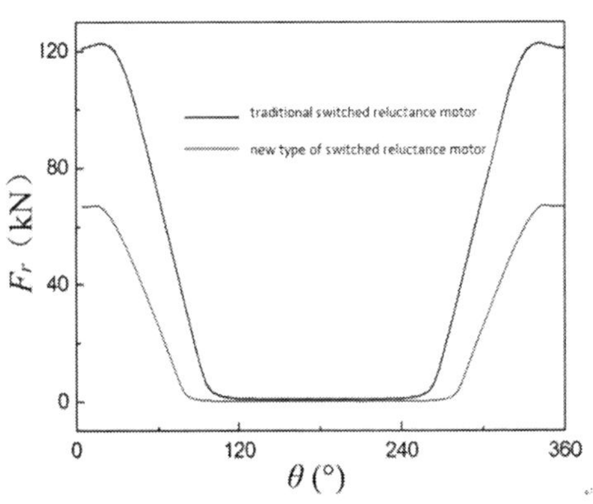

Fig.6. Radial electromagnetic force distribution of SRM in an electrical cycle

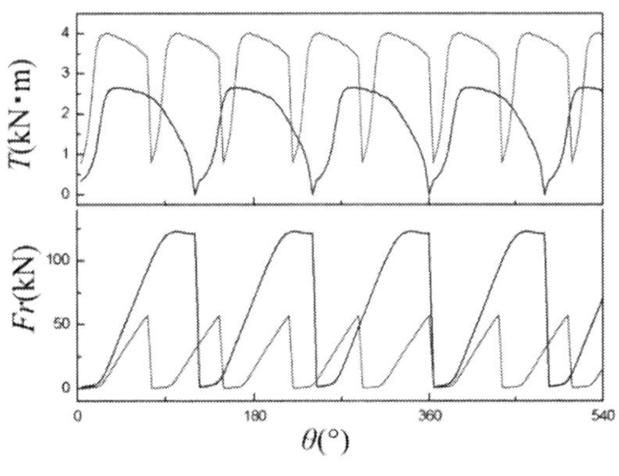

Fig.7. The contrast of two kinds of SRM electromagnetic torque and radial electromagnetic force

IOP Conf. Series: Materials Science and Engineering **207** (2017) 012080 doi:10.1088/1757-899X/207/1/012080

4. Response analysis under excitation of radial electromagnetic force

In order to verify the effect of the new SRM on vibration and noise reduction, we compare the traditional motor with the new motor and analyze the response of the traditional motor and the new motor under excitation of radial electromagnetic force. In the analysis, the radial electromagnetic force obtained from the above analysis is applied to the stator pole, and only a set of excitation windings are energized when the motor run normally. Fig.8 shows the response of traditional motors and new motors under excitation of radial electromagnetic force. For traditional switched reluctance motor, the maximum displacement of first-order, second-order , third-order response were 2.92×10-2mm,4.83×10-2mm,8.41×10-1mm respectively; For new switched reluctance motor, the maximum displacement of first-order, second-order , third-order response were 1.30×10-2mm,2.89×10-2mm,1.02×10-1mm.Taking the most easily occurring first-order response as an example, the maximum displacement of new switched reluctance motor is 55.48% less than traditional switched reluctance motor. The results have verified the structure of new switched reluctance motor on vibration and noise reduction.

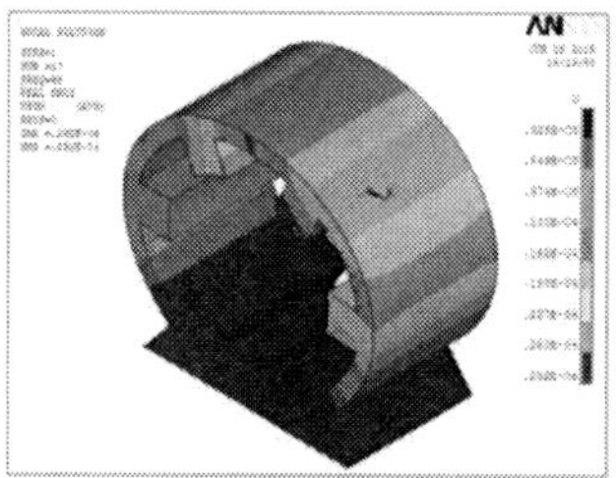

A traditional motor response

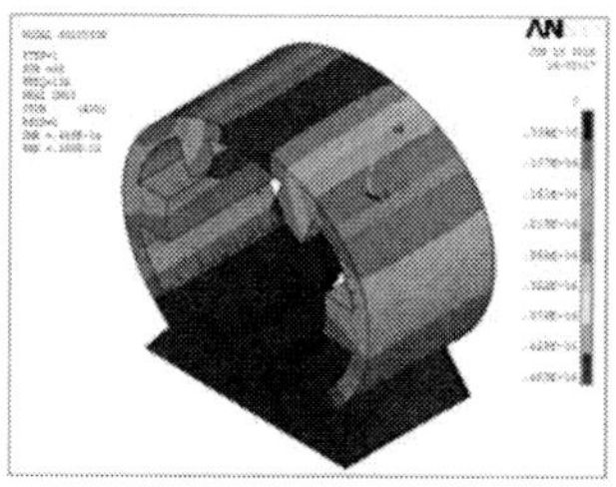

The second order response of traditional motor

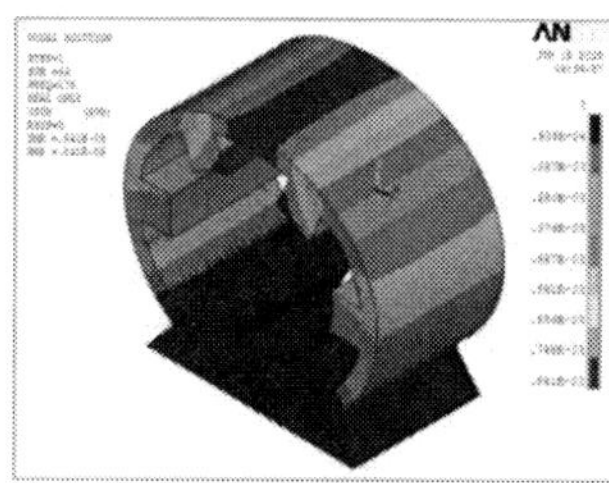

Traditional motor third-order responses

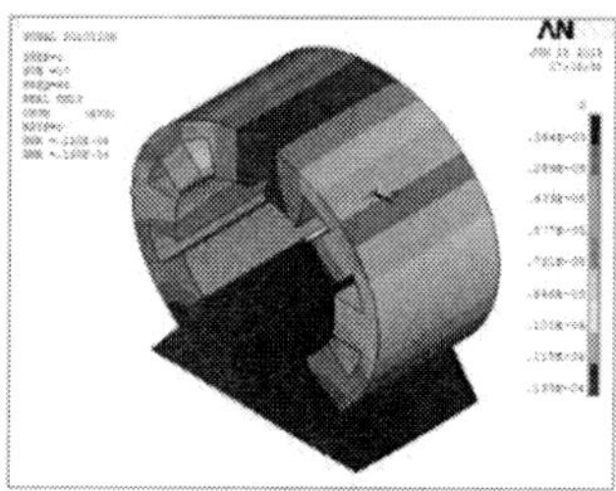

A new type motor response

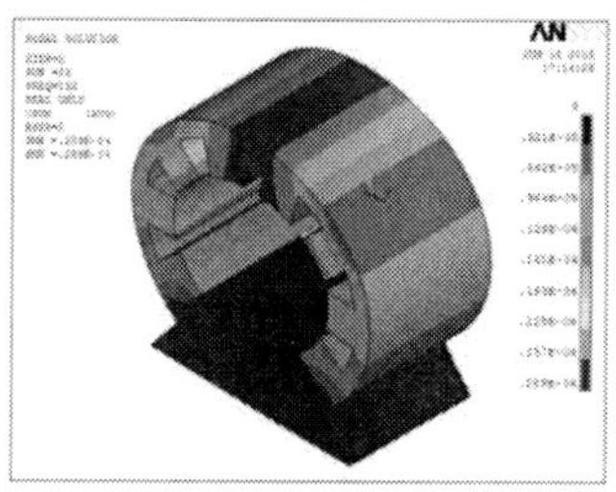

New type of motor second order response

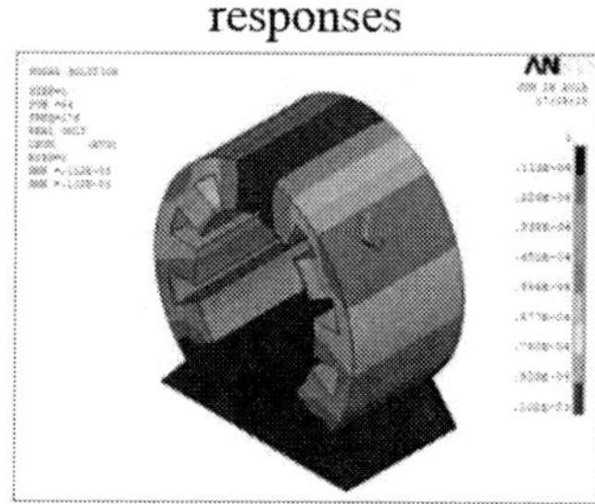

Three new motor response

Fig.8. traditional motor and a new type under the radial electromagnetic force excitation response analysis.

5. Conclusion

In this paper, the finite element method, the contrast analysis of the new type 10/4 switched reluctance motor with traditional 6/4 switched reluctance electric magnetic force distribution. Results show that under the condition of the same excitation, a new electromagnetic torque of switched reluctance motor than traditional electromagnetic torque of switched reluctance motor increases by 51.7%, while the radial electromagnetic force is decreased by 45.3%. Finally by comparing the analysis of the vibration of the traditional motor with new motor, taking the most easily occurring first-order response as an example, the maximum displacement of new switched reluctance motor is 55.48% less than traditional switched reluctance motor, proving the effect of damping noise reduction of the new switched reluctance motor. In general, the new type of switched reluctance motor has a larger electromagnetic torque output capacity and smaller radial electromagnetic force, can effectively achieve the purpose of damping noise reduction.

References

[1] Qionghua Zhan,Switched Reluctance Motor[M]. Version 1. Wuhan:Huazhong University of Science&Technology Press, 1992.
[2] Chen Yongxiao, Zhu Ziqiang, Ying Shancheng. Motor noise analysis and control[M]. Hangzhou:ZHEJIANG UNIVERSITY PRESS, 1987.
[3] Wang Honghua. Switch reluctance motor speed control technology[M]. Beijing:China Machine PRESS, 1995.
[4] Zhang Xin, Wang Xiuhe, Yang Yubo, Wei Bei.Based on the rotor teeth on both sides of the slot of the switched reluctance motor vibration suppression method research[J].Proceedings of the CSEE.2015(6):1508-1515.
[5] Sun Jianbo,Zhan Qionghua, Wang Shuanghong, Switched reluctance motor vibration noise and low torque ripple control strategy[J]. Proceedings of the CSEE, 2008, (12):134-138.
[6] Tan Ping, Qu Suichun, Ma Shiwei. Switched reluctance motor vibration noise reduction technology research[J]. Journal of Hunan University of Technology, 2010, 24:66-69.
[7] Materu P N, Krishnan R. Steady-state analysis of the variable-speed switched-reluctance motor drive[J]. IEEE Transactions on Industrial Electronics, 1989, 36(4):523-529.
[8] Matsui R, Nakao N, Akatsu K. Torque/current ratio improvement and vibration reduction of switched reluctance motors using multi-stage structure[C]// Power Electronics Conference (IPEC-Hiroshima 2014 - ECCE-ASIA), 2014 International. IEEE, 2014:1128-1134.
[9] Sun Jianbo, Zhan Qionghua, Huang Jin. Switched reluctance motor stator vibration modal analysis[J]. Proceedings of the CSEE, 2005, 25(22):148-152.
[10] Liang X, Li G, Ojeda J, et al. Comparative Study of Classical and Mutually Coupled Switched Reluctance Motors Using Multiphysics Finite-Element Modeling[J]. IEEE Transactions on Industrial Electronics, 2014, 61(9):5066-5074.
[11] Sun W, Li Y, Huang J, et al. Vibration effect and control of In-Wheel Switched Reluctance Motor for electric vehicle[J]. Journal of Sound & Vibration, 2015, 338:105–120.
[12] Cameron, D.E.;Lang, J.H.;Umans, S.D..The origin and reduction of acoustic noise in doubly salient variable-reluctance motors[J].IEEE Transactions on Industry Applications. 1992, 28(6):1250~1255.
[13] Anwar M N, Husain I. Radial force calculation and acoustic noise prediction in switched reluctance machines[J]. Industry Applications IEEE Transactions on, 2000, 36(6):1589-1597.
[14] Abbasian M, Moallem M, Fahimi B. Double-Stator Switched Reluctance Machines (DSSRM): Fundamentals and Magnetic Force Analysis[J]. IEEE Transactions on Energy Conversion, 2010, 25(3):589-597.
[15] Fiedler J O, Kasper K /, De Doncker R W. Calculation Of The Acoustic Noise Spectrum Of Srm Using Modal Superposition[J]. Industrial Electronics IEEE Transactions on, 2010, 57(9):2939 - 2945.
[16] Salvado J A, Calado M R, Espirito Santo A E. A Simple yet Effective Semi-Anechoic Chamber for the Acoustic Characterization of an LSRA[J]. Latin America Transactions IEEE, 2013, 11(5):1190 - 1200.
[17] Zou Y, Cheng K W E, Cheung N C, et al. Deformation and Noise Mitigation for the Linear Switched Reluctance Motor With Skewed Teeth Structure[J]. IEEE Transactions on Magnetics, 2014, 50:1-4.
[18] Li J, Sun H, Liu Y. New rotor structure mitigating vibration and noise in switched reluctance motor[C]// Information Networking and Automation (ICINA), 2010 International Conference on. IEEE, 2010:V2-80-V2-84.
[19] Hasegawa Y, Nakamura K, Ichinokura O. A Novel Switched Reluctance Motor With the Auxiliary Windings and Permanent Magnets[J]. Magnetics IEEE Transactions on, 2012,

48(11):3855-3858.
[20] Arihara H, Akatsu K. Basic Properties of an Axial-Type Switched Reluctance Motor[J]. IEEE Transactions on Industry Applications, 2013, 49(1):59 - 65

IOP Conf. Series: Materials Science and Engineering **207** (2017) 012081 doi:10.1088/1757-899X/207/1/012081

Study of the anti-sand sucker rod pump

Hangxin Wei*, Bingxin Lv ,Wenkui Xi ,and Peng Yi

School of mechanical engineering, Xi'an Shiyou University, Xi'an, China

*Corresponding author e-mail: whx0604@163.com

Abstract. In order to solve the problem of sand stuck in the sucker rod pump, an anti-sand sucker rod pump is designed. The anti-sand sucker rod pump includes the conventional sucker rod pump and the swirl flow device. The sand particles can be separated from the oil in the swirl flow device, so the plunger of the sucker rod pump cannot be stuck. The motion equation of the sand particles in oil is deduced. The virtual model of the swirl flow device is built in GAMBIT software. And simulation of solid-liquid two phase flow is simulated in software FLUENT. The simulation results show that the swirl flow device can realize the sand particles separation from the oil completely. So the pump can have the effect of anti-sands.

1. Introduction

The sucker rod pump is one of the important equipment's in the oil production [1]. However, the extracted fluid in many oil wells contains an amount of solid particles (sands). So the conventional sucker rod pump has some problem in the sandy wells. For example, the sands suspended in the oil will cause the wear of the plunger of the sucker rod pump if the sands fall into the clearance between the plunger and the barrel. In the worst case, the pump even is stuck and become failure.

Up to now, some anti-sand sucker rod pumps have been developed to prevent the pump from sand damage [2, 3]. However, these anti-sand sucker rod pumps cannot have a good effect [4]. In recent years, the swirl flow technology has been a research topic in the oil and gas industry [5]. So a new type of sucker rod pump based on the swirl flow is designed in this paper. The principle of the new sucker rod pump is that the swirl flow device in the sucker rod pump can separate the sand from the oil because of the centrifugal force. So the sand cannot fall into the clearance between the plunger and the barrel of the pump.

2. Structure and principle of the anti-sand sucker rod pump

The structure of the anti-sand sucker rod pump is shown in Fig.1. It includes the pump barrel, plunger, standing valve assembly, traveling valve assembly and the swirl flow device. When the sucker rod pump is used to lift the oil in the oil wells, the oil flows into the bottom of the pump through the screen pipe. Although the screen pipe can filter some sands for the pump, there are still some sands which can enter the pump.

The plunger is moved up and down reciprocated by traction of the rod string. When the plunger is moved up, the space between the plunger and the standing valve becomes larger. So the pressure is decreased and the standing valve is opened. The oil is sucked into the pump barrel. In addition, the oil pressure on the traveling valve is increased, so the valve is closed. The oil is lifted into the swirl flow device with the movement of the plunger. On the other hand, when the plunger is moved down, the space between the pump barrel and the standing valve becomes smaller, the oil pressure increases and the standing is closed. At the same time, the traveling valve is opened and the oil in the barrel is also lifted into the swirl flow device.

For the conventional pump, when there are sand particles in the oil fluid, some sands may fall down due to the gravity. If the sands enter into the clearance between the plunger and the barrel, the plunger will be abrasive, or even cause the sand plug afterward. However, if the sand sucker rod pump in this paper is used, the oil will be moved along the screw line when the oil enters into the swirl flow device. Since the density of the sand is bigger than that of the oil, the sands are centrifugally thrown toward the inside wall of the tube and the oil flow inward of the tube. Next, the sand may move up to the well head, or falling down the annular space, as is shown in Fig.1. In this way, the sand cannot go into the clearance between the plunger and the barrel. So the pump can have the anti-sands effect.

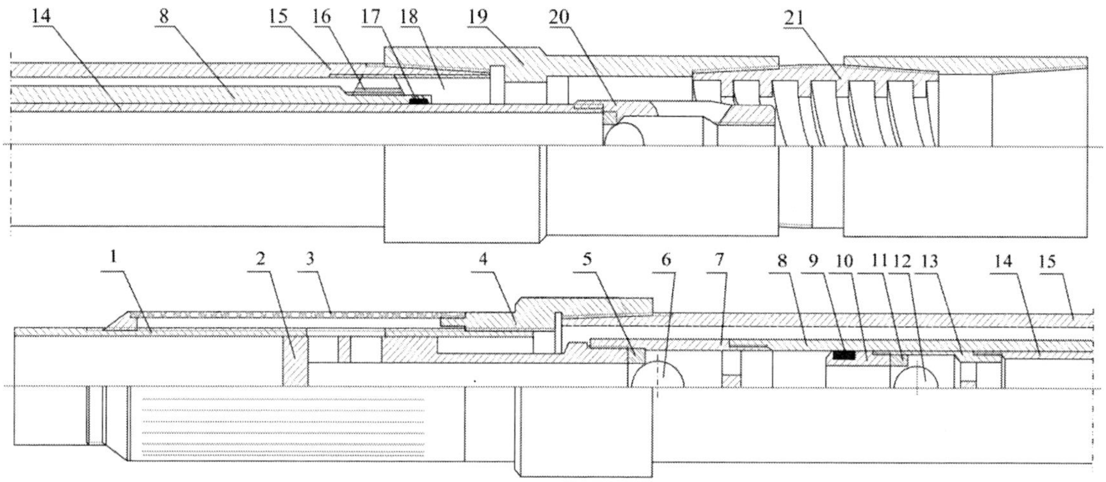

Figure 1. Structure of the sand rod sucker rod pump

1-bridge pipe, 2-bridge nipple, 3-screem pipe, 4-coupling, 5-seat, 6-ball, 7-standing valve, 8-middle barrel, 9-bottom sand abrasive block, 10-plug, 11-seat, 12-ball, 13-traveling seat, 14-plunger, 15-outer barrel, 16-connecting part, 17-upper sand abrasive sand block, 18- annular space for storage sands, 19-nipple, 20-travling valve assembly, 21-swirl flow device

3. Sand particles flow equation in the swirl flow device

The motion of the solid (sand) particles in fluid can be described in Lagrange equation. When a solid particle moves in the solid-liquid two phase fluid field, it is applied the following forces: gravity, additional mass inertia force, pressure gradient force, Magnus force and Saffman force. So the dynamical equation of a solid particle can be written as:

$$m_{\mathrm{p}} \frac{\mathrm{d}u_{\mathrm{p}}}{\mathrm{d}t} = m_{\mathrm{p}} g + F_{\mathrm{VM}} + F_{\mathrm{P}} + F_{\mathrm{B}} + F_{\mathrm{ML}} + F_{\mathrm{SL}} + F_{\mathrm{X}} \tag{1}$$

In Eq.(1), m_{p} is the mass of a particle; up is the velocity of the particle; g is the gravity acceleration; FVM is the additional mass inertia force; F_{p} is the pressure gradient force; F_{B} is the Basset force; F_{ML} is the Magnus force; F_{SL} is the Saffman force; F_{X} is the buoyancy. These forces can be written as follows:

$$F_{\mathrm{VM}} = \frac{4}{3} K_{\mathrm{m}} \rho_{\mathrm{ff}} \pi \, r_{\mathrm{p}}^{3} \frac{\mathrm{d}}{\mathrm{d}t} \left(u_{\mathrm{f}} - u_{\mathrm{p}} \right) \tag{2}$$

$$F_{\mathrm{p}} = -\frac{4\pi\, r_{\mathrm{p}}^{3} \nabla p_{\mathrm{f}}}{3} \tag{3}$$

$$F_{\mathrm{B}} = K_{\mathrm{B}} \frac{1}{\rho_{\mathrm{pp}} r_{\mathrm{p}}} \frac{3}{4} \sqrt{\frac{\mu_{\mathrm{f}} \rho_{\mathrm{ff}}}{\pi}} \int_0^{t} \frac{1}{\sqrt{t-\tau}} \frac{\mathrm{d}}{\mathrm{d}t} \left(u_{\mathrm{f}} - u_{\mathrm{p}}\right) \mathrm{d}\tau \tag{4}$$

$$F_{\mathrm{ML}} = \pi\, r_{\mathrm{p}}^{3} \rho_{\mathrm{ff}} \omega \left(u_{\mathrm{f}} - u_{\mathrm{p}}\right) \tag{5}$$

$$F_{\mathrm{SL}} = K_{\mathrm{L}} \sqrt{\rho_{\mathrm{ff}} \mu_{\mathrm{f}}}\, r_{\mathrm{p}}^{2} \sqrt{\left|\nabla u_{\mathrm{p}}\right|} \left|u_{\mathrm{f}} - u_{\mathrm{p}}\right| \tag{6}$$

Where, K_{m} and K_{L} are the coefficient; ρ_{ff} is the density of the oil; r_{p} is the radius of a solid particle; u_{f} is the slip velocity of the oil; u_{p} is the slip velocity of the solid particle; p_{f} is the pressure of the oil; ρ_{pp} is the density of the solid particle; t is time; ω is the angular velocity of the solid particle. Because the average flow rate of the oil in the sucker rod pump is low and the diameter of a sand particle is very small, F_{VM}, F_{B} and F_{X} are very small. So these forces can be omitted here.

4. Simulation of the solid-liquid two phase motion in swirl flow device

We used the CFD software to simulate the movement of solid-liquid two phase fluid in the swirl flow device. The software are GAMBIT and FLUENT. The GAMBIT software is used to build the geometry model of the swirl flow device. The FLUENT software is used to simulate the movement of the solid-liquid two phase fluid. Firstly, the swirl flow device is modeled in GAMBIT software. The model is shown in Fig.2. The swirl flow device contains two departments: swirl segment and annual passage. Then, the model is meshed, which the type of mesh is TGIRD and the space is 2mm. The inlet for motion simulation is the inlet of the screw slot and the type of simulation is Velocity. The outlet for simulation is the out of the screw slot and the type of simulation is Outflow.

When we simulated the solid-liquid two phase fluid in FLUENT, the RSM turbulent model is used because the movement of the fluid is swirl flow [6]. The algorithm of numerical solution in FLUENT is SIMPLEC. The inlet velocity of the fluid is 1.2m/s. The length of swirl segment is 250mm. The whole length of the swirl device is 600mm.

The simulation results are shown in Fig.3 and Fig.4. The solid particle trace is shown in Fig.3. We can see that the solid particles can realize the swirl flow movement. Fig.4 shows the separation process between the oil and the solid particles. There are all six cross section of the swirl device. The inlet of the swirl device is at y=0mm, and the outlet is at y=600mm. In the annular area, the blue color means that the portion of the sand particles is 0% and the red color means that the portion of the sand particles is 100%. In the first picture of Fig.4, the content of sand particles is 10% and is distributed uniform in the oil. Then the sand particles are separated gradually from the oil, which we can see it from the next five pictures. In the last picture, the sand particles are moved in the margin of the annular area and the oil is located in the center of the annular area. So it means the sand particles are separated completely from the oil.

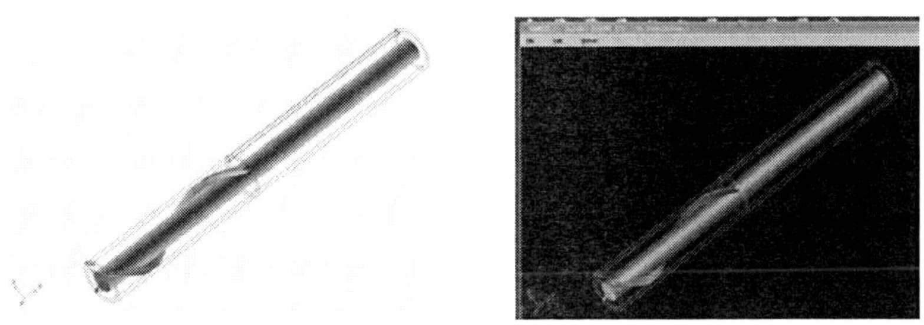

Figure 2. Virtual model of swirl flow device in GAMBIT

IOP Conf. Series: Materials Science and Engineering **207** (2017) 012081 doi:10.1088/1757-899X/207/1/012081

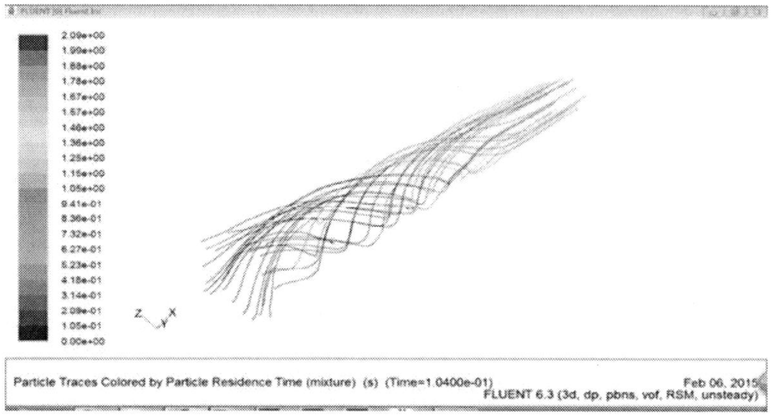

Figure. 3 Particles trace in the swirl flow device

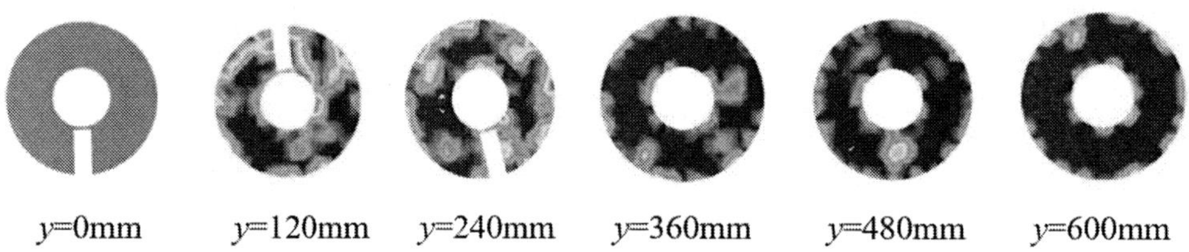

Figure 4. Distribution of the sands in the oil

5. Conclusion

Sands in the oil well can make the sucker rod pump failure. So the anti-sands sucker rod pump is designed in this paper. The principle of the sucker rod pump is based on the swirl flow. When sands in the oil pass the swirl flow device, sands will be separated from the oil because of the centrifugal force. Then sands cannot fall into the clearance between the plunger and the barrel. The swirl flow device is modeled in GAMBIT and the solid-liquid two phase flow is simulated in FLUENT software. The simulation result shows that the sand is separated from the oil completely.

Acknowledgments

This work was financially supported by the National Natural Science Foundation of China (No.51405385) and the Scientific Research Project of Shanxi Province, China (No. 2014K07-20).

References

[1] G. TAKACS, Sucker Rod Pumping Handbook, first ed., MA Elsevier Press, Holland, 2015.

[2] J. Stachowiak, "New Sucker Rod Pump Technology with Integral Screen for Sandy Wells," in SPE Production and Operation Symposium, SPE Conference Proceedings, edited by SPE. Society of Petroleum Engineers, Oklahoma, USA, 2015, pp. 390-395.

[3] A.A. Isaev, V.I. Malykhin and A.A. Shariffullin, "Investigation of Effect of Suspended Solid in Extracted Fluid on Pump Performance and Development of Means for Catching Them," in SPE Russian Petroleum Technology Conference, edited by SPE. Society of Petroleum Engineers, Moscow, Russia, 2015, pp. 1-21.

[4] G. Jinglai, L. Wei, Y. Zhaoju, Development and application of sand setting sucker rod pump, China Petroleum Machinery. 28 (2010) 47-49.

[5] S. Meher, F. Gioia, T. Catalin, Investigation of swirl flows applied to the oil and gas industry, SPE Projects, Facilities & Construction. 3(2009) 1-6.

[6] L. Jun, L. Nan, L. Jiayi, Numerical simulation research on eddy current drainage gas recovery technology, Oil Drilling & Production Technology. 35 (2013) 65-68.

Horizontal dual-point excitation and fatigue test of full-scale wind turbine blade

Zujin Pan[1, *], Jianzhong Wu[1], Xinhua Zhao[3], Yan Xie[2]

[1]School of Mechanical Engineering, Tongji University, Shanghai, China
[2]Shanghai Aeolian Wind Energy Technology Development (Group) Co, Ltd., Shanghai, China
[3]Ningbo Intelligent Manufacturing Industry Research Institute, Ningbo, China

*Corresponding author e-mail: pzj5275259@163.com

Abstract. In order to meet the requirements of large-blade fatigue testing, a horizontal dual-point linear exciter loading program is developed. In this paper, Lagrange's equation and FEM method are used to simulate the dual-point excitation theory model. The horizontal dual-point excitation characteristics of the blade are validated by the experimental method. The relationship among the dual-point exciting force, cycle time and bending moment during blade fatigue test in flap wise direction is studied. The advantages of horizontal dual-point excitation in blade fatigue test are analysed. Compared with vertical single-point excitation when the equal bending moment is reached, the exciting force and energy consumption required for dual-point excitation are reduced by 60.0% and 96.8%. Resonant frequency of the test system is improved by reducing the dead weight because of dual-point horizontal excitation. The entire blade fatigue test time is shortened by 7.4%. The use of horizontal dual-point excitation could save energy and shorten the entire fatigue test time.

1. Introduction

The reliability of wind turbine blades should be validated through fatigue test [1-4]. The single-point fatigue test of the large-blade requires high power for the exciter, moreover, bending moment on the blade applied by the single-point exciter could not fit well with target value in the blade long direction [5-6]. The dead weight is used to adjust the local bending moment on the blade during fatigue test, which will reduce the blade natural frequency and the resonant excitation force that lead to the entire fatigue tests need more energy consumption and test time. In order to meet the requirements of large-blade fatigue test, a horizontal dual-point linear exciter loading program is developed [7-8]. By changing the position and force of the two exciter, the bending moment of the blade is adjusted to fit the target value. The test equipment fatigue loss and spare parts replacement decrease because of the excitation force of each exciter is reduced by replacing the single exciter cycle loading on the blade with two exciters.Xin Lai's[9-10]study showed that there is problem of electromechanical coupling that influencing the control of vibration hammers. In this experiment, the resonant loading was carried out by two linear motor exciters. Compared with the vibration hammer, the linear motor has advantage of the orientation of loading force is in one direction, which eliminates the influence of additional torsional force on the coupling between the two linear motor exciters. The loading cycle time precision of the exciter has a great influence on the blade vibration bending moment. The cycle time deviation of linear motor loading system could control in less than 1ms. At present, the excitation force cycle loading direction during the blade fatigue test in the flap wise direction is perpendicular to the ground[11-14].The linear motor's axis force should reduce the gravity of dynamic mass, spring force, friction and others, of which the rest is the exciting force. The gravity of dynamic mass is huge in flap wise direction. If the blade is turn to 90° and exciter is horizontal, the linear exciter's axis force input is direct to exciting force. The efficiency of linear exciter will increase so much. From the current domestic and foreign documents, the way that full-scale blade cycle loading is applied by horizontal dual-point exciters during fatigue test in flap wise direction has not studied yet.

In this paper, the numerical model of fatigue loading system is established on the basis of the horizontal dual-point excitation test method during blade fatigue test in flap wise direction. The vibration characteristics of the system are simulated by finite element method. Horizontal dual-point excitation test is carried out on 40.3m blade mounted on the test bench. The results of the finite element simulation model are validated through the test, which provides a theoretical basis for precise control of dual-point fatigue loading of large-blades and optimal design of blades.

2. Theory modal of horizontal dual-point excitation

2.1. Numerical modal

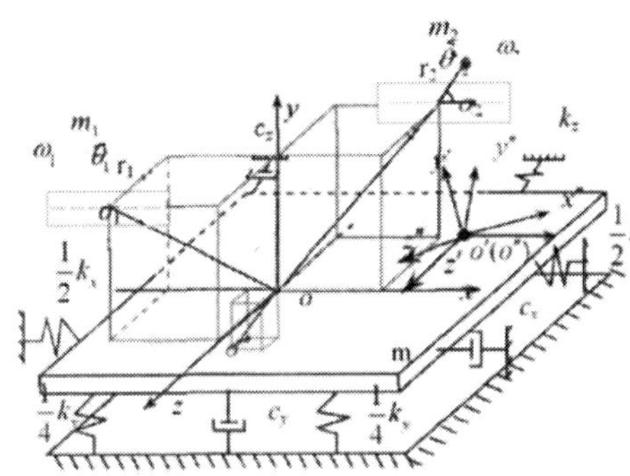

Fig 1. Dynamic model of dual-point horizontal loading system of wind turbine blade fatigue test

$$
\begin{aligned}
&(m+\sum_{i=1}^{2}m_i)\ddot{x}+k_x x+c_x\dot{x}=0\\
&(m+\sum_{i=1}^{2}m_i)\ddot{y}+\sum_{i=1}^{2}m_i\ddot{\theta}_i r_i\cos\theta_i+k_y y+c_y\dot{y}=0\\
&(m+\sum_{i=1}^{2}m_i)\ddot{z}+k_z z+c_z\dot{z}=0\\
&(ml_{01}^2+\sum_{i=1}^{2}m_i l_{i1}^2)\ddot{\varphi}_1+\\
&(ml_{01}l_{02}\sin\beta_{01}\sin\beta_{02}-\sum_{i=1}^{2}m_i l_{i1}l_{i2}\cos\beta_{i1}\cos\beta_{i2})\ddot{\varphi}_2+k_{\varphi_1}\varphi_1+c_{\varphi_1}\dot{\varphi}_2\\
&=\sum_{i=1}^{2}m_i l_{i1}\cos\beta_{i1}(\ddot{\theta}_i r_i\cos\theta_i-\dot{\theta}_i^2 r_i\sin\theta_i)\\
&(ml_{02}^2+\sum_{i=1}^{2}m_i l_{i2}^2)\ddot{\varphi}_2+\\
&(ml_{01}l_{02}\sin\beta_{01}\sin\beta_{02}-\sum_{i=1}^{2}m_i l_{i1}l_{i2}\cos\beta_{i1}\cos\beta_{i2})\ddot{\varphi}_1+k_{\varphi_2}\varphi_2+c_{\varphi_2}\dot{\varphi}_2\\
&=\sum_{i=1}^{2}m_i l_{i2}\cos\beta_{i2}(\ddot{\theta}_i r_i\cos\theta_i-\dot{\theta}_i^2 r_i\sin\theta_i)
\end{aligned}
\quad (1)
$$

The dynamic model of the dual-point fatigue loading test system is shown in Fig.1.In order to simplify the system, here it is assumed that the damping force and elastic force of the blade in each direction are the linear functions of velocity and stroke in the direction. As fig.1 shows, $oxyz$ is absolute coordinate system ; $o'x'y'z'$ and $o''x''y''z''$ are dynamic coordinate systems; o is the center of mass of entire system(include blade and two exciters); o_1 and o_2 are the center point position of the two masses respectively. m Is total mass include blade and two exciters; m_i is the mass of exciter i; r_i is mass stroke of exciter i; ω_i is the frequency of exciter i, θi is the phase of exciter i(i = 1, 2).

The dynamic equation of the vibration system is established according to the Varangian equation, x, y, z, φ_i, θ_i(i = 1, 2) as generalized coordinates.In equation: φ_i is the angle of $oxyz$ and two dynamic coordinate system respectively (i = 1, 2); l_{i1} is oo_i projection in the yoz coordinate plane; l_{i2} is oo_i projection in the xoy coordinate plane; l_{01} and l_{02} is oo''' projection in the yoz and xoy coordinate plane respectively; β_{i1} is the angle of l_{i1} and z-axis positive direction; β_{i2} is the angle of l_{i2} and x-axis positive direction; β_{0i} is the angle of l_{0i} and y-axis negative direction; k_x, k_y, k_z, $k_{\varphi i}$ and c_x, c_y, c_z, $c_{\phi i}$ are the blade stiffness and damping coefficient in the direction of x, y, z, φ_i ,respectively.

2.2. *Results of FEM analysis*

The stiffness and mass of the blade are assigned to the beam element properties in finite element model, and excitation forces of 1250N and 350N are applied at 16 m and 24 m away from blade root, respectively. The resonant frequency of the blade is 0.610Hz, and the amplitude of the 24m position is 1000mm. It can be seen from the phase-frequency diagram that the phases of difference positions on the blade are substantially coincident and the phase transitions from 0 ° to 180 ° at the resonant frequency.

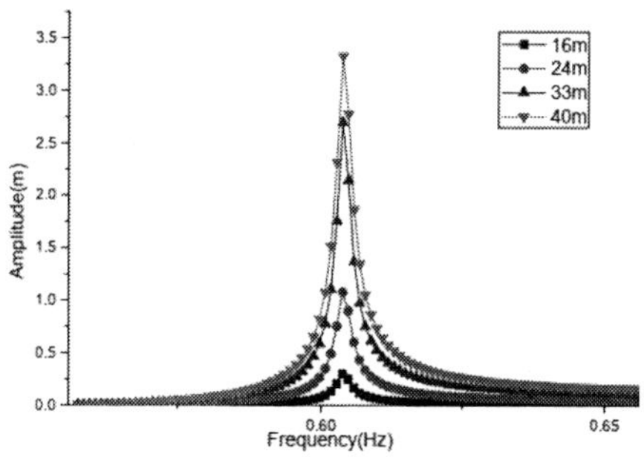

Fig2. Blade Amplitude-frequency curves of dual-point horizontal loading system

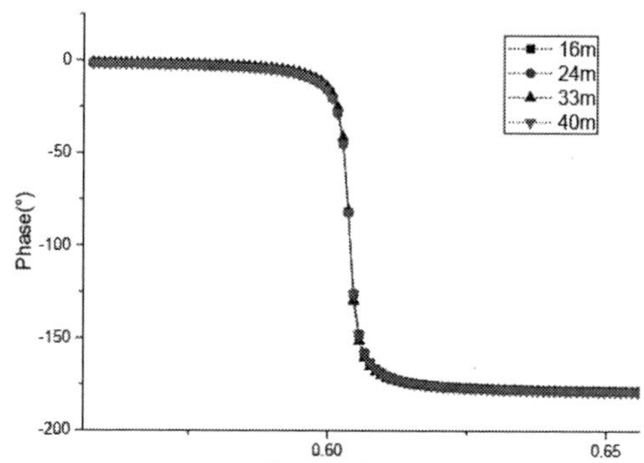

Fig3. Blade phase-frequency curves of dual-point horizontal loading system

3. Experiment results and discussion

The 40m blade with the leading edge facing up and the trailing edge facing down was mounted on the test bench during the fatigue test in the flapwise direction. The two exciters applied a horizontal excitation force on the blades by reciprocating the dynamic mass. Exciter A and B were installed at 16m and 24m away from the root. The clamp weights were 800Kg, 900Kg, 1300Kg and 300Kg at the distance of 12m, 16m, 24m and 32m from blade root, respectively.The exciting force was changed by adjusting the number of dynamic mass and stroke. The stroke ratio of the two exciters was obtained by separately exciting the blades to the target bending moment with the exciters A and B respectively, and then set the two linear motor exciters in accordance with the ratio of the stroke exciting the blade to the target bending moment.

Fig.4(a) Horizontal dual-point exciter mounted on the blade

Fig.4(b) Vertical single-point exciter mounted on the blade

3.1. Comparison of bending moment

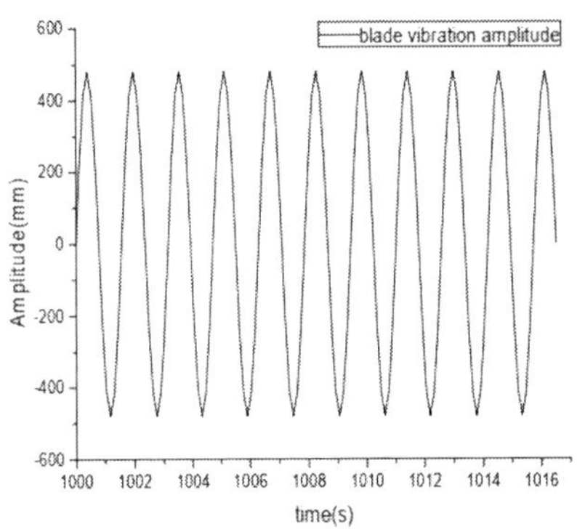

Fig.5 Amplitude-time curves in the blade 24m position

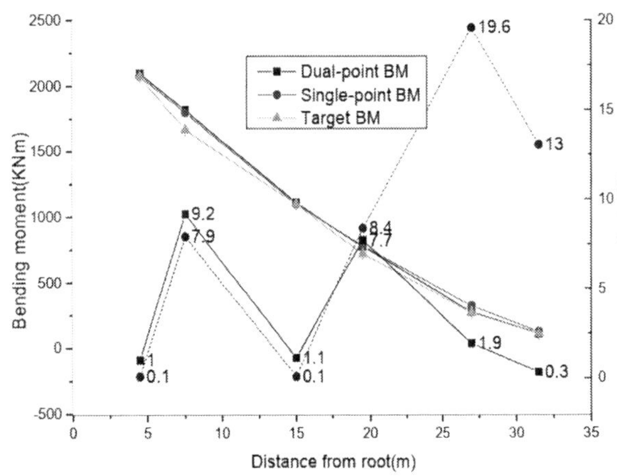

Fig.6 Comparison of single-point, dual-point exciting and target bending moment

The vibration amplitude of the blade at 24m is 960mm when blade is exciting to target bending moment. Compared with the theoretical value of 1000mm, the deviation is 40mm. The stability of the amplitude data among cycles indicates that the two exciters can co-load well. The ratio of strain to bending moment per cross section is obtained by static calibration of the blade, through which the peak-to-peak value obtained by the vibration of the blade is converted into the bending moment. The difference of bending moments among single - point excitation, two - point excitation and target bending moment is compared.

As we can see from Fig.6, the target bending moment can be included both by single-point and dual-point excitation curves. The round dots on the graph represent the difference between the single point and target bending moment, and the square dots represent the difference between the dual-point and target bending moment. The fitting deviation of single-point excitation and target bending moment reach 19.6%, while the dual-point excitation can be controlled within 9.2%. It shows that the bending moment distribution of the dual-point excitation on the blade can fit the target bending moment better.

3.2. Exciting force and energy consumption comparison

When the same bending moment on the blade is reached, the maximum exciting force of dual-point horizontal excitation and single-point vertical excitation are 1250N and 3112N respectively, the excitation force of dual-point is 60.0% less than that of single-point. Reduced excitation force is conducive to improving the life of test equipment and reducing the replacement of spare parts. Fatigue test for large-blade can be done only by adding exciter modules.

The energy consumption of one cycle of the horizontal exciter can be divided into four 1/4 cycles to calculate. The energy in the horizontal direction is the loading force and displacement integral. The energy consumption of the vertical exciter is a displacement integral that overcomes the mass gravity and exciting force. When the exciter is moving upward, force is the gravity of the mass plus the exciting force. When the exciter is moving downward, the force is the gravity of the mass minus the exciting force. In one cycle, the exciting force acts as a neutralizing force in the exciter energy consumption. The energy consumption for the exciter is equal to the integration of the mass gravity and displacement.

$$F = 8\pi sm\omega^2 \tag{2}$$

$$w_h = 4\int_0^s Fds = 4\int_0^S 8\pi sm\omega^2 ds = 16\pi m\omega^2 s^2 \tag{3}$$

$$w_v = \int_{-s}^{s} (F+G)ds + \int_{s}^{-s} -(G-F)ds = 4GS \tag{4}$$

F is linear motor exciting force;s is the dynamic mass stroke;m is mass; ω is frequency;G is the gravity of masss.

When the blade fatigue test is changed to horizontal dual-point excitation, the energy consumption is only 3.22% of the vertical single-point excitation. The linear motor needn't to work to overcome the mass of gravity in the horizontal direction. So the axial force of linear motor output is almost direct to the excitation force of the blade.

Table1. Comparison between single and dual points exciting

Item No.	Exciting direction	Mass Stroke (mm)	Dynamic mass (kg)	Frequency (Hz)	Exciting forces (N)	Single cycle work (J)	2E6cycls electricity consumption (KWh)
1	Vertical single	210	1860	0.563	3112	15312	8506
2	Horizontal A	168	780	0.616	1250	420	233
	Horizontal A	53	780	0.616	394	42	23

3.3. Resonant frequency and test cycle time

As the use of dual-point excitation, bending moment distribution on the blade can be adjusted through changing two exciting force and position, so you can reduce the use of deadweight. Blade fatigue test includes flapwise and edgewise directions, which need at least 2 million cycle times in each direction. So the resonant frequency directly affects the entire test time. Fatigue test time become a critical bottleneck in test efficiency due to the fact that the blade need to be mounted on the test bench during fatigue testing. The frequency of the system include blade and equipment is 0.616 Hz. Compared with the theoretical value of 0.610Hz, the deviation is 0.006Hz.The dual-point test saves 3 days in one direction compared with the single-point test, which reduces the entire fatigue test cycle time by 7.4%.

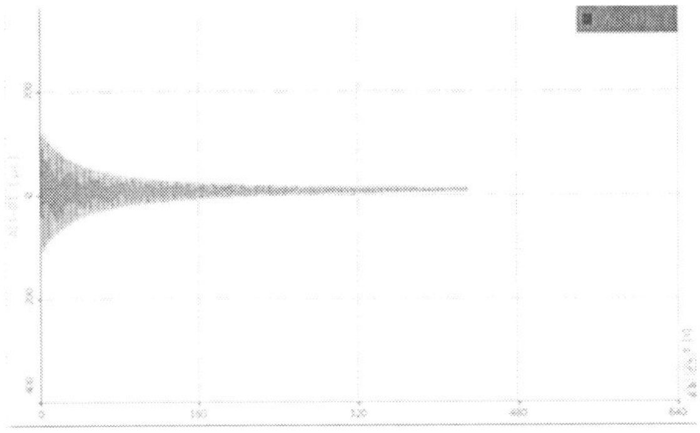

Fig.7(a) Amplitude-time diagram of dual-point exciting

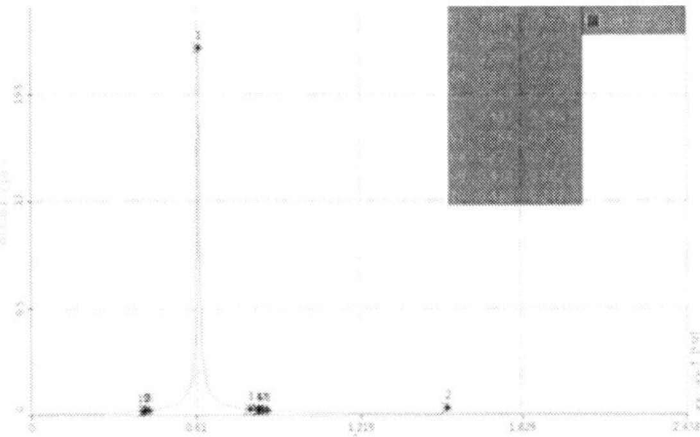

Fig.7(b) First Flap frequency of dual-point exciting

Table2. Cycle time of fatigue test

Item No.	Case	Frequency(Hz)	Cycle Time (s)	2E6cycles test time(day)
1	blade	0.853	1.172	27
2	Single-point	0.563	1.776	41
3	Dual-point	0.616	1.623	38

4 Conclusion

The horizontal dual-point excitation is used to carry out the continuous cyclic loading on the blades. The exciting forces is saved by 60.0% and the energy consumption is reduced by 96.8% during the whole fatigue test period.

The whole blade fatigue test cycle is shortened by 7.4% because of the resonant frequency increases due to the dead weight decrease during the dual-point exciting loading system.

By comparing that the frequency and 24m position amplitude deviation is 0.006Hz and 0.04m respectively, the deviation of experimental results and theoretical calculation is in line with regulation requirement.

Acknowledgments

This work was financially supported by Shanghai Aeolian Wind Energy Technology Development (Group) Co, Ltd.

References

[1] Bin Yang, Dongbai Sun.Testing, inspecting and monitoring technologies for wind turbine blades:A survey [J]. Renewable and Sustainable Energy Reviews, 2013, 22: 515–526.

[2] H.F. Zhou, H.Y. Dou,L.Z. Qin, et al. A review of full-scale structural testing of wind turbine blades [J]. Renewable and Sustainable Energy Reviews, 2014, 33: 177–187.

[3] J. Marin, Barroso.Paris, J. Canas.Study of fatigue damage in wind turbine blades [J]. Eng Fail Anal, 2009,16:656-68.

[4] HG Lee, MG Kang, J. Park. Fatigue failure of a composite wind turbine blade at its root end [J]. Compos Struct,2015,133:878-85.

[5] Lei-an Zhang, Zhongbin Wang, Weisheng Liu, et al.Vibration self-synchronous features of a wind turbine blade single-point fatigue loading test system [J]. Journal of vibration and shock, 2015, 34(13): 42-47.

[6] Lei-an Zhang,Xuemei Huang.Study of wind turbine blade vibration characteristc under single-point fatigue load driven [J]. Acta energiae solaris sinica, 2015, 36(5): 1112-1116.

[7] Gaohua Liao, Jianzhong Wu, Xin Lai. Experimental study and coupling characteristics of double hammer excitation fatigue loading system [J]. Journal of Central South University (Science and Technology), 2016, 47(11); 3692-3699.
[8] Lei-an Zhang, Xue-mei Huang, Jin-kai Yao. Vibration coupling features for a dual-point fatigue loading system of wind turbine blades [J]. Journal vibration and shock, 2015, 34(17): 82-87.
[9] Xin Lai, Jianzhong Wu, Zhen Zhang. Synchronous vibration system of multi-hammer and synchronous control strategy[J].Journal vibration and shock, 2012, 31(3); 147-152.
[10] Xin Lai, Jianzhong Wu, Bo Ruan, et al.Numerical simulation and experiments on electromechanical coupling characteristics of pile hammer synchronous vibration system [J]. Journal of Vibration Engineering, 2012, 25(2); 167; 173.
[11] HG Lee, J. Park.Optimization of resonance-type fatigue testing of a full-scale wind turbine blade [J]. Wind Energy, 2015
[12] N. Dervilis, M. Choi, I. Antoniadou.Novelty detection applied to vibration data from a CX-100 wind turbine blade under fatigue loading [J]. Modern Practice in Stress and Vibration Analysis 2012, 2012, 1:6.
[13] M.A.Eder, F.Belloni, A.Tesauro, T.Hanis.A multi-frequency fatigue testing method for wind turbine rotor blades[J].Journal of Sound and Vibration, 2017, 388: 123 - 140.
[14] H. Hosseini-Toudeshky, M. Jahanmardi, M.S. Goodarzi.Progressive debonding analysis of composite blade root joint of wind turbines under fatigue loading [J]. Composite Structures, 2015, 120: 417–427.

Effects of aerodynamic fairing on full scale blade fatigue test

Zujin Pan[1], Jianzhong Wu[1], Yuanrong Sun[2], Liu Jian[2]

[1]School of Mechanical Engineering, Tongji University, Shanghai, China
[2]Shanghai Aeolon Wind Energy Technology Development (Group) Co.,Ltd., Shanghai, China

*Corresponding author e-mail: pzj5275259@163.com

Abstract. The reliability of large blades should be verified by means of full scale fatigue test. In order to solve the problem of lack of exciting force during fatigue test in the flap wise direction, the program that aerodynamic fairing is installed in the tip of blade to reduce the air resistance is proposed. The numerical model of blade vibration and damping ratio calculation is established. The relationship between damping ratio, exciting force and amplitude is constructed by finite element method respectively. The difference of the exciting bending moment of blade and the damping ratio before and after the installation of aerodynamic fairing is compared respectively. The results show that damping ratio decreased by 27.9%. When the vibration of the blade reaches the target bending moment, the exciting force of the equipment decreases by 45.4%. It is an effective way to reduce the exciting force.

1. Introduction

In accordance with the wind turbine certification requirements, the reliability of blade should be verified through fatigue test of full scale blade [1].The cycle of dynamic load applied through fatigue test is to monitor the blade operating state. Defects in the blade during the process of production would be exposed through fatigue test, such as wrinkle, lace of adhesive paste, rich resin [2-6]. The state-of-the-art testing method for rotor blades in industry is based on resonance excitation where typically a dynamic mass excites the blade close to its first natural frequency [7-10].Due to lack of exciting force during fatigue test in flap wise direction, the blade is difficult to achieve the target bending moment. The air resistance has great impact on the damping ratio during blade fatigue test in the flap wise direction [1]. Fabio [11] et al. studied that macro-composites with star-shaped inclusions for vibration damping in wind turbine blades. Zhang [12-13] et al. investigated that tuned liquid dampers (TLDs) for vibration control and damping edgewise vibrations of wind turbine blade. In this paper, in order to solve the problem of insufficient exciting force, the program that aerodynamic fairing installed in the tip to reduce the air resistance during the blade vibration is proposed.

The relationship between damping ratio, exciting force and amplitude is established by means of the finite element model in this paper, which is the theoretical basis to adjust the excitation force. Differences of bending moment, damping ratio and exciting force are compared between before and after the installation of the aerodynamic fairing by means of the fatigue test of full scale blade. This study provides an effective way to reduce the exciting force and fatigue loss of testing equipment.

2. Numerical model of Resonance and damping ratio calculation

2.1. Vibration equation of blade

It is assumed that the wind turbine blade test system is composed of the object of mass m and light spring. The test system is subjected to the elastic force of the spring ($F = -kx$), the air resistance ($R = -cv$) which is proportional to the speed. The periodic exciting force ($H = F_0 \cos\omega t$, $F_0 = ml\omega^2$) under the action of forced vibration. According to Newton's second law to get the test system's motion equation.

$$m\ddot{x} + c\dot{x} + kx = F_0 \cos\omega t \tag{1}$$

Set $\omega_n{}^2 = \frac{k}{m}$, $2\xi = \frac{c}{\sqrt{mk}} = \frac{c}{\omega_n m}$, substituting the above equation,

$$\frac{d^2x}{dt^2} + 2\xi\omega_n\frac{dx}{dt} + \omega_n{}^2x = \frac{F_0}{m}\cos\omega t \tag{2}$$

The general solution of the second order constant coefficient linear nonhomogeneous differential equation is,

$$x = \frac{F_0}{\omega_n^2}\frac{1}{2\xi}\cos\left(\omega t - \frac{\pi}{2}\right) \tag{3}$$

The amplitude is proportional to the exciting force and inversely proportional to the damping ratio.

2.2. Damping ratio solving equation

From the vibration response signal of the blade, the modal parameters reflecting the blade are obtained. Damping ratio defines the characteristic that the energy gradually reduced impacted by the resistance of the blade in the course of movement. The blade amplitude decays exponentially during the free vibration attenuation. The attenuation response can be expressed as,

$$\begin{aligned} x(t) &= Ae^{-\zeta\omega_n t}\left[\frac{1-2\zeta^2}{1-\zeta^2}\sin(\omega_d t+\varphi) + \frac{2\zeta}{\sqrt{1-\zeta^2}}\cos(\omega_d t+\varphi)\right] \\ \omega_d &= \omega_n\sqrt{1-\zeta^2} \end{aligned} \tag{4}$$

Set the time before and after the adjacent vibration periods T are t_1 and t_2, respectively. The ζ is tiny, so we can derive from eq. (8)

$$\frac{x_1(t)}{x_2(t)} = \frac{Ae^{-\zeta\omega_n t_1 \sin(\omega_d t_1+\varphi)}}{Ae^{-\zeta\omega_n t_2 \sin(\omega_d t_2+\varphi))}} \tag{5}$$

Set $t_2 = t_1 + T, T = \frac{2\pi}{\omega_d}$

$$\sin(\omega_d t_1 + \varphi) = \sin(\omega_d t_2 + \varphi) \tag{6}$$

The ratio of the amplitude of the vibration of two adjacent periods is called the attenuation coefficient.

$$\eta = \frac{x_1(t)}{x_2(t)} = \frac{x_n(t)}{x_{n+1}(t)} = \frac{Ae^{-\zeta\omega_n t_1}}{Ae^{-\zeta\omega_n(t_1+T)}} = e^{\zeta\omega_n T} \tag{7}$$

In Eq. (11), $x_k(t)$, $x_{k+n}(t)$ is the ratio of the amplitudes of adjacent n periods, damping ratio can be expressed as,

$$\zeta = \frac{1}{2n\pi}\ln\frac{x_k(t)}{x_{k+n}(t)} \tag{8}$$

The damping ratio of the blade can be determined from the amplitude attenuation of multiple cycles.

3. Results of FEM

3.1. The target amplitude converted from the blade target bending moment.
The length of blade is 56.6m. The blade tip is to be cut off at a radius of 48m in order to reduce the air resistance. Figure 1 shows the deflection of blade applied by gravity, dead weight and fatigue target bending moment, respectively. When the blade reaches the target bending moment, the tip deflection is 6.3m. The deflection which calculated from the FEM applied by target bending moment of fatigue is set as target amplitude of blade vibration.

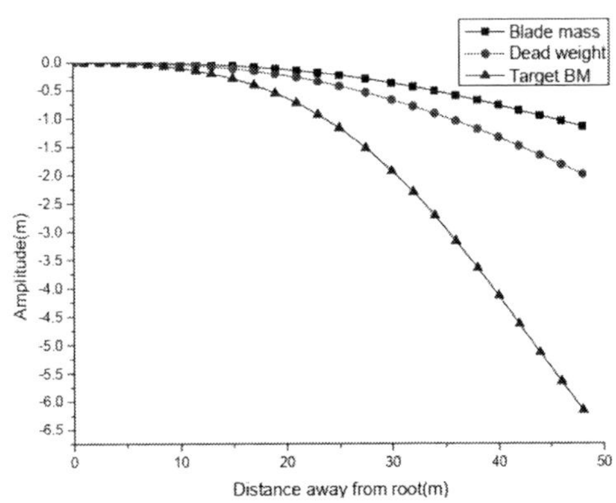

Figure 1. Deflection of blade in different cases

IOP Conf. Series: Materials Science and Engineering **207** (2017) 012083 doi:10.1088/1757-899X/207/1/012083

3.2. Effects of exciting force and damping ratio on vibration amplitude of blade

Figure 2 shows the linear relationship with a slope of 1.9E-3 between the different exciting forces and blade amplitudes at the damping ratio of 0.217% during fatigue test in the flap wise direction. When the tip amplitude reaches 6.3m, the exciting force is 3300N. The phase transitions from 0 To 180 in the resonant frequency of 0.485 Hz.

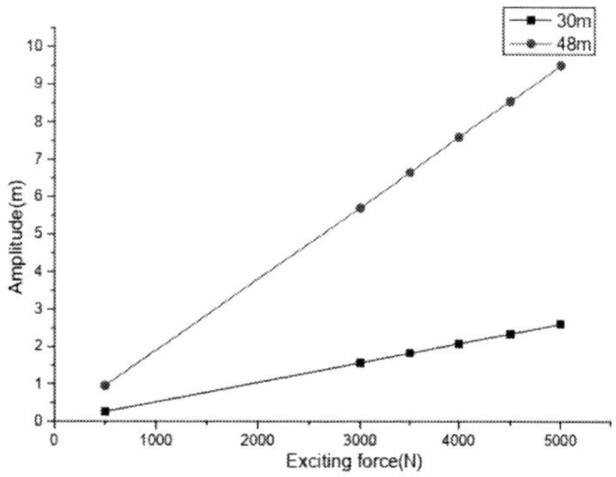

Figure 2. Amplitude vary exciting force curves

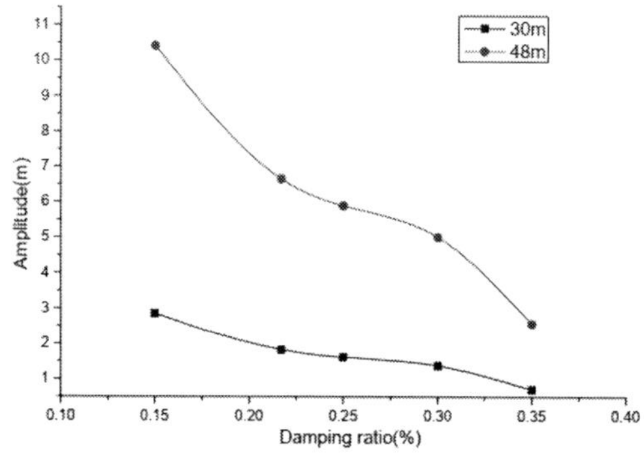

Figure 3. Amplitude vary damping ratio curves

The aerodynamic fairing installed in the tip of blade to reduce air resistance has great impact on the damping ratio during blade fatigue test in the flapwise direction. The stiffness and mass distribution of blade, shape of aerofoil, air temperature and humidity have a great influence on damping ratio, which will change during the test cycle. The relationship between the vibration amplitude of the blade under different damping ratio conditions is established at the exciting force of 3500N in figure 3, which provides theoretical basis for the adjustment of the exciting force and ensures that the blades are loaded cyclically on the target bending moment. The amplitude of the tip is 6.3m when the exciting force is 3500N and the damping ratio is 0.230%.

4. Results of experiments

The effects of aerodynamic fairing on the vibration bending moment are verified by fatigue test of full scale blade. The blade is mounted in the flapwise direction with the pressure side facing up and the suction side facing down. Exciter installed in the 25m away from the root. The aerodynamic fairing is installed in the tip from 38m to 48m areas. The strain gauges are bonded to the center of spar cap of each section of the blade. The force is applied at the tip of blade to produce bending moments in the each blade section. The coefficient that is the ratio of the bending moment to the strain in the blade section, by which the dynamic strain is transformed into the bending moment.

4.1. Vibration bending moment

The vibration bending moment in flap wise direction is the mean of the bending moment converted from strains on spar cap in the pressure side and the suction side. The target bending moments are enveloped by the vibration bending moments of the blade through adjusting the exciting force and the dead weight. Comparing the vibration bending moments before and after the installation of the aerodynamic fairing with the target bending moments in the area of the blade 4m to 36m fatigue test verification respectively, both which can envelop the target bending moments and meet the fatigue test specification requirements. Because of the weight of aerodynamic fairing, the ratio of vibration bending moments of the blade to target bending moments in the area of tip is higher than before, while in the area of root is lower than before.

Figure 4. Aerodynamic fairing and blade mounted in flapwise direction

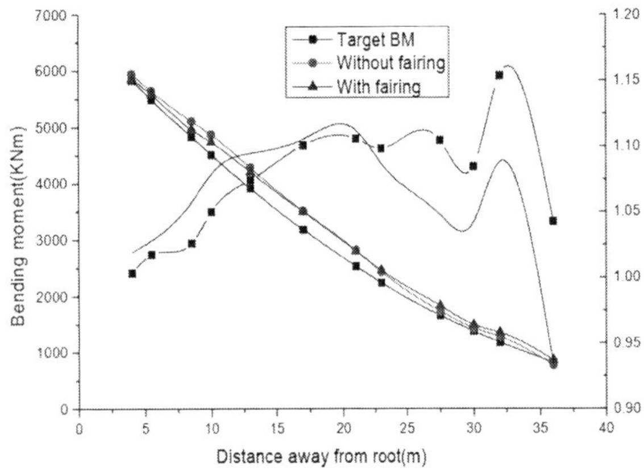

Figure 5. blade resonant bending moment with target value

4.2. Damping ratio and exciting force before and after installation of aerodynamic fairing

The damping of the blade is test in the flapwise direction, which include the structural damping and aerodynamic damping. The tip is released suddenly when the blade reach a certain amplitude, and then let the blade free vibration attenuation to within 1mm amplitude. The amplitude of the blade is acquired at a sampling frequency of 50 Hz. The damping ratio is obtained by the theoretical calculation formula (8).

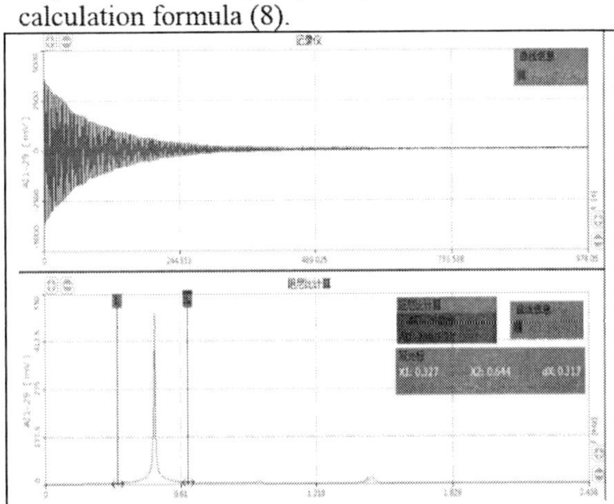

Figure 6.(a)damping ratio before installation of aerodynamic fairing

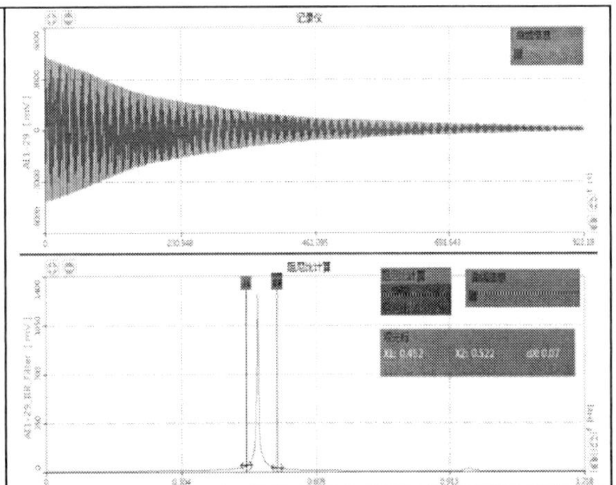

Figure 6.(b)damping ratio after installation of aerodynamic fairing

Table 1. Frequency and exciting force in flap wise direction

Case	Frequency (Hz)	Damping (%)	Exciting Position (m)	Mass(Kg)	Stroke (mm)	Exciting force (N)	Theory prediction (N)
1	0.491	0.300%	25	1936	330	6077	4185
2	0.479	0.217%	25	1890	195	3316	3300

Case 1 and Case 2 in Table 1 are the damping ratio change of the blade and the exciting force required to achieve the target bending moments before and after the installation of the aerodynamic fairing, respectively. Compared with the theoretical calculation, the excitation frequency of the test system is 0.491Hz before the installation of the aerodynamic fairing, and the exciting force is proportional to the square of the frequency. The dynamic mass stroke of the exciter is 330 mm, which is the maximum stroke the exciter can achieve. The friction that exciter has to overcome increases rapidly, so the vibration efficiency of the equipment dropped significantly. These two factors cause the actual exciting force higher than the theoretical excitation force.

The theoretical exciting force 3300N and the actual 3316N are basically consistent after the installation of the aerodynamic fairing. The mass of blade in the tip increases because of the weight of fairing is very sensitive to frequency, which lead to a decrease in blade frequency from 0.491 Hz to 0.479 Hz. Compared with the theoretical calculation of 0.485Hz, the deviation is 1.25%. The damping ratio decreased from 0.301% to 0.217%, the damping ratio decreased by 27.9%, While the exciting force decreased by 45.4%. The exciting force of the blade test system is not simply the output force of the exciter, but the result of the coupling of the exciting force of the equipment and the motion of the blade. The decrease of damping ratio after installation of the aerodynamic fairing can effectively reduce the exciting force at the same bending moment.

5. Conclusion

The vibration bending moments of the blade after the installation of the aerodynamic fairing can envelop the target bending moments, which are in the range of 100% to 115% compared with the target, and meet the requirements of the test specification.

The aerodynamic fairing does not change the structure damping of the blade, which reduces the exciting force by decreasing the vibration of the aerodynamic resistance. The damping ratio of the blade after installation of the aerodynamic fairing is reduced by 27.9%. The exciting force can be reduced by 43.3% when the target bending moments are reached. It is an effective way to reduce the exciting force.

The exciting force of the equipment and damping ratio of test system after the installation of the aerodynamic fairing are basically the same as the theoretical calculation.

Acknowledgments

This work was financially supported by Shanghai Aeolon Wind Energy Technology Development (Group) Co, Ltd.

References

[1] DNV-GL Renewables Certification. Rotor blades for wind turbines[S].DNVGL-ST-0376,2015. http://www.dnvgl.com is the officially binding version.

[2] H. Hosseini-Toudeshky, M. Jahanmardi, M.S. Goodarzi. Progressive debonding analysis of composite blade root joint of wind turbines under fatigue loading[J]. Composite Structures, 2015,(120):417–427.

[3] N. Dervilis, M. Choi, I. Antoniadou. Novelty detection applied to vibration data from a CX-100 wind turbine blade under fatigue loading[J]. Modern Practice in Stress and Vibration Analysis,2012,1:6.

[4] M.A.Eder, F.Belloni, A.Tesauro, T.Hanis. A multi-frequency fatigue testing method for wind turbine rotor blades[J]. Journal of Sound and Vibration 2017,(388):123–140.

[5] Kezhong Shi, Xiaolu Zhao, Jianzhong Xu. Research on fatigue test of large horizontal axis wind turbine blade[J].Acta Energiae Solaris Sinica, 2011, 32(8):1264-1268.

[6] Hui Li, Tong-guang Wang, Cheng chen, et,al. Structure design of wind turbine blade[D].Science press,2015.

[7] Lei-an Zhang, Xuemei Huang, Xiuting Wei. Energy consumption analysis of large wind turbine

blade fatigue loading mode[J].Renewable Energy Resources,2012,30(8):50-52.
[8] Xiaowei Cui. Fan blade fatigue testing machine design simulation research[D]. Guangzhou: East china university of science and technology,2014:49-67.
[9] Lei-an Zhang, Zhongbin Wang,Weisheng Liu,et al.Vibration self-synchronous features of a wind turbine blade single-point fatigue loading test system[J]. Journal of Vibration and Shock,2015,34(13):42-47.
[10] Gaohua Liao, Jianzhong Wu. Vibration analysis of a dual-axial fatigue loading system for wind turbine's blades test. Noise and Vibration Control, 2014,34(5):114-116.
[11] Fabio Agnese, Fabrizio Scarpa. Macro-composites with star-shaped inclusions for vibration damping in wind turbine blades[J]. Composite Structures, 2014,108(2): 978-986.
[12] Zili Zhang, Andrea Staino, Biswajit Basu,et al. Performance evaluation of full-scale tuned liquid dampers (TLDs) for vibration control of large wind turbines using real-time hybrid testing [J].Engineering Structures, 2016,126(11): 417-431.
[13] Zili Zhang, Soren R.K. Nielsen, Biswajit Basu,et al.Nonlinear modeling of tuned liquid dampers (TLDs) in rotating wind turbine blades for damping edgewise vibrations[J]. Journal of Fluids and Structures, 2015,59(11):252-269.

IOP Conf. Series: Materials Science and Engineering **207** (2017) 012084 doi:10.1088/1757-899X/207/1/012084

First-principles study of electronic, mechanical and optical properties of mixed valence SmB_6

Lihua Xiao [1, 2, 3*], **Yuchang Su**[2] **Ping Peng**[4] **and Dongsheng Tang**[1]

[1]College of Physics and Information Science, Hunan Normal University, Changsha, China
[2]School of Materials Science and Engineering, Central South University, Changsha, China
[3]School of Materials and Metallurgical Engineering, Guizhou Institute of Technology, Guiyang, China
[4]School of Materials Science and Engineering, Hunan University, Changsha, China

*Corresponding author e-mail: xiaolihua@git.edu.cn, dstang@hunnu.edu.cn

Abstract. The mechanical and optical properties, electronic structure, and theoretical hardness of mixed valence SmB_6 are calculated from first principles using density functional theory. The calculated results are in excellent agreement with previously reported experiments and theory. The band structures of SmB_6 reveal that this material has the qualities of a semiconductor with a minimum gap. The elastic constants, bulk modulus, shear modulus and Young's moduli of SmB_6 are obtained. The calculated results indicate that SmB_6 is a brittle material. The calculated theoretical hardness is 24.00 GPa. The optical properties of SmB6 are discussed in detail. It is shown that SmB_6 absorbs in the near infrared and visible light range. Therefore, SmB_6 has the potential to be used as a heat-absorbing coating in order to shield objects from solar heat radiation.

1. Introduction

Rare earth hexaborides (RB_6) have been widely studied because of their low densities, low thermal expansion coefficients, high melting points, high harnesses, high chemical stabilities, superconductivity capabilities, [1-4] magnetic properties,[5,6] high efficiency thermionic emission abilities,[7-10] narrow band gap semiconductivities,[11] and others.[12,13] Recently, these materials have been used in transparent heat-insulating windows[14-17] by taking advantage of their strong near-infrared radiation absorption and their high visible light transmittance that arise from the free electron plasmatic behaviors. The unusual properties of rare earth hexaborides are ascribed to the interaction of the 5*d* and 4*f* electrons of R with the 2*p* conduction electrons of B. These interactions seem to play an important role in determining the electrical, optical and mechanical properties of these materials.

The mixed valence compound SmB_6 belongs to a family of extremely hard and stable metal hexaborides. SmB_6 has been well studied because of its topological Kondo insulator properties which have intrigued physicists for many years. There have been significant theoretical and experimental

Efforts to understand the physical properties of SmB_6. Recent theoretical predictions have suggested that the mixed valence compound SmB_6 exhibits a topological Kondo insulator phase.[18-21]

Researchers have experimentally probed the topological Kondo insulator properties of SmB_6 by performing transport measurements,[22-25] torque magnetometer measurements,[26] neutron measurements,[27,28] various angle-resolved photoemission spectroscopy,[29-35] and scanning tunneling microscopy measurements.[36-38] The mechanical and optical properties of SmB_6 are very important owing to its applications as photoelectric functional material. At present, only a few reports have performed to investigate the mechanical and optical properties for SmB_6 and the results also provided some useful information. [16, 17, 39-42] However, the systematical information of mechanical and optical properties is still lack, which hinders the better understandings and applications of SmB_6. A systematic compilation of these properties is crucial for further material design and better understanding how to utilize SmB_6. Therefore, a systematic study of the electronic, mechanical and optical properties of SmB_6 must be performed to provide supplements to these shortages using the present first principles calculations.

In the present work, we investigated the electronic, mechanical and optical properties of SmB_6 using the first-principles calculations. This paper is organized as follows. First, the computational details based on first-principles calculations are depicted. Then, the structural properties, electronic structures, elastic constants, mechanical and optical properties for SmB_6 are presented. Finally, the conclusions of the present work are given.

2. Computational details

SmB_6 is ferromagnetic at low temperatures. Therefore, the spin polarized calculations, including spin-orbit coupling, were performed using density-functional theory (DFT) [43] as implemented in the Cambridge sequential total energy package (CASTEP) code. [44] The Vanderbilt-type ultra-soft pseudopotential was used to describe the interactions between ions and electrons. The exchange-correlation energy was described by a generalized gradient approximation (GGA) of the Pardew-Burke-Ernzerhof (PBE) [45] parameterization. The ultra-soft pseudo-potentials in this work were performed with $4f^6 5p^6 6s^2$ and $2s^2 2p^1$ valence-electron configurations for Sm and B atoms, respectively. The change of total energy during the optimization was converged to 1×10^{-6} eV and the forces per atom were reduced to 0.02 eV/Å. The Brayden-Fletcher-Goldfarb-Shannon (BFGS) algorithm was applied to relax the crystal structure to reach the ground state where both the cell parameters and the fractional coordinates of the atoms were simultaneously optimized. Brillouin zone (BZ) integrations were performed using Monk horst-Pack [46] *k*-point meshed. The *k* point separation in the Brillouin zone of the reciprocal space was $10\times10\times10$ for SmB_6. The cut-off energy for plane wave expansions was determined to be 400 eV. The separation of the reciprocal space was around 0.01 $Å^{-1}$ and the self-consistent field (SCF) tolerance was set to 5.0×10^{-7} eV/atom. The tolerance for elastic constants was set to within 1.0×10^{-6} eV/atom, a maximum force within 0.002 eV/ Å, and a maximum strain amplitude within 0.003 GPa.

3. Results and discussion

3.1. Structural properties and elastic constants

At ambient pressure, SmB_6 crystallizes in the cubic CsCl lattice structure and with a $Pm3m$ space group in which the hexaboride cluster (B_6) form a three-dimensional network with the octahedral structure. The Sm atom and B_6 octahedron occupy the site of cesium and chlorine, respectively. The crystal structure is shown in Fig. 1. Sm is located at the Wyckoff position 1*a* (0, 0, 0) and the B atoms at the 6f (1/2, 1/2, *z*) site (Fig. 1a). The boron-framework consists of octahedral B_6 groups joined together as shown in Fig. 1b. The framework has two kinds of boron-boron bonds where $B\text{-}B_{in}$ is on the edge of the octahedron and $B\text{-}B_{out}$ is between the octahedral. The $B\text{-}B_{in}$ distance is larger than the $B\text{-}B_{out}$ distance in SmB_6, which shows that the large octahedron keeps a large distance between neighboring Sm atoms.

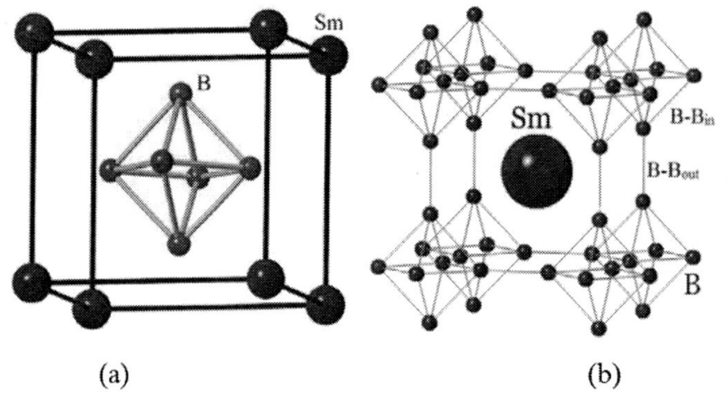

Figure 1. The crystal structure of SmB6. (a) Primitive cell and (b) Occupied sites of Sm atoms'–Bin and B–Bout are the inter-octahedral and intra-octahedral boron bond lengths, respectively.

The calculated lattice parameter (a_0) and internal parameter (z) of SmB_6 are listed in Table 1 along with available experimental and theoretical values. The calculated parameters are slightly larger than the experimental ones due to the inherent nature of the GGA approximation. The results of our calculations are in good agreement with previous experimental and theoretical data.

Table 1. Calculated and experimental structural parameters and the plasma energy (ωp) of SmB_6

	a_0(Å)	z	B-B$_{in}$ (Å)	B-B$_{out}$ (Å)	M-B(Å)	ω_p(eV)
Calc.[a]	4.1704	0.2018	1.7588	1.6831		1.34
	4.133[42]					
Expt.	4.1346[47]	0.2018[47]	1.7438[47]	1.6688[47]	3.0381[47]	
	4.131[39]					
	4.133[48]					

[a]This work.

In order to study the mechanical properties of SmB_6, its Elastic constant C_{ij}, bulk modulus B, shear modulus G, Young's modulus E and Poisson's ratio σ were calculated. The experimental data from the literature are shown in Table 2. For the cubic crystal, the mechanical stability leads to restrictions on the elastic constants as follows: $C_{44} > 0$ $C_{11} > C_{12}$, $C_{11} + 2C_{12} > 0$. The elastic constant C_{11} represents the elasticity in length while C_{12} and C_{44} are related to the elasticity in shape. Table 2 shows that the elastic constants of SmB_6 satisfy all of these criteria. It is important to note that the calculated elastic properties of SmB_6 agree well with the available experimental data. [41] Unfortunately, to our knowledge, there are no theoretical data in which to compare our elastic constants of SmB_6. Therefore, our results were compared to the experimental and theoretical elastic constants of other hexaborides, such as RB_6 (R=Y, La, Eu and Gd) [49] and MB_6 (M=Ca, Sr and Ba). [50] When comparing to these data, our calculated results are reasonable. From the ratio of B/G and Poisson's ratio σ, one can judge whether a material is ductile or brittle. The threshold of B/G and Poisson's ratio σ was found to be around 1.75 and 0.26, respectively. When $B/G > 1.75$ and $\sigma > 0.26$, the material behaves in a ductile manner, whereas it behaves in a brittle manner with values lower than 1.75 and 0.26, respectively. It is obvious from Table 2 that SmB_6 is brittle in nature.

Table 2. Elastic constants Cij(GPa), bulk modulus B(GPa), shear modulus G(GPa),Young's modulus E(GPa), B/G, and Poisson's ratio σ of SmB_6

	C_{11}	C_{12}	C_{44}	B	G_V	G_R	G	B/G	E	σ
Calc.[a]	415.89	9.88	75.69	145.22	126.61	101.03	113.82	1.28	415.43	0.023
	465[40]	29[40]	56[40]	174.1[40]						
Expt.	417[41]	-67[41]	78[41]	166.0[41]						

3.2. Electronic structure, magnetic properties and theoretical hardness

The energy band of the mixed valence compound SmB_6 was investigated to compare the exchange interaction. The energy band corresponding to the spin-up and spin-down split apart is due to the exchange coupling between the electrons. The spin energy band structures of SmB_6 are show in Fig. 2. The low energy band structures of SmB_6 have characteristics of semiconductors in that they have with a minimum gap of about 130 meV along the X-R direction in the majority spins (spin up) states and about 200 meV in the minority spins (spin down) states. The Sm-4*f* orbitals form narrow bands under the Fermi level in the majority spins (spin up) states. Our calculated energy band structures are in good agreement with results produced by the LDA + Gutzwiller method of Lu *et al* [51] and with the experimental results from transport and optical measurements which revealed the formation of a small gap only at temperatures under 50 K. [52-54]

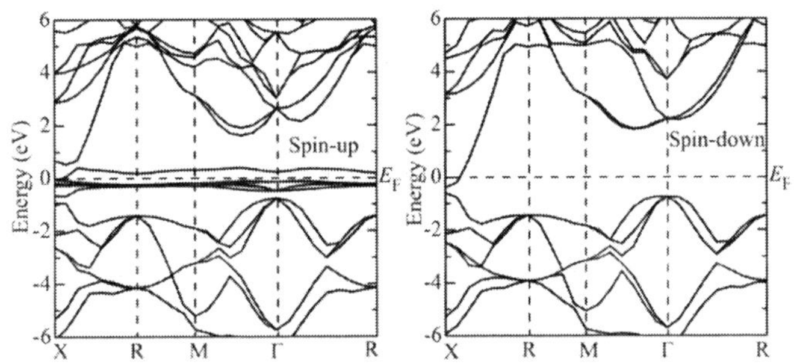

Figure 2. The spin energy band structure of SmB_6

The origin of the magnetic behaviors of SmB_6 was studied by calculating and analyzing the spin-projected total and partial density of states (DOS) for magnetic spin-states (Fig. 3). The spin-projected density of states of SmB_6 consists of four parts. The first part of bands consists of Sm 5*p* and B 2*s*2*p* states (-20 to -12 eV). The second part of bands (-10 to -1 eV) is characterized by B 2*p* states and some amount of Sm 5*s*5*p* states in order to make the distribution of spin-projected total density of states overall symmetric. It is interesting to note that 4*f* states cross the Fermi energy E_F, and hybridize with 5*d* states near the E_F, which is consistent with various experimental results. [55-57] The last group of bands (above E_F) is mainly made up of Sm 5*d*4*f* states. The Sm 5*d* band has a higher energy than La 5*d* in the lanthanum hexaboride due to the mixed valence character of SmB_6. The bottom of the conduction band and the top of the valence band are mainly formed from hybridized Sm 4*f* states and from B 2*p* states; there is also some amount of Sm 5*d* states and B 2*p* states. Figure 3 shows that orbital polarization is negligible for SmB_6.

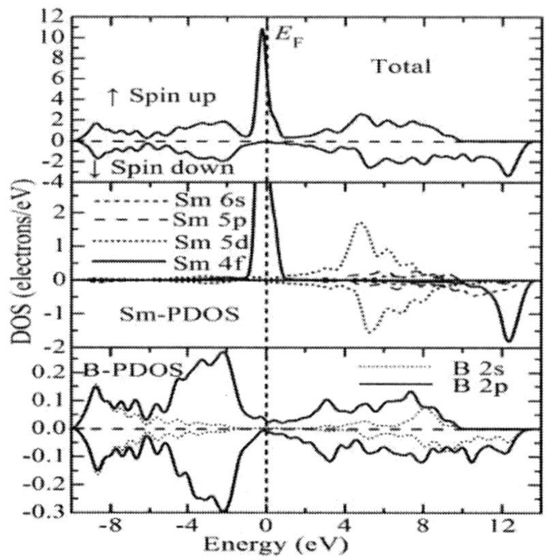

Figure 3. Spin-projected densities of states of SmB6

The Mullikan charge and overlap population are useful in evaluating the covalent, ionic, or metallic nature of bonds in a compound. Therefore, the Mullikan charge population for SmB_6 was calculated in order to better understand bonding in this compound. The Mullikan population results are given in Table 3. The calculated charge transfer from Sm to B is about 2.29 *e*. This value indicated that the bonding behaviour in SmB_6 is a combination of covalent and ionic. The Mullikan charge populations for SmB_6 were also calculated in order to obtain the intrinsic hardness of SmB_6 on the basis of Gao's model.[54] The intrinsic hardness of each chemical bond was calculated using the data in Table 4. Previous work has suggested that the hardness of complex compounds can be expressed as a geometric average of the hardness of all binary systems in the solid. This rule of considering the joint contributions of different bindings was applied to our calculations. [50] The calculated hardness values listed in Table 4 generally agree with previous experimental results. [39]

Table 3. Mullikan charge population of SmB6

Species	*s*	*p*	*d*	f	Total (*e*)	Charge (*e*)
B	0.85	2.53	0.00	0.00	-0.38	
Sm	1.81	5.17	0.81	5.92	13.71	2.29

Table 4. Calculated Mullikan population analysis and hardness (GPa) of SmB6.

	Atom	Charge (e)	Bond	Population	Length (Å)	N^{μ}	Vcell ($Å^3$)	v^{μ}	H^{μ}	H	$H_{Exp.}$ [39]
SmB_6	Sm_{1a}	2.29	$B\text{-}B_{in}$	0.58	1.6831	3	72.53	4.49	35.12	24.00	17.8±0.065
	B_{6f}	-0.38	$B\text{-}B_{out}$	0.45	1.7587	12		5.13	21.82		

3.3. Optical properties

The optical properties of a solid material are determined by the frequency-dependent dielectric function $\varepsilon(\omega) = \varepsilon_1(\omega) + \varepsilon_2(\omega)$, which characterizes the linear response of the material to an electromagnetic radiation (e.g. Governs the propagation behavior of radiation in a medium). The imaginary part of the dielectric function $\varepsilon2(\omega)$ is calculated from the momentum matrix elements between the occupied and unoccupied electronic states. The real part $\varepsilon_1(\omega)$ is derived from the imaginary part $\varepsilon_2(\omega)$ by the Kramers-Kronig transformation. All other optical constants, such as reflectivity, the absorption value, and the electron energy-loss function value are derived from ε_1 (ω) and ε_2 (ω).[58]

The complex dielectric function, reflectivity, the absorption spectrum, and the electron energy-loss function as a function of photon energy are presented in Fig. 4-7. The calculated static dielectric constant ε_1 (ω) was found to be 36.67. The peaks of the imaginary part of the dielectric function ε_2 (ω) are related to the electron excitation. These peaks which have a relatively strong intensity are located at 0.47 eV (A), 4.09 eV (B), and 6.98 eV (C). Peak A is ascribed to the transition of the inner electron excitation of Sm 4*f* states. Peak B and peak Care ascribed to the transition between B 2*p* states in the valence bands and Sm 5*d* states in the conduction bands.

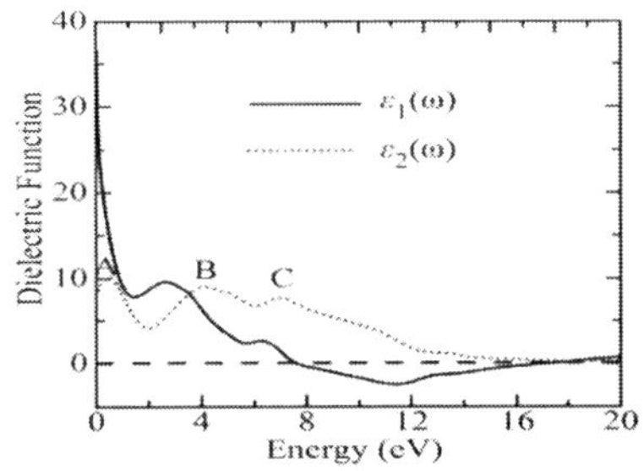

Figure 4. (Color online) Dielectric function of SmB6.

The reflectivity spectra were derived from the calculated dielectric functions for SmB_6 and are shown in Fig. 5. The calculated reflectivity was compared to the experimental data from Kimura *et al.* [59] Evidently, The experimental value shows a large value of reflectivity in the low energy range. Our calculated reflectivity also shows similar characteristics. In addition, the calculated result shows a dip below 1.78 eV in the visible light range. This is in good agreement with the experimental data from Kimura *et al.* (1.8 eV). [60]

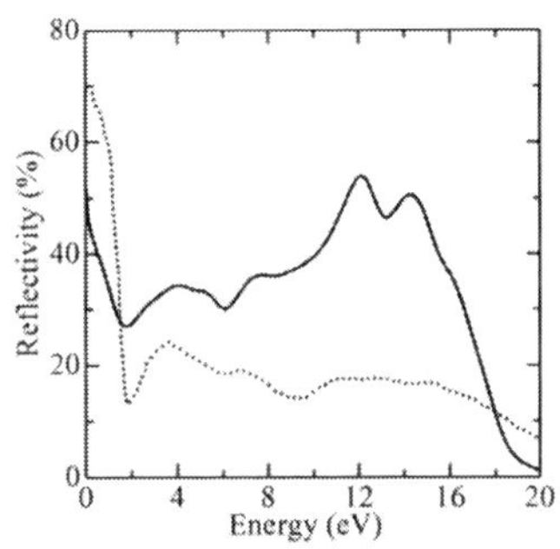

Figure 5. Reflectivity of SmB6. The dotted line stands for experimental data of SmB6. [59]

By analyzing the transmittance profiles of nanoparticle dispersions of rare-earth hexaborides, Takeda et *al.* concluded that hexaboride nanoparticles have potential uses as coatings for solar control windows to shield against solar heating. These results inspired us to study the absorption spectrum of SmB_6. Fig. 6 shows the calculated absorption spectrum of SmB_6. The strong absorption coefficient above 10^5 cm^{-1} occurs in the near infrared and visible light range because it is a metallic compound. The absorption spectrum has a sharp increase of about 3 eV in the high-energy region. Most significantly, there are no significant absorption peaks in the near infrared and visible light ranges. Therefore, we conclude that SmB_6 is a candidate for use in heat-absorbing coatings to shield solar heating.

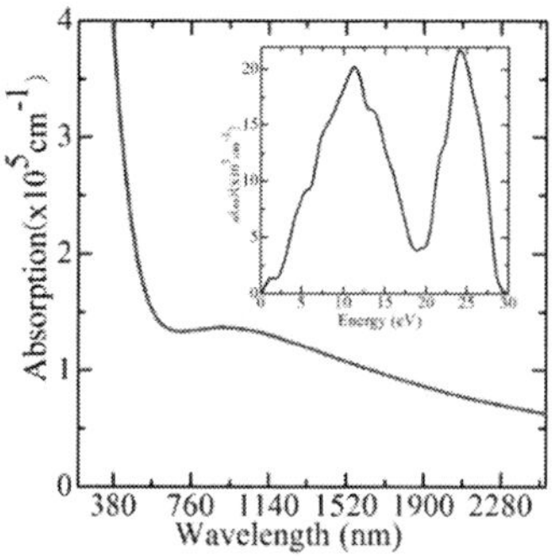

Figure 6. Absorption spectrum of SmB6.

The electron energy-loss function describes the energy loss of electrons that quickly traverse materials. The peak in the low-loss region characterizes the plasma resonance and the peak position corresponds to the relevant plasma frequency ω_p. Fig. 7 shows the calculated electron energy-loss spectrum. According to Kimura *et al.*, [59] the peak at 1.34 eV is derived from the excitation of the plasmon of the conduction electron. Additionally, the peak at 17.4 eV can be attributed to the excitation of the plasmon of the valence band which consists of the boron 2s and 2p bonding states and the $4f_{5/2}$ state of Sm atom. The peak at 28.0 eV shows the transition or excitation from 5p states. By comparing this information with other literature, [16] we conclude that the very weak peak at 1.34 eV, which is the plasma energy $\hbar\omega_p$ of SmB_6, shows a plasma resonance in the low-energy scale. This corresponds to an abrupt descent in reflectivity.

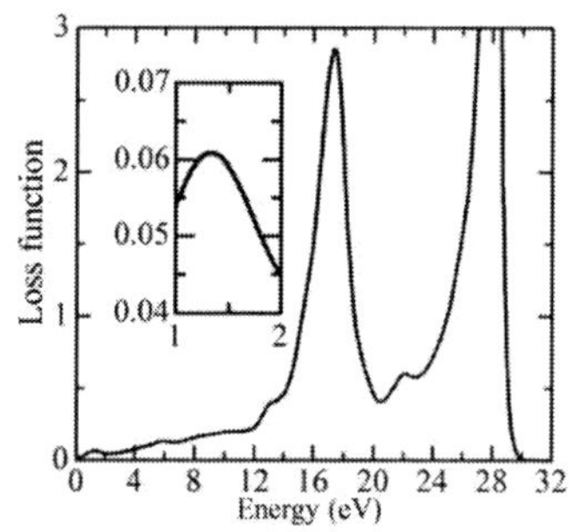

Figure 7. Energy-loss spectrum of SmB6.

4. Conclusion

In summary, first-principles calculations were used to investigate the structural parameters, mechanical properties, electronic structure, theoretical hardness and optical properties of SmB_6. The band structures of SmB_6 showed that this material has properties of a semiconductor (i.e. a minimum gap). The elastic constants, bulk modulus, shear modulus and Young's moduli of SmB_6 were calculated. These results showed that SmB_6 is brittle in nature. A theoretical hardness of 24.00 GPa was calculated. Finally, the dielectric function, reflectivity, absorption spectrum and electron energy-loss spectrum were obtained and discussed in detail. Our calculated reflectivity spectrum is in good agreement with the experimental literature. The absorption spectrum of SmB_6 shows that it has the potential for use as a heat-absorbing coating in order to shield solar radiation. We hope that our calculated results provide the theoretical data that are necessary for further material design and development of applications of the topological Kondo insulator SmB_6.

Acknowledgments
This work was financially supported by the National Natural Science Fund (No. 11574081), Guizhou Province Science and Technology Department-Guizhou Institute of Technology Joint Fund (Guizhou Science and Technology Agency LH [2014] 7357) and the Natural Science Foundation of Guizhou Province (Guizhou Science and Technology Agency J [2014]2086).

References
[1]. Matthias B T, Geballe T H, Andres K, et al. Superconductivity and ant ferromagnetism in boron-rich lattices. Science. 159 (1968).
[2]. Hiebl K, Sienko M J. Chemical control of superconductivity in the hexaborides. Inorg Chem. 19 (1980) 2179-2180.
[3]. Bat'ko I, Bat'kova M, Flachbart K, et al. Electrical resistivity and superconductivity of LaB6 and LuB12. J. Alloys Compd. 217 (1995) L1-L3.
[4]. Schell G, Winter H, Rietschel H, et al. Electronic structure and superconductivity in metal hexaborides. Phys. Rev. B 25 (1982) 1589.
[5]. Shannon J R, Sienko M J. Long-range magnetic interactions in europium hexaborides. Inorg Chem. 11 (1972) 904-906.
[6]. Segawa K, Tomita A, Iwashita K, et al. Electronic and magnetic properties of heavy rare-earth hexaboride single crystals. J Magn Magn Mater. 104 (1992) 1233-1234.
[7]. Zhang H, Zhang Q, Zhao G, et al. Single-crystalline GdB6 nanowire field emitters. J. Am. Chem. Soc. 127 (2005) 13120-13121.
[8]. Zhang H, Tang J, Zhang Q, et al. Field emission of electrons from single LaB6 nanowires. Adv. Mater. 18 (2006) 87-91.
[9]. Xu J, Hou G, Li H, et al. Fabrication of vertically aligned single-crystalline lanthanum hexaboride nanowire arrays and investigation of their field emission. NPG Asia Mater. 5 (2013) e53.
[10]. Xu J, Hou G, Mori T, et al. Excellent Field-Emission Performances of Neodymium Hexaboride (NdB6) Nano needles with Ultra-Low Work Functions. Adv. Funct. Mater. 23 (2013) 5038-5048.
[11]. Ishii M, Aono M, Muranaka S, et al. Raman spectra of metallic and semiconducting metal hexaborides (MB6). Solid State Commun. 20 (1976) 437-440.
[12]. Kerley E L, Hanson C D, Russell D H. Lanthanum hexaboride electron emitter for electron impact and electron-induced dissociation Fourier transform ion cyclotron resonance spectrometry. Anal. Chcm. 62 (1990) 409-411.
[13]. Takeda M, Fukuda T, Domingo F, et al. Thermoelectric properties of some metal borides. J Solid State Chem. 177 (2004) 471-475.
[14]. Schelm S, Smith G B. Dilute LaB6 nanoparticles in polymer as optimized clear solar control glazing. Appl. Phys. Lett. 82 (2003) 4346-4348.
[15]. Takeda H, Kuno H, Adachi K. Solar Control Dispersions and Coatings with Rare-Earth Hexaboride Nanoparticles. J. Am. Chem. Soc. 91 (2008) 2897-2902.
[16]. Xiao L, Su Y, Zhou X, et al. Origins of high visible light transparency and solar heat-shielding performance in LaB6. Appl. Phys. Lett. 101 (2012) 041913.
[17]. Xiao L, Su Y, Chen H, et al. Study on the electronic structure and the optical performance of YB6 by the first-principles calculations. AIP Advances, 1 (2012) 022140.
[18]. Dzero M, Sun K, Galitski V, et al. Topological Kondo insulators. Phys. Rev. Lett. 104 (2010) 106408.
[19]. Dzero M, Sun K, Coleman P, et al. Theory of topological Kondo insulators. Phys. Rev. B 85 (2012) 045130.
[20]. Miyazaki H, Hajiri T, Ito T, et al. Momentum-dependent hybridization gap and dispersive in-gap state of the Kondo semiconductor SmB6. Phys. Rev. B 86 (2012) 075105.
[21]. Lu F, Zhao J Z, Weng H, et al. Correlated topological insulators with mixed valence. Phys. Rev.

Lett. 110 (2013) 096401.
[22]. Wolgast S, et al. Low temperature surface conduction in the Kondo insulator SmB6. Phys. Rev. B 88 (2013) 180405(R).
[23]. Kim D J, et al. Surface hall effect and nonlocal transport in SmB6: evidence for surface conduction. Sci. Rep. 3 (2013) 3150.
[24]. Kim D J, Xia J. et al. Topological surface state in the Kondo insulator samarium hexaboride. Nat. Mater. 13 (2014) 466.
[25]. Zhang X H. et al. Hybridization, inter-ion correlation, and surface states in the Kondo insulator SmB6. Phys. Rev. X 3 (2013) 011011.
[26]. Li G, Xiang Z, Yu F, et al. Quantum oscillations in Kondo Insulator SmB6 . arXiv preprint arXiv:1306 (2013) 5221.
[27]. Fuhrman W T, Leiner J, Nikolić P, et al. Spin-exciton and topology in SmB6. arXiv preprint arXiv:1407 (2013) 2647.
[28]. Fuhrman W T, Leiner J, Nikolić P, et al. Interaction Driven Subgap Spin Exciton in the Kondo Insulator SmB6. Phys. Rev. Lett. 114 (2015) 036401.
[29]. Jiang J, Li S, Zhang T, et al. Observation of possible topological in-gap surface states in the Kondo insulator SmB6 by photoemission. Nat. Commun. 4 (2013) 3010.
[30]. Neupane M, Alidoust N, Xu S Y, et al. Surface electronic structure of the topological Kondo-insulator candidate correlated electron system SmB6. Nat. Commun. 4, (2013) 2991.
[31]. Zhu Z H, Nicolaou A, Levy G, et al. Polarity-driven surface metallicity in SmB6. Phys. Rev. Lett. 111 (2013) 216402.
[32]. Xu N, Shi X, Biswas P K, et al. Surface and bulk electronic structure of the strongly correlated system SmB6 and implications for a topological Kondo insulator. Phys. Rev. B 88 (2013) 121102(R).
[33]. Frantzeskakis E, de Jong N, Zwartsenberg B, et al. Kondo hybridization and the origin of metallic states at the (001) surface of SmB6. Phys. Rev. X 3 (2013) 041024.
[34]. Xu N, Biswas P K, Dil J H, et al. Direct observation of the spin texture in SmB6 as evidence of the topological Kondo insulator. Nat. Commun. 5 (2014) 4566.
[35]. Xu N, Matt C E, Pomjakushina E, et al. Exotic Kondo crossover in a wide temperature region in the topological Kondo insulator SmB6 revealed by high-resolution ARPES. Phys. Rev. B 90 (2014) 085148.
[36]. Ruan W, Ye C, Guo M, et al. Emergence of a Coherent In-Gap State in the SmB6 Kondo Insulator Revealed by Scanning Tunneling Spectroscopy. Phys. Rev. Lett. 112 (2014) 136401.
[37]. Yee M M, He Y, Soumyanarayanan A, et al. Imaging the Kondo insulating gap on SmB6. arXiv preprint arXiv:1308 (2013) 1085
[38]. Rößler S, Jang T H, Kim D J, et al. Hybridization gap and Fano resonance in SmB6. Proc Natl Acad Sci. 111 (2014) 4798-4802.
[39]. Futamoto M, Aita T, Kawabe U. Microhardness of hexaboride single crystals. Mater Res Bull. 14 (1979) 1329-1334.
[40]. Jie D, Tong Z, Li Z, et al. Elastic properties and electronic structures of lanthanide hexaborides. Chin. Phys. B. 2015, 24(9): 096201.
[41]. Smith H G, Dolling G, Kunii S, et al. Experimental study of lattice dynamics in LaB6 and YbB6. Solid State Commun. 53 (1985) 15-19.
[42]. Singh N, Saini S M, Nautiyal T, et al. Electronic structure and optical properties of rare earth hexaborides RB6 (R= La, Ce, Pr, Nd, Sm, Eu, Gd). J. Phys.: Condens. Matter 19 (2007) 346226.
[43]. Kohn W, Sham L J. Self-consistent equations including exchange and correlation effects. Phys. Rev. 140 (1965) 1133.
[44]. Segall M D, Lindan P J D, Probert M J, et al. First-principles simulation: ideas, illustrations and the CASTEP code. J. Phys. Condens. Matter 14 (2002) 2717.

[45]. Perdew J P, Burke K, Wang Y. Generalized gradient approximation for the exchange-correlation hole of a many-electron system. Phys. Rev. B 54 (1996) 16533.
[46]. Monkhorst H J, Pack J D. Special points for Brillouin-zone integrations. Phys. Rev. B 13 (1976) 5188.
[47]. Funahashi S, Tanaka K, Iga F. X-ray atomic orbital analysis of 4f and 5d electron configuration of SmB6 at 100, 165, 230 and 298 K. Acta Crystallogr., Sect. B: Struct. Sci 66 (2010) 292-306.
[48]. Zhang M, Jia Y, Xu G, et al. Mg-Assisted Autoclave Synthesis of RB6 (R= Sm, Eu, Gd, and Tb) Submicron Cubes and SmB6 Submicron Rods. Eur J Inorga Chem. 2010 (2010) 1289-1294.
[49]. Grechnev G E, Baranovskiy A E, Fil V D, et al. Electronic structure and bulk properties of MB6 and MB12 borides. Low Temp Phys 34 (2008) 921.
[50]. Huang B, Duan Y H, Sun Y, et al. Electronic structures, mechanical and thermodynamic properties of cubic alkaline-earth hexaborides from first principles calculations. J. Alloys Compd. 635 (2015) 213-224.
[51]. Menth A, Buehler E, Geballe T. Magnetic and semiconducting properties of SmB6. Phys. Rev. Lett. 22 (1969) 295–297.
[52]. Allen J, Batlogg B, Wachter P. Large low temperature Hall effect and resistivity in mixed valent SmB6. Phys. Rev. B 20 (1979) 4807–4813.
[53]. Cooley J, Aronson M, Fisk Z, et al. SmB6: Kondo insulator or exotic metal? Phys. Rev. Lett. 74 (1995) 1629.
[54]. Gao F. Theoretical model of intrinsic hardness. Phys. Rev. B 73 (2006) 132104.
[55]. Chazalviel J N, Campagna M, Wertheim G K, et al. Study of valence mixing in SmB6 by x-ray photoelectron spectroscopy. Phys. Rev. B 14(1976) 4586.
[56]. Allen J W, Johansson L I, Lindau I, et al. Surface mixed valence in Sm and SmB6. Phys. Rev. B 21(1980)1335.
[57]. Beaurepaire E, Kappler J P and Krill G. X-ray-absorption near-edge structure study in mixed-valent samarium systems. Phys. Rev. B 41(1990) 6768.
[58]. Li C, Wang B, Li Y, et al. First-principles study of electronic structure, mechanical and optical properties of V4AlC3. J. Phys. D: Appl. Phys. 42 (2009) 065407.
[59]. Kimura S, Nanba T, Tomikawa M, et al. Electronic structure of rare-earth hexaborides. Phys. Rev. B 46 (1992) 12196.
[60]. Kimura S, Nanba T, Kunii S, et al. Interband optical spectra of rare-earth hexaborides. J Phys Soc Jpn. 59 (1990) 3388-3392.

Numerical simulation of VAWT on the effects of rotation cylinder

Shuda Xing [1 2,*], **Yang Cao** [1 2], **Fuji Ren** [2]

[1]School of Mechanical Engineering, Nantong University, Nantong, Jiangsu, China
[2] School of Engineering, Tokushima University, Tokushima, Japan

*Corresponding author e-mail: 1286287339@qq.com

Abstract. Based on Finite Element Analysis Method, studying on Vertical Axis Wind Turbine (VAWT) which is added rotating cylinder in front of its air foils, especially focusing on the analysis of NACA6 series air foils about variation of lift to drag ratio. Choosing the most suitable blades with rotary cylinder added on leading edge. Analysis indicates that the front rotating cylinders on the VAWT is benefit to lift rise and drag fall. The most suitable air foil whose design lift coefficient is 0.8, the blades relative thickness is 20%, and the optimistic tip speed ratio is about 7.

1. Introduction

Due to the energy crisis and the rising costs of oil products, wind turbines were recognized and developed for its potential in power generation. Compared to wind-oriented horizontal axis turbine, as is shown in Figure 1, vertical axis turbines are affected less by wind direction and have no yaw systems. VAWTs have advantages in its quieter system and lower cost. What's more, Figure 2 shows that VAWTs makes it possible for Integration between wind power and constructions. In the near future, there will be more and more wind energy environmental protection buildings which combine wind energy with construction design.

Figure 1. Vertical axis wind turbine.

Figure 2. Integration of turbine and constructions.

In 1852, a physicist of German, Magnus [1], pointed out that the fluid exerts a force in vertical directions on the surface of rotating object. One and a half centuries later, Kozlov and Bychkov [2] applied Magnus effect to the development of new type of wind turbines, and it has been proved very effective. Thereafter, the University of British Columbia V.J.Modi [3] did research on a two-dimensional flat plate, airfoils, rectangular prism, even truck model with rotating cylinder added. They put forward the moving surface boundary layer control technology, and conducted a lot of experiments. The results show that the moving surface can increase the lift coefficient of airfoils. As a result of the relatively success of NACA6 series airfoils in laminar flow, this paper mainly focus on the numerical analysis of it.

2. Establishment of numerical model and parameter setting

To perform the VAWT simulations, we adopt the sliding mesh method. The flow field is calculated with an ideal incompressible gas, which is assumed valid for the present application because the Mach number is low. The sliding mesh method governing equations are posed as follows:

$$\frac{d}{d_t}\int_V \rho\phi dV + \int_{\partial V} \rho\phi(\vec{u} - \vec{u_g}).dA = \int_{\partial V} \Gamma\nabla\phi dA + \int_V s_\phi dV \tag{1}$$

Where is the density and $\vec{u}$ is velocity of fluid flow respectively, $\vec{u}_g$ is the velocity of moving grid, Γ is diffusion coefficient, s_ϕ and is the source. After the first order backward difference, the equation of time derivative term (1) can be described as (2):

$$\frac{d}{d_t}\int_V \rho\phi dV = \frac{(\rho\phi V)^{n+1} - (\rho\phi V)^n}{\Delta t} \tag{2}$$

Where n and n+1 expressed in the formula are respectively the current time and the next layer of time. In the calculations, the multi-block structure grid shown in Figure 3 was used. The grids around the airfoil nose is O grid and its trailing edge are topologically C grid. Due to rotation of cylinder, the grid quality near leading edge should be high.

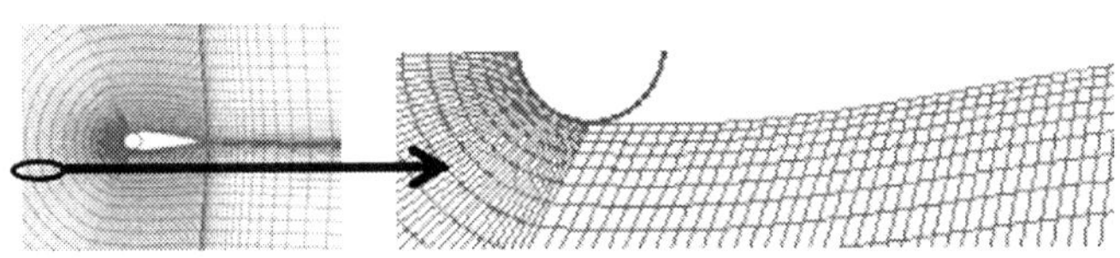

Figure 3. Computational grid of the leading edge rotating air foil and its magnification.

All the calculations were made for the Reynolds number Re = RU_∞/ν = 5×10^5, where $U\infty$ is the incident flow velocity and ν is the kinematic air viscosity, U = 10 m/s, and different design lift coefficient and relative thickness of the airfoils are distributed in table 1:

Table 1. Main parameters of new air foil.

parameters	values
gap value(mm)	0.003
chord(mm)	1000
lowest pressure points to the leading edge	30%*chord[4]
different design lift coefficient	0, 0.2, 0.4, 0.6, 0.8
relative thickness	12%, 14%, 16%, 18%, 20%

3. Numerical simulation analysis

The meaning of the curves in Figure 4 and Figure 5 are as follows: C indicates that the leading edge of airfoils with rotating cylinder added, the first number is the relative distance of minimum pressure point to the leading edge, the second number is the design lift coefficient, and the third number is the relative thickness. It can be seen in Figure 4 that the values of lift coefficient C_L are related to relative curvature. When the air flow over the surface of airfoils, the velocity upside of the airfoil is faster because of higher value of airfoil bending angle. But it is easy to result in the phenomenon of stall. For example, as we choose minimum pressure point whose relative distance to leading edge is 30% and lift coefficient is 0.4, the lift coefficient of new airfoil, leading edge rotation airfoil, is totally improved. The reason is that traditional airfoils are restricted by stall. However, leading edge rotation airfoils can solve this problem as the result of the existence of Magnus effect and be able to delay the stall. As shown in Figure 5, when relative curvature becomes larger, the lift to drag before 20 degree of attack angle increases more obvious. Nevertheless, after 20 degree of attack angle, lift to drag ratio C_L/C_D are less influenced, and the curves appear the phenomenon of cross. According to study on the lift and lift to drag ratio, the design lift coefficient 0.8 and the relative thickness 20% is superior. In contrast with the traditional airfoils, Leading edge rotating airfoils remarkably delay the phenomenon of stall.

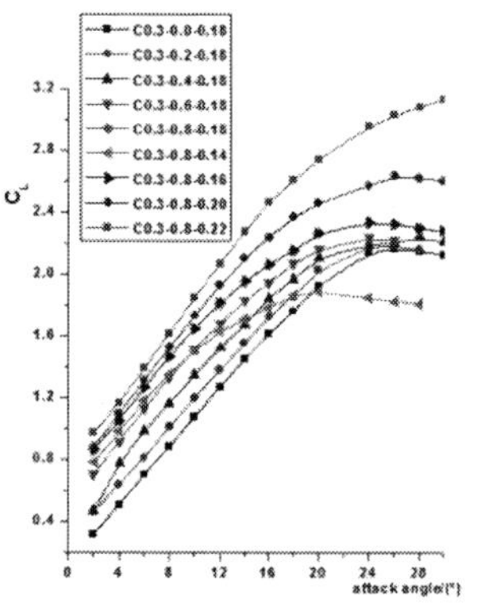

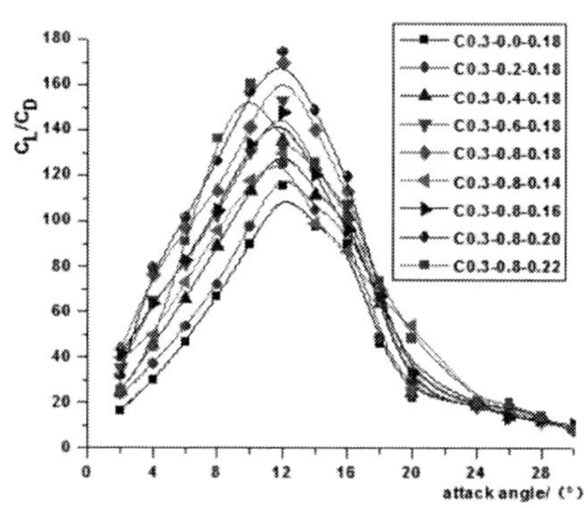

Figure 4. CL of different angles of attack. **Figure5.** CL/CD of Different angles of attack.

4. Construction of leading edge rotating VAWT model

4.1. Calculation method and parameter setting:

Leading edge rotating VAWT model consists of three blades. It includes two forms of rotational motion, the revolution of airfoils around the column and the rotation of cylinder. For double sliding grid calculation, the flow field is also calculated with an ideal incompressible gas. Moreover, k-e RNG turbulence model which mixes with the activation energy equation and the two order upwind difference discrete scheme based on pressure implicit SIMPLE algorithm are used.

Figure6 shows us that the grid is so organized that the cell dimensions increase with distance from the center. The wind wheel radius is 600mm, the shape of the wing chord length 140mm, blade combination installation angle of 2 degree [5]. The external boundary of the computation grid is a rectangle with a size of 15m*10m.The multi-block structure grid is a little complex for VAWT model. So it must take a lot of time. The inlet velocity is 10m/s, the rotation speed ratio of the cylinder is 4.

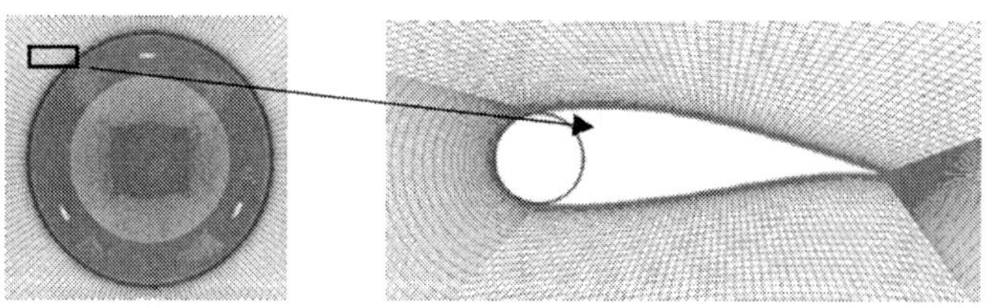

Figure6. Computational grid of vertical axis wind turbine model.

4.2. Results analysis

The Figure7 indicates the presence of vortex tubes that are created due to massive flow separation, and that quickly disintegrate and turn into further downstream. In addition to the wind turbines own energy consumption, the more obvious of vortex phenomenon, the greater consumption of whole VAWT. It's apparently to see Figure 7(b) which is the Leading edge rotation airfoil VAWT is better than Figure 7(a) that is the traditional VAWT because of rotation cylinder about vortex. Compared with traditional airfoils, rotating cylinder of new airfoils can adjust the directions of wind flow, so the downstream wake phenomenon of new VAWT has been significantly improved, and the vorticity produced from wind turbine farther.

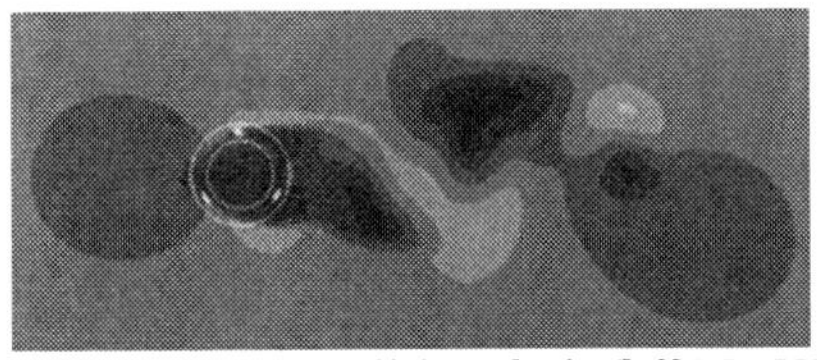

(a)Velocity contour of traditional air foil VAWT at t=2s

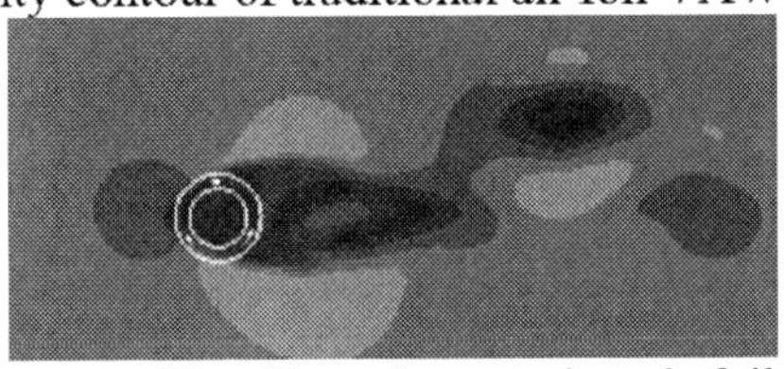

(b) Velocity contour of Leading edge rotation air foil VAWT at t=2s

Figure7. Velocity contour of VAWT model.

The torque which changes cyclically with time is shown in figure8. The negative value is the torque that drives the wind turbine to rotate and the positive torque is just the opposite, which will prevent vertical axis wind turbine model from rotating. In the whole, the wind turbine rotation torque is positive. It is just we want to see.

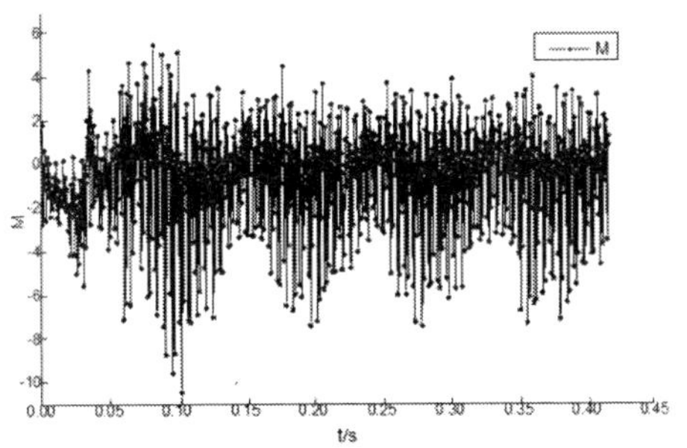

Figure8. Cp with tip speed ratio.

Figure9 shows that the maximum Cp is at the tip speed ratio (TSP) 7.1. When the speed is up to 140rad/s, the maximum power is 260w, as is shown in figure10. That is to say, when we give the speed of wind velocity is more than about 160rad/s, it will sharply decrease, and the rotating cylinder affect weaker on its improvement.

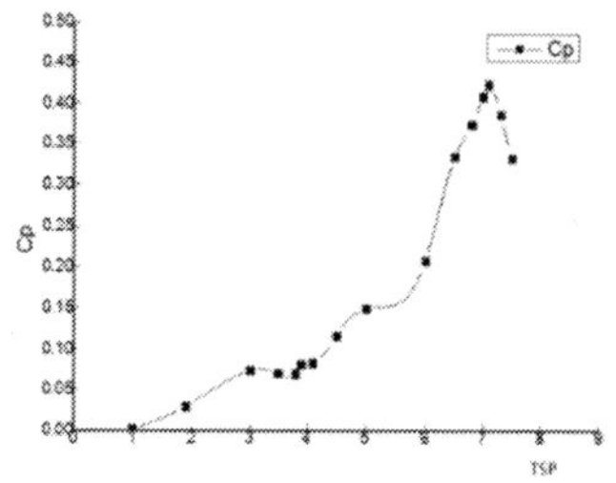

Figure9. Cp with tip speed ratio.

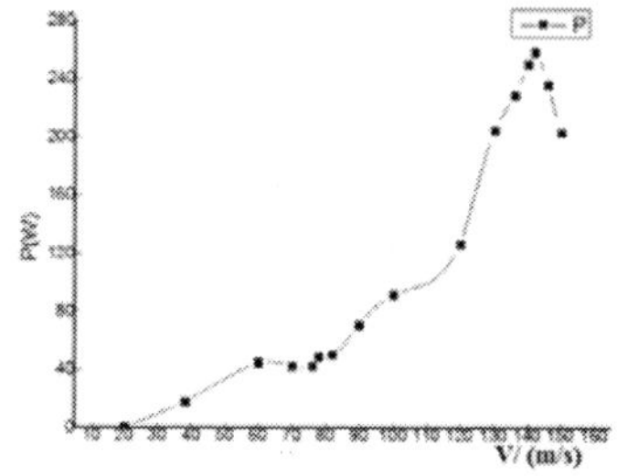

Figure.10 Power with the rotating speed.

5. Conclusions

The program shows that the leading-edge rotating cylinder is a successful device in increasing the sectional lift coefficient and lift-to-drag ratio at low angles of attack, hence reducing the need for higher angles of attack. Also, with high-speed rotation of the cylinder, the stall has been delayed. The increase in the lift coefficient and the delay in the stall angle of attack were about 10 degrees. The most suitable airfoil whose design lift coefficient is 0.8and the relative thickness is 20%, and the optimistic tip speed ratio is about 7. Selection methods of optimal airfoils and the calculation of the two-dimensional VAWT model provide a good reference for the construction of the three-dimensional model and other numerical analysis.

References

[1] L. Prandtl, Application of the "Magnus Effect" to the wind propulsion of ships[R]. NACA Technical Memorandum, 1926,13: 93–108.
[2] Bychkov N.M., Dovgal A.V., Kozlov V.V. Magnus wind turbines as an alternative to the blade ones[J]. Journal of Physics: Conference Series 75 (2007) 012004. 2007.
[3] Munshi S.R., Modi V.J., Yokomizo T. Fluid dynamics of flat plates and rectangular prisms in the presence of moving surface boundary-layer control[J]. Journal of Wind Engineering and Industrial Aerodynamics, 1999, 79(1-2):37-60.
[4] ZHUANG Yue-Qing, HUANG Dian Gui. Umerrical study on the aerodynamic characteristics theaerodynamic characteristics of the wind turbine airfoil with rotating cylinder[J],Journal of engineering thermal physics, 2011,32(1):43-46
[5] K.S.P.yankov, M.N.Toporkov. Mathematical Modeling of Flows in Wind Turbines with a Vertical Axis[J]. Fluid Dynamics, 2014, 249-25

Application test of a Detection Method for the Enclosed Turbine Runner Chamber

Yunlong Liu[1,2,*], Dingjie Shen[1,2], Yi Xie[1,2], Xiangwei Yang[1,2], Yi Long[1,2] and Wenbo Li[1,2]

[1]State Grid Hunan Electric Power Company Research Institute, Changsha, China
[2]Hunan Xiangdian Test Research Institute Co.Ltd. Changsha, China

*Corresponding author e-mail: 13874809964@163.com

Abstract. At present, for the existing problems of the testing methods for the key hidden metal components of the turbine runner chamber, such as the poor reliability, the inaccurate locating and the larger detection blind spots of the detection device, under the downtime without opening the cover of the hydropower turbine runner chamber, an automatic detection method based on real-time image acquisition and simulation comparison techniques was proposed. By using the permanent magnet wheel, the magnetic crawler which carry the real-time image acquisition device, could complete the crawling work on the inner surface of the enclosed chamber. Then the image acquisition device completed the real-time collection of the scene image of the enclosed chamber. According to the obtained location by using the positioning auxiliary device, the position of the real-time detection image in a virtual 3D model was calibrated. Through comparing of the real-time detection images and the computer simulation images, the defects or foreign matter fall into could be accurately positioning, so as to repair and clean up conveniently.

1. Introduction

A plurality of hydropower plants at home and abroad have been occurred serious accidents, such as cover rubbing and damaged fault [1-3]. To prevent major accidents of the turbine, it is necessary to carry out the technology research of the fault detection and protection monitoring for the key components under the turbine cover. However, it is a time-consuming and labor-intensive work to open the cover of the turbine runner chamber [4-6]. So the cover is just opened during A-type repair (usually every 5 to 8 years). When making short period B-type or C-type repair in hydropower plant, the endoscope is often used to extended into the chamber under the closed cover from the vacuum breaking valve hole above or the relief hole below to observe the key components such as the decompression panel and the water board in the enclosed turbine runner chamber, as shown in Fig. 1. Since the endoscope hose manual operation, the length and angle of the inserted probe cannot be controlled, it is difficult to determine the specific location of the inner chamber of the detection probe. Therefore, it is impossible to determine the position and properties accurately. Even if the defects or foreign matter is observed, it also cannot effectively guide the maintenance staff to repair. In addition, there is a big blind spot observation [7-10].

In this paper, an automatic detection method for the enclosed turbine runner chamber based on real-time image acquisition and simulation comparison techniques was proposed. The magnetic crawler with permanent magnet wheel could carry the real-time image acquisition device to complete the crawling work on the inner surface of the enclosed chamber. Then the image acquisition device completed the real-time collection of the scene image of the enclosed chamber. According to the obtained location by using the positioning auxiliary device, the position of the real-time detection image in a virtual 3-D model was calibrated. Through comparing of the real-time detection images and the computer simulation images, the defects or foreign matter fall into could be accurately positioning, so as to repair and clean up conveniently. It has important practical value to realize the automatic detection of the difficult demolition and complicated metal components of the enclosed turbine runner chamber, to solve the detection problems of common concern of hydropower manufacturing industry at home and abroad.

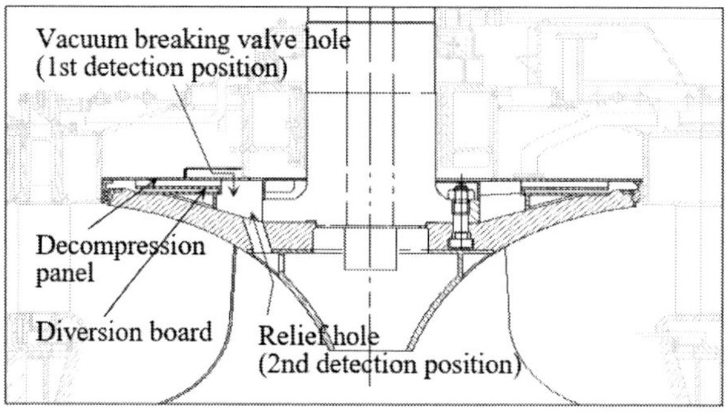

Figure 1. Detection position diagram of the turbine runner chamber

2. Design of detection system

The detection control system contains the motor control system and the data acquisition and processing system, which could realize the coordinated control of the motors and complete interactive communication with data acquisition and processing module. The detection control system contains 4 motors, and a half of them are used to control the motion of the magnetic crawler, and the others are used to achieve the perspective control of the image acquisition and illumination device.

2.1. Motion control

The magnetic crawler has 2 driving wheels (or rear wheels) and 2 driven wheels (or front wheels). The driving wheel and the driven wheel on the same side are connected with the belt and the 2 driving wheels are controlled by 2 motors respectively. Through coordinated control of the 2 motors, it could be realized to control the magnetic crawler to forward, backward and steering.

2.2. Viewing angle control

The steering and transmission device contains 2 motors. One of them is used to achieve the 360 degree rotation around the z-axis direction, and the other is used to achieve the 180 degree rotation around the x-axis direction. It can be realized to achieve the real-time image acquisition of 180 degree viewing angle through coordinated control of two motors.

2.3. Comparison

The simulation comparison system is used to establish the 3-D structural model of the enclosed turbine runner chamber. By using the positioning and distance measuring device, the characteristic parameters which contain the real-time position coordinates of the magnetic crawler and the acquired image could be obtained. Then the observation point and the position of the image could be calibrated in the corresponding 3-D simulation model, to realize the split-screen display of the real-time acquired image and the simulation image.

IOP Conf. Series: Materials Science and Engineering **207** (2017) 012086 doi:10.1088/1757-899X/207/1/012086

2.4. Split-screen display

According to the positioning and distance measuring device, the real-time acquisition feature image could be positioned. Then the position of the corresponding feature image in the virtual model could be calibrated and the real-time acquisition feature image and the corresponding feature image in the virtual model could be split-screen display in the computer control interface. As a result, the corresponding virtual image with the same size virtual foreign matter would be displayed on another screen at the same time.

3. Application test

3.1. Design of magnetic crawler

During the whole detection system, the magnetic crawler is one of the most important components, whose stability and reliability have direct influence on the scope and detection performance of the system. The schematic diagram of the magnetic crawler and the transmission device is shown in Fig. 2.

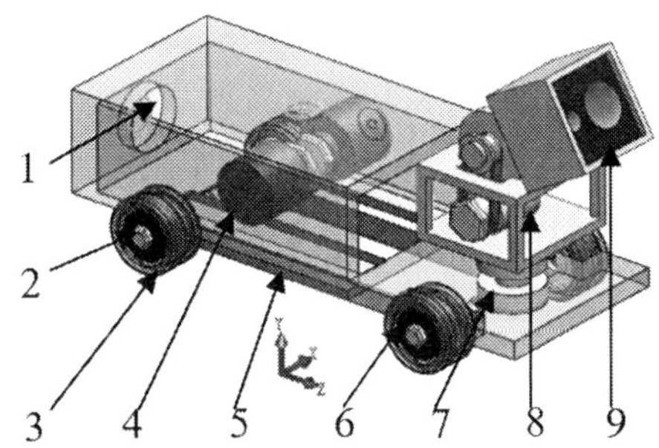

Figure 2. Schematic diagram of magnetic crawler and transmission device: 1- Cable interface, 2- rear wheels(drive wheels), 3- wear-resistant rubber wheel, 4- drive motor, 5- belt, 6- front heels(driven wheels), 7- motor around y-axis, 8- motor around x-axis, 8- image acquisition device.

The magnetic crawler contains the front wheels (driven wheels), the rear wheels (drive wheels), the drive motor, the belt and the wear-resistant rubber wheel. All the wheels are made by the permanent magnet, and the magnetic crawler which carries the real-time image acquisition device could complete the crawling work on the ferromagnetic surface of inclination angle of less than 55 degrees.

3.2. Processing magnetic crawler

The processing physical photo of the magnetic crawler is shown in Fig. 4. The cables are used to supply energy and achieve communication connection with computer control system. The magnetic crawler has compact and airtight structure and be able to meet the testing requirements of the hydropower plant turbine runner chamber.

3.3. Detection of hydropower plants turbine runner chamber

Taking the 6th turbine of a hydropower plant as a testing example, the relief hole below of the turbine runner chamber is as shown in Fig. 3.First of all, the virtual 3-D model of the turbine runner chamber must be established due to the actual structure and size parameters. And then, putting the magnetic crawler into the relief hole below of the turbine runner chamber and connecting with the computer control system, the magnetic crawler was controlled to walk foreword. Meanwhile, the machine perspective in virtual model followed up according to the acquired real-time position parameters of the magnetic crawler. As a result, it was realized that the real-time acquisition of images and the virtual extract images split displayed comparably, as shown in Fig. 4.

The test results show that the magnetic crawler has stable, reliable and well real-time operation, and the detection system has clear image, stable signal and well synchronization transmission, to be suit for internal inspection of the turbine runner chamber without opening the cover of the hydropower turbine.

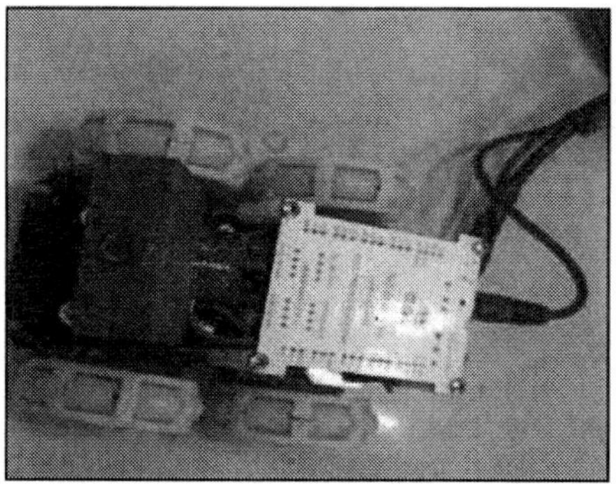

Figure 3. Photo of relief hole below

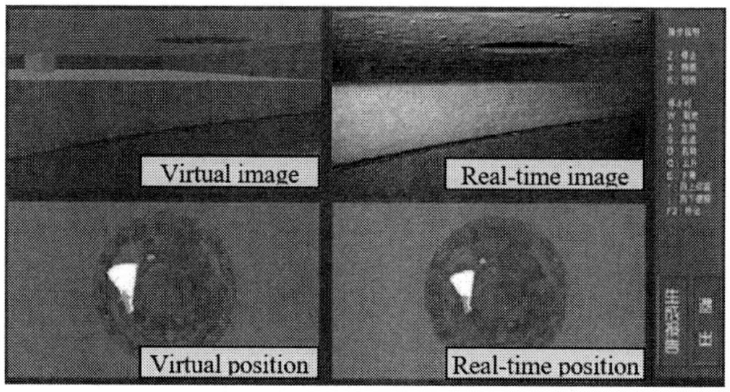

Figure 4. Split screen display interface of system

4. Conclusion

According to the designed detection device, the problems of the existing testing methods for the key hidden metal components were improved significantly. The application test showed that the designed automatic detection device could calibrate accurately and clean up the defects or foreign matter conveniently without opening the cover of the enclosed hydropower turbine runner chamber. According to the existing testing technical problems for the key hidden metal components of the turbine runner chamber, such as the poor reliability and the larger detection blind spots, it is proved to be effective of the proposed automatic detection method by using the simulation comparison system.

Acknowledgments

This research was financially supported by the State Grid Hunan Electric Power Company (NO. 5216A514004A) and the Hunan Xiangdian Test Research Institute Co.Ltd. (NO.18).

References

[1] HE Cheng-bing, JIANG Di, ZHANG Ya-fei, et al. Analysis and Process on Rub-impact Fault of Large Turbine Generator Units, Turbine Technology. 1(2013) 47-50.
[2] ZHANG Liang. Study on troubleshooting approach of turbine during running, China Hi-tech Enterprises. 0(2015) 61-62.
[3] YANG Jiandong, ZHAO Kun, LI Ling, et al. Analysis on the causes of units 7 and 9 accidents at Sayano-Shushenskaya hydropower station, Journal of Hydroelectric. 30(2011) 226-234.
[4] HE Chengbing, JIANG Di, ZHANG Yafei, et al. Analysis and process on rub-impact fault of large turbine generator units, Turbine Technology. 1(2013) 47-50.
[5] LU Xunyuan and TONG Xiaozhong. Vibration characteristics, analysis and diagnosis for rubbing between shaft/rotor and starts of turbine-generator in the field, Power Engineering. 22(2002) 2020- 2024.
[6] TANG Hua. Rebirth of Russia Sayano-Shushenskaya hydroelectric power station, Energy. 12(2014) 20-23.
[7] ZHANG Liang. Analysis on Fault Handling Methods in Turbine Operation, China High Technology Enterprises. 32 (2015) 61-62
[8] QU Zhao, ZHU Jianlin and LAI Xu. Study on hydroelectric generating unit on-line state monitoring system based on virtual instruments, Electric Power Construction. 2 (2005) 61-64.
[9] DONG Baoguo. Detection of Transformer Oil Leakage Based on Image Processing, Electric Power Construction. 11 (2013) 121-124.
[10] GAO Qiang and WEN Yao. Sag Calculation of Transmission Lines Based on Segmentation Image Analysis, Electric Power Construction. 11 (2010) 38-42.

IOP Conf. Series: Materials Science and Engineering **207** (2017) 012087 doi:10.1088/1757-899X/207/1/012087

Manganese oxides-based composite electrodes for supercapacitors

Dongyun Su[1, 3], Jun Ma[2], Mingyu Huang[3], Feng Liu[2], Taizhou Chen[2], Chao Liu[2], Hongjun Ni[3],*

[1] Nantong Vocational University, Nantong, Jiangsu Province 226007, China
[2] Nantong Science and Technology College, Nantong, Jiangsu Province 226007, China
[3] Nantong University, Nantong, Jiangsu Province 226007, China

*Corresponding author e-mail: ni.hj@ntu.edu.cn

Abstract. In recent, nanostructured transition metal oxides as a new class of energy storage materials have widely attracted attention due to its excellent electrochemical performance for supercapacitors. The MnO2 based transition metal oxides and their composite electrode materials were focused in the review for supercapacitor applications. The researches on different nanostructures of manganese oxides such as Nano rods, Nano sheets, nanowires, nanotubes and so on have been discovered in recent years, together with brief explanations of their properties. Research on enhancing materials' properties by designing combination of different materials on the micron or Nano scale is too limited, and therefore we discuss the effects of different components' sizes and their synergy on the performance. Moreover, the low-cost and large-scale fabrication of flexible supercapacitors with high performance (high energy density and cycle stability) have been pointed out and studied.

1. Introduction

Supercapacitors are regarded as a new promising type of device for energy storage and conversion, due to its high power density, rapid charging and discharging, and long cycle life. In the actual research, however, those problems such as low energy density and low capacitance is the bottleneck, limiting its development. Therefore, it's essential to develop new type of supercapacitor electrode materials to resolve these problems [1-3]. Transition metal oxides (e.g. MnO2) are regarded as the most promising supercapacitor materials due to their high theoretical capacitances, excellent cycling performance, abundance, low cost and environment friendliness. But weak conductivity and low specific surface area of the materials make their capacities far less than the theoretical values. Usually designing a unique morphology is used to increase the materials' specific surface area, and inducing other highly conductive materials is used to enhance the conductivity [4-6]. However, most of the reports only focus on the influence of one of these two factors. These researches are rare that enhance materials' property by designing the combination between capacitor materials with different activity under the micro or Nano scale, therefore it's still not clear about the synergy effect of different components neither size effect on the performance[7-8]. It is a great challenge to realize low-cost and large-scale fabrication of flexible SCs with high performance.

2. MnO_2 and MnO_2-based composite electrode materials

2.1. MnO2 electrode material

A great deal of studies has been attained in developing of fabrication different types of MnO_2 nanostructures. The key factors influencing the electrochemical capacitance are conductivity and specific surface area. However, MnO_2 has poor conductivity which couldn't be changed. So designing different nanostructures of MnO_2 electrode materials is the key to improve the capacitance. A summary of pure MnO_2 obtained with different nanostructures and their supercapacitor performances are discussed in Table 1.

As show in Table 1, MnO_2 can be synthesized in dozens of crystalline and disordered forms with variety structures, which were determined by the synthesis methods and nanostructures. Various MnO_2 with different crystalline structures, morphologies, pore structures, microstructures, and unique architectures have been synthesized by changing the parameters or using different reactions. Also, the specific capacitance of MnO_2 could be affected by its specific surface area, such as, Nano flake, particle size, urchin-like nanostructures with different specific surface area, contributing to various specific capacitances.

Table 1. Methods and performances of MnO2 electrodes

Synthesis methods	Nanostructure	Capacitance	Ref.
Carboxylic acid-mediated	Hollow sphere amorphous	281 F g^{-1} at 0.5 mA cm^{-2}	6
Template engaged redox	Hollow structures	366 F g^{-1} at 5 mV s^{-1}	7
Hydrothermal	Hierarchical Nano flower	347 F g^{-1} at 5 mV s^{-1}	8
Hydrothermal route	Crystalline porous	275 mAh g^{-1} at 40 mA g^{-1}	9
Chemical deposition	Mesoporous	173 F g^{-1} at 0.25 A g^{-1}	10
Microwave-assisted reflux	Urchin-like nanostructures	311 F g^{-1} at 0.2 A g^{-1}	11
hydrothermal method	Nano flake films	2.3 mF cm^{-2} at 0.025 mA cm^{-2}	12

2.2. MnO_2-based composite electrode materials

According to the categories of electrode materials, different types like carbon materials, transition metal oxides and conductive polymers also can be used as the doping materials with MnO_2, which has been widely reported, as shown in Table 2.

In addition, there are also many reports about some other metal oxides doped with MnO_2, as shown in Figure 1. Zhu et al [21] reported that hierarchically porous yet densely packed MnO_2 microspheres doped with Fe_3O_4 nanoparticles via a one-step and low-cost ultrasound assisted method. The results show that single-crystalline Fe_3O_4 particles of 3-5 nm in diameter are homogeneously distributed throughout the spheres and none are on the surface. The specific capacitance is optimized at an Fe/Mn atomic ratio of r = 0.075 to be 448 F/g at a scan rate of 5 mV/s, which is nearly 1.5 times that of the extremely high reported value for MnO_2 nanostructures. MnO_2 based with polymer such as PPy also has been studied. The composites formed by electrochemical polymerization of pyrrole deposited onto MnO_2 particles, have a big specific surface area for excellent performance.

In terms of materials design and preparation of MnO_2, the high specific surface area and good conductivity should be taken into the consideration. However, many reports only focus on the influence of one of these two factors, which improves the capacitance in limitation. Therefore, the choice of doped materials and design of nanostructure are very important for the capacitance improvement.

In addition, electrolyte optimization has been emphasized consistently for enhancing capacitance. The choice of electrolyte is much close to the increase of potential window, which can achieve highly desirable resulting in higher energy density and power density. Up to now, the electrolyte becomes the major cost factor and limitation of the next generation of supercapacitors.

In summary, choosing suitable component and reaction methods to build composite structures with rational design has a pivotal role in electrochemical properties. The synergistic effect and unique structure of composites should be taken full advantages. Due to the pure MnO_2-based pseudo capacitor with poor conductivity and cycle performance, supercapacitor materials should be designed basing on synergistic effect between the electrical double layer capacitor component and pseudo-capacitor component.

Table 2. Methods and performances of MnO2-based composites electrodes

Synthesis methods	Nanostructure	Composite materials	Capacitance	Ref.
Vacuum filtration	Ultrathin Nano sheets	MnO_2 /Graphene	267 F/g at 0.2 A/g	13
Hydrothermal	Core-shell	PPy/MnO_2	141.6 F g-1 at 2 mA cm-2	14
Wet chemical	Core-Shell Nanowire	α-Fe_2O_3 /MnO_2	838 F g-1 2mV s−1	15
Chemical agents	Nano needle structure	Graphene/MnO_2	327.5 F g-1 at 10 mV s−1	16
Hydrothermal	Core/Shell Nano sheet	Co_3O_4 Nanowire@MnO_2	480 F g-1 at 2.67 A g-1	17
Hydrothermal	Core/shell Nano sphere	Carbon@MnO_2	175 F g-1 at 100 mV s−1	18
Hydrothermal	Core/porous	MnO_2 /CNT	205 F/g at 2 mV/s	19
Hydrothermal	Nano architectures	Co_3O_4@MnO_2	1224 F/g at 5 mV/s	20

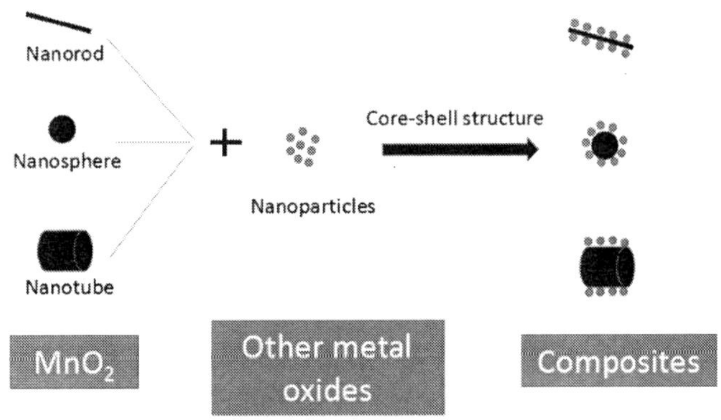

Figure 1. Scheme of the composites of MnO2 and other metal oxides

3. Summary

In this review, we have systematically pointed out that the recent researches and progress on MnO_2 and MnO_2-based composites as energy storage materials for supercapacitors. And the important of synergistic effect and size effect have been given an explanation. It is crucial to synthesize materials with suitable morphology and good conductivity since that have a profound influence on the storage capacity. So, the further detail explanations for the relationship between different materials should be researched in future.

Acknowledgments

This work was jointly supported by the Qing Lan Project, the Advanced Access Engineers for Higher Vocational Colleges Teachers of Jiangsu Province (2015FG032), the Natural Science Foundation of Jiangsu Province (BK20161289), the Research Innovation Program for College Graduates and Students of Jiangsu Province (KYZZ15_0043), the Foundation of Nantong Vocational University

(1512102), the College Students Innovation and Entrepreneurship Training Program of Jiangsu Province, Nantong Research Project (GY12015020).

References

[1] J. Liu, D. Xue, Thermal Oxidation Strategy towards Porous Metal Oxide Hollow Architectures, Adv. Mater. 20 (2008) 2622–2627.
[2] M. Olivares-Marín, J.A. Fernández, M. Lázaro, C. Fernández-González, A. Macías-García, V. Gómez-Serrano, F. Stoeckli, T.A. Centeno, Cherry Stones as Precursor of Activated Carbons for Supercapacitors, Mater. Chem. Phys. 114 (2009) 323–327.
[3] M.T. Lee, J.K. Chang, Y.T. Hsieh, W.T. Tsai, Annealed Mn-Fe Binary Oxides for Supercapacitor Applications, J. Power Sources 185 (2008) 1550–1556.
[4] T. Lu, Y.P. Zhang, H.B. Li, L.K. Pan, Y.L. Li, Z. Sun, Electrochemical Behaviors of Graphene-ZnO and Graphene-SnO2 Composite Films for Supercapacitors, Electrochim. Acta 55 (2010) 4170–4173.
[5] K.R. Prasad, N. Miura, Electrochemically Synthesized MnO2-Based Mixed Oxides for High Performance Redox Supercapacitors, Electrochem. Commun. 6 (2004) 1004–1008.
[6] J. Ma, H.J. Ni, D.Y. Su, X.K. Meng, Bioelectricity generation from pig farm wastewater in microbial fuel cell using carbon brush as electrode, International Journal of Hydrogen Energy. 41 (2016) 16191–16195.
[7] S. Dhibar, C.K. Das, Silver Nanoparticles Decorated Polyaniline/Multiwalled Carbon Nanotubes Nanocomposite for High-Performance Supercapacitor Electrode, Ind. Eng. Chem. Res. 53 (2014) 3495–3508.
[8] J. Ma, S.C. Tang, J.A. Syed, X.K. Meng, Asymmetric hybrid capacitors based on novel bearded carbon fiber cloth–pinhole polyaniline, RSC Adv. 6 (2016) 82995–83002.
[9] Y.M. Dai, S.C. Tang, S. Vongehr, X.K. Meng, Silver Nanoparticle-Induced Growth of Nanowire-Covered Porous MnO2 Spheres with Superior Supercapacitance, ACS Sustainable Chem. Eng. 2 (2014) 692–698.
[10] Y. Munaiah, B. Gnana Sundara Raj, Facile synthesis of hollow sphere amorphous MnO2: the formation mechanism, morphology and effect of a bivalent cation-containing electrolyte on its supercapacitive behavior, J. Mater. Chem. A. 1 (2013) 4300-4306.
[11] G. Yan, J.W. Cai, Preparation and capacitance behavior of manganese oxide hollow structures with different morphologies via template-engaged redox etching, Journal of Power Sources. 239 (2013) 347-355.
[12] G. Zhu, L.J. Deng, Hydrothermal preparation and the capacitance of hierarchical MnO2 nanoflower, Physicochem. Eng. Aspects. 434 (2013) 42-48.
[13] L.L. Peng, X. Peng, Ultrathin two-dimensional MnO2 /Graphene hybrid nanostructures for high-performance flexible planar supercapacitors, Nano Lett. 13 (2013) 2151−2157.
[14] A. Bahloul, B. Nessark, Polypyrrole-covered MnO2 as electrode material for supercapacitor, Journal of Power Sources. 240 (2013) 267-272.
[15] S. Debasish, G.K. Gobinda, High-Performance Pseudocapacitor Electrodes Based on α−Fe2O3/MnO2 Core−Shell Nanowire Heterostructure Arrays, J. Phys. Chem. C. 117 (2013) 15523−15531.
[16] K. Myeongjin, H. Yongseon, Graphene/MnO2-based composites reduced via different chemical agents for supercapacitors, Journal of Power Sources. 239 (2013) 225-233.
[17] J.P. Liu, J. Jiang, Co3O4 Nanowire@MnO2 Ultrathin Nanosheet Core/Shell Arrays: A New Class of High-Performance Pseudocapacitive Materials, Adv. Mater. 23 (2011) 2076-2081.
[18] Y. Zhao, Y. Meng, Carbon@MnO2 coreeshell nanospheres for flexible high-performance supercapacitor electrode materials, Journal of Power Sources. 259 (2014) 219-226.
[19] H. Xia, Y. Wang, J.Lin Hydrothermal synthesis of MnO2/CNT nanocomposite with a CNT core/porous MnO2 sheath hierarchy architecture for supercapacitors, Nanoscale Research Letters. 6 (2011) 1-10.

[20] J. Ma, H.J. Ni, D.Y. Su, M.Y. Huang, X.X. Wang, The research status of Nafion ternary composite membrane, International Journal of Hydrogen Energy. 37 (2012) 13185–13190.
[21] J. Zhu, S.C. Tang, Hierarchically Porous MnO2 Microspheres Doped with Homogeneously Distributed Fe3O4 Nanoparticles for Supercapacitors, ACS Appl. Mater. Interfaces. 6 (2014) 17637-17646.

Application of copper-based heterogeneous catalysts in organic wastewater treatment

Lingfeng He[1, a], Yuan Chen[2, b], Liang Shi[3, c], Yongli Zhang[1, *]

[1]School of Environmental and Chemical Engineering, Foshan University, Foshan 528000, China
[2]School of Foreign Studies, South China Normal University, Guangzhou 510000, China
[3]Environmental Protection Engineering of Sikete, Foshan University, Foshan 528000, China

[*]Corresponding author e-mail: 670511266@qq.com, [a]331538792@qq.com, [b]891532849@qq.com, [c]402225029@qq.com

Abstract. Copper-based heterogeneous catalyst is a kind of catalyst which is created by the effect that copper, ferrum, palladium, lanthanum together with their corresponding oxides have on the carriers such asγ-Al_2O_3. With the development of related research in recent years, its applications and theories are becoming more and more abundant. The properties, structures and designs of experiments of copper-based heterogeneous catalysts are reviewed and the establishment of applied kinetics equations is briefly introduced in this paper.

1. Introduction

The copper-based catalyst has many unique properties and quite a few catalysts with different catalytic properties could be made based on their different types of metals as well as oxides and component proportions. Oxides made from the dopant of noble metals and their oxides can exert better catalytic effect than those from the single component. This catalytic method is called catalytic wet air oxidation (CWAO for short). CWAO has been extensively studied since the 1950s, and in recent years, it has attracted much international attention. Catalytic oxidation of organic wastewater by the modification of the properties of copper-based catalysts has become another research area for environmental engineering. Zhou Yan-bo et al. established the applied kinetics model according to catalytic oxidation of Cu-Fe-Pd-La catalyst's reaction with methyl orange but he didn't build any models for other factors [1]. In this paper, the kinetic model of catalytic oxidation is established according to several influential factors: the concentration of methyl orange, pH, rational speed, and the catalyst particle diameter, the reuse of catalyst, dissolved oxygen, oxygen ion concentration, turbulence and jetting.

Table 1. Original Parameter of Water Sample of Methyl Orange

Concentration of methyl orange (mg/L)	COD (mg/L)	Turbidity (NTU)	Absorbance	PH
2238.5	3775	83.76	40.8	7.04

Table 2. Catalyst preparation

Name	Copper - based heterogeneous catalyst
Preparation method	Take transitional metals Cu, Fe, Co salt, rare metals Ce, La salt and noble metals Pt, Ru, Pd salt as the source of metal and make 20 mL6 wt% nitrate solution according to a certain metal proportion. Then immerse 10 g γ-Al_2O_3 carrier particles in the prepared nitrate solution at room temperature for 12 h, and later dry it for another 12 h in an oven at 105 ° C. After that bake it in a muffle furnace (heating rate 8 ° /min) for 4 h at the setting temperature and after cooling it down at a constant rate (cooling rate 10°/min), the needed catalyst would be obtained

2. Copper - based heterogeneous catalyst

Copper is a mixture of copper, ferrum, palladium, lanthanum and their oxides. Heterogeneous refers to the liquid phase, solid phase and etc. When these mixtures are supported on the pores of γ-Al_2O_3, the catalyst and carrier are formed. The catalyst is obtained through the metal salt being immersed in γ-Al_2O_3, dried and calcined. Compared with the traditional wet air oxidation (WAO) method, this method is easier in the scale-up treatment of organic wastewater.

Copper based heterogeneous catalysts can reduce the total cost due to the advantages of less investment, higher treatment efficiency and lower operating cost.

The heterogeneous catalyst has good recyclability and high catalytic efficiency. During the reaction, it is in solid form and after that process, it becomes easier to detach from reaction solution. The components of catalyst consist of transitional metals, rare earth metals and noble metals and so on. Transitional metal, a kind of homogeneous catalysis, is easy to dissolve but it owns high catalytic activity. The limitation of rare earth metal lies in catalytic inactivity, but this kind of metal can change catalyst structure so that it can increase the possibility of contacting with oxygen, and produce more free radicals and improve the overall performance. As for noble metal, its drawbacks are high price and limited application but it possesses good catalytic effect and strong stability.

3. Research status of kinetics

The study of kinetics is helpful to clarify and reflect mechanism. Relevant scholars have already done some related researches.

Liu Shao-lei has studied the kinetics by studying Pd-Fe/Graphene catalytic cathode and electrolytic degradation of 4-Chlorphenol [2].

Table 3. Original Parameter of the Water Sample

Drug Name	Reagent Grade	Pharmaceutical Manufacturers
4-chlorophenol	Analytical Reagent	Shanghai Feixiang Chemical Factory

Table 4. Catalyst preparation

Name	Pd-Fe/Graphene multifunctional catalytic cathode
Preparation method	Step 1: Prepare graphite oxide by adopting modified Hummer method; Step 2: Restore graphite oxide by adopting photo-reduction method and complete making the metals load on the surface of the catalyst.

Guo Min analyzed the reaction kinetics by removing nitrate from Nano-FeO/Pd/Cu particles (water sample and catalyst can be seen in Table 1.5 and 1.6) and concluded that the reduction of nitrate by Nano-FeO/Pd/Cu particles was in accordance with the pseudo-first-order reaction kinetics [3]. The factors that affect the removal rate are as follows: Nano-FeO/Pd/Cu dosage, initial DO, stirring speed, coexisting oxygen anion, nitrate initial concentration, initial pH, etc. All are different affecting factors from those studied by the aforementioned scholars. After learning from the experience and making analogy, it is found that the subject can also be studied around those several factors.

Table 5. Original Parameter of Water Sample

Drug Name	Preparation Method
Micro-polluted acid salt	As experimental request, Wahaha pure water and potassium nitrate need to prepare the solutions of different concentration gradients and simultaneously take into account the influence of other oxygen anions and organics. On such basis, add materials such as sodium nitrite, potassium chlorate, potassium bromate, potassium dihydrogen phosphate, sulfuric acid Potassium, sodium bicarbonate, humic acid (HA) into the potassium nitrate solution and then prepare expected mixed solution according to a certain proportion.

Table 6. Catalyst performance

	Nano-FeO/Pd/Cu
Color	Black
Properties	Agglomeration: slight suspense occurs in the solution; Good settle ability: generally settlement is completed within 20 min or so. It is stored in deoxidized anhydrous ethanol and it is wet added quantitatively during reduction reaction.

Li Chang-fang used nano Pd/Fe catalytic to reduce degradation BDE-47 (water sample and catalyst seen in Table 1.7 and 1.8) The reacting rate of BDE-47 increases with the increase of Pd loading rate [4].

Table 7. Original Parameter of Water Sample

Drug Name	Reagent Grade	Pharmaceutical Manufacturers
2,2',4,4'-Tetrabromodiphenyl ether	99%	American Chemical Services
2,2',4,4'-Tetrabromodiphenyl ether	Standard sample	American Accu Standard Company

Table 8. Catalyst Preparation

Name	Nano Pd/Fe
Method	NaBH4 is used as reduction agent to reduce Fe^{2+}.Nano zero valent Fe particles are obtained. Then $PdCl_2$ is used as Pd agent and the Pd is loaded on the surface of zero valent Fe.

Tang Di studied the treatment of PVA by the pure iron-carbon system under acidic conditions (water sample and catalyst are shown in Table 1.9 and 1.10) and concluded the kinetics equation V1=1.03x10-3p1.59, and the kinetics equation of the treatment with iron-carbon system by rare earth La V2=1.81x10-2P1.105M0.323[5].

Table 9. Original Parameters of Water Samples

Drug Name	Preparation Method
PVA Simulated wastewater	Calculate and weigh a certain amount of PVA that has been dried in the blast drying oven of 110 degrees Celsius; Stir and at the same time add PVA slowly into the cold water of about 25 ℃ to make it swell and disperse fully; 10 min later, put it into the water bath kettle of 90℃ below to dissolve (during this process of dissolution, stir it properly); In about 10 h when it is completely dissolved, cool it down to the room temperature; Filtering so as to remove impurities from the solution and dilute the concentration of filtered solution to about 1500 mg/L

Table 10. Parameters of Catalyst

System	Content of Iron/g	Content of Coke/g	Concentration of Lanthanum Nitrate /g/L
Rare Earth La- iron Carbon System	100	100	0.4

In the Fe/Fenton system, Wang Jing discovered that the degradation of methylene blue in 0-45 min accords with the pseudo-first-order reaction kinetics equation.(water sample and catalyst are seen in Table 1.9 and 1.10) The activation energy was 82.708KJ/mol. The study found that a new kind of methylene blue decomposes intermediate products of benzothiazole, and derived methylene blue in Fe/Fenton system degrades along three paths [6]. The previous research of this subject is two-step reaction. Therefore, we can assume that there are also intermediate products in the subject, and they also degrade along three paths.

Table 11. The Original Parameters of Water Sample

Drug Name	Reagent Grade	Pharmaceutical Manufacturers
Methylthionine Chloride	BS	Shanghai Jingchun reagent Co.,Ltd

Table 12. Catalyst preparation

Catalyst	Method
Ferric iron Hydrotalcites	Weigh a certain amount of $FeSO_4 \cdot 7H_2O$ and Fe2$(SO_4)_3$ and put them in a three-necked flask; Use the approved deionized water to prepare the solution, to achieve the equation $[Fe^{2+}]+[Fe^{3+}]=1.2$ mol•L-1; Under the protection of $[Fe^{2+}]/[Fe^{3+}]=2:1$, 1:1 and N2 and the condition of $[Fe^{2+}]/[Fe^{3+}]=1:0$ and air atmosphere, in the water-bath of 40°C, add 2 mol/L NaOH solution to the solution until the pH of the solution is constant; After the reaction is completed, locate the three-necked flask in ice-bath for 4 h, filter, washed with ice water through N2 and at last wash twice with ethanol of 0°C; Store it in the refrigerator finally in after dry it at room temperature.
Ferrocene modified resin	Weigh a certain amount of styrene cation exchange resin and put it in a conical flask; Immerse HC1: HNO_3 = 1: 1 mixed acid to remove impurities on its surface; Place the pretreated resin in 0.5 mol-L-1 ferrocene-ethanol solution and stirred for 7 h; After the reaction, filter the mixture and rinse with cthanol to remove excessive ferrocene; Dry it at the room temperature for 2 h and seal it for preservation.

Acknowledgments
This work was supported by Water Pollution Control Engineering Technology Research Center of Foshan (2016GA10159), 2016 Foshan Science and Technology Project.

References

[1] Zhou Yan-bo. Preparation of multi component supported catalysts and properties of catalytic wet air oxidation (MS., Liaoning University of Science and Technology, China 2016)
[2] Liu Shao-lei. Study on Pd-Fe/Graphene Multifunctional Catalytic Cathode and Degradation of 4-Chlorophenol by Multi-electrodes System (MS., Beijing Forestry University, China 2015)
[3] Gou Min. Study on the Removal of Nitrate by Nanometer Fe°/Pd/Cu Composite Materials in Drinking Water (MS., Zhejiang University of Technology, China 2013)
[4] Li Chang-fang. Study on Catalytic Debromination of 2, 2', 4, 4'- Tetrabromodiphenyl Ether by Nanoscale Pd/Fe Bimetallic Particles (MS., South China University of Technology, China 2012)
[5] Tang Di. Study on Rare Earthla-Iron and Coke Micro-Electrolysis Used for PVA Waste-Water

(MS., Donghua University, China 2016)
[6] Wang Jing. Mechanisms and application of heterogeneous Fenton systems catalyzed by Ferrocene and FeFe-LDH in the treatment of Methylene Blue (Ph.D., Kunming University of Science and Technology, China 2014)

Effect of humic acid on transformation of soil heavy metals

Shengzhe Wu, Rui Li, Shuyang Peng, Qiuyong Liu and Xi Zhu*

School of Information and Safety Engineering, Zhongnan University of Economics and Law, Wuhan 430073, China

*Corresponding author e-mail: zhuxi@zuel.edu.cn

Abstract. The pattern of transformation among different fractions of heavy metals in soil draws great attention. This article investigated the effect of humic acid on soil heavy metal under different pH and temperature. The results showed that with the increasing of the concentration of humic acid, the concentration of available Cu, Pb decreased greatly, though the decrease of available Cd was light. Also, pH of soil had certain impact on the concentration of available Cu, Pb, Cd while the influence of environmental temperature was minor. The removal efficiency of contaminated soil was 57.283% (Cu), 2.645% (Cd) and 15.485% (Pb) while it could reach more than 98% in contaminated solution.

1. Introduction

At present, China is paying close attention to the heavy metal contaminated soil [1]. The heavy metal pollution of soil is persistent and difficult to transform, which affects the soil fertility and biochemical activity, and enters the food chain through plant's absorption, causing an adverse impact on people's health. The existence of heavy metals in the soil can be divided into six forms: water-soluble, active, carbonate, organic, iron-manganese, and silicic acid. Among them, the water-soluble and active state are the effective modes of plant, which are easy to be absorbed by plants, while the rest are not[2].A sequence of factors like soil pH, oxidation-reduction potential, cation exchange capacity, soluble organic matter and so on will affect the solubility and mobility of heavy metals in the soil [2,3]. Udom [4] found that the complexation of organic matter and heavy metals can significantly reduce the migration and bioavailability of pollutants in the soil. Zhang Yali [5] also pointed out that organic fertilizers have the effect of reducing the bioavailability of heavy metals in soil and changing the morphology of Cd.

Humic acids (HAs) have a unique structure and are widely distributed in the soil system. The application of HAs to heavy metal remediation in the soil has been extensively studied. Bruna Alice Gomes De Melo et al. [6] found that the cationic metal interacts with the phenolic and carboxyl groups on the molecular chain of HAs to form spherical compounds, which claimed the soil remediation mechanism of HAs. Applying HAs to the soil can reduce the content of soluble heavy metals and the content of the organic form of Cu^{2+}, Cd^{2+}, Zn^{2+} and Pb^{2+}, so as to the activity, toxicity and bioactivity of Cu^{2+}, Cd^{2+}, Zn^{2+}, Pb^{2+} in the soil [7]. Yates & von Wandruszka's [8] experiments also showed that HAs had the greatest affinity with Pb^{2+}, Cu^{2+}.

However, when the external environment changes, there is less concern about the effect of HAs applied to heavy metals contaminated soil. In this study, we applied different contents of HAs to the soil that was in various conditions (pH and temperature) contaminated by Cu, Pb and Cd, in order to explore the effect and law adsorbed by HAs, which has great practical significance.

2. Materials and methods

2.1. Experimental material and instrument

The experimental sample was collected from 10-20cm surface soil layer of uncontaminated plant-covered area in Zhongnan University of Economics and Law. Remove the impurities and pack with polyethylene bag. Air and mix the sample after filtrating by 10-mesh sifter. Immersed 3kg sample with 1L liquor (30mg/L cadmium nitrate, 1500mg/L cupric nitrate, 3000mgL lead nitrate). Settle the sample for 30 days and filtrate with 100-mesh sifter.

The sample was preprocessed by the Jena Top Wave microwave digestion instrument. The heavy metal was measured by Jena Znieet 700P atomic absorption spectrophotometer.

2.2. Experimental approach

The effects of exchangeable heavy metal were measured in different percentage of humic acid dosage (0.2%, 0.5%, 1%, 3%, 5%), temperature (10℃, 20℃, 30℃, 40℃, 50℃) and pH (3, 5, 7, 9, 11). The measurement referred to Tessier's five-step approach.

Put 2.50g experimental sample in 50ml centrifuge tube. Add 10ml distilled water and humic acid into the centrifuge tube. Adjust the pH by 0.1mol/L NaOH. Shake the centrifuge tube for 18 hours at the setting temperature. Took down the tube after shaking and added 10ml MgCl2 (pH=7,2mol/L) in it. Shake the tube for another hour. Save the sample after filtrating it by 0.22μm filter membrane.

Take 0.125g humic acid and 20ml simulative solution (cupric nitrate 12mg/L, lead nitrate 68mg/L, cadmium nitrate 0.76ml/L) in tube. Shake it at 20℃ for 18h. Filtrate with 0.22μm filter membrane.

3. Results and discussion

3.1. Leaching pattern in man-made contaminated soil

As for man-made contaminated soil, fractions of heavy metal in soil transform when time elapsing, thus making it important to study leaching pattern in man-made contaminated soil. Fig.1 shows that pH and temperature both have an obvious impact on available state of heavy metals. For available Pb and Cu, their concentrations declined while temperature declined or pH rose. The influence of temperature is mild within the range 20°C~40°C, and there is a sharp drop when pH gets 11; for available Cd, the leaching amount is comparatively low when the pH ranging 3~11 and will increase slightly in strong alkaline environment. Also, there's great difference in leaching amount under different temperatures.

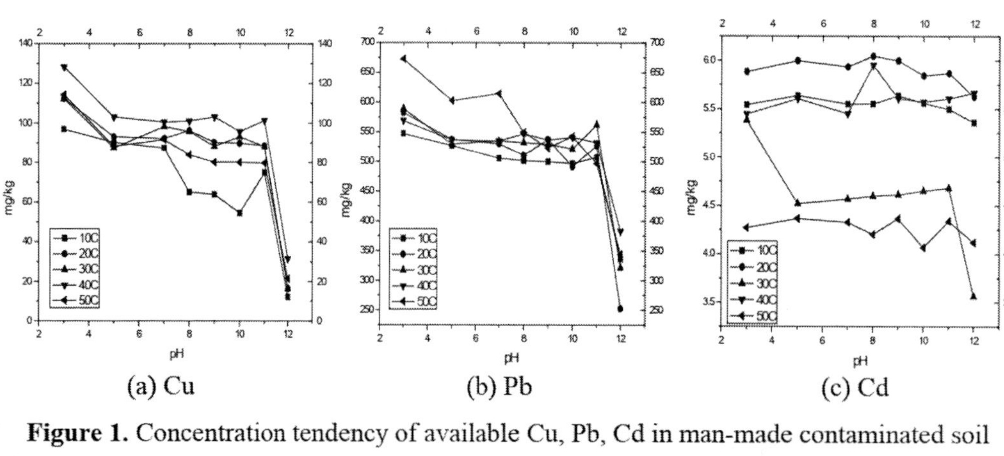

(a) Cu (b) Pb (c) Cd

Figure 1. Concentration tendency of available Cu, Pb, Cd in man-made contaminated soil

3.2. The impact of humic acid on different heavy metals

3.2.1. The impact of humic acid on Cu. The ratio of humic acid, pH, temperature all have an obvious impact on the concentration of available Cu in soil. As is seen from fig.2, the concentration of available Cu in soil declined sharply when the ratio of humic acid or pH rose, while temperature showed little influence on it under alkaline environment. The declining trend is comparatively smooth when the ratio of humic acid varied between 0 and 1.00%, however, it become somewhat steep after continuing increasing.

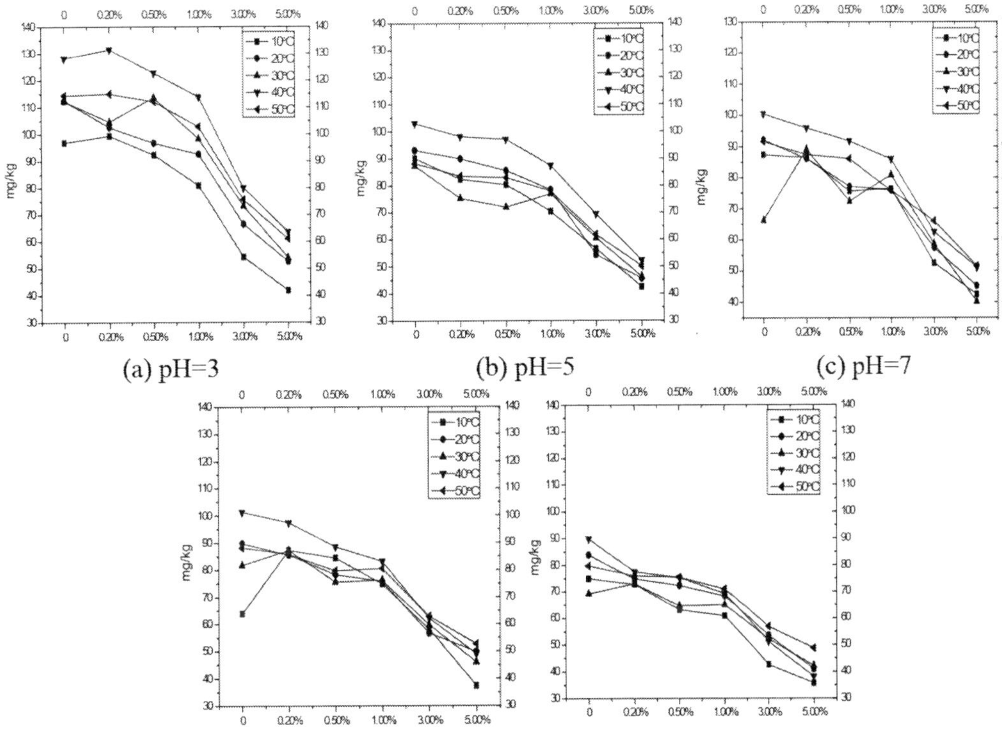

(a) pH=3 (b) pH=5 (c) pH=7

(d) pH=9 (e) pH=11

Figure 2. Concentration tendency of available Cu

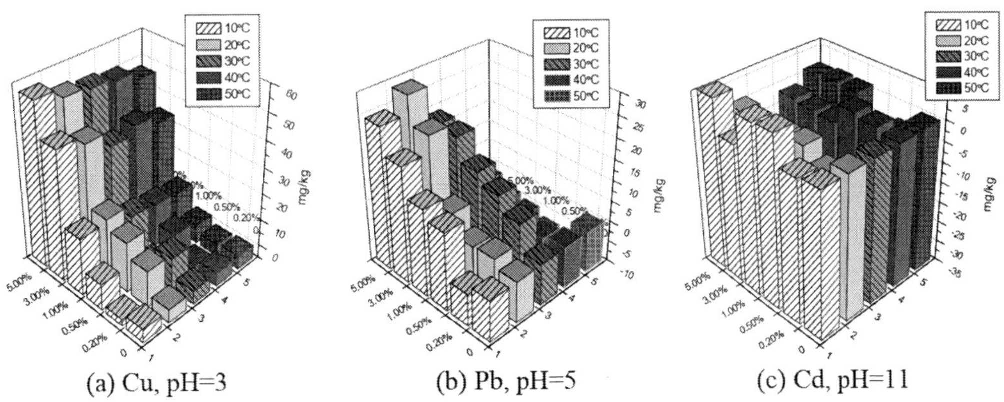

(a) Cu, pH=3 (b) Pb, pH=5 (c) Cd, pH=11

Figure 3. The influence of humic acid on available

The ratio of humic acid and temperature both have an obvious impact on the removal of available Cu in soil. Fig.3(a) shows that the removal efficiency was around 0~55% basically. This ose with more humic acid added, though it won't be much affected by pH. Under acidic environment, its negative correlation with temperature was obvious, and the removal efficiency worsened under 30°C with the increasing of pH.

From fig.2 and fig.3(a), the more the humic acid added, the less the concentration of available Cu in soil solution was. Humic acid is a kind of organic matter, whose inner structure contains not only abundant benzene ring but also some complex functional groups such as hydroxyl group (-COOH) which can provide electrons to coordinate with heavy metals to form into complex compounds or chelates. The coordination mentioned above is positively related to the concentration of humic acid. Meanwhile, humic acid can easily be adsorbed in the surface of soil colloid. Therefore, adsorption sites in soil particle will increase. It is the combined effects of coordination and adsorption that makes the concentration of available heavy metals decrease in a comparatively large scale [9].

3.2.2. The impact of humic acid on Pb. The ratio of humic acid, pH, temperature all have an obvious impact on the concentration of available Pb in soil. From fig.4, the concentration of available Pb in soil declined when the ratio of humic acid, temperature or pH rose. And, the declining trend was obvious when the ratio of humic acid varied between 3.00% and 5.00%. But, it reached a balance state on the whole under alkaline and high temperature environment.

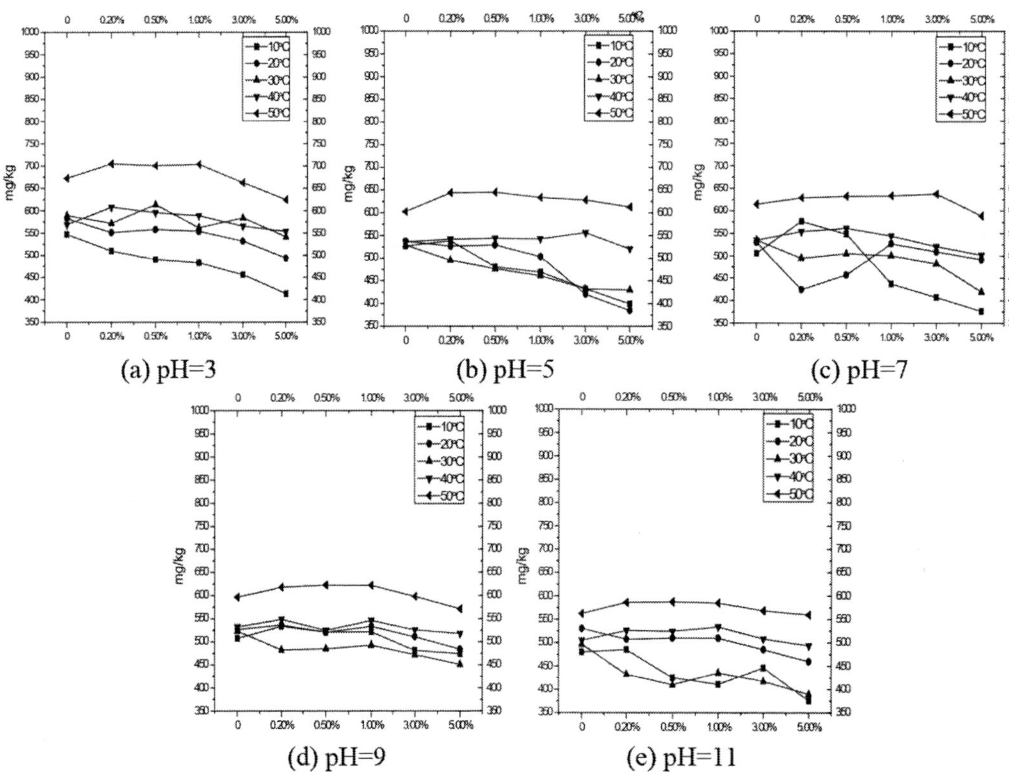

(a) pH=3 (b) pH=5 (c) pH=7

(d) pH=9 (e) pH=11

Figure 4. Concentration tendency of available Pb

The ratio of humic acid, pH and temperature both have a certain impact on the removal of available Pb in soil. As is seen from fig.3(b), the removal efficiency was around -10%~30% basically. This rose with more humic acid, lower temperature and lower pH. The best results appeared in 20°C (under acidic or neutral environment), or 30°C (under alkaline environment).

IOP Conf. Series: Materials Science and Engineering **207** (2017) 012089 doi:10.1088/1757-899X/207/1/012089

Fig. 3(b) and fig.4 show that, humic acid can efficiently remove Pb in soil, and its mechanism is quite similar to Cu. With the increasing of pH in soil solution, the concentration of available heavy metals droped. The influence of pH towards the adsorption efficiency of humic acid can be explained in three aspects listed below. Firstly, under lower pH, the dissolution-precipitation equilibrium of Cu, Pb, Cd is destroyed, thus freeing the electrons. In contrast, new dissolution-precipitation equilibrium comes into being under higher pH, thus making the concentration of available heavy metals drop. Secondly, causes of the dropping lie in the tendency to form more carbonate complexes under alkaline environment. Thirdly, more and more humic acid dissolve in soil solution when pH rises, which results in the rising of conditional stability constants between humic acid and Cu, Pb, Cd complexes. This, furtherly results in the reinforcement of their complexing coordination [10].

3.2.3. The impact of humic acid on Cd. The ratio of humic acid and temperature both have a certain impact on the concentration of available Cd in soil while pH shows little influence. Fig.5 shows that on condition that other things being the same, the concentration of available Cd increased when the ratio of humic acid ranging 0~0.20%, however, it droped when more humic acid added (the ratio gets 5.00%).

As is seen from fig.5, there is no direct relationship between the concentration of humic acid and its adsorption on available Cd. On one hand, the existence of metals ions such as Pb will prevent Cd from binding with adsorption sites of humic acid in soil particle, which is resulted from two main reasons. Firstly, the adsorptive dynamics of soil confirm that the latter adsorbed Pb will replace the former adsorbed Cd. Secondly, law of mass action also may account for this antagonistic action. For the concentration of Pb is much higher than Cd's in mixed contaminated solution, so soil tends to adsorb the higher one, namely, Pb [11]. On the other hand, because the formation of Cd and humic acid is a kind of weak coordination compounds (whose stability constants are smaller than of Pb and Cu), this makes it doesn't have much affinity with solid surface [12].

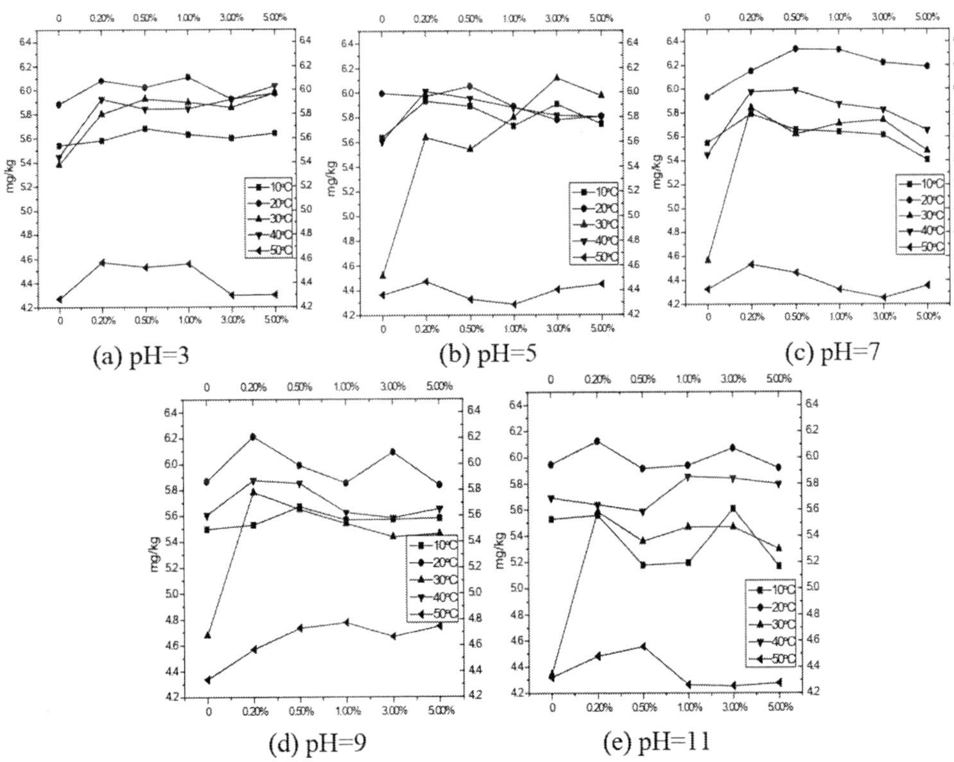

Figure 5. Concentration tendency of available Cd

The removal efficiency of humic acid on available Cd in soil was not ideal. Fig.3(c) shows that humic acid could barely remove available Cd for the highest removal efficiency was just about 5% only in strong alkaline and low temperature environment.

For Cu, it's concentration of available fraction reached its peak at 40℃; for Pb, that number was 50℃; and for Cd, it was 20℃. Comparatively speaking, unlike pH and the ratio of humic acid, there is no such thing as regular pattern of the impact of temperature on available heavy metals. This may be explained with the different influence of different temperatures on different heavy metals' activities and the stability of their complexes, which needs a further study to be explained more specific.

3.3. The impact of humic acid on migration rate of Cu, Pb, Cd in contaminated solution

Humic acid has better removal efficiency on heavy metals in contaminated solution. As fig.6 shows, the removal efficiency of humic acid aimed at Cu and Cd could be over 98%. The removal efficiency of Pb even reached 99.996%. However, the counterparts in contaminated soil were 57.283% (Cu), 2.645% (Cd), 15.485% (Pb) respectively. There are competitive absorption and antagonistic action among various heavy metals in contaminated soil while no outside factors will affect the removal efficiency for there only exists single heavy metal in contaminated solution. The existence of competitive ions may influence the adsorption efficiency of humic acid toward Cu^{2+}, Pb^{2+}, Cd^{2+}: the competitive ions are likely to lower the adsorption quantity of themselves firstly, then it is the competition that results in different adsorption amount of each ions. On the contrary, high concentration of Cd^{2+} is able to promote the adsorption of Cu^{2+} instead, which is likely to be the result of the existence of adsorption sites in competition. Therefore, this is shown as preferential adsorption of different adsorbents toward these three ions [13]. Besides, the removal efficiency can be influenced by quite a lot factors (the physicochemical property and composition of soil, the characteristics of heavy metals, etc.), because different property and composition of soil may lead to different adsorption mechanism and form different heavy metal compounds. In the process of adsorptive dynamics, the adsorption of Pb in soil may mainly be the result of ion-exchange reaction and precipitation reaction. However, precipitation reaction may play a major role in the adsorption of Cd [14].

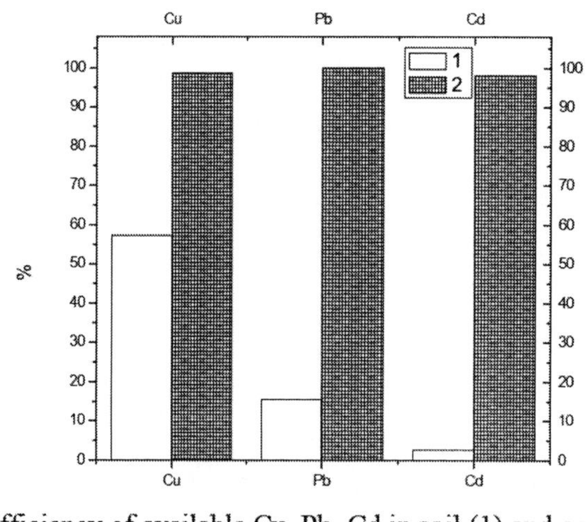

Figure 6. Removal efficiency of available Cu, Pb, Cd in soil (1) and contaminated solution (2)

4. Conclusion

1) Humic acid can effectively lower the concentration of available state of Cu and Pb but slightly influence Cd. The maximal removal efficiency of Cu can reach 57.480%, but 28.372% for Pb and only 6.528% for Cd.

2) pH plays an important role in the stabilizing process of humic acid toward Cu, Pb and Cd in the soil. The rise of pH will cause an obvious decline in the concentration of available heavy metals.

3) Temperature has a certain impact on the stabilizing process of the availability of Cu, Pb and Cd in soil. For instance, when pH=5 and the ratio of humic acid is 5.00%,Cu and Cd get to their maximal removal efficiencies at 10°C, while for Pb it is at 20°C. These differences are possibly caused by different activity and stability of products under different conditions, whose specific mechanisms still need to be further studied.

Acknowledgments
This work was financially supported by the Youth Foundation of Zhongnan University of Economics and Law (31541311313) and the found of 2016 SITP.

References

[1] Ministry of Environmental Protection. Ministry of Land and Resources. The investigation of nationwide soil[R]. 2014-4-17. (in Chinese)
[2] Zhong-qiu Zhao, Yong-guan Zhu, Yun-long Cai. The transformation and influence factors of Cd in the soil-botanic system [J]. Ecology and Environment. 2015, 14(2): 282-286. (in Chinese)
[3] Nadeem Sarwar a, Muhammad Imran,et al. Phytoremediation strategies for soils contaminated with heavy metals: Modifications and future perspectives[J]. Chemosphere, 171 (2017) 710-721.
[4] Udom B E, Mbagwu J S C, Adesodun J K, et al. Distributions of zinc, copper, cadmium and lead in a tropical ultisol after long-termdisposal of sewage sludge [J]. Environment International, 2004, 30(4): 467-470.
[5] Ya-li Zhang, Qi-rong Shen. Yang Jiang. The ameliorative effect of organic fertilizer on the Cd contaminated soil [J].ACTA Pedologica sinica. 2001, 38(2), 212-218. (in Chinese)
[6] Bruna Alice Gomes de Melo, et al. Humic acids: Structural properties and multiple functionalities for noveltechnological developments [J]. Materials Science and Engineering C, 62 (2016) 967–974.
[7] Chen Jing, Huang Zhanbin. Effect of Humic Acid on Soil Restoration [J]. Humic Acid. 2014(4):30-34. (in Chinese)
[8] L.M. Yates, R.von Wandruszka. Decontamination of polluted water by treatment with a crude humic acid blend [J].Environmental Science & Technology, 1999, 33(12):2076-2080.
[9] Yu-feng Jiang, Jian-mei Yuan, Zi-yang Lu, Ai-ping2 Wang, Hui Chen. The effect of humic acid on species of Cu, Cd, Pb, Zn in sew age farm. Journal of Northwest Normal University (Na tur al Science), 2005,41(6):42-46.(in Chinese)
[10] YU Gui-Fen, JIAN G Xin, SUN Lei. A Review for Effect of Organic Substances on the Availability of Cadmium in Soils [J]. ACTA ECOLOGICA SINICA. 2002, 20(5): 770-775. (in Chinese)
[11] Fei Song, Yu-wen Guo, Xiao-yi Liu, The combined pollution of Cd, Pb, Zn in the soil [J]. ACTA scientiae circumstantiae.1996, 16(4). (in Chinese)
[12] Environmental Chemistry [M]. edited by Shu-gui Dai. Beijing. Higher Education Press. 1997: 155-157. (in Chinese)
[13] Shu-Yan Hu (2008). Adsorption and competitive adsorption of heavy metals on humic acid and fulvic acid (in Chinese)
[14] Peng Guo. Adsorption Kinetic Characteristics of Heavy Metals in Urban Soils and the Relationship with the Physicochemical Properties of Soils [J]. Environmental Protection Science. 2008 34(6):23-26. (in Chinese)

IOP Conf. Series: Materials Science and Engineering **207** (2017) 012090 doi:10.1088/1757-899X/207/1/012090

Detection Mechanism of Parallel Defect using Scanning Inductive Thermography

Xianzhang Zuo[1], Benchu Song[1], Yongjiang Hu[1], Yunze He[2,*]

[1]Department of UAV Engineering. Ordnance Engineering College, Shijiazhuang 050003, China

[2]College of Electrical and Information Engineering, Hunan University, Changsha, 410082, China

*Corresponding author e-mail: hejicker@163.com

Abstract. Aiming at the requirement of workpiece integrity for parts processing line, on-line detection using inductive heating thermography for the moving workpieces on the assembly line is studied. In this paper, the detection mechanism of pulsed eddy current thermography for moving workpieces defects is analysed. A two-dimensional model of a magnetic material (45 steel), on which there is a crack parallel to the coil is established by the finite element software named COMSOL 5.2. By analysing the changes of the temperature curves, normalized curves and the temperature difference curves, the optimal detection area for parallel cracks is proposed. The consistency of the conclusions is verified by the experimental platform. The paper can provide a theoretical guidance for quantitative detection using eddy current thermography.

1. Introduction

With the development of astronautics and new energy technology, it is more and more important to ensure the quality of products and engineering by nondestructive testing technology. Eddy Current Pulsed Thermography (ECPT) [1, 2] is a very effective nondestructive testing method. Material failures like defect, fatigue, corrosion and residual stress, etc. have been detected and evaluated using ECPT [3, 4].

In present researches, most works related to ECPT are focused on static configuration [5, 6], which lead to a low detection efficiency and cannot be directly applied to industrial production. Today, with the development of IR camera, its view is large enough for some samples. There are three kinds of Scanning configuration under research. Scanning laser spot thermography is mainly used for crack detection in metal and semi-conductor based on the lateral heat conduction [7]. Scanning Induction Thermography (SIT), which heats the sample in linear scanning style in the camera view, is more efficient and reliable[8]. The dynamic configuration, used in this paper, called Line Scanning Thermography (LST) [9, 10] patented by NASA is shown in Fig. 1.

IOP Conf. Series: Materials Science and Engineering **207** (2017) 012090 doi:10.1088/1757-899X/207/1/012090

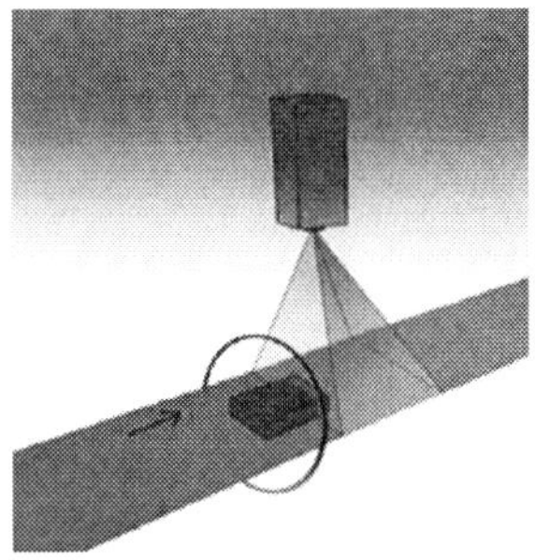

Fig. 1. Schematic of PECT dynamic detection

In this paper, a two-dimensional model is established by the finite element software. The temperature curves, normalized curves and differential curves of the defect and the reference point are given to analyze and study the detection mechanism of PECT dynamic detection for parallel cracks, which lay a foundation for the quantitative identification of surface defects on workpieces.

2. Detection Mechanism

For isotropic materials, on the surface of workpieces the temperatures of points on dotted lines should be equal on account of the same distance from the coil in the ideal case as shown in Fig. 2 (a). Therefore, cracks will be detected as temperatures of fixed pixel change when the coil and the thermal camera are relatively stationary while the workpieces move.

The detection mechanism for parallel defects (such as minor cracks) is shown in Fig. 2 (b). The side that the coil first passes is defined as the front side and the one the coil passes afterwards is defined as the rear side. The heat conduction is prevented near the parallel cracks on the surface of the workpiece. Therefore, the temperature of both sides of the cracks will change and appear as "hot zone" or "cold zone" on the camera. Because parallel cracks affect only the heat diffusion and have no effect on the eddy current, they will be detected in Area II III, and IV. But with the change of the relative position between the coil and the crack, the temperatures of two sides of the crack must be equal at some time. That is, Area III may have a blind spot. Therefore, it is important to determine the optimal detection area.

To explain the mechanism for parallel defects accurately, the model was introduced at first. Secondly, the simulation results using line source move in camera view [11, 12] were done. Next, the experimental results using LST were presented. Finally, conclusions were given.

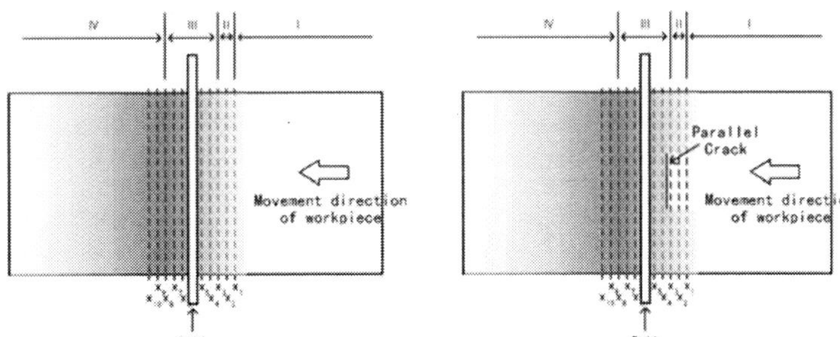

(a) Workpiece without Defect (b) Parallel Crack

Fig. 2. Top View of Dynamic Detection

3. Numerical Studies

3.1 Simulation Model

This paper uses the finite element simulator COMSOL5.2a for the two dimensional simulation of the eddy current field and the temperature field of 45 steel, a typical ferromagnetic material used in engineering. The temperature distribution near the crack is analyzed.

The workpiece is a 140mm × 20mm steel plate, as shown in Fig. 3. There is an artificial crack on the surface of the workpiece. The center of the crack, which is 1mm wide and 1mm deep is located at 0mm. Definition of sampling point: A is the front side of the crack that the coil passes first, B is the rear side that the coil passes afterwards, and X is the reference point representing areas with no defect. The coil is reduced to a long straight wire parallel to the crack and perpendicular to the screen. The excitation frequency is 256 kHz, and excitation current is 100A. The movement distance is 80mm, the coil moving from -40mm to 40mm at 20mm/s. The initial temperature is 293.15K. The electromagnetic and thermal parameters of the materials are shown in Table 1.

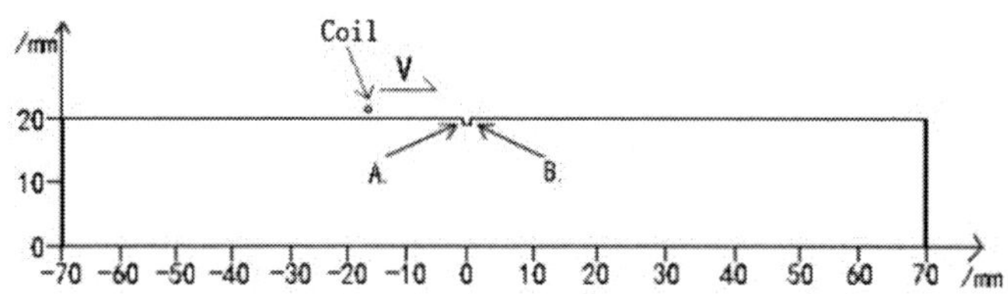

Fig. 3. Simulation Model

Table 1. The electromagnetic and thermal parameters of the materials

Parameters	Steel
$\sigma/(S\cdot m^{-1})$	5.4348×10^{6}
μ_r	200
α/K^{-1}	1.22×10^{-5}
$\rho/(kg\cdot m^{-3})$	7870
$C_p/(J\cdot kgK^{-1})$	440
$k/(W\cdot mK^{-1})$	44.5
$\delta/(m)$	3.328×10^{-5}

3.2 Analysis of Simulation

The simulation results are shown in Fig. 4 (a) when speed is 20mm/s and excitation current is 100A. The coil moving from -40mm to 40mm, the excitation time is 4s. The temperature curves change at -16 mm, so they are intercepted from -16mm to 24mm. As the distance between the coil and crack changes, three curves separate in different degrees. When the coil moves to -11mm, the temperature of A rises the fastest and becomes "hot zone"; X is the second; B rises the slowest and becomes "cool zone". Three temperature curves cross at -1mm. As the coil continues to advance, the temperature rise speed of A slows and A changes from "hot zone" to "cool zone"; B tends to be the same as X. The normalized curve is shown in Fig. 4 (b). During temperature rise, A reaches its maximum first, X is the second, and .the upward trend of B appears the latest, but the slope is maximum. During temperature drop period, A has the slowest descent rate while B and X tend to be consistent.

Compared with the non-defective areas, A cannot transfer heat to B due to the insulation of parallel cracks. A will accumulate more heat, thus presenting a higher temperature, while B presents a lower temperature because of a small amount of heat. At the end of heating period, the heat induced at B is blocked by the defect, so temperature B exceeds A's. In the process of cooling, A cannot accumulate the heat transferred from B, thus presenting a low temperature. X is a sampling point in areas with no defect. As the surface of the workpiece is smooth and continuous, the temperature curve of X is smooth too.

To eliminate the influence of ambient temperature, A, B and X are subtracted separately and the temperature change is studied as shown in Fig. 4 (c).

$$W1 = T_A - T_B \tag{1}$$

$$W2 = T_A - T_X \tag{2}$$

$$W3 = T_B - T_X \tag{3}$$

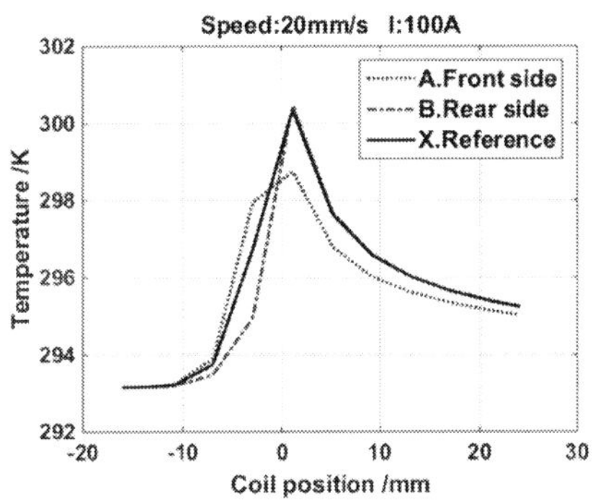

(a)Position of Coil - Temperature

Curves of Sampling Points

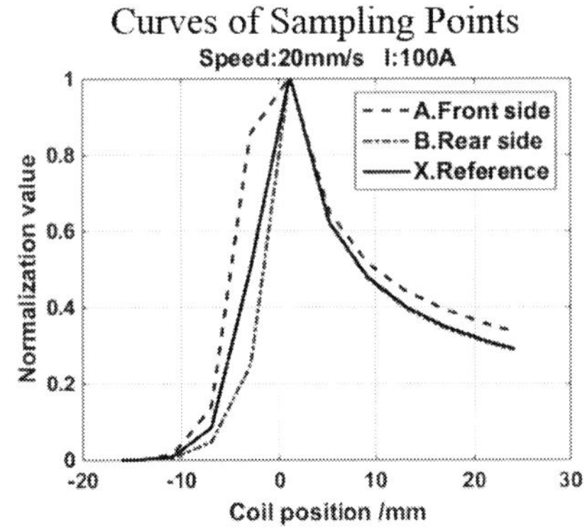

(b)Normalized Curve of Fig. 4 (a)

IOP Conf. Series: Materials Science and Engineering **207** (2017) 012090 doi:10.1088/1757-899X/207/1/012090

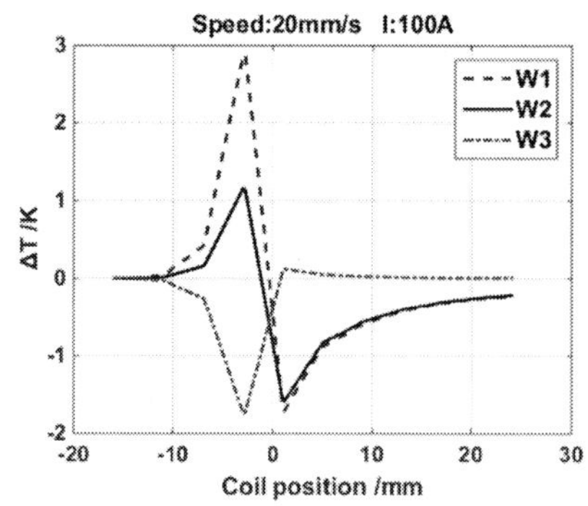

(c) Curves of Temperature Difference

Fig. 4. Curves of Sampling Point Temperature and Coil Position

It can be seen when the coil moves to -3mm (the defect is 3mm in front of the coil), W1, W2 and W3 peak, with steep and narrow rise and fall intervals, indicating that the contrast (W1) of "hot and cold" between the two sides of the defect is the most obvious. At the same time, the temperature difference between the defective and nondirective areas (W2, W3) is maximum, so the defect profile can be revealed most clearly. At 1 mm (the defect is 1mm behind the coil), W1 and W2 peak, the fall interval of which is more moderate and larger in width. The profile of the front side of defect is more obvious now. However, W3 is approximately zero, indicating that the rear side of the defect almost overlaps the non-defective areas.

Therefore, there are two best observation moments in the defect detection of uniform velocity motion. For the detection of parallel crack under the above excitation conditions, the best detection positions are at the points when the coil is 3mm on the front side from the defect and 1.5mm on rear side from the defect, because the differences of the sampling points of the defect are large here. The detection interval is short when the coil moving to 3mm on the front side of the defect and the defect shows "hot zone" and "cold zone". The detection interval at 1.5mm at the rear side of the coil is wide, but only "cold zone" appears on the front side of the defect.

4. Analysis of Experiment

As the coil will block the vision, the experiment, different from the simulation, cannot get the full-time temperature at sampling points. Therefore, the experimental setup is shown in the Fig. 5 (a), consistent with Fig.2. The thermal camera and the excitation source remain stationary and the workpiece is moving on the chassis pushed by the slipway. The excitation source is HB-X5K from WUXI HUABO GAOPIN TECHNOLOGY COMPANY, which has an excitation power range of 0.5-5kW and a excitation frequency of 1MHz.

Fig. 5 (b) shows a thermal image, a fuzzy defect profile can be seen. Temperature curves of sampling points (M1-M5 and L1-L5) are shown in Fig. 6. The sampling interval is 1.5mm. A parallel crack is observed on the left of the coil in the picture.

IOP Conf. Series: Materials Science and Engineering **207** (2017) 012090 doi:10.1088/1757-899X/207/1/012090

(a) Experimental Setup

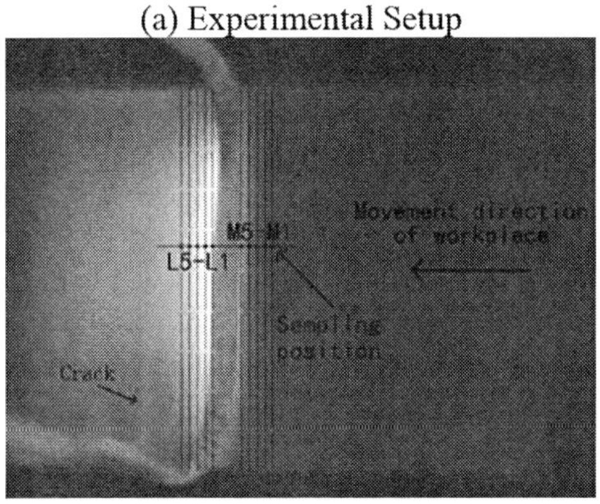

(b) Image of Detection

Fig. 5. Experimental Image

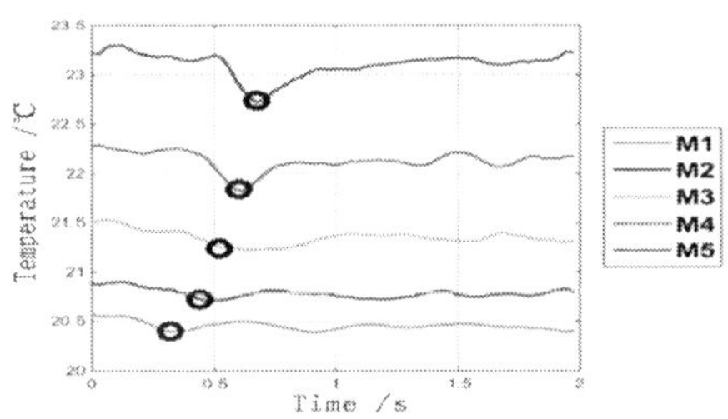

(a) Different Positions on the Front Side of the Coil

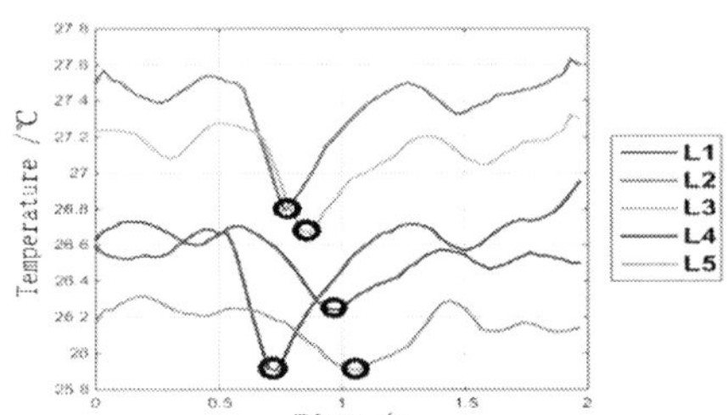

(b) Different Positions on the Rear Side of the Coil

Fig. 6. Temperature Curves of Sampling Points

Temperature curves of sampling points are shown in Fig. 6, when the moving speed is 20mm / s and excitation power is 3750W. The temperature of each sampling point is constant, unless the defect passes, which results in a low temperature. It is consistent with the results of Fig. 4 that B is less than A and X in the heating phase. On the front side of the coil, as the sampling points move closer to the coil, both the constant temperature and changes at the defect increase. On the other side of the coil, with the sampling points moving away from the coil, the constant temperature does not change monotonically. For example, the temperature of L2 is the highest but the temperature change at L1 is maximum.

In actual detection, the maximum temperature change of the workpiece is not directly below the coil. The point with the maximum temperature is not the best place to detect defects. In this experiment, defects can be identified on both sides of the coil, with the recognition capability greater on the rear side than on the front side.

5. Conclusions

The detection method is effective for the surface cracks on the moving workpieces. By analyzing the simulation and experimental data, fives are drawn:

(1) The temperature of the front side of the defect increases earlier and faster, the descending process is gentler and the absolute value of temperature change is smaller. While the temperature rise of the rear side of the defect is later and the absolute value of the temperature change is bigger.

(2) The maximum temperature of the workpiece is on the rear side of the coil. There are two optimal observation positions before and after the coil rather than below the coil. Experimental results show that the detection effect on the rear of the coil is better.

Acknowledgements

The work was supported by National Natural Science Foundation of China (Grant No. 61501483) and Natural Science Foundation of Hebei Province (Grant No. 51307183).

References

[1] Wilson, J,Tian, G. Y,Abidin, I. Z., etc. Pulsed eddy current thermography: System development and evaluation [J]. Insight: Non-Destructive Testing and Condition Monitoring, 2010, 52 (2): 87-90.

[2] Al-Qubaa, A. R.,Tian, G. Y.,Wilson, J, etc. Feature extraction normalized cross-correlation for pulsed eddy current thermographic images [J]. Measurement Science and Technology, 2010, 21 115501.

[3] Gao B, Bai L, Woo WL, Tian G. Thermography pattern analysis and separation. Applied Physics Letters. 2014; 104:251902 - -5.

[4] Li K, Tian GY, Cheng L, Yin A. State Detection of Bond Wires in IGBT Modules Using Eddy Current Pulsed Thermography. IEEE Transactions on Power Electronics. 2014; 29:5000-9.

[5] Chang Dong, Xianzhang Zuo, Gui Yun Tian, Sumin Qian. Analysis of the Crack Orientation Effect on Pulsed Eddy Current Stimulated Thermography [J]. Journal of Shanghai Jiao Tong University, 2013(5):779~785. (in chinese)

[6] Ruizhen Yang, Yunze He, Bin Gao, Gui Yun Tian, Jianping Peng. Lateral heat conduction based eddy current thermography for detection of parallel cracks and rail tread oblique cracks [J]. Measurement 2015, (66):54~61.

[7] Schlichting J, Ziegler M, Maierhofer C, Kreutzbruck M. Flying laser spot thermography for the fast detection of surface breaking cracks. 18th World Conference on Nondestructive Testing. Durban, South Africa2012.

[8] Renil Thomas K, Krishnan Balasubramaniam. Scanning Induction Thermography (SIT) for Imaging Damages in Carbon-Fibre Reinforced Plastics (CFRP) Components [J]. AIP Conference Proceedings, 2015, (1650):306~313.

[9] Kaltmann D. Quantitative line-scan thermographic evaluation of composite structures. Melbourne:

RMIT University; 2008.
[10] Varis J, Rantala J, Hartikainen J. An infrared line scanning technique for detecting delaminations in carbon fibre tubes. NDT&E Int. 1996; 29:371-7.
[11] Hoshimiya T, Hoshimiya J, Tsuda M. Line-focus Beam-scan Time-domain Active Thermography with External Control. 11th International Conference on Quantitative InfraRed Thermography. Naples Italy2012.
[12] Hoshimiya J, Hoshimiya T. Whole-electronic line-focus light-scanner for active thermography. The 12th International Conference on Quantitative InfraRed Thermography. Bordeaux2014.

Elementary analysis on application of ANSYS/LS-DYNAD in the detection of integrate pile

Leting Zhang[1, a], **Yaping Wu**[2, b], **Mingqiang Wei**[3, c]

[1]Lanzhou JiaoTong University, Gansu Vocational and Technical College of Communications, Gansu Lanzhou 730030, China
[2]Lanzhou JiaoTong University, Gansu Lanzhou 730030, China
[3]Lanzhou JiaoTong University, Northwest University for Nationalities, Gansu Lanzhou 730030, China

[a]417363499@qq.com, [b]wypsw@163.com, [c]258664939@qq.com

Abstract. In this paper, by using the ANSYS/LS-DYNA nonlinear dynamic finite element analysis method. The numerical simulations for the integrity pile, and get the rules and characteristics of stress wave about complete pile. Verify the feasibility of the method of ANSYS/LS-DYNA nonlinear dynamic finite element analysis in the pile integrity detection applications. The research results can provide guidance for the pile integrity detection of different types. Then this paper simulated pile necking. And we get the patterns and characteristics of stress wave reflection about the necking pile, which show the method of ANSYS/LS-DYNA nonlinear dynamic finite element analysis can detect the type and location of pile foundation defect accurately.

1. Introduction

Among forms of foundation, pile foundation has high bearing capacity and smaller settlement. As the pile foundation is able to load transfer from the cover layer to the stable layer, it can achieve a safe and reliable requirements. With the high speed development of society and economy, the pile foundation has been widely used in many engineering fields such as transportation, civil and industrial building, port, bridge etc. However, it has strong concealment in the course of the construction and inevitably leads to various problems in the construction phase being difficult to detect. It is Necessary to accurately determine whether there are defects in the pile foundation. Therefore, in order to ensure the quality of construction projects and the degree of the specific location of the defect, the pile foundation must be strictly tested, to prevent accidents to taking remedial measures economical, efficient and easy to operate. Thus, only by improving the detection reliability of pile foundation, whole pile foundation construction quality and safety can be guaranteed. So, with the continuous progress of human science and technology, theoretical methods and engineering applications of pile foundation detection become the research areas of many scholars. For a long time, scholars have concluded that some of the more effective methods, such as method of reverberation-ray matrix, dynamic testing techniques based on the wave theory, etc.

In this paper, solving nonlinear analysis program of ANSYS/LS-DYNA conducts a preliminary numerical simulation to display the stress wave pile foundation, establishes the model of complete pile, and draws a waveform graph complete pile of post-processing program by Lsprepostd to compare with the theoretical curve by method of reverberation-ray matrix. This shows that the applications of ANSYS / LS-DYNA is feasible in the pile integrity testing, and studies further the stress wave propagation characteristics of the pile.

2. The introduction of ANSYS

ANSYS finite element analysis software is a multi-purpose computer design program, which can integrate all kinds of structures, fluids, thermal, acoustic, electromagnetic, and collision sports, so that it has been widely used in various engineering fields in recent years. The finite element method (the finite element method) is a numerical method used in the current engineering analysis most widely. It is highly valued in the engineering sector, because it has high versatility and effectiveness, etc. With the rapid development of computer technology and science, ANSYS finite element analysis software has now become an integral part of computer-aided design and aided manufacturing. In this paper, ANSYS only provides the necessary pre-processing and post-processing for LS-DYNA, and specific solving procedure is completed by the LS-DYNA solver.

LS-DYNA is a highly versatile display dynamic analysis package which has complete function. It can simulate reality in a variety of complex issues, particularly suitable for solving a variety of two-dimensional, three-dimensional geometric nonlinearity, material nonlinearity, non-contact linear, explosions and other problems; LS-DYNA is focusing on showing algorithm for solving, especially for analyzing the impact of various nonlinear structural dynamics.

3. Computational model of complete pile

3.1. The basic assumption

The pile foundation, the pile length is generally much larger than the cross-sectional diameter or side length of the pile. So we can treat it as a one-dimensional elastic body. When momentary vibration is applied to top of the pile, there will be elastic wave passing along the pile, According to the basic principles of stress wave reflection method. The propagation of elastic wave obeys theory of one dimensional wave equation, and meets the following basic assumptions:

(1) Top of the pile is applied to the momentary vibration in the elastic range.

(2) Pile material is homogeneous or segmented homogeneous and isotropic.

(3) Cross-section of the pile remains flat when it happens longitudinal vibration, and the stress is distributed uniformly in each of the cross-section, and they are the axial stress, and all the stress components are zero in the other direction.

(4) The length of the longitudinal wave is much larger than the cross-sectional size of the pile. So do not consider the lateral displacement will affect the longitudinal motion.

3.2. Computational models

First, select a single complete pile. Pile - Soil relations is soil around the pile but bottom soilless. The pile length is 8m, and pile sectional side length is 0.5m, and soil-sectional side length is 0.6m, and soil depth is 9m. The top of the pile is expressed with 1, and pile bottom with 2. According to the above parameters and using ANSYS/LS-DYNA, it can build a finite element model of complete pile, showing in Figure 1. Selecting the first 49 nodes at the top of the pile top as the research object, recording the coordinates ID displayed on the toolbar, it can be named that x equals -0.08333333582, y equals -0.08333333582, z equals 0. The wave curve is shown in Figure 2.

IOP Conf. Series: Materials Science and Engineering **207** (2017) 012091 doi:10.1088/1757-899X/207/1/012091

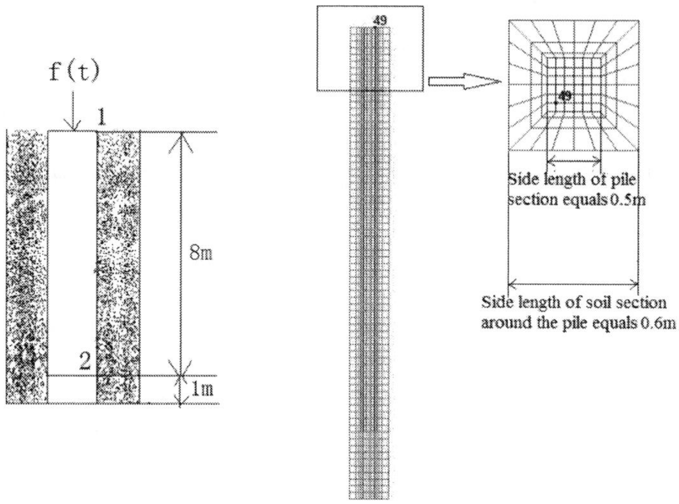

Fig. 1 Finite Element Simulation of relationship between complete pile and soil

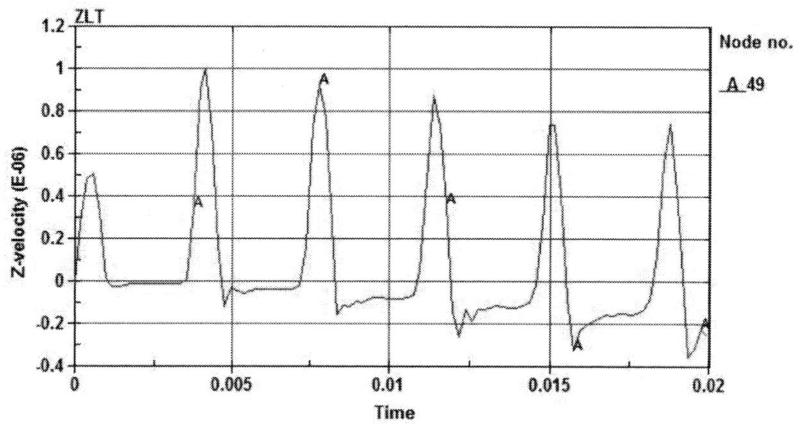

Fig. 2 Velocity waveform curve in the z-direction about the 49 node in the top of the complete pile

3.3. The analysis of the computation result

(1) Contrasted with the theoretical curve

Using the principle of reverberation ray matrix convention to building the MATLAB program, it will produce a complete pile top node velocity waveform curve in the Z direction, which is the theoretical curve. Figure 3 is comparison chart of a model curve and the theoretical curve.

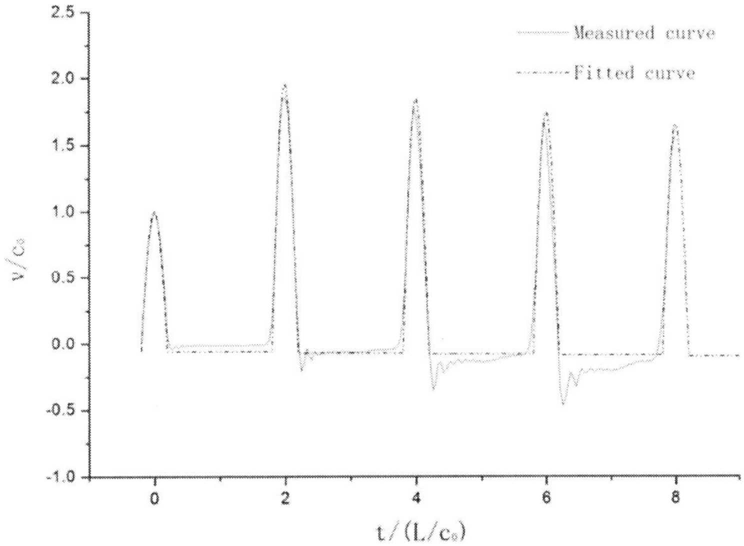

Fig. 3 Comparison between measured curve and fitted curve of complete pile

From the figure 3, the model curve fits well with the theoretical curve, indicating it is able to use ANSYS / LS-DYNA to establish complete pile model and get fine and available data.

(2) Reverse calculate model parameters

By velocity waveform curve of complete pile top the 49 node to reverse calculated model parameters, and it is shown in Figure 2. Elastic modulus of pile is E *equals* $2.7\times10^{10} N/m^2$, and the density is ρ *equlas* $2500 kg/m^3$.

Computational process is following:

① It can obtain the stress wave propagation velocity in the pile. c_0 *equals* $3286.34 m/s$.

② The incident peak time of incident wave is 0.5ms.

③ The first reflection wave's peak time is 5.393ms, and the propagation path is 121, and the total spreading length is 16m. Because of the stress wave propagate from the top of the pile reaches the bottom of the pile, and pile bottom is overhead. So elastic modulus of pile becomes smaller, which means impedance becomes smaller too, resulting the incident wave and the reflected wave are in the same direction. The difference time Δt equals 4.893ms, and get the L *equals* $\Delta tc_0/2$ *equals* $8.04m$, which is very close to the simulation of the pile length.

(3) Waveform characteristic curve analysis

① The complete pile has only two interfaces that top and bottom. Elastic modulus of pile becomes smaller when the stress wave reaches the top or bottom of the pile, which means impedance becomes smaller too, resulting the incident wave and the reflected wave are in the same direction. There is a relatively smooth transition between any two of the reflected wave of reach pile interface. The entire waveform curves no obvious glitches phenomenon, which is the typical characteristic of a complete pile different from defective pile.

② The reflected wave reaches the bottom of the pile in the correct position, and the peak of the peak is about twice the peak incident;

③ All the reflection wave peak appear the natural decay phenomenon, which indicate damping pile and the pile of subsoil body has consumed the role of stress wave signal.

(4) Comparison between measured values and fitted values of complete pile is shown in Table 1.

Tab. 1 Comparison between measured values and fitted values of complete pile

peak time of the reflected wave (ms)	measured values (m)	fitted values (m)	errors	propagate path
5.393	8.04	8	0.5%	121

4. Calculation model of necking pile

4.1. Causes of defects

Necking pile is the phenomenon that pile cross-sectional dimension of the pile local necking pile narrow after the molding. There are many causes constriction, for instance, in the usual place pile, the infusion finished after estuation cause excessive necking, as well as the absence amount of concrete in the tube, or concernment workability not meeting the requirements, will result in a corresponding reduction neck phenomenon. Furthermore, the soil around the pile itself has a more significant impact, because the greater the moisture content of water of the soil, the greater the viscosity, the easier soft soil necking occurs which is due to the higher moisture content of the clay, a common place pile produce construction process to the surrounding soil and squeeze more intense disturbance, and it will bring a greater excess pore water pressure, so after pulling out the tube, the water pressure will act directly on the body of the new cast concrete piles, if the water pressure is greater than the new cast concrete piles lateral pressure, it will produce shrink neck phenomenon.

If the constriction of the cross-sectional dimension of the pile is reduced, the strength will be decreased, and the bearing capacity will fall, so the effective and accurate method for detecting the position of the defective pile is necessary.

4.2. Computational models

Pile - Soil relations is the end of a pile has soil but pile has no soil. The pile's length is 8m, and pile's sectional side length is 0.6m, and the length of a partial sectional side in expanding part is 0.5m, and the length of soil-sectional side at bottom of the pile is 1.2m, and the soil height is 1m. The top of the pile is expressed with 1, and pile bottom with 2, and constricted portion starting with 3, and constricted portion of termination point with 4. According to the above parameters and using ANSYS/LS-DYNA, it can build a finite element model of necking pile, showing in Figure 4. Selecting the first 145 nodes at the top of the pile top as the research object, recording the coordinates ID displayed on the toolbar, it can be named that x equals -0.150000006, y equals -0.150000006, z equals 0. The wave curve is shown in Figure 5.

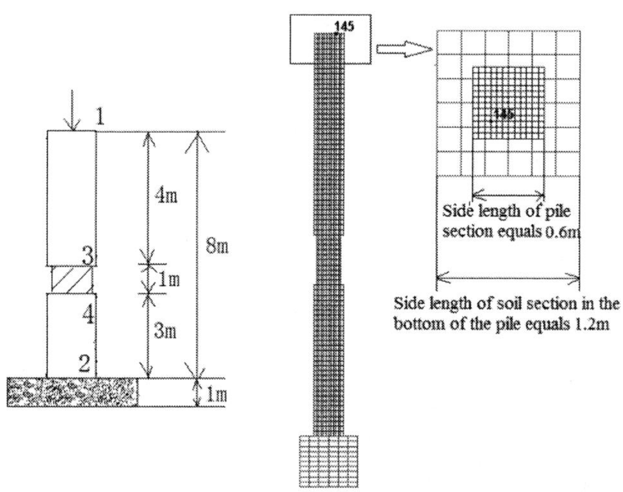

Fig. 4 Finite Element Simulation of relationship between necking pile and soil

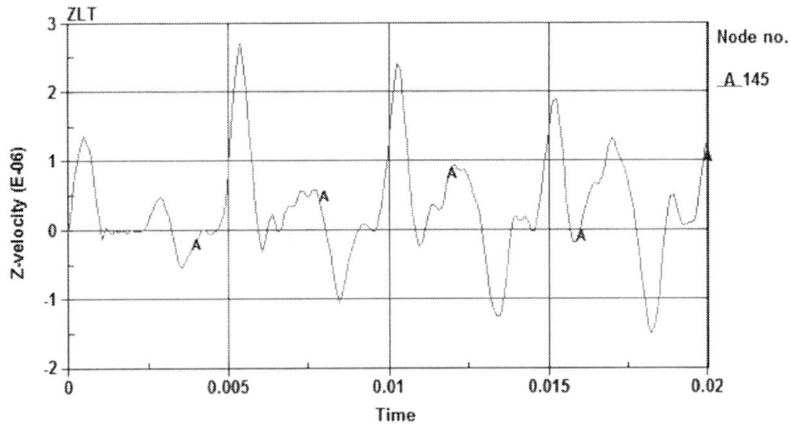

Fig. 5 Velocity waveform curve in the z-direction about the 145 node in the top of the necking pile

4.3. The analysis of the computation result

(1) Contrasted with the theoretical curve

Using the principle of reverberation ray matrix convention in the third chapter to building the MATLAB program, it will produce a complete pile top node velocity waveform curve in the Z direction, which is the theoretical curve. Figure 6 is comparison chart of a model curve and the theoretical curve.

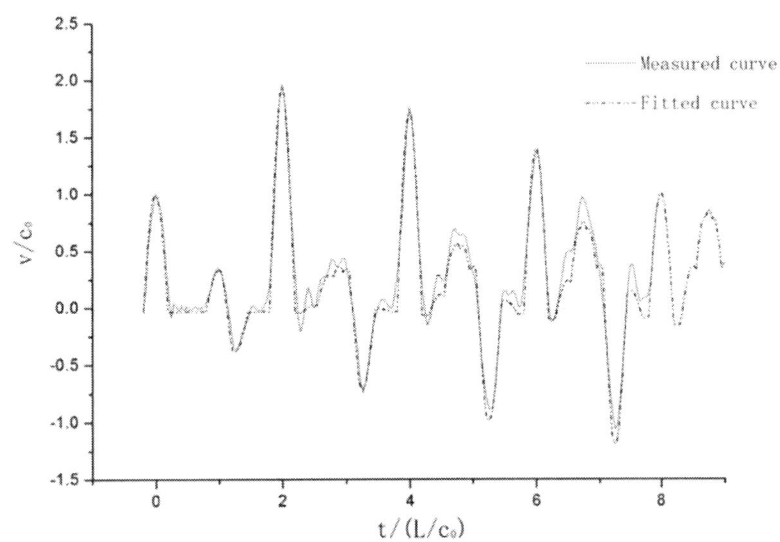

Fig. 6 Comparison between measured curve and fitted curve of necking pile

From the figure 6, the model curve fits well with the theoretical curve, indicating it is able to use ANSYS / LS-DYNA to establish necking pile model and get fine and available data.

(2) Reverse calculate model parameters

By velocity waveform curve of segregation pile top the 145 node to reverse calculated model parameters, it is shown in Figure 2. The Pile elastic modulus is E *equals* $2.7\times10^{10} N/m^2$, and the density is ρ *equlas* $2500 kg/m^3$.

Computational process is following:

① Obtaining the stress wave propagation velocity in the pile. c_0 *equals* $3286.34m/s$.

② The incident peak time of incident wave is 0.489ms.

③ The first reflection wave's peak time is 2.9ms, and the propagation path is 131, and the total spreading length is 8m. Since the stress wave propagation across the interface impedance when it reaches constriction, so cross-sectional area of pile becomes smaller, which means impedance becomes smaller too, resulting the incident wave and the reflected wave are in the same direction. The difference time Δt equals 2.471ms, and the L is the starting position of the constriction, which is very close to pile necking analog design flaw location.

④ The peak time of the second reflected wave is 3.499ms, and the propagation path is 13431, and the total length is 10m. On account of encountering the interface impedance after the stress wave propagates through the constriction, pile cross-sectional area enlarge and the impedance becomes larger, so the reflected wave occurs which is in reverse direction which the incident wave, and the time difference is 0.599ms. We know that L *equals* $\Delta tc_0/2$ *equals* $0.98m$, which is the length of the neck portion and is very close to the length of the construction defect pile analog design.

⑤ The third wave's reflection peak time is 5.395ms, and propagation path is 1342431, and the total propagation length is 16m, which is in the same direction with the incident wave, and the wave peak time difference between them is 4.906ms. We know that L *equals* $\Delta tc_0/2$ *equals* $8.06m$, which is the total length of the pile and is very close to the length of analog design.

⑥ The position of termination point of the pile neck is 4.05 plus 0.98 equals 5.03m, that is very close to 5m which is design position of a pile necking partial termination point.

⑦ The length of following pile under the pile body is 8.06 minus 5.03 equals 3.03m, which is very close to the length of 3m of analog design.

Tab. 2 Comparison between measured values and fitted values of necking pile

type of pile	peak time of the reflected wave (ms)	measured values (m)	fitted values (m)	errors	explanation	Propagate path
necking pile	2.9	4.05	4	1.2%	starting point of the necking portion	131
	3.499	5.03	5	0.6%	ending point of the necking portion	13431
	5.395	8.06	8	0.7%	pile bottom	1342431

(3) Waveform characteristic curve analysis

① In the initial interface and termination interface of the necking part, it will produce reflection wave which is in the same direction with the incident wave or in the reverse, which is caused by the change of impedance what is affected by the change in cross-sectional area of the pile. That means the cross-sectional area of the pile changes more, the wave impedance changes well, and the same direction or reverse reflection wave peaks will become increasingly prominent;

② If reflected wave reaches the bottom of the pile in the correct position, the peak of the peak is about twice the peak incident;

③ For the necking pile, the first reflected wave which will reach the starting point of the reduced-diameter portion and the incident wave in the same direction, and the reflected signal in the necking porting and the incident wave are the same direction too, which is typical characteristics between the pile necking and widening piles, piles segregation;

④ All the reflection wave peak appear the natural decay phenomenon, which indicate damping pile and the pile of subsoil body has consumed the role of stress wave signal.

5. Conclusion

In the paper, using ANSYS/LS-DYNA, it establishes finite element model for dynamic testing of reflection method which is aimed at complete pile. By analysing the final results can be seen the numerical simulation can effectively show that process, rules and features of stress wave in the pile. Besides, this paper also establishes finite element model for dynamic testing of reflection method which is aimed at pile necking. By analysing the final results, it can be seen that the model has fine simulating effect for necking piles and reflects rules and characteristics of the waveform curve of the pile necking.

However, this method is still in its infancy, and the future remains to be further developed so that it can detect the various types of defects of pile foundation.

ANSYS/LS-DYNAD can better simulate the stress wave under transient loads. It can also clearly see the waveform curve at the each location of pile. It can also display the stress state of each cell, and the dynamic stress of pile. These show that ANSYS/LS-DYNAD has the powerful capabilities to solve the problem of nonlinear simulation.

In short, testing the integrity of pile foundation is a complex problem, and hoping this article to provide a theoretical reference for other engineering personnel.

Acknowledgments

This research was supported by grants from National Science and Technology Support Program. The name is high-altitude and alpine region of highway construction technology (No. 2014BAG05B05), the Central Universities Fundamental Research (No.31920150072).

References

[1] YU Yun-yan, ZHAO De-an, PENG Dian-hua, MA Hui-ming, JIN Pei-jie, HOU Dian-ying.Experimental Research on Frost Heave and Salt Expansion of the Improved Saline Soil of Railway Roadbed Filling at South of Xinjiang [J]. Journal of Lanzhou Jiaotong University, 2009, (1).

[2] YU Yun-yan.Transient response of a damaged continuous beam [J]. Explosion and Shock Waves, 2009, (1).

[3] YU Yun-yan, YU Li-fen .Wave Responses of Integrate Pile Embedded Partially in Homogeneous Soil [J]. Journal of Lanzhou Jiaotong University, 2009, (4).

[4] Yun-yan, YU Li-fen.TRANSIENT WAVE RESPONSE OF AN INTEGRATED PILE PARTIALLY EMBEDDED IN LAYERED SOIL [J]. Acta Mechnica Solida Sinica, 2009, (6).

[5] YU Yun-yan, YU Li-fen. Transient wave of the pile with variable modulus embedded partially in homogeneous soil [J]. Journal of Vibration Engineering, 2010, (2).

[6] YU Yun-yan, YU Li-fen. Transient wave of the pile with variable sections embedded partially in homogeneous soil [J]. Chinese Journal of Computational Mechanics, 2010, (6).

[7] Yu Yunyan, Bao Yih-Hsing, Chen Yunmin.DISCUSSION ON QUALITY TESTING OF FRAME STRUCTURE EMBEDDED PARTIALLY IN SOIL [J]. chinese journal of theoretical and applied mechanics, 2006, (3).

[8] Chengcui. The Factors Influencing the Integrity detection Results of Pile Foundation Based on the Low-Strain Reflected Wave Method and Countermeasure. (Maritime University, Dalian, China)

[9] Yun-yan.Studies on wave responses of a defective frame structure embedded partially in soil and influence factors [J]. Journal of Vibration Engineering, 2007, (2).

[10] Yun-yan.Stress wave propagation and study on influencing factor in frame structure embedded partially in soil [J]. Chinese Journal of Computational Mechanics, 2007, (5).

IOP Conf. Series: Materials Science and Engineering **207** (2017) 012092 doi:10.1088/1757-899X/207/1/012092

The research macro-mechanical properties of rock based on discrete particle model

Luchen Zhang, Shuchen Li, Shisen Zhao, Qikai Liao

Geotechnical and Structural Engineering Research Center, Shandong University, Jinan, Shandong 250061, China

Abstract. The influences of different microscopic parameters of the macroscopic mechanical properties are not the same. Based on discrete element granular flow theory, particle contact stiffness, friction coefficient, connection strength and other microscopic parameters of the material properties of macroscopic mechanical properties are analyzed combined with laboratory experiments. Using triaxial test to study Ⅳ surrounding rock specimen under different mechanical properties. Based on discrete element granular flow theory, using PFC2D to analyze the relationship between different microscopic parameters and macroscopic mechanical properties. It is found that particle contact stiffness has great influence on the peak strength, macroscopic initial tangent modulus and elastic modulus; With the increase of the friction coefficient between the particles, the peak strength increases; connection strength is greater, the greater the material cohesion and Poisson's ratio is smaller. Research results may reflect the influences of microscopic parameters on the macroscopic mechanical properties of materials, and provide a basis for the selection of microscopic parameters in the numerical calculation.

1. Introduction

At present, domestic and foreign scholars have proposed many methods to simulate the mechanical properties of the material under external force based on different research purposes and requirements. Generally, the proposed methods are: continuous medium and discrete element. According to the function of discrete element method, it can be divided into macroscopic and microscopic discrete elements. The former focuses on solving problems (UDEC and 3DEC) caused by relatively large discontinuities (such as faults, joints and joints surface between structure and foundation, etc.); while the latter is mainly used for discontinuous contact surfaces or points (PFC, YADE) with large number, for example, the contact surfaces and points between the rupture surfaces of rock mass, or between the particles of the materials. A famous scholar, Peter Cundall, by using the theory of microscopic discrete elements, developed the discrete element numerical PFC commercial software [1-2], which is widely applied to solve various problems of research material microstructure.

A foreign scholar, Bagnold, was the first one to reveal the constitutive relation between the particles. In the 1950s, Bagnold conducted a coaxial shear rheological test for wax ball by using glycerin - alcohol - aqueous solution, and then proposed the famous Bagnold's law [3]. Then Ogawa introduced the concept of "granular temperature", and on this basis, the kinetic theory of granular flow model was rapidly established. The study on particle flow began to develop quickly after 1970s. In 1979, Cundall and Strack first developed a two-dimensional discrete element simulation program for studying the mechanical properties of particulate media, which were the beginning of discrete element simulation of particle motion [4-6].

At present, the discrete element PFC program referring to the field of rock engineering is mainly applied to the fracture brittle and mechanical properties simulation of hard rock mass, strength characteristics of jointed rock mass and rock engineering [7-9]. M. Doležalová et al. [10] studied the effect of stress path on the basic characteristics of the material by using PFC; for PFC applications, the Japanese scholars have been in a leading position, of which T. Hoda and I. Towhata have successfully solved the liquefaction and stability problems of the dam under seismic loading by using PFC[11]. Xu Jinming et al.[12] taking an example of limestone, obtained the mechanics data of rock through laboratory experiments, and linked the micromechanics and macroscopic parameters of the particles by using particle flow numerical method, then built a microstructure model of limestone. Based on the particle flow theory, J. W. Park and J. J. Song carried out a 3D simulation [13] for joint direct shear tests with PFC3D software through characterizing joint unit by weakening the strength of the particles. Zhu Huanchun [8] introduced the basic research ideas and work steps of PFC, the model building, the development of rock rupture during the test, changes and failure mode of the stress - strain during the engineering. Nevertheless, there is still no clear explanation for the microscopic parameter selected by the scholars, therefore, there has been a problem referring to the calculation of PFC.

Select the appropriate meso-mechanical parameters of particles is vital for macroscopic stress-strain curves of material simulation. The meso-mechanical parameters of different materials have different effects on the macroscopic mechanical properties. Currently, the tunnel rock mass conditions are not good; the study of the rock material with low strength is few. In this paper, taking an example of grade Ⅳ railway tunnel rock mass, we will analyze the effect of micromechanical parameters of particles, such as contact stiffness, friction coefficient and connection strength on the macroscopic mechanical properties of the material, combining with laboratory tests.

2. Laboratory Test

2.1. *Specimen Preparation*

Taking a certain grade Ⅳ railway tunnel rock mass for the study, and according to the " Railway Tunnel Design Specifications", the physical parameters of rock mass are as shown in Table 1. Preparation of model material has a similar relation with the values in the table. Materials are mainly composed of sand, fly ash, barite powder, lubricants and water. After conducting a series of trial ratio, that river sand and barite powder are aggregate, and fly ash and lubricants are cement materials can be determined, with the ratio of 1: 1: 0.32: 0.09. Formulation was conducted based on the material mixture ratio, and compacted in a mold by using a compaction meter to achieve the required density, as shown in Fig.1. Model samples are cylindrical, with a diameter of 39.5mm, and a height of 80mm, as shown in Fig. 2.

Table 1 Physico-mechanical indices of grade Ⅳ railway tunnel rock mass

Rock mass grade	Gravity γ (KN/m^3)	The elastic resistance coefficient K (MPa/m)	Deformation modulus E(GPa)	Poisson's ratio μ	Internal friction angle φ (°)	Cohesion c (MPa)	Calculated friction angle φ_c (°)
Ⅳ	20～23	200～500	1.3～6	0.3~0.35	27～39	0.2～0.7	50～60

Table 2 Physico-mechanical properties of samples

No.	Gravity(KN/m^3)	Elastic modulus (MPa)	Poisson's ratio	Flying shears		Triaxial		Rock mass grade
				c (kPa)	φ (°)	c (kPa)	φ (°)	
Model material	21.5	27.8	0.33	14.6	34.9	11.3	Model material	21.5

The untrained triaxial shear was applied to the experiment, with a strain rate of 1%. Through a series of experiments, the physical and mechanical parameters of the complete samples made by this material are shown in Table 2, and the apparatus used in the experiment are shown in Fig. 3.

Fig. 1 Compaction meter and molds

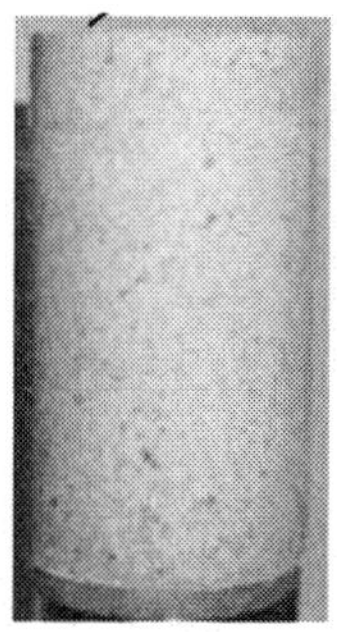

Fig.2 Specimen

Fig.3 Schemes of displacement sensor

2.2. Test Results

For each series of samples, the test was repeated exactly under the same experimental conditions. Three samples were made for each series, and intermediate values were selected as representative test results.

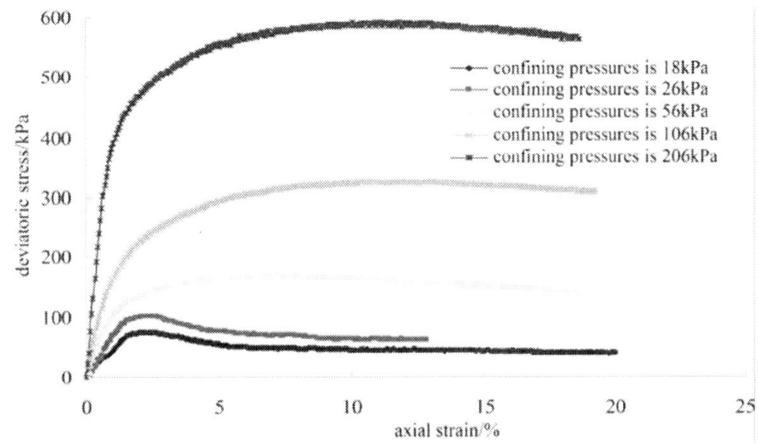

Fig.4 Stress - strain curve with different confining pressures

Under different confining pressures, the grade Ⅳ stress - strain curve of the rock specimen are shown in Fig.4, which shows that: with the increase of confining pressure, peak strength and residual strength increases; the larger the confining pressure, the more obvious the plasticity of the specimen; under low confining pressure (confining pressure of 18,26 KPa), the strength of the specimen increased to its peak and then the stress intensity ddeclined, representing as stress softening, which is the ideal model of elastic-brittle; under high confining pressure, the stress after peak represented as stress hardening, which is the ideal elastoplastic model. The stress - strain curves during cyclic loading and unloading are shown in Fig. 5, of which the fitting curve functions is σ_1=27.824ε_1-0.0475 .

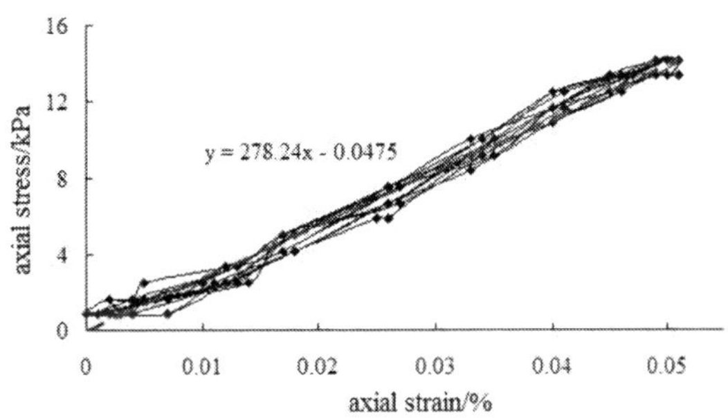

Fig.5 Stress - strain curves during cyclic loading and unloading

3. Research on macroscopic mechanical properties of materials

PFC2D calculation program uses micromechanics mechanics parameters to represent the particles of materials as well as the mechanical properties of their bonding. Before calculating the value of the model, we must first assume several microscopic parameters for the model, and compare the macroscopic mechanical properties and the results obtained in the laboratory. If the result is inappropriate, we need to adjust the parameters of micromechanics, until the calculated results are almost the same as the experimental one, then these parameters can be applied to the actual model calculations. Therefore, the adjustment of microscopic mechanics parameters is very important [14].

3.1. The establishment of specimen model

The sample size is the same as that of during the triaxial experiment of model material, which is 8.0 cm * 3.95cm, the radius of the particles is uniformly distributed within the range of 0.35mm ~ 0.8mm, and the particle density is 2200 kg / m^3. A sample contains a total of 3260 particles, and the sample parameters are shown in Table 3. For PFC2D, there are two kinds of connection models of particle: contact connection and parallel connection. During the destruction of material particles, the reduction of the stiffness can be modeled by a parallel connection model. As the parallel connection model is more suited for the simulation of rock materials, and it can reflect the cumulative process of material damage, the paper uses a parallel connection model. There are generally three steps for the experimental simulation: generating a sample, consolidation and loading. (as shown in Fig. 6).

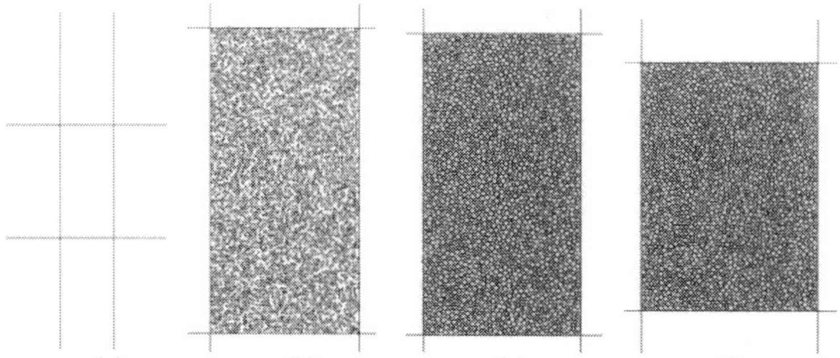
(a) (b) (c) (d)

Fig. 6 The particle flow processes of indoor samples

Table 3 Input parameters of PFC model

Density (kg/m3)	Sample size(cm)	Particle size(mm)	Original porosity	Coefficient of amplification	Normal contact stiffness(GPa)	Stiffness ratio	Friction coefficient	Connection stiffness(kPa)
2200	8.0×3.95	0.35~0.8	0.12	1.6	0.8~40	0.5~8.0	0.1~0.9	0.5~100

By using numerical servo system, PFC maintains a constant confining pressure [14] by the way of adding the loading through relative movement of the top and bottom of the wall while adjusting the displacement on both sides of the wall. Particle distribution after loading the model sample is shown in Fig.6 (d). Throughout the loading and unloading process, PFC program can automatically record the horizontal and vertical stress(σ_x , σ_y)and strain(ε_x , ε_y); wherein, σ_x and σ_y represent the contact area (the side area and top area of the particle sample) divided by the total contact force of the wall and particle.

3.2. Effects of different contact stiffness on macroscopic mechanical properties of materials

Under the conditions that particle friction coefficient of 0.5, the connection strength of 500Pa, with different contact stiffness (8e8, 2e9, 4e9, 8e9 and 4e10), and the calculated stress - strain curve of the model is shown in Fig. 7; and the relationship curve of particle contact stiffness and elastic modulus are shown in Fig.8. As shown in Fig. 7, with the increasing particle contact stiffness, the peak strength of the material also increased, as well as the macro initial tangent modulus, while the axial strain decreases when the shear is being damaged, this has a great influence on the stress - strain curve shape. As shown in Fig.8, the elastic modulus of the material increases along with the increase of the stiffness of particles, this demonstrates a substantial proportional relation.

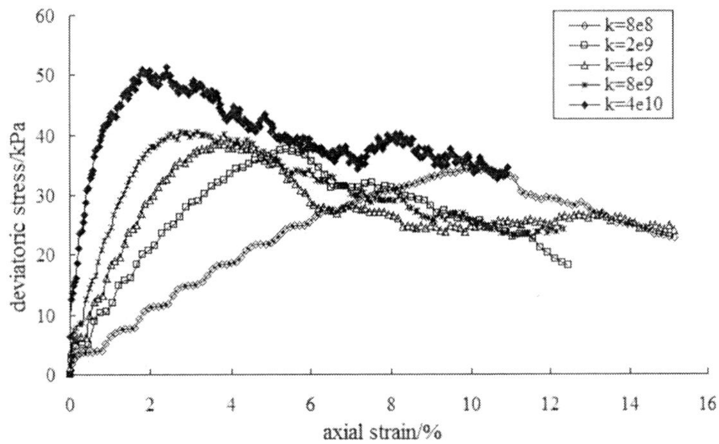

Fig.7 Stress - strain curve with different contact stiffness

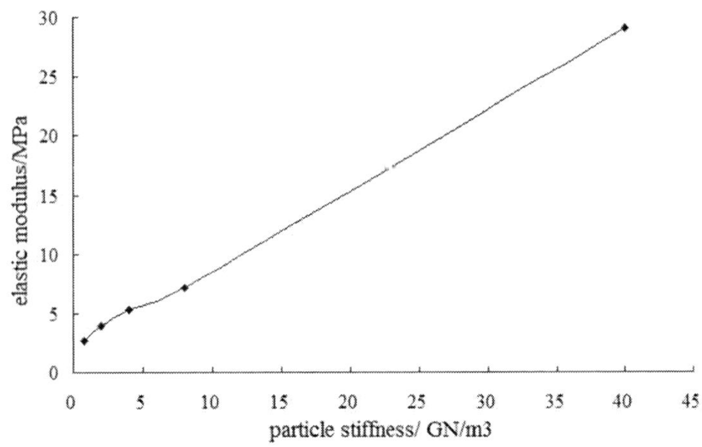

Fig. 8 The relationship of particle contact stiffness and elastic modulus

3.3. Effects of different stiffness ratio on macroscopic mechanical properties of materials

Under the conditions of different stiffness ratio (normal stiffness / shear stiffness), respectively 0.5,1,2,4 and 8, the deviatoric stress - strain curves of samples are shown in Fig.9. As can be seen, the stiffness ratio of the material has little effect on the stress - strain curve shape of the material, with little change of the peak strength and residual strength. The relation diagram of stiffness ratio and elastic modulus is shown in Fig. 10. As can be seen from the figure, with the increase of the stiffness ratio, elastic modulus decreases.

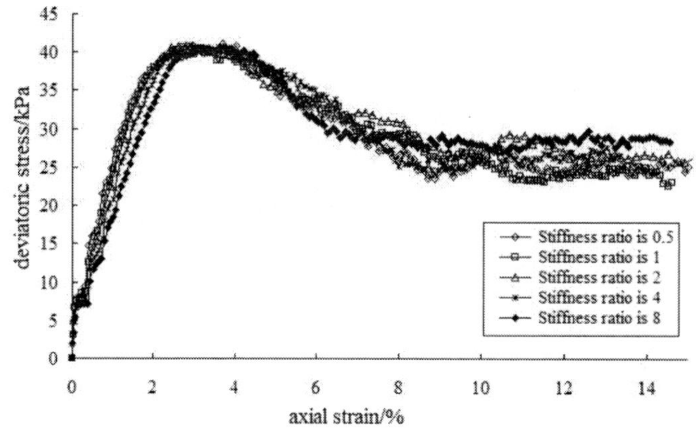

Fig. 9 The stress - strain curve with different stiffness ratio

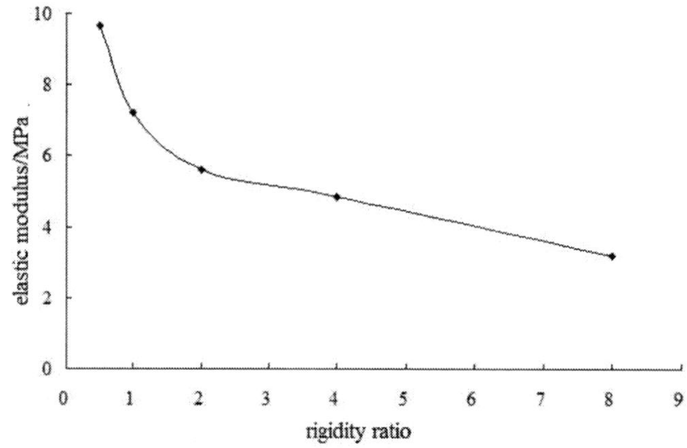

Fig.10 The relationship curve of elastic modulus and rigidity ratio

3.4. Effects of different friction coefficients on macroscopic mechanical properties of materials

Under certain conditions, when the friction coefficient changed from 0.1, 0.3, 0.5, 0.7 to 0.9, the relationship curves between deviatoric stress and axial strain of the sample are shown in Fig. 11, and the relationship curve of peak stress and the friction coefficient are shown in Fig. 12.

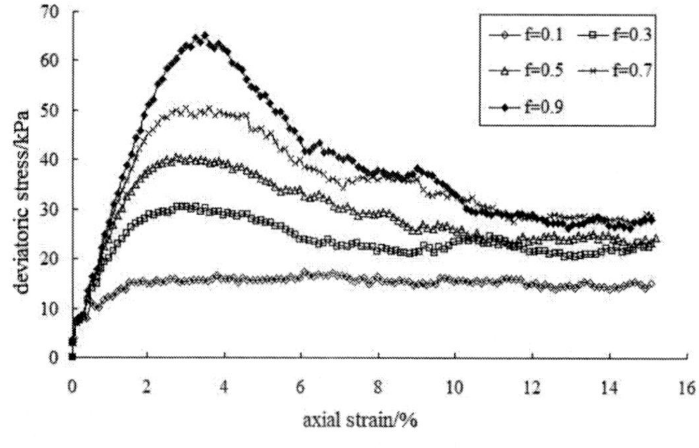

Figure 11 Deviatoric stress - strain curve with different friction coefficients

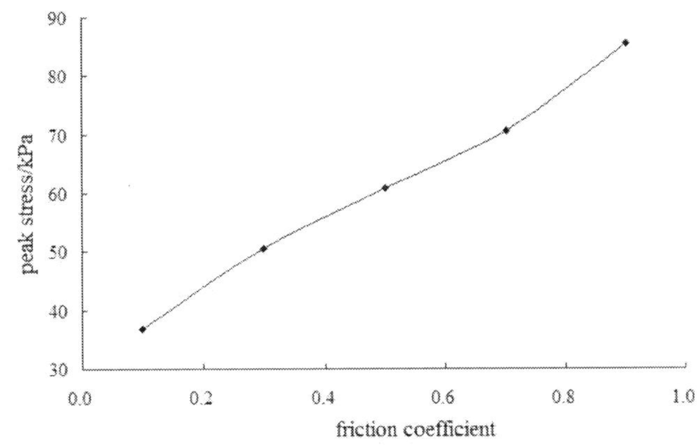

Fig. 12 The relationship curve of peak stress and the friction coefficient

As can be seen in Fig. 11 and 12, that: the peak strength of the sample increases with the increase of the friction coefficient between the particles, substantially representing a direct proportion relationship; when the friction coefficient is small, the material exhibits strain hardening characteristics. The larger the friction coefficient, the material will exhibit strain softening characteristics; residual strength of the material also increases as the friction coefficient increases, but the change is small. In addition, as can be seen in Fig. 11, the variation of friction coefficient has little effect on the initial tangent modulus of the material; this is because of the connection strength between the material particles. Only overcome the connection strength first, can the particles slide. Under the condition of small strain, the connection strength of the model can reach a maximum value; as the strain continues to increase, the model will soon get damaged; after the strain reached a certain value, the volume of the sample would stop increasing; when the stress-strain curve tends to level, only friction component provides strength.

3.5. Effects of different connection strength on macroscopic mechanical properties of materials

Under certain conditions, when the connection strength was taken 0.5 MPa, 5 MPa, 20 MPa, 50 MPa and 100 MPa respectively, the relationship curves of deviatoric stress - axial strain of the samples are shown in Fig. 13. Under different connection strengths, the relationship curves of volumetric strain - axial strain of the materials are shown in Fig. 14. Under different connection strengths, the relationship curves of the connection strength and Poisson's ratio are shown in Fig. 15. The relationship curves of the connection strength and cohesion are shown in Fig. 16.

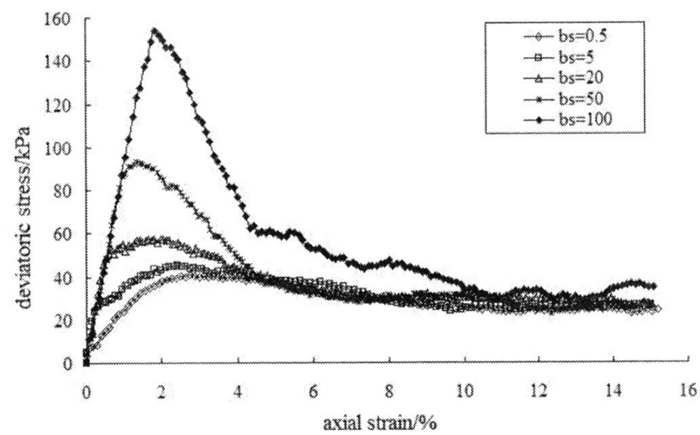

Fig. 13 Deviatoric stress - strain curve with different connection strength

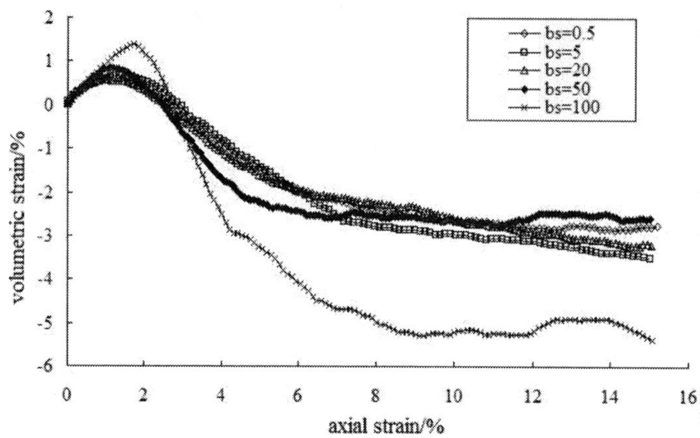

Fig. 14 Volumetric strain - axial strain curve with different connection strength

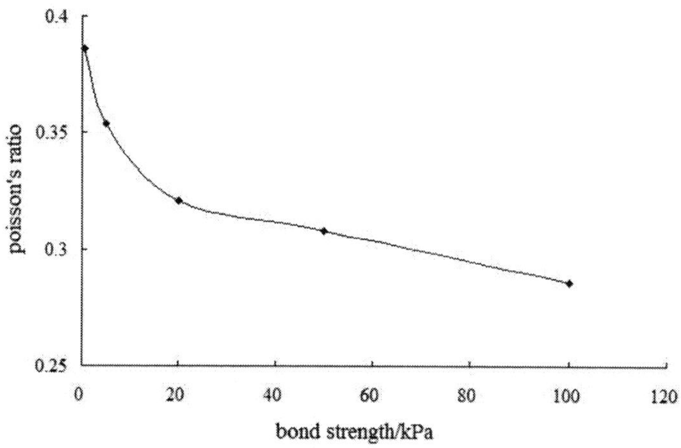

Fig. 15 Connecting the relationship between strength and Poisson's ratio

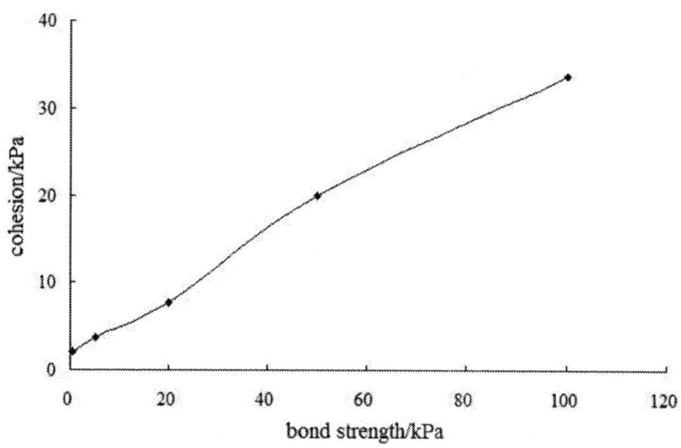

Fig. 16 The curve of Connection strength and cohesion

As it can be seen in Fig. 13 and 16, the greater the connection strength of particles, the greater the peak intensity of the sample; the greater the cohesion of the material, the smaller the Poisson's ratio, and more obvious the softening characteristics of the stress - strain curve will be. Under different connection strengths, there are little changes in the residual strength of the sample.

In summary, the rule of the influence of each micromechanics parameters of materials on the macroscopic mechanical properties can be obtained. According to the given material stress-strain curves, the microscopic parameters in the PFC calculation model can be adjusted, so as to match the stress-strain curves in the model test. After several adjustments, final parameters and results in the calculation model can be obtained, as shown in Table 4. Fig. 17 is a comparison chart of the test results and calculated results when the confining pressure of the sample is 10KPa.

Table 4 The basic parameters and calculated results of PFC model

Density (kg/m^3)	Particle size(mm)	Original porosity	Normal contact stiffness(GPa)	Stiffness ratio	Friction coefficient	Connection strength(kPa)	Calculated results			
							E (MPa)	μ	c (kPa)	φ (°)
2200	0.35~0.8	0.16	20	1	0.8	10	26.4	0.35	9.87	32.6

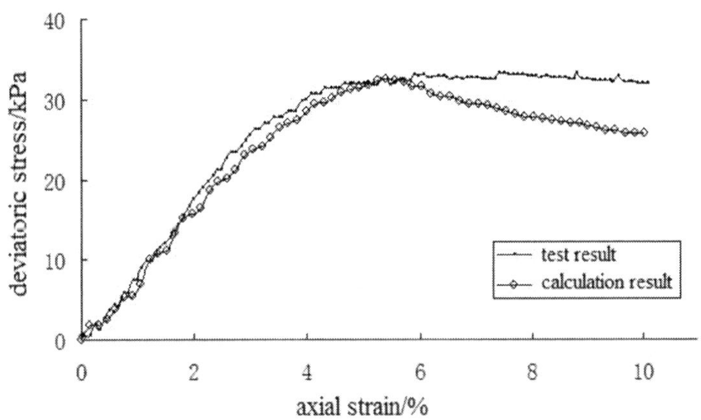

Fig. 17 Stress-strain curve of tests and PFC model

4. Conclusion

Taking the grade Ⅳ railway tunnel rock mass as an example, based on the discrete particle flow theory, combined with the results of laboratory and numerical calculation, we studied the effects of various microscopic parameters of particles including the contact stiffness, the friction coefficient, connection strength on the macroscopic mechanical properties of the materials. The following conclusions can be obtained:

(1) The entire physical mechanical parameters of the specimen similar to grade Ⅳ railway tunnel rock mass can be obtained through a series of laboratory tests.

(2) According to the calculated results, we found that the connection stiffness of particles has a great effect on the stress - strain curve of materials. As the connection stiffness of particles increases, the peak strength of the material increases, as well as the elastic modulus and the macro initial tangent modulus, while the axial strain decreases when the shear failed. And the effect of stiffness on the strain - stress of the material is little.

(3) The peak intensity of the sample will increase proportional with the increase of friction coefficients, and the effect of the friction coefficient on the initial tangent modulus of the material is little, mainly due to the connection strength between the particles of the materials.

(4) With the increase of the connection strength, both the peak strength and cohesion of the sample will increase, Poisson's ratio will decrease, and the more obvious softening characteristic the stress - strain curves of the sample will be. By adjusting the microscopic parameters in PFC calculations, it can well match the test results.

References

[1] Itasca Consulting Group Inc.. PFC-2D (particle flow code in 2dimensions) theory and background [R]. Minnesota, USA: Itasca Consulting Group Inc., 2002.

[2] WANG Chengbing. Study on the progressive failure mechanism of the surrounding rock of tunnel constructed in soft rock [D]. Doctoral dissertation, Tongji University, 2007.

[3] Bagnold R A. Experiments on a gravity-free dispersion of large solid particles in a Newtonian fluid under shear. Proc R Soc Lond A, 1954, 225:49-63.

[4] Strack O D L, Cundall P A. The distinct element method as a tool for research in granular media, Part Ⅰ.Report to the National Science Foundation, Minnesota: University of Minnesota, 1978.

[5] Cundall P A, Strack O D L. Particle flow code in 2 Dimensions [A].Itasca Consulting Group,Inc.,1999.

[6] Cundall P A, Strack O D L. A discrete numerical model for granular assemblies [J]. Geotechnique, 1979, 29(1): 47~65.

[7] LIU Ning, ZHANG Chunsheng, CHU Weijiang. Simulating time-dependent failure of deep marble with particle flow code [J]. Chinese Journal of Rock Mechanics and Engineering, 2011, 30(10): 1989-1 996.

[8] ZHU Huanchun. PFC and application case of caving study [J]. Chinese Journal of Rock Mechanics and Engineering, 2006, 25(9): 1 927–1 931.

[9] Billaux, D., F. Dedecker and P.Cundall. A Novel Approach to Studying Rock Damage: The Three Dimensional Adaptive Continuum / Discontinuum Code [J]. Rock Engineering, 2004: 723-728.

[10] Doležalová M, Czeme P, Havel E. Micromechanical modeling of stress path effects using PFC-2D code [A]. In: Konietzky H ed. Numerical Modeling in Micromechanics via Particle Methods [C]. [s. l.]: [s. n.], 2002. 173-182.

[11] Hoda T, Towhata I. Flow deformation of ground due to liquefaction during earthquake [A]. In: Konietzky H ed. Numerical Modeling in Micromechanics via Particle Methods[C]. [s. l.]: [s. n.], 2002. 141–150.

[12] XU Jinming, XIE Zhilei, JIA Haitao. Simulation of mesomechanical properties of limestone using particle flow code [J]. Rock and Soil Mechanics, 2010, 31(2): 390-395.

[13] PARK J W, SONG J J. Numerical simulation of a direct shear test on a rock joint using a bonded-particle model [J]. International Journal of Rock Mechanics and Mining Sciences, 2009, 46(8): 1315–1328.

[14] ZENG Yuan. Microscopic mechanics of soil failure and PFC numerical simulation [D]. Doctoral dissertation, Tongji University, 2006.

Analysis of influence of seepage on stability of foundation pit

Yongfu Li[a], Junrui Chai[b] and Zengguang Xu[c]
Xi`an University of Technology, Xi`an 710048, China

[a]674115503@qq.com, [b]jrchai@xaut.edu.cn, [c]xzg19821211@163.com

Abstract. Numerical analysis is executed to deep foundation pit by using the finite element software. In combination with the site data, the consequences with the consideration of groundwater seepage are contrasted to the consequences without the consideration of groundwater seepage including the stability of the foundation pit. The analysis indicates that the result with the consideration of seepage is relatively close to the site data. Compared with the result without consideration of groundwater seepage, the maximum horizontal offset of the pile increases approximate 5mm and the location of the maximum horizontal displacement moves 2m down. So seepage is harmful to the retaining structure of foundation pit and should be considered in the design of foundation pit.

1. Introduction
As China's real estate market continued to hot in recent ten years, in order to make full use of urban space and improve the carrying capacity of land resources, the three-dimensional development of city was more and more urgently needed [1,2], and people began to seek space for development in the air and underground[3]. Currently various types of underground space has been developed in the world's major cities[4], such as the multi-storey basement of high-rise building, underground railway and underground stations, underground parking garage, underground streets, underground shopping malls etc [5]. The common characteristic of developing underground engineering is to excavate the soil mass and result in a large number of foundation pit [6, 7]. Excavation and dewatering of deep foundation pit on pervious foundation will result in consolidation of soil and flow of pore water, water movement in pores tends to be strong because of difference of percolation water head [8], which affects internal force and deformation of enclosure system, the stability of the foundation pit and the surrounding structure. So it is necessary to consider the influence of seepage.

2. Engineering survey

2.1. Project introduction
According to the survey report, the foundation pit engineering ±0.00 is equivalent to the Yellow Sea elevation +4.65, base height is -12.30 m, design relative elevation of site natural floor is -1.15 m, and excavation depth of foundation pit is 11.15 m. The safety grade of foundation pit sidewall is two and importance coefficient is 1.0.

IOP Conf. Series: Materials Science and Engineering **207** (2017) 012093 doi:10.1088/1757-899X/207/1/012093

2.2. Engineering geology and hydrogeology condition

2.2.1. Engineering geological condition. The site is mainly alluvial formation of Quaternary Holocene strata, including sandy soil and cohesive soil. The distribution of soil layers is as follows:

(1) Plain fill: brown, tan and which is made up of silty clay. Its part is crushed stone and concrete pavement and a small amount of plant roots, loose structure~ slightly dense, thickness of the layer is generally 0.40~2.30 m.

(2) Silty clay: tan, grayish yellow, wet, plastic and soft plastic, relative elevation of floor is 0.06~2.12 m, thickness of the layer is generally 0.50~2.60 m.

(3) mucky silty clay: gray brown, brown grey, saturated, flow plastic, local soft plastic, local pinch a small amount of thin powder or silt, relative elevation of floor is -8.31~-5.10 m, thickness of the layer is generally 5.50~9.00 m.

(4) Silty clay soil: gray, brown gray, very wet, slightly dense, containing mica debris, medium uniform, shake response is rapid, initial and dry strength are low, relative elevation of floor is -10.20~-7.18 m, thickness of the layer is generally 7.80~11.60 m.

(5) Silty sand and silt: brown gray, gray, saturated, slightly dense, the mineral composition is W quartz, feldspar, relative elevation of floor is -18.43~-16.18 m, thickness of the layer is generally 0.80~4.00 m.

(6) Powder and sand: brown gray, gray, saturated, medium dense, a small amount of sand particles, medium uniformity, relative elevation of floor is -28.17~-24.18 m, thickness of the layer is generally6.00~11.50 m.

(7) Silty clay with fine sand and silt: powder, fine sand gray, saturated, slightly dense, particle composition uniformity, relative elevation of floor is -26.29~-25.93 m, thickness of the layer is generally0.7~1.2 m.

(8) Powder and fine sand: brown gray, bluish grey, saturated, medium dense, particle composition of medium uniform and partial clay or silty clay.

2.2.2. Hydro geological condition. This site groundwater of project is mainly the pore diving, there is weak confined water in the local and the bedrock contains a small amount of bedrock fissure water. Shallow groundwater belongs to seasonal changes. Geological data display that the average depth of groundwater in the site area is 0.20~2.40 m, the annual average stable water level of groundwater in the site area is 0.60~1.60 m.

2.3. Supporting scheme of foundation pit

As mentioned above, the groundwater level of the project is seriously affected by the surrounding water, and the ground water level is higher and larger precipitation depth. So pit dewatering and well dewatering combined with single row of soil cement mixing makeup water is used. In order to reduce the impact on the surrounding environment of the project design, focusing on the internal bracing wall type support, mainly in the following two forms: Filling pile with internal support and SMW with internal support.

2.4. Horizontal displacement monitoring

Taking measuring point 1 and 2 as an example, the article selected that pit excavation has been completed. The monitoring data is taken as reference when the displacement variation of pile body has stabilized. The monitoring depth is below pile top 10 m. The horizontal displacement of the pile is shown in Table 1. Table 1 shows that the horizontal displacement of the retaining structure increases from ten millimeters in the excavation to dozens of millimeters after the excavation. The maximum horizontal displacement is 24.73mm. Pile horizontal displacement increases with pile length and then decreases.

Table 1 The horizontal displacement of pile changes with depth

Distance (m)	Displacement(mm) Measuring point 1	Displacement(mm) Measuring point 2
0	18.37	17.43
1	19.78	18.74
2	20.83	20.11
3	21.94	21.42
4	23.01	22.60
5	23.96	23.82
6	24.72	24.71
7	24.24	24.73
8	23.98	23.75
9	22.18	23.09
10	21.64	21.56

3. Software modelling

3.1. Calculation profile
Considering the influence of foundation pit excavation on surrounding soil disturbance range and seepage effect on environment, the depth of 3D model is 40 m, Plane size is 100 m x 150 m. In the calculation process, the mixing piles and soil mass are analyzed as the whole. In the model of foundation pit, the soil mass is analyzed by Mohr-Coulomb model and other supporting structures are analyzed according to elastic body. The parameters required for this model are shown in Table 2. Parameters include thick layer, average severity, natural moisture content, void ratio, cohesion, friction angle, compression modulus.

Table 2 Physical and mechanical parameters of soil

Name of soillayer	1(m)	2(kN/m3)	3(%)	4(-)	5(KPa)	6(°)	7(MPa)
①Plain fill	0.8	19.0	-	-	-	-	-
②Silty clay	1.8	18.2	34.7	0.963	24	6.1	3.8
③Muddy silty clay	7	18.4	41.3	1.143	8	3.0	2.8
④Silty clay soil	2.9	18.3	32.4	0.908	17	18.1	4.6
⑤Silty sand clay	9.6	18.7	28.8	0.798	5	28.1	11.2
⑥Powder sand	8.7	18.9	26.1	0.741	6	30.0	11.8
⑦Silty~clay	1.3	18.6	28.9	0.839	5	27.6	11.5
⑧Powder~sand	7.9	19.1	25.6	0.755	6	28.0	12.6

3.2. Modelling calculation
With the existence of assumptions and parameters of each index, we have established three dimensional finite element models by software. The scope of the calculation model is from outside the pit edge line outward extending about 40 m and size is 100 m * 150 m * 40 m. The studied foundation pit is nearly symmetrical, the total length of X axis is 19.30 m and the total length of Y axis is 73.62 m. Limiting the bottom displacement of the model and the horizontal displacement of each side is 0, the upper surface is a free boundary without any constraints. The calculation model is shown in Figure 1.

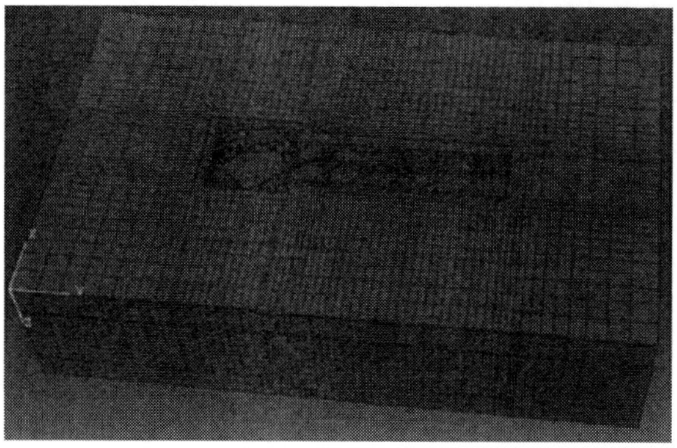

Fig. 1 The map of simulation grid

4. Calculation results and analysis

In order to study the influence of seepage on foundation pit deformation and stability, the following models are calculated by two conditions, one is to consider only the excavation and without considering changes in groundwater level, known as the working conditions I ; the other is to consider the excavation and precipitation, that is, the calculation of the use of soil dry weight and set the initial water table and precipitation wells, known as the working conditions II. With the development of the excavation and dewatering, firstly, unsteady seepage is formed in soil and after the excavation has been completed for some time, stable seepage field is formed in the soil. Through the comparison of two conditions, the influence of seepage on the deformation of foundation pit and retaining structure is analyzed.

Figure 2 gives the horizontal displacement nephogram of pile body about working conditions I and II when the final excavation depth is reached. From the picture we can see that deformation of pile tip is basically consistent about working conditions I and II, deformation is internal development of the pit and the horizontal displacement value displayed in the working conditions II is larger than the horizontal displacement value displayed in I.

(a) Working conditions I

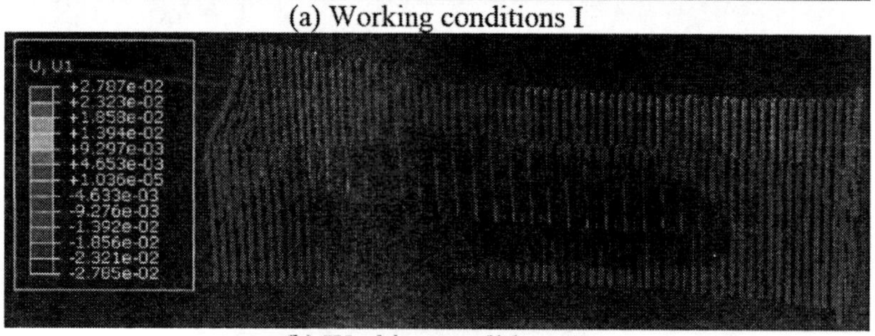

(b) Working conditions II

Fig. 2 Horizontal displacement of pile

Figure 3 gives the curves of horizontal displacement of pile with depth about inclined hole 1 and 2 when the final excavation depth is reached, including working conditions I, II and field monitoring data. From the picture we can see that① deformation of retaining pile is small at both ends and the middle is big, the upper part of the pile displacement and field monitoring conditions of II are close to the values, obvious deviation of central stiffness settings may be decided by the relatively small of the second supporting of the simulation;② the overall horizontal displacement of pile body under working condition II is larger than the horizontal displacement of pile body under I, the seepage makes the retaining piles and the soil around the pile move towards the pit. The maximum horizontal displacement of I occurs at a depth of about 8m, about 22mm, and the maximum horizontal displacement of II occurs at a depth of about 10m, about 27mm, they all occur in the bottom of the pit position. But the maximum horizontal displacement of II increases 5mm, and moves down 2m, and the deviation of horizontal displacement about working conditions I and II increases at first and then decreases. The variation trend of deviation and seepage force is same and the influence of seepage on horizontal displacement of pile body is larger.

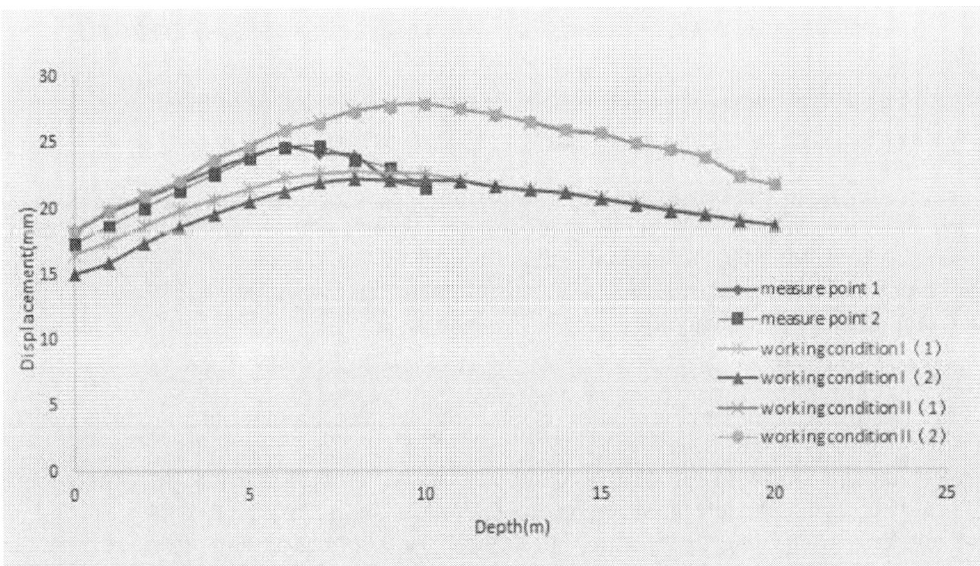

Fig. 3 The curves of horizontal displacement of pile with depth

Figure 4 gives the total displacement nephogram of the symmetrical surface of the foundation pit about I and II when the final excavation depth is reached. From the picture we can see that the total displacement of pit under working conditions I is obviously less than the total displacement of foundation pit under working conditions II. The deformation of foundation pit caused by excavation under working conditions I is limited in certain area of foundation pit and the deformation of foundation pit caused by excavation under working conditions II extends to the ground. The displacement nephogram is similar to the continuous sliding surface formed by the whole shear failure of foundation pit.

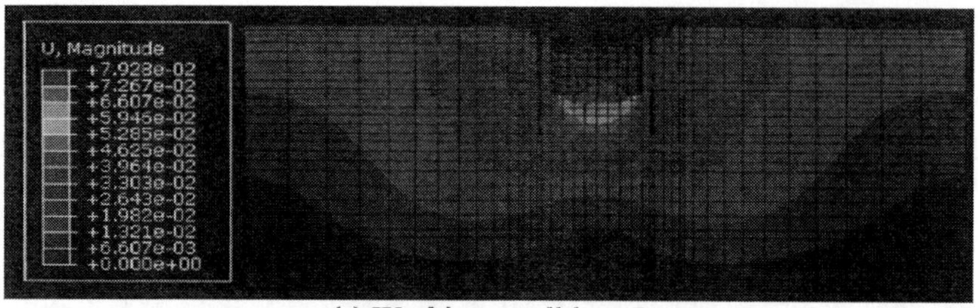

(a) Working conditions I

(b) Working conditions II

Fig. 4 Nephogram of total displacement of symmetric surface of foundation pit

5. Conclusion

From the results of two working conditions considering seepage and non-seepage we can see that the simulated data considering seepage flow are in agreement with field monitoring data and the horizontal displacement of retaining wall structure considering seepage are bigger than conditions without seepage. It shows seepage should be taken into account when the deformation of foundation pit and retaining structure is analyzed.

References

[1] Hou Xueyuan, Liu Guobin. Some views on the development of urban foundation pit engineering. Construction Technology. Vol.29 (2000) No.1, p.0005-03.

[2] Wei Junjun, Zhang Liwei. Seepage stability analysis of foundation pit. Journal of Water Resources and Architectural Engineering. Vol.10 (2012) No.3, p.0079-04.

[3] Yang Yue, Hou Gongyu. Influence of seepage on deformation and stress of foundation pit. Journal of Shenyang Jianzhu University (Natural Science). Vol.29 (2013) No.6, p.1053-06.

[4] Qian Deliang, Li Juncai. Deformation analysis of foundation pit under seepage. Journal of Nanjing University of Technology. Vol.37 (2015) No.1, p. 015.

[5] Chen Guoqing, Pan Yikai. Analysis and treatment of flooding accident in a deep foundation pit. Journal of Water Resources and Architectural Engineering. Vol.14 (2016) No.5, p. 044.

[6] Cao Liqiao. Finite element method analysis of bottom upheaval of deep foundation pits in soft-clay ground due to excavation. Chinese Journal of Geotechnical Engineering. Vol.35 (2013) No.2, p. 94-97.

[7] Ma Shaokun, Shao Yu. Deformation of deep foundation pits due to excavation considering change of void ratio and permeability coefficient with depth. Chinese Journal of Geotechnical Engineering.Vol.37 (2013) No. 2, p. 80-84.

IOP Conf. Series: Materials Science and Engineering **207** (2017) 012094 doi:10.1088/1757-899X/207/1/012094

Study on application of concrete sandwich insulation material in library building insulation

Zengzhang Yang*

Library of Shandong Women's University, Jinan 250300, China

*Corresponding author e-mail: wuyegl@163.com

Abstract: Energy shortage is the short slab that restricts the development of social economy, and the rational and effective use of energy is the principle of sustainable development. Building energy consumption accounts for about 30% of total social energy consumption, and this ratio has continued to rise, so the energy saving potential is great in the construction sector. In view of the building energy consumption problem, we produce green insulation building materials with the crop straw, and improve the construction of hot and humid environment. In this paper, we take concrete sandwich straw blocks in library building as the research object, through the experiment to test its winter heat consumption and summer power consumption indicators, carry out experimental study on thermal insulation performance, and explore the overall thermal and energy saving performance of concrete sandwich straw blocks in library building.

1. Introduction

According to statistics, China's per capita share of coal resources is 1/2 of the world average, and per capita oil occupancy is only 1/10 of the world average. However, the total consumption of coal resources accounts for more than 40% of the world total's, and oil consumption is also ranked in the world second. China's dependence on overseas energy is up to 50%. Energy shortage is the short slab that restricts the development of social economy, and the rational and effective use of energy is the principle of sustainable development. In the context of rapid economic development in China, the primary energy consumption has increased dramatically, while the building energy consumption accounts for around 30% of the social total energy consumption, and this proportion has a rising trend [1, 2]. Therefore, the energy saving potential of the construction industry is very big, which is of great significance to further accelerate the construction of a resource-saving and environment-friendly society and to ensure the energy security of our country.

Under the environment of building energy saving material, the concrete sandwich straw block as a new wall insulation materials, which is widely used in cold region and the hot summer and cold winter area, it meets the needs of energy-saving reform, and has broad market prospects[3]. A series of studies in this paper provide experimental basis for the extensive application of concrete sandwich stalks in buildings. The study of concrete sandwich stalks has great academic value, engineering significance and economic benefit.

2. The experimental procedure

2.1. Test equipment

The Laboratory has the equipment that the experiment requires; they are BES-C, SM1210B temperature and humidity data acquisition module, SLHT4 temperature and humidity sensors, heat sinks and so on.

1) Temperature and humidity acquisition module

SM1210B temperature and humidity data acquisition module supports 1-12 temperature and humidity sensors, and it can achieve low-cost temperature and humidity monitoring of the practical integration module.

This module can be used to: SMT industry temperature and humidity data monitoring; electronic equipment long temperature and humidity data monitoring; cold storage temperature and humidity monitoring; warehouse temperature and humidity monitoring; pharmaceutical GMP monitoring system; environmental temperature and humidity monitoring; telecommunications room temperature and humidity monitoring; and other various occasions that need to monitor the temperature and humidity.

Technical parameters and features: display temperature range: -40℃~123.8℃; sensor nominal temperature accuracy ± 0.5 ℃; display humidity range: 0-100RH; power consumption 2W; storage temperature -40 ℃ ~ 85 ℃.

2) The temperature and humidity sensor

The sensor consists of a capacitive polymer hygrometer and a temperature-sensitive element made of a gap-gap material, and on the same chip, it is seamlessly connected to the 14-bit A / D converter and the serial interface circuit. Therefore, the product has excellent quality, ultra-fast response, and anti-interference ability and so on. The temperature measurement accuracy is ± 0.4 ℃, and the humidity measurement accuracy is ± 3%.

3) BES-C portable microcomputer multi-channel data acquisition instrument

The instrument is mainly used for energy-saving building wall surface temperature, indoor and outdoor temperature, heat flux and other parameters of the field test. It integrates measurement, display, communication and other functions in one and has the characteristic of high precision measurement, full-featured, easy to use and so on. It is suitable for energy-saving building envelope thermal resistance, heat transfer coefficient of the field test, thermal testing of building materials and environmental monitoring and other occasions.

The technical parameters: Heat flow measurement channel: 16 road; range: 0 ~ ± 20mV; the measurement accuracy: ≤0.03mV; and the heat flow sensor use a plate heat flow meter. It is mainly used for wall heat flux measurement. The sampling period: 10 seconds to 24 hours (optional). Data storage capacity: 128,000 data can be stored.

2.2. Experimental program

This test is divided into two parts: winter test and summer test. In the concrete sandwich straw block building of library, we arrange two temperature and humidity collection points (1#, 2#) in the room. The distance between the collection point and the ground is about 70 cm, and the collection point not directly affected by the heat source. We arrange one temperature and humidity collection point (3#) in the outdoor, and this collection point is out of the sun. Therefore, we use the instrument to record the data of temperature and humidity collected by the three acquisition devices for 10 minutes.

In the winter experiment, we use the warm air blower to heat the library building of concrete sandwich straw block. In the summer experiment, we use the air conditioner to chill this building. Meanwhile, we use the ammeter to record the power consumption of the heater and air conditioner for seven days.

2.3. Experimental procedure

From February to March, the winter experiment is carried out in library building of concrete sandwich straw block. In the library, we use two rated power of 2500W heaters for winter heating. The first stage is constant power heating, and the second stage is constant temperature heating, set the temperature is 18 ℃. With the meter to collect heating power consumption during heating, and install the collection device of the temperature and humidity inside and outside, and then record the indoor and outdoor temperature and humidity changes during the test.

The summer experiment is carried out in August, in the library, use the air conditioning for the summer cooling of the house, and the cooling temperature is set at 26 degrees. We use the meter to record the power consumption during air conditioning and cooling, and use the temperature and humidity data acquisition instrument to record the indoor and outdoor temperature and humidity during the test.

2.4. DeST software

DeST software structure is composed of a number of separate modules connection, and it implements the concept of phased simulation [4, 5]. The software structure diagram is shown in Figure 1.

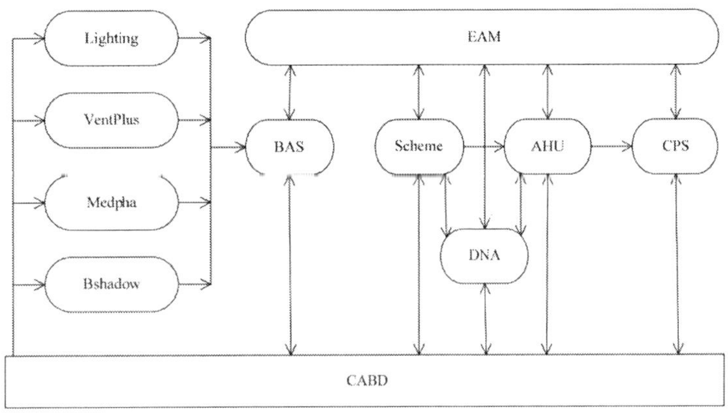

Figure 1. DeST simulation software structure

(1) Building shadow calculation module BShadow: use the geometric projection method to calculate the shadow details of the surface of the building at any time throughout the year.

(2) Outdoor meteorological parameters module Medpha: according to the design needs, we provide year-round meteorological data for the simulation analysis with extremely high temperatures, extremely low temperatures, very high solar radiation, minimal solar radiation and extremely high enthalpy.

(3) Natural lighting calculation module Lighting: responsible for indoor lighting calculation.

(4) Natural ventilation simulation module VentPlus: use the multi-regional network model to form a fluid network, and use the Newton method to obtain the air flow of each branch.

(5) Building thermal characteristics calculation module BAS: the core module of the building's thermal characteristics calculates the temperature and the load of the building.

(6) Air conditioning system design module Scheme: by simulating the performance of buildings under different air conditioning options, designers can explicitly use different solutions to produce results.

(7) Mechanical ventilation system analysis module DNA: mechanical ventilation system analysis module, which can complete the wind system design calculation and check the calculation.

(8) Air handling equipment AHU: through the simulation means of the designer's air treatment program to carry out annual verification, and it provides a quantified basis for the designer's program.

(9) Cold and heat source and water system simulation module CPS: after the designers determined the air treatment program, through the CPS simulation to analysis operating energy consumption of different cold and heat source system and water system, and provide designers with comparison data of different schemes.

(10) CABD: the graphical user interface, the designer can directly through the interface to carry out data design of various buildings (material, geometry, thermal interference).

(11) Economic evaluation module EAM: it can economically evaluate the life cycle costs of HVAC programs and provide economic basis of program options for HVAC designers and engineering construction departments.

3. Experimental results and analysis

3.1. Winter constant power test

In this experiment, the power offering is constant. Two heaters continue to run at constant power and measured the average daily power consumption is 94 degrees. The indoor temperature increased with the increase of the outdoor temperature, and the daily average temperature in the test stage reached 26.64 ℃. At this point, the indoor temperature is too high, and the humidity is relatively low. This brings discomfort to people's work and life, and the heaters constant power work consume a lot of energy, which is detrimental to energy saving and environmental protection. The winter constant power temperature chart is shown in Figure 2, and the winter constant power humidity chart is shown in Figure 3.

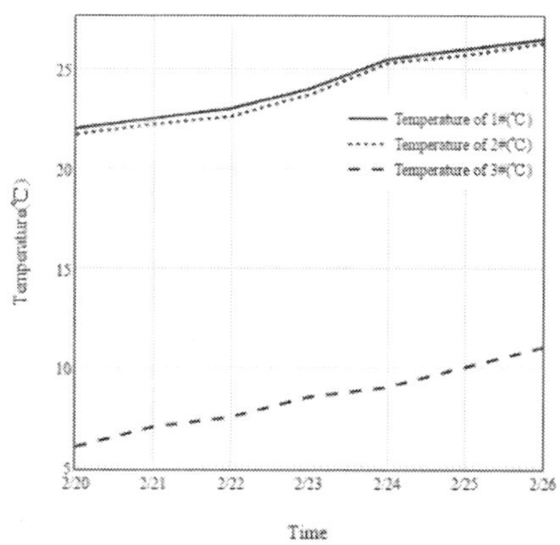

Figure 2. The winter constant power temperature chart

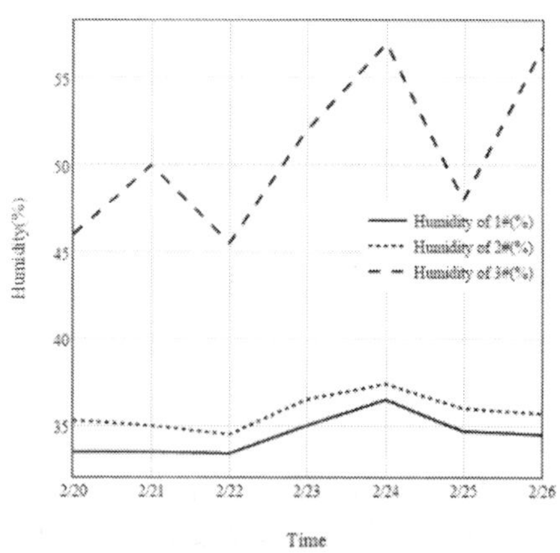

Figure 3. The winter constant power humidity chart

In the constant power test stage, the indoor humidity is relatively stable, about 35%, and the change of outdoor humidity is great. The indoor humidity is less affected by the change of the outdoor humidity.

3.2. Summer test

In this test, the daily average power consumption is 1.5°. The indoor temperature and humidity are relatively stable. The temperature is about 26℃, and the average humidity is about 70%. The changes of outdoor temperature and humidity are great. The indoor humidity is less affected by the change of the outdoor humidity. The summer temperature chart is shown in Figure 4, and the summer humidity chart is shown in Figure 5.

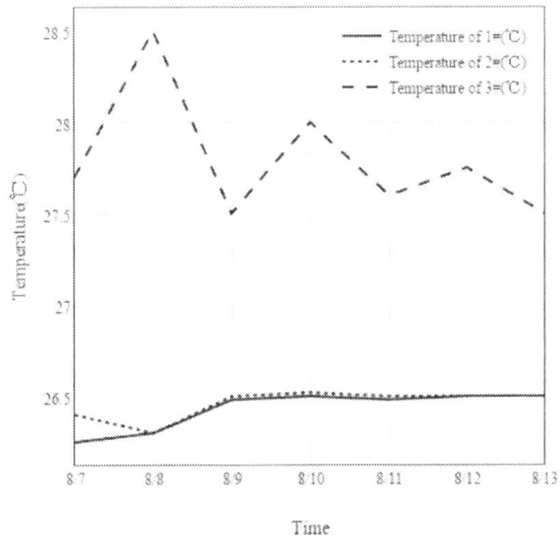

Figure 4. The summer temperature chart

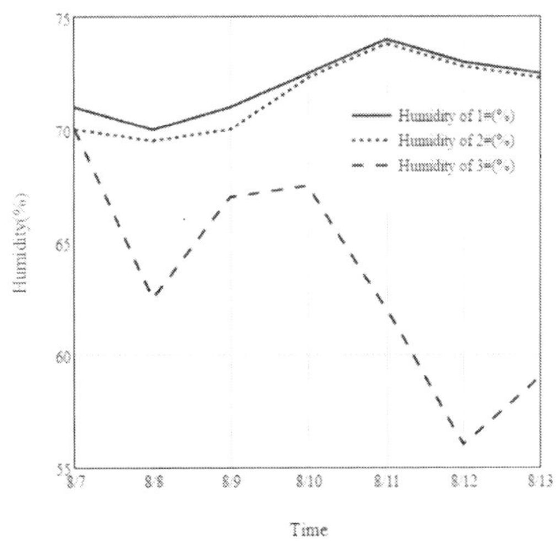

Figure 5. The summer humidity chart

3.3. Analysis of results

The humidity in the outdoor will affect the humidity inside the house based on the analysis of the relative humidity of the library building of the concrete sandwich straw block. But the hysteresis is existed in this process. In the short term, the indoor humidity environment is relatively stable. In the long-term outdoor environment, the indoor humidity will be changed by the outdoor humidity, and the trend is basically identical.

During the test, the temperature inside the building is mainly affected by summer cooling equipment and winter heating equipment. The outdoor temperature has a certain effect on the indoor temperature in the general trend. But the effect is far less than the effect of indoor insulation equipment. Therefore, the insulation function of the library building of the concrete sandwich straw block is effective.

1) Thermal insulation property in winter

According to *The Energy Saving Design Standard for Construction of Residence in Shandong Province DBJ 14-2012*, winter heat consumption index around the region of Jinan is 14.8W/m^2 .From the calculation by experiment, the heat consumption index of this building should be 9.35W/m^2 under the constant temperature condition setting as 18℃. Therefore, this experimental building has good heat insulation in winter which is designed in accordance with the energy saving standard.

2) Thermal insulation property in summer

We use DeST to simulate the summer energy consumption of ordinary concrete building. We also compare it with the measured situation of the library building of concrete sandwich straw block. In the summer experiment of concrete sandwich straw block building, measured power consumption is 1.5°per day, and simulated power consumption of DeST is 13.5° per day. Therefore, this experimental building has good heat insulation in summer, thereby reducing the power consumption.

4. Conclusion

Based on the experimental study on the thermal insulation performance of the concrete sandwich masonry block in library building, it is concluded that the calorific value measured in the winter test of the concrete sandwich block in library building is in accordance with the local regulations. The power consumption index of concrete sandwich straw block in library building measured in summer test, which has the advantages of good insulating property and distinctive effect of energy saving. In the short term, the change of outdoor humidity has little effect on the change of indoor humidity environment, and the humidity control performance of the straw compressed block is not significant. However, the overall humidity of indoor environment will change with the change of outdoor humidity environment, and the process is relatively slow.

References

[1] Ahmad, A. S., Hassan, M. Y., Abdullah, M. P., Rahman, H. A., Hussin, F., Abdullah, H., & Saidur, R. (2014). A review on applications of ANN and SVM for building electrical energy consumption forecasting. *Renewable and Sustainable Energy Reviews*, *33*, 102-109.

[2] Fan, C., Xiao, F., & Wang, S. (2014). Development of prediction models for next-day building energy consumption and peak power demand using data mining techniques. *Applied Energy*, *127*, 1-10.

[3] Castleton, H. F., Stovin, V., Beck, S. B., & Davison, J. B. (2010). Green roofs building energy savings and the potential for retrofit. *Energy and buildings*, *42*(10), 152-159.

[4] Tu, R., Liu, X. H., Li, Z., & Jiang, Y. (2011). Energy performance analysis on telecommunication base station. *Energy and Buildings*, *43*(2), 315-325.

[5] Zhou, F., Tian, X., & Ma, G. (2011). Investigation into the energy consumption of a data center with a thermosyphon heat exchanger. *Chinese Science Bulletin*, *56*(20), 185-190.

Exploration in the test of splitting resilient modulus of asphalt mixture

Peihong Li*

School of Traffic and Transportation Engineering, Changsha University of Science and Technology, Changsha, 410114, China

*Corresponding author e-mail: road-engineering@foxmail.com

Abstract. In terms of the non-correspondence about the bottom layer tensile stress provided by the compressive resilient modulus and the allowable tensile stress obtained by the splitting test, this paper puts forward the methods about splitting stepping testing by contrasting the compressive resilient modulus testing methods. Both the stepping load and resilient deformation are recorded. With the help of the concept of splitting stiffness modulus, the expression of splitting resilient modulus can be obtained to gain the splitting resilient testing flow. The results show that the resilient modulus which is measured by splitting stepping load can meet the design ranges of asphalt mixture and the stepping loading methods about splitting resilient modulus are worth popularizing.

1. Introduction

The structure design about asphalt pavement makes reference to the values about elastic modulus parameters of asphalt mixtures. The elastic modulus values in exiting specification adopt the compressive resilient modulus which can be measured from stepping loading. The stipulation in specification shows that the bottom layer tensile stress provided by the compressive resilient modulus must be less than or equal to the allowable tensile stress. Besides, the allowable tensile stress can be obtained by means that the splitting strength value is divided by the structure coefficient of tensile strength in splitting test. There are not correspondences between the compressive resilient modulus and the splitting strength value. This is a major deficiency about exiting specification of asphalt pavement and it is necessary to improve this weakness.

The elastic modulus of asphalt mixture has different values in different acting experiments and there are more differences about elastic modulus in different conditions. To correspond the pavement design parameters and improve the structure designs, the splitting resilient modulus is selected as the modulus parameter of asphalt mixture so that the splitting resilient modulus in calculating the bottom layer tensile stress can coincide with the allowable tensile stress. Nevertheless, the researches about elastic modulus in splitting condition are very few currently and the reason lies in lacking of equation about splitting modulus. Aiming at this puzzle, the article acquires the loading values and the resilient deformations in splitting stepping loading by contrasting with the testing method about compressive resilient modulus. By the lights of the concept about stiffness modulus of asphalt mixture, the mathematical expression about splitting resilient modulus is derived to obtain the splitting resilient testing method.

2. Defining methods of splitting resilient modulus

In the material mechanics method, the modulus is defined as the capacity to resist deformation and its modulus expression is $E=\sigma/\varepsilon$. The concept about σ is the stress in the range of elastic deformation and then the symbol about ε refers to the strain as the relative settlement. Asphalt mixture is a multivariate material and the external environment including temperature, traffic load and humidity have influences on its modulus. Therefore, the testing temperature should get command of fifteen degree centigrade in splitting strength and resilient modulus tests. In this temperature, the splitting resilient modulus can maintain stabilization relatively.

According to the uniaxial compression test in *Standard Test Methods of Bitumen and Bituminous Mixtures for Highway Engineering JTG E20-2011*, the resilient deformation values from stepping loading are measured to calculation the compressive resilient modulus. The splitting resilient modulus test can be put forward on the basis of compression method.

3. Testing methods of splitting resilient modulus

From the comparison of the method about compressive resilient modulus, the testing processes in AC-13 asphalt mixture splitting resilient modulus are described.

(1) The AC-13C Marshall specimens are shaped and the fixture containing 12.7mm wide and 50.8mm inner curvature is chosen. The specimens are kept by the temperature for more than 6 hours in fifteen degree centigrade.

(2) The process of splitting test are as follows. The loading rate in material machine is 50mm/min. The curves about load-time and vertical deformation-time are recalled in succession. Turn on the material machine, keep the load head on the fixture and adjust the data system. The specimens are loaded to destroy at a specified rate. The splitting test is in Figure 1:

Figure 1. Splitting test

(3) Many groups of corresponding splitting experiments are implemented to record the average of maximum splitting strength PT. The value about splitting strength is divided into ten ranks and then the first seventh are selected as stepping loading including 0.1P, 0.2P, 0.3P, 0.4P, 0.5P, 0.6P and 0.7P. The loading procedure is set as 2mm/min loading rate in material machine. The splitting resilient test is in Figure 2:

Figure 2. Splitting resilient test

(4) There has a compaction lasting one minute with the rate of 2mm/min to 0.2P. The material machine can read the vertical displacement automatically without using dial gauge to record the loading and unloading numbers.

(5) Being turned the first load 0.1P in the rate of 2mm/min and then discharged to zero, the specimen is waited for a bounce lasting for thirty seconds. The testing system can draw a deformation-time curve in this loading circulation by itself. Besides, the processes about loading and unloading including the second to seventh stage are carrying out. The resilient deformation is the difference between the deformation in certain load and the unloading deformation after thirty seconds resilience. The material properties appear not full elasticity but partial irrecoverable deformation. The stepping loading curve is in Figure 3 and the deformation curve is in Figure 4:

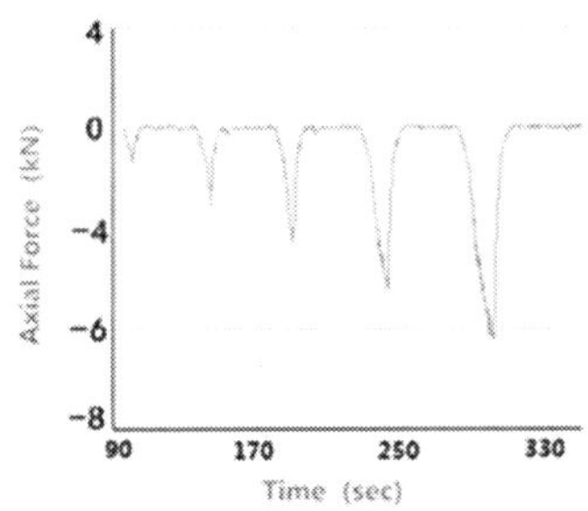

Figure 3. Stepping Load

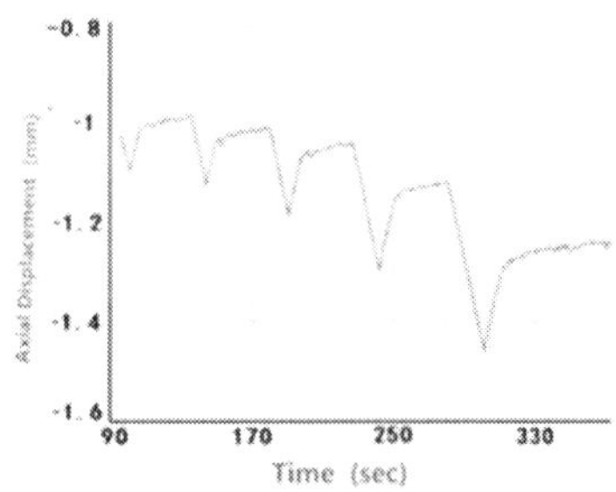

Figure 4. Stepping deformation

(6) Both the grade five load (N) and the resilient deformation (mm) are recorded.

According to the Standard Test Methods, the stiffness modulus expression in splitting condition is as follows:

$$ST=PT\times (0.27+\mu)/ (h\times XT) \quad (1)$$

This expression is applied to the splitting test in 50mm/min loading rate. The loading rate is too quick to reflect the plasticity and creep properties. The modulus status appear as elastic condition. The stiffness modulus is related to the loading time and testing temperature. The testing conditions include the fifteen degree centigrade about temperature and fifty millimeter per minute about loading rate. Every samples are being tested under the same conditions. P_T in stiffness modulus expression represents the maximum load in splitting destruction and X_T stands for the horizontal displacement which is correlated with Y_T vertical deformation. Similarly, P_T reflects the grade five load and Y_T is the grade five resilient deformation. In terms of the above relationships, ST can correspond to the splitting resilient modulus E in stepping load:

$$E=ST=PT\times (0.27+\mu)/ (h\times XT) \quad (2)$$

$$XT=YT\times (0.135+0.5\mu)/ (1.794+0.0314\mu) \quad (3)$$

The testing temperature should be strictly controlled in fifteen degree centigrade so that the Poisson's ratio of asphalt mixture is 0.3. The mathematical expression can be deduced:

$$E=3.608\times PT/ (h\times YT) \quad (4)$$

P_T is the grade five load (N); Y_T is the grade five resilient deformation (mm); h is specimen height (mm).

4. Results of splitting resilient modulus

In line with the above mentioned testing flow, the values about the grade five load and resilient deformation can be obtained. Contrasting the testing results and the design parameters about asphalt mixture compressive resilient modulus E is in Table 1.

Table 1. Comparison between testing results and design parameters

Asphalt Mixture	Specimen	Grade five load(N)	Grade five resilient deformation(mm)	Splitting resilient modulus(MPa)
AC-13	Sample1	6240	0.192	1847
	Sample2	6310	0.186	1928
	Sample3	6290	0.173	2066
	Sample4	6360	0.193	1872
Design parameter ranges of AC-13 asphalt mixture(15℃ tensile stress)				1800MPa-2200MPa

According to the Table 1 analysis, the results show that the splitting resilient modulus are in the context of tensile stress in fifteen degree centigrade and it is effective to measure the splitting resilient modulus by using stepping load method.

5. Conclusion

(1) Contrasting with testing methods of compressive resilient modulus, the methods of splitting resilient modulus are put forward tentatively. The splitting modulus mathematical expression -- $E=3.608\times P_T/ (h\times Y_T)$ can be deduced to calculate the values about splitting resilient modulus in stepping load. But both the loading rate and testing temperature should be strictly controlled when using this expression.

(2) The resilient modulus values which are measured by splitting resilient methods can meet the ranges of design parameters. This calculating methods about splitting testing can be available to compare the bottom layer tensile stress in splitting model with the allowable tensile stress in splitting test.

Acknowledgments

This work is supported by National Natural Science Foundation of China (NSFC) Grants (11072041), by Hunan province Natural Science Foundation Grants of China (10JJ6065) and by educational reform key project of Changsha University of Science and Technology (JG1102).

References

[1] JTG D50-2006, Specifications for Design of Highway Asphalt Pavement [S]. Beijing:China Communications Press, 2006: 30-35.
[2] LUO Zuo-fen,ZHENG Zhuan-chao.Research on the Resilient Modulus of the Asphalt Mixture under Splitting Loading [J]. JOURNAL OF HEBEI UNIVERSITY OF TECHNOLOGY, 2010, 39(3): 107-111.
[3] A.M.Hartman ect.Effect of Mixture Compaction on Indirect Tensile Stiffness and Fatigue [J]. Journal of Transportation Engineering, 2001, 9:370-378.
[4] LU Song-tao,CHEN Jie.Study on Stiffness Decay Regularity of Asphalt Mixture Based on

Accelerated Loading Test [J]. Journal of Highway and Transportation Research and Development, 2016, 33(5): 1-5.
[5] JTG E20-2011, Standard Test Methods of Bitumen and Bituminous Mixtures for Highway Engineering [S]. Beijing:China Communications Press, 2011: 253-258.
[6] TAN Zhi-ming,YU Xin-hua. Calculation of Sectional Flexural Stiffness Equivalent Modulus of Asphalt Pavement [J]. China Journal of Highway and Transport, 2012, 25(6): 37-42.
[7] SHRP Designation: M-009.Standard Method of Test for Determining the Fatigue Life of Compacted Bituminous Mixtures Subjected to Repeated Flexural Bending [M]. NCHRP. 1995.
[8] LI Yan-wei, MU Ke, SHI Xin, WANG Xuan-cang. Impact of Base-surface Contact Conditions on Mechanical Response of Asphalt Pavement [J]. Journal of Chang'an University (Natural Science Edition), 2014, 34(2): 38-44.
[9] Research Institute of Highway, Ministry of Communication Study of Dynamic Characteristic of Pavement Structure [R]. Beijing: Research Institute of Highway Ministry of Commications, 1999.

IOP Conf. Series: Materials Science and Engineering **207** (2017) 012096 doi:10.1088/1757-899X/207/1/012096

Optimal placement and sizing of wind / solar based DG sources in distribution system

Wanlin Guan*, Niao Guo, Chunlai Yu, Xiaoguang Chen, Haiyang Yu, Zhipeng Liu, Jiapeng Cui

State Grid Heilongjiang Electric Power Company Limited Electric Power Research Institute, Harbin, Heilongjiang, 150030, China

*Corresponding author e-mail: wanlinguan@163.com

Abstract. Proper placement and sizing of Distributed Generation (DG) in distribution system can obtain maximum potential benefits .This paper proposes quantum particle swarm algorithm (QPSO) based wind turbine generation unit (WTGU) and photovoltaic (PV) array placement and sizing approach for real power loss reduction and voltage stability improvement of distribution system. Performance modeling of wind and solar generation system are described and classified into PQ\PQ (V)\PI type models in power flow. Considering the WTGU and PV based DGs in distribution system is geographical restrictive, the optimal area and DG capacity limits of each bus in the setting area need to be set before optimization, the area optimization method is proposed . The method has been tested on IEEE 33-bus radial distribution systems to demonstrate the performance and effectiveness of the proposed method.

1. Introduction

Distributed generation (DG) in power system networks has rapidly increased as the demand of power system is growing exponentially. DG is small-scale power generation and usually located in distribution network [1]. DG units are mainly energized by wind, solar and fuel cell and have many advantages over centralized power generation. The optimum DG placement and sizing at planning stage of distribution system is necessary to achieve the reduction of power system losses and improved voltage profile [2].However, installing DG units at non-optimal place may get an opposite effect to what is desired. Selecting the best places for installing DG units and their preferable sizes in large distribution systems is a complex combinatorial optimization problem.

The optimal placement and sizing of DG in distribution networks has been continuously studied in order to achieve the maximum benefits. J.O. Kim [3] presented the improved Hereford Ranch algorithm to solve the DG allocation problem and getting significant effect. An analytical approach put forward by A Naresh to allocate DG for minimizing the total power losses in primary distribution network [4], DG is placed one by one in the network to obtain minimum power loss of the system. Authors in [5] considered a simple conventional iterative search technique along with NR (Newton Raphson) method of load flow study is implemented. But the analytical approach is not suitable for the large scale network for its large amount of calculation. GA [6] and PSO [7] algorithm, as the representative of the artificial intelligence algorithm, are applied to optimal the place and size of DG. Ref. [8] combined discrete enabled PSO and GA optimization methods to perform optimal location and DG size .However, in the later stage of evolution, the two algorithms are easy to fall into local optimal solution or stagnant state.

In [9], the voltage stability was considered as an objective function when dealing with optimum location of DG units for minimizing overall cost. Kumar and Banerjee [10] have studied about sizing and sitting of PV module and biomass gasifier based DG in isolated power system for distribution loss reduction only. The two objectives, minimization of total installation and operational costs, and minimization of the risk factor have been taken into account for system planning in Ref [11].Considering the interests of power supply and users, minimization of power losses and maximization of voltage stability due to finding weakest voltage bus as well as due to weakest link in the system are considered in fitness function [12].

However, in most of them, DGs are treated as PQ model type only (the real and reactive power output of DG is constant).In actually, power generation of wind unit and solar power need be detailed modeled based on their mechanism of generation. Wind and solar power DG have their operation modes and control characteristics, so handle DG as PQ model in power flow is not accurate.

Partha Kayal,et.[13] realized the sitting of the wind and solar based DG ,but the sizing optimum of DG was not considered. Though Ref. [14] presented an approach to find the optimum size of DG of three types at optimal power factor, the optimum placement of DG is also not solved .So optimal sitting and sizing of wind and solar based DG in distribution system simultaneously is necessary and impendency.

In this paper, a QPSO methodology is proposed to optimal placement and sizing of wind and solar based DG sources for real power loss reduction and voltage stability improvement of distribution system. The paper is organized as follows. In first section, double objective formulation is found as problem formulation for optimum DG placement and sizing. Second section is composed of performance modeling of wind and solar power generation system. In third section, area optimization method of DG units is proposed. Section fourth, basic mechanism of QPSO is presented, and also states the procedure to select appropriate locations and size for DG units using QPSO method. In fifth section, the method this paper put forward is tested in IEEE33-bus distribution systems .Finally some relevant conclusions are drawn.

2. Problem Formulation

Considering the interests of power supply and users, effectively combining the economy and safety, minimization of real power losses, maximization of voltage stability are considered as double objective formulation for optimum DG placement and sizing.

2.1. Real Power Losses Formulation

Considering N bus distribution system, the objective function for the minimization of real power loss is described as

$$f_{ploss} = \sum_{i=1}^{N-1} k_i r_i \frac{P_i^2 + Q_i^2}{U_i^2} \tag{1}$$

Where r_i is the resistance of the i_{th} branch; U_i is the voltage magnitude at bus i; P_i, Q_i are the active and reactive power inject bus i.

2.2. Voltage Stability Formulation

An improved distribution system voltage stability index is adopted. This index takes power flow solution as its criterion, the detailed description is as follows.

$$U_{stab,k} = 4\left[(XP_j - RQ_j)^2 + (XQ_j + RP_j)U_i^2\right]/U_i^4 \tag{2}$$

Where *Ustab, k* is the voltage stability index of line *k* (eg.bus *i* and *j* are start bus and end bus);*R* and *X*, are the resistance and reactance of line *k* respectively, ;P_j and Q_j are the active power and reactive power injected into bus *j* respectively; U_i is the voltage amplitude of bus *i*.

When the system is stable, all branch voltage stability of all branches values must be less than 1.0.However, when the system trend to crash, the crash bus must be the weakness point. So we can analyze the voltage stability of the system according to the distance between *Ustab* and 1.0.

The maximum voltage stability index value is the f_{Ustab} of the whole system and follows that

$$f_{Ustab} = \max\left\{U_{stab,1}, U_{stab,2}, \cdots, U_{stab,N}\right\} \tag{3}$$

Where *N* is the bus number of network; $U_{stab,1}, U_{stab,2}, \ldots, U_{stab,N}$ are the voltage stability of branch 1,2,…,*N* respectively .

2.3. Double objective formulation

Weight coefficient method is used to convert real power loss reduction and voltage stability improvement into a single Double objective formulation.

$$\min f = aw_1 \frac{f_{ploss}}{ploss_{init}} + bw_2 \frac{f_{ustab}}{Ustab_{init}} \tag{4}$$

Where *f* is the double objective optimization function; $ploss_{init}$, $Ustab_{init}$ is the initial real power loss and voltage stability index, by this time DG is not access to the network. *a* and *b* are the penalty factors of active loss and static voltage stability index respectively; w_1, w_2 are the weighting factors of active loss and static voltage stability index respectively ,however, $w_1+w_2=1$.If $w_1=1$, $w_2=0$, the single objective optimization is real power loss .Otherwise, the voltage stability index is treated as the single objective optimization.

a and *b* are defined as follows;

$$a = \begin{cases} N & \dfrac{f_{ploss}}{ploss_{init}} \geq 1 \\ 1 & \text{else} \end{cases} \tag{5}$$

$$b = \begin{cases} N & \dfrac{f_{Ustab}}{Ustab_{init}} \geq 1 \\ 1 & \text{else} \end{cases} \tag{6}$$

If the optimization of the real power losses greater than its initial (without DG) value, then *a*=*N*, *N* is *a* big positive value, which can increase the *f* value and give up this solution. According to this strategy, the objective function can toward to the optimal solution direction. Otherwise, *a* = 1. Update rule of *b* is same with *a*.

Subject to

(I) System power flow equations must be satisfied.

(II) Branch capacity and bus voltage constraints:

(III) DG capacity constraint

3. Performance modelling of wind and solar generation system

Power generation of Wind Turbine Generation Unit (WTGU) and PV array depends on their model and resource such as wind speed, solar radiation and ambient temperature. In this section modeling of WTGU and PV array is discussed.

3.1. Performance model of WTGU
Direct driven synchronous wind turbines are usually connected to grid by transducer which can control generator's active power output and reactive power respectively. So the direct driven synchronous and doubly fed wind turbines may be described as PQ model in flow calculations as well as power factor controlled combined heat and power generators. The type of DG is capable of injecting both real and reactive power.

Fixed speed and slip controlled asynchronous wind turbines which absorb reactive power from power system to build the magnetic field, don't have the ability of voltage regulation and this will lend to the increasing of network real power losses. Generally, compensative capacities are used to supply the reactive power which asynchronous generator need [15].

3.2. Performance model of PV array
Generally photovoltaic system is connected to power system by current controlled inverter. Photovoltaic system only supplies active power to power system. If photovoltaic system equipped with current inverter, the current output is constant. The corresponding compensative reactive power can be gotten by Eq. (7).

$$Q^{k+1} = \sqrt{|I_m|^2 (e^2_k + f^2_k) - P^2} \tag{7}$$

Where Q^{k+1} is the reactive power of DG in the $(k+1)$ iteration; e_k and f_k are the real part and imaginary part of voltage in the k iteration respectively; I_m is the current magnitude of DG; P is the active power of DG.

3.3. Power flow model type of WTGU and PV array
According to the Section 2.1 and 2.2, the WTGU and PV model DG will be classified into three type model in power flow, and described in Tab. 1.

Table 1. The classification of DG in power flow.

DG type	Classification	Model type in power flow
WTGU	Direct driven synchronous wind turbines	PQ
	Doubly fed wind turbines	PQ
	Fixed speed and slip controlled asynchronous wind turbines	PQ(V)
PV	photovoltaic system equipped with current inverter	PI

4. Area optimization method
Installing WTGU and PV based DGs in distribution system is geographical restrictive. As the storage of the local wind and solar power in distribution system are not identical, some areas are not suitable for installing WTGU or PV based DGs. So the optimal area and DG capacity limits of each bus in the setting area need to be set before optimization.

Based on this idea, this paper presents the area optimization method and describing as follows:

Two matrices are defined: Area optimization Matrix *NDG*, DG capacity limit Matrix *M*. The size of *NDG* is 1 * *N*, The size of *M* is 2 * *N*, and *N* is the bus number of the optimal area. *NDG* records the optional installation buses of the optimization area, but not mean all the buses need to be installed DG.*M* records the lower and upper limit of DG capacity in the optimal area. If the DG capacity less than the lower limit of DG, this bus will not be installed DG.

5. Optimal placement and sizing of DG

5.1. Overview of QPSO

Aiming at the shortcomings of PSO, Sun et al. proposed quantum particle swarm algorithm (QPSO) in 2004[16], QPSO improves the classic particle swarm optimization (PSO), combing the idea of quantum physics and key consideration of each particle's current local optimum information and global optimum location information. QPSO bases on DELTA, considering that the particles have quantum behavior. In quantum space, the particles make stochastic optimization search in the whole feasible solution space, so QPSO has stronger global optimization capability than standard PSO [17]. QPSO algorithm makes use of the wave function $\psi(X,t)$ to describe the state of the particles, obtains probability density function particles appear in a certain point by solving the Schrödinger Equation and gets particles position Eq. (8) as follows through Monte Carlo random simulation:

$$X(t) = p \pm \frac{L}{2}\ln(\frac{1}{u}) \quad (8)$$

Where

$$L(t+1) = 2 * \beta \left| mbest - X(t) \right| \quad (9)$$

QPSO algorithm evolutions are as follows:

$$mbest(t) = \frac{1}{M}\sum_{i=1}^{M} P_i(t) = [\frac{1}{M}\sum_{i=1}^{M} P_{i1}(t), \frac{1}{M}\sum_{i=1}^{M} P_{i2}(t), \cdots\cdots, \frac{1}{M}\sum_{i=1}^{M} P_{iD}(t)] \quad (10)$$

$$P_{id}(t) = \varphi * P_{id}(t) + (1-\varphi) * \mathrm{P}_{gd}(t) \quad (11)$$

$$X_{id}(t+1) = P_{id}(t) \pm \beta * \left| mbest(t) - X_{id}(t) \right| * \ln(\frac{1}{u}) \quad (12)$$

Where M is the number of particles in the population; D is the particle dimension; u and φ are the uniformly distributed random number in [0,1] respectively; P_{id} (t) is the current optimum position of particle i in the iteration t; P_{gd} (t) is the globally optimum position of particle i in the iteration $mbest$(t) is population average optimum position of all particles in the iteration i; P_{id} (t)is a random point between P_{id} (t) and P_{gd} (t). β is the contraction and expansion coefficient, it is an important parameter of QPSO. For better optimization effect, β dynamic changes with the number of iterations as follow:

$$\beta = m - (m-n) \times \frac{t}{MaxIters} \quad (13)$$

Where, $MaxIters$ is the maximum number of iterations. β increases from n to m with the iterations linearly, usually m=1,n=0.5.

5.2. The procedure for Optimaling placement and sizing of DGs

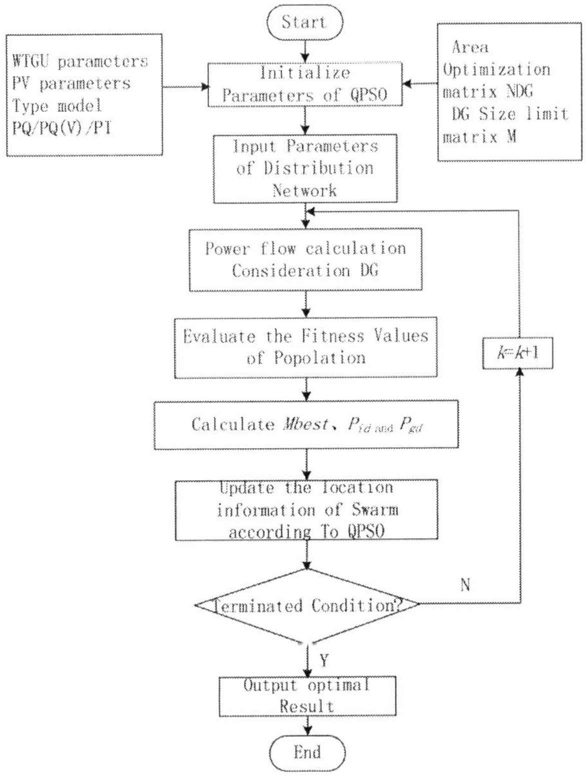

Figure 1. The flow chart of optimal placement and sizing of wind / solar based DG sources using QPSO.

6. Experiment results

The proposed method is tested on IEEE33-bus radial distribution systems, and results are obtained to evaluate its effectiveness. The substation voltage is considered as $1p.u$, Assume *Ploss* is the real power losses of network; *Ustab* is the voltage stability index .*Vmin* is lowest voltage of all buses. IEEE33 bus system is shown in Fig.2.

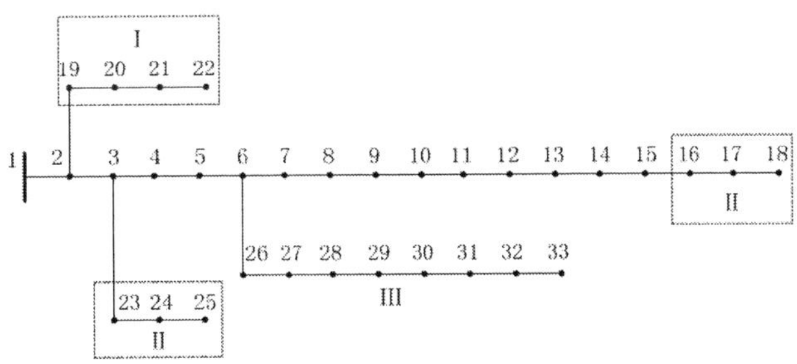

Figure 2. IEEE33 bus system

According to section 3, assume place DGs in three areas, these areas are defined as below. Area I:*NDG*=[19,20,21,22]; Area II:*NDG* =[16,17,18,23,24,25]; Area III:*NDG* =[25,26,27,28,29, 30,31,32]. The total real and reactive power loads of the system are 3715kW and 2300kvar respectively. The size of population is 50 and maximum number of iteration is 100 for all algorithms. Initial real power losses

(*Ploss*) is 202.6463kW, voltage stability index (*Ustab*) is 0.0792, and the lowest voltage of all buses (*Vmin*) is 0.9183pu.

Multiple asynchronous wind turbine (WTGU) and photovoltaic power generation (PV) type DGs will be placed in area I and II.

Install asynchronous wind turbines (WTGU) and photovoltaic system (PV) type DGs in area I and II. According to section 2, asynchronous wind turbines and photovoltaic power generation will be processed as PQ (V) model and PI model in power flow calculation respectively,.

This experiment is divided into six schemes, and defined as Table 1.

Table 2. Description of six schemes

Scheme Number	DG model	Upper limit/kW	Lower limit/kW
1	PQ(V)	2400	300
2	PQ(V)	2400	300
3	PQ(V)	600	300
4	PI	2400	300
5	PI	2400	300
6	PI	600	300

As WTGU and PV model DG have lower capacity limit, if the capacity of DG is less than the lower limit, the bus is not suitable for locating DG. The results are shown in Tab. 3. Fig.3 displays the convergence characteristics of scheme 2 and 5; Comparison voltage magnitude of scheme 2 and 5 is shown in Fig.4

Table 3. Optimal allocation results of WTGU and PV model DGs

Scheme	Area	DG type	Placement(Sizing/kW)	f	*Ploss*/kW	*Ustab*	*Vmin*/p.u.
1	I	PQ(V)	19(876.5);20(300);21(0);22(0)	0.9924	200.0077	0.0790	0.9139
2	II	PQ(V)	16(787.8);17(0);18(0); 23(825.2);24(0);25(422.7)	0.8175	158.1729	0.0677	0.9293
3	II	PQ(V)	16(404.4);17(371.2);18(0); 23(443.8);24(382.3);25(362.4)	0.8197	158.5534	0.0679	0.9290
4	I	PI	19(1089.5);20(0);21(0);22(0)	0.9851	197.0910	0.0790	0.9142
5	II	PI	16(501.3);17(0);18(0); 23(1080.8);24(0);25(557)	0.5842	88.4414	0.0580	0.9366
6	II	PI	16(500.6);17(0);18(0); 23(521.2);24(524.6);25(593.2)	0.5846	88.6169	0.0580	0.9365

Scheme1 plans to optimal placement and sizing of WTGU DG in area I. The optimization results show that bus 19 and 20 allocate DG of 876.5kW and 300kW respectively. However, bus 21 and 22 is not suitable for allocate DG. Assuming Rated power of WTGU is 300kw, so the bus 19 and 20 will be located 3units and 1 unit WTGU respectively.

Scheme2 plans to optimal placement and sizing of WTGU DG in area II. The optimization results show that bus 16 allocates DG of 787.8kW.

WTGU absorbs reactive power of 540.4kvar. According to the automatic switching program of parallel capacitor group of reactive power, 6 groups of parallel capacitor need to be accessed, the power factor of the WTGU increased from 0.8246 to 0.8246 and meets the demand of power factor. The actual compensation reactive power is 217.8kvar. However, WTGU also need absorbs 338.5kvar reactive power from the system.

After optimization, *Ploss* decrease from 202.6463kW to 158.1729kW, and reduction rate is 21.94%. *Ustab* decrease from 0.0792 to 0.0677, and reduction rate is 14.52%.

IOP Conf. Series: Materials Science and Engineering **207** (2017) 012096 doi:10.1088/1757-899X/207/1/012096

The difference of scheme 3 and 2 is the upper capacity limit. This scheme is set to analysis the capacity limit effect on the optimization results. The optimization results that the location of DG is different with scheme 2. Bus 18 is still not locate DG, the capacity of DG locating to the buses is between 300 and 600kw. As the upper limit of scheme 3 is less then scheme 2, the *Ploss* and *Ustab* are slightly bigger than scheme 2.

Scheme 4-6 plan to optimal placement and sizing of PV DG. As PV model DG do not need to absorb reactive power from the network, so *Ploss* and *Ustab* of scheme 4-6 is small then scheme 1-3. Compare with scheme 5 and 2, the placement of DG is same, it can be find that the DG model will not effect on the placement of DG so much. *Ploss* decrease from 158.1729kW (scheme 2) to 88.4414kW (scheme 5). *Ustab* decrease from 0.0677(scheme 2) to 0.0580(scheme 5).Scheme 6 and 3 also have the same rule.

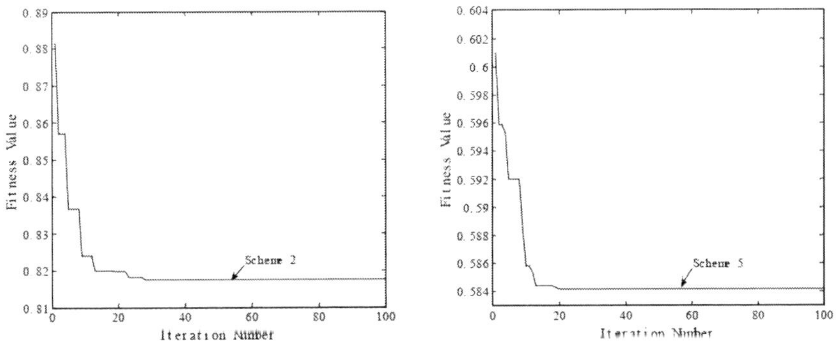

Figure 3. Convergence characteristics of scheme 2 and 5

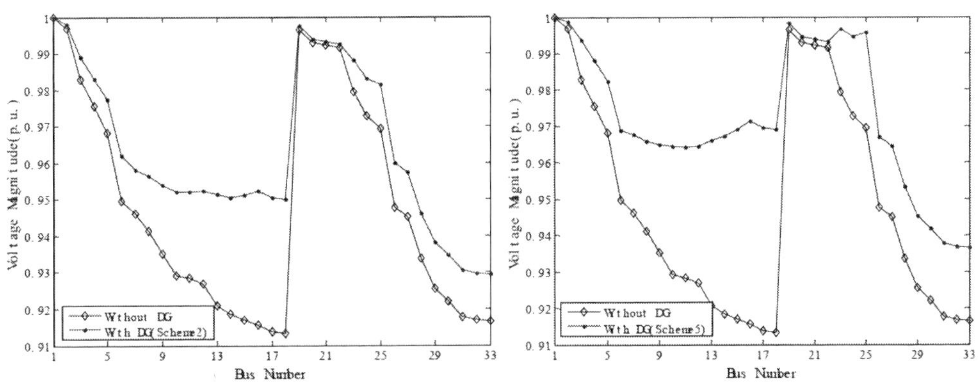

Figure 4. Comparison voltage magnitude of scheme 2 and 5 (With and without DG)

7. Conclusion

This paper has presented quantum particle swarm algorithm (QPSO) based method for optimal placement and sizing of wind and solar based Distributed Generation (DG) units in distribution system. Double objective formulation including real power loss reduction and voltage stability improvement is proposed. Utilizing performance modeling, optimal location and size selection of WTGU and PV array in distribution system are discussed. WTGU and PV array are classified into PQ\PQ (V)\PI type models in power flow. Considering the WTGU and PV based DGs in distribution system is geographical restrictive, some areas are not suitable for installing WTGU or PV based DGs. So the optimal area and DG capacity limits of each bus in the setting area need to be set before optimization, the area optimization method is proposed. Area optimization matrix (*NDG*) and DG capacity limit matrix (*M*) are defined. The area optimization method can help make the optimization more accurate. Several schemes have been experimented on IEEE33 bus radial distribution systems, the result show that proposed method is very helpful for optimizing placement and sizing of wind/solar based DG sources in distribution system.

References

[1] Ackermann T, Andersson G, Soder L. Distributed generation: a definition. Electric Power Syst Res 2001; 57: 195-204.
[2] AlRashidi MR, AlHajri MF. Optimal planning of multiple distributed generation sources in distribution networks: a new approach. Energy Convers Manage 2011; 52: 3301-8.
[3] Kim J O, Nam S W, Park S K, et al. Dispersed generation planning using improved Hereford ranch algorithm. Electric Power System Research 1998; 47(1): 47-55.
[4] Naresh A, Pukar M, Mithulananthan N. An analytical approach for DG allocation in primary distribution network. Elec Power Energy Syst 2006; 28: 669-78.
[5] Ghosh S, Ghoshal SP, Ghosh S. Optimal sizing and placement of distributed generation in a network system. Int J Elec Power Energy Syst 2010; 32(8): 849-54.
[6] Abou EI-Ela A A, Allam S M, Maximal optimal benefits of distributed generation using genetic algorithms. Electric Power Systems Research 2010, 80(7): 869-877.
[7] Prommee W, Ongsakul W. Optimal multiple distributed generation placement in microgrid system by improved reinitialized social structures particle swarm optimization. European Transactions on Electrical Power 2011, 21(1): 489-504.
[8] Moradi M H, Abedini M.A combination of genetic algorithm and particle swarm optimization for optimal DG location and sizing in distribution systems. International Journal of Electrical Power & Energy Systems 2012, 34(1): 66-74.
[9] Hemdan N G A, Kurrat M. Distributed generation location and capacity effect on voltage stability of distribution networks," in Proc. Annu. IEEE Conf., 2008, 1-5,
[10] Kumar MVM, Banerjee R. Analysis of isolated power systems for village electrification. Energy Sust Dev 2010;14: 213-22.
[11] Ganguly S, Sahoo NC, Das D. Multi-objective particle swarm optimization based on fuzzy-Pareto-dominance for possibilistic planning of electrical distribution systems incorporating distributed generation. Fuzzy Set Syst 2013; 213: 47-73.
[12] Aman MM, Jasmon GB, Bakar AHA, Mokhlis H. A new approach for optimum DG placement and sizing based on voltage stability maximization and minimization of power losses. J Energy Convers Manage 2013; 70: 202-10.
[13] Partha Kayal, C.K. Chanda, Placement of wind and solar based DGs in distribution system for power loss minimization and voltage stability improvement, Electrical Power and Energy Systems 2013; 53: 795-809.
[14] Kansal S, Kumar V, Tyagi B. Optimal placement of different type of DG sources in distribution networks. Interational Journal of Electrical Power Energy Systems 2013; 53: 752-760.
[15] Guan WL, Tan YH, Zhang HX, Song JL. Distribution system feeder reconfiguration considering different model of DG sources. International Journal of Electrical Power & Energy Systems. 2015; 68: 210-221.
[16] Sun J, Feng B, Xu W B. Particle swarm optimization with particles having quantum behavior. Proceedings of Congress on Evolutionary Computation. Portland: IEEE 2004: 325-331.
[17] Jamalipour M, Gharib M, Sayareh R, Khoshahval F. PWR power distribution flattening using quantum particle swarm intelligence. Annals of Nuclear Energy 2013; 56: 143-150.

IOP Conf. Series: Materials Science and Engineering **207** (2017) 012097 doi:10.1088/1757-899X/207/1/012097

Exergy analysis on industrial boiler energy conservation and emission evaluation applications

Henan Li*

Dalian Boiler and Pressure Vessel Inspection Institute, Dalian 116000, China

*Corresponding author e-mail: 172579206@qq.com

Abstract. Industrial boiler is one of the most energy-consuming equipments in china, the annual consumption of energy accounts for about one-third of the national energy consumption. Industrial boilers in service at present have several severe problems such as small capacity, low efficiency, high energy consumption and causing severe pollution on environment. In recent years, our country in the big scope, long time serious fog weather, with coal-fired industrial boilers is closely related to the regional characteristics of high strength and low emissions [1]. The energy-efficient and emission reducing of industry boiler is of great significance to improve China's energy usage efficiency and environmental protection. Difference in thermal equilibrium theory is widely used in boiler design, exergy analysis method is established on the basis of the first law and second law of thermodynamics, by studying the cycle of the effect of energy conversion and utilization, to analyze its influencing factors, to reveal the exergy loss of location, distribution and size, find out the weak links, and a method of mining system of the boiler energy saving potential. Exergy analysis method is used for layer combustion boiler efficiency and pollutant emission characteristics analysis and evaluation, and can more objectively and accurately the energy conserving potential of the mining system of the boiler, find out the weak link of energy consumption, and improve equipment performance to improve the industrial boiler environmental friendliness.

1. Description in Brief of Exergy Analysis

Characterization of material how much calories, one of the state parameter of enthalpy and enthalpy only expressed the unit mass material amount of calories, but did not show the merits of the heat quality. The quality of the energy is not the same, the same heat energy, its ability to effectively work is not the same. For example, although the quantity of the boiler flue gas is very big, but the heat is hard to use. Exergy is a quality material availability of calories, in the absence of other sources of heat, the steady flow of working medium from the initial state through reversible process changes to and environment medium in thermal equilibrium can be one of the biggest useful work, known as the working medium in the initial state of thermodynamic exergy. Thermodynamic exergy is one of the working medium state parameter, it corresponds to a certain temperature value is a single. In the process of actual heat, the sum of the entropy of the system of material is increasing, but the system is the sum of exergy is less. The exergy analysis for industrial boiler efficiency analysis, so as to improve the energy efficiency of industrial boiler is very timely and necessary [2].

Exergy analysis method is based on the first law of thermodynamics and the second law of exergy balance of thought, on the basis of the concept of its exergy balance for the tool, by studying the cycle of the effect of energy conversion and utilization, to analyze its influencing factors, to reveal exergy damage location, distribution and size, find out the weak links, and explore ways to improve the effect of energy conversion and utilization.

1.1. Exergy balance and exergy loss

Conservation of energy may be the most basic rule that scientists have yet discovered concerning the universe. Exergy energy use can be part of, it is not balance of payments, the actual transformation process, the part of the available can, will be transformed into unavailable exergy will reduce, called the exergy loss. This is not in violation of the law of conservation of energy and exergy balance is the sum of exergy and exergy loss in balance.

Exergy balance considering the number of energy, also take into account the power quality. exergy balance, the key is to write down all the exergy loss is needed to keep balance. Among them, the internal irreversible exergy loss of heat balance does not reflect. Two methods of analysis has a qualitative difference. Both and save the inner relationship between each other, exergy balance is based thermal equilibrium.

1.2. Exergy analysis and exergy efficiency

Usually heat balance and energy conversion efficiency and can't reflect the exergy utilization degree, we introduce the concept of exergy efficiency. exergy efficiency and the energy conversion efficiency is defined by a similar, the difference is that the exergy efficiency is income pay exergy and exergy ratio [3]. There is the concept of exergy efficiency, we can for a thermal system to establish exergy balance equation, and carries on the exergy analysis. Exergy analysis to quantitative calculation and loss of exergy balance, energy utilization. Balance is the foundation of energy flow to include relating to income and various losses, the allocation proportion can distinguish between the primary and secondary. Exergy analysis efficiency can be calculated, determine the degree of effect and the effective utilization of energy conversion. Exergy analysis can analyze the energy use of rationality, analysis of all kinds of damage and influence factors, puts forward improvement possibility and improving ways, and predict the improved energy saving effects.

2. Optimization and Application of Exergy Analysis

Exergy analysis of the development of the optimization can be divided into three levels: thermodynamic optimization; optimization of thermal economics; the combination of thermal economics and ecology. Thermodynamic analysis can only calculate the exergy loss and exergy loss, engineering decisions still need to consider factors such as investment, economic and environmental impact.

2.1. Thermodynamic optimization

The irreversible loss in the process of industrial boiler work with external loss of exergy analysis belongs to the thermodynamic optimization, it evaluated the combustion process in the effective measure of the fuel is converted to heat, corresponding combustion model and mathematical model is established, finish the calculation of exergy efficiency program, finally using numerical analysis has been influencing exergy efficiency in the working process of the main influencing factors.

2.2. Optimization of thermal economics

In recent years, with the method of exergy analysis and thermal economics are widely used in the energy and petrochemical industry in our country, energy conversion in the process of production, transfer, use, and damage are analyzed, reveals how much energy of each part, cause and position of the energy loss, to improve equipment, improve the energy utilization method and so on to find out.

Industrial boiler thermal system for the diagnosis and analysis of the main equipment of the system and thermodynamics of link integrity, point out the weak link in the system, and with the use of traditional heat balance method, the results were compared, for power plant operation optimization and energy-saving technological transformation provides a scientific basis [4].

2.3. The combination of thermal economics and ecology

When the further development of exergy analysis and thermal economics, combined with ecology gradually, it makes industrial system design and planning stage is considering the ecological environment protection, avoid environmental pollution in the process of production in the future.

Due to improve the boiler efficiency and reduce the effects of pollutants on the environment are interconnected, the boiler efficiency, improve the efficiency of combustion and heat transfer process, in particular, the effective use of resources and environment sustainable development will also bring new improvement. Reduce pollutants discharge, reducing the loss of the high grade energy, which improves the efficiency of the boiler. The combined realization of coal-fired boiler thermal efficiency and environmental protection will be the future boiler energy-saving reform of the economic operation and development trend

3. The Application of Exergy Analysis in Industrial Boiler

Industrial boilers in service at present have several severe problems such as small capacity, low efficiency, high energy consumption and causing severe pollution on environment. exergy analysis method for boiler equipment efficiency and pollutant emission characteristics of a detailed analysis and study is to establish a comprehensive consideration of the important methods for the evaluation system of the boiler thermal efficiency and environmental protection.

3.1. Exergy analysis in the application of boiler economic evaluation

By the exergy analysis, we can realize more profoundly the weak links of using energy systems and manage more rationally the energy process used. exergy analysis can get input materials such as boiler fuel and effective utilization and the balance of the relationship between the loss rate, so as to analyze the rationality of energy utilization. Furthermore, the method of exergy analysis may perfectly analyse the status of using energy of all kinds of boiler surface.

Boiler and heat transfer heat transfer process during burning process exergy efficiency and the exergy analysis can be concluded that fossil fuel energy use efficiency is equal to the loss and the ratio of input, so the efficiency of the boiler combustion is endowed with practical significance. the results show that the main loss process in the boiler for combustion and heat transfer process, the combustion process to reduce the power loss coefficient of air conditioning and inlet air temperature. Boiler exergy loss occurs mainly in the process of combustion chamber of a stove or furnace, this may be due to incomplete combustion, heat insulation effect is poorer, and the entropy production is bigger, rather than the calculation results show that the condenser heat loss is the main equipment, reveals the importance of exergy analysis from different angles [5].

In recent years, domestic scholars boiler parts a lot of calculation and analysis on modification of equipment and device, such as how to improve the feed temperature, at the same time in the boiler exhaust smoke under the premise of constant temperature appropriate increase secondary air temperature, etc. through the exergy analysis of boiler efficiency calculation and performance analysis it is concluded that the exergy efficiency of the boiler and its various parts, and the size of the exergy loss of each process, and the results with the boiler heat balance analysis results were compared, found that exergy loss mainly includes the combustion process of boiler exergy loss and exergy loss of heat transfer process, it is pointed out the direction of the in order to further improve the efficiency of the boiler. In combustion, heat transfer process as the breakthrough point, through oxygen-enriched combustion, improving steam initial parameter method to reduce the coal consumption of the boiler.

3.2. The application of exergy analysis in boiler pollutant emission characteristics

Exergy analysis can be used to improve the sustainable development of the society and nature, exergy is to explain the energy into the environmental impact of the best tools, exergy and the sustainable development of economy, and environment factors are closely linked. Energy due to the internal irreversibility, exergy loss may occur, so should try to reduce the loss to achieve sustainable development. Environmental impact and the relationship between the emissions and resource depletion can be created according to the physics principle based on the exergy of expression [6].

Many of the existing boiler emissions evaluation index, mostly for a number of pollutants are compared, and a single comprehensive evaluation of quantitative index of the performance of the boiler is less. For example, to calculate the desulfurization denitration, dust removal system, such as energy consumption and waste quantitative influence on the environment. exergy is measuring system suitable relationship between the loss and environmental loss, exergy analysis, for boiler can be caused by the pollutants into the environment influence quantitative evaluation analysis. Boiler system exergy magnitude depends on the state of the system and environment, only when the system and the environment balance, loss to zero.

Establish system discharge of wastes and waste heat exergy calculation model for description of the environmental pollution, In the boiler system, waste exergy is mainly refers to from the flue gas, solid waste burning, wastewater, and left in waste heat boiler of exergy when into the environment [7]. When calculating the overall boiler pollutants fire unavailable, need to consider different pollutant degree of damage to the environment, exergy of the potential difference is the measurement of environmental damage.

4. Conclusion

Exergy analysis to explain the reason of the loss of energy consumption, for the future energy efficiency are more accurate and scientific evaluation and prediction, will be important for the sustainable development of assessment tools. the exergy theory is applied to the actual boiler calculation of the ultimate goal is to draw a perfect unity of quantitative evaluation method of energy saving and emission reduction to find the lowest parts of the equipment or utilization in the process of running, and put forward concrete can reasonable transformation method. exergy calculation analysis of pollutant control device can further reveal the side of the boiler energy conservation and environmental protection work. exergy analysis to the boiler thermal efficiency and environmental protection of the establishment of a comprehensive quantitative evaluation method will play an important role.

Coal-fired boiler drawing surface and pollution control equipment system flow diagram, and calculating exergy analysis, establish the unity of the boiler economic performance and environmental performance evaluation index system, determine the best working condition of boiler operation, to avoid problems such as inefficient high emissions, industrial boiler is the urgent need of theoretical research and engineering application in the future. Based on the theory of exergy comprehensive consideration of economy and environmental protection, map corresponding with table, fire with fire, it is concluded that the two best integration performance. The establishment of the integrated model is of great significance for energy conservation and emissions reduction, and will become the evaluation of power plant boiler performance good or bad, to guide the actual modification, the importance of fault diagnosis and optimizing operation methods and indicators.

References

[1] Rosen M A. Second Law Analysis: Approaches and Implication [M].Int J Energy Res, 1999, (23): 410-450.

[2] Hamed O. Thermal performance of multi stage flash distillation plants in Saudi Arabia, Al Jubail plant and others. Desalination 2000, 128: 281–92. W. Strunk Jr., E.B. White, The Elements of Style, third ed., Macmillan, New York, 1979.

[3] Cerci Y. Exergy analysis of a reverse osmosis desalination plant in California. Desalination

2002; 142: 257–66
[4] Massoud M. Engineering thermofluids. Thermodynamics, fluid mechanics, and heat transfer; 2007. ISBN 10 3-540-22292-8, Springer.
[5] Saidi MH. Optimization of a combined heat and power PEFC by exergy analysis. J Power Sources 2005; 143: 179-84.
[6] Xie D, Wang Z, JinL, Zhang Y. Energy and exergy analysis of a fuel cell based micro combined heat and power cogeneration system. EnergyBuild2012; 50: 266-72.
[7] Dincer I, Rosen MA. Exergy. Energy, environmental and sustainable development. Elsevier; 2013. ISBN 978-0-08-097089-9.

Evaluation model of susceptibility to Cu hot shortness of Cu-containing LC steel

Yinghua Jiang[1, 2, *], Chunqian Xie[1, 2]

[1]Shougang Research Institute of Technology, Beijing 100043, China
[2]Beijing key Laboratory of Green Recyclable Process for Iron & steel Production Technology, Beijing 100043, China

[*]Corresponding author e-mail: yinghuajiang@163.com

Abstract. It is well known that copper causes the hot shortness problem when in the scrap based steelmaking process. Hot shortness occurs because of selective oxidation of the iron whereby the more noble copper is enriched at the steel-oxide interface and embrittlement take places during hot working. In this paper, in order to control the Cu hot shortness, the evaluation model of susceptibility to Cu hot shortness was investigated. We found that Cu hot shortness susceptibility is related to the oxidation rate, the copper equivalence and the scale-steel interface length ratio. Also, it is founded to exist a fairly nice correlation between the oxidation rate and other factors such as chemistry elements, oxidation conditions, especially the Ni content.

1. Introduction

Recycling of scraps in the EAF (Electric Arc Furnace) steelmaking process has become very important from the viewpoint of economic and environmental aspect [1]. Recycling of steel scrap accumulates undesirable impurity elements such as copper, nickel and tin which are difficult to remove in steel-making process. Copper is nobler than iron and it is enriched at the steel-oxide interface when iron is oxidized at high temperature. The copper-rich liquid penetrates with time into the steel along the austenite grain boundaries, weakens the bonds and induces the surface cracking during hot working. This cracking phenomenon is known as hot shortness [2]. The other impurity elements Sn and Ni affect the evolution of copper-rich liquid phase. Tin can increase the hot shortness by decreasing the solubility of copper in austenite. On the contrary, addition of Ni can reduce hot shortness. The beneficial effect of nickel is related to an increase in the solubility of copper in austenite and occlusion in the scales [3-5]. The effect of these elements on the solubility of copper in austenite can be expressed by means of Cueq parameter (Cueq=Cu+8Sn-Ni) [6]. Another beneficial effect of Ni is occlusion phenomenon. This process reduces the Cu hot shortness by involving metal enriched in copper and nickel to become entrapped in the iron oxide layer. A nickel content of >0.02wt has been found to promote occlusion. The occlusion of copper-rich material into the oxide layer has a requirement: the development of a rough oxide/metal interface [7]. In addition to the impurity elements, the hot shortness also depends on the oxidation condition such as oxidation pressure, temperature, time. In general, the Cu hot shortness is most severe at temperature around 1100℃ since the melting temperature of Cu is around 1089℃, resulting in the liquid Cu penetration into austenite grain boundaries [8].

In order to control and minimize the Cu hot shortness, it is necessary to understand the Cu hot shortness behavior during high temperature oxidation and establish a reliable evaluation method for Cu hot shortness quantitatively. The present work is undertaken to examine the effect of the important factors such as impurity elements and oxidation conditions on hot shortness quantitatively. Accordingly, on the basis of some theory reports and the experimental results, the evaluation model of susceptibility to Cu hot shortness is suggested.

2. Experimental Procedure

The chemical composition of the steels used in this study is listed in Table 1. They were melted in a vacuum furnace and the 30mm thickness ingots were hot forged. Specimens for oxidation experiment were cut to rectangular shapes of 20mm*15mm*3mm. Rectangular specimens were polished with #600 grit emery paper for the oxidation experiments. A steel specimen was first heated to 1010℃ and held at this temperature for ten minutes in flowing high purity Ar. Then, the inert gas was replaced by the selected oxidizing gas mixture (3%O2-9%CO2-17%H2O-N2). The specimens were isothermally oxidized for 4, 9 and 30minutes after heating to 1100℃~1200℃. Also, specimens for hot compression test with 10mm in diameter and 15mm in length were machined from the ingots. The specimen were then heated in the Gleeble at 1150℃, held for 9min to stabilize, and deformed at the highest axial speed resulting in a circumferential strain rate of 30mm/sec. The microstructures at the scale/steel interface after oxidation were closely observed by using scanning electron microscopy and energy dispersive X-ray spectroscopy in order to clarify the evolution of the Cu-rich phase. All images were taken in backscattered electron (BSE) mode. The Image J program was used to measure the scale-steel interface length, crack width and crack depth. The Crack index was given by the expression (1).

$$\text{crack index} = \text{crack width x crack depth x the number of crack} \quad (1)$$

Table 1. Chemical composition of steels. (wt. %)

Steel	C	Si	Mn	P	S	Cu	Sn	Ni	Cr	Al	N	Cueq
1	0.035	0.031	0.277	0.010	0.015	0.13	0.01	0.05	0.06	0.04	0.006	0.16
2	0.035	0.031	0.277	0.010	0.015	0.22	0.02	0.08	0.11	0.04	0.006	0.30
3	0.035	0.031	0.277	0.010	0.015	0.38	0.04	0.13	0.19	0.04	0.006	0.57

3. Results and Discussion

3.1. The effect of Ni content on the steel oxidation behaviour

As mentioned above, the Cu-liquid phase occurs during high temperature oxidation and is closely related to the chemistry parameters and oxidation conditions. The fairly nice correlation is founded that the oxidation rate and effective factors such as temperature [T], time [t], pressure [P] and Ni content influence each other and play a significant role in the oxidation behavior of steel surface. In order to investigate the relationship between oxidation rate and other factors in detail, a careful statistical analysis has been performed. Analysis result is expressed by using the following equation (2).

$$\Delta W/A = -0.184 + 0.000175\,[T] + 0.0006\,[t] + 0.00193\,[P] - 0.167\,[Ni] \quad (2)$$

From the expression, it is seen that the Ni content has a crucial influence on the oxidation rate. The relationship between Ni content and the oxidation rate is shown in Figure 1. In Figure 1 the oxidation rate reduces with increasing the Ni content. It has been reported that addition Ni make the scale-steel interface uneven and the uneven scale-steel interface provides a channel for the occlusion and further alleviate the Cu hot shortness. Figure 2 shows the relationship between Ni content and the scale-steel interface length, showing that the scale-steel interface length increase with an increase in Ni content.

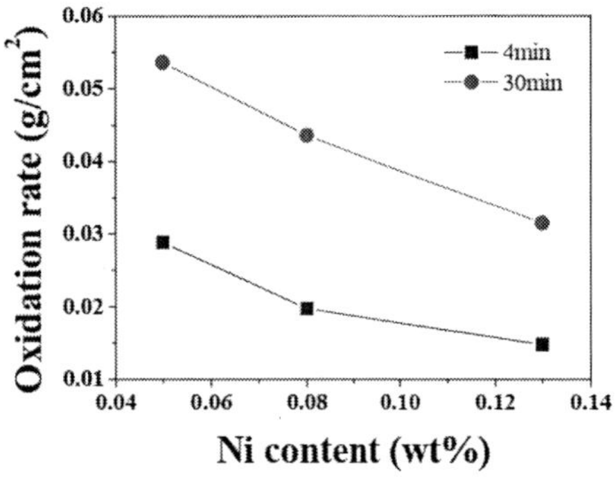

Figure 1. Effect of Ni content on the oxidation rate at 1150℃.

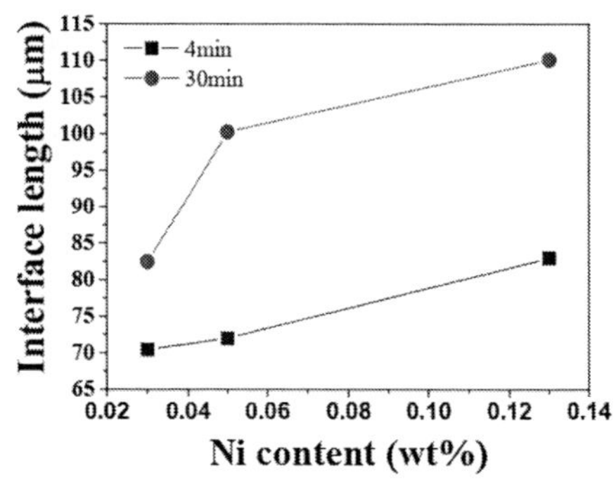

Figure 2. Effect of Ni content on the scale/steel interface length at 1150℃.

3.2. Model Development

On basis of given mechanism such as selective oxidation behavior, solubility and occlusion, the evaluation model of susceptibility to Cu hot shortness was investigated. Generally speaking, the presence of Cu-rich phase at the scale-steel interface was accompanied by surface cracking. Here, we try to evaluate the Cu hot shortness (surface cracking) by the degree of Cu enrichment at scale-steel interface, since the total enriched Cu amounts due to the oxidation should be the main origin of Cu penetration into grain boundaries. In order to measure the degree of Cu enrichment at scale-steel interface, the following operation is carried out. Firstly, the analysis describes the method that determines the maximum copper-rich amount on basis of oxidation behavior. There is the assumption that all copper is rejected from the scale and incorporated into the copper-rich layer, no diffusion of copper within the steel. The amount of copper accumulated within the copper-rich layer equals the amount of copper rejected due to scale formation during oxidation process with reference to Figure 3. The maximum amount of copper-rich can be expressed using equation (3):

$$W_{Cu\text{-}rich}^{max} = L \times C_0^{Cu} \tag{3}$$

Where L and C_0^{Cu} are the thickness of metal consumed and the initial concentration of copper, respectively. Secondly, with reference to [8], the scale thickness and the thickness of metal consumed can be calculated by equation (4) and (5), respectively.

$$X = \frac{\Delta W}{A} \times \frac{1}{K} \tag{4}$$

$$X = 1.84 \times L \tag{5}$$

Where X and $\Delta W/A$ are the scale thickness and oxidation rate, respectively. K is conversion factor (=1.279g/mm3). Then, according to the effect of the impurity elements on the solubility of copper, taking Ccu= Cueq=Cu+8Sn-Ni was calculated as the amount of copper-rich. Lastly, incorporating the effect of the uneven interface which is a channel for occlusion on Cu enrichment into the model, the equation (3) can be modified:

$$W_{Cu\text{-}rich} = \frac{\Delta W}{A} \times \frac{Cu_{eq}}{2.35} \times \frac{H_0}{H} \quad (6)$$

Where H_0, H and H_0/H are the initial scale-steel interface length (t=0) and the last scale-steel interface length (t=t) and interface length ratio respectively. Here, we use $W_{Cu\text{-}rich}$ to judge the degree of Cu enrichment at scale-steel interface.

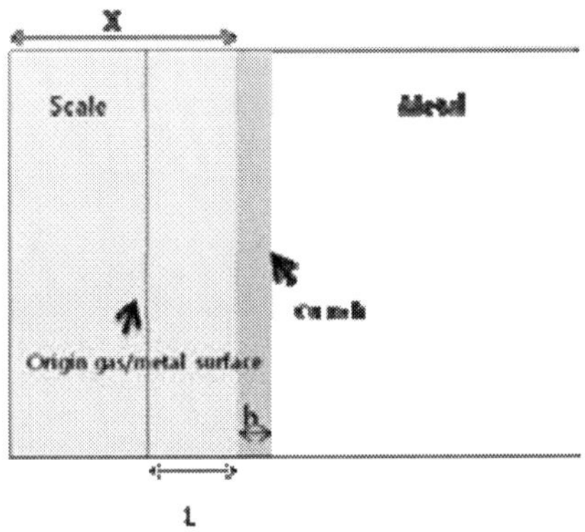

Figure 3. Schematic of arrangement of phases.

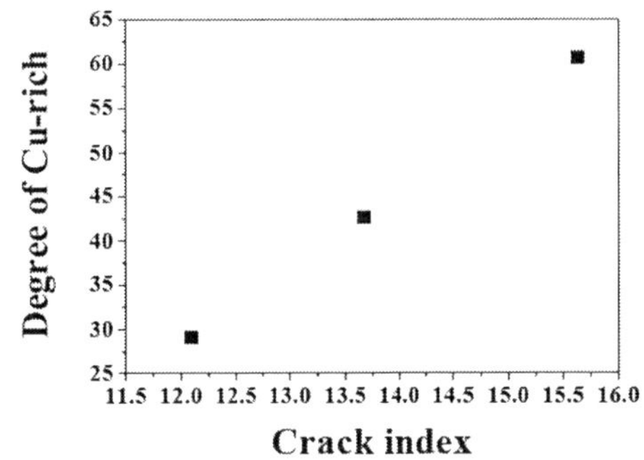

Figure 4. Effect of the degree of copper-rich (wcu-rich) on the crack index.

3.3. The effect of the degree of copper-rich on the crack index

In order to verify the model formula (6), some data input into the model formula to obtain the enriched amount of Cu. The calculation value matched with the crack index obtained from the experimental results. Figure 4 shows the relationship between the degree of Cu-rich and the crack index. With an increase in the degree of Cu-rich phase, the cracking index becomes large. This can be explained that the more Cu-rich phase enriches at the scale-steel interface, the more opportunities occur, allowing the liquid Cu penetration into grain boundaries.

4. Conclusion

The simple model for evaluating the susceptibility to Cu hot shortness of Cu-containing LC steel was investigated. Addition Ni results in a decrease in oxidation rate and an increase in the scale-steel interface length. The both effect are beneficial for reducing the Cu hot shortness. The oxidation rate is expressed by formula (2) by the mean of statistical analysis. There are the fairly nice correlations between the oxidation rate and important parameters such as Ni content and time etc. The evaluation model of susceptibility to Cu hot shortness is expressed in terms of Cu enrichment characteristics such as the oxidation rate, Cu equivalence, the scale-steel interface length ratio. From this model, the degree of Cu enrichment at scale-steel interface that can predict the degree of surface cracking has been successfully calculated.

References

[1] Miras. Djurovic, Bosko Perovic, Kata Kovacevic, The art of writing a scientific article, MATERIALI IN TEHNOLOGIJE, 36 (2002) 3-4.
[2] S. Akamatu, K. Koyama, CAMP-ISIJ, 7 (1994) 1712.
[3] Seok-Jong SEO, Kentaro ASAKURA, ISIJ International, 37 (1997) 232-239.
[4] Masaharu hatano, Kazutoshi Kunishige, Tetsu-to-Hagane, 88 (2002).

[5] N. Imai, N. Komatsubara, ISIJ International, 37 (1997) 217-223.
[6] A. Bergman, R. West, Scripta metallurgica, 22 (1988) 659-663.
[7] Bryan A. WEBLER, Seetharaman SRIDHAR, ISIJ International, 48 (2008) 1345-1353.
[8] Stephan Potschke, Achim R. Buchner, steel research int, 77 (2006).

IOP Conf. Series: Materials Science and Engineering **207** (2017) 012099 doi:10.1088/1757-899X/207/1/012099

Signal acquisition and scale calibration for beam power density distribution of electron beam welding

Yong Peng[1,*], Hongqiang Li[1], Chunlong Shen[2], Shun Guo[1], Qi Zhou[1], Kehong Wang[1]

[1]School of materials science and engineering, Nanjing University of Science and Technology, Nanjing 210094, China
[2]College of Marine and Electromechanical Engineering, Taizhou University, Taizhou 215300, China

[*]Corresponding author e-mail: ypeng@njust.edu.cn

Abstract. The power density distribution of electron beam welding (EBW) is a key factor to reflect the beam quality. The beam quality test system was designed for the actual beam power density distribution of high-voltage EBW. After the analysis of characteristics and phase relationship between the deflection control signal and the acquisition signal, the Post-Trigger mode was proposed for the signal acquisition meanwhile the same external clock source was shared by the control signal and the sampling clock. The power density distribution of beam cross-section was reconstructed using one-dimensional signal that was processed by median filtering, twice signal segmentation and spatial scale calibration. The diameter of beam cross-section was defined by amplitude method and integral method respectively. The measured diameter of integral definition is bigger than that of amplitude definition, but for the ideal distribution the former is smaller than the latter. The measured distribution without symmetrical shape is not concentrated compared to Gaussian distribution.

1. Introduction

Electron beam welding (EBW) is a high power density welding method that has many advantages over conventional techniques. The main advantages of this technique are large depth-to-width ratio, narrow heat-affected zones (HAZ), insignificant deformation and pure chemical composition of weld seam in a vacuum [1].

The process of EBW is influenced by a huge number of process parameters. The actual manufacture and design theory of an EBW electronic gun differ in equipment characteristics. The stability of an electron gun power supply system and vacuum changes could affect the electron beam diameter and power density distribution. The geometry and quality of the weld seam is strongly influenced by the process parameters such as accelerating voltage, beam current, welding speed and focus current. Therefore it is necessary to study the characteristics of the electron beam precisely and measure the electron beam quantity quantitatively, such as focus position, beam diameter and power density distribution [2, 3]. As the extremely high power density is about 10^7 W/cm^2 at the focus spot, it can melt any refractory material, which makes measurement very difficult. S. Rouquette built a

IOP Conf. Series: Materials Science and Engineering **207** (2017) 012099 doi:10.1088/1757-899X/207/1/012099

Gaussian source model in EBW and estimated the parameters of the energy dissipation in gas phase zone and liquid phase zone by Levenberg–Marquardt method [4]. Luo study the power density distribution of vacuum EBW under different welding conditions using the finite element model of rotary Gaussian body heat source [5]. The studies mentioned above are based on the ideal model.

In order to get the actual beam power density distribution for high-voltage EBW machine, we had developed the electron beam quality test system. With high-speed magnetic deflection technology, the system controlled the electron beam scanning on the Faraday cup sensor quickly, which solved the problem about high beam power density. As the electron beam welding environments with high electromagnetic interference, the collected signal is mixed with noise. In this paper, firstly the signal is processed by digital filter, and then the time-domain signal is calibrated corresponding to spatial scale, finally the power density distribution and beam spot diameter of electron beam cross-section is calculated by the special software.

2. Principle and Structure of the System

Fig. 1 represents a schematic diagram of the beam quality test system used for the electron beam measurements before welding process. Driven by the control system of the EBW machine, the electron beam deflected to the energy-absorbing device to wait for the test, which can prevent the high power electron beam from staying on the sensor and damaging the sensor. When the test began, the beam deflected periodically. The deflection process is shown in Fig.2, in which the deflection path is along the direction of the arrow. A tungsten film with a high melting point is used as the direct contacts with the electron beam, which installed on the surface of the Faraday cup sensor. There is a pinhole in the middle of the tungsten film. The pinhole size is Φ20mm. In the deflection process the sensor collected the electron entered into the pinhole, and then the current signal formed by the electron was converted to voltage signal which was collected by the high speed data acquisition card and stored in the industrial PC.

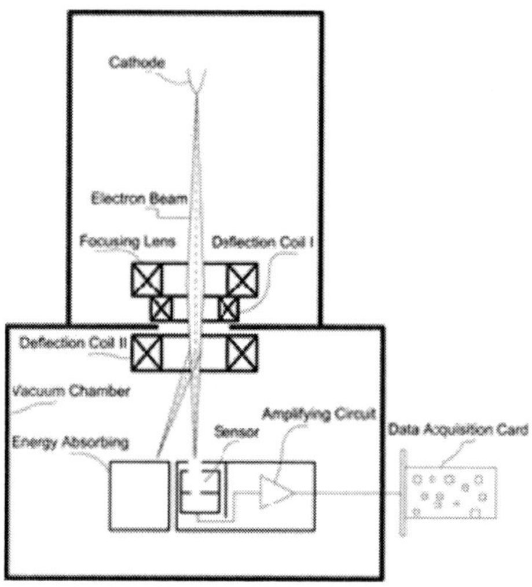

Fig.1. Schematic diagram of the beam quality test system.

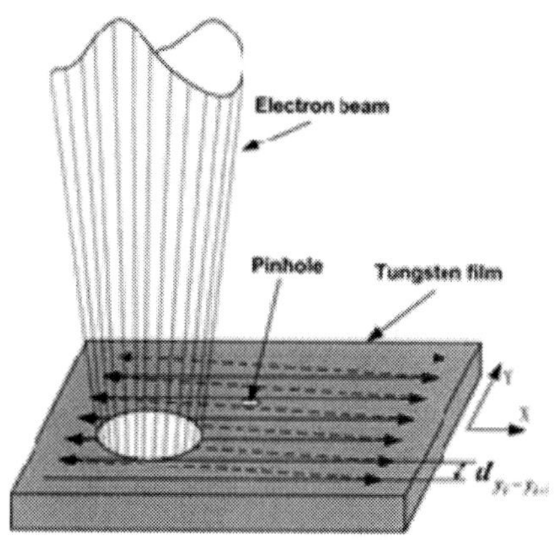

Fig.2. Sketch of the beam deflection.

3. Signal Acquisition

3.1 Acquisition mode and trigger source setting

The acquisition card PCI-1714 is set to post-trigger acquisition mode, which allows access to data based on a trigger event. The post-trigger acquisition starts when the PCI-1714 detects the trigger event and stops when the preset count of post-trigger samples has been acquired or when you stop the operation. For analog input operations, an external digital trigger event occurs when PCI-1714 cards detect either a rising or falling edge on the External A/D TTL trigger input signal which is from the beam quality test system. We defined the type of trigger source as rising-edge. Fig.3 shows that the PCI-1714 begins to collect data when it detects the rising edge of the digital pulse on the A-axis. At the same time, two control signals for electromagnetic deflection coil II in Fig.1 are generated by the beam quality test system. The particular phase relationship of the two signals shown on the B-axis and C-axis in Fig.3 can make the electron beam to deflect along the path shown in Fig.2.

3.2 Signal pre-segmentation and sampling frequency

In order to ensure the signal phase relationship, only the odd signal sequence corresponding to the solid line in Fig.3 can be used for the reconstruction of the electron beam power density distribution because of the same X-coordinate of starting point of beam deflection. If all of the collected signal were used to reconstruct the three-dimensional graphic of the electron beam power density distribution, that would lead to dislocation of the spatial data points.

The entire sampling signal collected by the PCI-1714 must be pre-segmented so as to obtain each section of the signal corresponding to each beam deflection (beam from one side to another side), and thus facilitate the subsequent data processing. As the sampling clock source of the data acquisition card PCI-1714 is the same as the digital clock source used for the generation of triangular control signal in the B-axis, which is the premise of the pre-segmentation. Only the frequency division coefficient of the two clock sources is different.

The pre-segmentation is described as following steps.

Step1: The frequency of the triangular control signal in the B-axis is ftriangle. The generation of the triangular control signal is based on the direct digital synthesis (DDS).

Step2: According to sampling theorem and signal reconstruction requirements, the sampling interval is the spatial distance corresponding to the two adjacent sampling data and must be approximately equal to the corresponding pinhole diameter 20um. If the beam deflection distance in the X-direction is 20mm, the resolution in the direction can be 1000. In the controller, the sample number N for each deflection process was set to 1024, which is equal to the pulse number inputted into acquisition card PCI-1714 within the time interval t_a.

Step3: After data collection the pre-segmentation is carried out every 1024 digital data and the pre-segmentation signal sequence is shown on the D-axis in Fig.3. After pre-segmentation it can be seen two cases of signal on the D-axis. One case is that the output signal amplitude is zero because the electron beam does not pass the pinhole in the corresponding deflection process and the electron is not collected by the sensor. The other case is that the output signal is nearly Gaussian distribution and the electron is collected by the sensor.

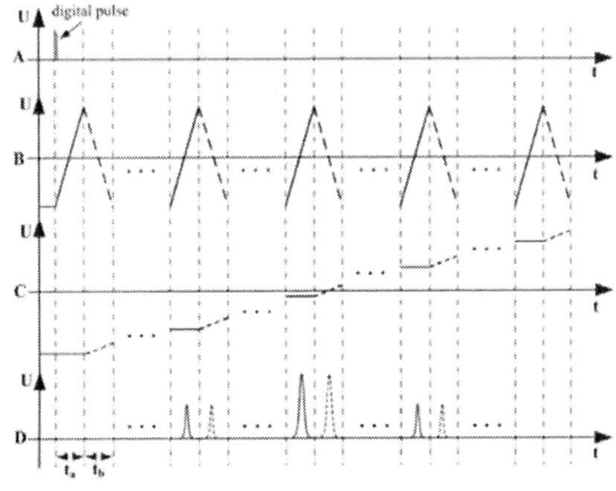

Fig.3. Diagram of the signal phase relationship.

3.3 Median filtering

The median filtering is a nonlinear digital signal processing technique based on order statistics theory. It can be used to filter noise effectively [6]. The basic principle of the technique is to run through the signal data by data, replacing each data with the median of its neighborhood.

The result of the median filtering is shown in Fig.4. By contrasting Fig.4(a) and Fig.4(b), we can see that the noise was removed significantly after median filtering.

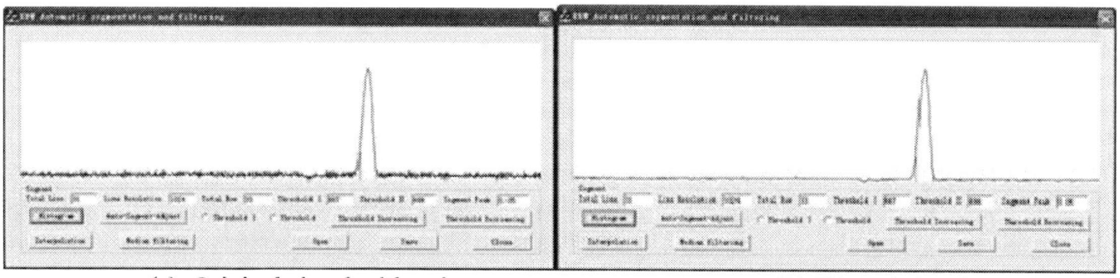

(a) Original signal with noise (b) Filtered signal

Fig.4. Median filtering.

3.4 Signal threshold segmentation

As shown in Fig.5, the three-dimensional graphics display of power density distribution of the electron beam cross-section was reconstructed with the filtered signal. It can be seen that the zero value signal occupy a large part of the signal range in the X-direction. In order to simplify the graphic display and calculation, we need to segment the signal again. The program interface for signal filtering and segmentation is shown in Fig.4(a). After pre-segmentation the signal sequence is projected onto the XOZ plane. With automatic segmentation and manual segmentation based on the signal strength threshold, we can guarantee the integrity of the signal segmentation.

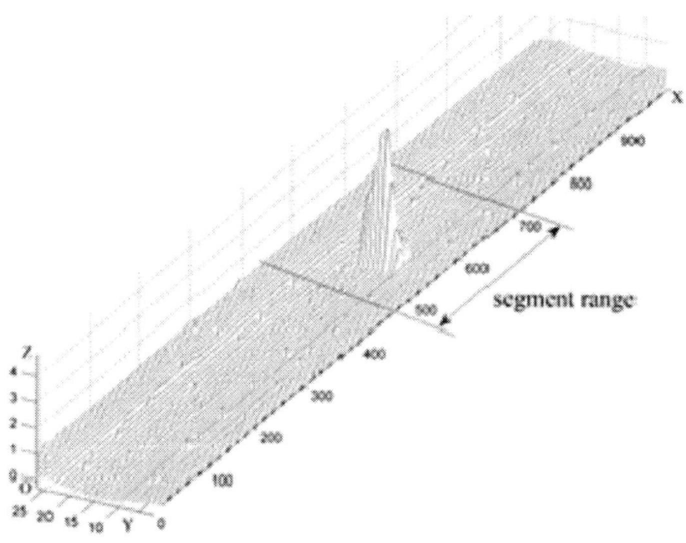

Fig.5. Signal threshold segmentation.

4. Signal calibration

4.1 Spatial scale calibration

If the electron beam deflection angle is small, the deflection of the electron beam is approximately linear with the control current intensity of the magnetic field.

For the X-direction and Y-direction of electron beam deflection, the deflection length on the sensor plane (tungsten film) corresponding to two adjacent collected data can be calculated by:

$$d_{x_i - x_i+1} = \frac{L_X}{N-1} = \frac{K_X H I_{X(\max)}}{N-1} \quad (1)$$

$$d_{y_i - y_{i+1}} = \frac{L_Y}{M} = \frac{K_Y H I_{Y(\max)}}{M} \quad (2)$$

Where L_X is the total deflection length in the X-direction, L_Y is the total deflection length in the Y-direction, H is the distance from the center of deflection coil to the sensor plane, I_X(max) is the maximum drive current in the X-direction, I_Y(max) is the maximum drive current in the Y-direction, N is the sample number in the X-direction, M is the stepper number in the Y-direction, K_X and K_Y are the constants which are related to the deflection coil and measured by the deflection experiment. Table 1 shows the experimental data for the determination of the deflection coil parameters.

Table 1. Experimental data of the beam deflection.

Deflection direction	H (mm)	I_X(max), I_Y(max) (A)	Starting point (mm)	End point (mm)	L_X, L_Y (mm)	K_X, K_Y (A-1)
X	330	1	252.532	272.493	19.961	0.060
Y	330	0.5	55.264	65.854	10.590	0.064

4.2 Definition of beam spot diameter

A. Amplitude definition

Firstly, taking the maximum amplitude of the beam power density as a reference data, the different percentages of the maximum amplitude ξ% (90%, 80%, etc.) correspond to a contour respectively. As shown in Fig.6, ideal cross-section distribution of electron beam power density was Gaussian with rotational symmetry. The corresponding contour is circular, and the beam spot diameter d_ξ can be defined as the diameter of the circle. This kind of definition is the most direct method. As the actual power density distribution influenced by the characteristics of the EBW machine is non-rotationally symmetric, we need an equivalent method to define the beam spot diameter. The area surrounded by the contour ξ % is defined as A_ξ, and the equivalent beam spot diameter d_ξ is defined as the diameter of equivalent circle of the area A_ξ in (3).

$$d_\xi = 2\sqrt{\frac{A_\xi}{\pi}} \quad (3)$$

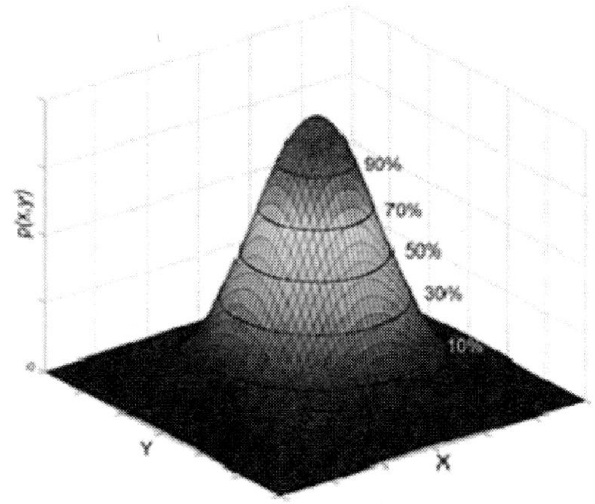

Fig.6. rotational symmetrical distribution

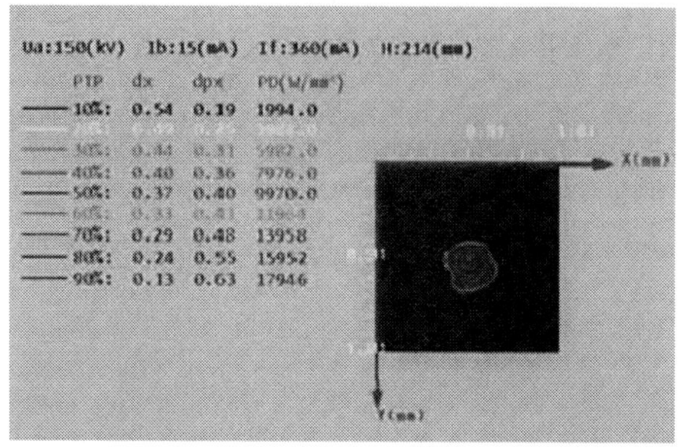

Fig.7. Contour plot of the power density distribution

B. Integral definition

The integral of beam power density with respect to beam cross-section area is calculated. If the integral value P_λ is equal to λ% of the total beam power P_{total}, the equivalent diameter is defined as $d_{P\lambda}$. The formulas are as follows.

$$P_\lambda = \lambda\% P_{\text{total}} = \int_{A_{P\lambda}} p(x,y)dA \tag{4}$$

$$d_{P\lambda} = 2\sqrt{\frac{A_{P\lambda}}{\pi}} \tag{5}$$

Where *p (x,y)* is the power density distribution, Ptotal is the total electron beam power.

5. Results and discussion

The beam power density distribution with the process parameters: accelerating voltage U_a=150kV, beam current I_b=15mA, focus current I_f=360mA, working distance H=214mm is shown in Fig. 7. The distribution graph is showed by the contour. The test results show the actual beam spot diameter by the two definition methods and the beam power density distribution. The beam spot diameters of different percentage are shown in Table 2.

Table 2. Beam spot diameter of different percentage.

ξ or λ(%)	10	20	30	40	50	60	70	80	90
d_ξ(mm)	0.54	0.49	0.44	0.40	0.37	0.33	0.29	0.24	0.13
$d_{P\lambda}$(mm)	0.19	0.25	0.31	0.36	0.40	0.43	0.48	0.55	0.63

The ideal power density distribution of the electron beam may take on a number of forms. The common form is the circular Gaussian distribution [7-10]. For a circular Gaussian distribution, the power density is given by,

$$p(x,y) = \frac{\Phi_b}{2\pi\sigma_b^2} \exp\left[-\left(x^2+y^2\right)/2\pi\sigma_b^2\right] \tag{6}$$

In this expression, Φ_b is the total beam power, σ_b (σ_b>0) is the standard deviation of the circular Gaussian beam, and the origin of the Cartesian coordinate is in the beam center.

Considering the importance of an accurate characterization of the beam geometry, it is suggested that the equivalent beam spot diameter is considered to be a value which ξ equals 10 or λ equals 90.

For the amplitude definition, the ideal beam spot diameter (ξ =10) can be calculated as in (8).

$$\begin{aligned} p(x_{10}, y_{10}) &= \frac{\Phi_b}{2\pi\sigma_b^2}\exp\left[-\left(x_{10}^2+y_{10}^2\right)/2\pi\sigma_b^2\right] \\ &= 10\% p(x,y)_{\max} = \frac{\Phi_b}{20\pi\sigma_b^2} \end{aligned} \tag{7}$$

$$d_{10} = 2\sqrt{x_{10}^2 + y_{10}^2} = 7.6\sigma_b \tag{8}$$

For the integral definition, the integrals over the two disks can easily be computed by switching from Cartesian coordinates to polar coordinates in (10) and the ideal beam spot diameter (λ=90) can be calculated as in (11).

$$\iint_D p(x,y)dxdy = \iint_T p(r\cos\theta, r\sin\theta)rdrd\theta \tag{9}$$

$$\begin{aligned} &\int_0^{2\pi} d\theta \int_0^{d_{P90}/2} p(r\cos\theta, r\sin\theta)rdr \\ &= \frac{\Phi_b}{2\pi\sigma_b^2}\int_0^{2\pi} d\theta \int_0^{d_{P90}/2} \exp(\frac{-r^2}{2\pi\sigma_b^2})rdr \\ &= \frac{\Phi_b}{2\pi\sigma_b^2} \bullet 2\pi \bullet \pi\sigma_b^2 \left(1-\exp(\frac{-d_{P90}^2}{8\pi\sigma_b^2})\right) = 90\%\Phi_b \end{aligned} \tag{10}$$

$$d_{P90} = 2.9\sigma_b \tag{11}$$

It can be seen that the diameter d_{P90} is bigger than d_{90} for the ideal power density distribution from (9) and (11). The measured results d_{10}=0.54mm, d_{P90}=0.63mm are not subject to the equations. The actual beam power disperse in a larger area compared to the Gaussian distribution and the measured electron beam power density distribution is not concentrated.

6. Conclusion

For the actual beam power density distribution of high-voltage electron beam welder, it is feasible to achieve the graphic reproduction of the beam power density distribution with the signal acquisition scheme and segmentation method. The noise in the signal of power density distribution can be removed effectively by median filtering. Experimental results show that with the process parameters U_a=150kV, beam current I_b=15mA, focus current I_f=360mA, working distance H=214mm, the measured diameter of integral definition is bigger than that of amplitude definition, but for the ideal distribution the former is smaller than the latter. The measured distribution without symmetrical shape is not concentrated compared to Gaussian distribution.

Acknowledgments
This work was financially supported by National Natural Science Foundation of China (Grant No. 51505226), National Natural Science Foundation of China (Grant No. 51375243), National Natural Science Foundation of Jiangsu (Grant No. BK20140784).

References
[1] Schultz H. Electron beam welding. Cambridge, Engl and: Abington Publising,1994.
[2] Palmer T.A., Elmer J.W. Improving process control in electron beam welding using the Enhanced Modified Faraday Cup. Journal of Manufacturing Science and Engineering,2008,130(4),041001- 041015.
[3] Dilthey U., Goumeniouk A., Nazarenko O.K, Akopjantz K.S. Mathematical simulation of the influence of ion-compensation, self-magnetic field and scattering on an electron beam during welding. Vacuum, 2001, 62 (2–3):87-96.
[4] Rouquette S., Guo J., Le Masson P. Estimation of the parameters of a Gaussian heat source by the Levenberg–Marquardt method: Application to the electron beam welding. International Journal of Thermal Sciences, 2007, 46 (2),128-138.
[5] Luo Y., Liu J., Ye H. An analytical model and tomographic calculation of vacuum electron beam welding heat source. Vacuum, 2010, 84 (6):857-863.
[6] Michaeli, L., Saliga, J, Sochova, L. Integral Nonlinearity Correction Algorithm Based on Error Table Optimizing and Noise Filtering. MEASUREMENT SCIENCE REVIEW, 2008, 8(2):29-32.
[7] Kawate, E., Hain, M.. New Scatterometer for Spatial Distribution Measurements. MEASUREMENT SCIENCE REVIEW, 2012, 12(2):56-61.
[8] Reddy D., Pratihar D.. Neural network-based expert systems for predictions of temperature distributions in electron beam welding process. The International Journal of Advanced Manufacturing Technology, 2011, 55(5), 535-548.
[9] Lacki P., Adamus K.. Numerical simulation of the electron beam welding process. Computers & Structures, 2011, 89(11–12), 977-985.
[10] Koleva E., Vutova K., Wojcicki S. G.. Use of radial distributions of the beam current density for evaluation of beam emittance and brightness. Vacuum, 2001, 62(2–3), 105-111.

IOP Conf. Series: Materials Science and Engineering **207** (2017) 012100 doi:10.1088/1757-899X/207/1/012100

Qualitative analysis of SBS modifier in asphalt pavements using field samples

Fengxia Chi[1, 2, a], **Zhifei Liu**[2, b]

[1]Zhejiang Institute of Communications, Hangzhou 311112, China
[2]Zhejiang Scientific Research Institute of Transport, Hangzhou 311305, China

[a]chifengxia@126.com, [b]whutlzf@126.com

Abstract. Series of tests are implemented to analysis the related characteristics of common asphalt and unknown asphalt mainly using Fourier Transform Infrared (FT-IR) and Dynamic Shear Rheometer (DSR) for chemical compositions and rheological properties of asphalt, respectively. In addition, a series of mechanical properties were performed on asphalt mixtures, including indirect tensile strength test and three point bending test at low temperature. Experimental results indicated that compared with common asphalt, the characteristic absorption peak of the unknown asphalt are appeared at 966cm-1and 699cm-1, which are accordant with the SBS modifier. The results of DSR indicated that the unknown asphalt's complex modulus is higher and the phase angle is lower. The mechanical tests indicated that some properties of the unknown mixture samples are increased by 24.7%~41.8% compared with common pavement sample, like the indirect tensile strength, the bending test at low temperature and indirect tensile resilient modulus. Comprehensive analysis indicates that SBS modifier is existed in the unknown asphalt pavement.

1. Introduction

Styrene-butadiene-styrene block copolymer (SBS) is a thermoplastic elastomer rubber that allows for greater performance than that possible with asphalt alone [1]. Due to this property, SBS can improve the asphalt in rutting resistance, thermal cracking, fatigue damage, stripping, and temperature susceptibility, which makes SBS one of the most promising and widely used polymer modifiers for asphalt. The content of SBS plays an important role on the performances of modified asphalt, so the content of SBS modifier is essential to quality control in construction.

In the last decades, several research efforts have investigated the use of SBS modifiers in asphalt. Some researches focused on the stability improvement of the SBS modified asphalt binder to ensure a good compatibility between SBS and asphalt [2] [3]. Some other researchers studied the rheological properties of SBS modified asphalt binder, performance of mixture and illustrated the mechanism of SBS modified asphalt [4] [5]. Because polymer content is an essential parameter of asphalt binder, some researchers investigated the method by quantification of polymer content in SBS modified asphalt [6] [7]. There are few studies on analysis of SBS modifier content for long service pavement.

This study is part of the research work of service pavement. The main objectives of this study are: i) to qualitatively analyze SBS modifier in asphalt pavements using field samples, and ii) to investigate the performance of asphalt mixture which was subjected to normal traffic for 8 years.

2. Experimental

2.1. Sample Preparation
Two types of samples asphalt mixture were obtained from Jinnan road in Zhejiang province for the performance tests: common asphalt mixture, which is pure asphalt cement without modifiers such as SBS, and unknown asphalt mixture, which is a need to determine whether there is SBS modifier. The core samples (100 mm in diameter) and slab samples (Long 400mm wide 400mm) were taken from the constructed wearing course. The field test section was subjected to normal traffic for 8 years. After the mechanical testing, they were further used for the extraction process to separate binder from aggregates.

2.2. FTIR
Fourier transform infrared spectroscopy (FTIR) was recorded on a TENSOR 27 spectrometer (Bruker, Germany) from 4000 to 400 cm-1 by KBr disk method, at a resolution of 4 cm-1.

2.3 Dynamic shear rheological characteristic
Dynamic shear properties were measured with MCR102 dynamic shear rheometer (MCR102, AntonPaar, Austria) in a parallel plate configuration with a gap width of 1mm. Measurements were performed at fixed frequencies 10 rad/s and temperature from 46 to 82 °C under controlled strain condition.

2.4 Indirect tensile test
The indirect tensile strength of specimens was measured at 25°C and at the loading rate 50 mm/min by using Universal Testing Machine (UTM-100). The peak load at failure is recorded and used to calculate the IDT Strength of the specimen

Indirect tensile resilient modulus was measured at 40°C which was according to ASTM D4123. The frequency of load application was 1 Hz with the load duration of 0.1 s in order to represent field conditions and the resting period of 0.9 s. Four samples were made for each of the two kinds of evaluated mixtures and their averages represented resilient modulus for each mixture.

2.5 Three point bending test
The slabs were cut firstly on highway pavement. Then the slabs sample were cut into the beam specimens with the length of 250±2.0mm, the height of 35±2.0mm and the breadth of 30±2.0mm in lab. The tests were carried on UTM-100 at the temperatures of -10 °C with a loading rate of 50mm/min. In order to reduce discreteness of mechanics properties of samples and ensure the reliability of the test results, there were six beams in every group.

3. Results and discussion

3.1 Chemical compositions analysis by FTIR
FTIR tests were conducted on the asphalt extraction from the common asphalt mixture and unknown asphalt mixture respectively. The testing results were shown in Fig.3. Two major bands at 2924cm-1 and 1460cm-1 were identified as the stretching vibration and deformation vibration for C-H bond of hydrocarbon. However, two major bands at 966 cm-1 and 699cm-1 were just observed for the unknown asphalt compared with absent peaks of common asphalt. The 966 cm-1 band corresponds to the C=C bond in butadiene while 699cm-1 band identifies the existence of styrene. Previous study [7] [8]on SBS polymer modified found that 1375cm-1, 810cm-1 of asphalt and 966cm-1,699cm-1 of SBS could be used for quantification of SBS content. Therefore, the two major bands, 966 cm-1 and 699cm-1, can be used to detect the presence of the SBS in the asphalt binder. It means there were SBS modifiers in the unknown asphalt.

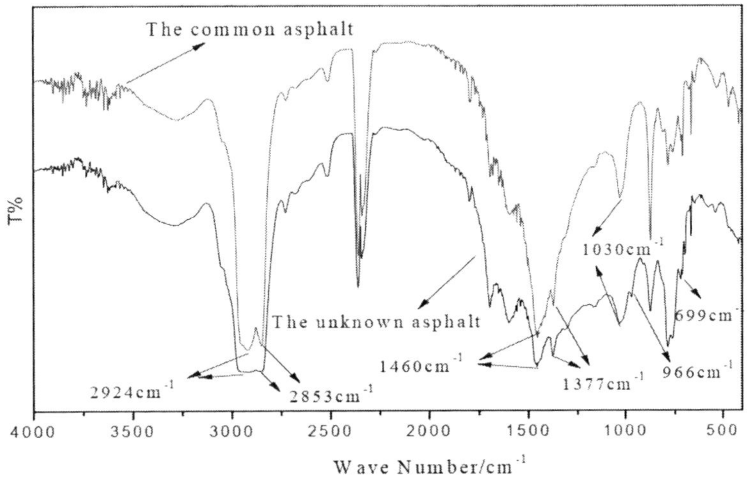

Figure 1. FTIR of asphalt extraction from mixtures

3.2 Dynamic shear rheological characteristics

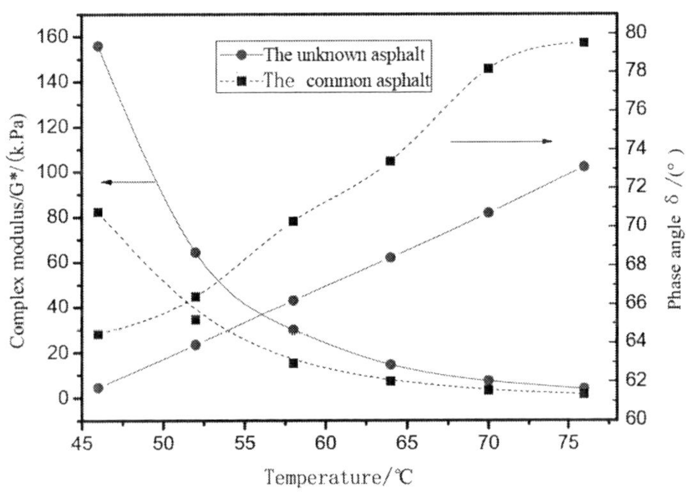

Figure 2. Rheological properties of asphalt using DSR

As a viscoelastic material, asphalt exhibits either elastic or viscous behavior. Dynamic shear test can be used to characterize the viscous and elastic behavior of the asphalt using complex modulus (G*) and the phase angle (δ) at different temperature. Fig.2 shows the temperature dependency of complex modulus G* and phase angle δ for two asphalts at range of temperature 46-76°C. The results also showed that the unknown asphalt binder has a lower phase angle and higher complex modulus at range of temperature 46-76°C compared to the common asphalt, especially at relatively high temperatures, This indicates that the common asphalt mixtures is more susceptible to rutting. The complex modulus of unknown asphalt at 46°C is 155.99kPa, which is comparable with common asphalt's 82.39kPa. The results above indicate that the deformation resistance of unknown asphalt is superior. Phase angle of unknown asphalt are increased from 61.7° to 73.1° at range of temperature 46-76°C, which is comparable with common asphalt's 64.5° to 79.5° . This means that the common asphalt binder has the higher phase angle significantly over temperature region. The increment of complex modulus and reduction of phase angle will contribute the resistance to the high temperature stability for asphalt binders and mixtures.

Measurements of δ are generally considered to be more sensitive to chemical structure than complex modulus [9]. The phase angle of unknown asphalt at 64°C is 68.4° . These data are in good agreement with other published results. Ma's study [10] indicated the phase angle of asphalt modified by SBS after pressure aging vessel test at 64°C was about 70.3°.The complex modulus and phase angle of unknown asphalt in Figure 2 are consistent with the rheological properties of SBS modified asphalt.

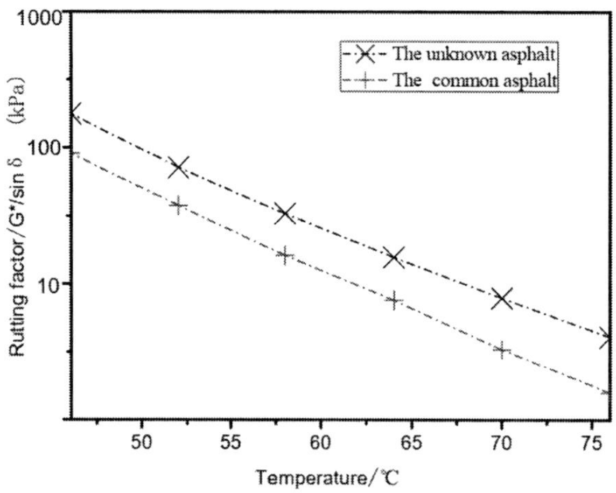

Figure 3. Rutting parameter of asphalt

A rheological parameter, G*/sinδ has been regarded as the rutting parameter of asphalt binder in the American Strategic Highway Research Program (SHRP). High G*/sinδ value was found to correlate with high rutting resistance. Figure 3 illustrates G*/sinδ measured at range of temperature 46-76°C. As has been indicated in this study, the modulus of unknown asphalt was higher and the phase angle was lower. The G*/sinδ values of the unknown asphalt binders were increased. The unknown asphalt results in the increase of rutting parameter (G*/sin δ) at high temperatures, indicated a higher rutting resistance of the binder.

3.3 Mechanical property

Table 1. Mechanical properties of asphalt mixtures

Property	Unknown asphalt mixture(a)	Common asphalt mixture (b)	(a-b)/b*100%
Indirect tensile strength at 25°C,MPa	1.97	1.58	24.7%
Bending strain at -10°C,με	1486.5	1109.4	34.0%
Indirect tensile resilient modulus at 40°C, MPa	5741	4050	41.8%

Tensile strength testing is used for evaluating the asphalt mixture's fatigue potential and moisture susceptibility. Previous studied have indicated that the tensile strength of asphalt mixture is related to cracking performance [11] [12]. Higher tensile strength means that asphalt pavement can tolerate higher strains before failing. As shown in Table 1, the average indirect tensile strength of the unknown asphalt mixture is about 24.7% higher than the common mixture. Here the unknown asphalt mixture shows 1.97 MPa and the common mixture shows 1.58 MPa. The unknown asphalt mixture exhibits a relatively higher indirect tensile strength, which indicates better cracking resistance compared to the common asphalt mixture.

Three point bending tests were performed to evaluate the low temperature cracking of the asphalt concretes. As shown in Table 1, the failure strain of the unknown asphalt mixture is about 34.0% higher than the common mixture. In general, asphalt mixtures that exhibit greater strain at failure in the low temperature range have better flexibility which leads to superior resistance to cracking.

Resilient modulus is a measure of materials responses to load and deformation. Generally, higher modulus indicates greater resistance to deformation. The results also show that the unknown asphalt mixture yields a higher resilient modulus value compared to the common mixture at 40°C. The resilient modulus is 5741MPa for the unknown asphalt mix and 4050MPa for the common mixture, which indicates greater resistance to deformation compared to the common asphalt mixture.

For comparing the mechanical properties obtained from different tests, the data from indirect tensile strength, bending strain, and resilient modulus for common and unknown asphalt mixtures, respectively, were ranked, as summarized in Table 1. Test results clearly identified that the unknown asphalt mixtures was better than common asphalt mixtures. These properties of unknown asphalt mixtures are consistent with the effects of SBS modified asphalt mixture.

4. Conclusion

The following results can be concluded:

(1) The chemical compositions analysis of the unknown asphalt Shows Two major bands at 966 cm-1 and 699cm-1 in FT-IR spectra compared to the common asphalt, indicated the unknown asphalt containing SBS modifier.

(2) The rheological behaviors of unknown asphalt investigated using Dynamic Shear Rheometer show significant high complex modulus and low phase angle compared to common asphalt resulting in increasing resistance to the high temperature stability.

(3) The mixture test results demonstrated up to 24.7% higher indirect tensile strength at 25°C, up to 34.0% higher bending strain, and 41.8% higher indirect tensile resilient modulus at 40℃ than the common asphalt. These properties of unknown asphalt mixtures are consistent with the effects of SBS modified asphalt mixture.

(4) Based on the analysis of the chemical composition of asphalt materials and the confirmation of rheological properties and mechanical properties, it is found that the asphalt used in unknown asphalt mixture section contains SBS modifier.

Acknowledgments

This research was supported by the Zhejiang Provincial Natural Science Foundation of China under Grant No. LY13E080006.

References

[1] Ouyang, C., Wang, S., Zhang, Y., & Zhang, Y. (2006). Improving the aging resistance of styrene–butadiene–styrene tri-block copolymer modified asphalt by addition of antioxidants. Polymer Degradation & Stability, 91(4), 795-804.

[2] Fu, H., Xie, L., Dou, D., Li, L., Yu, M., & Yao, S. (2007). Storage stability and compatibility of asphalt binder modified by sbs graft copolymer. Construction & Building Materials, 21(7), 1528-1533.

[3] Ou, Z. M., & Li, F. J. (2015). Studies the key-technologies of karamay matrix asphalt used in

sbs modified bitumen. Applied Mechanics & Materials, 744-746, 1308-1311.

[4] Hanyu, A., Ito, T., Kasahara, A., & Saito, K. (2004). Effect of the morphology of sbs modified asphalt on mechanical properties of binder and performance of mixture. Journal of the Eastern Asia Society for Transportation Studies, 9, 65-72.

[5] Zhao, X., Wang, S., Wang, Q., & Yao, H. (2016). Rheological and structural evolution of sbs modified asphalts under natural weathering. Fuel, 184, 242-247.

[6] Ye, Z. C., Chen, D., Ling, C., & Guan, F. Y. (2014). Quantifying sbs content in modified asphalt using fourier transform infrared spectroscopy. Advanced Materials Research, 1065-1069, 691-695.

[7] Sun, D. Q., Zhang, L. W., & Zhang, X. L. (2011). Quantification of sbs content in sbs polymer modified asphalt by ftir. Advanced Materials Research, 287-290, 953-960.

[8] Masson, J., Pelletier, L., & Collins, P. (2015). Rapid ftir method for quantification of styrene - butadiene type copolymers in bitumen. Journal of Applied Polymer Science, 79(6), 1034-1041.

[9] Larsen, D. O., Alessandrini, J. L., Bosch, A., & Cortizo, M. S. (2009). Micro-structural and rheological characteristics of sbs-asphalt blends during their manufacturing. Construction & Building Materials, 23(8), 2769-2774.

[10] Ma, T., & Yuan, H. (2013). Aging Behavior Characterization of SBS-Modified Asphalt for Recycling Purpose. International Conference on Sustainable Construction Materials: Design, Performance, and Application (pp.261-275).

[11] Katman, H. Y., Ibrahim, M. R., Matori, M. Y., Norhisham, S., Ismail, N., & Omar, R. C. (2012). Tensile strength of reclaimed asphalt pavement. International Journal of Civil & Environmental Engineering.

[12] Islam, M. R., Hossain, M. I., & Tarefder, R. A. (2015). A study of asphalt aging using indirect tensile strength test. Construction & Building Materials, 95, 218-223.

IOP Conf. Series: Materials Science and Engineering **207** (2017) 012101 doi:10.1088/1757-899X/207/1/012101

Corrosion behavior and microstructure of laser cladded bioactive coating on titanium alloy

Min Zheng[1,*], Liping Yang[1], Zhuo Wu[1], Jianbin Zhang[2] and Ding Fan[2]

[1]School of Mechanical and Electronical Engineering, Lanzhou University of Technology, Lanzhou 730050, China
[2]State Key Laboratory of Gansu Advanced Non-ferrous Metal Materials, Lanzhou University of Technology, Lanzhou 730050, China

*Corresponding author e-mail: zhminmin@sina.com

Abstract. Experimental results on bioactive modified treatment at titanium alloy was presented. A better metallurgical bonding of the single tracks with the substrate was obtained by the use of laser beam via an integral mirror. The samples were studied by Scanning Electron Microscope (SEM), Inductively Coupled Plasma (ICP), microhardness instrumentation and corrosion resistance testing. The facts show a significant influence of microhardness on the optimized experimental condition. Furthermore, it reveals that a significant influence of property, especially for corrosion resistance while the rare earth oxide ceria is mixed in the precursor powders. The dynamic factor is a possible influence of escape of granules. The laser-cladded bioceramic coating of ceria additive in pre-placed powders has more favourable corrosion resistance compared with the coating without rare earth oxide.

1. Introduction

Materials used in hard tissue are commonly exposed to very aggressive environments, such as excellent biocompatibility, high mechanical strength and toughness, extreme wearing and corrosion resistance, among others [1].Titanium and its alloys are widely used in biomedical devices owing to their exceptionally properties such as high specific strength, excellent corrosion and excellent biocompatibility. However, due to their poor tribological properties, the application of titanium and its alloys under severe wear and friction conditions is severely restricted [2-4]. But rather in the sense that surface engineering has emerged as an intense research area in the field of materials science [5].

Therefore, many methods have developed to deposit bioceramic materials on metallic implants, such as plasma spraying, ion-beam sputtering, electrophoretic deposition, and so on. But most of coatings made by these methods suffer from weak adherence. Functionally graded materials (FGMs) are advanced engineering materials designed for a specific performance or function in which a spatial gradation in structure and/or composition lend itself to tailored properties. This occurs by providing in-depth graded com-positions, microstructures and properties [6]. Laser cladding is one of the industrial preferred surface engineering techniques in which a laser is used as a heating source to overlay the precursor material with the substrate to form a sound interfacial bond, namely apply a chemically different material as a layer onto a given substrate [7-9]. Recently researchers have studied the influence of composition in fabrication of bioceramic structures using laser cladding. However, a systematic analysis of the effect of process parameters and corrosion behavior are not dealt with.

In this present work, we have analyzed the microstructure of coating as a function of the laser processing parameters. The improving corrosion behavior of the composite coating in human body fluid has also evaluated considering their biomedical applications.

2. Experimental work

The calcium phosphate bioceramic coating was fabricated on titanium alloy (Ti-6Al-4V) substrate by TJ-HL-T5000 CO_2 laser system equipped with integral mirror and processing lathe in argon shielding atmosphere. The preplaced powder was a mixture of calcium carbonate, calcium hydrogen phosphate, titanium powders and a little ceria. The weight per cent of calcium hydrogen phosphate and calcium carbonate is 81.1wt.% and 18.9wt.%, respectively. And 0.6 wt.%ceria was added in order to study the effect of rare earth oxide. To decrease thermal stress between coating and substrate during laser cladding, FGMs are advanced engineering materials designed for three-layer: (1) 80%Ti+20%M; (2) 40%Ti+60%M; (3) 100% M. Here, the "M" corresponds to the mixed powders mentioned above except titanium powders. Specimens with dimensions 50 mm×30 mm×5 mm. Synchronous powder feeding was not suitable for mixed cladding materials with density variation. Therefore, the gradient powders were mixed evenly, then pre-placed on the substrate and laser-cladded in sequence. The preplaced coating thickness is about 0.5mm.

The processing parameters of laser cladding were optimized at the experimental condition. In the condition of laser output power 2.5kW and laser beam size 15mm×1mm, laser scanning speed of 80, 100, 120, 140, 160mm/min, respectively.

Corrected SBF (c-SBF) with ion concentrations nearly equal to those that have almost similar compositions of inorganic ions to human blood plasma. The reagents and amount for preparation of each liter c-SBF were listed as follows: 8.035g NaCl, 0.355g $NaHCO_3$, 0.225g KCl, 0.231g $K_2HPO_4 \cdot 3H_2O$, 0.311g $MgCl_2 \cdot 6H_2O$, 0.292g $CaCl_2$, 0.072g Na_2SO_4, 6.118g $(HOCH_2)_3CNH_2$ (Tris). Tris and HCl serve as buffers to keep the PH value at 7.4. The preparation procedure was referenced by the works of T. Kokubo [10]. The laser-cladded specimen and the untreated substrate were immersed in SBF 7 days and 14 days.

The metallographic samples were prepared by standard mechanical polishing procedure and were chemically etched. The morphologies and the elemental distribution in different soak fluid of specimens were analyzed by scanning electron microscope (SEM, JEOL JSM-6700F, Japan) and Energy Dispersive X-ray analysis (EDX). The ion concentration soaked in SBF was tested by Inductively Coupled Plasma (ICP, Optima4300DV, Japan). Microhardness and corrosion resistance testing were also been evaluated. With a load of 300g and a loadtime of 10s, the distribution of microhardness was test by HX-1000TM Vickers microhardness tester.

3. Results and Discussion

The primary aim of this work is to study the effect of laser power and scan speed on microstructure, phase constituents and microhardness of laser cladded bioceramic coating on titanium alloy. The changes of laser scanning speed mean that the energy changes absorbed by the material surface, then directly affected the melting depth and width of molten pool. According to Tab. 1, the relation among scanning speed, melting depth and width of the laser cladded coating reveals that the melting depth and width of molten pool increased while the laser scanning speed reduced. The scanning speed should be selected higher than 100 mm/min because of the thickness of substrate. According to microstructure observation, microhardness and analytical calculations about heat transmission and mass transfer during laser cladding, the optimal processing parameters of laser cladding were as follows: laser output power 2.5kW, laser scanning speed 140mm/min, and laser beam size 15mm×1mm.

Table 1. Relation among scanning speed, melting depth and width.

scanning speed(mm/min)	80	100	120	140	160
melting depth	>5mm	>5mm	4.64	3.38	2.88
melting width	16.62	16.14	15.42	15.12	14.86

The microstructure of laser-cladded composite coating was investigated. The results indicated that the coating was metallurgically bonded to the titanium alloy substrate by the action of laser. The content of titanium was gradually decreased and the calcium and phosphorus were gradually increased. The laser-cladded composite coating contained such bioactive phases as hydroxyapatite (HA), β-tricalcium phosphate (β-TCP) and calcium titanate (CaTiO3), etc. The surface morphology of the laser cladded coating is shown in Fig.1. The appearance of irregularity microstructure on coating and more contact area may be in favor of the osseous tissue to grow along.

Fig.1 The surface morphology of the laser cladded coating

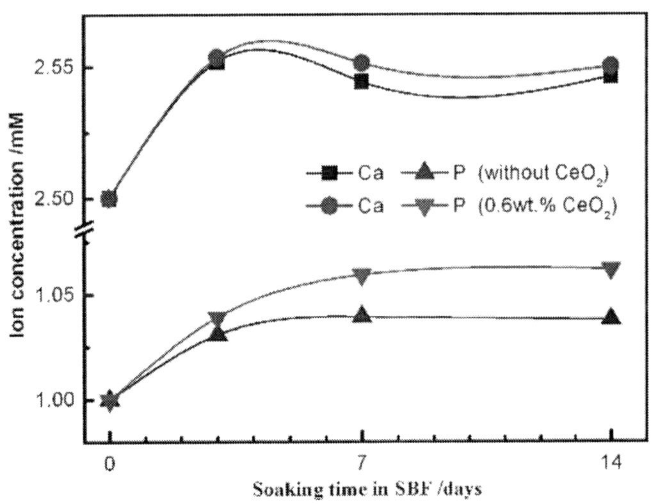

Fig.2 Distribution of ion concentrationsoakin SBF

Corrosion resistance testing in different soaked liquids indicated that the composite coating at the experimental condition was of bioactivity by SEM, EDX and ICP. The relationship between ion concentrations soaked in SBF at different periods is shown in Fig.2. As shown in the figure, ion concentration of phosphorus rises by the extension of soaking time. And the tends to be stabilization afte ra week later. The ion concentration of calcium rises in the beginning, and then appears a slight decline, and finally continue to rise. The ion concentration of calcium and phosphorus rise at preliminary stage, because the ions separate out from the surface of coating. As the extension of immersion time, this value of calcium and phosphorus easily to the equilibration in SBF. Apatite deposit is appeared on laser cladded coating. The ions of calcium and phosphorus should be consumed. While the addition of apatite deposit, ion concentration of calcium and phosphorus continued to increase and reached stabilization by the dissolution.

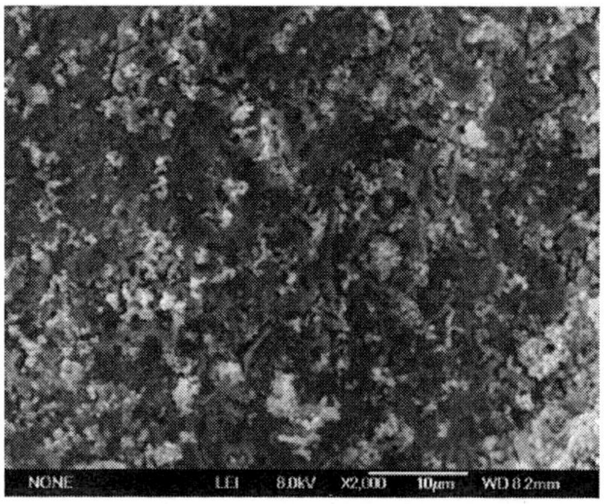

Fig.3 Morphology on surface soaked in SBF for 7 days **Fig.4** Morphology on surface soaked in SBF for 14 days

After soaking in SBF 7 (Fig.3) and 14 days (Fig.4), the flocculent precipitates were increased on the surface of coating and the apatite phases were gradually nucleated and grew. The morphology of precipitates was mainly composed of flake-like and the globular, flocculent and acicular morphology were also appeared. Moreover, the appearance of a small quantity of crystal whisker (diameter about 400nm, slenderness ratio about 40~100) in local zone sufficiently benefited intensify toughness. The diffraction peak of apatite extremely intensified and the CaO distinctly decreased when the coating soaked in SBF. It indicated that the coating had the ability of depositing apatite quickly. However, the untreated substrate surface was merely formed trace of soaking and some precipitation of salts.

4. Conclusion

The optimal processing parameters of laser cladding adopted according to microstructure, microhardness and analytical calculations about heat transmission and mass transfer. The results showed retention of large amount of calcium phosphate due to high cooling rates associated with laser processing. The corrosion resistance found to increase with laser cladded coating on titanium alloy substrate. The laser-cladded bioceramic coating of ceria additive in pre-placed powders has more favorable corrosion resistance compared with the coating without rare earth oxide. Under present experimental conditions, the laser cladding found to have strong influence on microstructure, phase constituents and corrosion resistance of composite coating.

Acknowledgments

This work was financially supported by Natural Science Foundation of Gansu Province (1508RJZA108) and Doctor Foundation of. Lanzhou University of Technology (0902ZX116).The authors gratefully acknowledge Professor Qibin Liu of Guizhou University in laser test.

References

[1] L. Bedian, M. Angel, V. Rodríguez, et al. Bio-based materials with novel characteristics for tissue engineering applications-A review, Int. J. Biol.Macromol. 98 (2017) 837-846.
[2] H.J. Rack, J.I. Qazi. Titanium alloys for biomedical applicationsMater. Sci. Eng. 26 (2006), 1269-1277.
[3] D.L. Shi: Introduction to Biomaterials (Tsinghua University Press, Beijing, China, 2005).
[4] Z.D. Dai, S.C. Pan, M. Wang, et al. Improving the fretting wear resistance of titanium alloy by laser beam quenching. Wear. 213(1997), 135-139.
[5] H. Assender, V. Bliznyuk, K. Porfyrakis, How surface topography relates to materials' properties.Science. 297 (2002) 973-976.
[6] N.Minoo,S.Kamyar. Functionally graded materials: A review of fabrication and properties. App. Mater. Today. 5(2016)223-245
[7] G.J. Cheng, D. Pirzada, M. Cai, et al. Bioceramic coating of hydroxyapatite on titanium substrate with Nd-YAG laser.Mater. Sci. Eng. 25 (2005)541-547.
[8] F. Lusquiños, J. Pou, M. Boutinguiza, et al. Main characteristics of calcium phosphate coatings obtained by laser cladding. Appl. Surf. Sci. 247 (2005) 486-492.
[9] F, Weng, C.Z. Chen, H.J. Yu. Research status of laser cladding on titanium and its alloys: A review,Mater. Des. 58(2014)412-425
[10] T. Kokubo, H. Takadama, How useful is SBF in predicting in vivo bone bioactivity. Biomaterials. 27 (2006) 2907-2915.

Experimental study of failure mechanism of thick pile caps

Tan Wang[1, 2], Lijun Dou[2] and Kun Yin[1]

[1]School of construction engineering, Jilin University, Changchun, China
[2]School of civil engineering, Changchun Institute of Technology, Changchun, China

Abstract. The failure mechanism of four-pile thick caps are analysed by experimental research. From the figure of crack development, analyse the impact of different reinforcement form of the bottom of the cap and different reinforcement form of horizontal mat. The analysis of the effect of concrete strength for loading capacity of the cap.

1. Introduction

From the research results of pile caps both at home and abroad the strut-and-tie theory can be used to simulate the stress state of thick pile cap. The strut-and-tie model of the thick pile caps refers to considering the concrete between the bottom of the column and the top of the pail as a compressed rod and considering reinforcement between the pile and pile as a pull rod. Based on this theory, the article put ward some design suggestions on pile caps by experiment and theoretical analysis.

2. Eexperimental research

The test adopted elector-hydraulic universal testing machine (WAW-2000) which is controlled by computer, the dimension of test pile cape as shown in Fig.1 and the reinforcement form of pile cap as shown in Fig.2.

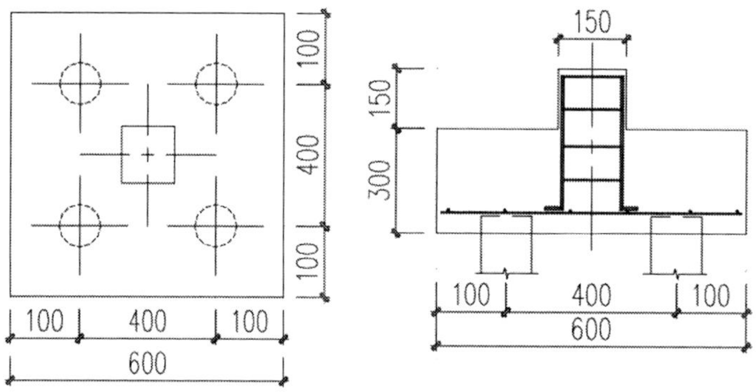

Fig.1 The test pile cap

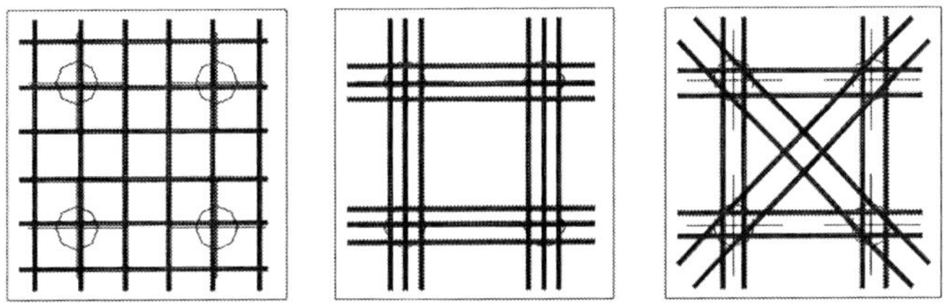

Fig.2 Reinforcement form at the bottom of the pile cap

3. Eexperimental research results

Synthesize the experimental results of the literature and we give out comparison shown as Tablet 1.

Table 1. Result of the experiment.

NO.	HRB335	concrete	reinforcement form	Cracking load	Ultimate load
ZJCT1	12Φ12	C40	A	620	810
ZJCT2	12Φ12	C40	C	630	850
CT1*	12Φ12	C30	A	450	700
CT2*	12Φ12	C30	B	450	750
CT4*	12Φ12	C30	C	400	780

Specimens ZJCT2 and CT4 adopted the same dimensions and the same reinforcement form, but the concrete strength level increased from C30 to C40, from the test result, cracking load enhanced conspicuously. The first reason is adopting higher mark of the concrete, the second reason is that the specimen ZJCT2 has been maintained indoor for over three months before being tested, but as a result, the strength of the concrete has enhanced and the concrete at surface of the cap have carbonized at the same time. But comparatively speaking, the enhanced of the ultimate load is inconspicuous. But its test value also exceed designed bearing capacity.

Therefore, during the designing, the concrete level can be enhanced properly, in order to enhance the cracking load. At the same time. The requirement in norm can be adopted, adopting 60d or 90d strength of the concrete to design the proportions of concrete mix, in order to save the concrete, under the premise of it also satisfy the bearing capacity requirement.

In order to show the pull rod effect the reinforcement should be placed between pile and pile. Combine with the result of test in chart 1 and analyze its placing form: Under the certain condition that the quantity of reinforcement is the same and the reinforcement is placed between pile and pile then the bearing capacity of the cap can be enhanced conspicuously. Placing reinforcement at slanting stitch cannot enhance the cracking load, it only can enhance the bearing capacity of the cap a little. From the figure of crack development (Fig.3), diagonal placement can retard the pile cap of the corner pile cracking firstly, because the reinforcement at slanting stitch work together with Two-way parallel reinforcement can strengthen the Anchoring effect of cap at corner. But which make the fissure are assembled at the bottom of the pile cap at the diagonal direction, because one side diagonal reinforcement parallel with the crack development direction, it has nothing to do with restraining the development of the fissure. Analyze from the above points, reinforcement form at Bottom, can still use the conventional two-way parallel arrangement, but, further study is needed, about arrangement above the pile top.

For the width of the reinforcement which is assembled at the top of the cap, the current research conclusions are not uniform, the literature [2] propose to arrange it within 1.5 times of the pile diameter. Based on the experimental results of numerous literature, even when the cap reaches the limit load, bottom reinforcement not yielding, it shows that even if the concentration of reinforcement in the pile diameter range is not fully be used. The breach of the cap depends on the breach of the concrete pull rod, so in order to achieve the purpose of optimizing the arrangement of reinforcement, we can take a small range, that is twice the diameter of the pile within the assemble of the centralized arrangement of reinforcement is enough. The other positions of the bottom of the pile cap should also be provided with the reinforcement rate should not less than 0.1%.

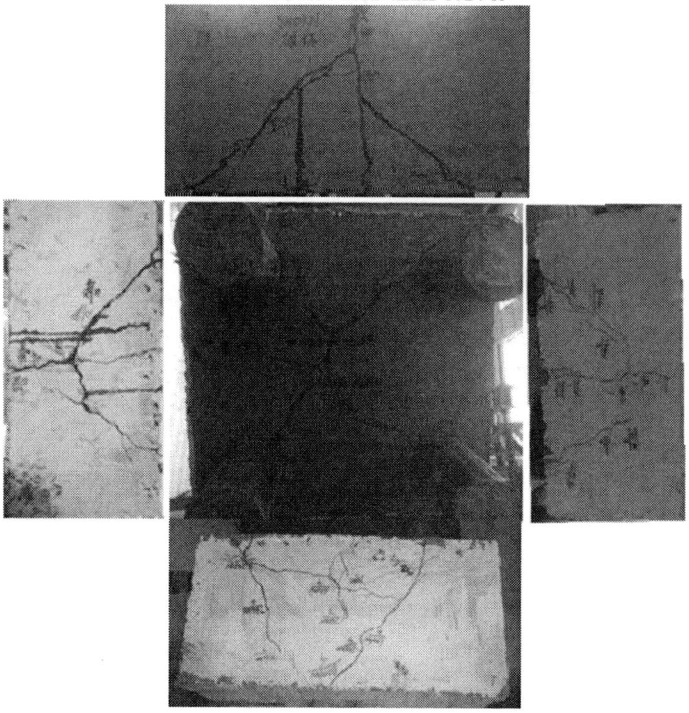

Fig.3 the figure of crack development

See from the experiment, the crack is developed firstly at the middle of the bottom side of the cap, this is the typical position of the bending crack, but, there was no significant increase in the mid-span fracture. When the limit load is reached, the crack is developed from the top of the pile to the bottom of the column, the cap is divided into four parts, and this is a typical form of punching destruction. Therefore, the destruction form of thick pile cap, obviously, it is neither the flexural destruction, nor the direct shear destruction, but the destruction of the diagonal tensile stress caused by shear stress and bending stress. For the whole cap, it can be regarded as bending shear destruction, but for each of the diagonal rod is a split destruction. In order to reduce the cracks caused by the tensile stress, the reinforcement mesh can be settled along the horizontal direction at the height of the cap,(Although there is no clear requirement in the code of the pile foundation, but in the concrete specification has been put forward, for thick reinforced concrete slabs, two way reinforcement mesh is settled at the middle part of the plate thickness, in order to improve the shear behavior. because, for the development of the inclined crack of pile foundation, Effect of horizontal reinforcement, not only reflected in the vertical direction of the crack, so as to improve the interlocking of the shearing force transfer of aggregate, also because of the role of the bolt can be provided to help shear transmission. The literature [3] also show that the horizontal reinforcement mesh can improve the bearing capacity of the cap. From the perspective of strut-and-tie model, it is to strengthen the anti-cracking ability of concrete diagonal rod.

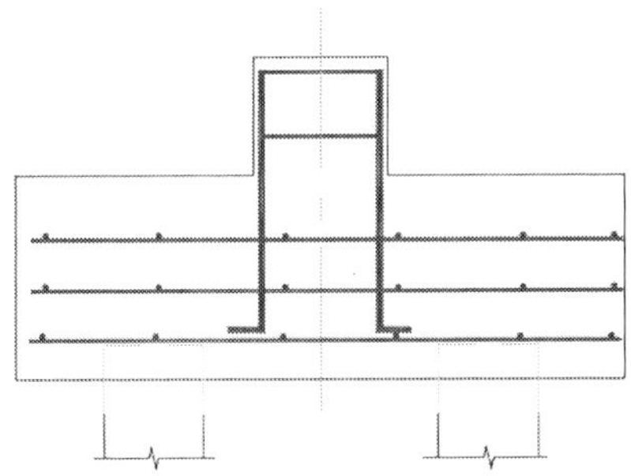

Fig.4 setting the horizontal reinforcement mesh

4. Conclusion

Based on the experimental study and theoretical analysis, the stress of pile cap is in accordance with the principle of strut-and-tie model. It can improve the bearing capacity of the pile cap by configuring the horizontal reinforcement mesh. Late strength of concrete can be used to optimize the mix proportion of concrete and reduce the amount of cement. Based on the strut-and-tie s model, the bearing capacity of the platform can be improved under the condition of using less reinforcement.

References

[1] Qian. Gu and Shuangshuang. Yang, Study on Form of Reinforcement Affecting Ultimate Loads of Thick Cap with Four Piles. J. of HUST. (Urban Science Edition).2016.23(4), pp 44~47.

[2] Robert L and Mokwa,J and Michael Duncan. Experimental Evaluation of Lateral-Load Resistance of Pile Caps[J]. Journal of Geotechnical and Geoenvironmental Engineering. 2001(2)Reference to a chapter in an edited book:

[3] You Sheng Deng,Jian Qiang Zheng,Chang Zhong Wan,Shao Hua He. Calculation for Settlement of High Cap Extra-Long Large Diameter Pile Groups[J]. Applied Mechanics and Materials. 2012(178).

Fatigue life analysis of cracked gas receiver of emergency cut-off system in gas gathering station

Junzhi Hu*, **Jiyong Zhou and Siyuan Li**

The second gas production plant of Petro China Changqing Oilfield Company, Shanxi, and China

*Corresponding author e-mail: hjzhi_cq@petrochina.com

Abstract. Small-scale air compressor and gas receiver are used as the driving gas of the emergency cut-off system in gas gathering station. Operation of block valve is ensured by starting and stopping compressor automatically. The frequent start-stop of compressor and the pressure fluctuation pose a threat to the service life of gas receiver, and then affect normal operation of the emergency cut-off system and security of gas gathering station. In this paper, the fatigue life of a pressure vessel with axial semi-elliptical surface crack in the inner wall is analyzed under the varying pressure by means of the theory of fracture mechanics. The influences of the amplitude of pressure fluctuation and the initial crack size on the residual life of gas receiver are discussed. It provides a basis for setting the working parameters of gas receiver of emergency cut-off system and determining the maintenance cycle.

1. Introduction

Emergency cut-off system is the most important guarantee for the safe operation of gas gathering station. For the gas field with gas collection technology of multi-well, high pressure and normal temperature, the emergency block valve is installed at the entrance of single well and exporting position in gas gathering station, and pressure monitoring points are located on the total throttle organ, separator and output position. When the pressure at the monitoring point exceeds the set value, block valves would cut off gas supply automatically to achieve overpressure protection of gas gathering station. In the early days, nitrogen gas that was supplied by nitrogen cylinders was used as gas supply and to drive emergency block valve on or off in the emergency cut-off system. Due to the large amount of nitrogen consumption, nitrogen cylinders needed to be replaced frequently. This leaded to overspending of manpower and material resource, and inconvenient management. Therefore, the air supply system, which is composed of small air compressor, gas receiver, dryer, filter screen, pressure regulator etc., is installed and used as gas supply of emergency block valve. The compressor can start and stop automatically, and avoid the emergency cut-off system in gas gathering station being influenced by climate and road. This new system could reduce demand of the material and financial resources due to the frequent replacement of nitrogen cylinders, abate operation cost of pneumatic valves, and ensure safe and stable operation of gas gathering station. But as gas supply of emergency cut-off system, gas receiver charges and discharges frequently and always stands the relatively large pressure difference. Accompanied by the fact that defects, such as corrosion, machining cracks and so on appear inevitably during service period, the safe operation of gas receiver is threatened [1].

In this paper, based on the theory of fracture mechanics, the fracture and fatigue properties of gas receiver with longitudinal surface cracks inside wall are analyzed, and the influences of the amplitude of pressure fluctuation and the initial crack size on the residual life of gas receiver are discussed. The results provide the basis for setting working parameters of gas receiver and determining the maintenance cycle properly.

2. Problem description

Gas receiver is generally made of steel Q345R, yield strength σ_s = 345MPa, and elastic modulus E = 210Gpa. The inner diameter of gas receiver D_i = 800mm, wall thickness t = 7mm, design pressure is 2.2MPa, and test pressure is 2.75MPa. When the supply gas pressure is lower than the set value (1MPa), compressor runs and inflates gas receiver; when the pressure exceeds the upper limit (2MPa), the compressor stops.

During the serving process of gas receiver, bulk defects and plane discontinuity may exist because of corrosion, incomplete welding, air gap and other reasons. Under the condition of internal pressure, the hoop stress of gas receiver is much higher than the axial stress (hoop stress $\sigma_\theta = pD_i / 2t$, axial stress $\sigma_z = pD_i / 4t$, where p is internal pressure), and the stress of the inner surface of the cylinder is higher than that of the outer surface. Therefore, the axial crack in the inner wall of the cylinder is more dangerous. Lin & Smith [2] and Carpinteri [3] had demonstrated that the initial surface defects with any shape would develop into a semi-elliptical crack with aspect ratio of 0.6 to 0.7 after several cycles of fatigue crack propagation. So, the axial crack in the inner wall of a tank is taken as a semi-elliptical surface crack, which can represent the actual crack properly, as shown in Fig.1.

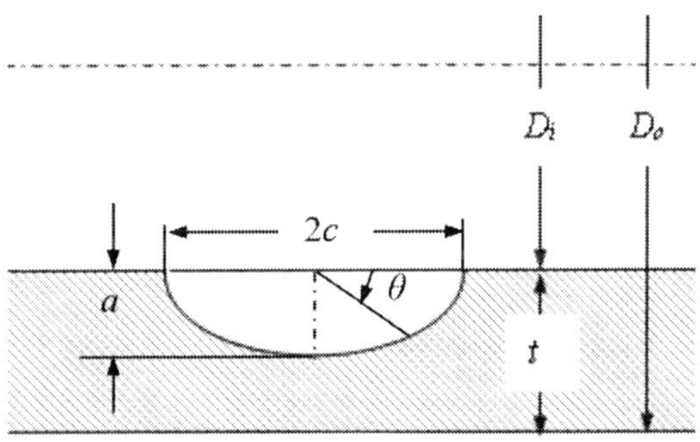

Figure 1 The sketch of semi-elliptical surface crack in the inner face of tank shell.

3. Fracture properties

Gas receiver that contains a longitudinal semi-elliptical surface crack is subjected to internal pressure. Considering the effect of air pressure on both the inner wall of tank and the crack surface, the stress intensity factor K_I of the crack can be expressed as [4]:

$$K_I = \frac{pD_i}{2t}\frac{\sqrt{\pi a}}{E(k)} \cdot 0.97 \cdot \left[M_1 + M_2\left(\frac{a}{t}\right)^2 + M_3\left(\frac{a}{t}\right)^4 \right] g \cdot f_\theta \cdot f_c \tag{1}$$

Where:

$$E(k)=\left[1+1.464\left(\frac{a}{c}\right)^{1.65}\right]^{1/2},\ M_1=1.13-0.09\frac{a}{c},$$

$$M_2=-0.54+\frac{0.89}{0.2+\frac{a}{c}},\ M_3=0.5-\frac{1}{0.65+\frac{a}{c}}+14\left(1-\frac{a}{c}\right)^{24},$$

$$g=1+\left[0.1+0.35\left(\frac{a}{t}\right)^2\right](1-\sin\theta)^2,\ f_\theta=\left[\sin^2\theta+\left(\frac{a}{c}\right)^2\cos^2\theta\right]^{\frac{1}{4}},$$

$$f_c=\left[\frac{D_0^2+D_i^2}{D_0^2-D_i^2}+1-0.5\sqrt{\frac{a}{t}}\right]\frac{2t}{D_i}.$$

In which a and c are the minor axis and major axis of the semi-elliptical crack, respectively, m. Here a/c is specified as 0.7; $E(k)$ is the complete elliptic integrals of the second kind; θ is circumferential angle of the crack front. The dimensionless stress intensity factor K_{non} is defined as:

$$K_{\text{non}}=\frac{K_I}{\frac{pD_i\sqrt{\pi a}}{2t}} \tag{2}$$

For the axial semi-elliptical surface cracks with a/t = 0.2, 0.4 and 0.6, the distributions of K_{non} along the crack front are calculated and shown in Fig.2. It can be seen that the K_{non} varies along the position of the crack front, and the larger K_{non} leads to faster growth rate of the crack. As a result, the shape of the crack front is flattened gradually as the crack propagating gradually. Based on the K_{non} at the deepest point of crack, the fatigue propagation process of crack is studied, and the fatigue life of gas receiver under different working parameters is predicted.

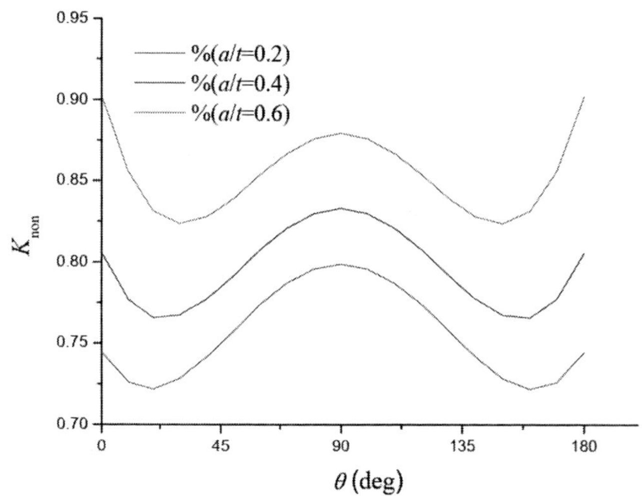

Figure 2 Distribution of K_{non} along crack front under different crack depth.

4. Fatigue life

By means of the formula derived by Paris, the crack growth rate in the tank inner wall could be estimated by Eq. (3).

$$\frac{\mathrm{d}a}{\mathrm{d}N} = C(\Delta K_I)^m \tag{3}$$

Where: N is the number of load cycles, ΔK_I is the range of stress intensity factor at the crack tip, MPa·m1/2; C and m are material constants related to working conditions, and chosen as $C = 7.97\times10^{-10}$ and m = 3.52 for steel Q345R. Fracture toughness at room temperature is K_{IC} =108 MPa·m$^{1/2}$. Substituting Eq. (1) into Eq. (3), and integrating from a_0 to a_c, the fatigue life of gas receiver could be obtained.

Here the surface crack with depth $a_0 = 0.1t$ is taken as an example, and the gas receiver is judged as failure when the deepest point of semi-elliptical surface crack penetrates tank wall, that means the critical crack size is $a_c = t$. The number of pressure cycles of the gas receiver varying along with the amplitude of the pressure fluctuation is shown in Fig.3. It can be seen that the fatigue life of gas receiver decreases sharply with the increase of the amplitude of pressure fluctuation. Under the condition that the compressor runs as $p <$ 0.6MPa and shuts down as $p >$ 2.2MPa, the fatigue life of gas receiver would be only 11751 cycles. If the gas pressure is adjusted so that the compressor runs below 1.2MPa and stops above 1.8MPa, the fatigue life of gas receiver would increase to 371107 times. Therefore, reducing the amplitude of pressure fluctuation can improve the service life of gas receiver greatly.

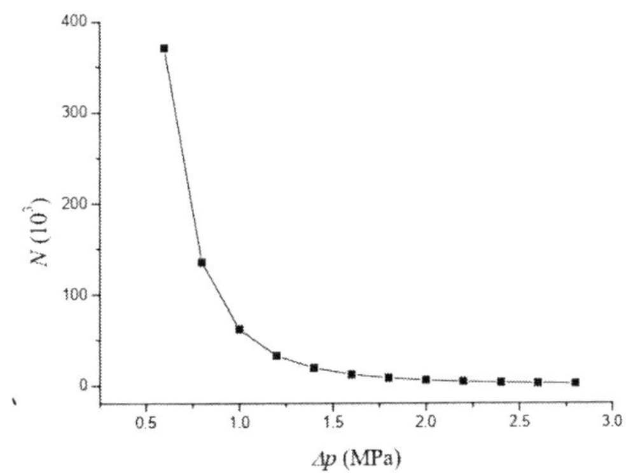

Figure 3 Fatigue life vs. the amplitude of internal pressure fluctuation.

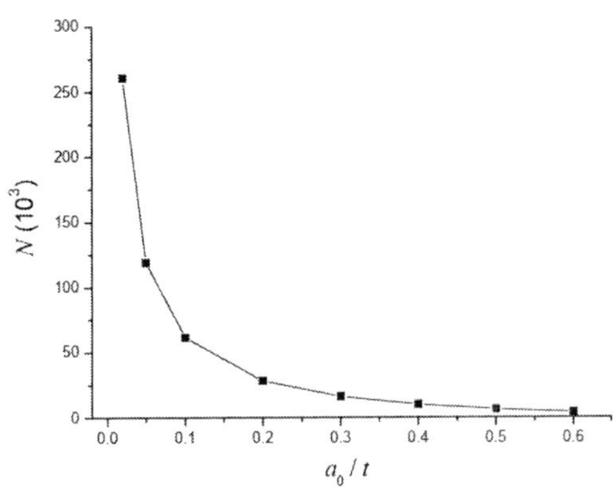

Figure 4 Fatigue life vs. initial crack depth.

Figure 4 shows the variation of the fatigue life of gas receiver with initial crack depth when the amplitude of pressure fluctuation is set as 0.6MPa. With the increase of the initial crack depth from 0.14mm (a/t = 0.02) to 4.2mm (a/t = 0.6), the number of residual cycles of gas receiver decreases from 260699 to 4112. With the increase of crack depth, the remaining life of gas receiver decreases sharply.

5. Conclusion

Based on the theory of fracture mechanics, the fracture properties of the axial semi-elliptical surface cracks in the inner wall of gas receiver under the action of internal pressure are studied. The influences of the amplitude of internal pressure fluctuation and initial crack depth on the remaining life of gas receiver are analyzed. It is found that the shape of the crack front is flattened gradually in the process of fatigue crack propagation. The residual life of gas receiver decreases sharply with the increase of either the amplitude of fluctuating internal pressure or the initial crack depth. Therefore, measures should be taken to ensuring safe operation of gas receiver as the following two aspects: improving the operating parameters to reduce the pressure fluctuation, and enhancing flaw detection to limit the defect size in the inner wall of gas receiver.

Acknowledgments

This work was supported by National Natural Science Foundation of China (11102173).

References

[1] GB/T 19624-2004, Safety assessment for in-service pressure vessels containing defects (In Chinese).
[2] X.B. Lin and R.A. Smith, Fatigue growth prediction of internal surface cracks in pressure vessels, J Pres Ves Technol, 120 (1998) 17-23.
[3] A. Carpinteri, Shape change of surface cracks in round bars under cyclic axial loading, Int J Fatigue, 15 (1993) 21-26.
[4] Chinese Aeronautical Establishment, Stress Intensity Factor Enchiridion, Beijing: Science Press, 1993(In Chinese).

IOP Conf. Series: Materials Science and Engineering **207** (2017) 012104 doi:10.1088/1757-899X/207/1/012104

Experimental Study on Color Durability of Color Asphalt Pavement

Shi Ning[1, 2*], Su Huan[3]

[1]Materials Science and Engineering, Chang'an University, Xi'an, China
[2]Transportation Research Institute of Jiangxi Province, Nan chang, China
[3]Jiangxi communications Career Technical College, Nan chang, China

*Corresponding author e-mail: 36517774@qq.com

Abstract. Aiming at the poor Color durability and the lack of research on Color asphalt pavement, spraying an anti-tire trace seal resin emulsion on the surface, a Color durable asphalt pavement was proposed. After long-term rolling and long-term aging test, the Color durability was evaluated by RGB function in Photoshop and trace residue rate formula. Test results proved that the Evaluation method was simple and effective. After long-term rolling, the Color of the road surface tends to a constant value. Spraying the emulsion on the road surface can resist tire traces. After long-term aging test, the resistance to tire traces was increased by 26.6% compared with the conventional type, while the former was 44.1% higher than the latter without long-term aging. The Color durable asphalt pavement can effectively improve the ability of Color asphalt pavement to resist tire traces, and significantly improve the Color durability of Color asphalt pavement.

1. Introduction

Color asphalt mixture is widely used in municipal roads parking lots and toll stations, the color pavement forms a beautiful landscape [1~2]. However, due to the repeated rolling of vehicle tires, the color of the pavement gradually fades.

The research on color pavement was mainly focused on the design of mixture, functional tests and construction technology [3~5], what makes the color pavement durable is keeping the color bright. So it is very important to research on color durability of the pavement.

A thermosetting acrylic resin is introduced in this paper, which is polymerized from acrylic monomer, initiator and distilled water. Spraying the thermosetting acrylic resin on the surface of color pavement, it can not only reduce the residual traces of tires, but also increase the surface water tightness, and improve the color durability of the pavement.

In this paper, tire rolling test is adopted to simulate the long-term vehicle rolling, 1000, 2000, 5000, 10000 and 20000 rolling times are chosen. Long-aging oven heating method is adopted to study the long-term aging performance of color.

IOP Conf. Series: Materials Science and Engineering **207** (2017) 012104 doi:10.1088/1757-899X/207/1/012104

2. Test conditions

2.1 Preparation for Color asphalt mixture

Aggregate gradation design for AC-13 modified asphalt mixture from technical specification for asphalt pavement construction in china the composition list in table 1. The asphalt-aggregate ratio 5.1% is determined by test.

Table 1 Gradation composition of AC-13 color asphalt mixture

Sieve size (mm)	16	13.2	9.5	4.75	2.36	1.18	0.6	0.3	0.15	0.075
Upper limit	100	100	85	68	50	38	28	20	15	8
lower limit	100	90	68	38	24	15	10	7	5	4
Optimum	100	96	77.5	42.5	25.5	21	15	12.5	8	4.5

2.2 Test on anti-sliding performance of Color asphalt mixture

Anti-sliding tests with and without protection film were carried out, the values of anti-sliding performance were measured by pendulum friction meter listing in table 2.

Table 2 Skid resistance value of color asphalt mixture

Indexes	Skid resistance value	Evaluation				
traditional	82	excellent	good	Medium	secondary	bad
spraying seal	74	≥42	37-42	32-37	27-32	<27

The anti-sliding performance with protection film is lightly lower than the traditional asphalt pavement, but the performance can meet china's requirements from specification.

3. Evaluation method

The original bright color of the pavement gradually turned black because of the tire trace residue between the pavement and the tires. The tire trace residual rate is evaluated by the changing rate of the color pavement the calculation of the color changing rate can be solved by the image processing technology in Photoshop software. RGB color space in three pixel combinations (R, G, B), different RGB pixel values can represent different depths of color, the original RGB value is(x, y, z), The RGB value after the test is defined as(x1, y1, z1), the value decrease in pace with the test times increasing.

The trace residual rate I is calculated by Formula 1.

$$i=\frac{x-x_1}{x+y+z}+\frac{y-y_1}{x+y+z}+\frac{z-z_1}{x+y+z} \tag{1}$$

The residual rate I value is between 0 and 1, the greater the value of me, the more serious traces residue. The wheel tracking test simulate vehicle loading, the following test parameters are simplified:

(1)The tire of rutting test machine simulate the tire of car,

(2)Constant tire speed,

(3)Test at room temperature only.

4. Test scheme

4.1 Specimens

The rutting test specimens were made according to the requirements of Chinese specification, the surface of the specimen was divided into two parts averagely the acrylic resin emulsion seal is sprayed on the lower half of the surface in the figure 1.

Eight points are located along the tire moving direction, marked on 2cm, 6cm, 10cm, 14cm, 17cm, 21cm, 25cm and 29cm. Every half part has 4 points. Take the marked points pictures using a HD digital camera every 1000 rolling times. The camera has 18 million pixels and the image resolution is 5184 x 3456.Every marked point RGB values were calculated in Photoshop software. It is noteworthy that backlight is useful for calculating RGB values when taking pictures.

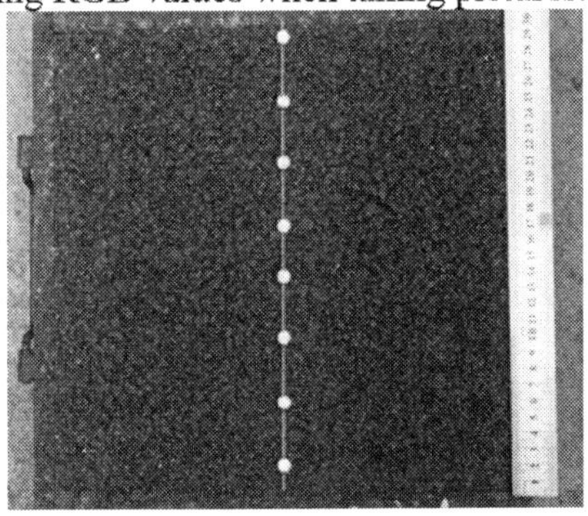

Fig.1 Trace residue mark

4.2 Test process

Three rutting test specimens of color pavement were made, spraying and marking on the specimen according to test scheme.

One of the specimens was subjected to wheel tracking test at ordinary temperature, another was heating in Long-aging oven 5 days at 85℃.The last one was keeping in ordinary temperature 5 days for comparison.

Taking the surface of each specimen pictures every 1000 rolling times. Each picture was processed by Photoshop software, the RGB values were calculated. The trace residual rate I for every picture was obtained.

5. Test results and analysis

5.1 Relationship between Colour durability and rolling times

The Figures 2 were taken after 1000, 2000, 5000 and 10000 rolling times. As can be seen from the pictures, with the increase rolling times, the surface of the red faded.

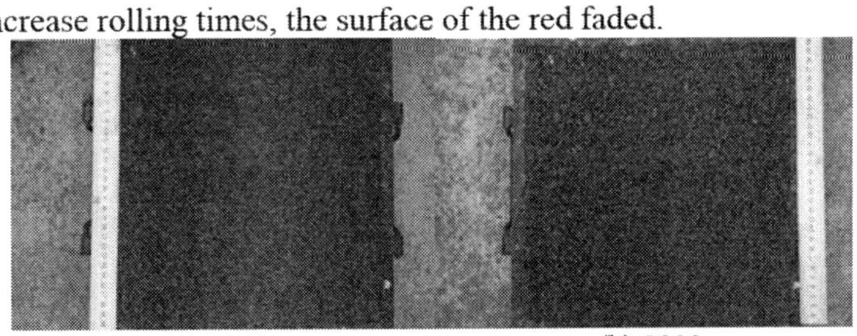

(a) 1000 (b) 2000

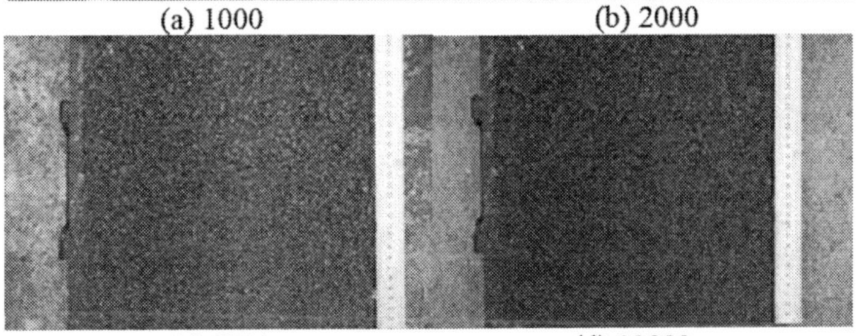

(c) 5000 (d) 10000

Fig.2 Tire trace residue degree

Table 3 Trace residue rate for every marking point under different rolling times

Trace residue rate / Types	Marked pints	1000	2000	5000	10000	20000
Durability	2cm	0.036	0.048	0.136	0.255	0.311
	6cm	0.040	0.054	0.104	0.202	0.303
	10cm	0.047	0.066	0.126	0.195	0.340
	14cm	0.033	0.064	0.137	0.221	0.328
Ordinary	17cm	0.089	0.166	0.265	0.368	0.473
	21cm	0.110	0.179	0.298	0.404	0.455
	25cm	0.133	0.181	0.284	0.375	0.462
	29cm	0.115	0.193	0.275	0.393	0.489

As seen in Table 3, the results show that trace residue rates increase with the adding times, but the spraying seal part has lesser trace residue rates. The emulsion seal can significantly improve the tire trace residue. When the number of rolling times up to 10000, the residual rate of traces of the two kinds of color asphalt mixture will increase sharply with the increase of rolling times, However, when the number is more than 10000 times, the increasing rate of trace residual rate of the two kinds of color asphalt mixture has slowed down, the curve is relatively stable, and the color durability is slightly higher than that of the ordinary type. When the color asphalt pavement with protection film reaches a certain residual rate, its surface color will tend to a constant value. Obviously, the color asphalt pavement with protection film to the constant value of the corresponding number of rolling times is much larger than ordinary color asphalt pavement. The acrylic resin emulsion seal can be more effective and long-term resistance to the traces of tire tracks, improve color asphalt pavement color durability.

5.2 Relationship between Colour durability and long-term aging

Two specimens of the red color mixture were prepared and tested according to the above methods. One of the specimens was put into the oven for long-term aging test. Another was keeping in ordinary temperature.

According to the formula of trace residual rate, the residual rates of color asphalt mixture were calculated under the different conditions, the relationship between two keeping conditions and residual rates of color asphalt mixture was drawn, as shown in Table4.

Table 4 Trace residue rate after long-term aging

Types	Trace residue rate (%)							
	2cm		6cm		10cm		14cm	
Durability	B	A	B	A	B	A	B	A
	0.069	0.061	0.072	0.067	0.076	0.063	0.086	0.084
	17cm		21cm		25cm		29cm	
Ordinary	B	A	B	A	B	A	B	A
	0.127	0.089	0.136	0.091	0.139	0.094	0.141	0.100

Type B is before long-term aging, Type A is after long-term aging.

As can be seen from Table4, whether or not after a long period of aging, the tire residue rate of color asphalt mixture spraying seal was significantly lower than the ordinary color asphalt pavement.

In the case of the ordinary color asphalt mixture, after the long-term aging, the residual rate of tire trace was significantly reduced. This may be due to the aging of the mixture resulting in a decrease in the color of the asphalt in the composition of the resin, and the color of the asphalt adhesion is also reduced, thereby reducing the black tire marks on the surface of the color asphalt adhesion.

In the case of the color asphalt mixture spraying seal, there was little change in tire trace residue rate before and after aging, the rates after aging were slightly lower than those before aging.

In conclusion, the color durability of the ordinary type of color asphalt mixture is influenced greater than that of mixture with protection under long-term aging. The color durability of asphalt pavement can effectively resist by the resin emulsion

6. Conclusions

In order to improve the color durability of asphalt pavement, a kind of thermosetting acrylic resin emulsion was sprayed on the surface of pavement to form a color durable asphalt pavement.

The residual traces of the specimen surface were processed by the Photoshop software, and the residual traces on the surface of the color asphalt pavement were quantified and evaluated by the RGB function and the residual rate formula. In addition, two kinds of methods of long-term aging were used to study the color durability of color asphalt pavement.

The following conclusions are obtained:

(1)A tire trace residual test method is proposed, which can be used to evaluate the color durability of the color asphalt pavement

(2)Trace residual tests under different rolling times compared the residual rates between color durability and ordinary of asphalt pavement, the color durability of the former is obviously higher than the latter.

(3)In addition, trace residual tests were carried out between color durability and ordinary of asphalt pavement after long-term aging, the color durability of the former compared to later type increased by 26.6%, the tests results without long-term aging show that the former resistance traces compared to latter one increased by 44.1%.It can be seen that the anti-tire resin seal coating emulsion can effectively improve the color durability of color pavement.

Reference

[1] Lee, H., and Y. Kim. Laboratory Evaluation of Color Polymer Concrete Pavement with Synthetic Resin Binder for Exclusive Bus Lanes. In Transportation Research Record: Journal of the Transportation Research Board, No. 1991, Transportation of Research Board of the NationalAcademies, Washington, D. C., 2007, pp. 124–132

[2] Liang, C., H. Wang, J. Lu, and H. Chen. The Investigationof ColoredNormal-Temperature Asphalt Concrete.Advanced Materials Research, Vol. 723, 2013, pp.686–693.

[3] Zhou, K. Introduction about Color Restoration Materials of Water-Based Color Pavement. Petroleum Asphalt, Vol. 26, No. 3, 2012, pp. 58–59.

[4] Lin, D., and H. Luo. Fading and Color Change sin Colored Asphalt Quantified by the Image Analysis Method. Construction and Building Materials, Vol. 18, No. 4, 2004, pp. 255–261.

[5] Wang,L., D. Han, and S. Chen. Overview of Colorful Road Pavement Technology and Its Development. Petroleum Asphalt, Vol. 23, No.1, 2009, pp. 68–71.

Influence of Cooling Duration on the PTCR Effect of Sm^{3+}-doped $BaTiO_3$-based Ceramics Sintered in a Reducing Atmosphere

Xuxin Cheng*, Xiaoxia L and Xiaoming Chen

College of Electronic Information and Mechatronic engineering, Zhaoqing University, Zhaoqing Road, Duanzhou District, Zhaoqing 526061, Guangdong, PR China

*Corresponding author e-mail: cxx101@126.com

Abstract. The influence of cooling duration on the electrical properties and the positive temperature coefficient of resistance (PTCR) characteristics of Sm-doped $BaTiO_3$ (BST) ceramics fired at 1200 °C for 30 min in a reducing atmosphere and deoxidized at 800 °C for 1 h were investigated. The cooling duration influenced the PTCR characteristics of the BST specimens. The room-temperature resistivity of the samples firstly reduced and then increased with the increase in the content of Sm_2O_3. A long cooling duration resulted in low resistivity of the BST ceramics. Meanwhile, the room-temperature resistivity of the BST ceramics decreased with an increase in cooling duration. Furthermore, the S2 samples exhibited a remarkable PTCR effect. The samples exhibited a resistance jump greater than 3.2 orders of magnitude and a low room-temperature resistivity of 228.4Ω·cm.

1. Introduction

It is well known that Barium titan ate ($BaTiO_3$) is a ferroelectric material, which transforms from a tetragonal perovskite structure at room temperature into a Para electric structure at the curie temperature [1]. Ti site or Ba site of $BaTiO_3$ are substituted by small quantities of pentavalent impurities (Nb^{5+}, Ta^{5+}, etc.) or trivalent donor (Y^{3+}, Sm^{3+}, etc.) becomes a semiconducting, respectively. Moreover they have the positive temperature coefficient of resistivity (PTCR) effect [2, 3], the phenomenon was explained by Heywang [4] and Jonker [5].

Semiconducting $BaTiO_3$-based PTCR ceramics are sintered by traditional method, which have a high room-temperature resistance. However, it is difficult to further decrease the room-temperature resistivity of the samples, and being attributed to the own characteristics of $BaTiO_3$-based materials. In order to further reduce room-temperature resistance, multilayer PTCR ceramics are prepared [6–8] by way of a reduction–reoxidation method [9]. Niimi et al. [10] recently reported that the ion radius of the donor dopant affects the PTCR characteristic of samples. Zubair M.A. et al. [11] found that a reduction in room-temperature resistance can be due to increase cooling rate. Chen H.P. [12] also indicated that PTC effect could be enhanced by a slower cooling rate. In addition, several researchers have proposed theory of grain-boundary oxygen adsorption and diffusion [13–15].

However, few studies have investigated the influence of cooling duration on electrical properties and PTCR effect of BST ceramics in a reducing atmosphere. Therefore, the PTCR characteristics of

$BaTiO_3$-based ceramics with doping Sm_2O_3 sintered in a reducing atmosphere was investigated in this paper.

2. Experimental Procedures

The starting materials were high-purity $BaCO_3$, TiO_2, SiO_2, and Sm_2O_3, and they were weighed according to the following formula: $(Ba_{1.022-x}Sm_x)$ TiO3 + 0.007 mol SiO_2 (BST). In this experiment, the components were mixed by wet ball milling for 4 h in deionized water using zirconia balls in a polyurethane jar and then calcined at 1150 °C for 2 h in air. After drying and sieving, the calcined powder was ground again by ball milling for 4 h. After mixing, the dried powder was mixed with dispersant, defoamer, solvent, and binder by ball milling for 18 h in a nylon pot and cast into green sheets of 55-μm thickness by the doctor-blade method. These sheets were stacked and laminated at 50 °C to form a ceramic block, which was cut into rectangular blocks (5.7 × 3.9 × 1 mm). Subsequently, the binder was removed by heating at 280 °C in air. Sintering was conducted in an aluminum tube at 1200°C for 30 min in a reducing atmosphere (3% H_2/N_2), with the heating and cooling rates being 5 and 3 °C/min, respectively, at a flow rate of 200 cm^3/min under a pressure of 1 atm. Three different cooling curves were used for our present experiment, as shown in Table 1. The sintered BST ceramics were deoxidized at 800 °C in air for 1 h, and the surfaces were rubbed with In–Ga alloy (60:40) to form electrodes. Resistance at room temperature was measured by a digital multimeter, and the temperature dependence of resistance was measured in a temperature-programmable furnace (ZWX-B, Huazhong University of Science and Technology, China) at a heating rate of 1.6 °C/min in the range of 25–250 °C.

Table 1. The various cooling curves during cooling.

The temperature range (°C)	Samples and the cooling time (min)		
	S1	S2	S3
1200→1150	17	17	FC[†]
1150→1150	30	0	0
1150→800	117	117	FC
800→Rt.	FC	FC	FC

†: Furnace Cooling

3. Results and Discussion

3.1. *Influence of cooling duration on electrical properties*

The dependence of the room-temperature resistivity of the BST based ceramics, deoxidized at 800 °C for 1 h after firing at 1200 °C for 30 minutes in a reducing atmosphere, on the dopant content is shown in Figure 1. From the figure, we can see that the room-temperature resistivity of the BST ceramics nearly decrease with increase in the concentration of doped Sm^{3+} for samples subjected to a long cooling time; however, it initially decreases and then increases as a function of the donor-doped content at a shot cooling time. It is indicated that the slower the cooling time, the smaller is the room-temperature resistivity of the specimens. Furthermore, the increase in resistance is attributed to the ionic compensation resulting from the formation of oxygen and cation vacancies. The decrease in resistance is generally due to the oxygen vacancies and the following electron compensation [16, 17]:

$$O_O \rightarrow V_O^{\bullet\bullet} + \tfrac{1}{2}O_2 + 2e' \qquad (1)$$

$$Sm_2O_3 \rightarrow 2Sm_{Ba}^{\bullet} + 2O_O + \tfrac{1}{2}O_2 + 2e' \qquad (2)$$

Moreover, the content of oxygen vacancies are increasingly higher at slower cooling rates. Thus, the concentration of the electrons plays a dominant role. So the room-temperature resistivity of the BST based ceramics is much less than that of the same samples sintered in air.

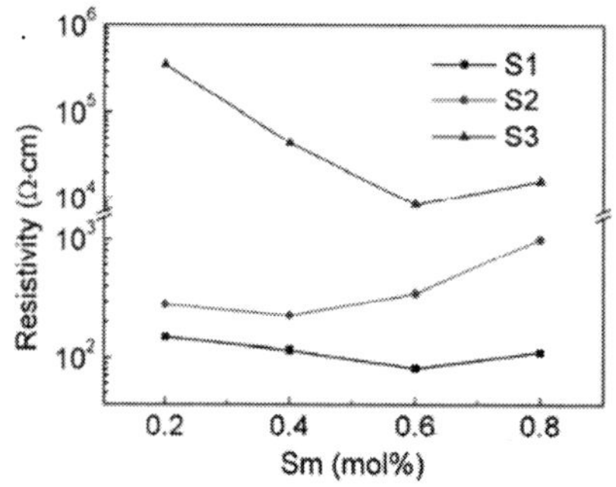

Figure 1. The room-temperature resistivity is as a function of the dopant concentration at different cooling methods.

3.2. Influence of the donor-doped content on PTCR effect

The temperature dependence of resistivity of the BST ceramics that is S2 samples as functions of the different Sm^{3+}-doped concentration of 0.2 mol% – 0.8 mol% are as shown in Figure 2. It is shown that the room-temperature resistivity of the ceramics that are deoxidized at 800 °C for 1 h after firing at 1200 °C for 30 min initially reduces and then increases with an increase in the dopant content from 0.2 mol% to 0.8 mol%. While, the critical donor-doped content for the S2 samples is 0.4 mol%. Moreover, the S2 ceramics doped Sm^{3+} of 0.4 mol% exhibit a significant PTCR effect, with a resistivity jump of 3.2 orders of magnitude, along with a low mean room-temperature resistivity of 228.4 Ω·cm. However, although the resistivity jump of the 0.2 mol% Sm^{3+}-doped BST ceramics is exhibited by 3.4 orders of magnitude, the room-temperature resistivity is slightly higher than that of the specimen with doping Sm^{3+} of 0.4 mol%. According to Heywang-Jonker model [4, 5], the PTC effect is the grain-boundaries effect, and a better PTCR effect may be attributed to the higher height of the potential Schottky barrier.

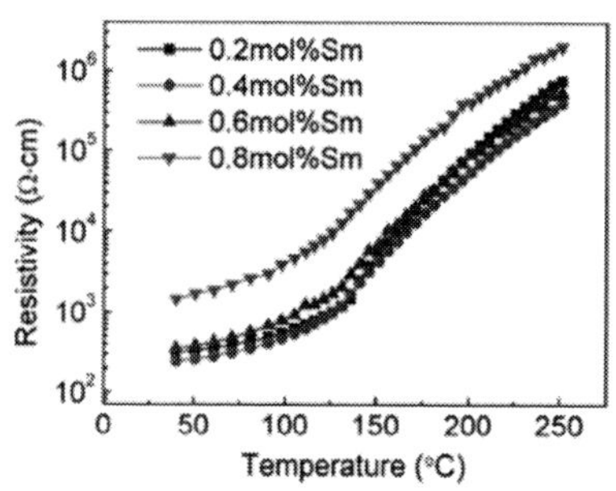

Figure 2. The temperature dependence of resistivity of the BST ceramics is as a function of the different Sm-doped contents.

3.3. Influence of cooling process on PTCR characteristics

The electrical properties and PTCR effect of the 0.4 mol% Sm-doped BST based ceramics for various cooling duration are as shown in Table 2. The ceramics are cooled at different cooling process after sintering at 1200 °C for 30 min in a reducing atmosphere and then deoxidized at 800 °C for 1 h. The cooling curves and their corresponding times are shown in Table 1. From the Table 2, we can see that the room-temperature resistivity of the S1, S2, and S3 samples are 115.8, 228.4, and 44056.6 Ω·cm, respectively. Moreover they show a remarkable PTCR effect with a resistance-jumping ratio of 3.0, 3.2, and 3.3, respectively. Furthermore, the temperature coefficient (α_T) of the BST ceramics is also observed, whereas α_T is the following formula: $\alpha_T = [(\ln R_{Tc+25°C} - \ln R_{Tc+10°C}) / 10°C] \times 100\%$, where Tc is the Curie temperature, $R_{Tc+25°C}$ is the resistance of the samples at a temperature of Tc+25 °C, so does $R_{Tc+10°C}$. On comparing the experimental results, the room-temperature resistivity of the S1 samples is found to be the lowest. This is because the content of oxygen vacancies in S1 specimens is the highest. According to Eq. (1), it is found that the longer the cooling duration, the higher is the concentration of electrons in the samples, thus resulting in a smaller room-temperature resistivity. This result indicates that the electrical properties of the BST samples are closely related to the cooling duration. In addition, we can see that the α_T is more than that of both S2 and S3 ceramics in Table 2. It is shown that the S1 samples are very sensitive to the temperature change. Therefore, we found that the S1 samples can exhibit significant PTCR characteristics.

Table 2. The PTCR characteristics of the BST ceramics with doping Sm3+ of 0.4 mol% for the different cooling duration.

Sample	ρ (Ω·cm)	Lg(R_{max}/R_{min})	α_T (%)
S1	115.8	3.0	2.9
S2	228.4	3.2	3.0
S3	44056.6	3.3	2.4

4. Conclusion

We investigated the influence of cooling duration on the PTCR characteristics and electrical properties of BST ceramics deoxidized at 800 °C for 1 h after firing at 1200 °C for 30 min in a reducing atmosphere. The room-temperature resistivity firstly decreased and then increased with of the increasing of the content of Sm_2O_3. Long cooling duration resulted in low room-temperature resistivity of the BST samples. Furthermore, the resistivity of the grain boundaries of the samples decreased with the increase in cooling duration. Lastly, the S2 samples deoxidized at 800 °C for 1 h after sintering at 1200 °C for 30 min in a reducing atmosphere exhibited a resistance jump greater than 3.2 orders of magnitude and a low RT resistivity of 228.4 Ω·cm.

Acknowledgments

This work was financially supported by the National Natural Science Foundation of China (No. 51402258), the Natural Science Foundation of Guangdong Province (No. 2015A030313706), the Science and Technology Planning Project of Guangdong Province (No. 2014A010105056), the Characteristic Creative Project of Department of Education of Guangdong Province (Nos. 2016KTSCX152, 2014KTSCX189), the Science and Technology Planning Project of Zhaoqing City (No. 2015B010402003).

References

[1] M.W. Mancini and P.I. Paulin Filho, Direct observation of potential barriers in semiconducting barium titanate by electric force microscopy, J. Appl. Phys. 100 (2006) 104501.
[2] W. Heywang, Resistance anomaly in doped barium titanate, J. Am. Ceram. Soc. 47 (1964) 484.
[3] J. Illingsworth, H.M. Ai-Allak, A.W. Brinkman, and J. Woods, The influence of Mn on the grain-boundary potential barrier characteristics of donor-doped $BaTiO_3$ ceramics, J. Appl. Phys. 67 (1990) 2088–2092.
[4] W. Heywang, Barium titanate as a semiconductor with blocking layers, Solid-State Electron. 3 (1961) 51.
[5] G.H. Jonker, Some aspects of semiconducting barium titanate, Solid-State Electron. 7 (1964) 895–903.

[6] X.X. Cheng, D.X. Zhou, Q.Y. Fu, S.P. Gong, D.C. Zhao, Influence of sintering conditions on the electrical properties and the PTCR effect of the multilayer $Ba_{1.005}(Ti_{1-x}Nb_x)O_3$ ceramics with Ni internal electrode, J. Mater. Sci.: Mater. Electron. 23(12) (2012) 2202–2209.

[7] H. Niimi, K. Mihara, Y. Sakabe and M. Kuwabara, Preparation of multilayer semiconducting $BaTiO_3$ ceramics Co-fired with Ni inner electrodes, Jpn. J. Appl. Phys. 46 (10A) (2007), 6715–6718.

[8] X.X. Cheng, D.X. Zhou, Q.Y. Fu, S.P. Gong, Y.X. Qin, Effect of reoxidation annealing on the PTCR behavior of multilayer Nb^{5+}-doped $BaTiO_3$ ceramics with Ni internal electrode, J. Phys. D: Appl. Phys. 45(38) (2012) 385306 (7pp).

[9] P.H. Xiang, H. Harinaka, H. Takeda, T. Nishida, K. Uchiyama and T. Shiosaki, Annealing effects on the characteristics of high T_c lead-free barium titanate-based positive temperature coefficient of resistance ceramics, J. Appl. Phys. (2008) 104 094108.

[10] H. Niimi, T. Ishikawa, K. Mihara, Y. Sakabe and M. Kuwabara, Effects of Ba/Ti ratio on positive temperature coefficient of resistivity characteristics of donor-doped $BaTiO_3$ fired in reducing atmosphere, Jpn. J. Appl. Phys. 46 (2) (2007) 675–680.

[11] M.A. Zubair and C. Leach, Modeling the effect of SiO_2 additions and cooling rate on the electrical behavior of donor-acceptor codoped positive temperature coefficient thermistors, J. Appl. Phys., 103 (2008) 123713.

[12] H.P. Chen, T.Y. Tseng, The effect of cooling rate on the positive temperature coefficent resistivity characteristics of lanthanum-doped $Ba_{0.8}Sr_{0.2}TiO_3$ ceramics, J. Mater. Sci. Lett. 8 (1989) 1483–1485.

[13] M. Kuwabara Determination of the potential barrier height in $BaTiO_3$ ceramics, Solid State Electron 27 (1984) 929–935.

[14] A. Hasegawa, S. Fujitsu, K. Koumoto et al., The enhanced penetration of oxygen along the grain boundary in semiconducting barium titanate, Jpn. J. Appl. Phys. 30(6) (1991) 1252–1255.

[15] D. C. Sinclair, A R.West, Use of succinic acid to test the stability of PTCR barium titanate ceramics under reducing conditions, J. Am. Ceram. Soc. 78(1) 1995 241–244.

[16] X.X. Cheng, D.X. Zhou, Q.Y. Fu. PTC effect of chip type $Ba_{1.005-x}Sm_xTiO_3$ ceramics prepared by reduction sintering-reoxidation method, Trans. Ind. Ceram. Soc. 71(4) (2012) 100–105.

[17] J. Qi, Z. Gui, Y. Wang, Q. Zhu, Y. Wu, L. Li, The PTCR effect in $BaTiO_3$ ceramics modified by donor dopant, Ceram. Int. 28 (2002) 141.

IOP Conf. Series: Materials Science and Engineering **207** (2017) 012106 doi:10.1088/1757-899X/207/1/012106

Study on reinforcement of soil for suppressing fugitive dust by bio-cementitious material

Qiwei Zhan[1, 2], Chunxiang Qian[1, 2, *]

[1]School of Materials Science and Engineering, Southeast University, Nanjing, China
[2]Research Institute of Green Construction Materials, Southeast University, Nanjing, China

*Corresponding author e-mail: 230149661@seu.edu.cn,zhanqiwei168@139.com

Abstract. Microbial-induced reinforcement of soil, as a new green and environmental-friendly method, is being paid extensive attention to in that it has low cost, simple operation and rapid effects. In this research, reinforcement of soil for suppressing fugitive dust by bio-cementitious material was investigated. Soil cemented by bio-cementitious material had superior mechanical properties, such as hardness, compressive strength, microstructure, wind-erosion resistance, rainfall-erosion resistance and freeze-thaw resistance. The average hardness of sandy soil, floury soil and clay soil is 18.9 °, 25.2 ° and 26.1 °, while average compressive strength of samples is 0.43 MPa, 0.54 MPa and 0.69 MPa, respectively; meanwhile, the average calcite content of samples is 6.85 %, 6.09 %, and 5.96 %, respectively. Compared with the original sandy soil, floury soil and clay soil, the porosity decreases by 38.5 %, 33.7 % and 29.2 %. When wind speed is 12 m/s, the mass loss of sandy soil, floury soil and clay soil cemented by bio-cementitious material are all less than 30 g/(m2·h). After three cycles of rainfall erosion of 2.5 mm/h, the mass loss are less than 25 g/(m2·h) and the compressive strength residual ratio are more than 98.0 %. Under 25 cycles of freeze-thaw, the mass loss ratio are less than 3.0 %.

1 Introduction

Fugitive dust is a significant contributor to air pollution with harmful effects to human health [1-3]. Of all air pollution by the pollutants, fugitive dust is particular serious in that fugitive dust are very difficult to collect, separate and degrade [4-5]. Control of fugitive dust is becoming a hot spot at home and abroad for a time [6-8]. Generally, two measures can be taken to control fugitive dust pollution—physical method and chemical method. Physical method mainly includes sprinkling water, covering dust-controlling nets and building fence [9-11], while chemical method contains the dust-depressor type of wetting, hygroscopic, bond and complex [12-16]. Due to high energy consumption, large investment, complex operation and likely secondary pollution to the environment, physical and chemical methods are relatively difficult to be applied to the control of fugitive dust in large areas [17-19].

Nevertheless, bio-cementitious material, as a new green and environmental material, has stable and reliable effects, no secondary pollution, which, therefore, has become the most promising method in the control of fugitive dust. Firstly, CO_2 was absorbed, transformed and produced bicarbonate ions under the enzymatic action of *Paenibacillus*. Secondly, bicarbonate ions were transformed into carbonate ions under the condition of alkaline environment. Meanwhile, calcium ions in the fugitive dust were attracted to the bacteria cell wall due to the negative charge of the latter and upon addition of substrate to the bacteria. Finally, calcium ions were mineralized and precipitated to carbonate particles at the cell surface serving as the nucleation site. Fugitive dust was cemented in the process of forming calcite, and calcite-consolidation-layer was prepared ultimately which had a certain mechanical properties. The mechanism of mineralization and cementation of fugitive dust by bio-cementitious material based on CO_2 capture and utilization could be explained from the following equations:

$$CO_2 + H_2O \longleftrightarrow HCO_3^- + H^+ \tag{1}$$

$$HCO_3^- + OH^- \longleftrightarrow CO_3^{2-} + H_2O \tag{2}$$

$$Ca^{2+} + Cell \longleftrightarrow Cell\text{-}Ca^{2+} \tag{3}$$

$$Cell\text{-}Ca^{2+} + CO_3^{2-} \longleftrightarrow Cell\text{-}CaCO_3 \tag{4}$$

This research, reinforcement of soil for suppressing fugitive dust by bio- cementitious material was investigated. The hardness and compressive strength of different type's soil were presented successively, and the calcite content and porosity of samples were also verified accurately. Meanwhile, the wind-erosion resistance and rainfall-erosion resistance of samples were measured in wind tunnel teat and artificial simulation of rainfall devices. Finally, the freeze-thaw resistance of samples was demonstrated in the case of high and low temperature cycle. The results of these studies proved the feasibility of developing more environmentally friendly bio-cementitious material and showed that different types soil have a strong influence on microstructure and properties of samples cemented by bio-cementitious material.

2 Materials and methods

2.1 Experimental materials

Bio-cementitious material consisted of two parts: bacteria powder and calcium source. Bacteria power is Paenibacillus, while the calcium source is calcium nitrate. Cultivation of the *Paenibacillus* was conducted in sucrose culture (10 g of sucrose and 3 g of sodium hydrogen phosphate were dissolved in deionized water to 1 L, and the pH value was adjusted to about 7.0) at 35 °C for 24 h. The harvested microorganisms were kept in a refrigerator at 4 °C for stock prior to use in spraying process, while bacteria powder was used directly without cultivation in blending process.

2.2 Treating process

Wind erosion plates (internal diameter 20.0 cm, height 2.0 cm) were used in this study, and they were packed in the following order: first a layer of approximately 0.2 cm of gauze was placed at bottom of the wind erosion plates, followed by 2.0 cm of fugitive dust. All experiments were conducted at room temperature of 25±2 °C. Then, bacteria solution was sprayed evenly on the surface of fugitive dust after dissolved calcium nitrate. Afterwards, the samples were dried at room temperature of 25±2 °C until they were dried. The drying is stopped when the weight of the sample remains constant, which is achieved when its variation does not exceed 0.01%. The samples were collected and characterized.

2.3 Analysis of samples

The calcite content of samples was measured by Thermogravimetric-differential scanning calorimetry (TG-DSC), and it was carried out on STA449 F3 thermogravimetric analyzer (Nietzsche, Germany).

The analyses were carried out simultaneously in a nitrogen atmosphere at a heating rate of 10 °C·min^{-1} between room temperature and 1000 °C.

Porosity was measured by the gravity method, using vacuum saturation. It consists of submitting the sample to a moderate oven-drying at a temperature of 60 ℃. The drying is stopped when the weight of the sample remains constant, which is achieved when its variation does not exceed 0.01%. The weight of the dried sample, denoted as m_d, is then measured with an accurate balance. Afterwards, the specimen is immersed under vacuum in a container filled with distilled water. The weight of the samples is measured at different times until it becomes stable; the sample is then considered as fully saturated with water and its weight denoted as *ms*. In addition, the volume of the sample (Vv) is precisely measured with a pycnometer. ρ_w is the volume density of water. Finally, porosity (p) is determined using the following formula:

$$P = \frac{m_s - m_d}{\rho_w Vv} \quad (5)$$

The wind-erosion resistance was conducted in wind tunnel test at 0 angle for 1 h, and the wind speed was 4, 6, 9, 12 m/s respectively. The wind-erosion resistance could be represented via the mass loss of unit area of samples. In other words, it could be represented by comparing the change of mass before and after experiment.

The rainfall-erosion resistance was conducted in the artificial simulation of rainfall devices. The rainfall duration on the surface of samples was 6 h, and the rainfall intensity was 2.5 mm/h, then dried at 60℃ in the oven for 24 h, and it was recorded as a cycle. The change of mass loss and compressive strength were obtained under the different cycles. According to the change, the rainfall-erosion resistance was evaluated accurately.

The freeze-thaw resistance was conducted in the constant temperature device. The non-dried sample first was in -20℃ for 6 hours, then was in 20℃ for 6 hours, and it was recorded as a cycle. The change of mass loss was obtained under the different cycles. According to the change, the freeze-thaw resistance was obtained.

3 Results and discussion

3.1 Hardness and compressive strength

The influence of different type's soil on hardness and compressive strength of samples cemented by bio-cementitious material are presented in Fig. 1. The experimental results indicate that different types soil significantly affect hardness and compressive strength of samples. It could be seen from Fig. 1 that average hardness of samples, cemented sandy soil, floury soil and clay soil is 18.9 °, 25.2 ° and 26.1 °, while average compressive strength of samples is 0.43 MPa, 0.54 MPa and 0.69 MPa, respectively. The average hardness and compressive strength of samples cemented clay soil are largest, while that of samples cemented sandy soil are smallest. The results show that the performance of the soil was improved by bio-cementitious material effectively, and different type's soil had a strong influence on hardness and compressive strength. The reasons might be that clay soil itself has a certain bond force, but the sand soil is without bond force.

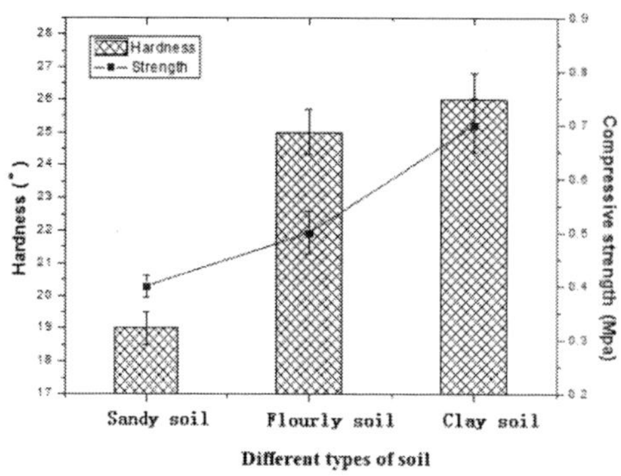

Fig.1 The hardness and compressive strength of samples cemented by bio-cementitious material

3.2 Calcite content

The influence of different type's soil on calcite content of samples cemented by bio-cementitious material are shown in Fig. 2. As Fig. 2 showed, average calcite content of samples, cemented sandy soil, floury soil and clay soil is 6.85 %, 6.09 %, and 5.96 %, respectively. Original calcite content in sandy soil, floury soil and clay soil is 0.21 %, 0.32 % and 0.28 %. However, the average calcite content of samples cemented sandy soil is larger than that others. The major causes for this end, among others, are the differences between particle size and pores of soil. Larger particle size and pores, which are conducive to the transfer, enrichment and transformation of carbon dioxide, result in more calcite.

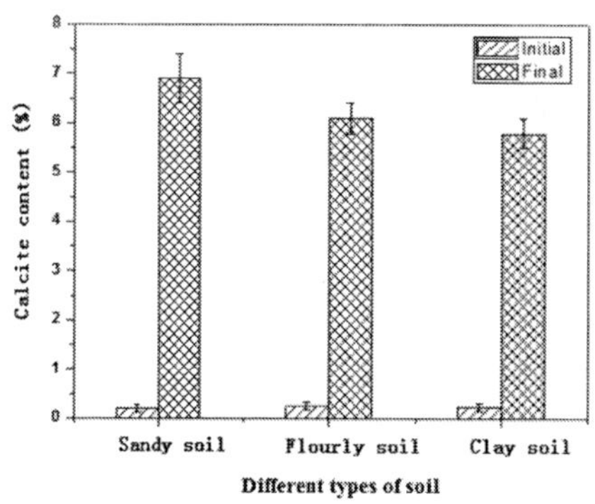

Fig. 2 The calcite content of samples cemented by bio-cementitious material

3.3 Porosity

The influence of different type's soil on porosity of samples cemented by bio-cementitious material are presented in Fig. 3. The results show that the initial porosity of samples, cemented sandy soil, floury soil and clay soil inclines to decrease, but the reduced range is different. Prudent observations of Fig. 3 indicates that the porosity of samples, cemented sandy soil, floury soil and clay soil effectively decreases by 38.5 %, 33.7 % and 29.2 %, respectively. The results are in keeping with the calcite content results, and more calcite filled with more pores in sandy soil.

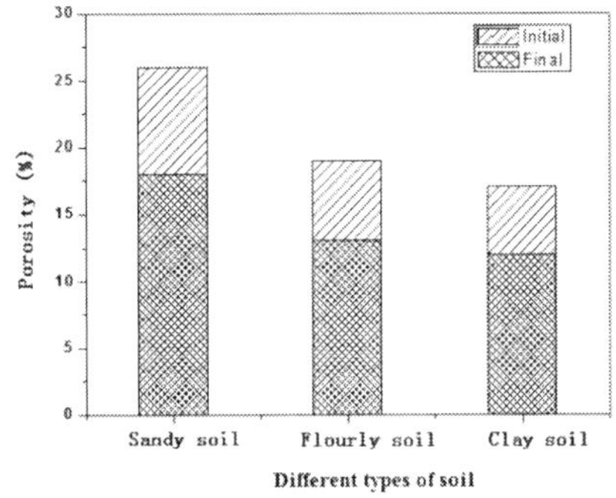

Fig. 3 The porosity of samples cemented by bio-cementitious material

3.4 Wind erosion resistance

The influence of different type's soil on wind erosion resistance of samples cemented by bio-cementitious material are shown in Fig. 4. Fig.4 showed, under different wind speed, the mass loss of original sandy soil is about 282, 1051, 1642 and 2189 g/ ($m^2 \cdot h$) respectively, the mass loss of original floury soil is about 312, 1223, 1805 and 2276 g/ ($m^2 \cdot h$) respectively, and the mass loss of original clay soil is about 504, 1415, 1981 and 2477 g/($m^2 \cdot h$) respectively. With the increase of wind speed, the mass loss is significantly increased.

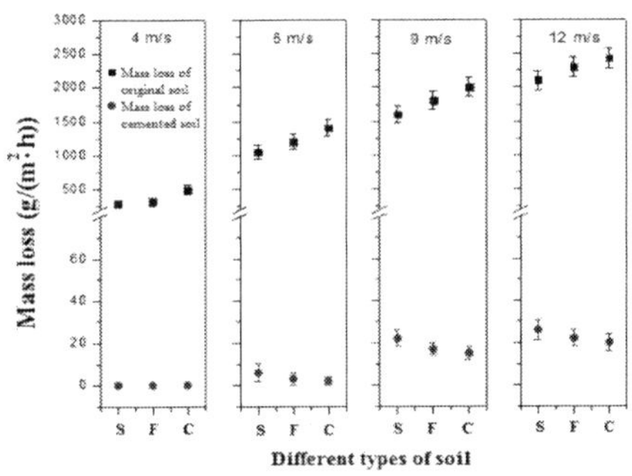

Fig.4 The wind-erosion resistance of samples cemented by bio-activated cementitious material

(S: sandy soil, F: floury soil, C: clay soil)

However, the mass loss of sandy soil, floury soil and clay soil cemented by bio-cementitious material are all less than 30 g/($m^2 \cdot h$), and no matter how much the wind speed, the change of the mass loss is not obvious. When the wind speed is 4 m/s, different type's soil samples cemented by bio-cementitious material could be achieved without obvious loss. As can be seen from Fig. 4, different type's soil has a certain influence on wind erosion resistance. In the same case, the wind erosion resistance of clay soil is better than that of sandy soil and floury soil.

3.5 Rainfall-erosion resistance

The influence of different type's soil on rainfall-erosion resistance of samples cemented by bio-cementitious material are shown in Fig. 5. Fig.5 showed, under the different number of cycle, the mass loss of original sandy soil, floury soil and clay soil is all more than 600 g/($m^2 \cdot h$), and the rainfall-erosion resistance is relatively poor.

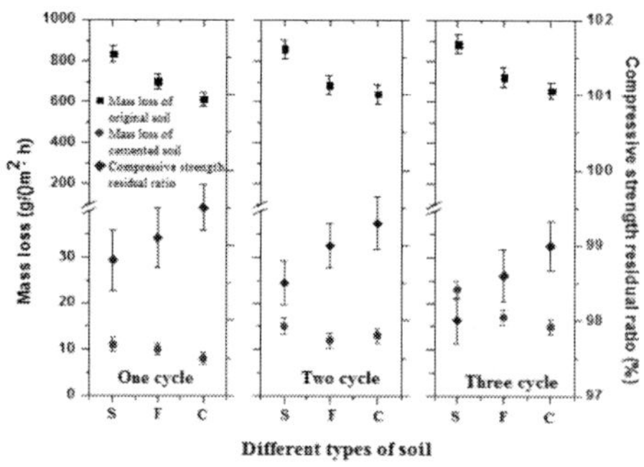

Fig.5 The rainfall-erosion resistance of samples cemented by bio- cementitious material

(S: sandy soil, F: floury soil, C: clay soil)

Nevertheless, in the one cycle, the mass loss of sandy soil, floury soil and clay soil cemented by bio-cementitious material are all less than 15 g/(m^2·h), then the change of the compressive strength was little and the compressive strength residual ratio are more than 99.0 %. Compared with the one cycle, the mass loss are all slightly increased and the compressive strength residual ratio are appreciably decreased in the two cycle. In spite of this, the mass loss are less than 25 g/(m^2·h) and the compressive strength residual ratio are more than 98.0 % in the three cycle. With the increase of number of cycle, the change of mass loss and compressive strength residual ratio are almost non-existent. It indicates that sandy soil, floury soil and clay soil cemented by bio-cementitious material have a good property of rainfall-erosion resistance.

3.6 Freeze-thaw resistance

Experimental data and numerical correlation on freeze-thaw resistance of samples cemented by bio-cementitious material are shown in Fig.6. Fig. 6 showed that the mass loss ratio of sandy soil cemented by bio-cementitious material at 1, 5, 10, and 15, 20 and 25 cycle is 0.13 %, 0.20 %, 0.58 %, 1.18 %, 1.89 % and 2.94 %. Then the mass loss ratio of floury soil cemented by bio-cementitious material at 1, 5, 10, and 15, 20 and 25 cycle is 0.08 %, 0.11 %, 0.20 %, 0.43 %, 0.71 % and 1.21 %, while the mass loss ratio of clay soil cemented by bio-cementitious material at 1, 5, 10, 15, 20 and 25 cycle is 0.05 %, 0.09 %, 0.15 %, 0.31 %, 0.59 % and 0.92 %. The experimental results show that the freeze-thaw resistance of sandy soil, floury soil and clay soil cemented by bio-cementitious material is apparently different. With the increase of number of cycle, the mass loss ratio is significantly increased, in particular in sandy soil. The freeze-thaw resistance of clay soil is better than that of sandy soil and floury soil.

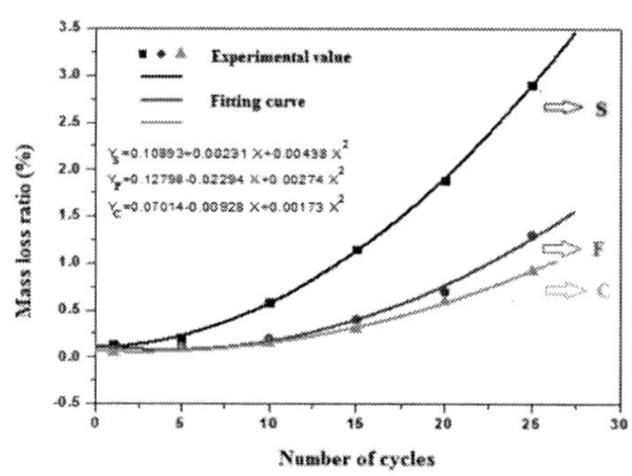

Fig. 6 The freeze-thaw resistance of samples cemented by bio-cementitious material

(S: sandy soil, F: floury soil, C: clay soil)

Basis on experimental data, numerical correlation of the freeze-thaw resistance at different number of cycle are carried out, and the equation could be found in Fig.6. The experimental value are in agreement with the simulated values, and the correlation curve could be effectively used to forecast and analyze the freeze-thaw resistance of samples.

4 Conclusions

This research, reinforcement of soil for suppressing fugitive dust by bio-cementitious material was investigated. The results obtained in this work can be summarized as follows: sandy soil, floury soil and clay soil cemented by bio-cementitious material have superior mechanical properties; furthermore, different type's soil have a certain influence on the mechanical properties. Using bio-cementitious material, the average hardness of sandy soil, floury soil and clay soil is 18.9 °, 25.2 ° and 26.1 °, while average compressive strength of samples is 0.43 MPa, 0.54 MPa and 0.69 MPa, respectively; Meanwhile, the average calcite content of samples is 6.85 %, 6.09 %, and 5.96 %, respectively. Compared with the original sandy soil, floury soil and clay soil, the porosity decreases by 38.5 %, 33.7 % and 29.2 %. With the increase of wind speed, the change of the mass loss is not obvious; When wind speed is 12 m/s, the mass loss of sandy soil, floury soil and clay soil cemented by bio-cementitious material are all less than 30 g/($m^2 \cdot h$). At the same time, the rainfall-erosion resistance of samples cemented by bio-cementitious material is proved; After three cycles of rainfall erosion of 2.5 mm/h, the mass loss are less than 25 g/($m^2 \cdot h$) and the compressive strength residual ratio are more than 98.0 %. The experimental results show that the freeze-thaw resistance of sandy soil, floury soil and clay soil cemented by bio-cementitious material is apparently different. The freeze-thaw resistance of clay soil is better than that of sandy soil and floury soil. For different type's soil, different amounts of bio-cementitious material could be used to achieve the same effect.

Acknowledgements

This work was supported financially by the National Nature Science Foundation of China (Grant No. 51372038), Key Research and Development Program of Jiangsu Science and Technology Department (Grant No. BE2015678), Science and Technology Research Program of Jiangsu Transport Department (Grant No. 2015T32) and Scientific Research Foundation of Graduate School of Southeast University (YBJJ1566).

References

[1] Zhang zhihui, Wu Fan. Health impairment due to building construction dust pollution. J Tsinghua Univ (Sci &Tech), 2008, 48(6): 922-925.

[2] Zhao Pusheng, Feng Yinchang, Jin Jing, et al. Characteristics and control indicators of fugitive dust from building construction sites [J]. Acta Scientiae Circumstantiate, 2009, 29(8): 1618-1623.

[3] Ma Qaing, Wang Mei. Preparation and research on Dust Suppressant in Construction Site. Industrial safety and environmental protection, 2014, 40(8):61-63.

[4] Chen Li, Han Tingting, Li Tao, et al. Estimation of the effect derived from wind erosion of soil and dust emission in tianjin suburbs on the central district based on WEPS model. Chinese Journal of Environmental Science, 2012, 33(7): 2197-2203.

[5] He Jijun, Cai Qiangguo, Tang Zejun. Wind tunnel experimental study on soil wind erosion control with PAM. Journal of Soil and Water Conservation, 2007, 21(2): 12-15.

[6] Xu Xianying, Tang Jinnian, Jin Hongxi, et al. Study on sand-fixing benefits of three new-type chemical sand-fixing materials. Journal of Soil and Water Conservation, 2005, 19(3): 62-65.

[7] Han Fang, Zhu Jimao, Jin Longzhe. Research about the capability of new dust suppression medicament. Safety and Environmental Engineering, 2003, 10(4): 31-33.

[8] Wang Yinmei, Han Wenfeng, Chen Wenwu. Analysis of chemical sand-fixing materials for sand fixation in arid desert area. The Chinese Journal of Geological Hazard and Control, 2004,

15(2): 78-81.

[9] Zhang Huajun, Liu Fei,Zhang Xingkai,et al. The overview of road dust suppression agent. Science and technology of labour protection, 2000, 6: 40-42.

[10] Zhou Yufen. Effect of construction activities on particulate matter pollution along roadways. Journal of Southwest Agricultural University, 2004(4): 518-520.

[11] ZhaoXiuyong,Cheng Shuiyuan,Tian Gang,et al. Construction fugitive dust pollution and control in beijing. Journal of Beijing University of Technology, 2007(10): 1086-1090.

[12] Bai Xiangbing, Liu Jian, Yan Yingtao, et al. Fugitive dust pollution of city and situation and prospect of study on dust-depressor. Journal of Shaanxi University of Technology, 2005, 21(4): 43-46.

[13] Peng Xiaolan, Wu Chao. Progress of chemical dust suppressants and their application. Journal of Safety Science and Technology, 2005, 1(5): 44-47.

[14] Zhang Wen'an, Huo Leixia, Liu Hailong, et al. Study on preparation and properties of complex coal dust inhibitor. Coal Chemical Industry, 2009(5): 21-24.

[15] Amato F, Querol X, Johansson C, et al. A review on the effectiveness of street sweeping, washing and dust suppressants as urban PM control methods [J]. Science of the Total Environment, 2010, 408(16): 3070-3084

[16] Copeland C R, Eisele T C, Kawatra S K. Suppression of airborne particulates in iron ore processing facilities [J]. International Journal of Mineral Processing, 2009, 93(3-4): 232-238

[17] Karin E, Rolf M. Impact of fine materials content on the transport of dust suppressants in gravel road wearing courses [J]. Journal of Materials in Civil Engineering, 2011, 23(8): 1163-1170

[18] Medeiros M A, Leite C M M, Lago R M. Use of glycerol by-product of biodiesel to produce an efficient dust suppressant [J]. Chemical Engineering Journal, 2012, 180: 364-369

[19] Jin Longzhe,Yang Jixing,Ou Shengnan. Experimental study of wetting chemical dust-depressor. Journal of safety and environment,2007,7(6):109-112.

Review on fatigue behavior of high-strength concrete after high temperature

Dongfu Zhao, Penghe Jia*, Haijing Gao

Beijing University of Civil Engineering and Architecture, Beijing, China

*Corresponding author e-mail: 18800161811@163.com

Abstract. The fatigue of high-strength concrete after high temperature has begun to attract attention. But so far the researches work about the fatigue of high-strength concrete after high temperature have not been reported. This article based on a large number of literature. The research work about the fatigue of high-strength concrete after high temperature are reviewed, analysed and expected, which can provide some reference for the experimental study of fatigue damage analysis.

1. Introduction

In practice, HSC structures not only need to withstand static load, but also have to bear cyclic loads such as vehicle, wind, wave or others, and sometimes even suffer a fire or other high temperature history, which will cause fatigue damage. At present, the research on the fatigue properties of HSC after high temperature has become a central issue for scholars.

So far, there are many researches on high-strength concrete, but most researches [1-3] mainly focus on the basic properties of HSC, such as strength and durability. The studies on the fatigue damage of concrete after high temperature are few and only focus on axial compression fatigue of plain concrete after high temperature [4]. While, fatigue properties of HSC after high temperature have not been reported. In this paper, the research on the fatigue properties of HSC after high temperature is reviewed, analyzed and evaluated. It can provide reference for fatigue test and damage analysis of HSC after high temperature.

2. Research status of fatigue behaviour of HSC

Wu Peigang et al [5] studied the axial compression fatigue behavior of HSC under constant-amplitude and variable-amplitude repeated loads. Based on the tests, fatigue strength and fatigue deformation of HSC under constant-amplitude repeated loading were analyzed, the empirical formula of longitudinal total strain and residual strain of fatigue strength were given, and a formula for judging the fatigue failure by residual strain was put forward. Lu Xiaobin et al [3] pointed out that HSC with lateral confining pressure σ3 growth, the axial peak strain ε1c and transverse peak strain ε3c also significantly increased, and in the triaxial compression of different loading path seemed to limit strength σ1c and $\tau_{oct}-\gamma_{oct}$ curve of the status quo was minimal. Tian Yaogang et al [6] studied the introduction of the damper of concrete under the vibration load of bending fatigue performance. Compared with the same level of strength of HSC, the anti-fatigue performance of HSC with damping function was obviously improved. With the increase of vibration fatigue load, the service life of HSC with damping function was obviously reduced. On the basis of experiment, the vibration fatigue life prediction equation of HSC with damping function under different failure probabilities was established. He Zhenjun et al [7] conducted a high-strength high-performance concrete at different stress ratio of multi-axis tension and compression tests, which showed that the compressive strength and deformation of multi-axial tension and compression depend on the stress state and stress ratio, and the multi-axial tensile strength $\sigma 1f$ and $\sigma 3f$ were smaller than the uniaxial tensile and compressive strength ft and fc at all stress ratios. And the peak strain $\varepsilon 3p$ of the main pressure was the largest in uniaxial compression.

A dynamic triaxial compression test of HSC was carried out by JKGran et al [8]. It pointed that stress-strain relationships, stress and strain measurements in triaxle compression, which provided data and a reference for the complete nonlinear relationship between stress and strain tensors. Ertekin Aztecan et al [9] performed the Drucker-Prager yield criterion for triaxle compression tests of ordinary and HSC. The Drucker-Prager parameter value increased with the increase of the ordinary concrete strength, and the Drucker-Prager parameter equation and the correctness of the equation were proved by the data collected.

3. Research status of mechanical properties of HSC after high temperature

Li Lijuan et al [10] conducted a high-temperature test on HSC (100 MPa), which studied its appearance, compressive strength, flexural strength and splitting tensile strength after 500℃ and 800℃.HSC after high temperature could occurs bursting phenomenon, with the fire temperature increasing, compressive strength, flexural strength and tensile strength splitting gradually become smaller, microstructure gradually worse, Mainly as follows: the loss of crystal water, cement hydrate decomposition occurs when the fire temperature reaches 800℃, the crystal water all lost, cement hydrate all the decomposition, the structure becomes loose. He Zhenjun [11-12] conducted C50 and C6O of HSC under multi-axial stress state strength and deformation performance test after high temperature. And the corresponding failure modes and mechanisms, multiaxial strength, peak strain, stress-strain curves of the specimens under different stress states and different stress ratios after different high temperatures were analyzed, and the corresponding strength and deformation conclusions were given. The failure criterion formula of HSC under multi-axial stress condition was established. Zhao Dongfu et al [13] studied the microstructural changes of HSC after different temperature and different constant temperature time from different angles by means of ultrasonic, scanning electron microscopy, X-ray diffraction and so on. The physical and chemical changes and the resulting microstructure changes were analyzed, studies had shown that with the heating temperature and constant temperature time, the compressive strength of concrete showed a trend of decreasing.

Long T et al [14] studied the fire performance of HSC and compiled the fire test data. It indicated that HSC and plain concrete showed great difference in the temperature range of 20℃ to 400℃. Fu-Ping Cheng et al [15] studied the stress-strain curves of high-strength concrete at 20℃, 100℃, 200℃, 400℃, 600℃ and 800℃, and pointed out that the compressive strength of HSC increased with temperature. The compressive strength at 800℃ was about one quarter of its initial strength. Masood Ghandehari et al [16] tested compressive strength, splitting tensile strength and corresponding ultrasonic pulse velocity of high-strength concrete after 100℃, 200℃, 300℃ and 600℃. It was found that with the increase of temperature, HSC was higher than the compressive strength loss rate and the residual strength measured at the ultrasonic pulse velocity was slightly lower than the value directly exposed to more than 200°C.

4. Conclusions and outlook

By summarizing and analysing the research on the fatigue properties of HSC after high temperature in recent years. It is found that there are many researches focused on the mechanical properties but little research focused on the fatigue properties. What's more, many researchers considered the fatigue, complex environment and other factors separately, which made those studies don't meet the objective engineering practice well and can't correctly determine the damage process and extent of HSC.

Because of different research purposes and angles, the studies on the fatigue of HSC after high temperature didn't connect the temperature history well with macro-mechanical properties, and cannot reveal its evolution mechanism under high temperature and complex load. The relationships between temperature, length of time and macro-mechanical properties of HSC have not been established yet. The research needs to be studied further and more systemic, which will be the main direction to study the fatigue performance of HSC.

Acknowledgements
This work was financially supported by the National Natural Science Foundation of China (51378045), Beijing Collaborative Innovation Center of Energy Conservation and Emission Reduction Technology (Beijing University of Civil Engineering and Architecture).

References
[1] He Zhenjun, Song Yupu. Multiaxial tensile-compressive mechanical behaviour of plain high-strength high-performance concrete [J].Engineering mechanical,2010, 27(10):190-195.
[2] Gyu-Yong KIM,Young-Sun KIM,Tae-Gyu LEE.Mechanical properties of high-strength concrete subjected to high temperature by stressed test [J].Transactions of Nonferrous Metals Society of China,2009,19:128-133.
[3] Lu Xiao-bin, Thomas Hsu Cheng-Tzu. Behavior of high strength concrete with and without steel fiber reinforcement in triaxial compression [J].Journal of Materials in Civil Engineering, 2007,19 (3):261-268.
[4] Zhou Xingang. Preliminary research on fatigue behavior of concrete after exposed to high temperature [J].Industrial Construction,1996,26(5),33-36.
[5] Wu Peigang,Zhao Guangyi,Bai Liming.Fatigue behavior of high strength concrete under compressive cyclic loading[J].China Civil Engineering Journal,1994,27 (3):33-40.
[6] Tian Yaogang, Hu Shuguang, Wang Fazhou, Ding Qingjun. Analysis of Fatigue Properties of Damping and High Strength Concrete[J].Journal of Building Materials,2009,12(3):328-331.
[7] Zhenjun He, Wanlin Cao, Jiaxing Zhang. Multiaxial mechanical properties of plain recycled aggregate concrete [J].Magazine of Concrete Research,2014,67(8):401-413.
[8] By J.K.Gran,A.L.Florence,J.D.Colton. Dynamic triaxial tests of high-strength concrete [J].Civil Engineering and Architecture,1989,115(5):891-904.
[9] Ertekin Öztekin,Selim Pul,Metin Hüsem. Experimental determination of Drucker-Prager yield criterion parameters for normal and high strength concretes under triaxial compression [J]. Construction and Building Materials,2016,112:725-732.
[10] Li Lijuan,Xie Weifeng et al. Performance of 100MPa high strength concrete (HSC) after high temperature treatment [J].Journal of Building Materials,2008,11(1):100-104.
[11] He Zhenjun. Experimental study on the multiaxial meehanieal behavior of plain high-strength concrete before and after high temperatures [D].Dalian: Dalian University of Technology,2008.
[12] He Zhenjun, Song Yupu. Strength and failure criterion of plain high-strength high-performance concrete under biaxial compression surviving high temperatures [J].Engineering Mechanics, 2009,26(2):124-130.
[13] Zhao Dongfu, Liu Mei. Experimental study on residual strength and nondestructive testing of high strength concrete after high temperature[J].Journal of Building Structures,2015,36:365-372.
[14] Long T.Phan,Nicholas J.Carino.Fire performance of high strength concrete: research needs[C]. Advanced technology in structural engineering. structures congress,2000:1-8.
[15] Fu-Ping Cheng,V.K. R.Kodur,Tien-Chih Wang. Stress-strain curves for high strength concrete at elevated temperatures. Journal of Materials in Civil Engineering,2004,16(1):84-90.
[16] Masoud Ghandehari,Ali Behnood, Mostafa Khanzadi. Residual mechanical properties of high-strength concretes after exposure to elevated temperatures [J].Journal of Materials in Civil

Engineering,2010,22(1):59-64.
[17] Jianzhuang Xiao, Zhiwei Li, Qinghai Xie, Luming Shen. Effect of strain rate on compressive behaviour of high-strength concrete after exposure to elevated temperatures[J].Fire Safety Journal,2016,83:25-37.
[18] Nadja Oneschkow. Fatigue behaviour of high-strength concrete with respect to strain and stiffness [J].International Journal of Fatigue,2016,87:38-49.
[19] Zhao Dongfu. Study on fatigue criterion of plain concrete under multiaxial fatigue loading [D]. Dalian: Dalian University of Technology,2002.
[20] Cao Wei,Song Yupu,Liu Haicheng. Fatigue properties of plain concrete under triaxial variable-amplitude tension-compression cyclic loading [J].Engineering Mechanics,2006,23(3):111-117.

IOP Conf. Series: Materials Science and Engineering **207** (2017) 012108 doi:10.1088/1757-899X/207/1/012108

Research on provenance of MVT Pb-Zn deposits

Huiguo Liu

Guizhou Vocational Technology Institute, Guiyang, Guizhou 550023, China.

*Corresponding author e-mail:huiguo_liu@outlook.com

Abstract. Based on previous research materials, provenance of MVT Pb-Zn deposits are explored. The results suggest that in China, MVT Pb-Zn ores form when deep met allogenic materials are transported to shallow zones through large deep zones and taken away after they are leached by fluids. All typical MVT Pb-Zn ores present features of crust sources all over the world.

1. Introduction

Produced in carbonatite and with evident epigenetic characteristics, MVT Pb-Zn deposits are well-known for their typical development in Mississippi River, the United States. This category of deposits has been explored for over a century (Garven, 1985; Ge, 1992; Appold, 1999; Leanch et al., 2005). Over the past few years, people have paid more attention to research on MVT Pb-Zn deposits with the growing demands for lead and zinc in the world. Based on previous research, the author analyzes material sources o MVT Pb-Zn deposits, in order to make basic materials available for future research on this kind of deposits.

2. Provenance of MVT Pb-Zn deposits

Concerning MVT deposits, people have acknowledged that metals of fluids originate from sedimentation basins or effects of leaching produced by brine in passing by the water bearing bed. It has been recognized that solubility of metals is impacted by temperature, pressure and fluid composition, particularly by fluid ligands of basins (Basukli et al, 2004). Many scholars consider that joint effects of content, temperature, pH value and element activity are decisive factors for formation of ligands and capacities of metal ions in solutions (Leach et al, 2005) Although both sulphides and organic metal complex compounds are likely to carry metal ions, the solubility of sulphides is low and organic matter is not common in MVT deposits. Owing to high content of chlorides in fluids, complex compounds of metal chlorides are most likely to transport metals of fluids. Since ore-forming fluids are cryogenic (<200℃) with a narrow range of changes, pH value of fluids is influenced by carbonatite of wall rocks, and the effects of temperature and pH value upon Pb and Zn solubility of ore-forming fluids are far weaker than reduction of sulphur activity (Plumlee et al, 1994), such reduction is highly effective for controlling content of Pb and Zn in ore-forming fluids.

Nowadays, content of metals is extremely high in brine of oilfields, where pH value of brine is rarely below 4 but mostly above 4.5 to 6 (Emsbo, 2000). In such brine, the content of Pb and Zn can be a high date, negatively correlated to content of reduced sulfur. In line with characteristics of MVT ore-forming fluids, oilfield brine has become convincing evidences for brine metallogenesis of MVT Pb-Zn deposits. However, it is still difficult to clarify how to transport ore-forming fluids and contributions of intruded or surrounding strata.

Sangster et al (1990) have summarized characteristics of lead isotopes in 17 MVT Pb-Zn deposits and mines. In the 207Pb/204Pb versus 06Pb/204Pb dot plot (Fig.1), lead isotope values make up a narrow band, which may be represented by a straight line with a slope of 0.096. Such distribution is considered to be the result to the mixing between leached lead from Precambrian basements and lead from secondary radioactivity of overlaying sediments.

Apart from Pine Point, composition of lead isotopes is associated with age differences between crystalline basements and met allogenic adjacent rocks in most MVT Pb-Zn deposits. In case of greater age differences, there will be more lead isotopes in radioactive sources that enter the deposits. The extent to which radioactive lead is blended into the deposits is dependent upon geological and geochemical changes to ore-forming materials during their transport and deposition. For instance, homogeneous composition of lead isotopes in Pine Point suggests that these isotopes would be the single sources of lead, or multi-source lead is homogenized in the process of extracting or transporting ore-forming fluids. The ratio of lead isotopes from radioactive sources varies so greatly that its dots form a line. Such changes indicate that lead comes from multiple sources, including further alternation of lead-included adjacent rocks on impervious strata, multi-source rocks or multiple sources of brines.

After statistically analyzing test results of 570 lead glance and sphalerite (Fig 2), Leach et al. (2005) have reported that lead tends to be from basements in many MVT Pb-Zn deposits (or ore concentration areas). For instance, the value of μ is relatively high in Bushy Park, Pering, Gays River, and Midlands Region, Lennard Shelf, Upper Mississippi Valley, Southeast Missouri regions and Tri-State regions, where lead is supposed to be from upper crusts.

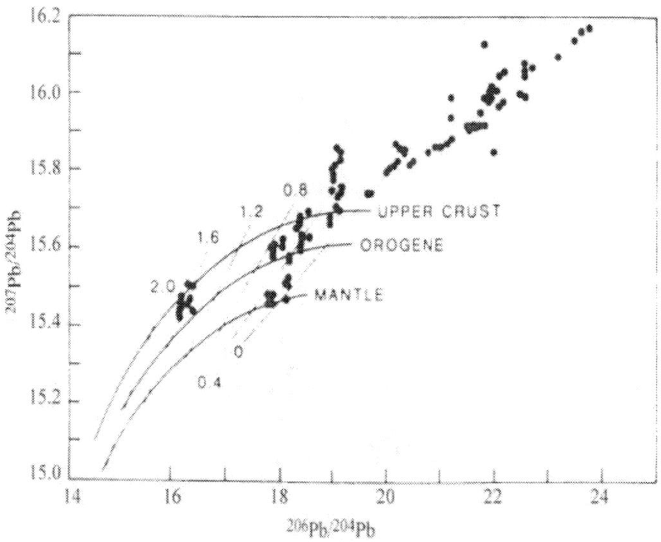

Fig 1 Growth of Lead Isotopes in 17 MVT Deposits and Mines [Sangster et al (1990)]

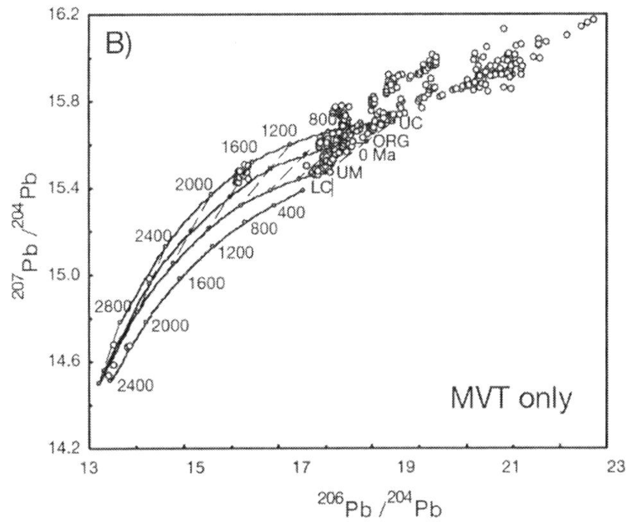

Fig. 2 Composition of Lead Isotopes in 30 MVT Deposits [Leach et al. (2005)]

Lead isotopes of ores and deposits in Sichuan, Yunnan and Guizhou (including Maozu, Qilin Mines, Tianbao Mountain and Dalianis) are mapped upon the 207Pb/204Pb versus 206Pb/204Pb. Some of them fall upon the average evolution belt of upper crusts or include lead from crusts, or those under the evolved orogenic belt are partially from the earth's mantle (Zhang et al., 2005). The projection points are relatively scattered in ore concentration areas of northwestern Hunan Province such as Huayuan, Fenghuang and Longshan, mostly distributed in evolved belts of upper crusts and orogenic belts, and partially allocated in the earth's mantle and lower crusts (Zhong et al., 2007). As a whole, ore-forming material sources of Chinese MVT Pb-Zn deposits mostly come from basement strata above orogenic belts, and deep met allogenic materials are possibly leached by fluids to form ores when they are transported to shallow zones through large deep fault zones.

3. Feature of S isotope

MVT Pb-Zn deposits are formed by a wide range of sulfur isotopes, which are different from sulfur isotopes of magma. Characteristic distribution of δ34S has been detected in all areas, which apparently reflects formation mechanisms of sulphur sources and H2S. The distribution range of sulphur elements in MVT Pb-Zn deposits is consistent with sulphur deviation of various crust sources, including sulfur-containing organic matters, H2S gas, and sulfate-containing evaporate, basin brine, inter-particle seawater and par genetic sulphate. In general, δ34S of sulfide in MVT Pb-Zn deposits is similar to rock isotopes near the deposits. However, the ratio of such δ34S is lower than that of contemporaneous seawater isotopes in adjacent rocks. Hence, there must be a crust of rocks as a direct source of sulphur for a MVT Pb-Zn deposit, where the sulfur of rocks is available from the seawater. Such typical light sulphur isotopes related to seawater of the depositional period of adjacent rocks shall be attributed to fractionation of sulphur isotopes in the process of reducing various sulphate rocks.

In some areas, the composition of δ34S covers a relatively wide range, which suggests mixing of sulphur from different sources, fractionation of isotopes with the same function as a type of minerals, imbalance between mineral pairs in terms of temperature or chemical environment and kinetic effects. Generally, biological reduction of sulphate is considered to result from a wide range of δ34S. Nevertheless, deposition temperature is mostly higher than that necessary for growth of bacteria, the biological reduction of sulphate is separated from its precipitation in time and space when these compounds deposit.

Although the ratio of sulphur isotopes greatly varies in typical worldwide MVT deposits, such isotopes have been mostly discovered to come from crusts. For a single deposit or area, sulphur would be from one or more sources such as sulfate-containing evaporite, syngeneic seawater, diagenetic sulfate, sulphur-containing organic matters, and H2S reservoirs and reduced sulfur of anaerobic water in basins. However, sulfate of seawater (wrapped inside various minerals by sediments) or syngeneic seawater is possibly still the major source of sulphur, which is subsequently reduced. Sulphate is reduced by three different mechanisms, including bacterial sulfate reduction (BSR), thermal degradation of sulphur in organic matters and thermochemical sulfate reduction (TSR).

Based on materials about δ34S, it may be known that sulphur of MVT Pb-Zn deposits is mostly from seawater sulphate, mixed with fluids and closely related to reduction with organic matters (Zhou et al., 2011).

4. Conclusions

After a comprehensive analysis, several conclusions are reached as follows: 1) Ore-forming material sources of Chinese MVT Pb-Zn deposits mostly come from basement strata above orogenic belts, and deep met allogenic materials are possibly leached by fluids to form ores when they are transported to shallow zones through large deep fault zones. 2) Sulphur isotopes of typical worldwide MVT deposits are generally from crusts in spite of significant changes to their ratio.

References

[1] Garven G. The role of regional fluid inclusions: a new tool in constraining source regimes and its implications for the genesis of Mississippi Valley-type deposits. Economic Geology, 1985, 80: 307-324.

[2] Ge S and Garven G. Hydro mechanical modeling of tectonically-driven groundwater flow with application to the Arkoma basin. Geophys. Res., 1992, 97: 9119-9144.

[3] Appold M S and Garven G. The hydrology of ore formation in the Southeast Missouri district: numerical models of topography-driven fluid flow during the Ouachita orogen . Economic Geology, 1999, 94: 913-936.

[4] Leach D L, Sangster D F, Kelly K D. Sediment-hosted lead-zinc deposit: A global perspective. Economic Geology, 2005, Economic Geology 100th Anniversary Volume: 307-324.

[5] Basuki N I and Spooner E T C. A review of fluid inclusion temperatures and salinities in Mississippi Valley-type Zn-Pb deposits: Identifying thresholds for metal transport . Exploration and Mining Geology, 2004, 11(1-4): 1-17.

[6] Plumlee G S, Leach D L, Hofstra A H. Chemical reaction path modeling of ore deposition in Mississippi Valley-type deposits of the Ozark region, U.S Midcontinent. Economic Geology, 1994, 89: 1361-1383.

[7] Emsbo P. Gold in Sedex deposits . Society of Economic Geologists review, 2000, 13: 427-437.

[8] Sangster D F. Mississippi Valley-type and Sedex lead-zinc deposits: A comparative examination. Transactions of the Institution of Mining and Metallurgy, Sec. B, 1990, 99: 21-42.

[9] Zhang Changqing. Distribution, characteristics and genesis of mississippi Valley-type lead-zinc deposits in the triangle area of Sichuan-Yunnan-Guizhou Provinces. China university of Geosciences, 2005, 73-77.

[10] Zhong Jiusi, Mao Changming. Discussion on Metallogenic characteristics and Metallogenic mechanism of MVT type Pb-Zn deposit in northwestern Hunan. Land resources, 2007, 4(6):52-56.

[11] Zhou Lingjie, Zhang Zhengwei, Zhang Zhongshan et al. Study on metallogenic systems and geochemistry methodology of MVT lead-Zinc deposits, Bulletin of mineralogy, petrology and geochemistry , 2011,30(2):223-233.

Institute of Physics
Dirac House, Temple Back
Bristol BS1 6BE UK

ISSN: 1757-8981
ISBN 978-1-5108-4265-6

9 781510 842656

CURRAN ASSOCIATES INC.
proceedings
.com